Conductor–Insulator Quantum Phase Transitions

Conductor–Insulator Quantum Phase Transitions

Edited by

Vladimir Dobrosavljević, Nandini Trivedi,
James M. Valles, Jr.

OXFORD
UNIVERSITY PRESS

Great Clarendon Street, Oxford, OX2 6DP,
United Kingdom

Oxford University Press is a department of the University of Oxford. If furthers the University's objective of excellence in research, scholarship, and education by publishing worldwide. Oxford is a registered trade mark of Oxford University Press in the UK and in certain other countries

First Edition published in 2012

Impression: 1

British Library Cataloguing in Publication Data

Data available

Library of Congress Cataloging in Publication Data

Data available

ISBN 978–0–19–959259–3

Printed in China on acid-free paper by
CC Offset Printing Co. Ltd

Preface

Conductor–Insulator Quantum Phase Transitions was born out of a very successful conference held at Ohio State University from January 9–11, 2008 with the same title. The thrust of the workshop was to highlight the current experimental and theoretical progress and fundamental open questions on quantum phase transitions between conducting and insulating phases in low-dimensional systems, specifically in the areas of:

- superconductor–insulator transitions
- metal–insulator transitions
- pseudogap behavior in underdoped high-T_c cuprates.

In this book we also include a weblink to the highly prized evening lecture by P.W. Anderson in celebration of "50 years of Anderson Localization".[1] Anderson talked about his motivation to look at disordered systems because of Feher's experiments on silicon doped with phosphorus in the 1950s. It was uncanny just how this topic has reemerged today in the context of single qubits!

It is clear that the words "Anderson model" are written all over the book and the conference, starting with the 1958 paper "Absence of diffusion in certain random lattices",[2] the pairing of exact eigenstates idea for disordered superconductors,[3] and the seminal paper on the RVB theory of high-T_c superconductivity.[4] But then one can also see how the field has moved forward through the integrated efforts of theory, experiment, and computation.

What was unique about this workshop were the special discussion sessions every afternoon and during dinner on critical issues and open questions in each of the above topics. Following the workshop we felt that it would be useful for the broader community to not only be made aware of the recent developments in this field but also the open questions highlighting the spirit of the discussions. Hence this multi-author book on *Conductor–Insulator Quantum Phase Transitions* (CIQPT) with both pedagogical and current research content.

The conceptual understanding of such phase transitions has proved difficult and challenging, since the problem typically cannot be reduced to standard paradigms, suggesting the need for conceptually new theoretical ideas. Over the last few years

[1] On a Windows machine, the link is: mms://streaming1.osu.edu/media2/physics/philanderson.wmv. On a Mac or other operating system, it is necessary first to download the plugin at http://www.microsoft.com/download/en/details.aspx?displaylang=en&id=9442, then use link mms://streaming1.osu.edu/media2/physics/philanderson.wmv.

[2] *Phys. Rev.* 109, 1492 (1958).

[3] *J. Phys. Chem. Solids* 11, 26 (1959).

[4] *Science* 237, 1196 (1987).

fascinating new results have started to emerge in this field, both on the experimental and theoretical fronts. Despite much recent progress, there are no other accessible and pedagogical texts on the subject, making it difficult for students and other young researchers to access the wealth of new information. We hope CIQPT fills exactly this gap.

Each article has been written with a three-pronged purpose: (a) a pedagogical component, a review, accessible to a typical graduate student or junior researcher—as such this portion addresses the lacunae between traditional condensed matter courses where disorder is dealt with in a perfunctory manner and research in metal–insulator transitions; (b) a research component that highlights new paradigms and new theoretical developments that provide an understanding of strongly interacting and disordered systems and novel high-precision local scanning probes that reveal spectral maps in a disordered system; and (c) an open questions component that points to research directions for future generations of researchers.

This monograph is a unique contribution in the field of condensed matter physics, touching on topics of quantum phase transitions, disorder and strong correlation effects, and novel local imaging and spectroscopic studies to characterize the system as it evolves across the transition. Starting with P.W. Anderson's historical perspective on the localization problem documented in his classic paper 50 years ago, in which he showed the existence of localized states, the first part of the book deals with the metal–insulator transition. Starting with an introduction by one of the editors (Dobrosavljević), the subsequent articles will span the evolution of non-interacting electrons in random media (Slevin and Ohtsuki) to the metal–insulator transition in the two-dimensional electron gas (Kravchenko for experiments and Punnoose and Finkelshtein for theory) as well as in other systems driven by a combination of both disorder and strong correlations (Miranda and Dobrosavljević). The theoretical articles are well balanced with experimental probes that visualize the critical fluctuations at the transition (Yazdani). Other probes of the transition have been local optical conductivity, imaging of the local compressibility, and noise fluctuations, and glassy behavior (Popovic). Insights from quantum Monte Carlo simulation methods (Scalettar and collaborators) are also included.

The second part of the book focuses on the superconductor–insulator transition (SIT). The introduction by one of the editors (Trivedi) lays the framework for the present status and open questions in the field of SIT. It provides an overview on the effect of disorder and magnetic field on superconductors, the role of amplitude and phase fluctuations, and the nature of the insulating state. The subsequent articles develop the theme further, with an experimental overview on the quantum phase transition driven by tuning the thickness, electric field, and gating (Goldman), a novel insulating phase at high magnetic fields (Sambandamurthy and Armitage), evidence of Cooper pairs in the insulating phase through signatures in the Bohm–Aronov effect (Stewart and Valles, Jr.), magnetic field and magnetic impurity effects (Xiong), and spin effects across the SIT (Adams). These are followed by imaging and scanning tunneling spectroscopy revealing the effects of disorder in high-T_c superconductors (Davis and collaborators), theoretical investigations of the SIT (Loh and Trivedi) and vortices in a disordered superconductor (Michaeli and Finkelshtein).

We acknowledge funding from ICAM (Institute for Complex Adaptive Matter), National High Field Magnet Laboratory, and The Ohio State University. These funds allowed us to have strong participation from younger faculty and students, and keep the field vibrant and alive.

Vlad Dobrosavljević, Nandini Trivedi, and James Valles

Contents

List of abbreviations

AHH	Anderson–Hubbard Hamiltonian
ARPES	angle-resolved photoemission
AT	Almeida–Thouless
BCS	Bardeen–Cooper–Schrieffer
BEC	Bose–Einstein condensation
BFK	Bose–Fermi Kondo
CDW	charge-density wave
CMR	colossal magnetoresistance
CPA	coherent potential approximation
DFT	density-functional electronic structure methods
DMFT	dynamical mean-field theory
DOS	density of states
DQMC	determinant quantum Monte Carlo
EDMFT	extended DMFT
FFLO	Fulde–Ferrell–Larkin–Ovchinnikov
FM	ferromagnetic
FOV	fields of view
HH	Hubbard Hamiltonian
JJAs	Josephson junction arrays
KTB	Kosterlitz–Thouless–Berezinskii
LCD	lower critical dimension
LDOS	local density of states
LE	Lyapunov exponent
MBE	molecular-beam epitaxy
MC	magnetoconductance
MEM	maximum entropy method
MFT	mean-field theory
MIT	metal–insulator transition
MOSFET	metal-oxide-semiconductor field-effect transistor
MR	magnetoresonance
NHC	nano-honeycomb
NRG	numerical renormalization group
NS	neutron scattering
OC	orthogonality catastrophe
PDF	probability density function
PG	pseudogap
PHS	Particle–hole symmetry

PLNS	Pauli-limited normal state
PM	paramagnetic
PR	pairing resonance
QCP	quantum critical point
QMC	quantum Monte Carlo
QPI	quasiparticle interference
QPT	quantum phase transition
RFIM	random-field Ising model
RG	renormalization group
RKKY	Ruderman–Kittel–Kasuya–Yosida
RS	replica symmetric
RSB	replica symmetry-breaking
SCHA	self-consistent harmonic approximation
SI-STM	spectroscopic imaging scanning tunneling microscopy
SIT	superconductor–insulator transition
SOP	superconducting order parameter
S-OS	spin–orbit scattering
SOS	spin orthogonal stripe
SRS	spin-rotation symmetry
STM	scanning tunneling microscope
TCR	temperature coefficient of the resistivity
TDOS	tunneling density of states
TFIM	Ising model in a transverse field
TMT	typical medium theory
TRS	time-reversal symmetry
VRH	variable-range hopping
WL	worldline
XPS	X-ray photoelectron spectroscopy
ZBA	zero-bias anomaly
ZFC	zero-field cooled

List of contributors

P.W. Adams
Gonzalo Alvarez
N. Peter Armitage
P. Chakraborty
S. Chiesa
Elbio Dagotto
J. C. Davis
P.J.H. Denteneer
V. Dobrosavljević
Shuai Dong
H. Eisaki
Alexander M. Finkel'stein
K. Fujita
A.M. Goldman
S.V. Kravchenko
E.A. Kim
M.J. Lawler
D.H. Lee
Yen Lee Loh
K. Michaeli
E. Miranda
Adriana Moreo
Tomi Ohtsuki
T. Paiva
Dragana Popovic
Alexander Punnoose
G. Sambandamurthy
R.T. Scalettar
A.R. Schmidt
Cengiz Sen
Keith Slevin
M.D. Stewart, Jr.
S. Story
Nandini Trivedi
S. Uchida
James M. Valles, Jr.
Peng Xiong
Ali Yazdani
Rong Yu

Part I

Metal-Insulator Transitions

1
Introduction to Metal–Insulator Transitions

V. Dobrosavljević

Department of Physics and National High Magnetic Field Laboratory
Florida State University,Tallahassee, FL 32306, USA

1.1 Why study metal–insulator transitions?

The metal–insulator transition (MIT) is one of the oldest, yet one of the fundamentally least understood problems in condensed matter physics. Materials which we understand well include good insulators such as silicon and germanium, and good metals such as silver and gold. Remarkably simple theories (Ashcroft and Mermin, 1976) have been successful in describing these limiting situations: in both cases low-temperature dynamics can be well described through a dilute set of elementary excitations. Unfortunately, this simplicity comes at a price: the physical properties of such materials are extremely *stable*. They prove to be very difficult to manipulate or modify in order to the meet the needs modern technology, or simply to explore novel and interesting phenomena.

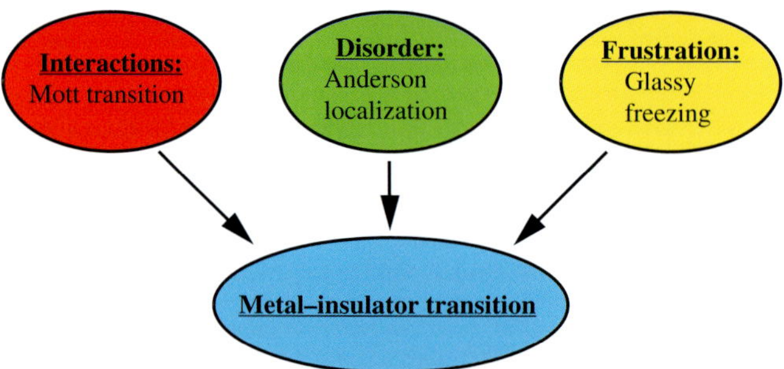

Fig. 1.1 Three basic mechanism for electron localization.

The situation is more promising in more complicated materials, where relatively few charge carriers are introduced in an otherwise insulating host. Several such systems have been fabricated, even years ago, and some have, in fact, served as basic building blocks of modern information technology. The most familiar are, of course, the doped semiconductors, which led to the discovery of the transistor. More recent efforts drifted to structures of reduced dimensionality and devices, such as silicon MOSFETs (metal-oxide-semiconductor field-effect transistors), which can be found in any integrated circuit.

1.1.1 Why is the metal–insulator transition an important problem?

In contrast to elemental materials, in systems close to the MIT the physical properties change dramatically with the variation of control parameters such as the carrier concentration, the temperature, or the external magnetic field. Such sensitivity to small changes is, indeed, quite common in any material close to a phase transition. In doped insulators this sensitivity follows from the vicinity to the metal-insulator transition. The sharp critical behavior is seen here only at the lowest accessible temperatures, because a qualitative distinction between a metal and an insulator exists only at $T = 0$ (Fig. 1.2). Since the basic degrees of freedom controlling the electrical transport properties are electrons, and the transition is found at $T = 0$, quantum fluctuations dominate the critical behavior. The MIT should therefore be viewed as

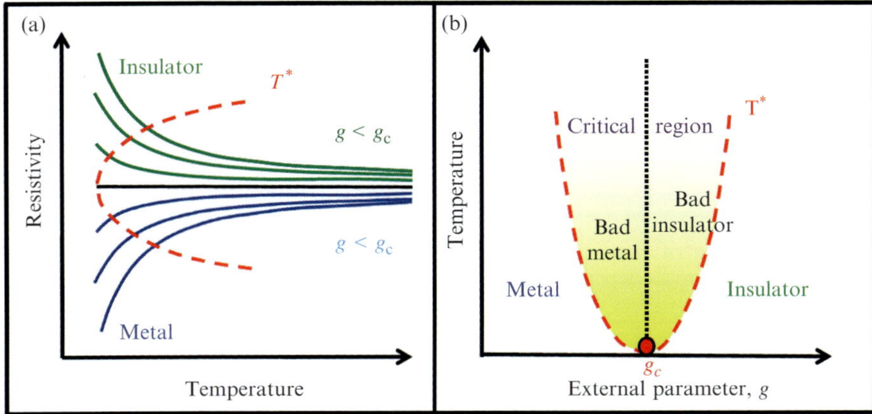

Fig. 1.2 Quantum critical behavior near a metal–insulator transition. Temperature dependence of the resistance for different carrier concentrations is shown schematically in (a). Well-defined metallic or insulating behavior is observed only at temperatures lower than a characteristic temperature $T < T^*$ that vanishes at the transition. At $T < T^*$, the system is in the "quantum critical region," as shown in (b). As the system crosses over from metal to insulator, the temperature dependence of the resistivity changes slope from positive to negative.

perhaps the best example of a quantum critical point, a subject that has attracted much of the physicists' fancy and imagination in recent years (Sachdev, 2011). As near other quantum critical points, one expects the qualitative behavior here to display a degree of universality, allowing an understanding based on simple yet fundamental physical pictures and concepts. Before we understand the basic mechanisms and process that control this regime (Fig. 1.1) one can hardly hope to have control over material properties even in very simple situations.

1.1.2 Why is the metal–insulator transition a difficult problem?

From the theoretical point of view the problem at hand is extremely difficult for reasons that are easy to guess. The two limits, that of a good metal and that of a good insulator, are very different physical systems, which can be characterized by very different elementary excitations. For metals, these are fermionic quasiparticles corresponding to electrons excited above the Fermi sea. For insulators, in contrast, these are long-lived bosonic (collective) excitations such as phonons and spin waves. In the intermediate regime of the MIT, both types of excitations coexist, and simple theoretical tools prove of little help. Some of these conceptual difficulties are a general feature of quantum critical points. For quantum critical points involving spin or charge ordering, the critical behavior can be described by examining the order parameter fluctuations associated with an appropriate symmetry-breaking. In contrast, the MIT is more appropriately described as a dynamical transition, and an obvious order-parameter theory is not available. For all these reasons, the intermediate regime between the metal and the insulator has remained difficult to understand both from the practical and the conceptual point of view.

1.1.3 Disorder and complexity

After more then sixty years of study, the subject of MITs has become a very wide and complicated field of research. In many complex materials, particularly in transition-metal oxides and other strongly correlated systems, the emergence of the MIT is often accompanied by changes in magnetic or structural symmetry. In many such cases the transition is dominated by material-specific details, and it is often of first order, not a critical point. Somewhat surprisingly, the situation is in fact simpler in presence of sufficient amounts of disorder, where the metal and the insulator have the same symmetries and the transition is not associated with any uniform ordering. In such cases, strong evidence exists indicating that the transition is a genuine quantum critical point of a fundamentally new variety, with properties that most likely dominate the behavior of many materials.

At first glance disorder may be viewed simply as a nuisance and an inessential complication. In recent years, though, fascinating new evidence has emerged (Fig. 1.3) revealing that genuinely new phenomena (Miranda and Dobrosavljević, 2005) arise in presence of disorder and impurities. Put quite simply, a given configuration of impurities may locally favor one or another of several competing phases of matter. In many instances this gives rise to strongly inhomogeneous states that feature an enormous number of low-lying metastable states—giving rise to quantitatively new excitation, slow relaxation, and glassy fluctuations and response. These phenomena often dominate the observable properties in many systems, ranging from colossal magneto-resistance manganites and cuprates to diluted magnetic semiconductors, and even Kondo alloys. Even more surprisingly, recent work suggests that a plethora of intermediate heterogenous phases may emerge between the metal and the insulator,

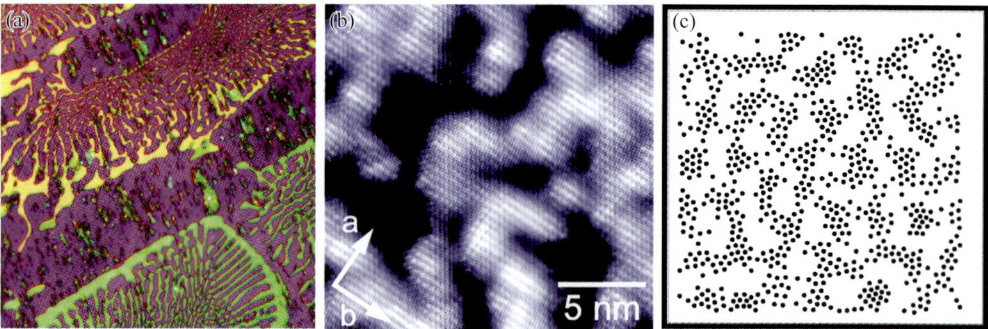

Fig. 1.3 In the last few years, fascinating examples of complex ordering around the MIT have started to emerge, due to advances of both the experimental probes and the theoretical tools available. (a) percolative conduction in the half-metallic ferromagnetic and ferroelectric mixture (La,Sr)MnO_3 (Park *et al.*, 2004). (b) inhomogeneous charge distribution revealed by scanning tunneling microscope (STM) spectroscopy (Kohsaka *et al.*, 2004) on underdoped cuprate $Ca_{2-x}Na_xCuO_2Cl_2$. (c) strikingly similar "stripe glass" ordering is observed in a computer simulation of an appropriate model (Reichhardt *et al.*, 2005).

possibly even in the absence of disorder. Such *complexity* emerges as a new paradigm (Dagotto, 2005) of the MIT region, most likely requiring a description in terms of probability distribution function functions rather then simple minded order parameters.

1.1.4 Metal–insulator transitions in the strong correlation era

The subject of MITs came to a renewed focus in the last two decades, following the discovery of high-temperature superconductivity, which triggered much activity in the study of "bad metals" (for a recent perspective, see Basov and Chubukov (2011)). Many of the materials in this family consist of transition metal or even rare earth elements, corresponding to compounds that are essentially on the brink of magnetism. Here, conventional approaches proved of little help, but recent research has lead to a veritable avalanche of new and exciting ideas and techniques both on the experimental and the theoretical fronts. In many ways, these developments have changed our perspective on the general problem of the MIT, emphasizing the deep significance of the physics of strong correlation. A common theme for all these systems seems to be the transmutation of conduction electrons into localized magnetic moments (Anderson, 1978), a feature deeply connected to the modification of the fundamental nature of elementary excitations as the MIT is crossed.

1.1.5 The scope of this overview

In this overview we will not discuss all the possible examples of MITs, but will focus on the basic physical mechanisms that can localize the electrons in the absence of magnetic or charge ordering, and produce well-defined quantum critical behavior. Our emphasis will be on results providing evidence that strong-correlation physics dominates such quantum critical points, where physical pictures based on weak-coupling approaches prove insufficient or even misleading.

1.2 Basic mechanisms of metal–insulator transitions

What determines whether a material is a metal or an insulator? In most cases the answer is provided by simply examining the electronic band structure (Ashcroft and Mermin, 1976) and the pattern of chemical bonding of a given compound. At present, solid state physicists and quantum chemists have an impressive toolbox of theoretical methods to determine the band structure with sometimes surprising accuracy. Quite generally, one calculates all the accessible electronic levels (Slater, 1934) for the valence electrons in a solid and populates them according to the Pauli principle. If the highest occupied electronic state—the Fermi energy—is within a band gap, then the material is an insulator, since it takes a finite (generally large) energy to excite the electron to the lowest accessible state in order to carry electrical current. Otherwise, when electronic bands are partially filled, then we expect metallic behavior.

1.2.1 Band transitions

Can one induce a MIT within this band theory picture? This is possible if the gap can be induced to open at the Fermi surface by rearranging the charge or the spin density of the electrons in the ground state, i.e. when the system undergoes an ordering transition. Typically, this corresponds to some kind of Fermi surface instability where a charge or spin-density wave (Gruner, 2000) formation leads to unit-cell doubling. An important early example was the Slater theory (Slater, 1951) of itinerant antiferromagnets, where the gap opens due to magnetic ordering. Such instabilities are also common in low-dimensional solids such as organic charge-transfer salts (Gruner, 2000), leading to rich phase diagrams with many exotic properties. Such ordering transitions typically take place at finite temperature, and can often be successfully described using conventional approaches based on the band-theory picture. These situations have attracted considerable attention in recent years, but we will not them discuss them further in this overview.

When does band theory work?

The success of the band-theory approach was so impressive that already by the 1930s Slater had announced (Slater, 1934) that solid-state physics was a solved problem, and that we only needed fast computers to accurately predict physical properties of any material. In the last few decades computers have become amazingly fast, but in many interesting cases the band-structure approach hs proved insufficient. When does that happen? To understand this important issue from a general point of view we need to recall the implicit assumptions of the band-structure approach.

The band-theory picture describes the dynamics of one electron moving through a solid, while the effects of all the other electrons is approximated by modifying the effective potential energy surface—the pseudopotential—on which it moves. This approximation is generally expected to be valid whenever the kinetic energy of the electrons is dominant over the other energy scales in the problem. A most naive estimate would involve simply calculating the so-called r_s-number: $r_\mathrm{s} = E_c/E_F$, where E_c is the average Coulomb energy per particle and E_F is the Fermi energy. The r_s-number ranges between three and five even in good metals, and thus one would naively think that band theory should never work. However, one has to keep in mind the following important facts (Ashcroft and Mermin, 1976) that minimize the role of interactions.

- In metals screening reduces the magnitude of electron–electron and electron–impurity interactions to a significant degree.
- The largest part of the Coulomb energy (Hartree and exchange terms) contributes to redefining the pseudopotential, and only the "correlation" energy gives rise to many-body effects.
- The Pauli principle considerably restricts the phase space for electron–electron scattering, as described by Fermi-liquid renormalizations.

As a result, the excitations in an electron gas can be viewed as a dilute collection of quasiparticle excitations, as described by the Landau Fermi liquid theory (Landau, 1957, 1959). In a nutshell, a Fermi liquid is "protected" by a large kinetic energy scale of the electrons in their ground state—a direct result of their Fermi statistics. In good metals, the Fermi energy is typically in the electronvolt range and the effects of electron correlations and impurity scattering can be treated as small perturbations (Abrikosov *et al.*, 1975).

And... when does it fail?

In materials close to the MIT the situation is very different. Here, the Fermi energy is typically small, and the "quantum protectorate" of the Pauli principle starts to weaken. This situation is found, for example, in:

- narrow-band materials such as the transition-metal oxide V_2O_3.
- doped semiconductors such as Si:P or diluted two-dimensional electron gases.
- doped magnetic (Mott) insulators, e.g. the famous high-T_c cuprate $La_{2-x}Sr_xCuO_4$.

In all these cases the potential energy terms coming from either residual electron–electron interactions or due to disorder (electron–impurity interaction) become comparable to the Fermi energy, and the ground state of the system can undergo a sudden and dramatic change—the electrons become bound or "localized." The material ceases to conduct, although band theory does not predict any gap at the Fermi surface. In the following we briefly describe the early ideas on how this might take place and discuss some general features of such quantum critical points.

1.2.2 Interactions: the Mott transition

Many insulating materials have an odd number of electrons per unit cell, and band theory would thus predict them to be metals, in contrast to experiments. Such compounds (e.g. transition-metal oxides) often have antiferromagnetic ground states, leading Slater to propose that spin-density wave formation (Slater, 1951) was probably at the origin of the insulating behavior. This mechanism does not require any substantial modification of the band-theory picture, since the insulating state is viewed as a consequence of a band gap opening at the Fermi surface.

According to Slater (1951), such insulating behavior should disappear above the Neel temperature, which is typically in the 10^2-K range. Most remarkably, in most antiferromagnetic oxides, clear signatures of insulating behavior persist at temperatures well above any magnetic ordering, essentially ruling out Slater's weak-coupling picture.

What goes on is such cases was first clarified in early works by Mott (1949) and Hubbard (1963), tracing the insulating behavior to strong Coulomb repulsion between electrons occupying the same orbital. Within this picture, which is appropriate for narrow-band systems (Mott, 1990), the electrons tunnel between weakly hybridized atomic orbitals, as described by a Hubbard Hamiltonian

$$H_{\mathrm{HUB}} = -\sum_{\langle ij\rangle\sigma}\left(tc^{\dagger}_{i\sigma}c_{j\sigma} + \mathrm{h.\,c.}\right) + \sum_{j\sigma}\epsilon_j c^{\dagger}_{j\sigma}c_{j\sigma} + U\sum_{j} c^{\dagger}_{j\uparrow}c_{j\uparrow}c^{\dagger}_{j\downarrow}c_{j\downarrow}. \tag{1.1}$$

Here, the operator $c^{\dagger}_{i\sigma}$ creates an electron of spin σ in the ith orbital, t is the tunneling element describing the inter-orbital hybridization, ϵ_j represents the corresponding site energy, and U describes the on-site Coulomb repulsion.

When the lattice has integer filling per unit cell, then electrons can be mobile only if they have enough kinetic energy ($E_K \sim t$) to overcome the Coulomb energy U. In the narrow band limit of $t \ll U$, the electrons do not have enough kinetic energy, and a gap opens in the single-particle excitation spectrum, leading to Mott-insulating behavior. This gap $E_g \approx U - B$ (here $B \approx 2zt$ is the electronic bandwidth; z being the lattice coordination number) is the energy an electron has to pay to overcome the Coulomb repulsion and leave the lattice site. In the ground state, each lattice site is singly occupied, and the electron occupying it behaves as a spin-$\frac{1}{2}$ local magnetic moment. These local moments typically interact through magnetic superexchange interactions (Anderson, 1959) of the order $J \sim U/t_{ij}^2$, leading to magnetic ordering at temperatures of order $T_J \sim J$. The insulating behavior, however, is not caused by magnetic ordering, and will persist all the way to temperatures $T_{\mathrm{Mott}} \sim E_g \gg T_J$. In oxides, $T_{\mathrm{Mott}} \sim E_g \sim 10^3$–$10^4$ K is typically on the atomic (eV) scale, while magnetic ordering emerges at temperatures roughly an order of magnitude lower $T_J \sim 100$–300 K.

We note that Mott's simple argument for the stability of the interaction-driven insulator does not directly rely on a periodicity of a lattice. Even if the site energies ε_i or the hopping elements t_{ij} are random variables of moderate variance, the Mott gap will persist, provided that the on-site repulsion U is large enough as compared to the typical kinetic energy. This is true, since the Mott gap essentially measures the energy for an electron to hop to the nearest lattice site—an inherently local process which does not depend much on long-range periodicity of the lattice. This behavior is precisely what takes place in a doped semiconductor deep in the insulating phase, which should be viewed as a strongly disordered Mott insulator. Clear evidence for the correctness of this picture is provided by optical experiments (Thomas *et al.*, 1981), which provide unambiguous evidence of the coexistence of the Mott gap at sufficiently low doping levels.

When the kinetic energy and the Coulomb interaction are comparable, the system finds itself in the vicinity of the Mott transition (Fig. 1.4). Experimentally, the bandwidth can often be controlled by modifying the orbital overlap t, and thus the electronic bandwidth. In several transition-metal oxides, for example, this is possible by applying external hydrostatic pressure. From the theoretical perspective, describing the vicinity of the Mott transition proves quite difficult due to the lack of a small parameter characterizing this non-perturbative regime. Still, after more then thirty years of work on the problem, several theoretical approaches have emerged, which provide a physical picture of the transition region. The early arguments of Mott and Hubbard make it clear that the gap will close for $U \lesssim B$, but the precise form of the critical behavior remained elusive.

Correlated metallic state close to the Mott transition

An important step in elucidating the approach to the Mott transition from the metallic side was provided by the pioneering work of Brinkman and Rice (1970). This work, which was motivated by experiments on the normal phase of ^{3}He (for a review, see Vollhardt (1984)), predicted a strong effective mass enhancement close to the Mott transition. In the original formulation, as well as in its subsequent elaborations

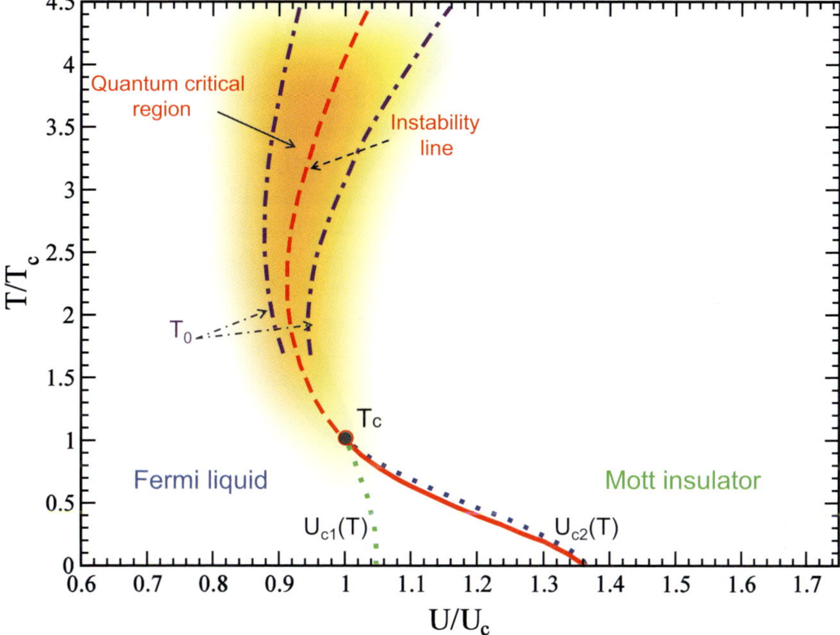

Fig. 1.4 Phase diagram for a fully frustrated half-filled Hubbard model calculated from DMFT theory. At low temperatures the Fermi liquid and the Mott insulating phases are separated by a first-order transition line, and the associated coexistence region. Very recent work Terletska *et al.* (2011) established that at $T > T_c$ the intermediate metal–insulator crossover region show all the features expected for the quantum critical regime, including the characteristic scaling behavior for the family of resistivity curves.

(slave-boson mean-field theory (Kotliar and Ruckenstein, 1986) and dynamical mean-field theory (DMFT) (Georges *et al.*, 1996)), the effective mass is predicted to continuously diverge as the Mott transition is approached from the metallic side

$$\frac{m^*}{m} \sim (U_c - U)^{-1}. \tag{1.2}$$

A corresponding coherence (effective Fermi) temperature

$$T^* \sim T_F/m^* \tag{1.3}$$

is predicted above which the quasiparticles are destroyed by thermal fluctuations. As a result, one expects a large resistivity increase around the coherence temperature, and a crossover to insulating (activated) behavior at higher temperatures. Because the low-temperature Fermi liquid is a spin singlet state, a modest magnetic field of the order

$$B^* \sim T^* \sim (m^*)^{-1} \tag{1.4}$$

is expected to also destabilize such a Fermi liquid and lead to large and positive magnetoresistance.

The physical picture of Brinkmann and Rice seems to suggest that the Mott transition should be viewed as a quantum critical point, where power-law behavior of characteristic crossover scales is expected as the transition is approached. On the other hand, the Mott transition discussed here describes the opening of a correlation-induced spectral gap in the absence of magnetic ordering, i.e. within the paramagnetic phase. Such a phase transition is, therefore, not associated with spontaneous symmetry-breaking associated with any static order parameter. Why should the phase transition then have any second-order (continuous) character at all?

The answer is that... in fact it does not! Later work (Moeller *et al.*, 1999; Park *et al.*, 2008), which presented better descriptions of inter-site correlations, suggested that such a transition should generically have a (weakly) first order character, in agreement with early ideas of Mott (1949). The effective mass, even if it does not exactly diverge at the transition, is still expected to be significantly enhanced in its close vicinity. Such behavior is indeed seen in various Mott systems (Georges *et al.*, 1996) that have been studied, including bulk and two-dimensional He^3 liquids (Fig. 1.5), transition-metal oxides, and organic charge-transfer salts.

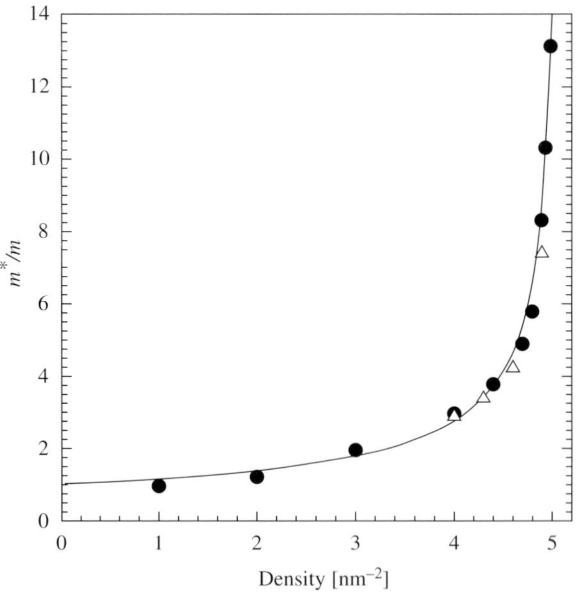

Fig. 1.5 Clear evidence of strong mass enhancements can be seen in experiments on monolayer He^3 films on graphite (Casey *et al.*, 2003) . In this system, the solid phase (Mott insulator) can be approached when the density is increased by the application of hydrostatic pressure.

Physical content of the effective mass enhancement

How should we physically interpret the large effective mass enhancement that is seen in all these systems? What determines its magnitude if it does not actually diverge at the transition? An answer to this important question can be given using a simple thermodynamic argument, which does not rely on any particular microscopic theory or specific model. In the following, we present this simple argument for the case of a clean Fermi liquid, although its physical context is, of course, much more general.

In any clean Fermi liquid (Abrikosov *et al.*, 1975) the low-temperature specific heat assumes the leading form

$$C(T) = \gamma T + \cdots , \tag{1.5}$$

where the Sommerfeld coefficient

$$\gamma \sim m^*. \tag{1.6}$$

In the strongly correlated limit ($m^*/m \gg 1$) this behavior is expected only at $T \lesssim T^* \sim (m^*)^{-1}$, while the specific heat should drop to much smaller values at higher temperatures where the quasiparticles are destroyed. Such behavior is indeed observed in many systems showing appreciable mass enhancements.

On the other hand, from general thermodynamic principles, we can express the entropy as

$$S(T) = \int_0^T dT \frac{C(T)}{T}. \tag{1.7}$$

Using the above expressions for the specific heat, we can estimate the entropy around the coherence temperature

$$S(T^*) \approx \gamma T^* \sim O(1). \tag{1.8}$$

The leading effective mass dependence of the Sommerfeld coefficient γ and that of the coherence temperature T^* cancel out!

Let us now explore the consequences of the assumed (or approximate) effective mass divergence at the Mott transition. As $m^* \longrightarrow \infty$, the coherence temperature $T^* \longrightarrow 0+$, resulting in large residual entropy

$$S(T \longrightarrow 0+) \sim O(1). \tag{1.9}$$

We conclude that the effective mass divergence indicates the approach to a phase with finite residual entropy!

Does not this result violate the third law of thermodynamics? And how can it be related to the physical picture of the Mott transition? The answer is, in fact, very simple. Within the Mott insulating phase the Coulomb repulsion confines the electrons to individual lattice sites, turning them into spin-$\frac{1}{2}$ localized magnetic moments. To the extent that we can ignore the exchange interactions between these spins, the

Mott insulator can be viewed as a collection of free spins with large residual entropy $S(0+) = R \ln 2$. This is precisely what happens within the Brinkmann–Rice picture; similar results are obtained from DMFT, a result that proves exact in the limit of large lattice coordination (Georges *et al.*, 1996).

In reality, the exchange interactions between localized spins always exist, and they generally lift the ground-state degeneracy, restoring the third law. This happens below a low temperature scale T_J, which measures the effective dispersion of inter-site magnetic correlations (Moeller *et al.*, 1999; Park *et al.*, 2008) emerging from such exchange interactions. In practice, this correlation temperature T_J can be very low, either due to the effects of geometric frustration, or additional ring-exchange processes, which lead to competing magnetic interactions.

We conclude that the effective mass enhancement, whenever observed in experiment, indicates the approach to a phase where large amounts of entropy persist down to very low temperatures. Such situations very naturally occur in the vicinity of the Mott transition, since the formation of local magnetic moments on the insulating side gives rise to large amounts of spin entropy being released at very modest temperatures. A similar situation is routinely found (Flouquet, 2005; Hewson, 1993; Stewart, 1984) in the so-called "heavy fermion compounds" (e.g. rare-earth intermetallics) featuring huge effective mass enhancements. Here, local magnetic moments coexist with conduction electrons giving rise to the Kondo effect, which sets the scale for the Fermi-liquid coherence temperature $T^* \sim 1/m^*$, above which the entire free spin entropy $S(T^*) \sim R \ln 2$ is recovered (Fig. 1.6). This entropic argument is, in turn, used to experimentally prove the existence of localized magnetic moments within the metallic host.

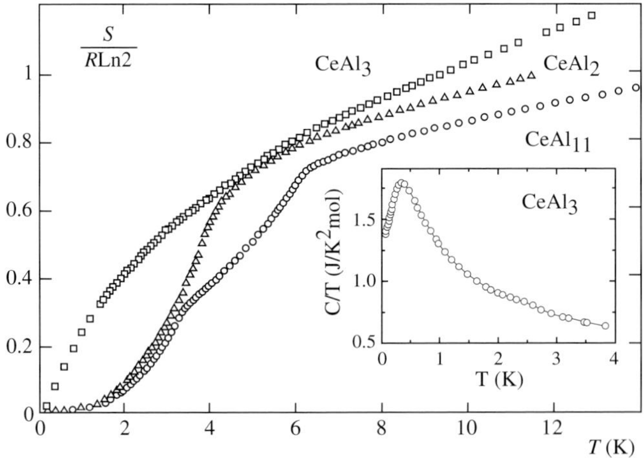

Fig. 1.6 Temperature dependence of entropy extracted from specific heat (inset) experiments (Flouquet, 2005) on several heavy-fermion materials. Essentially the entire doublet entropy $S = R \ln 2$ is recovered by the time the temperature has reached $T^* \approx 10$ K, consistent with a large mass enhancement $m^* \sim 1/T^*$.

We should mention that other mechanisms of effective mass enhancement have also been considered. General arguments (Millis, 1993) indicate that m^* can diverge when approaching a quantum critical point corresponding to some (magnetically or charge) long-range-ordered state. This effect is, however, expected only below an appropriate upper critical dimension (Sachdev, 2011), reflecting an anomalous dimension of the incipient ordered state. In addition, this mechanism is produced by long-wavelength order-parameter fluctuations, and is thus expected to contribute only a small amount of entropy per degree of freedom, in contrast to local moment formation. It is interesting to note that weak-coupling approaches, such as the popular "on-shell" interpretation of the random-phase approximation (Ting *et al.*, 1975), often predict inaccurate or even misleading predictions (Zhang and Das Sarma, 2005) for the effective mass enhancement behavior. As discussed in Chapter 6, more accurate modern theories such as DMFT can be used to benchmark these and other weak-coupling theories, and reveal the origin of some of the pathologies found when they are applied to strong coupling.

Finite temperature metal–insulator coexistence region

Since the Fermi-liquid metal and the paramagnetic Mott insulator do not differ on symmetry grounds, there is no reason why these two phases cannot coexist in a finite range of parameter space. Indeed, recent theories (Georges *et al.*, 1996) as well as several experimental studies (Limelette *et al.*, 2003*a*,*b*) in clean Mott systems find such a metal–insulator coexistence region (Fig. 1.7) leading to a finite-temperature first-order phase transition line. Experimentally, an appreciable drop of resistivity is seen as the system is driven though such a finite-temperature Mott transition, which separates the Mott-insulating state and the metallic (Fermi-liquid) state. Similar to standard liquid–gas systems, the coexistence region and the associated first-order line terminate at the critical end-point at $T = T_c$. The corresponding critical behavior has been carefully studied in recent experiments (Limelette *et al.*, 2003*a*) on chromium-doped V_2O_3, and was found to belong to the standard Ising universality class.

Weak disorder near Mott transitions

All quantities display a discontinuity across any first-order phase transition, but this jump can be reduced in the presence of impurities or disorder, which tend to locally favor one or the other phase. Well-known droplet arguments (Imry and Ma, 1975) then suggest that sufficiently strong disorder can completely suppress such a first-order transition (Berker, 1991), eliminating the finite-temperature metal–insulator coexistence region. To illustrate this argument, consider an uncompensated doped semiconductor (Paalanen and Bhatt, 1991), where the bandwidth of the impurity band can be tuned by varying the donor concentration n. Assume that n is chosen to lie just below its critical value $n_c(T)$, so that even a small increase in n would drive the system through a Mott transition, leading to a large resistivity drop. Here, the reduced donor density $\delta n(T) = (n - n_c(T))/n_c(T)$ plays the role of the magnetic field in an ordinary ferromagnet, which can be used for $T < T_c$ to drive the system through a first-order transition where the magnetization jumps. A similar first-order

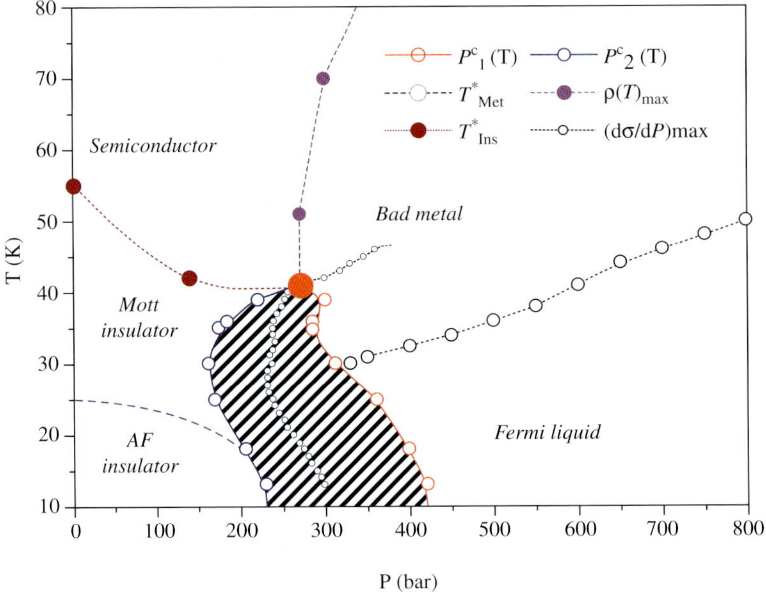

Fig. 1.7 Phase diagram of the organic salt $\kappa-(BETD\text{-}TTF)_2Cu[N(CN)_2]Cl$ (Limelette *et al.*, 2003*b*). In this material, increasing hydrostatic pressure broadens the electronic bandwith, favoring the metallic state. The first-order Mott transition extends at finite temperature up to the critical end-point at $T_c \approx 40$ K. The corresponding coexistence region (shaded) displays hysteresis in transport.

Mott metal–insulator transition is precisely what one would expect in our case as well, if only the donors were ordered with perfect periodicity. In fact, the original work of Mott (1949) considered precisely this scenario, where one imagines varying the donor concentration while ignoring the local density fluctuations.

Unfortunately, in a real material, donor ions assume random positions within the semiconducting host. A given region of size L can have a local concentration $n(L) \gtrsim n_c$, favoring the formation of a metallic "droplet." For $\delta n_L = n(L) - n_c > 0$, the energy of the metallic phase will be lower by the amount

$$\Delta E_{\mathrm{M-I}}(L) = \varepsilon \delta n_L, \tag{1.10}$$

where ε is a constant measuring the density-dependent (free) energy difference between the metal and the insulator. Creating such a droplet will create a domain wall, which costs surface energy

$$E_{\mathrm{s}} = \sigma L^{d-1}, \tag{1.11}$$

where σ is proportional to the surface tension of the droplet. The droplet will be formed only if

$$\Delta E_{\mathrm{M-I}}(L) > \sigma L^{d-1}. \tag{1.12}$$

Note, however, that $\Delta E_{\mathrm{M-I}}(L) \sim \delta n_L$, which is a random quantity. To calculate the probability that a droplet of size L will be formed, we need to calculate the probability of a density fluctuation

$$\delta n_L > \sigma L^{d-1}/\varepsilon. \tag{1.13}$$

Assuming that the donor density fluctuations are uncorrelated on large enough scales, the probability distribution is given by the central limit theorem

$$P(\delta n_L) \sim \exp\left\{-\frac{1}{2}\frac{\delta n_L^2}{L^d W^2}\right\}, \tag{1.14}$$

where W is a constant measuring density fluctuations (disorder strength) on the microscopic scale. The probability that the droplet of size L will for will, therefore, be of order

$$P(L) \sim \exp\left\{-\frac{1}{2}\frac{\sigma^2 L^{2d-2}}{\varepsilon^2 L^d W^2}\right\} = \exp\left\{-\frac{1}{2}\frac{\sigma^2 L^{d-2}}{\varepsilon^2 W^2}\right\}. \tag{1.15}$$

As we can see, for $d > 2$, large droplets are exponentially suppressed, i.e. their concentration is exponentially small, and for sufficiently weak disorder ($W \ll \sigma/\varepsilon$), even the very small droplets are exponentially rare. The first-order transition remains sharp, at least at low enough temperatures.

But what happens when the temperature is increased and we approach the critical end-point at $T = T_c$? Here, we expect the droplet surface tension to decrease as a power of the correlation length $\xi \sim (T_c - T)^{-\nu}$ (in mean-field theory $\nu = \frac{1}{2}$), so even weak disorder starts to have an appreciable effect. More precisely, here $\sigma(T) \sim \xi^{-3}$ and $\varepsilon(T) \sim \xi^{-1}$ (see, for example Goldenfeld (1992)), and even small droplets start to proliferate. The transition is then smeared down to temperatures such that $\varepsilon(T)W/\sigma(T) \sim O(1)$, i.e. $W \sim \xi^{-2} \sim (T_c - T)^{-2\nu}$. We conclude that in the presence of weak disorder, the critical temperature is depressed by

$$\delta T_c(W) = T_c(0) - T_c(W) \sim W^{1/2\nu}. \tag{1.16}$$

When disorder is sufficiently strong ($W \sim \sigma(0)/\varepsilon(0)$), the first-order jump is completely eliminated at finite temperature, and only a smooth metal–insulator crossover remains. Such behavior is clearly seen in all standard doped semiconductors, where the positional disorder in donor or acceptor ions is so strong that no evidence of a finite-temperature Mott transition can be seen. A sharp distinction between the metal and the insulator is then found only at $T = 0$, where the transition reduces to the conventional quantum critical point. Of course, the resulting critical behavior is completely different from that of a clean system, where all the conduction electrons simultaneously turn into local magnetic moments at a well-defined critical concentration n_c.

In the random case the local density undergoes strong spatial fluctuations. As a result, many regions form where the local density is much lower then the average. Here, one expects the electrons to undergo local Mott localization. In the remaining regions the local density is higher than average and the electrons remain itinerant. This simple physical picture thus suggests a two-fluid behavior (Paalanen *et al.*, 1998) of conduction electrons and local magnetic moments—a situation very different to what one expects in a weakly disordered metal. Describing such disorder-enhanced strong correlation effects theoretically is extremely difficult, since the theory must account for the effective interaction between such disorder-induced local moments and the remaining itinerant electrons. Developing such a theory should include an appropriate description of the corresponding Kondo screening processes in a disordered environment (Dobrosavljević *et al.*, 1992). This remains one of the most challenging open problems in this field, although important advances have recently been accomplished (Miranda and Dobrosavljević, 2005) based on dynamical mean-field approaches.

An especially interesting situation is found in $d = 2$, where both the bulk energy gain $\Delta E_{\mathrm{M-I}}(L) = \varepsilon\delta n_L \sim LW$, and the surface energy $E_{\mathrm{s}} = \sigma L$ both scale linearly with the droplet size L, allowing for *arbitrarily large droplets*. The typical droplet size thus diverges, and the first-order transition is suppressed for arbitrarily weak disorder. This situation may be relevant for high-mobility (weak-disorder) two-dimensional electron gases (Abrahams *et al.*, 2001). Here, behavior reminiscent of a Mott transition ($m^* \sim (n - n_{\mathrm{c}})^{-1}$) seems to emerge (Kravchenko and Sarachik, 2004) only at $T = 0$, while only a smooth metal–insulator crossover persists at finite temperatures. We should note, however, that the above expressions are valid only for sufficiently large droplets containing many impurities, such that Gaussian statistics applies. At weak disorder, the average distance between impurities ℓ is large, and the disorder can produce only droplets larger then a certain minimum size $L_{\min} \gg \ell$. We thus expect that reasonably large-scale inhomogeneities should emerge (Fig. 1.8) when weak disorder is introduced near first-order MITs in $d = 2$.

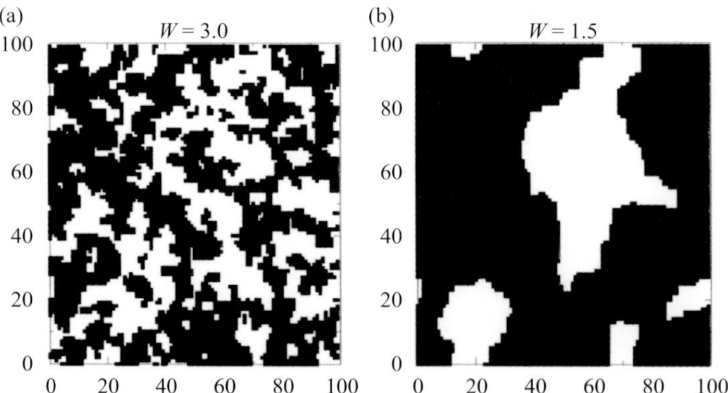

Fig. 1.8 The simplest model for disorder-induced cluster states near first-order phase transitions is provided by the random-field Ising model (Imry and Ma, 1975). An illustration is provided by recent simulation results (Moreo *et al.*, 2000), which show how in d=2, stronger disorder ($W = 3$; (a)) creates many small size clusters, while only few large ones remain for sufficiently weak disorder ($W = 1.5$; (b)). Similar behavior is expected (Imry and Ma, 1975) near any disorder-smeared first-order phase transition.

Is Wigner crystallization a Mott transition in disguise?

The original ideas of Mott (1949), who thought about doped semiconductors, envisioned electrons hopping between well-localized atomic orbitals corresponding to donor ions. In other Mott systems, such as transition-metal oxides, the electrons travels between the atomic orbitals of the appropriate transition-metal ions. In all these cases, the Coulomb repulsion restricts the occupation of such localized orbitals, leading to the Mott-insulating state, but it does not provide the essential mechanism for the formation of such tightly bound electronic states. The atomic orbitals in all these examples result from the (partially screened) ionic potential within the crystal lattice.

The situation is more interesting if one considers an idealized situation describing an interacting electron gas in the absence of any periodic (or random) lattice potential due to ions. Such a physical situation is achieved, for example, when dilute carriers are injected into a semiconductor quantum well (Ando *et al.*, 1982), where all the effects of the crystal lattice can be treated within the effective mass approximation (Ashcroft and Mermin, 1976). This picture is valid if the Fermi wavelength of the electron is much longer then the lattice spacing, and the quantum mechanical dynamics of the Bloch electron can be reduced to that of a free itinerant particle with a band mass m_b. In such situations, the only potential energy in the problem corresponds to the Coulomb repulsion E_C between the electrons, which is the dominant energy scale in low carrier density systems. At the lowest densities, $E_C \gg E_F$, and the electrons form a Wigner crystal lattice (Wigner, 1934) to minimize the Coulomb repulsion.

Here, each electron is confined not by an ionic potential, but due to the formation of a deep potential well produced by repulsion from other electrons. The same mechanism prevents double occupation of such localized orbitals, and each electron in the Wigner lattice reduces to a localized $S = \frac{1}{2}$ localized magnetic moment. A Wigner crystal is therefore nothing but a magnetic insulator: a Mott insulator in disguise. At higher densities, the Fermi energy becomes sufficiently large to overcome the Coulomb repulsion, and the Wigner lattice melts (Tanatar and Ceperley, 1989). The electrons then form a Fermi liquid. The quantum melting of a Wigner crystal is therefore an MIT, perhaps in many ways similar to a conventional Mott transition. What kind of phase transition is this? Despite years of effort, this important question is still not fully resolved.

What degrees of freedom play the leading role in destabilizing the Wigner crystal as it melts? Even in the absence of an accepted and detailed theoretical picture describing this transition, we may immediately identify two possible classes of elementary excitations that potentially contribute to melting, as follows.

- *Collective charge excitations* ("elastic" deformations) of the Wigner crystal. In the quantum limit, these excitations have a bosonic character, but they persist and play an important role even in the semi-classical ($k_B T \gg E_F$) limit, where they contribute to the thermal melting of the Wigner lattice (Thouless, 1978). These excitations clearly dominate in high magnetic field (Chen *et al.*, 2006), where both the spin degrees of freedom and the kinetic energy are suppressed due to Landau quantization.

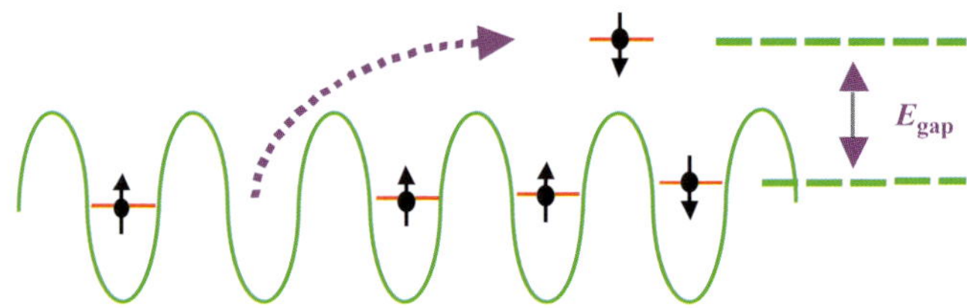

Fig. 1.9 In a Wigner crystal, each electron is confined to a potential well produced by Coulomb repulsion from neighboring electrons, forming a spin-$\frac{1}{2}$ local moment. The lowest-energy particle–hole excitation creates a vacancy–interstitial pair (Cândido *et al.*, 2001), which costs an energy E_{gap} comparable to the Coulomb repulsion.

- *Particle–hole excitations* leading to vacancy–interstitial pair formation (Fig. 1.9). These excitations have a fermionic character, where the spin degrees of freedom play an important role. Virtual excitations of this type give rise to superexchange processes that produce magnetic correlations (Cândido *et al.*, 2004) within the Wigner crystal. These excitations would exist even if dynamic deformations of the Wigner lattice were suppressed, for example by impurity pinning. Recent quantum Monte Carlo simulations indicate (Cândido *et al.*, 2001) that the effective gap for vacancy–interstitial pair formation seems to collapse precisely around the melting of the Wigner crystal. If these excitations dominate, then the melting of the Wigner crystal is a process very similar to the Mott MIT, and may be expected to produce a strongly correlated ($m^*/m \gg 1$) Fermi liquid on the metallic side. Behavior consistent with this possibility has recently been documented (Kravchenko and Sarachik, 2004) in several two-dimensional electron systems. A microscopic theory for a simplified model of such Wigner–Mott transitions has recently been solved (Amaricci *et al.*, 2010; Camjayi *et al.*, 2008), clarifying the mechanism for the effective mass enhancement on the conducting side.

We should mention, however, that the Wigner crystal melting in zero magnetic field is believed (Tanatar and Ceperley, 1989) to be a weakly first-order phase transition. Conventional (e.g. liquid–gas or liquid–crystal) first-order transitions are normally associated with a density discontinuity and global phase separation within the coexistence dome. For charged systems, however, global phase separation is precluded by charge neutrality (Gor'kov and Sokol, 1987). In this case, one may expect the emergence of various modulated intermediate phases, leading to bubble or stripe (Jamei *et al.*, 2005), or possibly even "stripe glass" (Schmalian and Wolynes, 2000) order. While convincing evidence for the relevance of such "nano-scale phase separation" has been identified (Terletska and Dobrosavljević, 2011) in certain systems (Jaroszyński *et al.*, 2007), recent work seems to indicate (Clark *et al.*, 2009; Waintal, 2006) that such effects may be negligibly small for Wigner crystal melting.

1.2.3 Localization by disorder

A small concentration of impurities or defects simply produces random scattering of mobile electrons. In ordinary metals, the kinetic energy of electrons is generally so large that the random potential due to impurities can be treated as a small perturbation. In this case, the Drude theory (Ashcroft and Mermin, 1976) applies, where the conductivity takes the form

$$\sigma \approx \sigma_o = \frac{ne^2\tau_{\rm tr}}{m}, \tag{1.17}$$

where n is the carrier concentration, e the electron charge and m its band mass. According to Matthiessen's rule (Ashcroft and Mermin, 1976), the transport scattering rate takes additive contributions from different scattering channels, namely

$$\tau_{\rm tr}^{-1} = \tau_{\rm el}^{-1} + \tau_{\rm ee}^{-1}(T) + \tau_{\rm ep}^{-1}(T) + \cdots . \tag{1.18}$$

Here, $\tau_{\rm el}^{-1}$ is the elastic scattering rate (describing impurity scattering), and $\tau_{\rm ee}^{-1}(T)$, $\tau_{\rm ep}^{-1}(T), \ldots,$ describe inelastic scattering processes from electrons, phonons, etc. It is important to note that in this picture the resistivity $\rho = \sigma^{-1}$ is strictly a monotonically increasing function of temperature

$$\rho(T) \approx \rho_o + AT^n, \tag{1.19}$$

where $A > 0$, and the exponent n depends on the scattering process ($n = 1$ for electron-phonon scattering; $n = 2$ for electron–electron scattering, etc.). The residual resistivity $\rho_o = \sigma^{-1}(T = 0)$ is thus viewed as a measure of impurity (elastic) scattering.

In contrast, in low-carrier-density systems, the impurity potential is comparable or larger then the Fermi energy, and the electrons can get trapped, i.e. "localized", by the impurities. Of course, this process generally leads to a sharp MIT only at $T = 0$, since at finite temperature the electrons can overcome the impurity binding potential through thermal activation. In the low-temperature limit, a continuous metal—insulator transition is typically found (Rosenbaum *et al.*, 1980), where the conductivity decreases in a power-law fashion

$$\sigma(T = 0) \sim (n - n_{\rm c})^{\mu},$$

where the conductivity exponent μ characterizes the critical point. The precise value of the critical exponent μ, as well as other aspects of the transition depend strongly on the specific form of disorder, and especially on the characteristic length scale for disorder.

Percolation transition

The simplest situation where disorder can cause electron localization is found in cases where the random potential is sufficiently "smooth." More precisely, consider the situation where the spatial correlation length for the random potential is much larger then the phase coherence length L_ϕ (Lee and Ramakrishnan, 1985).

In this regime, the influence of the random potential can be described in a semi-classical picture, where quantum interference processes can largely be ignored, and the calculation of the resistivity reduces to solving a percolation problem (Stauffer and Aharoni, 1994) describing a random resistor network.

As the electrons are added to the system, the electron liquid will first fill the deepest potential wells, thus forming small metallic "puddles." At low temperature, regions where the random potential $V(\mathbf{r}) > E_F$ is essentially free of electrons and therefore represent insulating areas. When the electron density (and thus E_F) increases, the metallic puddles grow and eventually connect at the percolation threshold; the system becomes metallic.

The critical behavior of the conductivity within such a percolation approach has been studied in detail (Stauffer and Aharoni, 1994), giving

$$\mu \approx \begin{cases} 2.0 \text{ for } d = 3, \\ 1.3 \text{ for } d = 2. \end{cases} \tag{1.20}$$

Note (see Fig. 1.10) that such dimensionality-dependent critical behavior is seen only within a narrow critical region. For $d = 3$, for example, the conductance behavior is rougly linear, in agreement with the prediction of the effective-medium theory (Kirkpatrick, 1973), which does not capture such dimensionality dependence. This theory, ilke any mean-field description, works equally well in any dimension—everywhere except within a narrow critical region, where long-wavelength fluctuations produce dimensionality-dependent behavior.

The percolation theory predictions can be directly compared to experiments, and agreement is found in a number of systems such as granular metals (Beloborodov

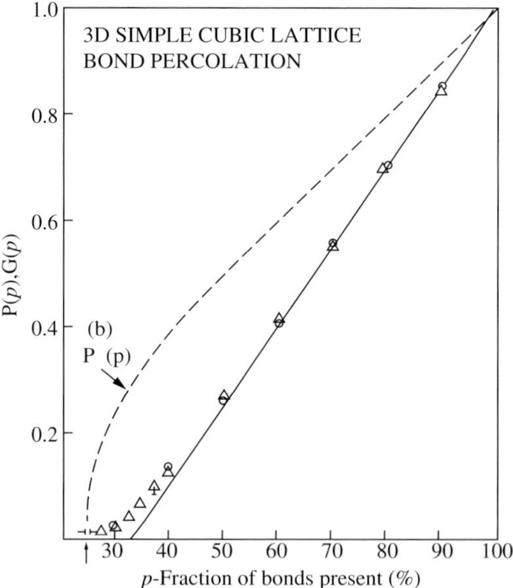

Fig. 1.10 Critical behavior at the bond percolation for a three-dimensional cubic lattice, following (Kirkpatrick, 1973). Percolation probability(dashed line) $P(p)$ and conductance (symbols) $G(p)$ are shown as a function of the bond probability p. The solid line is the prediction of an (approximate) effective medium theory.

et al., 2007), where the characteristic inhomogeneities scale is sufficiently large. The percolation scenario seems also to apply (Shahar *et al.*, 1997) to certain experiments showing apparent violation of the expected quantum critical scaling in quantum Hall plateau transitions. Similar behavior is found, for example, in manganese oxide materials showing colossal magnetoresistance, where disorder induces nano-scale phase separation (Dagotto, 2002), but the transport behavior can be well described using an effective random resistor model and the underlying percolation processes. Here the droplets sizes are not necessarily very large, but the temperatures are sufficiently elevated to produce sufficiently short L_ϕ. Such percolative phase coexistence has very recently been observed by nano-scale X-ray imaging on a thermally-driven Mott transition in VO_2 (Qazilbash *et al.*, 2011).

In other systems, most notably uncompensated doped semiconductors, low-temperature studies find $\mu \approx 0.5$, indicating drastic departure from the percolation picture. Can a better description of quantum effects provide the answer? Or does one have to include strong correlation effects as well? To address this question, it is useful to first discuss the prediction of a full quantum theory of localization for non-interacting electrons in the presence of disorder.

Tunneling versus localization

The semiclassical percolation picture assumes that in the insulating phase the electrons are confined to potential wells. It ignores the possibility of tunneling through barriers separating the wells—a process that is allowed by quantum mechanics. Although the tunneling probability is exponentially small with the barrier size, it always remains finite. Naively, one would expect that the tunneling processes would lead to an exponentially small but finite conductivity even in the regime where classical percolation would predict insulating behavior. According to this argument, quantum mechanical tunneling would smear the metal–insulator transition, and no true insulating behavior would be possible, even at $T = 0$. This argument is, of course, incorrect because is based on an incomplete description of quantum mechanics. It does take into account the tunneling effect, but it ignores the crucial interference processes without which electronic bound states could not be formed. To fully appreciate this, we should recall that interference processes are what gives rise to the formation of quantized electronic orbits even for simple bound states within atoms or molecules.

Anderson localization

The possibility that true electronic bound states can be formed in the presence of a random potential was first discussed by Anderson in 1958 (Anderson, 1958). This pioneering work argued that sufficiently strong randomness will localize all the electronic states within a given band, leading to a sharp metal–insulator the transition at $T = 0$. Anderson's original argument takes the local point of view, which provides a very transparent physical picture of how localization can occur, as follows.

Suppose that an experiment or a computer simulation can examine only local quantities associated with a particular lattice site. Can such a study determine whether the material is a metal or an Anderson insulator? The answer (Anderson, 1958) is, somewhat surprisingly, yes. One simply has to determine the escape rate $\hbar/\tau_{\text{esc}}$ from the local orbital. The recipe how to do this has a long history, and is provided by Fermi's Golden Rule:

$$\frac{\hbar}{\tau_{\text{esc}}} = t^2 \rho_{\text{loc}}(\varepsilon). \tag{1.21}$$

where t is the appropriate matrix element between the local orbital and its environment. The local density of available states $\rho_{\text{loc}}(\varepsilon)$, to which an electron of energy ε can escape, is proportional to the wavefunction amplitude on this site

$$\rho_{\text{loc}}(\varepsilon) \sim |\psi_i(\varepsilon)|^2. \tag{1.22}$$

If electronic states in the relevant energy range are all localized (i.e. bound), then only a small number of such states have appreciable overlap with the given site. Therefore, in the case of the Anderson insulator, the local density of states (LDOS) will consist of only a few discrete δ-function peaks (Fig. 1.11) with appreciable weight. These are very improbable at the Fermi energy.

At strong disorder $W \gg B$, the typical density of available states and thus the *typical escape rate* from a given site can be shown to vanish (Anderson, 1958), and the electron remains localized. In the Anderson insulator, the spectrum of the local environment of any given site has a gap with a certain typical size measuring the localization strength. According to this point of view, the ultimate localization mechanism in both the Mott and the Anderson insulator is somewhat similar: the electron cannot find levels to which it can escape.

Anderson's original arguments demonstrated that sufficiently strong disorder is able to localize all electronic states within a narrow band of electronic states. At

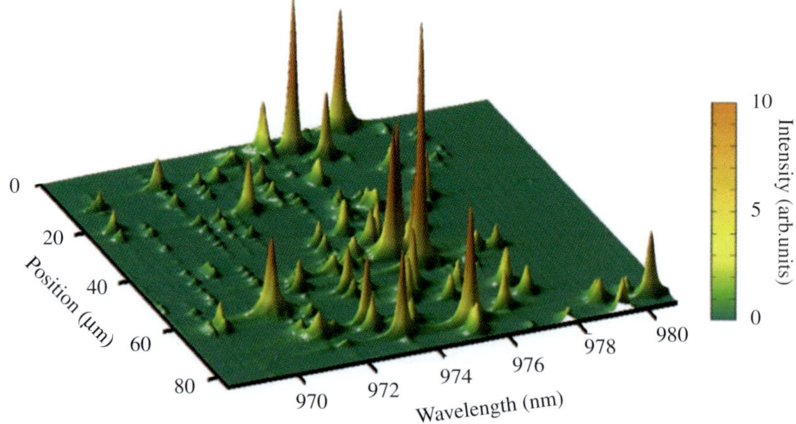

Fig. 1.11 Wavefunction amplitude $|\psi_i|^2$ of strongly localized states. Since the concept of Anderson localization applies to any wave in random media, recent efforts have documented localization of light in disordered optical media. Shown here are spectra experimentally observed in photonic crystals (Sapienza *et al.*, 2010).

weaker disorder, only the states near the band edge are expected to localize, but other states at energies $E > E_c$—the so-called "mobility edge"—remain extended. As the Fermi energy is increased (for example by carrier doping), the system undergoes an Anderson MIT. Following the development of the scaling theories of localization (Abrahams *et al.*, 1979), and especially due to recent progress in numerical studies of the problem, the corresponding critical behavior is now well understood for non-interacting electrons. These studies established (see also Chapter 3) that for non-interacting electrons at $T = 0$, all the electronic states remain localized for dimensions $d \leq 2$, while a continuous MIT is found in higher dimensions. According to most recent estimates, the corresponding critical exponent

$$\mu \approx 1.58 \tag{1.23}$$

for $d = 3$, and becomes even larger in higher dimensions (see below).

It is interesting to note that the conductivity exponent at the $d = 3$ Anderson transition is not too different from that of percolation theory ($\mu_{\text{perc}} \approx 2$). In most experimental systems where the MIT has been studied, the observed exponent is much smaller ($\mu \lesssim 1$). This indicates that disorder-driven mechanisms that ignore electron–electron interactions cannot hope to explain the behavior at the MIT.

1.2.4 Basic interaction effects in disordered systems

In most realistic systems both the disordered strength and the electron–electron interactions have comparable magnitudes. In such cases, the localization mechanisms of Mott and Anderson cannot be considered separately, since each will influence and affect the other. Despite recent advances (see Chapter 6), a complete theory of such a Mott–Anderson transition remains incomplete. Nevertheless, the early physical arguments of Mott (1990) already indicate how new and more complicated behavior must emerge when both mechanisms are at play.

The Mott–Anderson transition

To see this most simply, consider a disordered Hubbard model in the strongly localized (atomic) limit. In the absence of disorder, each site has two energy levels, $\varepsilon_0 = 0$ and $\varepsilon_1 = U$, where U is the on-site interaction potential. If the system is half-filled, then each site is singly occupied; the levels ε_1 remain empty. We have one local magnetic moment at each site, and a gap equal to U to charge excitations.

When disorder is added, each of these energy levels is shifted by a randomly fluctuating site energy $-W/2 < \varepsilon_i < W/2$. The situation remains unchanged for $W < U$, as all the levels $\varepsilon_1'(i) = U + \varepsilon_i$ remain empty (for half-filling the chemical potential is $\mu = E_F = U/2$). For larger disorder, those sites with $\varepsilon_i > U/2$ have the level $\varepsilon_0'(i) = 0 + \varepsilon_i > \mu$ and are empty. Similarly, those sites with $\varepsilon_i < -U/2$ have the excited level $\varepsilon_1'(i) = U + \varepsilon_i < \mu$ and are doubly occupied. Thus for $W > U$ a fraction of the sites are either doubly occupied or empty. The Mott gap is now closed, although a fraction of the sites still remain as localized magnetic moments. We can describe this state as an inhomogeneous mixture of a Mott and an Anderson insulator (as shown in

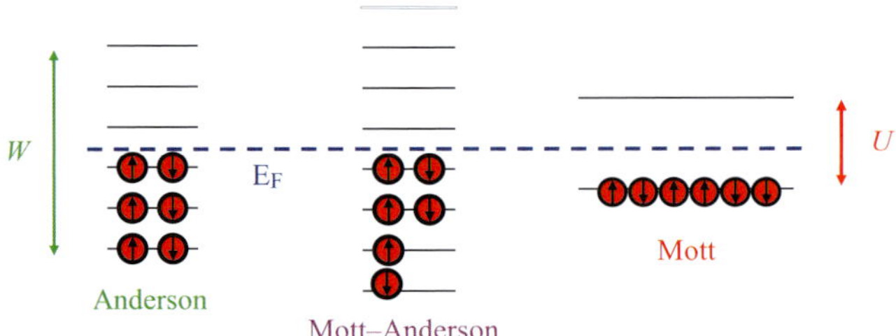

Fig. 1.12 Energy level occupation in the strongly localized (atomic) limit for an Anderson (left), a Mott (right), and a Mott–Anderson (center) insulator. In a Mott–Anderson insulator, the disorder strength W is comparable to the Coulomb repulsion U, and a two-fluid behavior emerges. Here, a fraction of localized states are doubly occupied or empty as in an Anderson insulator. Coexisting with these, other states remain singly occupied, forming local magnetic moments, as in a Mott insulator. Note that the spins of the local moments may be randomly oriented indicating the absence of magnetic ordering. The chemical potential is represented by the dashed line.

the center of Fig. 1.12). However, the empty and doubly occupied sites have succeeded in completely filling the gap in the average single-particle density of states (DOS).

This physical picture of Mott, which is schematically represented in Fig. 1.12, is very transparent and intuitive. The non-trivial question is how the strongly localized (atomic) limit is approached as one crosses the MIT from the metallic side. To address this question one needs a more detailed theory for the MIT region, which was not available when the questions posed by Mott and Anderson were put forward.

Coulomb gap

Local moment formation leading to the Mott or the Mott–Anderson insulating state is most important for narrow bands, where the on-site Coulomb repulsion ("Hubbard U") dominates. This mechanism is most effective close to half-filling, since local moment formation requires exactly one electron per orbital. In this regime the long-range (inter-site) component of the Coulomb interaction plays a secondary role, because on-site repulsion opposes charge rearrangement. Such a situation is found in narrow impurity bands (deeply insulating regime) of uncompensated doped semiconductors such as Si:P. Deep in the insulating regime, each electron forms a hydrogenic bound state with exactly one phosphorus ion, forming a spin $S = \frac{1}{2}$ local magnetic moment, and charge rearrangements are suppressed.

A more complicated situation is found (Shklovskii and Efros, 1984) away from half filling, which can be realized, for example, in partially compensated Si:P,B. Here the electrons can occupy different localized states, and many charge rearrangements are possible. This is the regime considered by the well-known theory of Efros and Shklovskii (1975; 1984), which focuses on a classical model of spinless electrons distributed among strongly localized states, as given by the Hamiltonian

$$H = \sum_{j \neq i} \frac{e^2}{\kappa |\mathbf{r_i} - \mathbf{r_j}|}(n_j - \overline{n})(n_j - \overline{n}) + \sum_{j \neq i} \varepsilon_i n_i. \tag{1.24}$$

Here $n_j = 0, 1$ is the occupation number of the remaining localized states with (bare) energy ε_i at position $\mathbf{r}_i$, $\overline{n}$ is the average occupation per site, and κ is the dielectric constant of the insulator. For localized electrons, the single-particle (tunneling) DOS is then simply the probability distribution of the local energy levels

$$N(\varepsilon) =< \delta(\varepsilon - \varepsilon_i) > \tag{1.25}$$

of an electron occupying a localized state shifted (renormalized) by the electrostaic potential produced by the electrons on all remaining sites.

$$\varepsilon_i \to \varepsilon_i^R = \varepsilon_i + \sum_{j \neq i} \frac{e^2}{\kappa |\mathbf{r_i} - \mathbf{r_j}|}(n_j - n). \tag{1.26}$$

This electrostatic shift depends, of course, on the precise electronic configuration, favoring those charge configurations that lower the energies of the occupied states. In absence of disorder, this effect leads to charge ordering—Wigner crystallization—opening a hard gap at the Fermi energy.

When sufficiently strong disorder is present, the Wigner gap is smeared, leading to the soft "Coulomb gap" (Fig. 1.13). The original argument of Efros and Shklovskii (1975) rested on a stability argument for the ground state with respect to any single-particle displacement from a given occupied site i to an empty site j. Stability requires every such excitation to cost positive energy, giving

$$\varepsilon_j - \varepsilon_i - \frac{e^2}{\kappa |\mathbf{r_i} - \mathbf{r_j}|} > 0. \tag{1.27}$$

If the states i and j are within energy ε from the Fermi level, i.e. $|\varepsilon_i - \varepsilon_{\mathrm{F}}| < \varepsilon$, then their typical separation in space is $r(\varepsilon) = e^2/\kappa\varepsilon$ is large if ε is small. The DOS $N(\varepsilon) \sim \frac{d}{d\varepsilon}[r^{-3}(\varepsilon)]$ then has a soft gap around the Fermi energy. In three dimensions the result is

$$N(\varepsilon) = \frac{3\kappa^3}{\pi e^6}(\varepsilon - \varepsilon_{\mathrm{F}})^2. \tag{1.28}$$

More generally, for interactions of the form $V(r) \sim r^{-\alpha}$, with $\alpha < d$ (in d spatial dimensions), a generalization of this argument gives (Pankov and Dobrosavljević, 2005) $N(\varepsilon) \sim (\varepsilon - \varepsilon_{\mathrm{F}})^{\beta}$, with $\beta = (d - \alpha)/\alpha$. Arguing that incomplete (scale-dependent) screening modifies the form of the Coulomb interaction within the quantum critical region of the MIT, recent work suggested (Lee *et al.*, 1999) that in this case $\alpha = 2$, leading to $\beta = \frac{1}{2}$ in $d = 3$. While such critical scaling seems consistent with the experimental results on Si:B, a more precise treatment of quantum effects will be needed for a more convincing theory of the Coulomb gap in the quantum critical region.

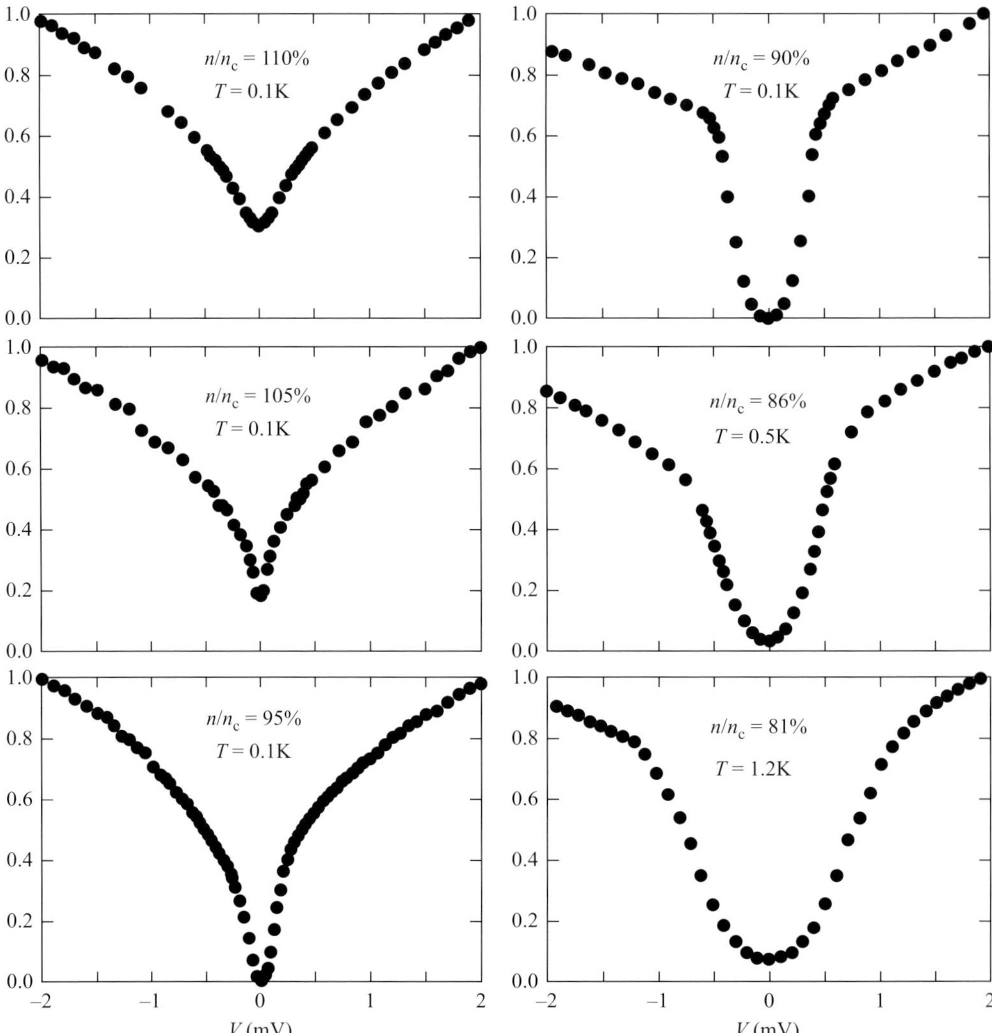

Fig. 1.13 Tunneling density of states spectra observed (Lee *et al.*, 1999) across the metal–insulator transition in Si:B. The Coulomb gap is seen to gradually close and change shape as the transition is crossed. This crossover behavior was interpreted to reflect the emergence of screening as one approaches the metallic phase.

It is very important to reiterate that the Anderson localization mechanism by itself does not lead to opening of any kind of gap at the transition. This scenario, which provided a popular and attractive scenario for interpreting many transport experiments, rests on the concept of a "mobility edge"—the energy separating the extended from the localized states. The critical concentration then obtains when all the localized states are filled up and the Fermi energy reaches the mobility edge. Precisely at the critical point one expects (Mott, 1990) the electronic states "above" the Fermi energy to be extended, while those "below" the mobility edge to remain localized. A test of these ideas has very recently become possible through high-resolution scanning tunneling microscopy (STM) experiments (Richardella *et al.*, 2010) which can directly determine the degree of wavefunction localization in an energy-resolved fashion. In contrast to the conventional mobility edge scenario, these experiments provided striking evidence that, close to the critical concentration, the electronic states precisely at the Fermi energy are the ones most strongly localized. This experiment, which will be discussed in more detail in Chapter 7, provided direct evidence that interaction effects cannot be neglected near the MIT, where pseudogap opening plays a key role in controlling the localization of electronic states. Is the relevant interaction mechanism directly related to Coulomb glass phenomena? Only time will tell. Still, all these experiments make it clear that the fundamental—but yet unresolved—physics questions posed by the early work of Mott cannot be ignored.

Coulomb glass

The long-range Coulomb interactions produce, however, another important effect. Because Coulomb repulsion favors a uniform charge configuration while disorder opposes it, these competing interactions give rise to frustration and the emergence of many meta-stable electronic states. As described in Chapter 8, this typically leads to gradual freezing of electrons, and to slow relaxation and aging, in a fashion surprisingly similar to glassy phenomena in spin glasses or super-cooled liquids.

The precise relation of the Coulomb gap formation and the glassy freezing has, however, long remained controversial and ill-understood (Grannan and Yu, 1993). On the one hand, even the early arguments of Efros and Shklovskii indicated that the long-range nature of the Coulomb interaction was a key feature for the formation of the Coulomb gap. On the other, thermal and/or quantum fluctuations can allow charge rearrangements, generically leading to the phenomenon of screening, practically eliminating the long-range part of the interaction. In contrast, if the electrons are (partially or fully) frozen in a glassy state, then screening may remain incomplete, allowing the long-range nature of the Coulomb interaction to manifest itself. This physically plausible idea (Lee and Ramakrishnan, 1985) has found support in very recent theoretical work (Pankov and Dobrosavljević, 2005; Pastor and Dobrosavljević, 1999; Pastor *et al.*, 2001), which argues that the two phenomena typically go hand-in-hand.

Both the formation of the Coulomb gap and the emergence of glassy features are, at this time, well established features in strongly disordered insulators. But how should this influence the approach to the MIT? Recent theoretical (Dobrosavljević *et al.*, 2003*b*; Pastor and Dobrosavljević, 1999) and experimental (Bogdanovich and Popović, 2002; Jaroszynski *et al.*, 2004) work has suggested that in some cases it may dramatically affect the critical region, perhaps even leading to an intermediate metallic-glass phase with unusual transport properties (Dalidovich and Dobrosavljević, 2002). This important question remains far from settled. Still, its fundamental importance was recognized early on by Mott (1990), who noted that the phenomenon of screening must be dramatically modified as one crosses from a metal to an insulator. In short, the localization of electrons immediately produces the demise of screening, so the two phenomena must be profoundly linked. Mott's dream was to understand how this "unscreening" occurs at the MIT, a physical question of basic importance, but one that is typically not addressed by most conventional theories, thus remaining a major challenge for future work.

1.3 Current theories of the metal–insulator transition

1.3.1 The metal–insulator transition as a critical point

Absence of minimum metallic conductivity

The early ideas of Mott and Anderson identified the basic mechanism for the MIT, but did not provide specific and detailed predictions for the critical behavior, or even the precise nature of this phase transition. In fact, up to the late 1970s, Mott's arguments (Mott, 1990) suggested that the transition was discontinuous, where a *minimum metallic conductivity* should exist on the metallic side even at $T = 0$. Mott's early argument examined the transport behavior based on Drude's picture, where increasing disorder simply reduces the elastic mean-free path ℓ. Since the scattering rate $\tau_{\rm tr}^{-1}$ from any impurity assumes a maximum possible value that is finite (the so-called "unitarity limit"), the corresponding mean-free path $\ell = v_{\rm F}\tau_{\rm tr}$ cannot be shorter then a microscopic lower cutoff a of the order of the lattice spacing. Therefore, it was argued, the conductivity of any metal is bounded from below by the "Mott limit"

$$\sigma \geq \sigma_{\rm min} = \frac{ne^2 a}{m v_{\rm F}}.$$

Early low-temperature experiments on many materials seemed to confirm these predictions by only finding metallic conductivities in excess of $\sigma_{\rm min}$. The MIT in disordered systems was thus assumed to have a first-order character, similar to the Mott transition in clean systems. Similarly, high-temperature behavior in a number of metallic systems was found to display "resistivity saturation," where the Mott limit is approached due to incoherent (inelastic) scattering.

With the development of more advance cryogenic techniques, lower temperatures and more precise measurements became available. Our perspective on the fundamental nature of the transition has been deeply influenced by the ground-breaking experiments on Si:P in early 1980s (Paalanen *et al.*, 1982; Rosenbaum *et al.*, 1980). These experiments provided evidence (Fig. 1.14) of metallic conductivities as much as two orders of magnitude smaller then $\sigma_{\rm min}$. They also made it clear that the MIT in doped semiconductors is a continuous (second-order) phase transition (Paalanen *et al.*, 1982), which bears many similarities to conventional critical phenomena. This important observation has sparked a veritable avalanche of experimental (Paalanen and Bhatt, 1991; Sarachik, 1995) and theoretical (Abrahams *et al.*, 1979; Schaffer and Wegner, 1980; Wegner, 1976, 1979) studies, most of which have borrowed ideas from studies of second-order phase transitions. Indeed, many experimental results were interpreted

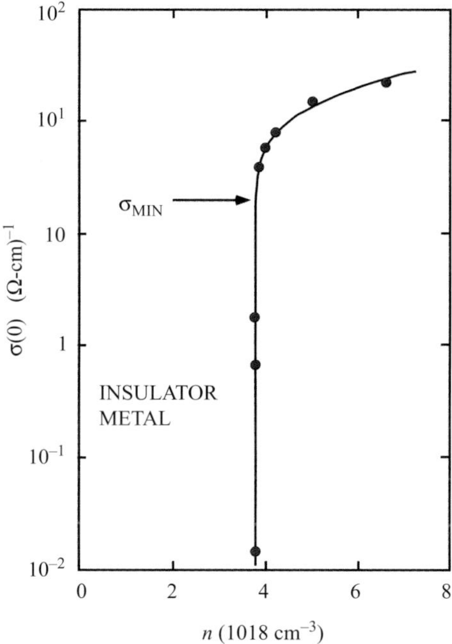

Fig. 1.14 Critical behavior of the conductivity extrapolated to $T \to 0$ for uncompensated Si:P (Rosenbaum *et al.*, 1980). Sharp power-law behavior with critical exponent $\mu \approx \frac{1}{2}$ extends over a surprisingly large concentration range. Finite values of the conductivity much smaller then σ_{M} (shown by arrow) are observed close to the transition.

using scaling concepts (Lee and Ramakrishnan, 1985), culminating in the famed scaling theory of localization (Abrahams *et al.*, 1979) and the subsequent extensions to incorporate the interaction effects (Belitz and Kirkpatrick, 1994; Castellani *et al.*, 1984; Finkel'stein, 1983, 1984).

Phenomenological scaling formulation

Microscopic theories describing the MIT remain controversial and somewhat incomplete. We should stress, however, that scaling behavior near second-order phase transitions is a much more robust and general property then any particular approximation scheme or microscopic model. Historically, the scaling hypothesis of Widom (1965) was put forward for conventional (classical) critical phenomena long before the microscopic theory of Wilson and Kogut (1975) became available. It has provided crucial guidance for experimentalists to systematically analyze the experimental data, and has provided a framework and direction for the development of microscopic theories.

A phenomenological scaling hypothesis can be formulated for quantum criticality (Sachdev, 2011) as well, in direct analogy to conventional critical phenomena. In particular, if the scaling description is valid, then a single correlation length $\xi \sim \delta n^{-\nu}$ exists, characterizing the system and the corresponding timescale $\tau_\xi \sim \xi^z$, both of which diverge in a power-law fashion at the critical point. Here, $\delta n = (n - n_c)/n_c$ is the dimensionless distance from the transition, and we have introduced the correlation length exponent ν and the "dynamical exponent" z. Because of the Heisenberg uncertainty principle, the corresponding energy (temperature) scale

$$T^* \sim \frac{\hbar}{\tau_\xi} \sim \delta n^{\nu z}$$

vanishes as the critical point is approached.

For the MIT the conductivity plays a role similar to an order parameter, as it vanishes at $T \longrightarrow 0$ in the insulating phase. Its sharp critical behavior at $T = 0$ is rounded at finite temperature, suggesting a scaling behavior similar to that of a ferromagnet in an external (symmetry-breaking) field. The conductivity can therefore be written in a scale-invariant form as

$$\sigma(\delta n, T) = b^{-\mu/\nu} f_\sigma \left(b^{1/\nu}\delta n, b^z T\right).$$

Here, T is the temperature, b is the length rescaling factor, and μ is the conductivity exponent.

The $T = 0$ behavior $\sigma\,(T = 0) \sim \delta n^\mu$ can be obtained by working at low temperatures and choosing $b = \delta n^{-\nu} \sim \xi$. We obtain the scaling form

$$\sigma(T) = \delta n^\mu \widetilde{\phi}_\sigma(T/\delta n^{\nu z}), \tag{1.29}$$

where $\widetilde{\phi}_\sigma(y) = f_\sigma(1, y)$. Finite-temperature corrections in the metallic phase are obtained by expanding

$$\widetilde{\phi}_\sigma(y) \approx 1 + a y^\alpha, \tag{1.30}$$

giving the low-temperature conductivity of the form

$$\sigma(\delta n, T) \approx \sigma_o(\delta n) + m_\sigma(\delta n) T^\alpha. \tag{1.31}$$

Here, $\sigma_o(\delta n) \sim \delta n^\mu$ and $m_\sigma(\delta n) \sim \delta n^{\mu - \alpha\nu z}$. Since the *form* of the scaling function $\phi_\sigma(y)$ is independent of the distance to the transition δn, the exponent α must take an universal value in the entire metallic phase, and therefore can be calculated by perturbation theory at weak disorder. For example, the interaction corrections in $d = 3$ (Lee and Ramakrishnan, 1985) lead to $\alpha = \frac{1}{2}$.

This scaling argument provides a formal justification for using the predictions from perturbative quantum corrections as giving the leading low-temperature dependence in the entire metallic phase. Note, however, that the prefactor $m_\sigma(\delta n)$ (i.e. its dependence on δn) is not correctly predicted by perturbative calculations, since it undergoes Fermi-liquid renormalizations that can acquire a singular form in the critical region near the MIT.

The temperature dependence at the critical point (in the critical region) can be obtained if we put $\delta n = 0$, and choose $b = T^{-1/z}$, giving

$$\sigma_c(T) = \sigma(\delta n = 0, T) \sim T^{\mu/\nu z}. \tag{1.32}$$

We can also write

$$\sigma(n,T) = T^{\mu/\nu z}\phi_\sigma(T/T_o(n)), \tag{1.33}$$

where $\phi_\sigma(x) = x^{\mu/\nu z}\widetilde{\phi}_\sigma(x)$, and the crossover temperature $T_o(\delta n) \sim \delta n^{\nu z}$.

How to experimentally find quantum criticality?

To demonstrate quantum critical scaling around a MIT one must adopts the following systematic procedure in analyzing experimental data:

1. Plot $\sigma(\delta n, T)$ vs T for several carrier concentrations n. Simple power-law behavior $\sigma_c \sim T^x$ is expected only at the critical point, and we can identify the critical concentration ($n = n_c$) as the only curve that looks like a straight line when the data are plotted on a log–log scale.
2. From the slope of the $\sigma_c(T)$ we find the critical exponent $x = \mu/\nu z$.
3. Having determined $\sigma_c(T) = \sigma(0,T)$ we can now plot $\phi_\sigma(T/T_o(\delta n)) = \sigma(\delta n, T)/\sigma_c(T)$ as a function of $T/T_o(\delta n)$. The crossover temperature is determined for each concentration n in order to collapse all the curves on two branches (metallic and insulating) of the scaling function $\phi_\sigma(y)$. This procedure does not assume any particular functional form (density dependence) for the crossover scale $T_o(\delta n)$.
4. Next, we plot $T_o(\delta n)$ as a function of δn on a log–log scale to determine the corresponding exponent νz. If scaling works, then $T_o(\delta n)$ should vanish as the transition is approached, and we expect to find the same exponent from both sides. The conductivity exponent is then obtained from $\mu = x\nu z$.
5. A crosscheck can be obtained from extrapolating the metallic curves $\sigma(\delta n, T) \longrightarrow \sigma_o(\delta n)$ to $T \longrightarrow 0$, and by determining the exponent μ from the relation $\sigma_o(n) \sim \delta n^{\mu}$.

All the above expressions are quite general, and can be considered to be a phenomenological description of the MIT. This is of particular importance in instances where a scaling approach is utilized to systematically analyze the experimental data in instances where an accepted microscopic theory is not available. Such a situation is found in several two-dimensional systems, where beautiful and convincing scaling behavior is observed, providing evidence that the MIT is a well-defined quantum critical point.

How to theoretically approach quantum criticality?

To better understand the *physical content* of the scaling approach to the MIT, we should contrast it to standard approaches to critical phenomena, which are by now perfectly well understood. In most cases, the theory is built based on the emergence of spontaneous symmetry-breaking within the low-temperature (ordered) phase. Based on identifying the appropriate order parameter φ describing such symmetry-breaking, one typically proceeds in the following steps:

1. Formulate an appropriate Landau theory, which defines how the free energy $F[\varphi]$ depends on the (spatially fluctuating) order parameter $\varphi(\mathbf{x})$.

2. Mean-field theory (MFT) is obtained by minimizing $F[\varphi]$ and ignoring the spatial fluctuations of $\varphi(\mathbf{x})$.
3. Examine the effects of long-wavelength spatial fluctuations of the order parameter beyond MFT.
4. The most singular effects of spatial fluctuations are found close to critical points, and are re-summed using renormalization group (RG) methods.

When this program is implemented in practice, one finds that the mean-field description (steps 1 and 2) suffices everywhere except in a very narrow interval around the critical point. In fact, it is precisely by examining the leading corrections to MFT one is able to theoretically estimate the size of the so-called "Ginzburg" region where non-MFT behavior can be observed. Accounting for non-MFT behavior within such a critical region is much more difficult, and requires the powerful arsenal of RG methods (Goldenfeld, 1992). The practical calculations simplify considerably near the upper critical dimension d_{uc}, where the fluctuations corrections are logaritmically weak, and can be effectively re-summed by a perturbative RG methods, using $\varepsilon = d_{\text{uc}} - d$ as a small parameter in the theory (Wilson and Fisher, 1972). For standard magnetic critical phenomena $d_{\text{uc}} = 4$, this RG program has been effectively implemented, and all the appropriate critical exponents calculated to leading order in $\varepsilon = 4 - d$. When the results are extrapolated to $d = 3$, very impressive agreement with both experiments and numerical simulation results has obtained. The theory of conventional (thermal) critical phenomena can therefore be considered a closed book.

What to do if an order-parameter description is not available?

We emphasize that the above "standard" approach to criticality relies on being able to identify an appropriate symmetry-breaking scheme. For phenomena such as magnetic- and charge-ordering or superconductivity, this program can be straightforwardly extended to the quantum domain, as first discussed by Hertz (1976). In such cases, especially for insulating magnets, the familiar approach of Landau theory and weak-coupling RG methods has been studied in detail, and met some success in describing quantum criticality (Sachdev, 2011).

When it comes to the MIT, the situation is more complicated. Here, despite convincing experimental evidence for criticality, the conventional approach cannot be applied directly. The fundamental difficulty lies in the absence of an obvious symmetry-breaking scheme needed to build a Landau theory. For this reason, a simplistic mean-field description of the MIT is not readily available, and one is forced to look for alternative approaches.

A clue on how this may be possible is again found by analogy to conventional critical phenomena, where one generally expects the fluctuations to increase in importance in low dimensions. Their effects are particularly strong near the lower critical dimension (LCD) d_{lc}, where they are able to completely suppress the ordering. These phenomena allow for a particularly elegant approach in systems with continuous broken symmetry where $d_{\text{lc}} = 2$. Here, an alternative perturbative RG treatment based on an $\varepsilon = d - 2$ expansion has been developed following early ideas of Polyakov (1975).

The simplest example is the behavior of a Heisenberg magnet near $d = 2$, where one examines the low-temperature spin-wave corrections to the spin stiffness. Finite corrections are found for $d > 2$, while logarithmic singularities of the form $T \ln L$ (T is the temperature and L is the system size) arise precisely at $d = 2$. This result indicates the instability of the ordered phase due to infinitesimal thermal fluctuations indicating that $d_{\rm lc} = 2$, and allows for a perturbative RG treatment based on expanding around two dimensions (Nelson and Pelcovits, 1977).

In the case of disorder-driven MITs, one should examine the effects of weak disorder on the stability of the metallic phase. If similar singular corrections arise in low dimensions, then one can not only hope to identify the LCD, but it should also be possible to develop a perturbative RG scheme. This elegant approach does not require developing an appropriate symmetry-breaking scheme, thus bypassing the essential stumbling block in the theory for the MIT. Instead, it focuses on those physical processes that describe how weak disorder modifies a clean Fermi liquid. In those instances where the system is close to instabilities of the clean system, this approach may be insufficient because new types of low-energy excitation may become important. Such a situation may be found if the clean system is sufficiently close to Mott or Wigner transitions, in which case strong correlation effects may require a different theoretical framework and approach. Independent of these issues, the physical content of theories describing perturbative disorder effects within conventional Fermi liquids is sufficiently rich and non-trivial, and in the following we discuss its basic ideas and results.

1.3.2 Scaling theories of disorder-driven transitions

Phenomenological β-function

An elegant and compact description of the scaling behavior around a critical point is provided by the β-function formulation. What we want to emphasize here is that the β-function description is simply a alternative language one can use, rather then a microscopic theory. However, its straightforward application to the MIT is based on several implicit assumptions that allow for a simplified phenomenology, and which we discuss as follows.

In its original formulation as presented by the "gang of four" (Abrahams *et al.*, 1979), the β-function describes how the conductance of the system changes with the (effective) system size. In principle, one could imagine taking a finite-size chunk of disordered metal, and attaching it to contacts. If the experiment is repeated for different sample sizes, then one could experimentally determine how the conductance depends on the system size.

Why use the conductance and what do we expect to find? The essential physics is easy to see by thinking about what happens far from the MIT. In a good metal disorder is weak, and the conductivity σ is large and finite even at $T = 0$, and the standard Drude theory applies. From Ohm's law, the conductance g then scales with the system size L as

$$g_{\rm met}(L) = \sigma L^{d-2}.$$

For increasing system sizes (in $d > 2$), the conductance grows as L^{d-2}. In the opposite limit of very strong disorder, we expect all the electrons to form bound (localized) states with impurities. If ξ is the characteristic (localization) length of these bound states, then the conductance is expected to decrease exponentially

$$g_{\text{ins}}(L) \sim \exp\{-L/\xi\}.$$

More generally, we may expect that $g(L)$ increase with L in the metal and decrease in the insulator. What is not *a priori* obvious is how $g(L)$ behaves around the transition. The seminal work on the scaling theory for Anderson localization concentrated on the logarithmic rate of change of the conductance with length scale, by defining the β-function

$$\beta(g) = \frac{d(\ln g)}{d \ln L}. \tag{1.34}$$

This quantity is expected to be positive in a metal and negative in the insulator. Its precise form, or a possible dependence on the sample size L, is not *a priori* clear.

To make more specific predictions on the critical regime, the "gang of four" made two key assumptions, as follows:

1. $\beta(g)$ is a function of g only, but does not depend explicitly on L.
2. $\beta(g)$ is a smooth (analytic) function near the transition.

In particular, in an ohmic metal, we find

$$\beta_{\text{met}} = d - 2, \tag{1.35}$$

while in the localized insulator

$$\beta_{\text{ins}} = \ln g. \tag{1.36}$$

Since $\beta_{\text{met}} > 0$, and $\beta_{\text{ins}} < 0$, assumptions (1) and (2) then suggest that $\beta(g)$ has to change sign at some *finite* value of the conductance $g = g_{\text{c}}$, and we can write

$$\beta(g) \approx s \ln(g/g_{\text{c}}), \tag{1.37}$$

where $s = \beta'(g_{\text{c}})$ is the critical slope of the β-function. Defining the logarithmic variable $t = \ln(g/g_{\text{c}})$, we can now integrate the β-function equation

$$\frac{d(\ln t)}{d \ln L} \approx s,$$

from the microscopic cutoff ℓ to the sample size L, and write

$$t(L) = t_o(L/\ell)^s.$$

This integration has to be carried out up to the length scale (i.e. the correlation length) $L = \xi$ such that the renormalized distance to the transition $\delta g(L) = (g(L) - g_{\text{c}})/g_{\text{c}} \approx$

$t(L) \sim O(1)$. Since at short scales $t_o = t(\ell) \sim \delta g_o \sim \delta n$, we find

$$\xi \sim t_o^{1/s} \sim \delta n^{1/s}.$$

Thus, the correlation length exponent

$$\nu = 1/s.$$

At this scale the conductivity is expected to saturate to a (size-independent) macroscopic value

$$\sigma \approx g_c \xi^{d-2} \sim \delta n^{(d-2)\nu}.$$

We conclude that the assumptions implied by the β-function formulation, i.e. the scaling theory of localization, predict the validity of "Wegner scaling" for the conductivity exponent

$$\mu = (d-2)\nu.$$

It is worth emphasizing that postulating a particular form for the β-function is completely equivalent to formulating a scaling hypothesis for the critical behavior, *provided that the conductance is assumed to be finite at the transition.* Under these conditions, the scaling hypothesis states that the $T = 0$ conductance of a finite-size system takes a scaling form

$$g(\delta n, L) = f_\sigma\,(b^{1/\nu}\delta n, b/L).$$

Choosing $b = \delta n^{-\nu} \sim \xi$, we can write $g(\delta n, L) = \psi(\xi/L)$, where $\psi(x) = f_\sigma(1, x)$. The condition that the conductance is finite at the transition is equivalent to requiring that $\psi(0) = g_c$ is a finite constant. Using the definition of the β-function, we then find

$$\beta(g) = x^{-1}\psi'(x)/\psi(x),$$

where $x = \psi^{-1}(g)$. The form of the β-function can be directly extracted from the experimental data, provided that an appropriate scaling behavior is found.

Is there a β-function for percolation?

The β-function formulation of the "gang of four" (Abrahams *et al.*, 1979) may be viewed as a convenient phenomenological description of how the conductance depends on length scale. While its implicit assumptions prove correct for the problem of Anderson localization of non-interacting electrons, one may ask the same question for other models of the MIT. In particular, the percolation problem represents a consistent description of the transition in the semiclassical limit. Since percolation (Stauffer and Aharoni, 1994) is a well-characterized (classical) critical phenomenon, power-law scaling of all quantities is still valid, and many of the same questions raised by the "gang of four" can be again posed.

Since $d = 2$ remains above the lower critical dimension for percolation, Wegner scaling cannot be valid in this case. How is the scaling behavior modified in this case, and what would the β-function look like? Can it even be defined? How is the finite-temperature scaling modified, and what form does it take? These questions are important, since the precise answer allows us to distinguish experimental systems where percolation behavior dominates from those where genuine quantum critical behavior is at play. In the case of the percolation transition, we can give precise and rigorous answers to all these questions, as we discuss in the following.

To discuss transport behavior near the percolation transition, consider a resistor network corresponding to a random mixture of a metallic and and insulating component. Let the conductivities of the respective components be $\sigma_{\rm M}(T)$ and $\sigma_{\rm I}(T)$, both of which remain finite at $T \neq 0$, but with $\sigma_{\rm M}(0) = \sigma_o \neq o$, while $\sigma_{\rm I}(0) = 0$. As the relative fraction x of the insulating component increases past the percolation concentration $x_{\rm c}$, the overall conductivity of the network behaves as

$$\sigma(x, T = 0) \sim (x_{\rm c} - x)^{\mu}, \tag{1.38}$$

but such a sharp critical behavior emerges only at $T = 0$. At any finite temperature the sharp percolation transition is smeared, similarly as when a symmetry-breaking field is turned on in conventional critical phenomena. The family of conductivity curves generated by varying the percolation concentration x and temperature T at first glance looks very similar to those expected at any quantum localization transition, where sharp critical behavior also emerges only at $T = 0$.

How can we distinguish the two phenomena? The simplest way to do so is by focusing on the behavior at the the critical concentration, where the finite temperature conductivity is determined (Straley, 1977) by the conductivity of the insulating components

$$\sigma(x_{c}, T) \sim (\sigma_{\rm I}(T))^{u}. \tag{1.39}$$

The corresponding critical exponent $u = \frac{1}{2}$ in $d = 2$, and $u \approx 0.7$ in $d = 3$ (Straley, 1977). Therefore, one may first determine the location of the critical point by extrapolating the conductivity to $T = 0$ from the metallic side. One should then plot the temperature dependence at the critical point. At any quantum critical point, we generally expect a power-law temperature dependence, $\sigma_{\rm c}(T) \sim T^{(d-2)/x}$. In contrast, the conductivity of any insulator typically takes a exponential form

$$\sigma_{\rm I}(T) \sim \exp\{-(T_o/T)^{\alpha}\} \tag{1.40}$$

where $\alpha = 1$ for simple activated behavior, $\alpha = 1/(d+1)$ for Mott variable-range hopping (Mott, 1990), or $\alpha = \frac{1}{2}$ for Efros–Shklovskii hopping (Shklovskii and Efros, 1984). Since the measured conductance is expected to be a power of $\sigma_{\rm I}$, we conclude that in the presence of percolation, the conductivity will assume an insulating-like (exponential) temperature dependence only in the insulating phase, but even at the critical point. Thus, when the conductivity is plotted as a function of temperature on a log–log scale, one will find a single straight line (indicated power-law behavior at the critical concentration) in the case of quantum criticality, but no such curve will be found in case of percolation.

One can be even more precise, and specify the precise scaling behavior around such a percolation transition, which can be used to perform an alternative scaling analysis and collapse the experimental curves in presence of percolation. In this case, we expect (Stauffer and Aharoni, 1994) that the conductivity should assume the following scaling form

$$\sigma_{\text{perc}}(\delta x, h, L) = b^{-(d-2+\zeta)} f_{\text{perc}}(b^{1/\nu}\delta x, hb^{z'}, b/L). \tag{1.41}$$

Here, $\delta x = (x_{\text{c}} - x)$ is the distance to the critical point and $h = \sigma_{\text{I}}/\sigma_{\text{M}}$ plays the role of the symmetry-breaking field. Choosing $b^{1/\nu}\delta x = 1$ and taking $L \to \infty$, we can write

$$\sigma_{\text{perc}}(\delta x, h) = \delta x^{\mu} \phi_{\text{perc}}(h/\delta x^{\nu z'}). \tag{1.42}$$

The conductivity exponent

$$\mu = (d - 2 + \zeta)\nu, \tag{1.43}$$

which replaces Wegner scaling of the quantum case. Thus to collapse all the experimental curves, the argument of the scaling function should contain $h \sim \sigma_{\text{I}}(T)$, and not the temperature T. Working at the critical point ($\delta x = 0$), we similarly find the exponential relation

$$u = \frac{d - 2 + \zeta}{z'}. \tag{1.44}$$

Finally, let us examine the finite-size scaling behavior at $h = 0$. Working again at the critical point, we find

$$\sigma_{\text{c}}^{\text{perc}}(L) \sim L^{-(d-2+\zeta)}, \tag{1.45}$$

so that the critical conductance

$$g_{\text{c}}^{\text{perc}}(L) \sim L^{-\zeta}, \tag{1.46}$$

revealing the physical meaning of the anomalous dimension ζ, which describes transport on the critical percolation cluster. As we can see, the critical conductance vanishes in the case of percolation, in contrast to the scaling theory of localization. We conclude that although percolation does display conventional power-law finite-size scaling at the critical point, a β-function description is not possible. This behavior reflects the fact that $d = 2$ is not the lower critical dimension for percolation. The anomalous dimension ζ is very generally expected to vanish at ordinary quantum critical points describing conductor–insulator transitions. This result can be shown (Wen, 1992) to very generally follow from charge conservation for any quantum criticality displaying simple single-parameter scaling behavior. Although more complicated quantum scenarios are in principle possible, no microscopic quantum model has been identified to date showing $\zeta \neq 0$, or equivalently a lower critical dimension $d_{\text{lc}} < 2$.

Perturbative quantum corrections in disordered metals

What is the physical mechanism that invalidates Mott's bound on impurity scattering? To understand this, recall that Mott used Drude's picture, where scattering processes from each impurity or defect are assumed to independent and uncorrelated. This assumption is indeed justified at weak enough disorder, where the Drude prediction is recovered as a leading order contribution. For stronger disorder, multiple-scattering processes cannot be ignored, and they provide the so-called "quantum corrections" to Drude theory. In good metals the magnitude of the quantum corrections is generally small, modifying the conductivity by typically only a fraction of a percent. In this regime, the quantum corrections can be systematically obtained as next-to-leading corrections within weak-disorder perturbation theory for impurity scattering.

Detailed calculations and classification of all such perturbative quantum corrections were carried out in the late 1970s and early 1980s, and are by now well understood (Lee and Ramakrishnan, 1985). They consist of several additive terms,

$$\sigma = \sigma_o + \delta\sigma_{\mathrm{wl}} + \delta\sigma_{\mathrm{int}},$$

corresponding to the so-called "weak localization" and "interaction" corrections. These "hydrodynamic" corrections are dominated by infrared singularities, i.e., they acquire non-analytic contributions from small momenta or equivalently large distances. Specifically, the weak localization corrections take the form

$$\delta\sigma_{\mathrm{wl}} = \frac{e^2}{\pi^d}\left[l^{-(d-2)} - L_{\mathrm{Th}}^{-(d-2)}\right], \qquad (1.47)$$

where $l = v_F\tau$ is the mean free path, d is the dimension of the system, and L_{Th} is the length scale over which the wave functions are coherent. This effective system size is generally assumed to be a function of temperature of the form $L_{\mathrm{Th}} \sim T^{p/2}$, where the exponent p depends on the dominant source of decoherence through inelastic scattering.

The situation is simpler in the presence of a weak magnetic field or magnetic impurities. Here the weak-localization corrections are suppressed and the leading dependence comes from the interaction corrections first discovered by Altshuler and Aronov (1979)

$$\delta\sigma_{\mathrm{int}} = \frac{e^2}{\hbar}(c_1 - c_2\widetilde{F}_\sigma)(T\tau)^{(d-2)/2}. \qquad (1.48)$$

Here, c_1 and c_2 are constants, and $\widetilde{F}_\sigma$ is an interaction amplitude.

In $d = 3$, this leads to a square-root singularity $\delta\sigma_{\mathrm{int}} \sim \sqrt{T}$ and to a more singular logarithmic divergence $\delta\sigma_{\mathrm{int}} \sim \ln(T\tau)$ in $d = 2$. These corrections are generally more singular than the temperature dependence of the Drude term, and thus they are easily identified experimentally at the lowest temperatures. Indeed, the $T^{1/2}$ law is commonly observed (Lee and Ramakrishnan, 1985) in transport experiments in many disordered metals at the lowest temperatures, typically below 500 mK.

Similar corrections have been predicted for other physical quantities, such as the tunneling density of states and, more importantly, for thermodynamic response functions. As in Drude theory, these quantities are not expected to be appreciably affected by non-interacting localization processes, but singular contributions are predicted from interaction corrections. In particular, corrections to both the spin susceptibility χ, and the specific heat coefficient $\gamma = C_{\rm V}/T$ were expected to take the general forms

$$\delta\chi \sim \delta\gamma \sim T^{(d-2)/2},$$

again leading to logarithmic corrections in $d = 2$.

As in conventional Fermi liquid theories, these corrections emerged already when the interactions were treated at the lowest, Hartree–Fock level, as done in the approach of Altshuler and Aronov (1979). Higher order corrections in the interaction amplitude were first incorporated by Finkel'stein (1983; 1984), demonstrating that the predictions remained essentially unaltered, at least within the regime of weak disorder. In this sense, Fermi liquid theory has been generalized to weakly disordered metals, where its predictions have been confirmed in numerous materials (Lee and Ramakrishnan, 1985).

Anderson transition in $2 + \varepsilon$ *dimensions*

The essential idea of these approaches focuses on the fact that a weak, logarithmic instability of the clean Fermi liquid arises in two dimensions, suggesting that $d = 2$ corresponds to the lower critical dimension of the problem. In conventional critical phenomena, such logarithmic corrections at the lower critical dimension typically emerge due to long-wavelength fluctuations associated with spontaneously broken continuous symmetry. Indeed, the early work of Wegner (1979) emphasized the analogy between the localization transition and the critical behavior of Heisenberg magnets. It mapped the problem onto a field theoretical non-linear σ-model and identified the hydrodynamic modes leading to singular corrections in $d = 2$. Since the ordered (metallic) phase is only marginally unstable in two dimensions, the critical behavior in $d > 2$ can be investigated by expanding around two dimensions. Technically, this is facilitated by the fact that in dimension $d = 2 + \varepsilon$ the critical value of disorder W for the MIT is very small ($W_{\rm c} \sim \varepsilon$), and thus can be accessed using perturbative RG approaches in direct analogy to the procedures developed for Heisenberg magnets. In this approach (Abrahams *et al.*, 1979; Schaffer and Wegner, 1980), conductance is identified as the fundamental scaling variable associated with the critical point, which is an unstable fixed point of the RG flows. This RG calculation provides a scaling description predicting how the conductance depends on the system size, and thus produces the desired β-function in $d = 2 + \varepsilon$ dimensions. To leading order in ε the resulting critical exponent ν is predicted (Abrahams *et al.*, 1979; Schaffer and Wegner, 1980) to be

$$\nu^{-1} = \varepsilon + O(\varepsilon^2). \tag{1.49}$$

When this result is extrapolated to three dimensions ($\varepsilon = 1$), the conductivity exponent

$$\mu = 1 + O(\varepsilon). \tag{1.50}$$

This early prediction was initially widely acclaimed as a plausible theoretical explanation for the critical behavior commonly observed in several systems (e.g. compensated doped semiconductors, see below), where $\mu \approx 1$. From the theoretical side, this result was believed to be exact from more then ten years, since very tedious subsequent work established that higher-order corrections to ν, when evaluated to second and even third order in ε all vanished! Very surprisingly, a tour-de-force calculation by Hikami (to $O(\varepsilon^5)$) succeeded in finding a non-vanishing correction to this exponent, giving an estimate

$$\mu \approx 0.67 \tag{1.51}$$

in three dimensions. This calculation demonstrated that the exact value ($\mu \approx 1.58$ for the "orthogonal ensemble" describing ordinary potential scattering) lies hopelessly far from the estimates based on the ε-expansion.

This important result indicates a potentially serious shortcoming of any such perturbative RG schemes. As in ordinary (magnetic) critical phenomena, the ε-expansion around $d = 2$ seems to have extremely bad convergence properties (Castilla and Chakravarty, 1993) when extrapolated to $d = 3$, making it a virtually useless theoretical tool for making quantitative estimates for the critical exponents. The reason for generally poor convergence of $d = 2 + \varepsilon$ expansions is at present not fully understood, although related work (Kamal and Murthy, 1993) suggested that it reflects an inability to incorporate topological excitations (the "hedgehog" configurations for the $d = 3$ Heisenberg model studied) within any perturbative scheme. This behavior should be contrasted to that familiar from the $4 - \varepsilon$ RG approaches in standard critical phenomena. Here, a consistent description of the critical point is found even at the mean-field level, which becomes exact for $d > d_{\rm uc} = 4$. The fluctuation effects described by RG flows only provide the corrections to mean-field values for the critical exponents, and in general have proved to have much superior convergence properties, often allowing surprisingly accurate estimates when extrapolated to $d = 3$. Interestingly, the complete lack of spatial correlations within MFT allows for *all spin configurations* to be considered on an equal footing within the $4 - \varepsilon$ scheme , including both the smooth configurations described by perturbative methods, *and* the topological defects which they ignore.

Returning to MITs, one may speculate that the difficulties with perturbative approaches reflect that here the very existence of the phase transition emerges only due to fluctuation corrections; no transition is found at the mean-field (saddle-point) level describing Drude–Boltzmann theory. Developing a more appropriate mean-field description of the MIT thus appears to be one of the most promising directions for future work.

Generalized Fermi liquid theory for disordered electrons

In the mid 1980s, these ideas were extended with a great deal of effort in the formulation of a scaling theory of interacting disordered electrons by Finkel'stein (1983) and many followers (Belitz and Kirkpatrick, 1994; Castellani *et al.*, 1984). While initially shrouded by a veil of quantum field theory, these theories were later given a simple physical interpretation in terms of Fermi-liquid ideas (Castellani *et al.*, 1987; Kotliar, 1987) for disordered electrons. Technical details of these theories are of considerable complexity, and the interested reader is referred to the original literature (Belitz and Kirkpatrick, 1994; Castellani *et al.*, 1984; Finkel'stein, 1983, 1984). Here we just summarize the principal results, in order to clarify the constraints imposed by these Fermi-liquid approaches.

Within the Fermi liquid theory for disordered systems (Castellani *et al.*, 1987; Kotliar, 1987), the low-energy (low-temperature) behavior of the system is characterized by a small number of effective parameters, which include the diffusion constant D, the frequency renormalization factor Z, and the singlet and triplet interaction amplitudes γ_{s} and γ_{t}. These quantities can also be related to the corresponding quasiparticle parameters, which include the quasiparticle density of states

$$\rho_{\mathrm{Q}} = Z\rho_o, \tag{1.52}$$

and the quasiparticle diffusion constant

$$D_{\mathrm{Q}} = D/Z \sim D/\rho_{\mathrm{Q}}. \tag{1.53}$$

Here, ρ_o is the "bare" density of states which describes the noninteracting electrons. In the absence of interactions, the single-particle density of states is only weakly modified by disorder and remains noncritical (finite) at the transition (Schaffer and Wegner, 1980).

Using these parameters, we can now express the thermodynamic response functions as follows. We can write the compressibility

$$\chi_{\mathrm{c}} = \frac{dn}{d\mu} = \rho_{\mathrm{Q}}[1 - 2\gamma_{\mathrm{s}}], \tag{1.54}$$

the spin susceptibility

$$\chi_{\mathrm{s}} = \mu_{\mathrm{B}}^2 \rho_{\mathrm{Q}}[1 - 2\gamma_{\mathrm{t}}], \tag{1.55}$$

and the specific heat

$$C_{\mathrm{V}} = 2\pi^2 \rho_{\mathrm{Q}} T/3. \tag{1.56}$$

In addition, we can use the same parameters to express transport properties such as the conductivity

$$\sigma = \frac{dn}{d\mu} D_{\mathrm{c}} = \rho_{\mathrm{Q}} D_{\mathrm{Q}}, \tag{1.57}$$

as well as the density–density and spin–spin correlation functions

$$\pi(q,\omega) = \frac{dn}{d\mu}\frac{D_c q^2}{D_c q^2 - i\omega} \qquad \chi_s(q,\omega) = \chi_s \frac{D_s q^2}{D_s q^2 - i\omega}. \tag{1.58}$$

Here, we have expressed these properties in terms of the spin and charge diffusion constants, which are defined as

$$D_c = \frac{D}{Z(1-2\gamma_s)}; \quad D_s = \frac{D}{Z(1-2\gamma_t)}. \tag{1.59}$$

Note that the quantity D is *not* the charge diffusion constant D_c that enters the Einstein relation, eqn (1.57). As we can write $\sigma = \rho_o D$, and since ρ_o is not critical at any type of transition, the quantity D (also called the "renormalized diffusion constant") has a critical behavior identical to that of the conductivity σ. We should also mention that the quasiparticle diffusion constant $D_Q = D/Z$ has been physically interpreted as the heat diffusion constant.

Weak-coupling renormalization group approach of Finkel'stein

The Fermi-liquid relations provide constraints relating different Landau parameters, but they cannot predict their specific values, or how should they behave in the vicinity of the MIT. In practical applications of these approaches to disordered systems one assumes that the Landau parameters characterizing the clean Fermi liquid are known (say from experiments), and then examines how they are modified when impurities are added to the system. Precise prediction for these impurity effects have been obtained in the limit of weak disorder, where perturbative "quantum corrections" to all quantities have been calculated (Lee and Ramakrishnan, 1985). As in the non-interacting limit, singular logarithmic corrections are found in $d = 2$, signaling that a systematic weak-coupling approach can be developed, to investigate the instability of the metallic phase under impurity scattering. While conceptually simple and transparent, practical calculations to implement this program have proved of considerable complexity, even within the framework of weak-coupling approaches.

Perturbative renormalization group calculations (Belitz and Kirkpatrick, 1994; Castellani *et al.*, 1984; Finkel'stein, 1983, 1984) based on the $2+\varepsilon$ expansion have been used to make explicit predictions for the values of the critical exponents for different universality classes. The details of this formulation will not be elaborated here, as it has been discussed in great detail in several excellent reviews. Instead, we comment on the physical content of these theories, and indicate what aspects of the problem may and which may not be addressed using this framework.

Physical content of Finkel'stein's theory

The mathematical complexity of the formalism of Finkel'stein seems, at first sight, to shroud its physical content with a veil of mystery, and render it difficult to comprehend to all but the very few specialists working in the field. Later works (Castellani *et al.*, 1984, 1987; Zala *et al.*, 2001), however, succeeded to re-derive most of Finkel'stein's results using standard (albeit much less elegant) diagrammatic approaches, allowing for a simple and transparent physical interpretation. The main points include:

- This theory describes a disordered Fermi liquid, specifically the scale-dependence of the disorder-modified Landau parameters.
- As in the original scaling theory of localization, the transition is identified as an unstable fixed point for the conductance. If the interaction amplitudes retain finite (albeit renormalized) values at the fixed point, then the scaling scenario is essentially of the same form as for non-interacting localization. This behavior is found (Belitz and Kirkpatrick, 1994) for all the universality classes with broken time-reversal invariance: in presence of an external magnetic field, magnetic impurity scattering, or the presence of spin–orbit scattering (S-OS).
- The metallic phase, as in any Fermi liquid, retains the same qualitative behavior at $T = 0$ as a weakly disordered systems of non-interacting electrons. In particular, this means that all thermodynamic quantities (e.g. the spin susceptibility χ or the Sommerfeld coefficient $\gamma = C/T$) remain finite away from the transition.
- The scaling hypothesis, which is built into this formalism, guarantees that the low-temperature behavior of all quantities is qualitatively identical within the each (metallic or insulating) phase. For example, the leading low-temperature $\sqrt{T}$ corrections, first predicted by using the weak-disorder perturbation theory of Altshuler and Aronov, should retain their form throughout the metallic phase.

These results may lead—at least in principle—to a mathematically consistent description of the MIT in $d = 3$. In two dimensions, where Anderson localization is expected to destroy any metallic phase, the situation is more subtle. An especially interesting and still controversial situation is found for the "generic" model, i.e. in the absence of broken time-reversal invariance perturbations. Here, a qualitatively new behavior is found due to interaction effects. The triplet interaction amplitude γ_t is found to grow without limits under rescaling (i.e. as T is lowered), even at infinitesimal disorder (Finkel'stein, 1983, 1984). As a result, the "interaction" correction for the conductivity (which to leading order is proportional to γ_t) is found to produce an "antilocalization" effect, which is predicted to "overcome" the usual weak-localization term responsible for the instability of the $d = 2$ metallic phase. From this point of view, Finkelstein's results open a consistent possibility that interactions may be able to stabilize the metallic phase at sufficiently weak disorder, thus leading to a sharp MIT in $d = 2$.

Most remarkably, recent experiments on a variety two-dimensional electron gases have revealed behavior quite suggestive of the existence of precisely such a transition (Abrahams *et al.*, 2001). This controversial phenomenon, first discovered by Kravchenko *et al.* (1995), has reignited interest in this fundamental question. Its many facets continue to fascinate theorists and experimentalists alike, but much still remains to be understood, as described in more detail in Chapter 2. Even from the purely theoretical point of view, however, the physical interpretation of what precisely happens in a two-dimensional weakly disordered Fermi liquid still remains puzzling and controversial. The main issue is how to understand the singular behavior of the triplet interaction amplitude γ_t, which is found to grow without bounds at low temperature. In a disordered Fermi liquid it indicates enhanced spin susceptibility

$\chi \sim \gamma_t$, so this result may signal some kind of magnetic instability of the metallic phase. Various interpretations have been proposed, ranging from disorder-induced ferromagnetism (Belitz and Kirkpatrick, 1994), to disorder-induced local-moment formation (Castellani *et al.*, 1984; Dobrosavljević *et al.*, 1992; Milovanović *et al.*, 1989; Paalanen *et al.*, 1998). While some progress has recently been made by Punnoose and Finkel'stein (see Chapter 4), using large-N methods, the issue remains controversial and challenging.

Difficulties with runaway RG flows

From a more general perspective, any solution with such "runaway" coupling constant immediately brings into question the credibility of perturbative RG approaches. It is interesting to recall that similar situations have been encountered in several other problems, where disorder effects in interacting systems have proved difficult to describe using perturbative RG methods. The first such example is the famed random-field Ising model (RFIM) (Imry and Ma, 1975), where a perturbative arguments of Parisi and Sourlas (1979) suggested that the critical behavior of the disorder problem can be mapped to the corresponding clean problem in $d-2$ dimensions. Because the clean Ising model does not order in $d=1$, this work suggested that infinitesimal disorder would be sufficient to destroy the ordered phase even in $d=3$. Later rigorous work by Imbrie (1984) rigorously showed that a finite temperature ordering survives sufficiently weak disorder in $d=3$, thus invalidating the Parisi and Sourlas mechanism, which was claimed to hold to all orders in perturbation theory. The puzzle was soon resolved by more careful functional RG methods of Fisher (1985), who showed that the perturbative fixed point of Parisi and Sourlas proves unstable, leading to "runaway" RG flows. Subsequent work (De Dominicis *et al.*, 1995; Mezard and Monasson, 1994; Mezard and Young, 1992) found evidence that this reflects the proliferation of metastable states in the RFIM, a phenomenon which proves "invisible" in perturbative treatments. Interestingly, much later work (Belitz and Kirkpatrick, 1995; Kirkpatrick and Belitz, 1994) attempted a non-perturbative (albeit uncontrolled) solution of Finkel'stein's non-linear σ-model by proposing a new saddle-point solution, and an RG approach based on a $6-\varepsilon$ expansion, finding RG flows of a structure very similar to the Parisi and Sourlas theory for the RFIM. The authors interpreted these results as evidence that metastability and non-perturbative glassy effects cannot be ignored within the Finkel'stein model any more then they could for the RFIM.

Yet another example of non-perturbative disorder effects has been found in recent studies of Hertz–Millis (Hertz, 1976; Millis, 1993; Sachdev, 2011) models for itinerant quantum criticality. Here, early work attempted a perturbative RG treatment for weak disorder, only to find runaway flows. Later works by Hoyos *et al.* (2007), which used a complementary "strong-disorder renormalization group" methods, demonstrated that for these models disorder completely changes the physical nature of quantum criticality, leading to so-called "infinite-randomness fixed point" behavior (Miranda and Dobrosavljević, 2005; Vojta, 2006). In such cases, the behavior of the system is dominated by rare disorder realizations, leading to the formation of quantum Griffiths phases surrounding the critical point.

We should stress that these non-perturbative disorder effects do not bring into question the general ideas of scaling, but they they do seem to raise concerns about the application of weak-coupling RG methods. In cases such as the infinite-randomness fixed point universality class, the renormalized effective action describing the critical point assumes a qualitatively different form than that of the clean system. This simple fact makes it abundantly clear how and why perturbative methods may fail in a most dramatic fashion. Whether a similar mechanism resolves the puzzles identified by the perturbative solution of the Finkel'stein model remains to be established by future work.

Conceptual limitations of the disordered Fermi-liquid picture

Despite a great deal of effort invested in such calculations, the predictions of the perturbative RG approaches have met only limited success in explaining the experimental data in the critical region of the MIT. We should emphasize, though, that limitations associated with these weak-coupling theories do not invalidate the potential applicability of Fermi-liquid ideas *per se*. Having this in mind, it is also important to understand the fundamental limitations of the disordered Fermi-liquid picture that is the underpinning of Finkel'stein's field theory. The important physical conditions implicitly assumed by this formulation include the following.

- It describes leading low-temperature excitations, which are adiabatically connected to a noninteracting (but disordered) electronic system. These excitations, therefore, assume fermionic character. Other (spin or charge) collective excitations are assumed to play a subleading role in this temperature regime.
- The fermionic excitations are assumed to be sufficiently dilute that inelastic electron–electron scattering processes can be neglected. The leading low-temperature corrections then reflect elastic scattering processes on a static, but temperature-dependent renormalized disorder potential. Upon disorder averaging, this can produce different temperature dependence in the diffusive and ballistic regimes. As explained in recently by Zala *et al.* (2001), however, the basic physical mechanism for all these corrections is one and the same (Fig. 1.15). We should emphasize that these processes dominate only at sufficiently low temperatures, typically only a small fraction of the Fermi temperature. In presence of strong electronic correlations, the "coherence temperature" T^*, below which the Fermi-liquid picture applies, may be very low; a good example are so-called "heavy-fermion" systems, where $T^* \ll T_\mathrm{F}$ and a very broad incoherent metallic regime is found. In the presence of disorder, the coherence temperature T^* may be further reduced or even driven to zero (Miranda and Dobrosavljević, 2005). This situation is illustrated in certain exactly solvable models with strong correlation and disorder (Dobrosavljević and Kotliar, 1993; Tanasković *et al.*, 2004), which can be exactly solved in the limit of large coordination. Here strong disorder fluctuations stabilize "non-Fermi-liquid" metallic behavior at any non-zero temperature.
- The standard formulation of the disordered Fermi-liquid theory is derived at weak disorder, where the electrons are well delocalized across the entire sample. The Landau-interaction amplitudes are assumed to be self-averaging, and are thus

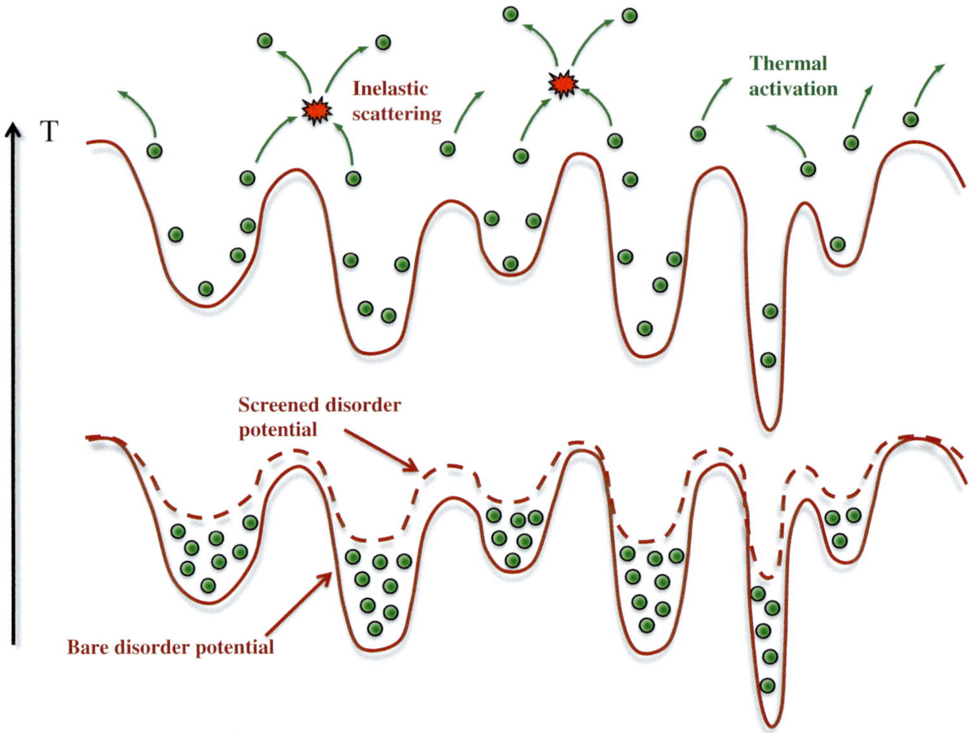

Fig. 1.15 In the disordered Fermi-liquid picture, the leading low-temperaure dependence of transport reflects elastic scattering off a renormalized, but temperature-dependent random potential (dashed line). At low temperatures (bottom), the potential wells "fill-up" with electrons; in the presence of repulsive (Coulomb) interactions, the screened (renormalized) potential has reduced amplitude (dashed line), leading to effectively weaker disorder. As the temperature increases (top), electrons thermally activate (shown by arrows) out of the potential wells, reducing the screening effect. This physical mechanism, which operates both in the ballistic and in the diffusive regime (Zala *et al.*, 2001), is at the origin of all "quantum corrections" found within the Fermi-liquid picture. It is dominant, provided that inelastic electron–electron scattering can be ignored. While this approximation is well justified in good metals, inelastic scattering (star symbol) is considerably enhanced in presence of strong correlation effects, often leading to disorder-driven non-Fermi liquid behavior (Miranda and Dobrosavljević, 2005) and electronic Griffiths phases (Andrade *et al.*, 2009; Miranda and Dobrosavljević, 2001; Tanaskovic *et al.*, 2005).

replaced by their averaged values. At stronger disorder, very strong spatial fluctuations may emerge (Miranda and Dobrosavljević, 2005), so that the interaction effects may be much more pronounced in certain regions of the sample. These effects are most important in strongly correlated electronic systems, where the assumption of self-averaging may completely break down, leading to the formation of "electronic Griffiths phases" (Andrade *et al.*, 2009; Miranda and Dobrosavljević, 2001; Tanaskovic *et al.*, 2005). If this happens, then the simplified version of disordered Fermi-liquid theory may prove insufficient or even misleading.

- The Fermi-liquid pictures assumes a unique ground state, implicitly ignoring the possibility metastable states resulting from the competition of disorder and the long-range Coulomb interactions. The associated electron glass behavior (Pastor and Dobrosavljević, 1999) is a well established feature of disordered insulators, but its possible role in the critical regime (Dobrosavljević *et al.*, 2003*b*) has been ignored by the weak coupling approaches.

1.3.3 Order-parameter approaches to interaction localization

Need for an order-parameter theory: experimental clues

In conventional critical phenomena, simple mean-field approaches such as the Bragg–Williams theory of magnetism, or the van der Waals theory for liquids and gases work remarkably well—everywhere except in a very narrow critical region (Goldenfeld, 1992). Here, effects of long-wavelength fluctuations emerge that modify the critical behavior, and its description requires more sophisticated theoretical tools, based on RG methods. A basic question then emerges when looking at experiments: is a given phenomenon a manifestation of some underlying mean-field (local) physics, or is it dominated by long-distance correlations, thus requiring an RG description? The answer for conventional criticality is well known, but what about MITs? Here the experimental evidence is much more limited, but we would like to emphasize a few well-documented examples which stand out.

- *Doped semiconductors such as Si:P.* These are the most carefully studied (Shklovskii and Efros, 1984) examples of MIT critical behavior. Here the density-dependent conductivity extrapolated to $T = 0$ shows sharp critical behavior (Paalanen and Bhatt, 1991) of the form $\sigma \sim (n - n_c)^{\mu}$, where the critical exponent $\mu \approx \frac{1}{2}$ for uncompensated samples (half-filled impurity band), while dramatically different $\mu \approx 1$ is found (Fig. 1.16) for heavily compensated samples of Si:P,B, or in presence of strong magnetic fields. Most remarkably, the dramatic differences between these cases is seen over an extremely broad concentration range, roughly up to several times the critical density. Such robust behavior, together with simple apparent values for the critical exponents, seems reminiscent of standard mean-field behavior in ordinary criticality.
- *Two-dimensional MITs.* Signatures of a remarkably sharp MIT have also been observed (Abrahams *et al.*, 2001; Kravchenko *et al.*, 1995; Popović *et al.*, 1997) in several examples of two-dimensional electron gases such as silicon MOSFETs. While some controversy regarding the nature or even the driving force for this transition remains a subject of intense debate, several experimental features seem robust properties common to most studied samples and materials. In particular, various experimental groups have demonstrated (Kravchenko *et al.*, 1995; Popović *et al.*, 1997) striking scaling of the resistivity curves in the critical region, (Fig. 1.17) which seems to display (Simonian *et al.*, 1997) remarkable mirror symmetry ("duality")

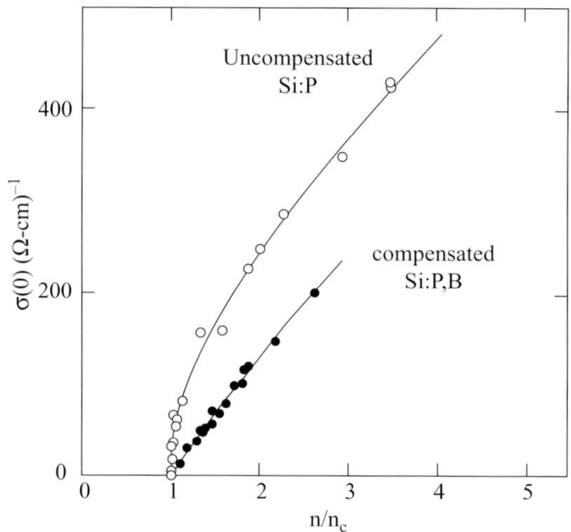

Fig. 1.16 Critical behavior of the conductivity for uncompensated $Si:P$ and compensated Si:P,B (Paalanen and Bhatt, 1991). The conductivity exponent $\mu \approx \frac{1}{2}$ in the absence of compensation, while $\mu \approx 1$ in its presence. Clearly distinct behavior is observed in a surprisingly broad range of densities, suggesting mean-field scaling. Since compensation essentially corresponds to carrier doping away from a half-filled impurity band (Shklovskii and Efros, 1984), it has been suggested (Lee and Ramakrishnan, 1985) that the difference between the two cases may reflect the role of strong correlations.

(Dobrosavljević *et al.*, 1997) over a surprisingly broad interval of parameters. In addition, the characteristic behavior extends to remarkably high temperatures, which are typically *comparable the Fermi temperature* (Abrahams *et al.*, 2001). One generally does not expect a Fermi-liquid picture of diluted quasiparticles to apply at such "high energies," or any correlation length associated with quantum criticality to remain long.

- *High temperature anomalies: Mooij correlation.* Surprisingly similar metal–insulator crossover is seen at elevated temperatures in several systems (Lee and Ramakrishnan, 1985) where the MIT is driven by increasing the level of disorder at fixed electron density. This behavior, first identified by Mooij (1973), has recently been studied in phase change materials (Siegrist *et al.*, 2011). Here, the temperature coefficient of the resistivity (TCR) is found to change sign, indicating the crossover from metallic ($d\rho/dT > 0$) to insulating(-like) ($d\rho/dT > 0$) transport around the values of the resistivity close to the "Mott limit" $\rho_c = 1/\sigma_{\min}$ (Fig. 1.18). This behavior is consistent with the early ideas of Mott, suggesting that in metals, as temperature is increased, the resistivity should display *resistivity saturation* (Fisk and Webb, 1976) as soon as the mean-free path approaches the atomic scale. Although reminiscent of what is seen in ($d = 2$) silicon MOSFETs, this behavior observed in bulk ($d = 3$) systems is inconsistent with expectations based on the scaling theories of localization (e.g. a divergent critical resistivity $\rho_c(T) \sim 1/T^{\frac{d-2}{z}}$).

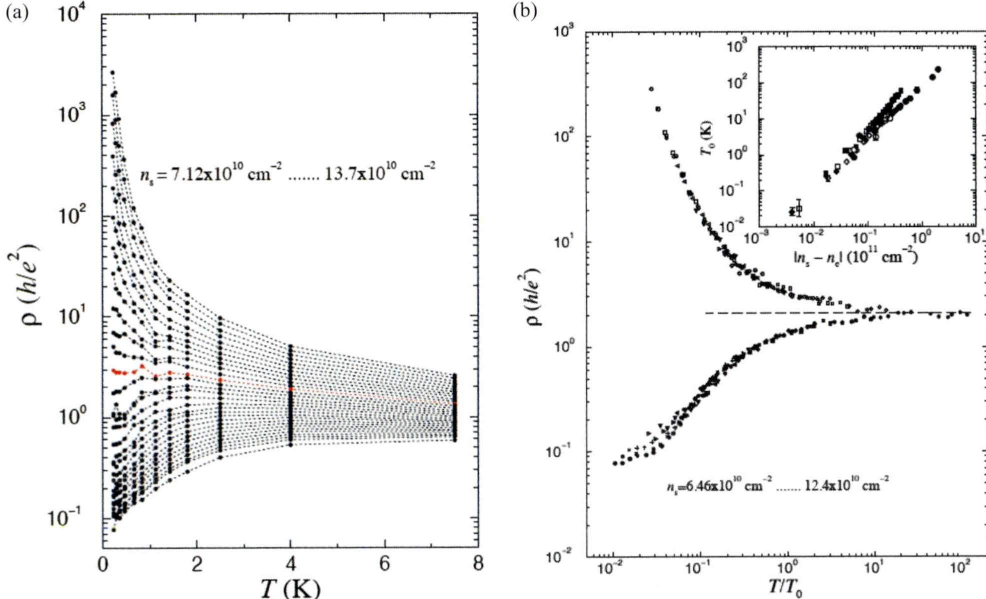

Fig. 1.17 The resistivity curves (a) for a two-dimensional electron system in silicon (Kravchenko *et al.*, 1995) show a dramatic metal–insulator crossover as the density is reduced below $n_c \sim 10^{11}$ cm^{-2}. Note that the system has "made up its mind" whether to be a metal or an insulator even at surprisingly high temperatures $T \sim T_F \approx 10$ K. (b) displays the scaling behavior, which seems to hold over a comparable temperature range. The remarkable "mirror symmetry" (Simonian *et al.*, 1997) of the scaling curves seems to hold over more then an order of magnitude for the resistivity ratio. This surprising behavior has been interpreted (Dobrosavljević *et al.*, 1997) as evidence that the transition region is dominated by strong-coupling effects characterizing the insulating phase.

Since any conceivable coherence length must be short at such elevated temperatures, local (incoherent) scattering processes are likely to be at the origin of this puzzling behavior. This should be contrasted to the well-known weak localization and interaction corrections in a disordered Fermi liquid, processes dominated by coherent multiple-scattering processes at long length scales.

- *High temperature violations of the Mott limit.* Examples of metallic ($d\rho/dT > 0$) transport with resistivities dramatically exceeding the Mott limit have been reported only in sufficiently clean systems not too far from the Mott-insulating state. This curious behavior was first noted shortly after the discovery of cuprate superconductors (Hussey *et al.*, 2004), and was quickly interpreted as a "smoking gun" of non-Fermi-liquid physics and strong correlation effects. Further examples have been documented (Fig. 1.19) in organic Mott systems (Limelette *et al.*, 2003*b*) and transition metal oxides such as V_2O_3 (Limelette *et al.*, 2003*a*). In all these cases, the violation of the Mott limit was found at temperatures exceeding the Fermi-liquid coherence scale T^*, and was observed to coincide with the suppression of the corresponding

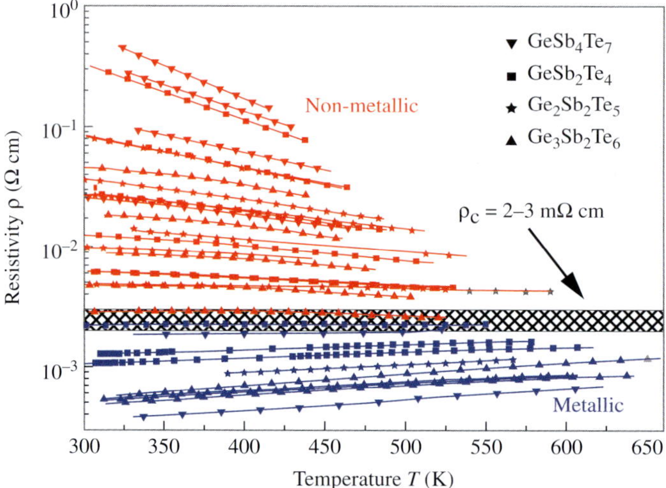

Fig. 1.18 High-temperature transport in the disorder-driven metal–insulator transition region in recently discovered "phase change" materials (Siegrist *et al.*, 2011). Here disorder is tuned at fixed carrier concentration by partially annealing the parent amorphous insulator. The change of the sign of TCR is found to occur precisely at the "separatrix" coinciding with the estimated Mott limit.

Drude peak in the optical conductivity. Similarly to resistivity saturation, such "bad metal" behavior seems to emerge only in the high-temperature incoherent regime, where local scattering processes dominate.

All these experiments taken together provide strong hints that in many systems of current interest an appropriate mean-field description is what is needed. It should provide the equivalent of a van der Waals equation of state, for the MIT problem of disordered interacting electrons. Unfortunately, such a theory has long been elusive, primarily due to a lack of a simple order-parameter formulation for this problem. Very recently, an alternative approach to the problem of disordered interacting electrons has been developed, based on dynamical mean-field (DMFT) methods (Georges *et al.*, 1996). This formulation is largely complementary to the scaling approach, and has already resulting in several striking predictions. In the following, we briefly describe this method, and summarize the main results that have been obtained so far.

The DMFT physical picture

The main idea of the DMFT approach is in principle very close to the original Bragg–Williams mean-field theories of magnetism (Goldenfeld, 1992). It focuses on a single lattice site, but replaces (Georges *et al.*, 1996) its environment by a self-consistently determined "effective medium," as shown in Fig. 1.20.

In contrast to the Bragg–Williams theory, the environment cannot be represented by a static external field, but instead must contain the information about the dynamics of an electron moving in or out of the given site. Such a description can be made precise by formally integrating out all the degrees of freedom on other lattice sites. In presence of electron–electron interactions, the resulting local effective action has an arbitrarily complicated form. Within DMFT (Georges *et al.*, 1996), the situation simplifies, and all the information about the environment is contained in the local single-particle spectral function $\Delta_i(\omega)$. The calculation then reduces to solving an appropriate quantum impurity problem supplemented by an additional self-consistency condition that determines this "cavity function" $\Delta_i(\omega)$.

The precise form of the DMFT equations depends on the particular model of interacting electrons and/or the form of disorder, but most applications (Georges *et al.*, 1996) to this date have focused on Hubbard and Anderson lattice models. The approach has been very successful in examining the vicinity of the Mott transition in clean systems, and it has met spectacular successes in elucidating various properties of several transition-metal oxides, heavy fermion systems and Kondo insulators (Kotliar *et al.*, 2006).

The central quantity in the DMFT approach is the local "cavity" spectral function $\Delta_i(\omega)$. From the physical point of view, this object essentially represents the *available electronic states* to which an electron can "jump" on its way out of a given lattice site.

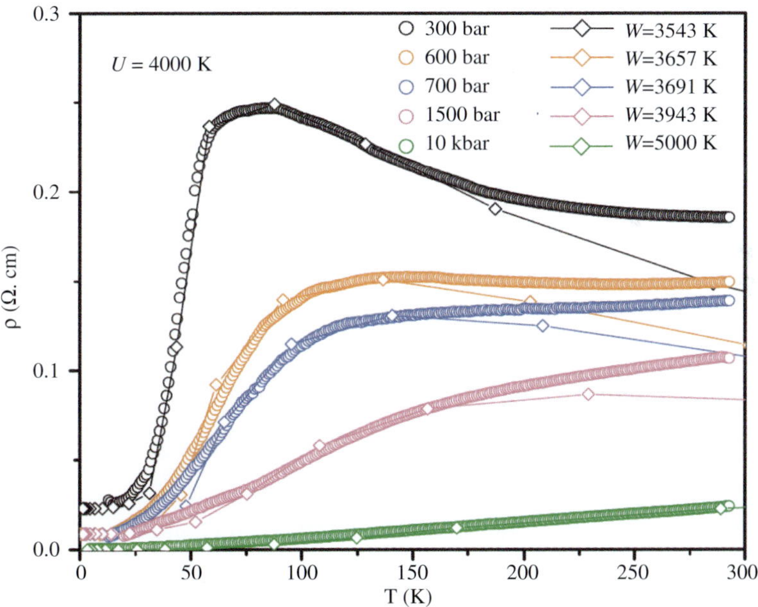

Fig. 1.19 Transport behavior in $\kappa-(\text{BETD-TTF})_2\text{Cu}[\text{N(CN)}_2]\text{Cl}$, a strongly correlated metal approaching the pressure-driven Mott transition. Here, Fermi-liquid transport ($\rho \sim T^2$) is found only at the lowest temperatures. The unusual transport behavior at higher temperatures, displaying pronounced resistivity maxima well above the Mott limit, has been quantitatively explained using DMFT theories (Limelette *et al.*, 2003*b*; Radonjic *et al.*, 2010).

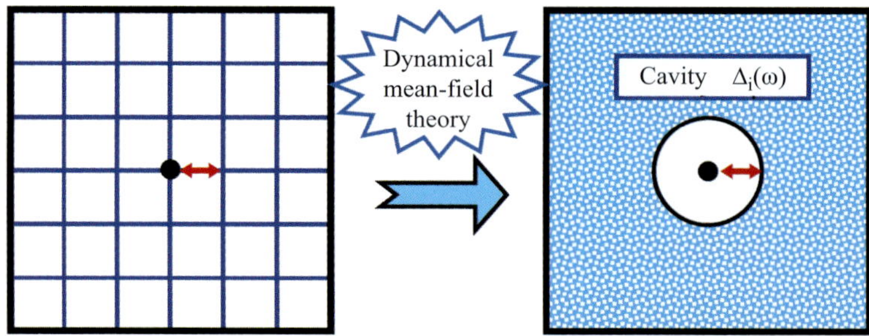

Fig. 1.20 In dynamical mean-field theory, the environment of a given site is represented by an effective medium, represented by its "cavity spectral function" $\Delta_i(\omega)$. In a *disordered* system, $\Delta_i(\omega)$ for different sites can be very different, reflecting Anderson localization effects (Anderson, 1958).

As such, it provides a natural order parameter description for the MIT (Dobrosavljević and Kotliar, 1998). Of course, its form can be substantially modified by either the electron–electron interactions or disorder, reflecting the corresponding modifications of the electron dynamics. According to Fermi's Golden Rule, the transition rate to a neighboring site is proportional to the density of final states, leading to insulating behavior whenever $\Delta_i(\omega)$ has a gap at the Fermi energy. In the case of a Mott transition in the absence of disorder, such a gap is a direct consequence of the strong on-site Coulomb repulsion, and is the same for every lattice site.

The situation is more subtle in the case of disorder-induced localization, as first noted in the pioneering work of Anderson (1958). Here, the *average* value of $\Delta_i(\omega)$ has no gap and thus cannot serve as an order parameter. However, as Anderson noted a long time ago, "...no real atom is an average atom..." (Anderson, 1978). Indeed, in an Anderson insulator, the environment "seen" by an electron on a given site can be very different from its average value. In this case, the *typical* "cavity" spectral function $\Delta_i(\omega)$ consists of several delta-function (sharp) peaks, reflecting the existence of localized (bound) electronic states (Fig. 1.21). Thus a *typical* site is embedded in an environment that has a *gap* at the Fermi energy—resulting in insulating behavior. We emphasize that the location and width of these gaps strongly vary from site to site. These strong fluctuations of the local spectral functions persist on the metallic side of the transition, where the typical spectral density $\Delta_{typ} = \exp < \ln(\Delta_i) >$ can be much smaller than its average value (Dobrosavljević *et al.*, 2003*a*). Clearly, a full *distribution function* is needed (Anderson, 1978) to characterize the system. The situation is similar as in other disordered systems, such as spin glasses (Mézard *et al.*, 1986). Instead of simple averages, here the entire distribution function plays a role of an order parameter, and undergoes a qualitative change at the phase transition. The DMFT formulation thus naturally introduces self-consistently defined *dynamical* order parameters, that can be utilized to characterize the qualitative differences between various phases. In presence of disorder, these order parameters have a character of distribution functions, which change their qualitative form as we go from the normal metal to the non-Fermi-liquid metal, to the insulator.

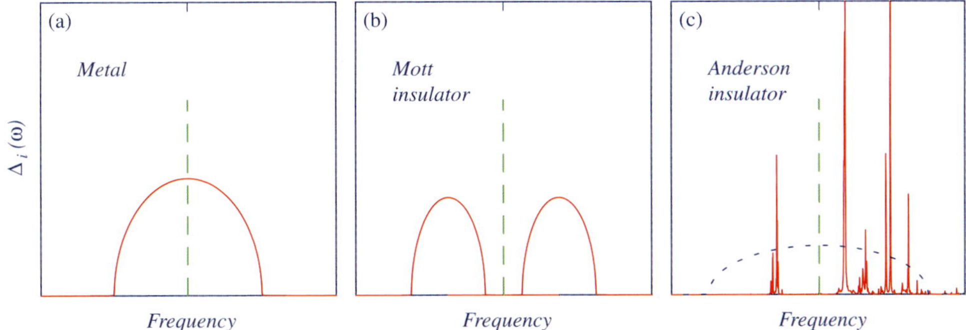

Fig. 1.21 The local cavity spectral function $\Delta_i(\omega)$ as the order parameter for the MIT. In a metal (a) there are available electronic states near the Fermi level (dashed line) to which an electron from a given site can delocalize. Both for a Mott insulator (b) and the Anderson insulator (c) the Fermi level is in the gap, and the electron cannot leave the site. Note that the *averaged* spectral function (dotted line in (c)) has no gap for the Anderson insulator, and thus cannot serve as an order parameter.

Physical content of DMFT approaches

DMFT and its various generalizations are designed to capture strong but local correlation effects. Here we will not discuss the technical details on these approaches, which will be discussed, in some detail, in Chapter 6. Instead, we give a brief summary of the advantages and the limitations of the existing DMFT theories, in applications to strongly correlated electronic systems with and without disorder.

- The approach is formally exact in the limit of large coordination, but in general it represents a "conserving" approximation scheme (Kotliar *et al.*, 2006), in the sense of Baym and Kadanoff. Although it reproduces Fermi-liquid behavior of metals at low temperature, it is not restricted to this regime. In fact, DMFT is *more accurate (essentially exact) in the high-temperature incoherent regime*, because this is where any spatial correlations ignored by DMFT become negligible.
- The local correlation effects are described through the local self-energies $\Sigma_i(\omega)$, defining both the quasiparticle effective mass $m^* = 1 - (\partial \Sigma_i / \partial \omega)_{\omega=0}$, and the inelastic scattering rate $\hbar \tau_{in}^{-1}(\omega) = -Im\Sigma_i(\omega = 0)$. In presence of disorder these quantities display spatial fluctuations (Miranda and Dobrosavljević, 2005), and need to be characterized with appropriate distribution functions.
- DMFT cannot be used to properly describe those phenomena, which are dominated by long-wavelength spatial fluctuations. It cannot provide a description of anomalous dimensions of various physical quantities within the narrow "Ginzburg" region around a critical point, at least for ordinary criticality. On the other hand, recent work suggested the possibility of "local quantum criticality" (Si *et al.*, 2001), where DMFT-like approaches may represent the proper description.

We should stress that the local dynamical description description provided by DMFT is a feature built-in from the start. The range of parameters and models where this approximation is valid is not obvious or precisely known at this time, despite the many successful applications (Kotliar *et al.*, 2006) of DMFT to various experimental systems. Interestingly, very recent work based on "holographic duality" (McGreevy, 2010; Sachdev, 2010), an idea borrowed from super-string theory, suggests that certain strong-coupling classes of quantum critical points may exist that have strictly local character. If these arguments prove correct, they may provide fundamental insight into why and where the local ideas implied by DMFT can be expected to apply, and in which cases they do not.

Applications and extensions of DMFT

The first practical application of DMFT has focused at the Mott transition in a single-band Hubbard model, shedding new light on systems ranging from transition-metal oxides to organic Mott systems. Even in this simplest "single-site" version, the method can be easily extended to multi-band and multi-orbital models, features that are of key importance in materials such as rare-earth intermetallics (heavy-fermion systems) and the recetnly discovered family of iron pnictides (Basov and Chubukov, 2011). The technical difficulties in applying the methods to multi-orbital systems has pleagued early studies, essentially due to lack of a reliable "quantum impurity solver" for DMFT equations. Very recent progress based on "continuous time quantum Monte Carlo" methods (Gull *et al.*, 2011) has provided a practical and efficient solution in many cases, opening avenues for applications to many materials.

Other complications have surrounded efforts to dovetail the DMFT methods for strong correlation with first-principle density-functional electronic structure methods (DFT). Considerable efforts have been invested in this program (Kotliar *et al.*, 2006), and over the last few years combined density-functional and DMFT studies have bloomed into a veritable industry, resulting in impressive and accurate first-principles descriptions of many strongly correlated materials. Finally, effects of inter-site correlations have also been a focus of much recent work, leading to various "cluster" generalizations of DMFT, resulting in substantial new insight (Haule and Kotliar, 2007; Yang *et al.*, 2011) in systems such as high-T_c cuprate superconductors. Different aspects of these theories have been discussed in detail in several recent reviews (Kotliar *et al.*, 2006; Maier *et al.*, 2005). In the following we focus on applications of DMFT to disordered systems (Miranda and Dobrosavljević, 2005), and list the physical phenomena and regimes that have found a natural description within the DMFT framework, but which remain difficult to address using complementary approaches.

The simplest DMFT theories focus on the dynamical effects of local interactions, while describing the environment of a given site as an average "effective medium." In presence of disorder, this formulation reduces (Dobrosavljević and Kotliar, 1994) to the well-known "coherent potential approximation" (CPA) in the noninteracting limit. This theory, while providing a reasonable description of disorder-averaged single-particle spectral functions, is unable to capture Anderson localization, as well as other imporant disorder effects such as the formation of inhomogeneous or glassy phases. Recent efforts, however, have resulted in various extensions of DMFT, which have succeeded to incorporate various aspects of such disorder-driven phenomena, as follows.

- *Statistical DMFT theories* (Dobrosavljević and Kotliar, 1997), which calculate the local site-dependent self-energies for a given fixed realization of disorder. Here the strong-correlation effects are described in DMFT fashion, while disorder fluctuations are treated in an exact numerical approach. In a sense, this method can be regarded as a quantum generalization of the TAP equation (Thouless *et al.*, 1977) approach to classical spin glasses. In applications to disordered Hubbard and Anderson lattice models, it led to the discovery of disorder-driven non-Fermi-liquid behavior (Miranda and Dobrosavljević, 2005) and electronic Griffiths phases (Andrade *et al.*, 2009; Miranda and Dobrosavljević, 2001; Tanasković *et al.*, 2004).
- *Typical medium theories* (Dobrosavljević *et al.*, 2003*a*), which treat both the interactions and disorder fluctuations in a DMFT-like self-consistent fashion. This method is the simplest order-parameter theory of Mott–Anderson localization.
- *Extended DMFT (EDMFT)* theories describing the effect of the bosonic collective modes due to inter-site interactions. The EDMFT approach, which was much utilized in recent work on quantum criticality in Kondo systems (Smith and Si, 2000), also proved very effective in describing quantum spin glasses (Sachdev *et al.*, 1995) and especially in describing the effects of the long-range Coulomb interaction (Chitra and Kotliar, 2000). It provided a quantitatively accurate description of the pseudogap phenomena in Coulomb systems (Pankov and Dobrosavljević, 2005) and clarified the relation between glassy freezing and formation (Pastor and Dobrosavljević, 1999) of the universal Efros–Shklovskii Coulomb gap in the presence of disorder.

1.3.4 Conclusions and outlook

Recent years have seen considerable progress in finding fascinating but often baffling examples of materials that seem to belong to the MIT region. In many of these examples, the electron localization does not appear to have a conventional "band transition" character, where some kind of uniform ordering leads to a band gap opening at the Fermi surface. Instead, mechanisms such as Mott and Anderson localization are invoked, processes that simply do not fit the cherished mold of spontaneous symmetry-breaking. Nevertheless, in systems ranging from heavy doped semiconductors, dilute two-dimensional electron gases, to organic charge-transfer salts, to cuprate-oxide and iron-pnictide materials, the metal-insulator transition region typically assumes the form expected of quantum criticality.

The traditional approaches to the MIT, dating to the 1980s, tried to circumvent the absence of an obvious order-parameter description by systematically examining the stability of the metallic (Fermi-liquid) phase to weak disorder. This approach, despite its formal elegance and conceptual simplicity, did not find many convincing applications, most likely because of its inherent inability to describe genuine strong-correlation effects—the role of "Mottness." Indeed, much of the effort of the theoretical condensed matter community over the last twenty years has focused on Mott physics in its many forms.

One central physics question has emerged from all these studies. This issue, first emphasized in the pioneering works of Mott and Anderson, relates to the question of what are the relevant length scales that dominate the "bad-metal" regime. The perspective provided by the Fermi-liquid picture of metals and the symmetry-breaking paradigm of conventional criticality both suggest that long length scales should hold the key. However, as hinted by the title of Phillip Anderson's 1978 Nobel Prize lecture "Local Moments and Localized States" (Anderson, 1978), sufficiently close to insulating states, the real-space representation may offer a better starting point.

Over the last twenty years, these local ideas acquired a precise and systematic language with the rise of DMFT approaches. This provided a natural *dynamical* order-parameter description for strongly correlated systems with and without disorder. The progress made by the various forms and extensions of DMFT is very encouraging, since they proved capable of incorporating all the basic mechanisms for electron localization. Many of these phenomena, such as the emergence of strongly inhomogeneous phases, or the description of glassy dynamics, proved to be beyond the scope of conventional Fermi-liquid-based theories of interaction–localization. In all fairness, however, the problem remains remains far from being resolved, and much more focused effort will be necessary to combine all the facets of these fascinating theories in a comprehensive and well established picture of the MIT region.

Acknowledgements

This work was supported by the NSF through grant DMR-1005751, and by the National High Magnetic Field Laboratory.

References

Abrahams, E., Anderson, P.W., Licciardello, D.C., and Ramakrishnan, T.V. (1979). *Phys. Rev. Lett.*, **42**, 673–676.

Abrahams, E., Kravchenko, S.V., and Sarachik, M.P. (2001). *Rev. Mod. Phys.*, **73**, 251–266.

Abrikosov, A.A., Gor'kov, L.P., and Dzyaloshinskii, I.E. (1975). *Methods of Quantum Field Theory in Statistical Physics.* Dover, New York.

Altshuler, B.L. and Aronov, A.B. (1979). *JETP Lett.*, **50**, 968.

Amaricci, A., Camjayi, A., Haule, K., Kotliar, G., Tanasković, D., and Dobrosavljević, V. (2010). *Phys. Rev. B*, **82**(15), 155102.

Anderson, P.W. (1958). *Phys. Rev.*, **109**, 1492–1505.

Anderson, P.W. (1959). *Phys. Rev.*, **115**(1), 2.

Anderson, P.W. (1978). *Rev. Mod. Phys.*, **50**(2), 191.

Ando, T., Fowler, A.B., and Stern, F. (1982). *Rev. Mod. Phys.*, **54**, 437–672.

Andrade, E.C., Miranda, E., and Dobrosavljević, V. (2009). *Phys. Rev. Lett.*, **102**, 206403.

Ashcroft, N. W. and Mermin, D. (1976). *Solid State Physics.* Thomson Learning, Toronto.

Basov, D. and Chubukov, A. (2011). *Nature Physics*, **7**, 272–276.

Belitz, D. and Kirkpatrick, T.R. (1994). *Rev. Mod. Phys.*, **66**, 261.

Belitz, D. and Kirkpatrick, T.R. (1995). *Phys. Rev. B*, **52**, 13922.

Beloborodov, I.S., Lopatin, A.V., Vinokur, V.M., and Efetov, K.B. (2007). *Rev. Mod. Phys.*, **79**(2), 469–518.

Berker, A.N. (1991). *J. Appl. Phys.*, **70**, 5941.

Bogdanovich, S. and Popović, D. (2002). *Phys. Rev. Lett.*, **88**, 236401.

Brinkman, W.F. and Rice, T.M. (1970). *Phys. Rev. B*, **2**, 4302.

Camjayi, A., Haule, K., Dobrosavljević, V., and Kotliar, G. (2008). *Nature Physics*, **4**, 932.

Cândido, Ladir, Bernu, B., and Ceperley, D.M. (2004). *Phys. Rev. B*, **70**, 094413.

Cândido, Ladir, Phillips, Philip, and Ceperley, D.M. (2001). *Phys. Rev. Lett.*, **86**(3), 492–495.

Casey, A., Patel, H., Nyéki, J., Cowan, B.P., and Saunders, J. (2003). *Phys. Rev. Lett*, **90**, 115301.

Castellani, C., Castro, C. Di, Lee, P.A., and Ma, M. (1984). *Phys. Rev. B*, **30**, 527.

Castellani, C., Kotliar, B.G., and Lee, P.A. (1987). *Phys. Rev. Lett.*, **56**, 1179.

Castilla, G.E. and Chakravarty, S. (1993). *Phys. Rev. Lett.*, **71**(3), 384–387.

Chen, Y.P., Sambandamurthy, G., Wang, Z.H., Lewis, R.M., Engel, L.W., Tsui, D.C., Ye, P.D., Pfeiffer, L.N., and West, K.W. (2006). *Nature Physics*, **2**, 452.

Chitra, R. and Kotliar, G. (2000). *Phys. Rev. Lett.*, **84**, 3678.

Clark, B.K., Casula, M., and Ceperley, D.M. (2009, Jul). *Phys. Rev. Lett.*, **103**(5), 055701.

Dagotto, E. (2002). *Nanoscale Phase Separation and Colossal Magnetoresistance*. Springer-Verlag, Berlin.

Dagotto, Elbio (2005). *Science*, **309**, 257.

Dalidovich, D. and Dobrosavljević, V. (2002). *Phys. Rev. B*, **66**, 081107.

De Dominicis, C., H.Orland, and T.Temisvari (1995). *J. de Physique I*, **5**, 987.

Dobrosavljević, V., Abrahams, E., Miranda, E., and Chakravarty, S. (1997). *Phys. Rev. Lett.*, **79**, 455.

Dobrosavljević, V., Kirkpatrick, T.R., and Kotliar, B.G. (1992). *Phys. Rev. Lett.*, **69**, 1113.

Dobrosavljević, V., Kirkpatrick, T.R., and Kotliar, G. (1992). *Phys. Rev. Lett.*, **69**, 1113.

Dobrosavljević, V. and Kotliar, G. (1993). *Phys. Rev. Lett.*, **71**, 3218.

Dobrosavljević, V. and Kotliar, G. (1994). *Phys. Rev. B*, **50**, 1430.

Dobrosavljević, V. and Kotliar, G. (1997). *Phys. Rev. Lett.*, **78**, 3943.

Dobrosavljević, V. and Kotliar, G. (1998). *Phil. Trans. R. Soc. Lond. A*, **356**, 1.

Dobrosavljević, V., Pastor, A.A., and Nikolić, Branislav K. (2003*a*). *Europhys. Lett.*, **62**, 76–82.

Dobrosavljević, V., Tanasković, D., and Pastor, A.A. (2003*b*). *Phys. Rev. Lett.*, **90**, 016402.

Efros, A.L. and Shklovskii, B.I. (1975). *J. Phys. C*, **8**, L49.

Finkel'stein, A.M. (1983). *Zh. Eksp. Teor. Fiz.*, **84**, 168. [Sov. Phys. JETP **57**, 97 (1983)].

Finkel'stein, A.M. (1984). *Zh. Eksp. Teor. Fiz.*, **86**, 367. [Sov. Phys. JETP **59**, 212 (1983)].

Fisher, Daniel S. (1985). *Phys. Rev. B*, **31**(11), 7233–7251.

Fisk, Z. and Webb, G.W. (1976). *Phys. Rev. Lett.*, **36**(18), 1084–1086.

Flouquet, J. (2005). Volume 15, Progress in Low Temperature Physics, pp. 139–281. Elsevier.

Georges, A., Kotliar, G., Krauth, W., and Rozenberg, M.J. (1996). *Rev. Mod. Phys.*, **68**, 13.

Goldenfeld, N. (1992). *Lectures on Phase Transitions and the Renormalization Group*. Addison-Wesley, Reading.

Gor'kov, L.P. and Sokol, A.V. (1987). *JETP Lett.*, **46**, 420.

Grannan, E.R. and Yu, C.C. (1993). *Phys. Rev. Lett.*, **71**(20), 3335–3338.

Gruner, G. (2000). *Density Waves in Solids* (1st edn). Perseus Books Group, New York.

Gull, E., Millis, A.J., Lichtenstein, A.I., Rubtsov, A.N., Troyer, M., and Werner, P. (2011). *Rev. Mod. Phys.*, **83**(2), 349–404.

Haule, K. and Kotliar, G. (2007). *Phys. Rev. B*, **76**(9), 092503.

Hertz, J.A. (1976). *Phys. Rev. B*, **14**, 1165.

Hewson, A.C. (1993). *The Kondo Problem to Heavy Fermions*. Cambrige University Press, Cambridge.

Hoyos, José A., Kotabage, Chetan, and Vojta, Thomas (2007, Dec). *Phys. Rev. Lett.*, **99**(23), 230601.

Hubbard, J. (1963). *Proc. R. Soc. (London) A*, **276**, 238.

Hussey, N.E., Takenaka, K., and Takagi, H. (2004). *Philos. Mag.*, **84**(27), 2847–2864.

Imbrie, J.Z. (1984). *Phys. Rev. Lett.*, **53**, 1747.

Imry, Y. and Ma, S.K. (1975). *Phys. Rev. Lett.*, **35**, 1399.

Jamei, R., Kivelson, S., and Spivak, B. (2005). *Phys. Rev. Lett.*, **94**, 056805.

Jaroszyński, J., Andrearczyk, T., Karczewski, G., Wróbel, J., Wojtowicz, T., Popović, Dragana, and Dietl, T. (2007). *Phys. Rev. B*, **76**(4), 045322.

Jaroszynski, J., Popovic, Dragana, and Klapwijk, T.M. (2004). *Phys. Rev. Lett.*, **92**, 226403.

Kamal, M. and Murthy, G. (1993). *Phys. Rev. Lett.*, **71**(12), 1911–1914.

Kirkpatrick, S. (1973). *Rev. Mod. Phys.*, **45**, 574.

Kirkpatrick, T.R. and Belitz, D. (1994). *Phys. Rev. Lett.*, **74**, 1178.

Kohsaka, Y., Iwaya, K., Satow, S., Hanaguri, T., Azuma, M., Takano, M., and Takagi, H. (2004). *Phys. Rev. Lett.*, **93**, 097004.

Kotliar, G. (1987). *Anderson Localization*. Springer, Berlin.

Kotliar, G. and Ruckenstein, A.E. (1986). *Phys. Rev. Lett.*, **57**(11), 1362.

Kotliar, G., Savrasov, S.Y., Haule, K., Oudovenko, V.S., Parcollet, O., and Marianetti, C.A. (2006). *Rev. Mod. Phys.*, **78**(3), 865–951.

Kravchenko, S.V., Mason, W.E., Bowker, G.E., Furneaux, J.E., Pudalov, V.M., and D'Iorio, M. (1995). *Phys. Rev. B*, **51**, 7038.

Kravchenko, S.V. and Sarachik, M.P. (2004). *Rep. Prog. Phys.*, **67**, 1.

Landau, L.D. (1957). *Sov. Phys. JETP*, **3**, 920.

Landau, L.D. (1959). *Sov. Phys. JETP*, **8**, 70.

Lee, Mark, Massey, J.G., Nguyen, V.L., and Shklovskii, B.I. (1999). *Phys. Rev. B*, **60**(3), 1582–1591.

Lee, P.A. and Ramakrishnan, T.V. (1985, 4). *Rev. Mod. Phys.*, **57**(2), 287.

Limelette, P., Georges, A., Jerome, D., Wzietek, P., Metcalf, P., and Honig, J.M. (2003*a*). *Science*, **302**, 89.

Limelette, P., Wzietek, P., Florens, S., Georges, A., Costi, T.A., Pasquier, C., Jerome, D., Meziere, C., and Batail, P. (2003*b*). *Phys. Rev. Lett.*, **91**, 016401.

Maier, Thomas, Jarrell, Mark, Pruschke, Thomas, and Hettler, Matthias H. (2005). *Rev. Mod. Phys.*, **77**(3), 1027–1080.

McGreevy, John (2010). *Physics*, **3**, 83.

Mezard, M. and Monasson, R. (1994). *Phys. Rev. B*, **50**, 7199–7202.

Mézard, M., Parisi, G., and Virasoro, M.A. (1986). *Spin Glass Theory and Beyond.* World Scientific, Singapore.

Mezard, M. and Young, A.P. (1992). *Europhys. Lett.*, **18**, 653–659.

Millis, A.J. (1993). *Phys. Rev. B*, **48**(10), 7183.

Milovanović, M., Sachdev, S., and Bhatt, R.N. (1989). *Phys. Rev. Lett.*, **63**, 82.

Miranda, E. and Dobrosavljević, V. (2001). *Phys. Rev. Lett.*, **86**, 264.

Miranda, E. and Dobrosavljević, V. (2005). *Rep. Prog. Phys.*, **68**, 2337.

Moeller, G., Dobrosavljević, V., and Ruckenstein, A.E. (1999). *Phys. Rev. B*, **59**(10), 6846–6854.

Mooij, J.H. (1973). *Phys. Status Solidi A*, **17**, 521–530.

Moreo, A., Mayr, M., Feiguin, A., Yunoki, S., and Dagotto, E. (2000). *Phys. Rev. Lett.*, **84**, 5568.

Mott, N.F. (1949). *Proc. Roy. Soc. (London)*, **A197**, 269.

Mott, N.F. (1990). *Metal-Insulator Transition.* Taylor & Francis, London.

Nelson, David R. and Pelcovits, Robert A. (1977). *Phys. Rev. B*, **16**(5), 2191–2199.

Paalanen, M.A. and Bhatt, R.N. (1991). *Physica B*, **169**, 231.

Paalanen, M.A., Graebner, J.E., Bhatt, R.N., and Sachdev, S. (1998). *Phys. Rev. Lett.*, **61**, 597.

Paalanen, M.A., Rosenbaum, T.F., Thomas, G.A., and Bhatt, R.N. (1982). *Phys. Rev. Lett.*, **48**, 1284.

Pankov, S. and Dobrosavljević, V. (2005). *Phys. Rev. Lett.*, **94**, 046402.

Parisi, G. and Sourlas, N. (1979). *Phys. Rev. Lett.*, **43**, 744.

Park, H., Haule, K., and Kotliar, G. (2008). *Phys. Rev. Lett.*, **101**(18), 186403.

Park, S., Hur, N., Guha, S., and Cheong, S.W. (2004). *Phys. Rev. Lett.*, **93**, 107207.

Pastor, A.A. and Dobrosavljević, V. (1999). *Phys. Rev. Lett.*, **83**, 4642.

Pastor, A.A., Dobrosavljević, V., and Horbach, M.L. (2001). *Phys. Rev. B*, **66**, 014413.

Polyakov, A.M. (1975). *Phys. Lett. B*, **59**(1), 79–81.

Popović, D., Fowler, A.B., and Washburn, S. (1997). *Phys. Rev. Lett.*, **79**, 1543.

Qazilbash, M.M., Tripathi, A., Schafgans, A.A., Kim, Bong-Jun, Kim, Hyun-Tak, Cai, Zhonghou, Holt, M.V., Maser, J.M., Keilmann, F., Shpyrko, O.G., and Basov, D.N. (2011). *Phys. Rev. B*, **83**(16), 165108.

Radonjic, Milos M., Tanaskovic, D., Dobrosavljević, V., and Haule, K. (2010). *Phys. Rev. B*, **81**(7), 075118.

Reichhardt, C., Reichhardt, C.J. Olson, and Bishop, A.R. (2005). *EPL (Europhysics Letters)*, **72**(3), 444.

Richardella, Anthony, Roushan, Pedram, Mack, Shawn, Zhou, Brian, Huse, David A., Awschalom, David D., and Yazdani, Ali (2010). *Science*, **327**(5966), 665–669.

Rosenbaum, T.F., Andres, K., Thomas, G.A., and Bhatt, R.N. (1980). *Phys. Rev. Lett.*, **45**, 1723.

Sachdev, S. (2010). *Phys. Rev. Lett.*, **105**(15), 151602.

Sachdev, S. (2011). *Quantum Phase Transitions* (2nd edn). Cambridge University Press, UK.

Sachdev, S., Read, N., and Oppermann, R. (1995). *Phys. Rev. B*, **52**, 10286.

Sapienza, L., Thyrrestrup, H., Stobbe, S., Garcia, P. David, Smolka, S., and Lodahl, P. (2010). *Science*, **327**(5971), 1352–1355.

Sarachik, M.P. (1995). In *Metal–insulator Transitions Revisited* (ed. P. Edwards and C. Rao). Taylor and Francis.

Schaffer, L. and Wegner, F. (1980). *Phys. Rev. B*, **38**, 113.

Schmalian, J. and Wolynes, P.G. (2000). *Phys. Rev. Lett.*, **85**, 836.

Shahar, D., Tsui, D.C., Shayegan, M., Shimshoni, E., and Sondhi, S.L. (1997). *Phys. Rev. Lett.*, **79**(3), 479–482.

Shklovskii, B.I. and Efros, A.L. (1984). *Electronic Properties of Doped Semiconductors.* Springer-Verlag.

Si, Q., Rabello, S., Ingersent, K., and Smith, J.L (2001). *Nature*, **413**(6858), 804–808.

Siegrist, T., Jost, P., Volker, H., Woda, M., Merkelbach, P., Schlockermann, C., and Wuttig, M. (2011, March). *Nat. Mater.*, **10**(3), 202–208.

Simonian, D., Kravchenko, S.V., and Sarachik, M.P. (1997). *Phys. Rev. B*, **55**(20), R13421–R13423.

Slater, J.C. (1934). *Rev. Mod. Phys.*, **6**, 209.

Slater, J.C. (1951). *Phys. Rev.*, **82**, 538.

Smith, J.L. and Si, Q. (2000). *Phys. Rev. B*, **61**, 5184.

Stauffer, D. and Aharoni, A. (1994). *Introduction to Percolation Theory* (2nd edn). Taylor & Francis, London.

Stewart, G.R. (1984). *Rev. Mod. Phys.*, **56**, 755.

Straley, J.P. (1977). *Phys. Rev. B*, **15**, 5733.

Tanaskovic, D., Dobrosavljević, V., and Miranda, E. (2005). *Phys. Rev. Lett.*, **95**, 167204.

Tanasković, D., Miranda, E., and Dobrosavljević, V. (2004). *Phys. Rev. B*, **70**, 205108.

Tanatar, B. and Ceperley, D.M. (1989). *Phys. Rev. B*, **39**(8), 5005–5016.

Terletska, H. and Dobrosavljević, V. (2011). *Phys. Rev. Lett.*, **106**(18), 186402.

Terletska, H., Vučičević, J., Tanasković, D., and Dobrosavljević, V. (2011). *Phys. Rev. Lett.*, **107**(2), 026401.

Thomas, G.A., Capizzi, M., DeRosa, F., Bhatt, R.N., and Rice, T.M. (1981). *Phys. Rev. B*, **23**(10), 5472–5494.

Thouless, D.J. (1978). *J. Phys. C: Solid State*, **11**(6), L189.

Thouless, D.J., Anderson, P.W., and Palmer, R.G. (1977). *Philos. Mag.*, **35**, 137.

Ting, C.S., Lee, T.K., and Quinn, J.J. (1975). *Phys. Rev. Lett.*, **34**(14), 870–874.

Vojta, T. (2006, 16 May). *J. Phys. A: Math. Gen.*, **39**, R143–R205.

Vollhardt, D. (1984). *Rev. Mod. Phys.*, **56**, 99.

Waintal, X. (2006). *Phys. Rev. B*, **73**(7), 075417.

Wegner, F.J. (1976). *Z. Phys. B Cond. Mat.*, **25**, 327–337.

Wegner, F. (1979). *Phys. Rev. B*, **35**, 783.

Wen, Xiao-Gang (1992). *Phys. Rev. B*, **46**(5), 2655–2662.
Widom, B. (1965). *J. Chem. Phys.*, **43**(11), 3898.
Wigner, E. (1934, Dec). *Phys. Rev.*, **46**(11), 1002–1011.
Wilson, K.G. and Fisher, M.E. (1972). *Phys. Rev. Lett.*, **28**(4), 240–243.
Wilson, K.G. and Kogut, J. (1975). *Physics Reports*, **12C**(2).
Yang, S.-X., Fotso, H., Su, S.-Q., Galanakis, D., Khatami, E., She, J.-H., Moreno, J., Zaanen, J., and Jarrell, M. (2011). *Phys. Rev. Lett.*, **106**(4), 047004.
Zala, G., Narozhny, B.N., and Aleiner, I.L. (2001). *Phys. Rev. B*, **64**, 214204(31).
Zhang, Y. and Das Sarma, S. (2005). *Phys. Rev. B*, **71**(4), 045322.

2
Metal–Insulator Transitions in Two-dimensional Electron Systems

S.V. Kravchenko

Physics Department, Northeastern University, Boston, Massachusetts 02115, USA,
and
Department of Physics, Loughborough University, Leicestershire, LE11 3TU, UK

2.1 Introduction

In two-dimensional (2D) electron systems—such as very thin metallic films or the active region of field-effect semiconductor transistors—the electrons are constrained to move in a plane. Three decades ago, a powerful renormalization group theory was put forward (Abrahams *et al.*, 1979) which indicated that at zero temperature, any amount of disorder, however small, would trap electrons, thus preventing conduction and the existence of a metallic state in two-dimensions (2D). The arguments were based on the quantum-wave nature of the electrons. A traveling electron wave can be scattered from impurities back to its starting point. If these returning waves interfere constructively, the electrons become localized within a certain range called the localization length, ξ. According to the renormalization group theory, the logarithmic derivative

$$\frac{\mathrm{d}\ln G}{\mathrm{d}\ln L} = \beta(G) \tag{2.1}$$

is a function of the conductance $G \equiv 1/R$ alone (here R is the resistance—not the resistivity—of the sample of size L). If the localization length is much bigger than the sample size (high G), the quantum interference effects can be disregarded and $G \propto L^{d-2}$, where d is the dimension of the sample: $d = 3$ for a three-dimensional (3D) sample, $d = 2$ for a 2D sample, and $d = 1$ for a one-dimensional (1D) sample. Therefore, according to eqn (2.1), $\beta(G \to \infty) = 1$, 0, and -1 in 3D, 2D, and 1D, correspondingly. If, however, the localization length is smaller than the sample size, the quantum interference becomes important and an additional factor $\exp(-L/\xi)$ appears in the equation for G:

$$G \propto L^{d-2}\, e^{-L/\xi}, \tag{2.2}$$

thus making $\beta(G)$ decrease with decreasing G, as schematically shown in Fig. 2.1.

As one sees, in a 3D case, $\beta(G)$ can be either negative or positive. When it is negative, the conductance decreases with the sample size, while for positive β, it

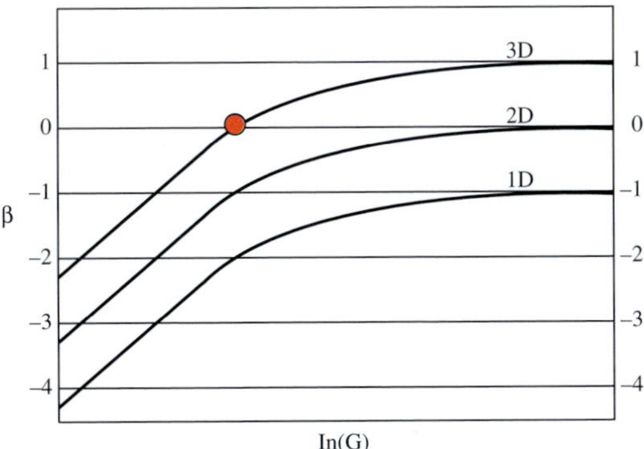

Fig. 2.1 $\beta(G)$ for 3D, 2D, and 1D cases. The dot indicates the point of the metal–insulator transition in 3D.

increases. Thus, infinitely big 3D sample will be an insulator for $\beta < 0$ and a conductor for $\beta > 0$. A solid circle on the figure at $\beta = 0$ indicates the point of the metal-to-insulator transition. However, in 2D and 1D cases, $\beta(G)$ is always negative meaning that an infinitely large (long) sample will always be an insulator.

The above considerations are true in the zero-temperature limit. At non-zero temperatures T, electron coherence breaks at a so-called phase-breaking time τ_ϕ, due to interactions with phonons. Therefore, if the corresponding phase-breaking length L_ϕ is smaller than the sample size, the latter becomes irrelevant and needs to be replaced with L_ϕ, which is roughly proportional to $1/T$. Therefore, eqn (2.1) can be replaced with

$$-\frac{\mathrm{d}\ln G}{\mathrm{d}\ln T} = \beta(G). \tag{2.3}$$

As the temperature is reduced, 3D system of an infinite size will become more and more insulating while $\beta < 0$ and more and more metallic when $\beta > 0$. However, although at higher temperatures 2D and 1D systems may appear metallic, they become more and more insulating as $T \to 0$, and an infinite 2D or 1D system will have zero conductance at $T = 0$. In the 2D case, detailed calculations yield logarithmic temperature dependence of the conductance:

$$G = G_{\mathrm{D}} + \frac{e^2}{h}\ln(T\tau/\hbar), \tag{2.4}$$

where G_{D} is the classical (Drude) conductance and τ is the mean-free time. Low-temperature experiments on both thin metal films (Dolan and Osheroff, 1979) and 2D systems of electrons in field-effect transistors (Bishop *et al.*, 1980; Uren *et al.*, 1980) confirmed these theoretical predictions.

The discovery of the quantum Hall effect by von Klitzing *et al.* (1980) challenged the scaling theory of localization. Obviously, under the conditions of the quantum Hall effect, 2D systems conducted electric current with the resistivity exponentially decreasing with decreasing temperature. Pruisken (1984) (see also Pruisken and Burmistrov (2005)) put forward the two-parameter scaling theory, according to which there are electronic states that are extended at the middle of each Landau level. Therefore, it was agreed that zero-temperature conductivity can exist in 2D in high magnetic fields $B_\perp$ perpendicular to the surface, but not at $B = 0$. The question of what happens to these extended states when the magnetic field approaches zero was addressed by Khmelnitskii (1984) and Laughlin (1984), who have shown that the energy of these states increases as $\omega_{\mathrm{c}} \equiv eB_\perp/m \to 0$, eventually lifting them above the Fermi energy and thus promoting an insulating state at $B = 0$ (see Fig. 2.2).

However, as 2D electron systems with increasingly high quality (low disorder) became available in a variety of devices, anomalous low-temperature transport properties were reported, seemingly inconsistent with the paradigm that the ground state of 2D systems must necessarily be insulating. These were strongly correlated 2D systems with a large ratio between the typical potential (Coulomb) and kinetic (Fermi) energies, E_{C} and E_{F}:

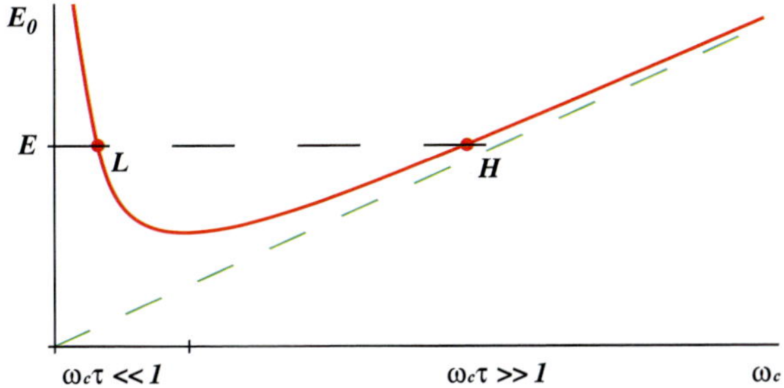

Fig. 2.2 Energy position of the lowest delocalized state vs magnetic field ω_c, as predicted by Khmelnitskii (1984) and Laughlin (1984). Below point L, the system is insulating. Adapted from Mkhitaryan *et al.* (2009).

$$r_s \equiv \frac{E_C}{E_F} = \frac{n_v\, m^*\, e^2}{\hbar^2 \epsilon \sqrt{\pi n_s}}, \tag{2.5}$$

where n_s is the areal density of electrons, ϵ is the dielectric constant, m^* is the effective mass, and n_v is the number of degenerate valleys in the spectrum. (Note that to reach high values of r_s, low electron densities are needed. The low-density regime can be reached only in low-disordered samples.) Clearly, a system with $r_s \gg 1$ cannot be considered non-interacting and, *a priori*, the conclusion of the scaling theory of localization is not valid in this regime. There were no theoretical predictions for the ground state of a strongly correlated 2D electron system except for the case of very high $r_s > r_s^* \approx 38$ at which the formation of Wigner crystal (Wigner, 1934) is expected (Attaccalite *et al.*, 2002; Tanatar and Ceperley, 1989). At intermediate strengths of interaction, $1 \ll r_s < r_s^*$, however, electrons are highly correlated but still not crystallized. This is the area on which I will focus in this chapter.

Below, I will present a summary of the experimental results on large r_s 2D electron and hole gases (2DEGs and 2DHGs) in clean semiconductor devices. I will particularly focus on experiments on the following systems: silicon metal-oxide-semiconductor field-effect transistors (MOSFETs), p-GaAs heterojunctions and quantum wells, n-GaAs heterojunctions, p- and n-SiGe quantum wells, and AlAs quantum wells. The typical values of r_s are $\sim$ 10–40, depending on the device. Note that the high r_s regime is easier to reach in Si MOSFETs than the much cleaner n-GaAs-based devices, due to relatively high effective mass ($0.19\, m_e$ compared with $0.07\, m_e$ in n-GaAs), lower average dielectric constant (7.8 compared to 12 in n-GaAs), and the existence of two degenerate valleys in the spectrum, which further increase r_s by a factor of 2 (see eqn 2.5). As a result, to reach the same interaction strengths, electron densities two orders of magnitude lower are needed in n-GaAs heterojunctions compared to those in Si MOSFETs.

The great similarities in the data obtained on different types of 2D system imply that the observed behaviors are robust, universal, and largely independent of details, in spite of the fact that in various 2D structures there are significant differences between the electronic spectrum and the nature of the disorder potential (short-range in Si MOSFETs and long-range in most of the other 2D systems considered here).

2.2 Transport properties

2.2.1 Metal–insulator transition

All low-disordered 2D systems, in which high-r_s regimes can be reached, exhibit a metal–insulator transition as a function of the electron density. This effect has been observed in Si MOSFETs, p- and n-GaAs quantum wells, p- and n- SiGe quantum wells, and n-AlAs quantum wells. Examples of the experimental data showing the resistivity ρ as a function of temperature T for different electron densities in various systems are given in Fig. 2.3. At low temperatures, the resistivity $\rho(T)$ exhibits a "metallic" temperature dependence ($\mathrm{d}\rho(T)/\mathrm{d}T > 0$) for electron densities $n_s > n_c$, where n_c is the well-defined critical value. Below this critical density, an insulating behavior ($\mathrm{d}\rho(T)/\mathrm{d}T < 0$) is observed. At the lowest temperatures $\rho \ll h/e^2$ for $n > n_c$, while $\rho \gg h/e^2$ for $n < n_c$. At the critical density, the resistivity is of the order of h/e^2. Note that if the Drude formula were valid here, $\rho = h/e^2$ would correspond to a mean-free-path ℓ equal to the Fermi wave-length: $\ell = 2\pi/k_F$. This is the well-known Ioffe–Regel criterion for the metal–insulator transition based on the uncertainty relation (Ioffe and Regel, 1960).

A quantum phase transition is defined, only at zero temperature, as the point at which the resistance changes abruptly from being finite (metallic) to infinite (insulating). At any non-zero temperature such a transition would manifest itself as a rapid crossover. Since all experiments are conducted at finite temperatures, it is impossible to tell unambiguously whether the transport data shown in Fig. 2.3 represent a quantum phase transition or a finite-temperature crossover. However, the resistivity does change by many orders of magnitude as n_s varies over a narrow range near n_c, and the lower the temperature is, the stronger the resistivity change. This strongly suggests that we are indeed dealing with a phase transition. It is interesting to compare the data for 2D and 3D metal–insulator transitions, the latter being shown in Fig. 2.3 (f). From the empirical viewpoint, the evidence for the metal–insulator transition in 3D, "allowed" by the scaling theory of localization, is no better than that in 2D.

2.2.2 Temperature dependence of $\rho(T)$ deep in the metallic regime

As shown in Fig. 2.4 and in the curves in Figs. 2.3 (b–e), in samples with large r_s and $\rho \ll h/e^2$, the resistance $\rho(T)$ increases with increasing temperature to a peak value $\rho(T_{max})$. In Si MOSFETs, where the effect is stronger than in other 2D systems, this increase may reach as much as an order of magnitude. At $T > T_{max}$, the resistance decreases with temperature by as much as a factor of two (see Fig. 2.4), before ultimately starting to increase again at higher temperatures, where phonon scattering starts to dominate. Still deeper in the metallic regime (i.e. at smaller r_s), the resistivity is a monotonically increasing function of the temperature (see the lowest curves in Fig. 2.3 (b-e)) over the whole temperature range.

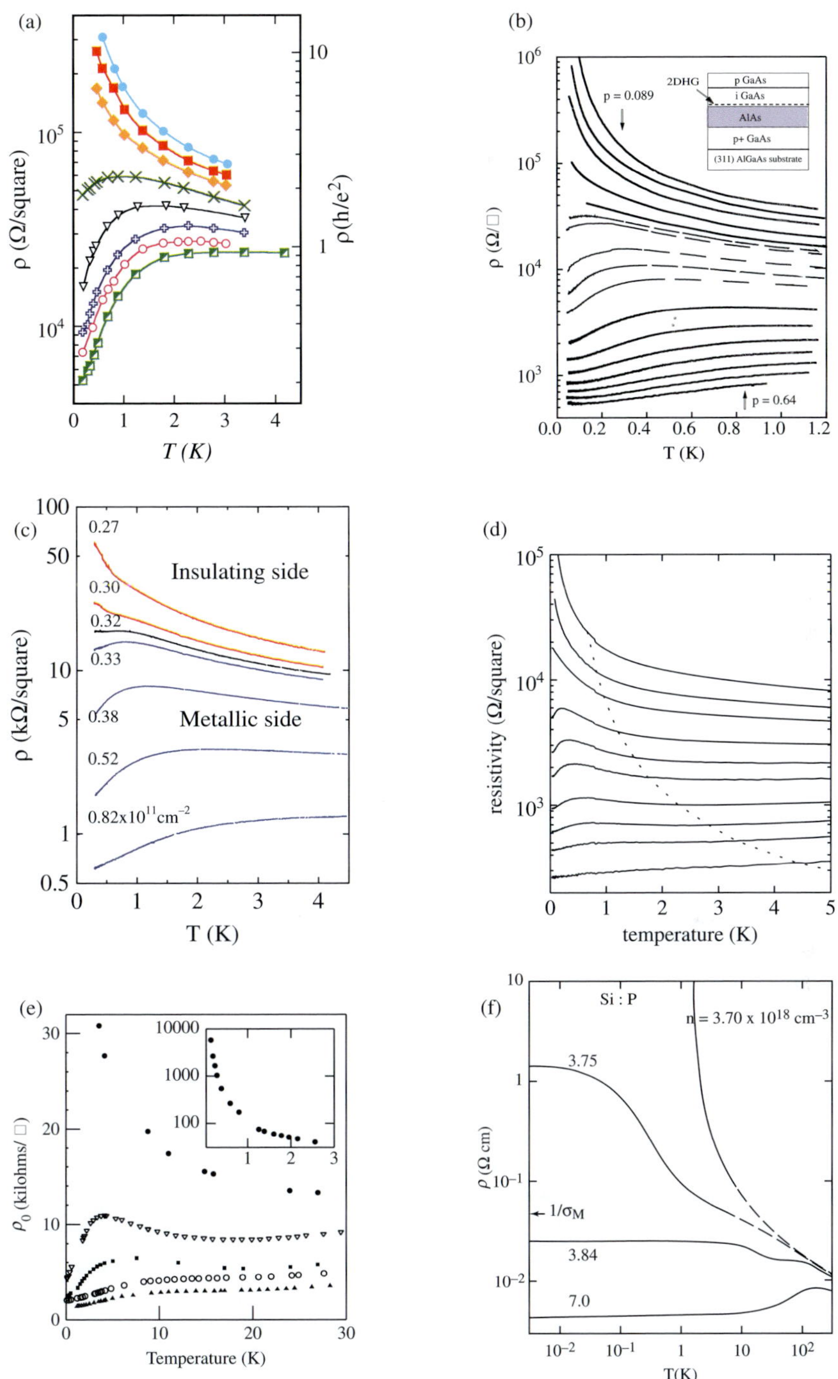

Fig. 2.3 Critical behavior of the resistivity near the 2D metal–insulator transition in: (a) Si MOSFET, (b) p-GaAs/AlGaAs heterostructure, (c) n-SiGe heterostructure, (d) n-GaAs/AlGaAs heterostructure, and (e) p-SiGe quantum well. For comparison, we also show in (f) the critical behavior of the resistivity in Si:P, a 3D system. Figures adapted from Anissimova *et al.* (2007), Hanein *et al.* (1998*a*), Lai *et al.* (2005), Lilly *et al.* (2003), Coleridge *et al.* (1997), and Rosenbaum *et al.* (1983), respectively.

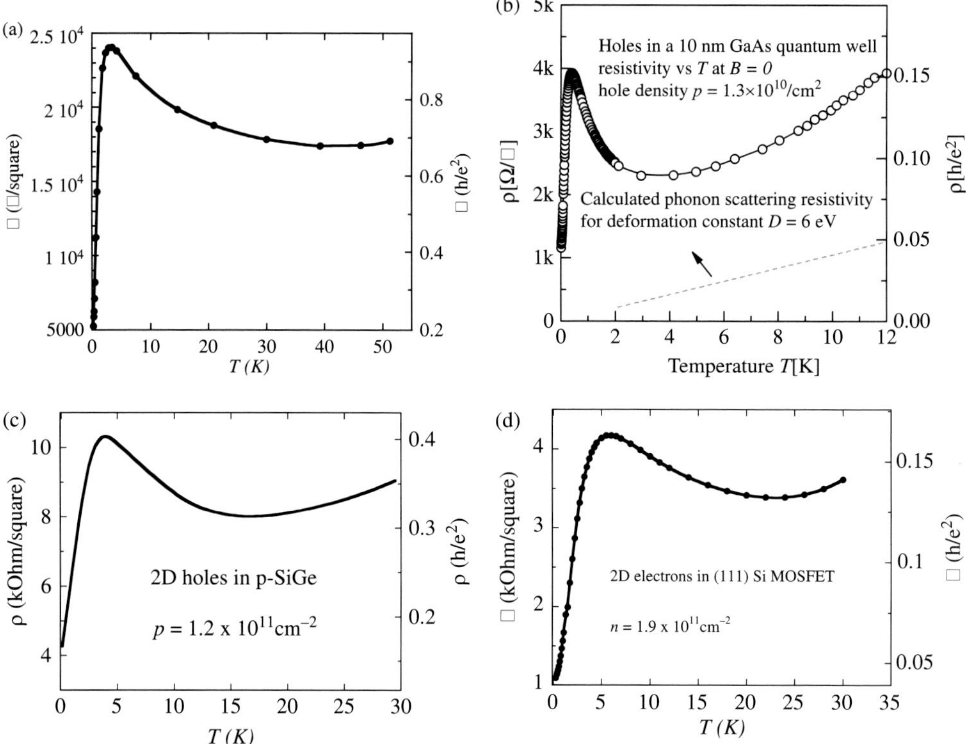

Fig. 2.4 Non-monotonic temperature dependence of the resistivity in (a) (100) Si MOSFET, (b) p-GaAs quantum well, (c) p-SiGe quantum well, and (d) (111) Si MOSFET deep in the metallic regime over an extended temperature range. Adapted from Mokashi and Kravchenko (2009) (a), Gao *et al.* (2005) (b), Coleridge (1997) (c) and McFarland *et al.* (2009) (d).

2.2.3 The effect of a parallel magnetic field

A magnetic field parallel to the 2D plane ($B_{\|}$) couples only to the electrons' spins. A sufficiently strong field completely spin-polarizes electrons when the Zeeman splitting exceeds the Fermi energy. As shown in Fig. 2.5, in Si MOSFETs, p-GaAs heterojunctions, n-AlAs quantum wells, and SiGe quantum wells with large r_{s}, the resistance exhibits a strongly positive magnetoresistance, increasing by as much as an order of magnitude as a function of $B_{\|}$, before eventually saturating for $B_{\|} > B^* \sim E_{\mathrm{F}}/g\mu_{\mathrm{B}}$. Here μ_{B} is the effective Bohr magneton and g is the Landé g-factor. It has been experimentally shown (Vitkalov *et al.*, 2001*a*) that the saturation field coincides with the field of the full-spin polarization to an accuracy of 5%. At temperatures below 50 mK, the magnetoresistance may exceed five orders of magnitude, as shown in Fig. 2.6.

At electron densities close to the point of the $B = 0$ metal–insulator transition, $B_{\|}$ can induce a metal–insulator transition as shown in Fig. 2.7. The critical behavior of

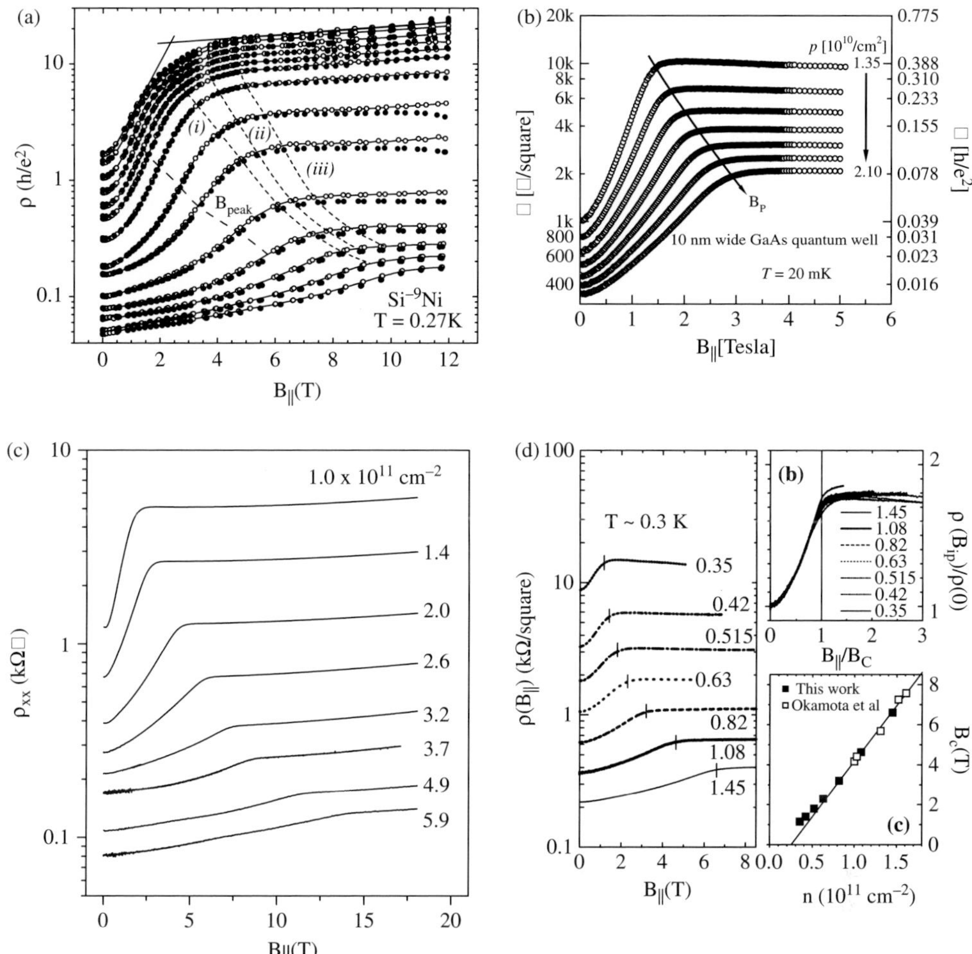

Fig. 2.5 Giant magnetoresistance in a parallel magnetic field in (a) strongly metallic Si MOSFET, (b) 10-nm wide p-GaAs quantum well, (c) n-AlAs quantum well, and (d) n-SiGe quantum well. Adapted from Pudalov *et al.* (1998), Gao *et al.* (2006), De Poortere *et al.* (2002) and Lai *et al.* (2005), respectively.

the resistance as a function of temperature, seen in Fig. 2.3 and also in Fig. 2.8 (a), is completely wiped away in a parallel magnetic field above B^* (see Fig. 2.8 (b)).

2.2.4 Absence of the "floating up" of the extended states as $B_\perp \to 0$

As has already been mentioned, a non-interacting 2D electron system is supposed to be insulating in the absence of a magnetic field, but at strong perpendicular magnetic fields $B_\perp$, it is conducting due to the existence of extended states in the middle of each Landau level. As $B_\perp \to 0$, these extended states should "float up" in

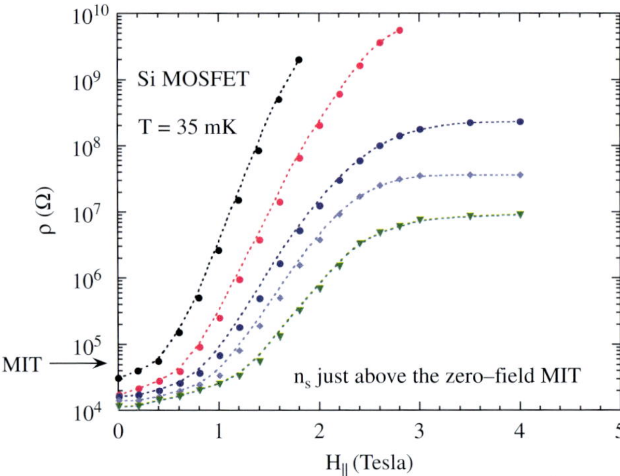

Fig. 2.6 Magnetoresistance of nearly-critical Si MOSFET at low temperatures. Adapted from Shashkin and Kravchenko (2000). MIT, metal–insulator transition.

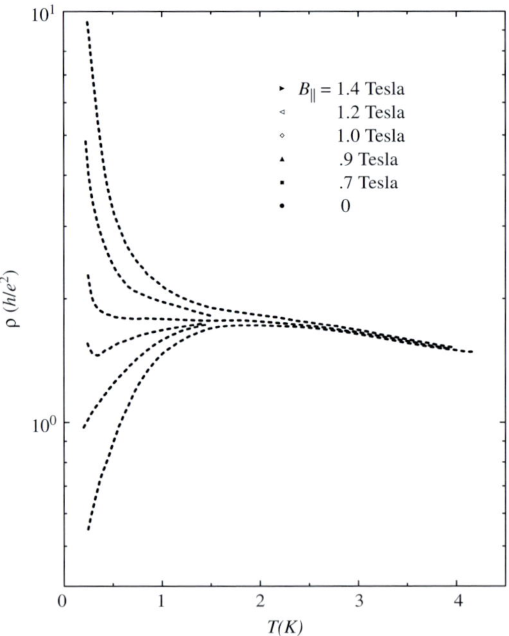

Fig. 2.7 Parallel magnetic-field-induced metal–insulator transition in a Si MOSFET. Adapted from Simonian *et al.* (1997).

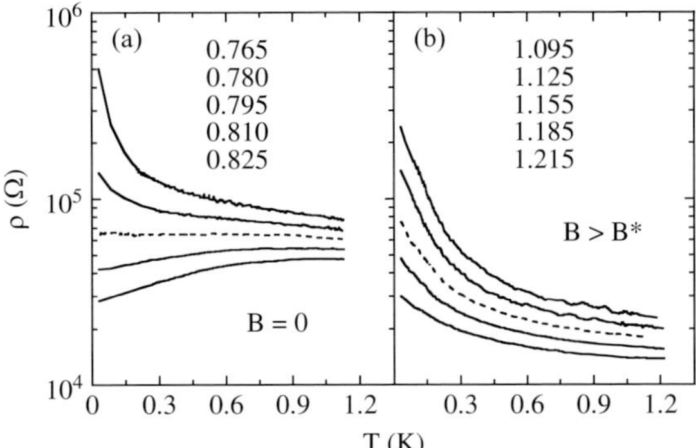

Fig. 2.8 Temperature dependence of the resistivity of an Si MOSFET at different electron densities near the met–insulator transition (a) in a zero magnetic field and (b) in a parallel magnetic field of 4 Tesla, which is above B^*. The electron densities are indicated in units of 10^{11} cm^{-2}. Adapted from Shashkin *et al.* (2001).

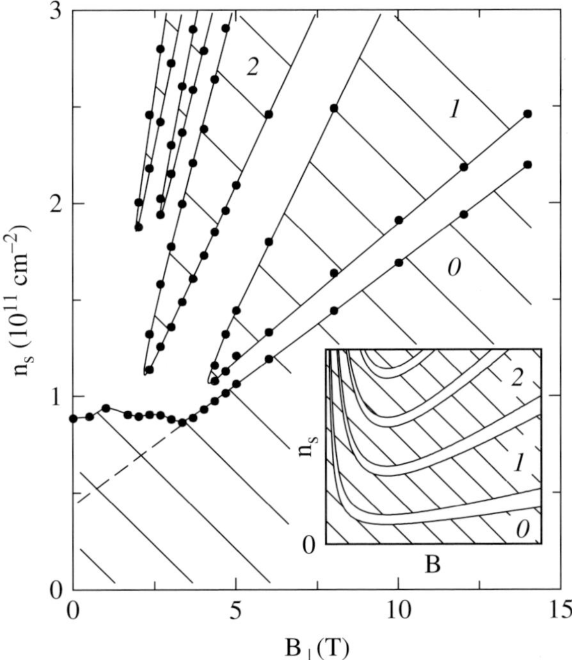

Fig. 2.9 Bands of the extended states (white areas) in a dilute Si MOSFET. The shaded area corresponds to localized states. The inset shows the expected "floating" of the extended states (Khmelnitskii, 1984; Laughlin, 1984). Adapted from Shashkin *et al.* (1993).

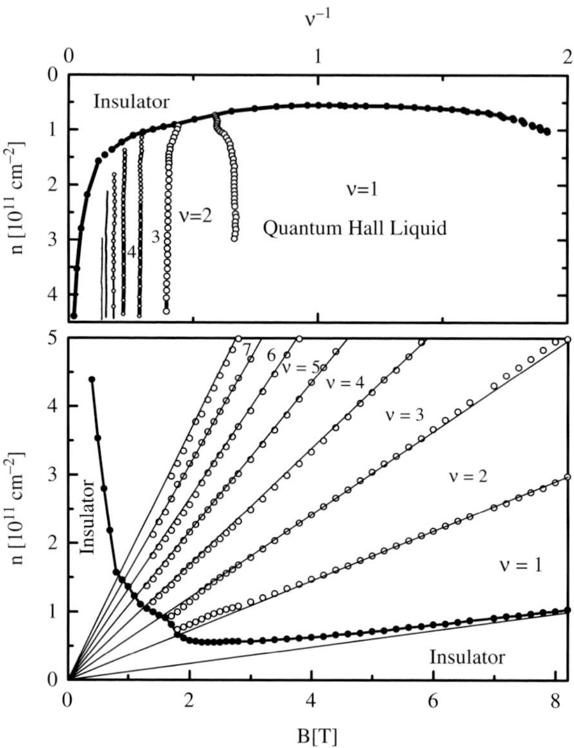

Fig. 2.10 A map of the extended states for a highly-disordered 2D hole system in a Ge/SiGe quantum well. The open circles represent the positions of the extended state in the quantum Hall effect regime. The solid circles correspond to the position of the lowest extended state. Numbers show the value of $\rho_{xy}h/e^2$. Adapted from Hilke *et al.* (2000).

energy thus making a system an insulator (Khmelnitskii, 1984; Laughlin, 1984), as schematically shown in Fig. 2.2 and also in the inset to Fig. 2.9. Indeed, such floating up of the extended states has been experimentally observed in strongly disordered, weakly interacting 2D systems, as shown in Fig. 2.10.

In contrast, in a strongly correlated 2D electron system in clean Si MOSFET samples with large r_s, as $B_\perp \to 0$ the extended states are observed to extrapolate to roughly the same value n_c that marks the zero-field metal–insulator transition in the same device (Shashkin *et al.*, 1993), as shown in Fig. 2.9. This further corroborates the identification of n_c as a critical point.

2.3 Magnetic properties of the electron liquid at large r_s

2.3.1 Spin susceptibility

Compared to the transport properties, the thermodynamic properties of strongly correlated 2D electron systems are less well explored. Most of the data on spin magnetization were obtained by analyzing the parallel-field magnetoresistance. Direct measurements of the thermodynamic magnetization have so far been performed only on one type of 2D systems, silicon MOSFETs (Prus *et al.*, 2003; Shashkin *et al.*, 2006).

The easiest way to estimate spin magnetization of 2D electrons (or holes) is to measure the magnetic field B^* at which the magnetoresistance saturates (and thus full spin polarization is reached) for different electron densities. For non-interacting electrons, the saturation field is proportional to the electron density:

$$B^* = \frac{\pi\hbar^2 n_s}{gm\mu_B},$$

where g is the Lande g-factor, m is the effective mass, and μ_B is the Bohr magneton. It turns out, however, that in strongly correlated 2D systems in Si MOSFETs, the $B^*(n_s)$ dependence extrapolates to a non-zero electron density n_χ, as shown in Fig. 2.11 (for comparison, the calculated $B^*(n_s)$ dependence is also shown, as a dashed line). The fact that the measured B^* lies significantly lower than the calculated value indicates that either g or m (or both) are larger than their band values. Moreover, if the extrapolation of B^* to zero is valid and the field of complete spin polarization indeed vanishes at $n_s = n_\chi$, then it indicates the occurrence of a spontaneous spin polarization at $n_s = n_\chi$. Many Si MOSFET samples of different quality have been tested, and it turned out that $n_\chi \approx 8 \times 10^{10}$ cm^2 is disorder-independent. In the highest-quality samples, n_χ was found to be within a few percent of the critical density for the metal–insulator transition n_c (but consistently below).

It is easy to recalculate the renormalized spin susceptibility using the data for $B^*(n_s)$:

$$\frac{\chi}{\chi_0} = \frac{n_s}{n_s - n_\chi},$$

where χ_0 is the "non-interacting" value of the spin susceptibility. Such critical behavior in a thermodynamic parameter usually indicates that a system is approaching a phase

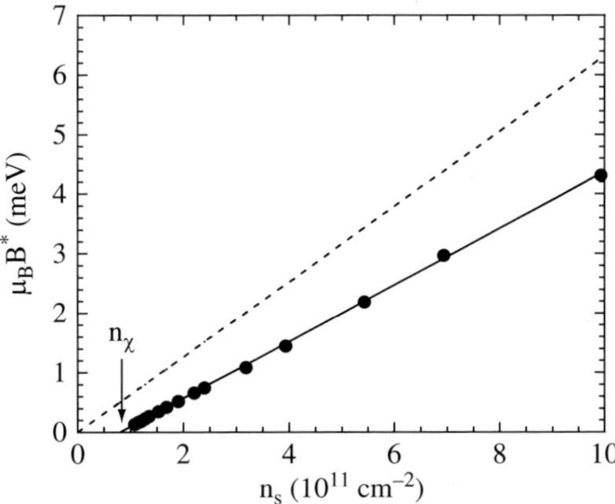

Fig. 2.11 Magnetic field of complete spin polarization vs electron density (solid line). The dashed line depicts B^*, calculated assuming that g and m are not renormalized.

transition, possibly of magnetic origin. However, evidence for the phase transition can only be obtained by thermodynamic measuring techniques. Given the tiny number of electrons in the 2D layer, magnetic measurements are rather hard to perform. A clever technique was designed and implemented by Prus *et al.* (2003) and Shashkin *et al.* (2006). The idea was to modulate the parallel magnetic field with a small AC field B_{mod} and to measure the tiny induced current between the gate and the two-dimensional electron system. The imaginary (out-of-phase) current component is proportional to $\mathrm{d}\mu/\mathrm{d}B$, where μ is the chemical potential of the 2D gas. By applying the Maxwell relation $\mathrm{d}M/\mathrm{d}n_{\mathrm{s}} = -\mathrm{d}\mu/\mathrm{d}B$, one can obtain the magnetization M from the measured current. Full spin polarization corresponds to $\mathrm{d}M/\mathrm{d}n_{\mathrm{s}} = 0$. Yet another way of finding the density of the complete spin polarization involves measurements of the thermodynamic density of states of the 2D system by measuring the capacitance of a MOSFET (Shashkin *et al.*, 2006): the thermodynamic density of states abruptly changes when the electrons' spins become completely polarized.

The results obtained for the spin susceptibility in this way are shown in Fig. 2.12. One can see that upon approaching the critical density of the metal–insulator transition, the spin susceptibility grows by almost an order of magnitude relative to its "non-interacting" value. This favors the occurrence of a spontaneous spin polarization (either Wigner crystal or ferromagnetic liquid) at low n_{s}, although in currently available samples the formation of the band tail of localized electrons at $n_{\mathrm{s}} < n_{\mathrm{c}}$

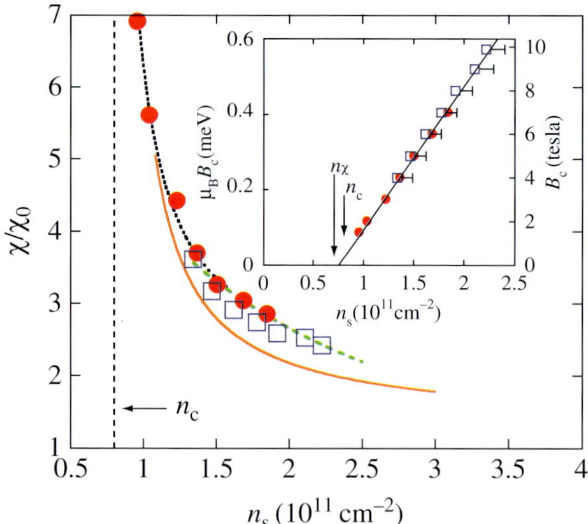

Fig. 2.12 Dependence of the Pauli spin susceptibility on electron density obtained by thermodynamic methods: direct measurements of the spin magnetization (dashed line), $\mathrm{d}\mu/\mathrm{d}B = 0$ (circles), and density of states (squares). The dotted line is a guide to the eye. Also shown by a solid line is the transport data of Shashkin *et al.* (2001). Inset: polarization field as a function of the electron density determined from the magnetization (circles) and magnetocapacitance (squares) data. The data for B_{c} are described by a linear fit which extrapolates to a density n_χ close to the critical density n_{c} for the $B = 0$ metal–insulator transition.

conceals the origin of the low-density phase. In other words, so far one can only reach an incipient transition to a new phase.

Strong dependences of the magnetization on n_s have also been seen (Lu *et al.*, 2008; Pudalov *et al.*, 2002; Vakili *et al.*, 2004; Vitkalov *et al.*, 2001*b*; Zhu *et al.*, 2003) in other types of device with n_s near the critical density for the metal–insulator transition.

2.3.2 Effective mass or g-factor?

In principle, the strong increase of the Pauli spin susceptibility at low electron densities can be due to either the increase of the effective mass or the Lande g-factor (or both). The effective mass was measured by several groups employing different methods (Anissimova *et al.*, 2006; Shashkin *et al.*, 2002, 2003), which gave quantitatively similar results. The values g/g_0 and m/m_b as a function of the electron density are shown in Fig. 2.13 (here $g_0 = 2$ is the g-factor in bulk silicon, m_b is the band mass equal to $0.19\,m_e$, and m_e is the free electron mass). In the high-n_s region (relatively weak interactions), the enhancement of both g and m is relatively small, both values slightly increasing with decreasing electron density in agreement with earlier data (Ando *et al.*,

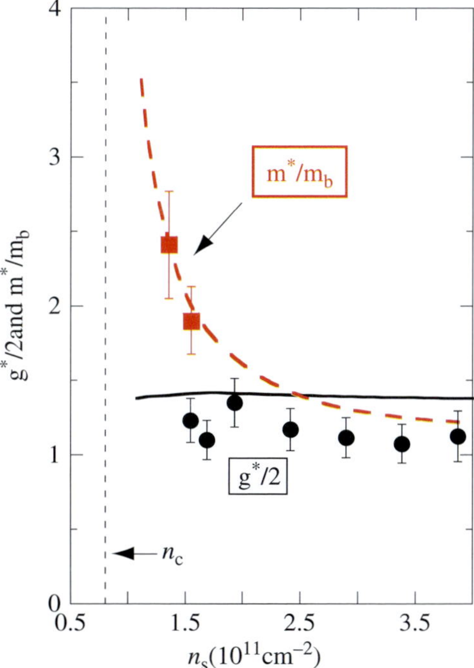

Fig. 2.13 g-factor and effective mass as a function of the electron density, obtained from transport measurements (Shashkin *et al.*, 2002) (solid and dashed lines, respectively). Also shown are the effective g-factor (circles) and the cyclotron mass (squares) obtained by measurements of thermodynamic magnetization (Anissimova *et al.*, 2006). The critical density n_c for the metal–insulator transition is indicated by the arrow.

1982). Also, the renormalization of the g-factor is dominant compared to that of the effective mass, which is consistent with theoretical studies (Chen and Raikh, 1999; Iwamoto, 1991; Kwon *et al.*, 1994).

In contrast, the renormalization at low n_s (near the critical region), where $r_s \gg 1$, is much more striking. As the electron density is decreased, the renormalization of the effective mass overshoots abruptly while that of the g factor remains relatively small, $g \approx g_0$, without tending to increase. Hence, it is the effective mass, rather than the g factor, that is responsible for the drastically enhanced spin susceptibility near the metal–insulator transition.

2.3.3 Effective mass as a function of r_s

The effective mass has also been measured in a dilute 2D electron system in (111)-silicon. This material system is interesting because the band electron mass $m_b = 0.36\, m_e$ is larger by a factor of about two than that for (100)-silicon. Also, (111)-samples used in this experiment have a much higher level of disorder than (100)-ones. Remarkably, it has turned out that the relative enhancement of the effective mass in this material system, i.e. m^*/m_b, is practically the same function of the interaction parameter r_s as in (100)-Si MOSFETs. In Fig. 2.14, the effective mass in units m_b is plotted as a function of $(1/r_s)^2 \propto n_s$. As seen in the figure, the mass renormalization m/m_b versus the interaction parameter r_s is coincident within the experimental uncertainty with that found in (100)-Si MOSFETs. Thus, one may conclude that the relative mass enhancement is determined solely by the strength of the electron–electron interactions, being independent of a 2D electron system.

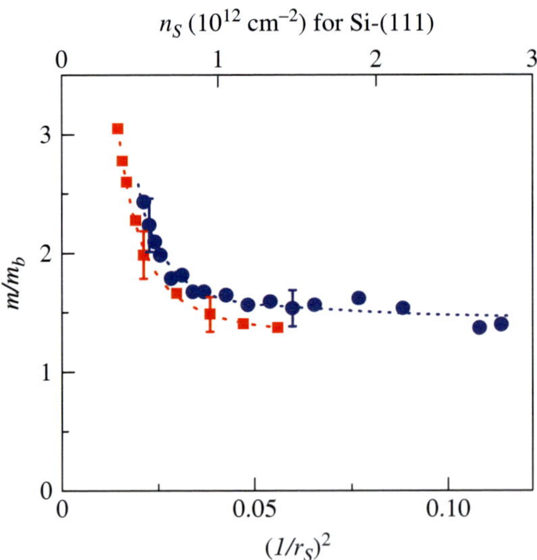

Fig. 2.14 The effective mass (dots) in units of m_b as a function of $(1/r_s)^2 \propto n_s$. Also shown by squares is the data obtained in (100)-Si MOSFETs (Shashkin *et al.*, 2003). The dashed lines are guides to the eye. From Shashkin *et al.* (2007).

Finally, I would like to note that moderate enhancements of the effective mass $m \approx 1.5\,m_b$ have been determined in 2D electron systems of AlAs quantum wells and GaAs/AlGaAs heterostructures. This is because the lowest accessible densities are still too high (Tan *et al.*, 2005; Vakili *et al.*, 2004).

2.4 Comparison with theory

Many attempts have been made to interpret the experimental observations summarized above. I will briefly review only theories making quantitative predictions, which can be compared with the experiment.

2.4.1 Ballistic regime ($T\tau/\hbar \gg 1$)

When the resistivity of a sample is much smaller than h/e^2, the electrons are in a ballistic regime at not-too-low temperatures ($T > h/k_B\tau$). In "screening" theories by Stern and Das Sarma (1985), Gold and Dolgopolov (1986), Dolgopolov and Gold (2000), Das Sarma and Hwang (1999; 2000; 2003; 2004), attempts have been made to explain the transport results in this regime using classical formulas for the resistivity, valid at $r_s < 1$, and extrapolating them to the case $r_s \gg 1$. Quantitatively successful comparisons between this theory and experiment have been reported by Das Sarma and Hwang (1999; 2000; 2003; 2004). However, at large r_s, the screening length λ_{sc} obtained in random-phase approximation is parametrically smaller than the spacing between electrons: $\lambda_{sc}\sqrt{\pi n} = (1/4)r_s^{-1} \ll 1$ (Spivak and Kivelson, 2006). Screening lengths smaller than the distance between electrons are clearly unphysical.

This approach was corrected by Zala *et al.* (2001*a*; 2001*b*), where the contribution due to the scattering from the Friedel oscillations induced by impurities has been considered. At small r_s, the "insulating-like" sign of $d\rho/dT$ has been obtained, which is indeed seen in experiments on samples with $r_s \sim 1$. However, when extrapolation is made to sufficiently large r_s, $d\rho/dT$ changes sign and $\rho(T)$ becomes a linearly increasing function of T. The same theory predicts complete suppression of the metallic behavior in parallel magnetic fields, strong enough to completely polarize spins, also in agreement with the experiments. However, one should keep in mind that this is a theory of *corrections* to the conductivity, which are small compared to the Drude conductivity. In experiments, changes of the resistivity by an order of magnitude are often observed.

2.4.2 Scaling theory of the metal–insulator transition in 2D: diffusive regime ($T\tau/\hbar \ll 1$)

The two-parameter scaling theory by Punnoose and Finkel'stein (2001; 2005) of quantum diffusion in an interacting disordered system is based on the scaling hypothesis that both the resistivity and the electron–electron scattering amplitudes, γ_2, become scale- (temperature-) dependent. The renormalization-group equations describing the evolution of the resistance and the scattering amplitude in 2D have the form (Punnoose and Finkelstein, 2001)

$$\frac{d\ln\rho^*}{d\xi} = \rho^* \left[n_v + 1 - (4n_v^2 - 1)\left(\frac{1+\gamma_2}{\gamma_2}\ln(1+\gamma_2) - 1 \right) \right], \tag{2.6}$$

$$\frac{d\gamma_2}{d\xi} = \rho^* \frac{(1+\gamma_2)^2}{2}, \tag{2.7}$$

where $\xi = -\ln(T\tau/\hbar)$, τ is the elastic scattering time, $\rho^* = (e^2/\pi h)\rho$, and n_v is the number of degenerate valleys in the spectrum. It turns out that $\rho(T)$ is a non-monotonic function of temperature, reaching a maximum value ρ_{max} at some temperature T_{max} and having metallic temperature dependence ($d\rho/dT > 0$) at $T < T_{max}$. Furthermore, $\rho(T)$ can be written as

$$\rho = \rho_{max}\, F\left(\rho_{max}\, \ln(T_{max}/T)\right), \tag{2.8}$$

where F is a *universal* function shown by a solid curve in Fig. 2.15 (a). The strength of spin-dependent interactions, γ_2, is also a universal function of $\ln(T_{max}/T)$, shown by a solid line in Fig. 2.15 (b).

Therefore, this theory accounts for large changes in resistivity, observed in experiment, and provides quantitative functions that can be directly compared with experimental data. Such a comparison was made by Anissimova *et al.* (2007). The interaction amplitude was extracted from the magnetoresistance. The results are presented in Figs. 2.15 (a) and (b). The agreement between theory and experiment is especially striking given that the theory has no adjustable parameters. Systematic deviations from the universal curves occur at lower densities, as higher-order correc-

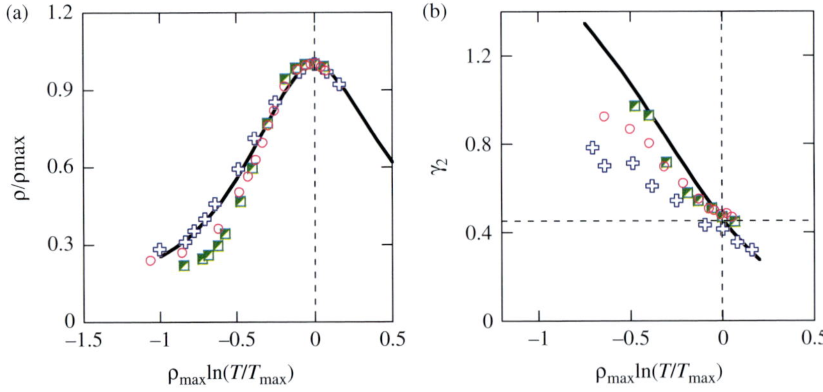

Fig. 2.15 Comparison between theory (lines) and experiment (symbols). (a) ρ/ρ_{max} as a function of $\rho_{max}\ \ln(T/T_{max})$, (b) γ_2 as a function of $\rho_{max}\ \ln(T/T_{max})$. Vertical dashed lines correspond to $T = T_{max}$, the temperature at which $\rho(T)$ reaches a maximum. Note that at this temperature, the interaction amplitude $\gamma_2 \approx 0.45$ (indicated by the horizontal dashed line in (b)), is in excellent agreement with theory. Electron densities are 9.87 (squares), 9.58 (circles), and 9.14×10^{10} cm^{-2} (crosses). Adapted from Anissimova *et al.* (2007).

tions in ρ become important. Furthermore, the resistivity reaches its maximum at $\gamma_2 \approx 0.45$ in excellent agreement with theory for $n_v = 2$.

The data for γ_2 allow calculation of the renormalized Landé g-factor $g^* = 2(1+\gamma_2)$. As shown in Fig. 2.16, at $n_s = 9.87 \times 10^{10}$ cm^{-2}, it grows from $g^* \approx 2.9$ at the highest temperature to $g^* \approx 4$ at the lowest. Therefore, the g-factor in the diffusive regime becomes temperature-dependent and increases with decreasing temperature, in agreement with Punnoose and Finkel'stein (2001; 2005). Note that at higher temperatures it is close to its temperature-independent "ballistic" value of about 2.8.

However, it should be noted that a theory based on electron interference effects can predict substantial changes of resistivity only in the near-vicinity of the critical point, where $\rho \sim h/e^2$. At much lower resistivities $\rho \ll h/e^2$, only small logarithmic corrections are possible, while the experiments demonstrate very large changes in resistivity even deep in the metallic region (see Figs. 2.3–2.7). Therefore, although theory by Punnoose and Finkel'stein (2001; 2005) quantitatively describes experimental data in the close vicinity of the transition, it cannot explain large effects far from the transition.

2.4.3 Spin susceptibility and the effective mass enhancement

There are several mechanisms that could lead to the strong enhancement of the effective mass at low carrier densities (high r_s). In the Fermi liquid theory, the enhancement of g and m is due to spin-exchange effects. The Fermi-liquid model has been extended to relatively large r_s (Chen and Raikh, 1999; Iwamoto, 1991; Kwon *et al.*, 1994), which is problematic. The main outcome is that the renormalization of g is large compared to that of m. In the limiting case of high r_s, one may expect a divergence of the g-factor that corresponds to the Stoner instability. These predictions are in obvious contradiction to the experimental data. The divergence of the effective mass and spin susceptibility follows also from the Gutzwiller variational approach (Brinkman and

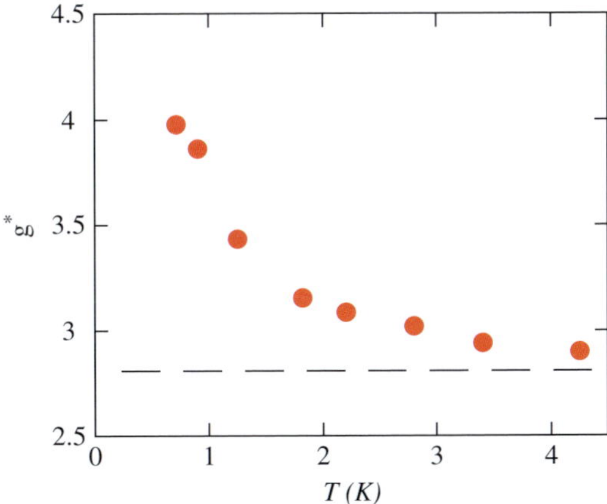

Fig. 2.16 g-factor vs temperature in the diffusive regime. The dashed line shows the value of the g-factor obtained in the ballistic regime. Adapted from Mokashi and Kravchenko (2009).

Rice (1970); see also Dolgopolov (2002)). The latest theoretical developments include the following. Using a renormalization group analysis for multi-valley 2D systems, it has been found that the spin susceptibility dramatically increases at disorder-dependent density for the metal–insulator transition while the g factor remains nearly intact (Punnoose and Finkelstein, 2005). However, this prediction is made for the diffusive regime $T\tau/\hbar \ll 1$ only, while the spin-susceptibility enhancement has been observed well into the ballistic regime $T\tau/\hbar \gg 1$. In the Fermi-liquid-based model of Khodel *et al.* (2005), a flattening at the Fermi energy in the spectrum that leads to a diverging effective mass has been predicted, which is in qualitative agreement with the experiment, although a detailed microscopic theory is needed before conclusions can be made. The strong increase of the effective mass has been also obtained (in the absence of the disorder) by solving an extended Hubbard model using dynamical mean-field theory (Camjayi *et al.*, 2008; Pankov and Dobrosavljević, 2008). This is consistent with experiment, especially taking into account that the relative mass enhancement has been experimentally found to be independent of the level of the disorder (Shashkin *et al.*, 2007).

2.5 Summary

Currently, there is no microscopic theory that can explain the whole variety of the observed phenomena, although the behavior in the close vicinity of the transition is quantitatively described by the renormalization group theory of Punnoose and Finkelstein (2005) without any fitting parameters. Deeper into the metallic regime, the origin of the large changes in the resistance remains unclear, with suggested explanations ranging from temperature-dependent screening (Das Sarma and Hwang 1999; 2000; 2003; 2004) to an analog of the Pomeranchuk effect (Spivak 2003; Spivak and Kivelson 2004; 2006). From the empirical perspective, numerous experiments on various strongly correlated 2D electron and hole systems strongly suggest the existence of the metal–insulator transition in 2D, contrary to the opinion that such a transition is impossible, which prevailed until recently.

Acknowledgements

This work was supported by DOE Grant DE-FG02-84ER45153 and by BSF grant #2006375.

References

Abrahams, E., Anderson, P.W., Licciardello, D.C., and Ramakrishnan, T.V. (1979). *Phys. Rev. Lett.*, **42**, 673.

Ando, T., Fowler, A.B., and Stern, F. (1982). *Rev. Mod. Phys.*, **54**, 437.

Anissimova, S., Kravchenko, S.V., Punnoose, A., Finkel'stein, A.M., and Klapwijk, T.M. (2007). *Nature Phys.*, **3**, 707.

Anissimova, S., Venkatesan, A., Shashkin, A.A., Sakr, M.R., Kravchenko, S.V., and Klapwijk, T.M. (2006). *Phys. Rev. Lett.*, **96**, 046409.

Attaccalite, C., Moroni, S., Gori-Giorgi, P., and Bachelet, G.B. (2002). *Phys. Rev. Lett.*, **88**, 256601.

Bishop, D.J., Tsui, D.C., and Dynes, R.C. (1980). *Phys. Rev. Lett.*, **44**, 1153.

Brinkman, W.F. and Rice, T.M. (1970). *Phys. Rev. B*, **2**, 4302.

Camjayi, A., Haule, K., Dobrosavljević, V., and Kotliar, G. (2008). *Nature Phys.*, **4**, 932.

Chen, G.-H. and Raikh, M.E. (1999). *Phys. Rev. B*, **60**, 4826.

Coleridge, P.T. (1997). *unpublished.*

Coleridge, P.T., Williams, R.L., Feng, Y., and Zawadzki, P. (1997). *Phys. Rev. B*, **56**, R12764.

Das Sarma, S. and Hwang, E.H. (1999). *Phys. Rev. Lett.*, **83**, 164.

Das Sarma, S. and Hwang, E.H. (2000). *Phys. Rev. B*, **61**, R7838.

Das Sarma, S. and Hwang, E.H. (2003). *Phys. Rev. B*, **68**, 195315.

Das Sarma, S. and Hwang, E.H. (2004). *Phys. Rev. B*, **69**, 195305.

Das Sarma, S., Lilly, M.P., Hwang, E.H., Pfeiffer, L.N., West, K.W., and Reno, J.L. (2005). *Phys. Rev. Lett.*, **94**, 136401.

De Poortere, E.P., Tutuc, E., Shkolnikov, Y.P., Vakili, K., and Shayegan, M. (2002). *Phys. Rev. B*, **66**, 161308(R).

Dolan, G.J. and Osheroff, D.D. (1979). *Phys. Rev. Lett.*, **43**, 721.

Dolgopolov, V.T. (2002). *JETP Lett.*, **76**, 377.

Dolgopolov, V.T. and Gold, A. (2000). *JETP Lett.*, **71**, 27.

Gao, X.P.A., Boebinger, G.S., Mills, Jr., A.P., Ramirez, A.P., Pfeiffer, L.N., and West, K.W. (2005). *Phys. Rev. Lett.*, **94**, 086402.

Gao, X.P.A., Boebinger, G.S., Mills, Jr., A.P., Ramirez, A.P., Pfeiffer, L.N., and West, K.W. (2006). *Phys. Rev. B*, **73**, 241315(R).

Gold, A. and Dolgopolov, V.T. (1986). *Phys. Rev. B*, **33**, 1076.

Hanein, Y., Meirav, U., Shahar, D., Li, C.C., Tsui, D.C., and Shtrikman, H. (1998*a*). *Phys. Rev. Lett.*, **80**, 1288.

Hanein, Y., Shahar, D., Yoon, J., Li, C.C., Tsui, D.C., and Shtrikman, H. (1998*b*). *Phys. Rev. B*, **58**, R13338.

Hilke, M., Shahar, D., Song, S.H., Tsui, D.C., and Xie, Y.H. (2000). *Phys. Rev. B*, **62**, 6940.

Ioffe, A.F. and Regel, A.R. (1960). *Prog. Semicond.*, **4**, 237.

Iwamoto, N. (1991). *Phys. Rev. B*, **43**, 2174.

Khmelnitskii, D.E. (1984). *Phys. Lett. A*, **106**, 182.

Khodel, V.A., Clark, J.W., and Zverev, M.V. (2005). *Europhys. Lett.*, **72**, 256.

Kwon, Y., Ceperley, D.M., and Martin, R.M. (1994). *Phys. Rev. B*, **50**, 1684.

Lai, K., Pan, W., Tsui, D.C., Lyon, S.A., Muhlberger, M., and Schäffler, F. (2005). *Phys. Rev. B*, **72**, 081313(R).

Laughlin, R.B. (1984). *Phys. Rev. Lett.*, **52**, 2304.

Lilly, M.P., Reno, J.L., Simmons, J.A., Spielman, I.B., Eisenstein, J.P., Pfeiffer, L.N., West, K.W., Hwang, E.H., and Das Sarma, S. (2003). *Phys. Rev. Lett.*, **90**, 056806.

Lu, T.M., Sun, L., Tsui, D.C., Lyon, S., Pan, W., Muhlberger, M., Schäffler, F., Liu, J., and Xie, Y.H. (2008). *Phys. Rev. B*, **78**, 233309.

McFarland, R.N., Kott, T.M., Sun, L., Eng, K., and Kane, B.E. (2009). *Phys. Rev. B*, **80**, 161310(R).

Mkhitaryan, V.V., Kagalovsky, V., and Raikh, M.E. (2009). *Phys. Rev. Lett.*, **103**, 066801.

Mokashi, A. and Kravchenko, S.V. (2009). *unpublished*.

Pankov, S. and Dobrosavljević, V. (2008). *Phys. Rev. B*, **77**, 085104.

Pruisken, A.M.M. (1984). *Nucl. Phys. B*, **235**, 277.

Pruisken, A.M.M. and Burmistrov, I.S. (2005). *Ann. Phys. (N.Y.)*, **316**, 285.

Prus, O., Yaish, Y., Reznikov, M., Sivan, U., and Pudalov, V.M. (2003). *Phys. Rev. B*, **67**, 205407.

Pudalov, V.M., Brunthaler, G., Prinz, A., and Bauer, G. (1998). *Physica B*, **249–251**, 697.

Pudalov, V.M., Gershenson, M.E., Kojima, H., Butch, N., Dizhur, E.M., Brunthaler, G., Prinz, A., and Bauer, G. (2002). *Phys. Rev. Lett.*, **88**, 196404.

Punnoose, A. and Finkelstein, A.M. (2001). *Phys. Rev. Lett.*, **88**, 016802.

Punnoose, A. and Finkelstein, A.M. (2005). *Science*, **310**, 289.

Rosenbaum, T.F., Milligan, R.F., Paalanen, M.A., Thomas, G.A., Bhatt, R.N., and Lin, W. (1983). *Phys. Rev. B*, **27**, 7509.

Shashkin, A.A., Anissimova, S., Sakr, M.R., Kravchenko, S.V., Dolgopolov, V.T., and Klapwijk, T.M. (2006). *Phys. Rev. Lett.*, **96**, 036403.

Shashkin, A.A., Kapustin, A.A., Deviatov, E.V., Dolgopolov, V.T., and Kvon, Z.D. (2007). *Phys. Rev. B*, **76**, 241302(R).

Shashkin, A.A., Kravchenko, G.V., and Dolgopolov, V.T. (1993). *JETP Lett.*, **58**, 220.

Shashkin, A.A. and Kravchenko, S.V. (2000). *unpublished*.

Shashkin, A.A., Kravchenko, S.V., Dolgopolov, V.T., and Klapwijk, T.M. (2002). *Phys. Rev. B*, **66**, 073303.

Shashkin, A.A., Kravchenko, S.V., and Klapwijk, T.M. (2001). *Phys. Rev. Lett.*, **87**, 266402.

Shashkin, A.A., Rahimi, M., Anissimova, S., Kravchenko, S.V., Dolgopolov, V.T., and Klapwijk, T.M. (2003). *Phys. Rev. Lett.*, **91**, 046403.

Simonian, D., Kravchenko, S.V., Sarachik, M.P., and Pudalov, V.M. (1997). *Phys. Rev. Lett.*, **79**, 2304.

Spivak, B. (2003). *Phys. Rev. B*, **67**, 125205.

Spivak, B. and Kivelson, S.A. (2004). *Phys. Rev. B*, **70**, 155114.

Spivak, B. and Kivelson, S.A. (2006). *Ann. Phys.*, **321**, 2071.

Stern, F. and Das Sarma, S. (1985). *Solid State Electronics*, **28**, 158.

Tan, Y.W., Zhu, J., Stormer, H.L., Pfeiffer, L.N., Baldwin, K.W., and West, K.W. (2005). *Phys. Rev. Lett.*, **94**, 016405.

Tanatar, B. and Ceperley, D.M. (1989). *Phys. Rev. B*, **39**, 5005.

Uren, M.J., Davies, R.A., and Pepper, M.J. (1980). *J. Phys. C*, **13**, L985.

Vakili, K., Shkolnikov, Y.P., Tutuc, E., De Poortere, E.P., and Shayegan, M. (2004). *Phys. Rev. Lett.*, **92**, 226401.

Vitkalov, S.A., Sarachik, M.P., and Klapwijk, T.M. (2001*a*). *Phys. Rev. B*, **64**, 073101.

Vitkalov, S.A., Zheng, H., Mertes, K.M., Sarachik, M.P., and Klapwijk, T.M. (2001*b*). *Phys. Rev. Lett.*, **87**, 086401.

von Klitzing, K., Dorda, G., and Pepper, M. (1980). *Phys. Rev. Lett.*, **45**, 494.

Wigner, E. (1934). *Phys. Rev.*, **46**, 1002.

Zala, G., Narozhny, B.N., and Aleiner, I.L. (2001*a*). *Phys. Rev. B*, **64**, 214204.

Zala, G., Narozhny, B.N., and Aleiner, I.L. (2001*b*). *Phys. Rev. B*, **65**, 020201(R).

Zhu, J., Stormer, H.L., Pfeiffer, L.N., Baldwin, K.W., and West, K.W. (2003). *Phys. Rev. Lett.*, **90**, 056805.

3
Anderson Localization

Keith SLEVIN[1] and Tomi OHTSUKI[2]

[1] Osaka University
[2] Sophia University

3.1 Introduction

According to the Einstein relation the DC conductivity of a degenerate conductor is proportional to the product of the density of states per unit volume at the Fermi energy ρ and the diffusion constant D of the electrons

$$\sigma = e^2 \rho D. \tag{3.1}$$

Here, $-e$ is the electron charge. From this formula there seem to be two reasons why a material would be an insulator. One is that the density of states at the Fermi level is zero. This is the case of a band insulator where the Fermi level lies in gap. It can also happen that the Coulomb interaction between the electrons can cause a gap to open at the Fermi level. This is the Mott transition that occurs in strongly correlated materials.

Another possibility is that the diffusion constant might be zero. At first sight, however, this seems unlikely. Treating the motion of the electrons classically, we can derive the following expression for the diffusion constant in terms of the mean-free path l, the Fermi velocity v_F and the dimensionality d of the conductor

$$D = \frac{1}{d} v_{\mathrm{F}} l. \tag{3.2}$$

In a strongly disordered conductor one expects that the mean-free path is small but not zero, so it seems we should be able to obtain a bad conductor but not an insulator in this way. The key insight of Philip Anderson in his seminal paper (Anderson, 1958), where the phenomenon of Anderson localization was first proposed, is that this picture can change drastically if the motion of the electrons is treated quantum mechanically. In the quantum case, interference between different trajectories can cause the diffusion constant to vanish even though the mean-free path remains finite. This is the phenomenon of Anderson localization.

In this short review, we shall not attempt to follow the historical development of the subject. Rather, we begin with the semiclassical theory of transport, and then describe the first step beyond that, which is the weak localization theory. We shall see that the weak localization theory raises serious questions concerning the validity of the semiclassical approach, especially in one and two dimensions. We investigate this in more detail in one dimension and then see how the results for all dimensions, and for weak and strong disorder, are combined in the scaling theory of localization. We describe the numerical approach to Anderson localization focusing in particular on the transfer matrix method and finite-size scaling. In closing, we describe some of the experimental signatures of Anderson localization.

3.2 Anderson localization

3.2.1 Semiclassical theory

In semiclassical theory (Ashcroft and Mermin, 1976) the motion of the electrons on atomic scales is treated quantum mechanically, while on the scale of the mean-free path l the motion is treated classically. The velocity v of an electron wavepacket formed from Bloch states in a metal that is subject to an external electric field E is determined by the equation

$$\frac{dv}{dt} = -\frac{eE}{m} - \frac{v}{\tau}. \tag{3.3}$$

Here, m is the effective electron mass and τ is a momentum relaxation time that characterizes the scattering of the electrons. We need to distinguish two sources of scattering. The first is scattering due to static disorder, i.e. defects and impurities in the crystal. The second is scattering due to dynamic disorder, i.e. other electrons and phonons. The latter depends strongly on temperature, while the former remains even at zero temperature.

We need not go into the details of the semiclassical theory here because we can arrive at the result we need directly from eqn (3.3). In a steady state, the derivative on the left-hand side of eqn (3.3) is zero and the velocity of the wavepacket is related to the electric field by

$$v = \frac{-eE\tau}{m}. \tag{3.4}$$

From this the conductivity is easily calculated

$$j = -env = \sigma E \quad \sigma = \frac{ne^2\tau}{m} \equiv \sigma_{\mathrm{D}}. \tag{3.5}$$

This is called the Drude conductivity. Here, n is the number of conduction electrons per unit volume. Noting that the mean-free time and the mean-free path are related by the Fermi velocity

$$l = v_{\mathrm{F}}\tau, \tag{3.6}$$

it is a straightforward but rather laborious exercise to verify that eqn (3.5) is consistent with the Einstein relation of eqn (3.1).

3.2.2 Weak localization

In the semiclassical theory, interference between scattering from different impurities is neglected. While at first sight, this would seem reasonable, if the disorder is weak, it is in fact highly questionable. The first step beyond this approximation is the theory of weak localization (Altshuler *et al.*, 1980*b*; Bergmann, 1984; Hikami *et al.*, 1980). In this theory, the temperature is assumed to be very low so that scattering from static disorder dominates and the scattering from other electrons and phonons can be neglected. The motion of the electron in the solid is described by the following Hamiltonian,

$$H = H_0 + V(\vec{x}) \quad H_0 = -\frac{\hbar^2}{2m}\nabla^2. \tag{3.7}$$

Here, V is a random potential describing scattering by defects and impurities. When the mean-free path is much longer than the Fermi wavelength,

$$k_F l \gg 1, \tag{3.8}$$

or equivalently when

$$\frac{\hbar}{\varepsilon_{\mathrm{F}} \tau} \ll 1, \tag{3.9}$$

where ε_F is the Fermi energy, we can use the free electron gas as a starting point, and take the scattering into account using diagrammatic perturbation theory. The theory is complicated but in systems with time-reversal symmetry an important contribution comes from the re-summation of a series of diagrams (the so-called maximally crossed diagrams) that describe quantum interference between time-reversed trajectories. It is found that constructive interference between these trajectories enhances back scattering and leads to a decrease in the conductivity. The correction to the Drude conductivity from these terms is called the weak-localization correction. The weak-localization corrections are (Hikami, 1981; Hikami *et al.*, 1980):

$$\sigma = \sigma_{\mathrm{D}} + \Delta\sigma \quad \Delta\sigma = \begin{cases} -\frac{2e^2}{\pi^2\hbar}(L-l) & 1D \\ -\frac{e^2}{\pi^2\hbar}\log\left(\frac{L}{l}\right) & 2D \\ \frac{e^2}{\pi^2\hbar}\left(\frac{1}{L}-\frac{1}{L}\right) & 3D \end{cases} \tag{3.10}$$

Here, L is the system size. There is something rather puzzling about these results, particularly in one and two dimensions. The calculation predicts that when the system size is increased the conductivity will tend to zero. In fact, this signals the breakdown of perturbation theory. In particular, for one dimension, the breakdown of perturbation theory seems to occur when the system size is comparable to the mean-free path. This calls into question the existence of any diffusive regime of normal transport in one dimension. We will investigate this further, using a completely different approach, the Landuer–Buttiker formalism, to calculate the conductance of a disordered one dimensional (1D) system.

3.2.3 One dimensional systems

According to what is referred to as the Landuer–Buttiker theory .()....().........(Datta 1995), the zero-temperature conductance

$$G = 2\frac{e^2}{h}g \tag{3.11}$$

(where the factor 2 takes account of spin degeneracy) of a 1D system in a two-terminal measuring geometry is given by

$$g = |t|^2. \tag{3.12}$$

Here, t is the transmission coefficient for an electron at the Fermi energy incident on the system and g is the dimensionless conductance defined by eqn (3.11). The reader will note that eqn (3.12) implies that the conductance of a perfect wire in the two-terminal measuring geometry is unity, not infinity. This is in agreement with experiment. This is attributed to an unavoidable contact resistance that is inherent in the two-terminal geometry (Datta, 1995).

Now let us suppose that this 1D system contains two impurities. The total transmission $t$can be expressed in terms of the transmission and reflection coefficients of the individual impurities

$$t = t_2 t_1 + t_2 r_1' r_2 t_1 + t_2 r_1' r_2 r_1' r_2 t_1 + \cdots = t_2 \frac{1}{1 - r_1' r_2} t_1 \tag{3.13}$$

Reference to Fig.3.1 should make this equation easy to understand. Let us now consider an ensemble of wires in which the distance between the impurities is uniformly distributed. The random distance between the impurities leads to a random phase in the product of the reflection coefficients, which in turn will clearly lead to large fluctuations of the conductance. This brings out an aspect of the problem that we have so far not discussed. The zero-temperature conductance is a quantity that fluctuates from sample to sample; it is not determined simply by the density of impurities or the mean scattering time. These mesoscopic fluctuations, which occur even in 3D disordered conductors (Altshuler and Khmel'nitskii, 1985; Lee and Stone, 1985), are especially strong in 1D systems. A complete analysis requires a rigorous calculation of the conductance distribution. For that we refer the reader to the review of Pendry (1994). Here, we shall not attempt to be rigorous.

Taking the logarithm of the modulus square of both sides of eqn (3.13) we have

$$\ln |t|^2 = \ln |t_1|^2 + \ln |t_2|^2 - \ln |1 - r_1' r_2|^2 . \tag{3.14}$$

Assuming a uniform distribution for the phase of the product of the reflection coefficients and averaging, we find

$$\langle \ln |t|^2 \rangle = \langle \ln |t_1|^2 \rangle + \langle \ln |t_2|^2 \rangle. \tag{3.15}$$

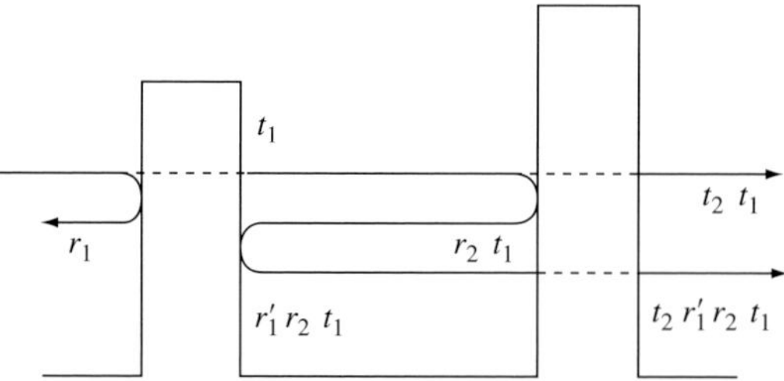

Fig. 3.1 Multiple scattering in system with two impurities.

If we now consider a long wire composed of many impurities, we are led to expect that

$$\langle \ln |t|^2 \rangle \approx -\alpha L. \tag{3.16}$$

Here, α is a positive constant and the approximation arises because we have neglected interference terms involving more than two impurities. While we have only considered the calculation of the mean, and not the full distribution of the conductance, these results suggest (by analogy with the central limit theorem) that the logarithm of the conductance will be normally distributed. It then follows that the typical conductance

$$g_{typ} \equiv \exp(\langle \ln g \rangle) \approx \exp(-\alpha L) \tag{3.17}$$

will decay exponentially with the size of the system.

Physical insight into the reason for the exponential decay of the conductance in one dimension is given by an argument due to (Thouless, 1977). In a wire of length L the average spacing between energy levels at the Fermi level is

$$\Delta = \frac{1}{\rho L} \tag{3.18}$$

If this wire is connected at the ends, each level will acquire a finite width, which we shall call the Thouless energy E_c. This energy is of the order

$$E_c = \frac{\hbar}{t_D} = \frac{\hbar D}{L^2}. \tag{3.19}$$

This formula is obtained by noting that a time of the order of t_D is required for an electron to diffuse to the ends of the sample. Identifying this as the lifetime of the state, we then use the energy–time uncertainty relation to obtain the broadening of the level. Now making use of the Einstein relation we see that the dimensionless conductance is, up to a constant, the ratio of these two energy scales

$$g = \frac{h}{2e^2} G = \frac{h}{2e^2} \frac{\sigma}{L} = \frac{h}{2e^2} e^2 \rho \, D \frac{1}{L} = \pi \frac{\hbar D}{L^2} \rho \, L = \pi \frac{E_c}{\Delta}. \tag{3.20}$$

The conductance on the left of this formula should be understood as some suitable average of the conductance distribution of an ensemble of wires, not the measured conductance of a single wire. This conductance is usually referred to as the Thouless conductance to make this clear. (Note also that the contact resistance that occurs in the two-terminal geometry is not present here since the measurement set up is not considered explicitly.)

Referring to eqn (3.20) we see that the conductance is the ratio of the broadening of each level to the spacing between adjacent levels. When the Thouless conductance is much larger than unity, we see that broadening of the levels is larger than the spacing between levels. If we put two such wires in series, levels in the two wires will overlap and there should be a high probability for an electron to tunnel through the wire.

If the Thouless conductance is much less than unity, levels in the two wires will not overlap and tunnelling probability can be expected to decrease exponentially.

3.2.4 The scaling theory of localization

Perhaps the single most important breakthrough in the theory of Anderson localization was the proposal of the scaling theory of localization (Abrahams *et al.*, 1979). In this theory the ideas of the renormalization group, which were developed to describe the critical phenomena of thermal phase transitions, and also the Kondo problem, are applied to the localization problem.

Consider a d-dimensional hyper cube of side L. Inspired by the renormalization group theory of phase transitions, let us suppose that the dimensionless conductance obeys a scaling law

$$\frac{d \ln g}{d \ln L} \equiv \beta(g). \tag{3.21}$$

The dimensionless conductance g that appears in this expression should be understood as a suitable average, such as the Thouless conductance or the typical conductance. To be clear, it is being assumed in eqn (3.21) that the derivative on the left is a function of g only. This function is called the beta function. This hypothesis is certainly correct in the metallic limit where

$$g = \sigma L^{d-2}. \tag{3.22}$$

and we immediately obtain

$$\beta(g) = d - 2 \quad g \to \infty. \tag{3.23}$$

It remains true, if we include the weak localization corrections

$$\beta(g) = d - 2 + \frac{a_d}{g} + \cdots \quad a_d < 0. \tag{3.24}$$

In the opposite limit of very small conductance, let us suppose that the 1D result of an exponentially decaying conductance applies in all dimensions

$$\beta(g) = \ln\left(\frac{g}{g_0}\right) \quad g \to 0. \tag{3.25}$$

Here, g_0 is some constant. Assuming a monotonic interpolation between these limits, we draw the schematic of the beta function shown in Fig. 3.2.

From the behavior of the beta function, some very important conclusions can be drawn:

1. In one and two dimensions, the beta function is always negative. This means that the conductance will vanish exponentially for large enough system size, and we conclude that all electronic states are localized in one and two dimensions. Note that in two dimensions the situation is marginal, and this conclusion depends crucially on the assumed monotonic behavior of the beta function. In fact, when

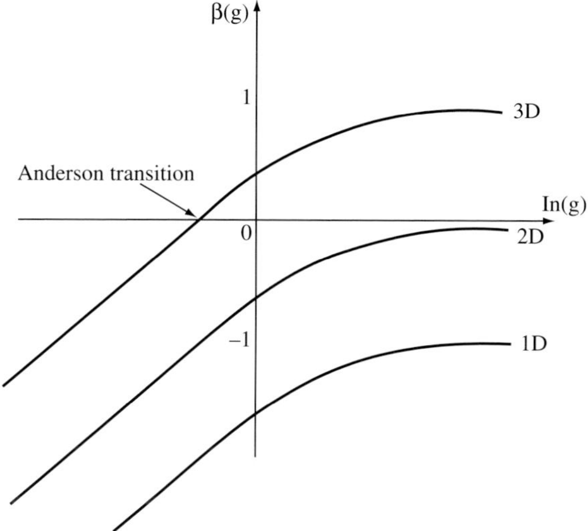

Fig. 3.2 Schematic of the beta function for one, two and three dimensions.

the spin–orbit interaction is considered, the beta function can become non-monotonic and a transition is possible in two dimensions in systems where the spin–orbit interaction is sufficiently strong.

2. In three dimensions there is a value of the conductance where the beta function is zero. This corresponds to a phase transition between an insulating and a metallic phase. This transition is called the Anderson transition.

It is helpful to formulate the scaling hypothesis in a slightly different way that brings out its physical content more clearly. In principle, we expect that the conductance of the hyper-cube is a function of the system size and microscopic length scales such as the Fermi wave number λ_{F}, mean-free path l, etc,

$$g = g(L, \lambda_{\mathrm{F}}, l, \cdots). \tag{3.26}$$

The scaling hypothesis is equivalent to the statement that once the system is much longer than any of these microscopic length scales, these microscopic length scales do not enter the formulae for the conductance explicitly but only through a correlation length ξ in a relation of the form

$$g = F(L/\xi). \tag{3.27}$$

The function F is called a scaling function. Since all the microscopic lengths are subsumed into a single correlation length, this is referred to as single-parameter scaling.

3.2.5 The Anderson transition

It is clear from eqn (3.21) that at the transition the conductance is independent of the system size. Looking at eqn (3.27), we can see that this corresponds to a divergence of the correlation length at the transition. This divergence is described by a power law and an associated critical exponent (Lee and Ramakrishnan, 1985)

$$\xi \sim |x - x_c|^{-\nu}. \tag{3.28}$$

Here, $x$refers to a parameter such as carrier density or impurity concentration that is used to drive the system through the transition, which we suppose to occur at some critical value x_c.

In practice, in experiments, it is the behavior of the conductivity near the transition, rather than the correlation length, that is observed. The conductivity also has a power-law dependence near the transition

$$\sigma \sim (x - x_c)^s \quad x > x_c, \tag{3.29}$$

but the associated critical exponent s is different. (We have assumed here that $x > x_c$corresponds to the metallic phase.) However, this exponent is not independent of the critical exponent that appears in eqn (3.28). This can be seen from the following argument.

When the system is in the metallic phase, the conductivity may be expressed using the scaling law of eqn (3.27) as

$$\sigma = L^{2-d} F\left(\frac{L}{\xi}\right). \tag{3.30}$$

For small values of its argument we expect that the scaling function behaves as a power law

$$F(x) \sim x^\eta \quad x \to 0. \tag{3.31}$$

In order that the size dependence cancels in eqn (3.30) we must have

$$\eta = d - 2 \tag{3.32}$$

and hence we conclude that

$$s = (d - 2)\nu. \tag{3.33}$$

This is called Wegner's scaling relation (Wegner, 1979).

In two dimensions, Wegner's relation implies that $s = 0$, meaning that the conductivity remains finite at the Anderson transition, and drops to zero discontinuously in the insulating phase.

One of the important features of critical phenomena is that the values of the exponents exhibit universality. Their values depend only on the dimensionality and symmetry of the system. For the Anderson transition the relevant symmetries are time-reversal symmetry and spin-rotation symmetry (Hikami, 1981). In fact, there is an exact correspondence with the symmetry classes that occur in the theory of random matrices (Mehta, 2004): the "orthogonal" class with both time-reversal symmetry (TRS) and spin-rotation symmetry (SRS), the "unitary" class where TRS is broken, and the "symplectic" class with TRS but where SRS is broken.

It has not so far proved possible to calculate the precise values of the critical exponents analytically (but they have been estimated numerically: see below). However, it has been possible to derive a lower bound on the exponent (Chayes *et al.*, 1986; Kramer, 1993):

$$\nu > \frac{2}{d} \tag{3.34}$$

Precisely at the critical point it has been found that the distribution of the wave-function amplitude has a multi-fractal form (Castellani and Peliti, 1986; Rodriguez *et al.*, 2009)

$$P_L(\alpha) = P_L(\alpha_0) L^{f(\alpha)-d} \quad \alpha = -\frac{\ln|\psi|^2}{\ln L}. \tag{3.35}$$

This distribution is characterized by a function $f(\alpha)$ called the multi-fractal spectrum. The position of the maximum of this spectrum is usually denoted α_0.

The lowest dimension above in which a phase transition occurs is called the lower critical dimension d_{L}. From the behavior of the beta function (Figs 3.2 and 3.2), a zero of the beta function, and hence an Anderson transition, exists for $d > 2$. The lower critical dimension of the Anderson transition is thus

$$d_{\mathrm{L}} = 2. \tag{3.36}$$

Systems with symplectic symmetry (i.e. systems with strong spin–orbit scattering) may be an exception where $d_{\mathrm{L}} < 2$(Asada *et al.*, 2006).

The upper critical dimension d_{U} is the dimension above which the critical exponents no longer depend on the dimension. In the case of the Anderson transition it is believed that (Castellani *et al.*, 1986; Mirlin and Fyodorov, 1994)

$$d_{\mathrm{U}} = \infty. \tag{3.37}$$

3.2.6 Self-consistent theory

Staring from a systematic investigation of the infrared divergences in weak disorder perturbation theory, Vollhardt and Wolfle 1980*a*; 1980*b*; 1982 derived the following self-consistent equation for the diffusion constant,

$$D(\omega) = D_0 - \frac{k_{\mathrm{F}}{}^{2-d}}{\pi m} \sum_q \frac{1}{q^2 - i\omega/D(\omega)}, \tag{3.38}$$

Here, ω is the frequency and D_0 is the classical diffusion constant. The sum over q is cut off at the inverse of the mean-free path. The theory is quite involved and we refer the reader to the original papers for details. The main results are consistent with the scaling theory of localization. The self-consistent theory predicts that the lower critical dimension is 2 and that the critical exponent is

$$\nu = \begin{cases} 1/(d-2) & 2 < d < 4 \\ 1/2 & d \geq 4 \end{cases}. \tag{3.39}$$

However, these predictions are not in agreement with numerical estimates of the critical exponent (see below), and it is generally believed that the upper critical dimension for the Anderson transition is infinity and not 4.

An alternative approach (Dobrosavljević *et al.*, 2003) is to derive a self-consistent equation for the local density of states (LDOS) by employing typical medium theory. The typical value of the LDOS

$$\rho_{typ} = \exp(\langle \log \rho(\vec{x}) \rangle) \tag{3.40}$$

is non-zero in the diffusive phase and zero in the insulating phase. Again the qualitative description obtained is in agreement with the scaling theory but the quantitative predictions for the critical exponent are not in agreement with numerical analyses. Nevertheless, a very interesting aspect of this approach is that it provides a natural starting point for a theory of the Anderson localization of interacting electrons (Aguiar *et al.*, 2009).

3.3 Numerical simulations

3.3.1 The Anderson model

In numerical simulations the most commonly used model is the Anderson model

$$H = \sum_m W_m c_m^+ c_m - \sum_{\langle mn \rangle} c_m^+ c_n. \tag{3.41}$$

The first term is a sum over all the sites of a simple cubic lattice. The second term is a sum over nearest neighbours. For simplicity the hopping term is set to unity and this then becomes the unit of energy for the model. The on-site potential is often taken to be uniformly distributed

$$P(W) = p(W)dW \quad p(W) = \begin{cases} 1/W & |W| \leq W/2 \\ 0 & |W| > W/2 \end{cases}. \tag{3.42}$$

The physics is not materially altered when other distributions such as Gaussian or Cauchy are used (Slevin and Ohtsuki, 1999). The model can be generalized to include an applied magnetic field (Governale and Ungarelli, 1998) and also the spin–orbit interaction (Ando, 1989). In the former case the hopping becomes a position-dependent complex phase, and in the latter the wavefunction at each site is replaced by a spinor and the hopping elements become matrices taken from the group SU(2) (Asada *et al.*, 2002).

In the absence of disorder, all sites have the same energy, which we may set as zero, and the wavefunctions are Bloch states. For simplicity let us consider a 1D lattice with spacing a. In this case we can label the sites by their position $x = Xa$

and the wavefunction amplitudes at the sites of an eigenstate of energy E obey the following Schrödinger equation:

$$-\psi_{X+1} - \psi_{X-1} = E\psi_X. \tag{3.43}$$

The solutions of this equation are

$$\psi_X = \exp(iKX). \tag{3.44}$$

The energy and wavenumber K are related by

$$E = -2\cos(K). \tag{3.45}$$

In the presence of disorder the Schrodinger equation becomes

$$W_X\psi_X - \psi_{X+1} - \psi_{X-1} = E\psi_X. \tag{3.46}$$

and the wavefunction amplitudes can no longer be calculated analytically. One possibility is to truncate the Hamiltonian, imposing suitable boundary conditions at the ends, and diagonalize it numerically. Another is to rewrite the Schrödinger equation as a transfer matrix equation.

3.3.2 Transfer matrices

The real-space transfer matrix is obtained by taking the energy E as an arbitrary parameter and rewriting the Schrödinger equation in the form

$$\begin{pmatrix} \psi_{X+1} \\ \psi_X \end{pmatrix} = \begin{pmatrix} W_X - E & -1 \\ 1 & 0 \end{pmatrix} \begin{pmatrix} \psi_X \\ \psi_{X-1} \end{pmatrix}. \tag{3.47}$$

In the absence of disorder the transfer matrix is a constant matrix independent of x

$$M_0 = \begin{pmatrix} -E & -1 \\ 1 & 0 \end{pmatrix}. \tag{3.48}$$

The eigenvalues and eigenvectors of this transfer matrix are

$$\exp(+iK) \quad \begin{pmatrix} 1 \\ \exp(-iK) \end{pmatrix} \tag{3.49}$$

and

$$\exp(-iK) \quad \begin{pmatrix} 1 \\ \exp(+iK) \end{pmatrix}. \tag{3.50}$$

Thus, in the absence of disorder, we can find the Bloch states by diagonalizing the transfer matrix.

The transfer matrix is often employed to solve scattering problems. Let us consider a disordered sample connected to two semi-infinite leads at the left and right. We can model this situation by supposing that the on-site potential is zero when $X \leq 0$

and $X \geq L+1$, and distributed according to eqn (3.42) in between. Consider now a situation in which a Bloch wave is incoming from the right. The wavefunction at the right of the sample is

$$\psi_X = A' \exp(+iKX) + B' \exp(-iKX) \; X \geq L \tag{3.51}$$

and to the left of the sample is

$$\psi_X = \exp(-iKX) \; X \leq 1. \tag{3.52}$$

Note that we have set the coefficient of the outgoing wave, rather than the coefficient of the incoming wave, to unity to facilitate the application of the transfer matrix method. Let us define

$$M_X = \begin{pmatrix} W_X - E & -1 \\ 1 & 0 \end{pmatrix}. \tag{3.53}$$

We can relate the wavefunction amplitudes at the left to wavefunction amplitudes at the right

$$\begin{pmatrix} \psi_{L+1} \\ \psi_L \end{pmatrix} = M \begin{pmatrix} \psi_1 \\ \psi_0 \end{pmatrix}, \tag{3.54}$$

where

$$M = M_L \cdots M_1. \tag{3.55}$$

From this transfer matrix we can calculate the transmission and reflection coefficients describing the scattering of the Bloch state by the random potential of the sample. To do this we must first calculate the probability current carried by a Bloch wave. Making use of the time-dependent Schrodinger equation

$$W_X \psi_X - \psi_{X+1} - \psi_{X-1} = i \frac{d}{dt} \psi_X \tag{3.56}$$

(we use units where $\hbar = 1$) the time derivative of the probability of finding the electron at site X may be expressed in terms of difference of the probability flux incoming on site X from the left, which we shall denote as j_{X-1}, and the probability flux incoming on site $X+1$

$$\frac{d}{dt} {\psi_X}^* \psi_X = j_{X-1} - j_X, \tag{3.57}$$

where

$$j_X = i \left({\psi_X}^* \psi_{X-1} - \psi_X {\psi^*}_{X-1} \right). \tag{3.58}$$

For a Bloch state with wavenumber K we find that the probability current is everywhere equal to

$$j = 2 \sin(K). \tag{3.59}$$

The transmission coefficient is the ratio of the transmitted to incoming flux amplitudes

$$t' = \frac{1}{B'}. \tag{3.60}$$

Similarly, the reflection coefficient is the ratio of the reflected to the incoming flux amplitudes

$$r' = \frac{A'}{B'}. \tag{3.61}$$

These can be calculated from the transfer matrix

$$\begin{pmatrix} A' \\ B' \end{pmatrix} = \frac{1}{2i\sin(K)} \begin{pmatrix} \exp(-iKL) & -\exp(-ik(L+1)) \\ -\exp(iKL) & \exp(iK(L+1)) \end{pmatrix} M \begin{pmatrix} \exp(-iK) \\ 1 \end{pmatrix}. \tag{3.62}$$

Let us now consider a more general situation where waves are incoming from both the left and the right

$$\psi_X = A\exp(+iKX) + B\exp(-iKX) \;\; X \leq 1. \tag{3.63}$$

The transfer matrix can now be used to relate the coefficients of the wavefunction at the left to the coefficients at the right

$$\begin{pmatrix} A' \\ B' \end{pmatrix} = T \begin{pmatrix} A \\ B \end{pmatrix} \;\; T = U^{-1}(L)MU(0), \tag{3.64}$$

where

$$U(X) = \begin{pmatrix} \exp(iK(X+1)) & \exp(-iK(X+1)) \\ \exp(iKX) & \exp(-iKX) \end{pmatrix}. \tag{3.65}$$

Here, we have introduced a new transfer matrix T. This should be distinguished from M, which we will refer to as the real-space transfer matrix. Note that T is also equal to a product of transfer matrices

$$T = T_L \cdots T_1 \;\; T_X = U^{-1}(X)M_X U(X-1). \tag{3.66}$$

The incoming and outgoing flux amplitudes are related by the scattering matrix

$$\begin{pmatrix} B \\ A' \end{pmatrix} = S \begin{pmatrix} A \\ B' \end{pmatrix} \;\; S = \begin{pmatrix} r & t' \\ t & r' \end{pmatrix}. \tag{3.67}$$

By rearranging eqn (3.67) we can express the transfer matrix in the form

$$T = \begin{pmatrix} t - \frac{rr'}{t'} & \frac{r'}{t'} \\ \frac{-r}{t'} & \frac{1}{t'} \end{pmatrix}. \tag{3.68}$$

When performing numerical calculations, we recommend checking that the calculated scattering matrix is unitary

$$S^\dagger S = 1. \tag{3.69}$$

This corresponds to the conservation of probability current. Alternatively, one can check that the transfer matrix obeys

$$T^\dagger \sum_C T = \sum_C \sum_C = \begin{pmatrix} 1 & 0 \\ 0 & -1 \end{pmatrix}, \tag{3.70}$$

which is equivalent.

The Hamiltonian we are considering has time-reversal symmetry, i.e. it commutes with the complex conjugation operator. This means that if ψ is a solution of the scattering problem at a given energy then so is its complex conjugate. It follows that if eqn (3.67) holds then so must

$$\begin{pmatrix} A^* \\ B'^* \end{pmatrix} = S \begin{pmatrix} B^* \\ A'^* \end{pmatrix}. \tag{3.71}$$

Since this must hold for arbitrary values of B and A' we deduce that the transfer matrix is symmetric

$$S = S^T. \tag{3.72}$$

Similarly, time reversal symmetry imposes a constraint on the transfer matrix

$$T = \begin{pmatrix} 0 & 1 \\ 1 & 0 \end{pmatrix} T^* \begin{pmatrix} 0 & 1 \\ 1 & 0 \end{pmatrix}. \tag{3.73}$$

Once the transmission coefficient has been calculated the zero-temperature two-terminal conductance may be calculated by using the Landauer–Buttiker formula (Imry and Landauer, 1999).

3.3.3 Lyapunov exponents

We shall now consider what happens when the length of the system becomes extremely long. The transfer matrix will be a product of a large number of random matrices and it is worth considering whether or not the distribution of the real-space transfer matrix will tend to some limiting form. Before doing this let us consider a much simpler problem.

Suppose that $x_1, \cdots, x_L$ are independently and identically distributed random variables and that

$$x = x_L \cdots x_1. \tag{3.74}$$

Taking the logarithm of both sides and applying the central limit theorem we see immediately that the logarithm of x will be normally distributed. In this case x is said to be log-normally distributed.

The problem of finding the transfer matrix distribution is more complicated because, unlike scalar multiplication, matrix multiplication is not commutative. Consider the Hermitian matrix

$$\Omega = \ln M^\dagger M. \tag{3.75}$$

This has real eigenvalues, which as a consequence of the current conservation condition of eqn (3.70) occur as a pair with opposite sign

$$\{-\nu, +\nu\} \ \nu \geq 0. \tag{3.76}$$

The Lyapunov exponents (LEs) are defined as the values of the following limit

$$\gamma_i = \lim_{L\to\infty} \frac{\nu_i}{2L} \ \ i = 1, 2. \tag{3.77}$$

Since the eigenvalues occur as pair with opposite sign, so do the Les:

$$\{-\gamma, +\gamma\} \ \gamma \geq 0. \tag{3.78}$$

It has been shown that this limit always converges to the same value for almost all sequences of the transfer matrices, i.e. with probability one (Crisanti *et al.*, 1993). This result can be thought of as analogous to the law of large numbers. The natural question is then: is there a central limit theorem for the LE? This has been proven in certain cases that seem not to be directly applicable to the localization problem (Crisanti *et al.*, 1993). However, let us proceed on the assumption that a central limit theorem does hold, i.e. that for finite but sufficiently large L the limiting distribution of the LEs is normal, with a variance that decreases inversely with the length L. To see the consequence of this for the distribution of the two-terminal conductance we may use Pichard's formula (Imry, 1986) to relate the two-terminal conductance directly to the eigenvalues of eqn (3.75):

$$g = \frac{2}{1 + \cosh(\nu)}. \tag{3.79}$$

Provided that

$$\nu > 0, \tag{3.80}$$

we deduce that for lengths sufficiently long that

$$L \gg \xi \ \ \xi = \frac{1}{\gamma} \tag{3.81}$$

the conductance will be log-normally distributed with a mean value of the logarithm that decays as

$$\langle -\ln\ g \rangle \approx \frac{2L}{\xi}. \tag{3.82}$$

In fact, for some special cases we may have $\nu = 0$. However, this is very unusual and the generic behavior is that of eqn (3.80). This is a proof, at least for physicists, that all states are localized in one dimension and confirms the conclusions reached above using less rigorous arguments.

The alert reader will have noticed that since the conductance distribution is log-normal, and since such a distribution is determined by two parameters, we should expect a two-parameter scaling of the conductance distribution. However, detailed calculations show that the mean and variance of this distribution are related (Slevin and Pendry, 1990*a*,*b*)

$$\mathrm{var}(-\ln g) = 2\langle -\ln g\rangle, \tag{3.83}$$

so it turns out that the conductance distribution obeys a single-parameter scaling law (Slevin and Pendry, 1990*a*,*b*).

$$p_L(\ln g) = p\left(\ln g; \frac{L}{\xi}\right) \tag{3.84}$$

However, care should be exercised when using this distribution. It gives an accurate description of the bulk of the distribution but is not a good description of the tails of the distribution, i.e. rare events. For a detailed discussion we recommend that the reader consult the review by Pendry (Pendry, 1994).

3.3.4 Finitesize scaling

So far we have limited our discussion of the numerical simulations to strictly 1D systems. We will now broaden the scope of our discussion to include 2D and 3D systems. One way of drawing conclusions about what happens in two and three dimensions is to simulate quasi-one-dimensional (Q1D) systems and then use the finite-size scaling method to extrapolate to two and three dimensions. For those readers unfamiliar with the finite-size scaling method, which is an important topic in statistical physics, we recommend consulting the book by Amit and Martin-Mayor (2005).

Consider a system that is extended in the x direction and has a uniform cross-section in the y and z directions. The transfer matrix is easily generalized to deal with this geometry by replacing the wavefunction amplitude in eqn (3.47) with a vector

$$\psi_x = \begin{pmatrix} \vdots \\ \psi_{XYZ} \\ \vdots \end{pmatrix} \tag{3.85}$$

and the site energy W_X with a matrix H_X consisting of the matrix formed from the Hamiltonian matrix elements between sites on layer X. The size of the transfer matrix is increased from 2 to $2N$, where N is the number of sites in the cross section. When we diagonalize the matrix in eqn (3.75), we find $2N$ eigenvalues that again occur in pairs of opposite sign:

$$\{-v_1, \cdots, -v_N, +v_N, \cdots, +v_1\} \qquad v_1 \geq v_2 \geq \cdots v_N \geq 0. \tag{3.86}$$

Taking the limit of eqn (3.77) we obtain a spectrum of Lyapunov exponents with the same symmetry.

To proceed we need to generalize our previous scattering theory calculation to the Q1D geometry. Here, we encounter a minor complication. In the absence of disorder, when we diagonalize the transfer matrix we find $2N$ eigenvalues and eigenvectors. However, not all of these correspond to Bloch states; some correspond to evanescent modes with imaginary (more generally complex) wavenumbers K. To calculate the scattering matrix correctly it is necessary to take account of these modes correctly (Ando, 1991; Slevin and Nagao, 1994). Here, in this short review, we will ignore this complication and suppose (unrealistically) that all the eigenvectors have eigenvalues with real K. Pichard's formula for the two-terminal conductance is then applicable:

$$g = \sum_{i=1}^{N} \frac{u \equiv 2}{1 + \cosh(v_i)}. \tag{3.87}$$

When the system is sufficiently long we again find eqn (3.82), where the Q1D localization length is now given by the inverse of the smallest positive Lyapunov exponent

$$\xi^{\text{Q1D}} = \frac{1}{\gamma_N}, \tag{3.88}$$

We conclude that all states are localized in Q1D in agreement with the general argument of Thouless (1977).

We now turn to the problem of extrapolating to two and three dimensions. The first step is to calculate the Q1D localization length for a strip ($N = L$) or a bar ($N = L \times L$) of linear transverse size L. (Note that we are changing our notation here: L now denotes the transverse size not the length of the system.) The second step is to fit the dependence of Q1D localization length on the transverse size and other parameters, such as energy and disorder, to a finite-size scaling law. Performing this procedure for a strip permits an extrapolation to two dimensions, and on a bar to three dimensions.

The finite size scaling law has the form

$$\Gamma \equiv L\gamma_N = f(uL^{\alpha}). \tag{3.89}$$

Here, $f$is a *scaling function* and α is a *critical exponent*, and u is a *relevant scaling variable* that is a function of model parameters such as energy E and disorder W,

$$u \equiv u(E, W, \cdots) \tag{3.90}$$

Note that the left-hand side of eqn (3.89) is a dimensionless ratio.

Let us now consider what we expect to happen when we extrapolate $L \to \propto$. If the corresponding 2D or 3D system is in the localized phase we expect the Q1D localization length to converge to the 2D or 3D localization length, so that

$$\Gamma \propto L \qquad \text{(localised phase)}. \tag{3.91}$$

If the corresponding 2D or 3D system is in the delocalized phase we expect (MacKinnon and Kramer, 1983)

$$\Gamma \propto L^{2-d} \qquad \text{(delocalised phase).} \tag{3.92}$$

First let us focus attention on the 3D situation. If an Anderson transition occurs, we expect to see a change from Γ increasing with L to Γ decreasing with L as a function of disorder or energy. Exactly at the critical point, we expect that Γ will be independent of L, i.e. *scale invariant*. For this to be consistent with eqn (3.89) we must have

$$u = 0 \tag{3.93}$$

at the critical point. By writing

$$\xi = |u|^{-\upsilon} \quad \upsilon = \frac{1}{\alpha} \tag{3.94}$$

we see that eqn (3.89) is simply another way of writing the single-parameter scaling hypothesis

$$\Gamma = F_{\pm}\left(\frac{L}{\xi}\right). \tag{3.95}$$

The scaling function has two branches: a branch corresponding to the delocalized phase

$$F_{+}(x) = f(x^{\alpha}) \quad x > 0 \tag{3.96}$$

and a branch corresponding to the localized phase

$$F_{-}(x) = f\left(-|x|^{\alpha}\right) \quad x < 0. \tag{3.97}$$

In Fig. 3.3 we plot numerical data for the Anderson model. There is an approximate common crossing point of the curves for different transverse size L, which confirms the existence of the Anderson transition for the 3D Anderson model. The existence of the transition was first confirmed in precisely this way (MacKinnon and Kramer, 1981; Pichard and Sarma, 1981).

Unfortunately, it is not possible to fit the data in Fig. 3.3 to eqn (3.89). The common crossing is approximate, while eqn (3.89) is consistent only with an exact common crossing. To fit the data properly it is necessary to take account of corrections to the scaling law arising from non-linearity of the scaling variables and irrelevant scaling variables (Slevin and Ohtsuki, 1999). We refer the reader to our original articles and will be content here to summarize our published estimates for the critical exponent in Table 3.1.

Turning now to the 2D case, we find that all numerical data can be rescaled onto a single localized branch, i.e. the localization length is always finite and there is no Anderson transition. Strictly speaking the region of weak disorder in two dimensions

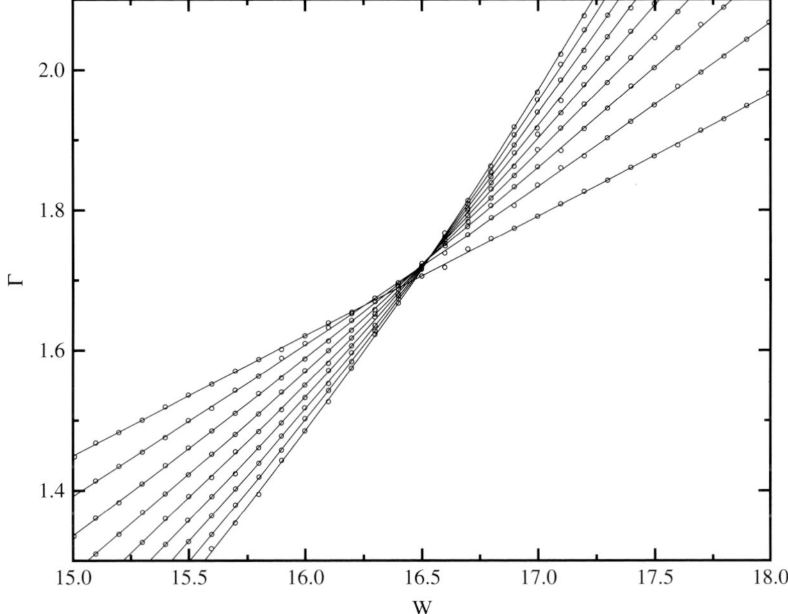

Fig. 3.3 The quantity Γ defined in eqn (3.89) for the Anderson model with a box distribution of the random potential of eqn (3.42) and Fermi energy at the band center. The numerical data have a precision of 0.1% and the lines are a finite-size scaling fit that includes corrections to scaling. The transverse size L=4, 6, 8, 10, 12, 14, 16, 18, and 20 (the slopes of the curves' increase are with L).

Table 3.1 Numerical estimates of the critical exponent of the Anderson transition. The precisions are 95% confidence intervals.

Critical exponent	**Universality class**	**Reference**
$\nu = 1.57 \pm 0.02$	3D orthogonal symmetry	Slevin and Ohtsuki (1999)
$\nu = 1.43 \pm 0.04$	3D unitary symmetry	Slevin and Ohtsuki (1997)
$\nu = 1.375 \pm 0.016$	3D symplectic symmetry	Asada *et al.* (2005)
$\nu = 2.73 \pm 0.02$	2D symplectic symmetry	Asada *et al.* (2002)
$\nu = 2.59 \pm 0.006$	Integer quantum Hall effect	Slevin and Ohtsuki (2009)

is not numerically accessible because the localization length becomes extremely long, making the finite-size scaling analysis extremely difficult. Nevertheless, the consensus is that the prediction of the scaling theory of localization, which is that all states are localized in two dimensions, is correct.

In fact, there are exceptions to this rule. One well known example is the extended state that appears at the centre of a Landau level in the integer quantum Hall effect

(Huckestein, 1995). Another is the extended phase that appears in systems with strong spin–orbit interaction (Ando, 1989; Asada *et al.*, 2002, 2004; Evangelou and Ziman, 1987).

3.3.5 Other approaches

In this section we will briefly mention other numerical approaches to the Anderson localization problem that we do not have space to cover in this brief review.

Instead of focusing on the Q1D localization length, it is possible to calculate the two-terminal conductance for a Q1D system using the formula

$$g = \mathrm{tr}(t^{\dagger} t). \tag{3.98}$$

This is the analogue of eqn (3.12) for a Q1D system in the two-terminal geometry. An exactly analogous finite-size scaling analysis of the moments of the conductance or its distribution can then be performed, yielding results entirely consistent with those for the Q1D localization length (Slevin *et al.*, 2001, 2003).

Another possibility is to diagonalize the Hamiltonian and examine either the statistics of the wavefunction or of the energy levels. In Fig. 3.4 we show a typical localized wavefunction, which was obtained using the JADAMILU sparse matrix diagonalization software (Bollhöfer and Notay, 2007). As can be seen from Fig. 3.4 most of the electron probability is supported on a small number of sites. The wavefunction has a very random structure with large intensity fluctuations near a "centre of localization" and exponential decay at large distances from this centre. At the critical point the distribution of wavefunction amplitudes is found to be multi-fractal and characterized by a multi-fractal spectrum (Evers and Mirlin, 2008).

Turning to the statistics of the energy levels, the distribution of the energy spacing between consecutive energy levels has a characteristic Wigner–Dyson form (the same as that found for Gaussian random matrix ensembles) in the delocalized phase and a Poisson form in the localized phase (Shklovskii *et al.*, 1993). Meanwhile, at the critical point the distribution is system-size independent.

Finally, the time-dependent Schrödinger equation can be solved numerically and the suppression of diffusion by disorder simulated directly. At the critical point anomalous diffusion is observed (Kawarabayashi and Ohtsuki, 1995; Ohtsuki and Kawarabayashi, 1997).

3.4 Experimental signatures of localization

In this section we shall briefly mention some the observable experimental consequences of Anderson localization.

One of the most well studied transport phenomena is weak localization, which can be regarded as precursor of Anderson localization, particularly in 2D systems such as MOSFETs, semiconductor hetero-structures and metallic films. In particular, weak localization is manifest in both the temperature dependence and the magnetic-field dependence of the conductivity.

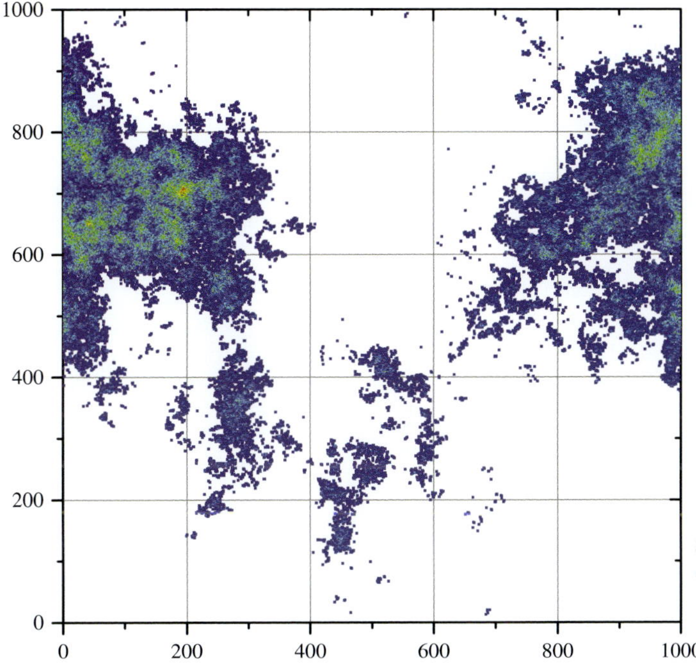

Fig. 3.4 A plot of the wavefunction intensity near the band centre in a realization of the 2D Anderson model with box distributed random potential (W = 5) and lattice size 1000 × 1000. The color indicates the logarithm of the wavefunction intensity (decreasing from red to blue). The plotted points support 99% of the electron probability. Approximately 80% of the electron probability is supported on the red–green points. On the remaining sites the intensity is negligible. Periodic boundary conditions have been used.

To understand the effect of weak localization in the temperature dependence, we must first note that the weak localization expressions of eqn (3.10) have been derived for zero temperature. At finite temperature, the interaction of the electron with other degrees of freedom (other electrons and phonons) can only be neglected on timescales less than the phase-coherence time and on length scales less than the related phase-coherence length. On longer time and length scales, single-particle interference phenomena are exponentially suppressed. In this case, we can continue to use weak localization formulae provided we replace the system size with the minimum of the system size and the phase-coherence length. We will give the explicit formulae for the 2D case only, since experimental investigations of the weak-localization correction are most often performed in 2D systems.

At room temperatures the phase-coherence length is comparable to the mean-free path, since the dominant scattering mechanism is electron–phonon scattering and the weak localization corrections (and all other interference effects) are negligible. (This is why the semiclassical theory works at room temperature.) At lower temperatures the phase-coherence length becomes longer than the mean-free path as scattering from static disorder such as defects and impurities becomes relatively more important. At very low temperatures it is usually found that the phase-coherence length has a power-law dependence on temperature

$$L_\varphi \propto T^{-p}, \tag{3.99}$$

with an exponent p the order of unity. In two dimensions, this leads to the following temperature dependence of the correction to the Drude conductivity:

$$\Delta\sigma_{2\mathrm{D}}(T) = \frac{e^2}{\pi^2\hbar} p \log(T). \tag{3.100}$$

Observation of this correction in experiments is complicated by the fact that an anomaly with the same temperature dependence arises from the Coulomb interaction between the electrons (Altshuler *et al.*, 1980*a*).

A clearer experimental signature of weak localization is observed in the magneto-conductance. In the presence of a magnetic field, time-reversal symmetry is broken and the quantum interference between time-reversed states is suppressed. If the magnetic field is weak, we may take account of this by replacing the system size in the weak-localization formulae with the magnetic length

$$l_B = \sqrt{\frac{\hbar}{eB}}. \tag{3.101}$$

In two dimensions, this leads to a positive magnetoconductance (negative magnetoresistance)

$$\Delta\sigma_{2\mathrm{D}}(B) = \frac{e^2}{2\pi^2\hbar} \ln B. \tag{3.102}$$

This has been observed in numerous experiments (Bergmann, 1984).

The signature of Anderson localization can be observed in the breakdown of Mathiessen's rule at low temperatures in strongly disordered materials. In a good conductor one expects that the contributions to the resistivity from disorder and temperature are additive. Thus, we expect that

$$\frac{d\,\rho}{dT} > 0 \qquad \text{(metal)}. \tag{3.103}$$

However, in strongly disordered materials this breaks down and it is found that at low temperatures the resistance increases as the temperature is lowered (Imry, 2002):

$$\frac{d\,\rho}{dT} < 0 \qquad \text{(insulator)}. \tag{3.104}$$

1D and 2D systems always seem to exhibit this behavior if the temperature is low enough.

In doped semiconductors, for example Si:P, Ge:Ga etc, a metal–insulator transition is observed as a function of the doping concentration (Itoh *et al.*, 2004). While this transition may be an Anderson transition, interpretation of the experiments is complicated by the fact that electrons interact via the Coulomb interaction.

Since Anderson localization is an interference phenomenon it can, in principle, be observed with other waves. While reports of the observation of an Anderson transition with light remain controversial (Aegerter *et al.*, 2006), a transition has been observed both for sound waves (Faez *et al.*, 2009) and for matter waves in cold atomic gases (Chabe *et al.*, 2008). These recent observations have given a renewed vigor and freshness to the field of Anderson localization.

References

Abrahams, E., Anderson, P.W., Licciardello, D.C., and Ramakrishnan, T.V. (1979). *Phys. Rev. Lett.*, **42**, 673–676.

Aegerter, C.M., Störzer, M., and Maret, G. (2006). *Europhys. Lett.*, **4**, 562.

Aguiar, M.C.O., Dobrosavljević, V., Abrahams, E., and Kotliar, G. (2009). *Phys. Rev. Lett.*, **102**, 156402–156404.

Altshuler, B.L., Aronov, A.G., and Lee, P.A. (1980*a*). *Phys. Rev. Lett.*, **44**, 1288.

Altshuler, B.L. and Khmel'nitskii, D.E. (1985). *JETP Lett.*, **42**, 359–363.

Altshuler, B.L., Khmel'nitzkii, D., Larkin, A.I., and Lee, P.A. (1980*b*). *Phys. Rev. B*, **22**, 5142.

Amit, D.J. and Martin-Mayor, V. (2005). *Field Theory, the Renormalization Group, and Critical Phenomena.* World Scientific, New Jersey, London.

Anderson, P.W. (1958). *Physical Review*, **109**, 1492.

Ando, T. (1989). *Phys. Rev. B*, **40**, 5325–5339.

Ando, T. (1991). *Phys. Rev. B*, **44**, 8017–8027.

Asada, Y., Slevin, K., and Ohtsuki, T. (2002). *Phys. Rev. Lett.*, **89**, 256601.

Asada, Y., Slevin, K., and Ohtsuki, T. (2004). *Phys. Rev. B*, **70**, 035115.

Asada, Y., Slevin, K., and Ohtsuki, T. (2005).

Asada, Y., Slevin, K., and Ohtsuki, T. (2006). *Phys. Rev. B*, **73**, 041102.

Ashcroft, N.W. and Mermin, N.D. (1976). *Solid State Physics.* Holt, New York.

Bergmann, G. (1984). *Phys. Reports*, **107**, 1–58.

Bollhöfer, M. and Notay, Y. (2007). *Comput. Phys. Commun.*, **177**, 951–964.

Castellani, C., Castro, C.D., and Peliti, L. (1986). *J. Phys. A*, **17**, L1099.

Castellani, C. and Peliti, L. (1986). *J. Phys. A.*, **8**, L429.

Chabe, J., Lemarié, G., Grmaud, B., Delande, D., Szriftgiser, P., and Garreau, J.C. (2008). *Phys. Rev. Lett.*, **101**, 255702–255704.

Chayes, J.T., Chayes, L., Fisher, D.S., and Spencer, T. (1986). *Phys. Rev. Lett.*, **57**, 2999.

Crisanti, A., Paladin, G., and Vulpiani, A. (1993). *Products of Random Matrices in Statistical Physics.* Springer, Berlin, New York.

Datta, S. (1995). *Electronic Transport in Mesoscopic Systems.* Cambridge University Press, Cambridge, New York.

Dobrosavljević, V., Pastor, A.A., and Nikolic, B.K. (2003). *Europhys. Lett.*, **62**, 76.

Evangelou, S.N. and Ziman, T. (1987). *J. Phys. C*, **20**, L235.

Evers, F. and Mirlin, A.D. (2008). *Rev. Mod. Phys.*, **80**, 1355–1363.

Faez, S., Strybulevych, A., Page, J.H., Lagendijk, A., and van Tiggelen, B.A. (2009). *Phys. Rev. Lett.*, **103**, 155703–155704.
Governale, M. and Ungarelli, C. (1998). *Phys. Rev. B*, **58**, 7816–7821.
Hikami, S. (1981). *Phys. Rev. B*, **24**, 2671–2679.
Hikami, S., Larkin, A.I., and Nagaoka, Y. (1980). *Prog. Theoret. Phys.*, **63**, 707–710.
Huckestein, B. (1995). *Rev. Mod. Phys.*, **67**, 357–396.
Imry, Y. (1986). *Europhys. Lett.*, **1**, 249.
Imry, Y. (2002). *Introduction to Mesoscopic Physics.* Oxford University Press, Oxford, New York.
Imry, Y. and Landauer, R. (1999). *Rev. Mod. Phys.*, **71**, S306.
Itoh, K.M., Watanabe, M., Ootuka, Y., Haller, E.E., and Ohtsuki, T. (2004). *J. Phys. Soc. Japan*, **73**, 173–183.
Kawarabayashi, T. and Ohtsuki, T. (1995). *Phys. Rev. B*, **51**, 10897–10904.
Kramer, B. (1993). *Phys. Rev. B*, **47**, 9888.
Lee, P.A. and Ramakrishnan, T.V. (1985). *Rev. Mod. Phys.*, **57**, 287.
Lee, P.A. and Stone, A.D. (1985). *Phys. Rev. Lett.*, **55**, 1622.
MacKinnon, A. and Kramer, B. (1981). *Phys. Rev. Lett.*, **47**, 1546–1549.
MacKinnon, A. and Kramer, B. (1983). *Z. Physik B*, **53**, 1–13.
Mehta, M.L. (2004). *Random Matrices.* New York.
Mirlin, A.D. and Fyodorov, Y.V. (1994). *Phys. Rev. Lett.*, **72**, 526.
Ohtsuki, T. and Kawarabayashi, T. (1997). *Journal of the Physical Society of Japan*, **66**, 314.
Pendry, J.B. (1994). *Advances in Physics*, **43**, 461–542.
Pichard, J.L. and Sarma, G. (1981). *J. Phys. C*, **6**, L127.
Rodriguez, A., Vasquez, L.J., and Romer, R.A. (2009). *Phys. Rev. Lett.*, **102**, 106406–106404.
Shklovskii, B.I., Shapiro, B., Sears, B.R., Lambrianides, P., and Shore, H.B. (1993). *Phys. Rev. B*, **47**, 11487–11490.
Slevin, K., Markos, P., and Ohtsuki, T. (2001). *Phys. Rev. Lett.*, **86**, 3594–3597.
Slevin, K., Markos, P., and Ohtsuki, T. (2003). *Phys. Rev. B*, **67**, 155106.
Slevin, K. and Nagao, T. (1994). *Phys. Rev. B*, **50**, 2380–2392.
Slevin, K. and Ohtsuki, T. (1997). *Phys. Rev. Lett.*, **78**, 4083–4086.
Slevin, K. and Ohtsuki, T. (1999). *Phys. Rev. Lett.*, **82**, 382–385.
Slevin, K. and Ohtsuki, T. (2009). *Phys. Rev. B*, **80**, 041304(R).
Slevin, K.M. and Pendry, J.B. (1990*a*). *Phys. Rev. B*, **41**, 10240–10242.
Slevin, K.M. and Pendry, J.B. (1990*b*). *J. Phys. C*, **2**, 2821.
Thouless, D.J. (1977). *Phys. Rev. Lett.*, **39**, 1167–1169.
Vollhardt, D. and Wolfle, P. (1980*a*). *Phys. Rev. Lett.*, **45**, 842.
Vollhardt, D. and Wolfle, P. (1980*b*). *Phys. Rev. B*, **22**, 4666.
Vollhardt, D. and Wolfle, P. (1982). *Phys. Rev. Lett.*, **48**, 699.
Wegner, F. (1979). *Z. Phys. B*, **36**, 1209.

4
Penultimate Fate of a Dirty-Fermi Liquid

Alexander Punnoose[1] and Alexander M. Finkel'stein[2]

[1] Physics Department, City College of the City University of New York, New York, New York 10031, USA
[2] Department of Physics, Texas A&M University, College Station, Texas 77843, USA

It is generally expected that in the absence of strong external fields and when lattice effects can be ignored, weakly or even moderately interacting systems will remain Fermi liquid at sufficiently low temperatures. This simple expectation, we now understand, is wrong in two dimensions as even the weakest disorder has been shown to destabilize the Fermi-liquid at very low temperatures (Castellani *et al.*, 1984; Finkel'stein, 1990), driving the system into a strongly-correlated unusual metal or into an insulator (Punnoose and Finkel'stein, 2005) depending on the initial (high-temperature) resistance of the system. Indications are that the the two phases are kept apart by a quantum critical point at a finite critical resistance (Punnoose and Finkel'stein, 2005). We argue that the observed metal–insulator transition (MIT) (Kravchenko *et al.*, 1994) in two-dimensional (2D) electronic systems is a manifestation of this break down (Anissimova *et al.*, 2007; Punnoose and Finkel'stein, 2001).

We are unfortunately very far from understanding the nature of the ultimate metallic state at absolute zero. We do, however, understand the process by which the Fermi liquid is destabilized. In the diffusion regime, i.e. for temperatures (frequencies) less than the elastic scattering rate, $T \lesssim 1/\tau$, diffusion modes and not quasiparticles are the low-lying propagating modes. Observable effects of quantum diffusion accumulate over long distances and long times and thereby develop only at much lower temperatures $T \lesssim \frac{1}{\tau}e^{-\sigma}$, where σ is the dimensionless conductance of the sample. Therefore, non-trivial effects are most clearly visible only in the vicinity of the transition where $\sigma \sim 1$. To follow the effects over a wide range of temperature requires a reasonable value for $1/\tau$, putting practical constraints on the observability of quantum diffusion effects. The scattering rates are much too low in ultra-high-mobility modulation doped or HIGFET structures ($1/\tau \ll 500$ mK); silicon MOSFETs, on the other hand, have mobilities that are not too high, leading to $1/\tau \sim 1$–2 K in the vicinity of the MIT, making them ideal playgrounds for studying quantum diffusion. It should be noted that in the absence of any other correlation-induced phase transition, all 2D systems will be destabilized at sufficiently low temperatures.

The theory of the dirty-Fermi liquid, which details the destruction of the Fermi-liquid phase up to and including the quantum critical point is reviewed here.

4.1 Experimental systems

The mobility of 2D quantum wells vary over orders of magnitude, ranging from $\sim 20,000$–$40,000\ \mathrm{cm}^2/\mathrm{Vs}$, as in typical n-type Si-MOSFETs, to hundreds of thousands and even millions, as in ultra-high-mobility p-type GaAs heterostructures. The source and therefore the nature of the impurity potentials vary greatly between these systems. It is, therefore, important to separate the effects of long- versus short-range random potential fluctuations on the scale of the interparticle distance, especially at very low carrier concentrations, when the wavelength of the carriers is considerably increased. Screening of the impurity potentials becomes difficult in these ultra-high-mobility systems; the resulting non-linear screening can induce non-trivial temperature dependencies in the transport properties of 2D systems and can in the extreme situation also lead to domain formations of high- and low-density regions (Fogler, 2004), preventing us from experimentally studying the fundamental problem of the effects of random quantum scattering in a Fermi liquid, whose full solution is unknown.

We argue below that MOSFETs provide an ideal environment to study the properties of a 2D dirty-Fermi liquid. The moderate values of the mobility of these systems results in an impurity elastic scattering rate slightly larger than a Kelvin. Comparing with the Drude conductance $\sigma = n\mu$, expressed in units of e^2/h, where n is the electron density and μ the mobility, moderately high mobility implies that the density in the vicinity of the quantum conductance is of the order of 10^{11} cm^{-2}. The corresponding Fermi temperature is of the order of 10 K, and therefore compared to holes in p-GaAs, the electrons in Si-MOSFETs are already degenerate at temperatures that are not particularly low. Also, because the electron–phonon coupling in silicon is weak due to the presence of inversion symmetry in the bulk, the contribution of electron–phonon scattering o the resistivity is negligible at low temperatures. Finally, another important characteristic of MOSFETs, in strong contrast to heterostructures, is that the elastic single-particle scattering rate is comparable to the transport scattering rate; they typically differ by a factor of ten in other systems. Quantum scattering is therefore the dominant scattering mechanism in MOSFETs.

The Coulomb interaction energy at the densities of interest are typically ten times larger than the Fermi energy, i.e. $r_{\rm s} = E_{\rm c}/E_{\rm F} \approx 10$, where $E_{\rm c}$ is the Coulomb energy of neighboring electrons. At these $r_{\rm s}$ values, detailed numerics indicate no phase transitions, at least up to $r_{\rm s} \approx 25$ where a paramagnet-to-ferromagnet transition is found to occur in a system with a single valley (no transition has been found in a two-valley system) (Marchi *et al.*, 2009; Palo *et al.*, 2005). At still-higher $r_{\rm s} \approx 35$, the electron liquid is expected to form a Wigner crystal, although none has been observed experimentally (Tanatar and Ceperley, 1989). These studies suggest a Fermi liquid with moderately strongly interacting quasiparticle scattering amplitudes around $r_{\rm s} \approx 10$.

MOSFETs have the additional advantage that they are a multi-valley system. The conduction band of an n-(001) silicon inversion layer has two almost degenerate valleys located close to the X-points in the Brillouin zone (Ando *et al.*, 1982). The presence of two equivalent valleys further enhances the correlation effects (Anissimova *et al.*, 2007; Punnoose and Finkel'stein, 2001) making MOSFETs ideal systems to study the effects of quantum disorder scattering in a Fermi liquid. In the following we consider the number of equivalent valleys to be equal to N.

4.2 Diffusion regime

Fermi-liquid renormalizations, like mass, g-factor, or equivalently, specific heat and spin susceptibility, occur on the scale $\lambda_{\rm F}$. As a result, the quasiparticle description remains valid when the elastic mean free path, $l \gg \lambda_{\rm F}$. This translates to $E_{\rm F} \gg \hbar/\tau$, where τ is the elastic scattering time related to l by the Fermi velocity. Multiple impurity scattering leads to the diffusive propagation of the quasiparticles for times $\gg \tau$ or frequencies $\omega \ll 1/\tau$, characterized by the quasiparticle diffusion constant, D. Fermi-liquid interactions between the quasiparticles, however, modify D in a scale-(frequency/temperature-) dependent way, making the conductivity σ, which is related to D through Einstein's relation $\sigma = (\partial n/\partial \mu)D$, temperature dependent. Altshuler, Aronov and Lee showed (Altshuler and Aronov, 1985; Altshuler *et al.*, 1980) that the corrections are localizing in the limit of weak quasiparticle interactions, so that the electronic system becomes an Anderson insulator as the temperature $T \to 0$. Here, $\partial n/\partial \mu$ is the thermodynamic density of states, which is not renormalized in the diffusion limit. (Experimentally, the compressibility is found to change sign in the strong-localization regime (Dultz and Jiang, 2000; Ilani *et al.*, 2001), which is outside the scope of the diffusion theory.)

In the discussion above, the Fermi-liquid scattering amplitudes were considered to be scale independent, as is the case in a clean Fermi liquid. It was shown, however, that the disorder-averaged amplitudes are in fact scale-dependent at low energies, $\omega \lesssim 1/\tau$, due to the diffusive propagation of the quasiparticles (Finkel'stein, 1983). For weak disorder (low resistance) the effective scattering amplitudes were shown to become strongly correlated over long distances, eventually diverging at a very small but finite scale (temperature) T^*. The enhanced correlation overcomes localization so that the resistance ρ^* at T^* is finite. This interplay of diffusion and interaction controls the physics of dirty-Fermi liquids down to the scale T^*, below which the liquid is no longer describable by diffusing Fermi-liquid quasiparticles.

In contrast to the weak disorder case, it was shown recently that the situation is very different when the disorder (resistance) is increased (Punnoose and Finkel'stein, 2005). For not-so-weak disorder the Fermi-liquid correlations were shown to be suppressed at long distance, so that the dirty-Fermi liquid becomes an Anderson-like insulator at low temperatures. It was also established in Punnoose and Finkel'stein (2005), in the framework of an effective large-N dirty-Fermi-liquid theory, that a quantum critical point at a finite critical resistance separates the metallic and the insulating phases, implying the existence of an MIT in two dimensions. The theoretical details of these results and their experimental tests are briefly discussed below.

4.3 Theory: background

The minimal model that describes the problem of quasiparticle diffusion in the field of impurities and interactions is the non-linear sigma model (Finkel'stein, 1983):

$$S[Q] = S_{\mathrm{dis}} + S_{\mathrm{int}}, \tag{4.1a}$$

$$S_{\mathrm{dis}} = -\frac{\pi\nu}{8}\int d^2\mathbf{r}\ \mathrm{Tr}\left[D(\nabla Q)^2 - 4z(\hat{\epsilon}Q)\right], \tag{4.1b}$$

$$S_{\mathrm{int}} = +\frac{\pi\nu}{8}\int d^2\mathbf{r}\ Q(\hat{\Gamma}_1 + \hat{\Gamma}_2 + \hat{\Gamma}_c)Q. \tag{4.1c}$$

The functional $S[Q]$ describes interacting quasiparticles with energies $\epsilon < 1/\tau$. ($\hat{\epsilon}$ is the energy matrix whose elements involve the energy of the quasiparticles.) Note that the region of higher energies $1/\tau < |\epsilon| \lesssim E_{\mathrm{F}}$, containing Fermi-liquid renormalizations of the clean liquid are incorporated as the input parameters into this theory. The Q-field here is an auxiliary matrix field describing the impurity scattering. It satisfies the constraints: $Q^2 = 1, Q = Q^\dagger$, and $\mathrm{Tr}\ Q = 0$. The fluctuations of Q in the particle-hole (diffusons) and particle–particle (cooperons) channels are described by the diffusive propagators $\mathcal{D}(q, \omega_n) \sim 1/(Dq^2 + z\omega_n)$ (we use Matsubara frequencies). These propagators capture the diffusive evolution of the charge and spin-density fluctuations at large times and length scales, i.e. small frequency and wavevectors. The parameter z is the frequency renormalization parameter (Finkel'stein, 1983, 1984), which can be interpreted as the renormalization of the density of states of the diffusion modes ($z = 1$ for non-interacting electrons), and as a result controls the thermodynamic properties of the dirty-Fermi liquid. The cooperons capture the effects of quantum interference, which at the perturbative level leads to the well known weak-localization corrections.

The quasiparticle interactions are characterized by the Fermi-liquid amplitude matrices: $\hat{\Gamma}_1$ and $\hat{\Gamma}_2$, in the diffuson channel and, $\hat{\Gamma}_c$, in the cooperon channel, respectively. The details of their matrix structure in the space of spin, frequency, and momentum are not shown here; they can be found in Finkel'stein (1990) and see Punnoose (2010) for their valley dependence. The magnitudes of the amplitudes, Γ_1 and Γ_2, are related to the standard zeroth order harmonics of the Fermi-liquid amplitudes $F_0^{\mathrm{s,a}}$ as $\Gamma_{1,2} = -F_0^{\mathrm{s,a}}/(1 + F_0^{\mathrm{s,a}})$.

The scaling of the parameters, D, z and the Γ_is ($i = 1, 2$), together describe the transport and thermodynamic properties of the dirty-Fermi liquid. Special care is, however, required when Coulomb interactions are present (Finkel'stein, 1983). The interaction amplitudes in this case include amplitudes of the kind that can be separated by cutting the statically screened long-range Coulomb line once. These amplitudes are usually denoted Γ_0. This distinction is important because the polarization operator that is irreducible to cutting a Coulomb line does not include Γ_0. Conservation laws then require that $z = 2N(\Gamma_0 + \Gamma_1) - \Gamma_2$, where N is the number of valley degrees of freedom (Finkel'stein, 1983; Punnoose, 2010).

As discussed in the last section, the diffusion constant and the Fermi-liquid scattering amplitudes become scale dependent in a dirty-Fermi liquid. It has been shown that renormalization group (RG) theory applied to the dirty-Fermi-liquid model given in eqn (4.1) is able to capture the scale dependence originating from the interplay of disorder and interactions to all orders in the interaction amplitude, making it the most promising analytical tool available to understand the physics of disordered systems. Pedagogical reviews of the RG theory can be found in Finkel'stein (1990) and Castellani *et al.* (1984). For more recent advances, see Baranov *et al.* (2002).

The scaling variables are described in terms of the dimensionless parameters, the resistance $\rho = 1/2N(2\pi^2\nu D)$, and the interaction Γ_2 and z. It follows, however, from the structure of the action in eqn (4.1) that the variables can be expressed as ρ and $\gamma_2 = \Gamma_2/z$, which does not contain z explicitly. In terms of these variables, the equations for ρ and γ_2 form a closed set of equations independent of z. In general, they take the form:

$$\frac{d\ln\rho}{d\xi} = \rho\beta_\rho(\rho, \gamma_2), \tag{4.2a}$$

$$\frac{d\gamma_2}{d\xi} = \rho\beta_{\gamma_2}(\rho, \gamma_2), \tag{4.2b}$$

$$\frac{d\ln z}{d\xi} = \rho\beta_z(\rho, \gamma_2). \tag{4.2c}$$

The scale $\xi = \log(1/T\tau)$, where to logarithmic accuracy $1/\tau$ is used as the upper cut-off. Note that the diffusive contribution involving the parameter γ_1 is universal when the dynamically screened Coulomb interaction is involved, i.e. when $\gamma_0 = \Gamma_0/z$ is included. Hence it does not appear as an explicit parameter in these equations (Altshuler and Aronov, 1985). It is a well established result with great importance for the general structure of the theory that the static amplitude Γ_0 does not acquire diffusion corrections and is therefore not renormalized. When combined with the conservation laws discussed above it follows that $d(2N\Gamma_1 - \Gamma_2) = dz$, from which one may extract the renormalization of Γ_1 (Finkel'stein, 1983). The above discussions imply that the electronic transport (ρ) and thermodynamic (involving z) properties are describable by a two-parameter scaling theory involving ρ and γ_2 only. (For repulsive interactions, the amplitude γ_c scales to a resistance-dependent value for finite N and hence is not considered here. The situation in the $N \to \infty$ limit is different and will be discussed later.)

The β-functions are multiplied by a factor ρ to emphasize that the sought after corrections appear as a result of the underlying diffusive propagators. The complete form of the β-functions is unknown. The general approach, however, is to expand the functions in a power series in ρ as $\beta(\rho, \gamma_2) \sim \beta_1(\gamma_2) + \rho\beta_2(\gamma_2) + \cdots$, such that for each power of ρ the full dependence on γ_2 is obtained. This is possible, in principle, because the maximum number of allowed interaction vertices are limited (before the ladder sums are included) by the number of momentum integrations involving the diffusive propagators, and since each propagator gives a factor of $1/D \sim \rho$, for a given order in ρ only a finite number of vertices contribute. So far, only the first-order (one-loop) results have been obtained exactly (Finkel'stein, 1983). Such a solution has the obvious limitation that it is applicable only when $\rho \ll 1$ and is not applicable near $\rho \sim 1$, where the MIT is experimentally found to be located (Kravchenko *et al.*, 1994). Therefore to approach the MIT, the disorder has to be treated beyond one-loop order (at least to two-loop order) while adequately retaining the effects of interaction. Exact results to order ρ^2 have not been obtained so far. It has been shown, however, that certain simplifications that come with standard large-N theories can be exploited to obtain the two-loop results (Punnoose and Finkel'stein, 2005). The RG equations to two-loop order within the large-N approximation showed, for the first time, the existence of a quantum critical point that describes the MIT separating the metallic from the insulating phase in 2D. Analysis of the quantum critical point revealed a non-Fermi-liquid fixed point with diverging specific heat and spin-susceptibility. The one- and two-loop RG equations are presented below and comparison with experiments are discussed.

4.4 Finite-N: one loop

The theoretical details of the evolution of ρ and γ_2 in this region with $\rho \ll 1$ were discussed in detail in Punnoose and Finkel'stein (2001). For completeness we give the equations and only discuss their salient features here. The RG equations generalized to include N equivalent valley degrees of freedom are:

$$\frac{d\ln\rho}{d\xi} = \rho\left[N+1-(4N^2-1)\Phi(\gamma_2)\right], \tag{4.3a}$$

$$\frac{d\gamma_2}{d\xi} = \frac{\rho}{2}(1+\gamma_2)^2, \tag{4.3b}$$

$$\frac{d\ln z}{d\xi} = \frac{\rho}{2}\left[-1+(4N^2-1)\gamma_2\right]. \tag{4.3c}$$

where $\Phi(\gamma_2) = (1+1/\gamma_2)\ln(1+\gamma_2) - 1$. The factor $(4N^2-1)$ corresponds to the number of spin-valley triplet channels and the factor N in eqn (4.3a) corresponds to the weak-localization corrections, which are enhanced due to the N extra degrees of freedom. The factor of one in (4.3a) is the contribution of the long-range Coulomb singlet amplitude, which is universal after dynamical screening; it should be emphasized that the factor of one arising in eqn (4.3b) for γ_2 has the same origin and, as a result, setting the initial value of γ_2 to zero does not switch the interactions off.

The following salient features are to be noted:

1. The amplitude γ_2 increases monotonically as the temperature is reduced.
2. The resistance, as a result, has a characteristic non-monotonic form changing from insulating behavior ($d\rho/dT < 0$) at high temperatures to metallic behavior ($d\rho/dT > 0$) at low temperatures. The change in slope occurs at a maximum value $\rho_{\max}$ at a temperature $T = T_{\max}$, neither of which are universal.
3. The corresponding value of the amplitude γ_2 at $T_{\max}$ is, however, universal at the one-loop order, depending only on N; for $N = 1$, it is 2.08, whereas for $N = 2$, it has the considerably lower value 0.45.
4. It follows from the form of the equations that the temperature dependence of $\rho(T)/\rho_{\max}$ and $\gamma_2(T)$ are universal when plotted as functions of $\rho_{\max}\ln(T/T_{\max})$.

As explained in the introduction, the theory described here is applicable only in the diffusive regime. It is critical that the diffusive regime (temperature and density range) is identified carefully before the scaling is tested. In Anissimova *et al.* (2007) a method that takes advantage of the B and T dependence of the magnetoconductance, $\sigma(B,T)$, in a parallel field was exploited to identify the diffusive regime. It is known that while $\sigma(B,T)$ scales as $(B/T)^2$ in the diffusive regime (Castellani *et al.*, 1998; Lee and Ramakrishnan, 1982), in the ballistic regime it scales as B^2/T (Zala *et al.*, 2001). This distinction was used to reliably identify the diffusive regime in a MOSFET sample. The experimental details and the results of the comparison with theory can be found in Anissimova *et al.* (2007), where for the first time the scaling of the interaction amplitude was established. Remarkably, the parameter γ_2 at $T = T_{\max}$ was found to correspond to 0.45 as predicted by theory for $N = 2$.

4.5 Two loops and the metal–insulator transition in the limit $N \to \infty$

It is instructive to analyze eqn (4.3) for $N \gg 1$. In this case it is easily seen that the value of the interaction amplitude γ_2 at $T = T_{\max}$, i.e. when $d\rho/dT = 0$ in eqn (4.3a), scales as $\gamma_2(T_{\max}; N) \approx \frac{1}{2N}$ for large N. Furthermore, because the presence of N-valleys increases the number of conducting channels, the resistance scales as $\rho \sim 1/N$. This can be explicitly taken into account by defining the maximum resistance, which is the only free parameter at the one-loop level, as $\rho_{\rm max} = 1/N$. Defining new variables $t = \rho(T)/\rho_{\rm max}$ and $\theta_2 = \gamma_2(T)/\gamma_2(T_{\rm max}; N)$, and keeping in mind that $N \gg 1$, eqn (4.3) can be expanded in powers of $1/N$:

$$\frac{d\ln t}{d\eta} = t\,[1 - \theta_2 + \mathcal{O}(\theta_2/N)]\,, \tag{4.4a}$$

$$\frac{d\theta_2}{d\eta} = t\,[1 + \mathcal{O}(\theta_2/N)]\,, \tag{4.4b}$$

$$\frac{d\ln z}{d\eta} = t\,[\theta_2 - \mathcal{O}(1/N)]\,. \tag{4.4c}$$

After rescaling, the temperature scale becomes $\eta = \ln(T_{\rm max}/T)$. It is now possible to take the limit $N \to \infty$, keeping t and θ_2 finite. Note that the parameter t corresponds to the resistance per valley, $t = 1/(2\pi)^2\nu D$, with ν being the density of states of a single spin and valley species. In the infinite-N limit eqn (4.4) reduces to

$$\frac{d\ln t}{d\eta} = t(1 - \theta_2), \tag{4.5a}$$

$$\frac{d\theta_2}{d\eta} = t, \tag{4.5b}$$

$$\frac{d\ln z}{d\eta} = t\theta_2. \tag{4.5c}$$

The form of eqn (4.5) illustrates the fact that in the $N \to \infty$ limit the maximum number of interaction amplitudes that can appear are limited by the power of t. This is to be compared with the function $\Phi(\gamma_2)$ appearing in eqn (4.3). Clearly, this is one of the main simplifications provided by taking the $N \to \infty$ limit.

Taking advantage of this simplification, the one-loop results above were extended to two loops, i.e. order t^2 in Punnoose and Finkel'stein (2005). It is easy to see that, unlike eqns (4.3), the solutions of eqns (4.5) are valid up to $T = 0$, at which point $\theta_2 \to \infty$ and $t \to 0$, providing an asymptotically exact theory all the way down to $T = 0$. (As discussed earlier, γ_2 for finite N diverges at a certain finite temperature T^* below which the RG equations are no longer valid.) It is therefore left to analyze the large t behavior at two-loop order.

The RG equations in the $N \to \infty$ limit are given as (Punnoose and Finkel'stein, 2005):

$$\frac{d\ln t}{d\eta} = t(\alpha - \Theta) + t^2(1 - \alpha - \alpha\Theta + c_t\Theta^2), \tag{4.6a}$$

$$\frac{d\Theta}{d\eta} = t(1 + \alpha + \alpha\Theta) - 4t^2\left[(1+\alpha)\Theta + \frac{\alpha}{2}\Theta^2 + c_\Theta\Theta^3\right], \tag{4.6b}$$

$$\frac{d\ln z}{d\eta} = t\Theta, \tag{4.6c}$$

where $c_t = (5 - \pi^2/3)/2$ and $c_\Theta = (1 - \pi^2/12)/2$. The parameter α takes values 0 and 1, depending on if the cooperon channel is absent or present, respectively. In the large-N limit, because rescattering in the cooperon channel is not accompanied by factors of N, such processes are irrelevant and hence $\theta_c = N\gamma_c$ appears on equal footing as θ_2. (Violation of time-reversal by a magnetic field, for example, will, however, suppress the contribution of γ_c.) The amplitudes θ_2 and θ_c always appear in the combination $\Theta = \theta_2 + \alpha\theta_c$.

It is clear from eqn (4.6b) that for large enough resistance $d\Theta/d\eta < 0$. Suppression of the Θ amplitude allows t to grow to large values (insulator) as the temperature is lowered. This is a clear departure from the metallic behavior observed for small t. The insulating and metallic phases are then separated by a fixed point given at (t_c, Θ_c), where $d\ln t/d\eta = 0$ and $d\Theta/d\eta = 0$. The quantum critical nature of the fixed point is apparent from the equation for z, which at (t_c, Θ_c) diverges as $z \sim 1/T^{t_c\Theta_c}$. The parameter z, which corresponds to the frequency renormalization, can be interpreted as the density of states of the underlying diffusion modes. Therefore it controls the behavior of the specific heat $C_v \sim (z\nu)T$. Hence, divergence in z implies a divergence in the specific heat constant C_v/T signaling the break down of the Fermi-liquid behavior at the critical point. Similar divergence is also expected in the Pauli spin susceptibility. Since the interaction parameter Θ is finite at the critical point, the divergence in the Pauli spin susceptibility is not related to any Stoner-like magnetic instability.

The feature that can be compared with experiments comes by observing in eqns (4.6) that, unlike in the metallic phase where $dt/d\eta < 0$ and $d\Theta/d\eta > 0$, in the insulating phase the opposite behavior is obtained, namely, $dt/d\eta > 0$ and $d\Theta/d\eta < 0$. In Anissimova *et al.* (2007), precisely such a scaling was confirmed experimentally.

4.6 Conclusions

We believe thereby that the two-parameter theory of the dirty-Fermi liquid reviewed here captures both quantitatively (for small disorder) and qualitatively (for large disorder) the physics of the MIT in two dimensions.

Acknowledgements

This work was partially supported by the US–Israel Binational Science Foundation grant 2006375. The work at City College was supported by DOE grant DOE-FG02-84-ER45153.

References

Altshuler, B.L. and Aronov, A.G. (1985). In *Modern Problems in Condensed Matter Physics*, p. 1. Elsevier, North Holland.

Altshuler, B.L., Aronov, A.G., and Lee, P.A. (1980). *Phys. Rev. Lett.*, **44**, 1288.

Ando, T., Fowler, A.B., and Stern, F. (1982). *Rev. Mod. Phys.*, **54**, 437.

Anissimova, S., Kravchenko, S.V., Punnoose, A., Finkel'stein, A.M., and Klapwijk, T.M. (2007). *Nature Physics*, **3**, 707.

Baranov, M.A., Burmistrov, I.S., and Pruisken, A.M.M. (2002). *Phys. Rev. B*, **66**, 075317.

Castellani, C., Di Castro, C., and Lee, P.A. (1998). *Phys. Rev. B*, **57**, R9381.

Castellani, C., Di Castro, C., Lee, P.A., and Ma, M. (1984). *Phys. Rev. B*, **30**, 527.

Dultz, S.C. and Jiang, H.W. (2000). *Phys. Rev. Lett*, **84**, 4689.

Finkel'stein, A.M. (1983). *Sov. Phys. JETP*, **57**, 97.

Finkel'stein, A.M. (1984). *Z. Phys. B*, **56**, 189.

Finkel'stein, A.M. (1990). *Sov. Sci. Rev., Sect. A*, **14**, 1.

Fogler, M.M. (2004). *Phys. Rev. B*, **69**, 121409(R).

Ilani, S., Yacoby, A., Mahalu, D., and Shtrikman, H. (2001). *Science*, **292**, 1354.

Kravchenko, S.V., Kravchenko, G.V., Furneaux, J.E., Pudalov, V.M., and D'Iorio, M. (1994). *Phys. Rev. B*, **50**, 8039.

Lee, P.A. and Ramakrishnan, T.V. (1982). *Phys. Rev. B*, **26**, 4009.

Marchi, M., De Palo, S., Moroni, S., and Senatore, G. (2009). *Phys. Rev. B*, **80**, 035103.

Palo, S., De Botti, M., Moroni, S., and Senatore, G. (2005). *Phys. Rev. Lett.*, **94**, 226405.

Punnoose, A. (2010). *Phys. Rev. B*, **81**, 035306.

Punnoose, A. and Finkel'stein, A.M. (2001). *Phys. Rev. Lett.*, **88**, 016802.

Punnoose, A. and Finkel'stein, A.M. (2005). *Science*, **310**, 289.

Tanatar, B. and Ceperley, D.M. (1989). *Phys. Rev. B*, **39**, 5005.

Zala, G., Narozhny, B.N., and Aleiner, I.L. (2001). *Phys. Rev. B*, **65**, 020201.

5
Numerical Studies of Metal–Insulator Transitions in Disordered Hubbard Models

S. Chiesa and R.T. Scalettar,[1]
P. Chakraborty,[2] P.J.H. Denteneer,[3]
T. Paiva[4] and S. Story[5]

[1] Physics Department, One Shields Ave., University of California, Davis, CA 95616
[2] Theoretical Physics III, Center for Electronic Correlations and Magnetism, Institute for Physics, University of Augsburg, D-86135, Augsburg, Germany
[3] Lorentz Institute, Leiden University, P.O. Box 9506, 2300 RA Leiden, The Netherlands
[4] Instituto de Física, Universidade Federal do Rio de Janeiro Cx.P. 68.528. 21941-972 Rio de Janeiro RJ, Brazil
[5] Physics Department, University of Arizona, Tucson, AZ 85721

5.1 Introduction

This introductory chapter presents a brief explanation for interest in numerical studies of disordered interacting lattice fermions, emphasizing new connections with optical lattice emulators (Feynman, 1982; Jaksch *et al.*, 1998; Lewenstein *et al.*, 2007), and a description of the DQMC method (Blankenbecler *et al.*, 1981). The focus will be on explaining the advantages and limitations of DQMC, as opposed to a detailed practical description of how to write a code. Several expositions with that emphasis can be found in the literature (Bai *et al.*, 2009; Gubernatis *et al.*, 1989; Loh *et al.*, 1990; Scalettar, 2003). Finally, we will mention alternative computational methods and some of their results.

5.1.1 Motivation

Tight-binding Hamiltonians for non-interacting electrons that are translationally invariant have plane-wave solutions in accordance with Bloch's theorem. The associated wave functions are delocalized throughout the lattice. Disorder breaks this translation invariance and gives rise to the possibility of eigenfunctions that can be concentrated about specific lattice sites, decaying exponentially as one moves away. Dimensionality affects how this phenomenon occurs in a fundamental way. In one dimension, an arbitrarily small amount of randomness robustly localizes all the eigenstates. While the same is true in two dimensions, the localization only barely happens (Abrahams *et al.*, 1979). Three-dimensional lattices have a richer array of possibilities, with the generic result being that a subset of the modes near the edges of the eigenvalue distribution is localized and another subset near the center is delocalized. A "mobility edge," whose location is a function of disorder strength, separates the two classes.

Given the marginal situation that occurs in two dimensions, it is natural to ask what the effect of interactions might be. Could they stop the localization and revive metallic behavior? Such two-dimensional metal–insulator transitions were extensively studied experimentally, first in the early 1980s when no metallic phase was found, and later in the 1990s when ultra-pure metal-oxide-semiconductor field-effect transistors were shown to exhibit conducting behavior. A large body of theoretical work, both analytic and numerical, has lent great insight into these experiments, without however, conclusively settling the issue. Useful references are Lee and Ramakrishnan (1985) and Belitz and Kirkpatrick (1994).

One of the commonly used theoretical models of this question is the "Anderson–Hubbard Hamiltonian" (AHH),

$$\hat{H} = -\sum_{<ij>,\sigma} t_{ij}\hat{c}^{\dagger}_{i\sigma}\hat{c}_{j\sigma} + \sum_{i\sigma}(\epsilon_i - \mu)\hat{n}_{i\sigma} + \sum_i U_i(\hat{n}_{i\uparrow} - \frac{1}{2})(\hat{n}_{i\downarrow} - \frac{1}{2}), \qquad (5.1)$$

which combines a spatially varying on-site Hubbard interaction $U_{\rm i}$ with randomness in the site energies ϵ_i and hopping matrix elements t_{ij}. Here $c^{\dagger}_{j\sigma}$ and $\hat{c}_{j\sigma}$ are creation and destruction operators for a fermion of spin σ on site j. $\hat{n}_{j\sigma}$ is the associated number operator. The results presented here will be entirely on two dimensional square lattices.

The original Hubbard Hamiltonian (HH) has uniform hoppings and site energies. As is well-known (Fazekas, 1999; Gebhard, 1997; Montorsi, 1992; Rasetti, 1991), at half filling, $\rho = \langle n_{i\uparrow} + n_{i\downarrow} \rangle = 1$, the ground state of the HH is a Mott insulator (MI) at sufficiently large U. This MI may be probed through the observation of a vanishing compressibility $\kappa = \partial n/\partial \mu$, or through the appearance of a gap in the spectral function $A(\omega)$. Away from half filling, $\rho \neq 1$, the HH is metallic. As described further below, DQMC simulations, among other methods, have established that this MI has long-range antiferromagnetic correlations (Hirsch and Tang, 1989; Varney *et al.*, 2009; White *et al.*, 1989). This is true on a square lattice at any U/t, owing to the perfect nesting of the Fermi surface. An open issue is the possibility that the metallic phase becomes superconducting as the temperature is lowered away from half filling. Our results will focus on the magnetic, spectral, and transport properties of the AHH to see how these properties, and analogous issues in the case of attractive interactions, are modified by disorder.

The experimental exploration of these issues may be poised to go through a new phase, with the advent of optical lattice emulators (Jaksch *et al.*, 1998). Such atomic systems, where interaction and disorder strength can be readily tuned, and which are in many ways more pure realizations of model Hamiltonians, have the potential to open a novel window in this long-standing question. At the same time, algorithm development and increases in computational power allow methods like DQMC to explore larger lattices than ever before and, indeed, lattice sizes that are comparable to those of optical lattice emulators in size.

Before concluding this section, we note a special "particle–hole" symmetry (PHS) of the HH on bipartite lattices, because it plays a key role in both the physics and the algorithm described here. Consider a canonical transformation that exchanges the fermion creation and destruction operators $c^{\dagger}_{i\sigma} \leftrightarrow (-1)^i c_{i\sigma}$. If the phase factor $(-1)^i$ is chosen to take opposite sign on the two sublattices of the bipartite lattice, then the HH is unchanged by the mapping as long as $\mu - \epsilon_i = 0$ on all sites. The presence of non-uniform hopping t_{ij} or interactions U_i does not affect PHS.

PHS has many interesting consequences for the physics of the HH. First, it is easy to see that the density on each site must be $\rho_i = 1$ (again under the condition $\epsilon_i - \mu = 0$). That is, the sites are half-filled for any value of temperature T, interaction strengths U_i, and even any pattern of hopping t_{ij}, including random ones. PHS turns out to have crucial implications for the appearance of a metal–insulator transition. This may seem to be a somewhat surprising result, since one often thinks the specific type of disorder is not relevant (e.g. in a renormalization-group sense in which one form of randomness will induce all others as low energy/length scales are integrated out). Despite this expectation, it turns out that systems with site and bond randomness behave in a fundamentally different way. As we shall see, PHS also plays a role in DQMC.

5.1.2 Determinant quantum Monte Carlo formalism

Most QMC methods proceed by writing down a path integral for the partition function, $Z = \mathrm{Tr}\, e^{-\beta \hat{H}}$, thereby mapping the problem onto a classical model in one higher dimension. As in Feynman's original prescription, in methods that utilize a discretized imaginary time $\beta = L\Delta\tau$, the Hamiltonian $\hat{H}$ is separated into kinetic ($\hat{K}$) and potential ($\hat{V}$) pieces, and Z is expressed as,

$$Z = \mathrm{Tr}\left[e^{\Delta\tau\hat{H}} e^{\Delta\tau\hat{H}} \cdots e^{\Delta\tau\hat{H}} \right] \approx \mathrm{Tr}\left[e^{\Delta\tau\hat{K}} e^{\Delta\tau\hat{V}} e^{\Delta\tau\hat{K}} e^{\Delta\tau\hat{V}} \cdots e^{\Delta\tau\hat{K}} e^{\Delta\tau\hat{V}} \right]. \quad (5.2)$$

In the AHH, $\hat{K}$ incorporates the hopping and site energy terms, which are quadratic in the fermion operators, and $\hat{V}$ is the quartic interaction term. Increasingly, methods that employ continuous time to avoid this discretization are being used (Beard and Wiese, 1996; Prokofév *et al.*, 1996; Rieger and Kawashima, 1999). We will, however, focus on the discrete version, as it is most relevant for our formulation of DQMC.

Beginning from the common representation of eqn (5.2), lattice QMC methods subsequently fall into two broad classes. In "worldline" (WL) approaches, complete sets of states are inserted between the incremental imaginary-time evolution operators, and Z is then expressed as a sum over the space and imaginary-time-dependent classical variables that label these eigenstates. The "weight" of a given configuration is given by the product of the matrix elements that arises in the construction.

WL methods (Hirsch *et al.*, 1982) for local Hamiltonians generally have a computation cost that scales linearly with the lattice size N and inverse temperature dimension $L = \beta/(\Delta\tau)$, since only a small, lattice-size-independent number of matrix elements are altered by a change to one of the classical variables, of which there are $o(NL)$. This linear scaling allows WL simulations, when they work, to be done on lattices of very large size. This has especially been the case over the last decade, when "loop" and "worm" algorithms have been developed which ensure that long autocorrelation times do not occur (Kawashima *et al.*, 1994; Sandvik, 1992, 1997; Sandvik and Kurkijarvi, 1991).

The key issue in the applicability of the WL approach is whether all the matrix elements have positive sign. If they do not, the method collapses spectacularly, and usually cannot give results even at high temperatures. Models that are safe from the sign problem include fermions in one dimension, and quantum spin Hamiltonians and the boson Hubbard model on bipartite lattices in any dimension. In particular, fermion models in two and higher dimensions (or even simply with a single impurity orbital) cannot be treated with WL techniques. This leads us to DQMC, which also has a sign problem, but one that is much less severe, and even absent entirely at certain special (PHS) points.

In DQMC, rather than inserting complete sets of states in eqn (5.2), the interaction exponential is converted from quartic to quadratic in the fermion operators through the introduction of an auxiliary Hubbard–Stratonovich field. In the work reported here, we use a discrete version of this transformation (Hirsch, 1985), which we write for the on-site Hubbard interaction,

$$e^{-U\Delta\tau(n_\uparrow - \frac{1}{2})(n_\downarrow - \frac{1}{2})} = \frac{1}{2} e^{-\frac{\Delta\tau U}{4}} \sum_{S=\pm 1} e^{\lambda S(n_\uparrow - n_\downarrow)}. \quad (5.3)$$

The parameter λ is given by $\cosh\lambda = e^{U\Delta\tau/2}$. We have suppressed the site index i in eqn (5.3) but evidently there is a Hubbard–Stratonovich variable $S_{i\tau}$ for each spatial site $i = 1, 2, \cdots N$ in eqn (5.1) and imaginary time $\tau = 1, 2, \cdots L$ in eqn (5.2).

The conversion of the quartic interaction term to a quadratic one allows the fermion trace to be performed analytically.

$$Z = \sum_S \det[I + B_{1\uparrow} B_{2\uparrow} \cdots B_{L\uparrow}] \det[I + B_{1\downarrow} B_{2\downarrow} \cdots B_{L\downarrow}]. \tag{5.4}$$

The matrices $B_{l\sigma}$ have dimension equal to the number of spatial lattice sites N and are functions of the Hubbard–Stratonovich field S. Their detailed structure will not be discussed here, but can be found in any of several review articles (Bai *et al.*, 2009; Gubernatis *et al.*, 1989; Loh *et al.*, 1990; Scalettar, 2003; White *et al.*, 1989).

DQMC then consists of sweeping through all the entries of the Hubbard–Stratonovich field, suggesting a change to each in turn. The ratio of the determinants before and after the change is used in a Metropolis accept/reject decision. In principle the CPU time in DQMC would scale as $N^4 L$, since there are NL elements $S_{i\tau}$ to update, and the computation of a determinant costs N^3. However, if only a single Hubbard–Stratonovich variable is altered, there is an $o(N^2)$ method to evaluate the determinant ratio. More precisely, if the "Green's functions"

$$G_\uparrow = [I + B_{1\uparrow} B_{2\uparrow} \cdots B_{L\uparrow}]^{-1} \qquad G_\downarrow = [I + B_{1\downarrow} B_{2\downarrow} \cdots B_{L\downarrow}]^{-1} \tag{5.5}$$

are known, the determinant ratio can be evaluated in just a few operations. Updating G after the change is $o(N^2)$. Again, the reader is referred to the literature for details (Bai *et al.*, 2009; Gubernatis *et al.*, 1989; Loh *et al.*, 1990; Scalettar, 2003; White *et al.*, 1989). The Green's function that is needed to do the Monte Carlo calculation also turns out to allow the rapid construction of almost any physical observable. For example, the estimators for the density and kinetic energy, are $\rho_i = 1 - G_{\rm ii}$ and $K = 4\sum_{\langle ij \rangle} t_{ij} G_{ij}$. Other formulae involving products of elements of G can be written for the double occupancy, spin, pair, and density correlation functions. Indeed, there are close comparisons possible between DQMC and traditional diagrammatic perturbation theory owing to the ability of DQMC to measure directly very general combinations of products of fermion operators and hence Green's functions, using an analog of Wick's theorem.

The fermion determinants in DQMC can go negative in the same way that the matrix elements change sign in the WL method. However, it turns out that this in general does not happen until the inverse temperature $T \leq t/4 = W/32$, where $W = 8t$ is the bandwidth of the two-dimensional model with uniform hopping. This is much colder than accessible to WL simulations. If one is describing a solid with bandwidth $W \sim 1\,\text{eV}$, then $T = t/4 = W/32 \sim 400$ K. One dramatic way to see the improved behavior of the sign in DQMC versus WL methods is to note that the WL method has a severe sign problem in two dimensions even at $U = 0$. DQMC, on the other hand, has no sign problem at any T or μ in the non-interacting limit. In fact, it solves the $U = 0$ problem exactly, G_σ is independent of S, and no Monte Carlo sampling is required whatsoever.

Although the precise onset of negative determinants depends on U and μ, the accessible temperatures are low enough to discern many interesting aspects of the physics of the AHH, including the renormalization of the non-interacting dispersion relation by U, the opening of the charge ("Mott") gap, and the beginning of the formation of magnetic order. In particular, as we shall see, T is low enough that the effects of U on the conductivity can be studied. Additionally, temperatures accessible to DQMC are well below those currently available in optical lattice emulators. Thus although superconducting transitions, if they occur, are well outside the range of DQMC, this technique has much to guide the optical lattice emulator community at present.

It is important to note that the special PHS discussed in Section 5.1.1 has a deep consequence for the sign problem. Specifically, the up and down determinants can be simply related to each other when $\mu - \epsilon_i = 0$, and always have the same sign. Their product is therefore positive, and there is no sign problem at any temperature. It is for this reason that DQMC was able successfully to answer the deep issue of the existence of long-range antiferromagnetic order in the two-dimensional HH at half filling, and can also be used to study variants such as random hopping and site dilution.

There are several refinements of this basic DQMC approach. Most importantly, great care must be taken to avoid round-off errors in the matrix multiplications that arise in the construction of G (Loh Jr. *et al.*, 1989; Sorella *et al.*, 1988; Sugiyama and Koonin, 1986; White *et al.*, 1989). Measurements can be extended to imaginary time-dependent quantities and, thereby, dynamical observables using the maximum entropy method (Gubernatis *et al.*, 1991). A ground-state, canonical-ensemble formulation is also possible (Furukawa and Imada, 1992; Sorella *et al.*, 1988). While noting these important details here, we refer the reader to the literature for their in-depth discussion.

5.1.3 Overview of other numerical approaches

DQMC provides one very powerful computational approach to the metal–insulator transition, and is the method that is the focus of the results presented in this article. However, there are, of course, many alternate numerical techniques which have provided great insights in the field. These include diffusion QMC, dynamical mean-field theory (DMFT), inhomogeneous Hartree-Fock approaches and exact diagonalization. We cannot review these methods and their results thoroughly here. What we will do in this section is briefly to point to these other approaches and some of the conclusions they have reached. At the end of this article, we will return to some of these methods and discuss how their results compare with our DQMC calculations. This discussion is of necessity somewhat idiosyncratic and certainly is not meant to imply that other numerical work that we have not mentioned has not had important impact on the field.

We first summarize the results of a study most closely related methodologically to the approach of this chapter. Caldara *et al.* (2000) and Srinivasan *et al.* (2003) investigated the disordered, repulsive HH in one and two dimensions using projector QMC to access the ground-state properties. Calculation of the ratio $r = \xi(U)/\xi(U = 0)$ of the inverse participation ratio ξ in the presence and absence of the on-site interaction

U indicates that electron–electron repulsion can cause significant delocalization, with values of $r \approx 1.5$–2, which increase as the system size becomes larger. While suggestive, these authors do not conclude that a metal–insulator transition occurs in the AHH.

DMFT and its extensions are powerful methods to treat correlated electron physics, which are now being applied to include the effects of disorder (Georges *et al.*, 1996; Hettler *et al.*, 1998; Vollhardt, 1993). Tanaskovic *et al.* (2003) apply DMFT to the question of disorder screening. They emphasize a fundamental question concerning the effectiveness of interactions in smoothing the random potential, namely that the screening length is inversely proportional to the compressibility κ, which, however, becomes small near an interaction-driven (Mott) insulating transition. In such a regime, therefore, one might think interactions are not so effective at screening the disorder. Their key conclusion, when the DMFT equations are treated with an accurate impurity solver, is that the strong enhancement of the effective mass can help counterbalance the effect of reduced κ and lead to a stabilization of metallic behavior. This effect is largest when interactions and disorder are of comparable magnitude, $U \sim \Delta_\mu$.

In related work, Aguiar *et al.* (2005) also employ DMFT, focussing on the effect of relatively weak disorder on the region of coexistence of metal and insulator, where both types of solution of the DMFT equations are found, at finite temperature. The coexistence region is reduced, but never entirely eliminated, as Δ_μ is ramped upwards.

Some of the most recent work of this group (Aguiar *et al.*, 2009) uses DMFT in a way that incorporates Anderson localization effects, the so-called "typical medium theory." Here the intriguing picture of a two-fluid model is presented, in which some electrons are Mott localized on certain sites, while others become Anderson localized. The local quasiparticle weights describe the tendency towards Mott physics, while the typical local density of states characterizes Anderson localization.

DMFT is also the methodology employed by Song *et al.* (2008) who examined the local density of states of the AHH to extract the inverse participation ratio and localization length. The localization length was found initially to increase as U was turned on, reaching a maximum, and then decreasing at yet stronger coupling. These DMFT calculations (Song *et al.*, 2008) differ from those of Tanaskovic *et al.* (2003) at several technical levels, e.g. in the use of an effective-medium approach versus an exact treatment of the disorder potential, and in the presence of a quasiparticle resonance at the Fermi surface. Differences in the conclusions regarding the degree of screening of the disorder potential by the interactions might originate in these technical details.

Radonjic *et al.* (2010) use DMFT to examine the conductivity of the half-filled AHH. Rather than starting with Anderson localization and examining the screening effect of U, here the starting point is the Mott insulator and the effect of Δ_μ in its disruption. A simple picture of this effect is a disorder-induced increase in the bandwidth and hence an increase in the critical value of interactions U_c for the Mott transition.

Finally, Byczuk *et al.* (2010) have recently used DMFT to evaluate the phase diagram of the half-filled AHH in the paramagnetic phase (i.e. ignoring the possibility of magnetic order). Starting from an Anderson insulator with $U = 0$ and $\Delta_\mu = 2W$ where W is the bandwidth, an insulator-to-metal transition is seen to occur at $U \sim W/2$.

As Δ_μ increases, the degree of correlation required to drive the metallic behavior also increases until, ultimately, when $\Delta_\mu > 3.5W$, no metallic phase is possible. When the possibility of antiferromagnetic order is included, the metallic region is substantially reduced in area but still occurs.

The effect of an in-plane (Zeeman) field, which is discussed here in Section 5.2.5, has been investigated by Berkovits and Kantelhardt (2002) using exact diagonalization for a second quantized Hamiltonian that includes both an on-site Hubbard interaction U_{H} and a long-range Coulomb interaction. While no metal–insulator transition is observed in the inverse participation ratio, it is shown that U_{H} tends to delocalize the electrons, and that in contrast a magnetic field increases the resistivity. Both trends are consistent with the results presented in this chapter. Interestingly, the long-range repulsion leads to increased resistivity. The exact diagonalization work of Kotlyar and Das Sarma (2001) focuses on the behavior of the charge stiffness (Drude weight) D_{c} as a function of the ratio of interaction to disorder strengths. Increasing U/Δ_μ does lead to a larger D_{c}, but only in the regime of large $\Delta_\mu/t \sim 3$, where these authors believe metallic behavior to be very unlikely. For small $\Delta_\mu/t \sim 0.5$ increasing U/Δ_μ barely affects D_{c} at all. Thus they conclude that a two-dimensional metal–insulator transition is unlikely to occur.

Herbut (2001) has also considered the effect of a parallel magnetic field within Hartree–Fock theory for the AHH. In agreement with the work discussed above, he finds that, at low temperatures and in the absence of a field, the interaction U screens the disorder and enhances the conductivity by a factor $(1 + UN(E_{\mathrm{F}})$, where $N(E_{\mathrm{F}})$ is the density of states at the Fermi energy. When the field is non-zero and the system becomes polarized the screening is reduced. The magnetoresistance can also be computed analytically via diagrammatic techniques or, equivalently, the quantum kinetic equation. The correction to the conductivity can be shown to be field dependent (Zala *et al.*, 2001).

While the subject of this chapter is second-quantized lattice Hamiltonians, it is worth noting interesting similarities with a range of quite different numerical models. Chui and Tanatar (1995) have used QMC simulations for continuum systems to show that impurities can shift the freezing transition of a pure electron gas of density n from $r_{\mathrm{s}} = 1/(\sqrt{\pi n}a_B)$ to $r_{\mathrm{s}} \approx 7.5$, in much better agreement with values observed experimentally for Si-MOSFETS (Pudalov *et al.*, 1993), and emphasizing the importance of the interplay of disorder and interactions. Their study also found interesting magnetic behavior, a feature that is also present in lattice models, especially near commensurate fillings. Another variant of the problem of the interplay of randomness and correlation is to consider disordered spinless fermion systems, where intersite interactions embody correlation effects. One example is the exact diagonalization work of Vojta *et al.* (1998), which shows that interactions can increase the conductivity in the strongly localized regime (large Δ_μ) but decrease σ for small Δ_μ.

We conclude this review by turning to studies of anomalies in the density of states, the topic of Section 5.3. The presence of zero-energy anomalies in the density of states is a widespread feature of interacting systems in the presence of disorder. In the case of completely localized charges on a disordered background potential, Efros and Shklovskii (1975) were able to rigorously prove that when the particles interact through an unscreened Coulomb potential, the density of states has to go to zero at the chemical potential at least as fast as ω^{D-1}. In the opposite case of weak interaction and disorder, Altshuler and Aronov (1979) employed the methods of many-body perturbation theory to show that in that case too an anomaly is present (although for different physical reasons) and that it is characterized by singular behavior as a function of ω as the chemical potential is approached. Wortis (pers. comm.) has shown that this phenomenon is present even in an ensemble of Hubbard dimers, after averaging over differences in the two site energies.

5.2 Repulsive Hubbard model: magnetism and transport

5.2.1 Introduction

This section contains a discussion of the results of determinant Quantum Monte Carlo simulations of the disordered repulsive HH on a square lattice. Initial considerations will address how site dilution affects the long-range magnetic order that occurs in the ground state of the clean system at half filling. Two types of dilution will be examined. In the first, the interaction is turned off by setting $U_i = 0$ on a randomly selected fraction f of sites. This has the effect of inhibiting moment formation by allowing double occupation. In the second, the site is actually removed from the lattice, so that it is not available as a location onto which fermions can hop. In this situation, magnetic order is inhibited by a reduction of neighbors. The next set of studies will be of random chemical potentials: the AHH. When the amplitude of ϵ_i becomes comparable to U, double occupancy increases and, as in the case of $U_i = 0$ impurities, magnetism is suppressed.

Following these studies of the interplay of magnetic order and randomness, we turn to the metal–insulator transition by looking at how increasing U affects the conductivity $\sigma_{\rm dc}$ of an Anderson insulating state arising from site disorder. We set the filling $\rho = \frac{1}{2}$ (quarter filling) to avoid the special Mott insulating and magnetic physics near $\rho = 1$ (half filling). The key conclusion is that $d\sigma_{\rm dc}/dT$ can change sign, becoming negative at low temperatures T, indicating metallic behavior. This phenomenon is seen to occur in a very similar way if the randomness is instead in the hopping.

Finally, a Zeeman magnetic field $B_\parallel$ is demonstrated to return the metallic phase created by non-zero U to insulating behavior. This effect can be ascribed to a reduction of the role of U when the fermions become spin polarized.

5.2.2 Effect of site dilution on magnetism at half filling

Cuprate (Scalapino, 1994) and iron-pnictide (Ishida *et al.*, 2009; Tesanovic, 2009) superconductors have layered structures for which two-dimensional models may be appropriate. In both sets of systems, superconductivity occurs in close proximity to a magnetic phase. In the cuprates the ordering is antiferromagnetic with wavevector $\mathbf{q}_* = (\pi, \pi)$. In the iron pnictides stripes occur with $\mathbf{q}_* = (\pi, 0)$, so the spins are ferromagnetically aligned in one direction. In both cases, changes in the density or chemical substitution of non-magnetic atoms rapidly suppress antiferromagnetic order and superconductivity arises. This same effect of doping was seen in the earliest DQMC simulations of the two-dimensional Hubbard model by Hirsch (Hirsch, 1985; Hirsch and Tang, 1989; White *et al.*, 1989). Indeed, he suggested that, unlike the mean-field phase diagram, which has a substantial region of antiferromagnetism around half filling, the true situation might be that long-range order exists only at precisely $\rho = 1$.

One can ask whether this sensitivity of magnetism to filling arises from the presence of a static set of sites that no longer have a moment, or whether the mobility of the defects plays a crucial role. This section first explores the effect of static $U_i = 0$ impurities on long-range magnetic order in the half-filled two-dimensional Hubbard model (Janis *et al.*, 1993; Ulmke *et al.*, 1998). Such sites would, for example, correspond to replacing Cu atoms in cuprate superconductors by Zn (as opposed to replacement of La by Sr, which dopes the lattice). This work is supplemented by examining a situation in which, rather than setting $U_i = 0$ but retaining the impurities as possible locations to which the electrons can hop, the sites are instead entirely clipped from the lattice by setting the hoppings to them to zero.

Consider the HH of eqn (5.1) with uniform hopping $t_{\rm ij} = t$, all site energies zero $\mu - \epsilon_i = 0$, and bimodal interaction strength: $U_i = 0$ with probability f, and $U_{\rm i} = U$ with probability $1 - f$. Hirsch and Tang (1989) used DQMC to demonstrate that at $T = 0$ and $f = 0$ there is long-range antiferromagnetic order. The way this result is established, using simulations on finite lattices, is through an appropriate scaling analysis in the system size. More specifically, the first step is the evaluation of the real-space spin–spin correlations and their Fourier transform, the magnetic structure factor,

$$c(j) = \langle (n_{i\uparrow} - n_{i\downarrow})(n_{i+j\uparrow} - n_{i+j\downarrow}) \rangle \qquad S(\mathbf{q}) = \sum_{\rm l} e^{i\mathbf{q}\cdot j}\, c(j)\ . \tag{5.6}$$

It is necessary to measure these quantities as a function of T and make sure T is small enough that the ground-state values of the correlation function have been reached on the lattice size in question. (See, for example, Fig. 5.3 (a)) Attention is focussed here on the ordering wavevector $\mathbf{q}_* = (\pi, \pi)$. However, note that $\beta S(\mathbf{q} = (0,0)\,)$ is equal to the uniform spin susceptibility, another quantity of considerable interest. Equation (5.6) employs the z component of spin but, owing to the spin–rotation invariance of the Hamiltonian, the x and y components are identical. Indeed, checking that measurements in different directions are consistent is a good way to verify ergodicity in the simulations and reduce error bars. If the spin–spin correlation function $c(j)$ is long ranged, the structure factor will grow linearly with the number of lattice sites $N = L \times L$, and hence $S(\pi, \pi)/N$ will approach a non-zero limit as $N \to \infty$.

Huse used spin-wave theory (Huse, 1988) to evaluate the way in which the thermodynamic limit is approached, and showed that the corrections both to the structure factor and to the spin–spin correlation function at maximal separation go inversely with the linear lattice size L. That is,

$$\frac{1}{N}\, S(\pi, \pi) = \frac{1}{3}\, m^2 + \frac{a}{L} \qquad c(\frac{L}{2}, \frac{L}{2}) = \frac{1}{3}\, m^2 + \frac{b}{L}, \tag{5.7}$$

where m is the antiferromagnetic order parameter and a and b are the coefficients of the linear corrections. Equation (5.7) allows for a reliable extrapolation to the thermodynamic limit from simulations on finite lattices: $S(\pi, \pi)/N$ and $c(L/2, L/2)$ are plotted versus $1/L$ for different lattice sizes, and the extrapolation $1/L \to 0$ (that is, $L \to \infty$) reveals the presence or absence of long-range order.

Here the $f = 0$ result of Hirsch and Tang is generalized by asking to what values of f long-range order persists. The finite-size scaling of the structure factor is shown in the inset of Fig. 5.1 (a) for $f = 0, \frac{1}{8}$ and $f = \frac{1}{2}$. The interaction strength $U = 8t$ and the lattice is half-filled, $\rho = 1$. The lower two values of f have a non-zero extrapolation of $S(\pi, \pi)$ to $1/L \to \infty$ and hence a non-zero order parameter m. The largest value, $f = \frac{1}{2}$ has $m = 0$. The main part of Fig. 5.1 (a) shows the extrapolated value m for different f. The critical f necessary to destroy magnetic order is $f_{\rm crit} \sim 0.45$. This conclusion is in sharp contrast to the much more dramatic effect of changing the density, where a shift of just a percent from half filling eliminates long-range magnetic order. The lattice sizes used in Fig. 5.1 were up to $N = 100$ sites, i.e. linear size $L = 10$. With currently available computers and algorithms, $N = 576$ ($L = 24$) can be simulated on a modest cluster of desktop computers. The U dependence of the order parameter of the $f = 0$ system was recently evaluated by Varney *et al.* (2009) on such large lattices.

An interesting feature of the $f \neq 0$ model is that the Mott-insulating phase does not occur at $\rho = 1$. Thus the antiferromagnetic and MI phases are separated, occuring

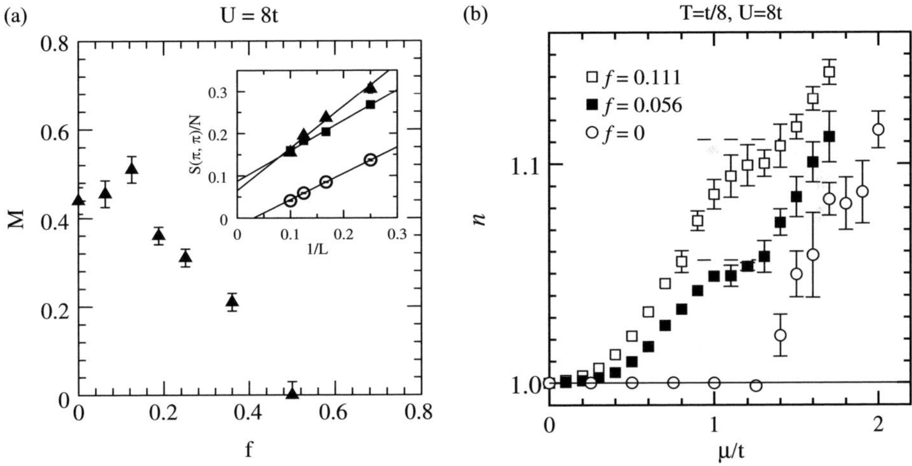

Fig. 5.1 (a) ground-state staggered magnetization m as a function of fraction f of $U_i = 0$ sites in a background of $U_i = 8t$ sites, and half filling $\rho = 1$. Inset: Finite-size scaling of the antiferromagnetic structure factor: $f = 0.0$, 0.125, (filled triangles and squares) are ordered, $f = 0.500$ (open circles) is not. For values of f that do not correspond to an integer number of defects, we have interpolated the results for the two bracketing concentrations. (b) density ρ vs μ for three fillings. The Mott plateau occurs at $\rho = 1 + f$ where f is the fraction of $U_i = 0$ sites.

at different fillings. This is illustrated in Fig. 5.1 (b) where the dependency of the density on chemical potential is shown. We see that the compressibility κ does not vanish at $\rho = 1$ but rather at $\rho = 1 + f$. A little thought reveals why this is the case. In the usual $f = 0$ model, extra fermions can be added to the lattice without causing double occupancy up to $\rho = 1$. But once the filling exceeds $\rho = 1$ the cost to add a particle μ must involve U. The jump in μ at $\rho = 1$ causes a horizontal plateau in $\rho(\mu)$ and vanishing compressibility. In the present model, when $f \neq 0$, there are additional sites that can become double occupied with low energy cost since the on-site U_i is zero. It is not until the filling reaches $1 + f$ that the chemical potential jumps and the Mott plateau in $\rho(\mu)$ occurs.

An important consequence of writing the interaction term in eqn (5.1) in particle–hole symmetric form is that setting $U_i = 0$ on some sites also means that the chemical potential varies from site-to-site. Indeed, this choice causes the Mott plateau of Fig. 5.1 (b) at $n = 1 + f$ to be accompanied by a plateau at $n = 1 - f$. This convention is frequently used, e.g. in studies of the periodic Anderson model, and is essential to control the sign problem in these DQMC simulations. This will be discussed further in Section 5.4.

Another way to observe the large-U Mott gap (or the Slater gap caused by antiferromagnetic order into which it crosses over at weak U) is through the density of states. As mentioned in the first chapter, the real-space many-body Green's function,

$$G_j(\tau) = \langle c_{i+j}(\tau) c_i^\dagger(0) \rangle \tag{5.8}$$

is simply related to the matrix inverses of eqn (5.5), which arise in the update procedure/determinant ratio in DQMC. The local density of states $A_{j=0}(\omega)$ is given by the inverse of the integral transform,

$$G_{j=0}(\tau) = \int d\omega \, \frac{A_{j=0}(\omega)}{e^{\beta\omega} + 1}. \tag{5.9}$$

For completeness we note that in momentum space $A_{\mathbf{p}}(\omega)$ (which is probed with angle-resolved photoemission spectroscopy) is given by,

$$G_{\mathbf{p}}(\tau) = \int d\omega \, \frac{A_{\mathbf{p}}(\omega)}{e^{\beta\omega} + 1} \qquad G_{\mathbf{p}}(\tau) = \sum_{\mathbf{p}} e^{i\mathbf{p}\cdot j} \, G_j(\tau) \, . \tag{5.10}$$

The full spectral function is $A(\omega) = \sum_{\mathbf{p}} A_{\mathbf{p}}(\omega)$. Inversion of eqns (5.9) and (5.10) requires an analytic continuation as discussed in Gubernatis *et al.* (1991).

In the pure Hubbard model, the vanishing of κ and the plateau in μ are reflected in a gap in $A(\omega)$. In this disordered system we instead focus on the local density of states $A_j(\omega)$ given by eqn (5.9). In Fig. 5.2 (a) results are shown for half filling and a dilution fraction $f = 0.11$ of $U_i = 0$ sites in a model with $U_i = 8$. The structure on the $U_i = 8$ sites is quite similar to the $f = 0$ case, with a well established Slater/Mott gap at $\omega = 0$. On the other hand, the spectral function on the $U_i = 0$ sites is completely different from what it would be for a homogeneous $U = 0$ system. There is a clear "induced" gap, which would not be present in the non-interacting system.

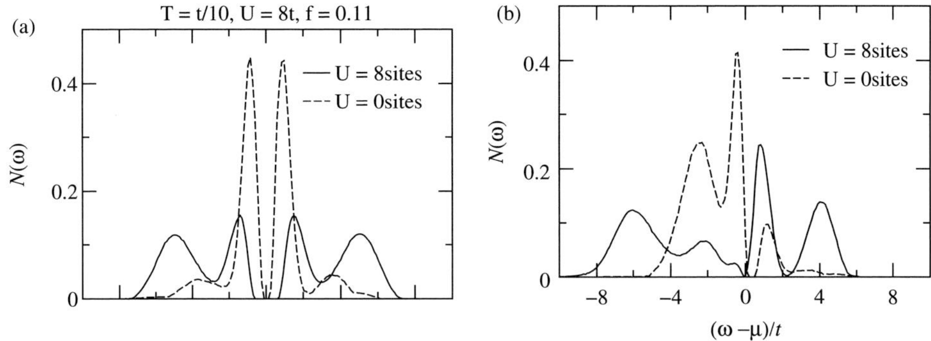

Fig. 5.2 (a) Density of states on $U = 0$ and $U = 8t$ sites at half filling. (b) $\rho = 1.11$. All spectra have a vanishing density of states at the Fermi energy.

It is worth noting that in discussions of the HH within the context of optical lattice emulators, where a smoothly varying external potential confines the atoms, a frequently used tool is the local density approximation, which asserts that the local properties of a model with a spatially varying chemical potential are very similar to those of a uniform system with the same local density. Clearly the results of Fig. 5.2 (a) imply that such a picture, where the properties on a given site are determined by the local energy scales alone, is invalid here. This is not too surprising, since the success of the LDA has been in situations with rather smoothly varying parameters, and here U_i changes very abruptly.

Figure 5.2 (b) examines density $\rho = 1 + f = 1.11$. As commented earlier, and shown from $n(\mu)$ in Fig. 5.2 (b), the charge gap of the Hubbard model with a fraction f of $U_i = 0$ sites is shifted to filling $1 + f$. Figure 5.2 (b) confirms this observation since $A_{\rm j}(\omega) = 0$ for both types of site.

A different realization of dilution is to remove sites from the lattice, as opposed to retaining them with a weakened interaction $U_i = 0$. The consequences for magnetism are shown in Fig. 5.3. Figure 5.3 (a) illustrates the analysis necessary to ensure that the temperature is low enough that the ground state is captured: β is increased until $S(\pi, \pi)$ saturates. Note that the low-T value grows markedly with lattice size, an indication that the order might be long ranged. (Similar plots were made to assess how big to take β in producing Figs. 5.1 and 5.2). From those $T = 0$ numbers, the scaling plot of the structure factor determines whether there is long-range order in the thermodynamic limit. The critical dilution fraction $f_{\rm crit} \sim 0.25$–0.30. This value is lower than in the previous case where correlated sites are replaced by uncorrelated ones, (see Fig. 5.1). It is not completely apparent why site removal is more hostile to $U_i = 0$ randomness. It might be that retaining sites allows for a remnant RKKY-like antiferromagnetic coupling even though the sites have only weak correlation energy.

The preceding section assessed the effect of impurities, which take the form of modifying the on-site repulsion on the magnetism and spectral function of the half-filled HH. A problem that is closely related to such site removal is the introduction of random chemical potentials (site energies). Since fermions are inhibited from hopping

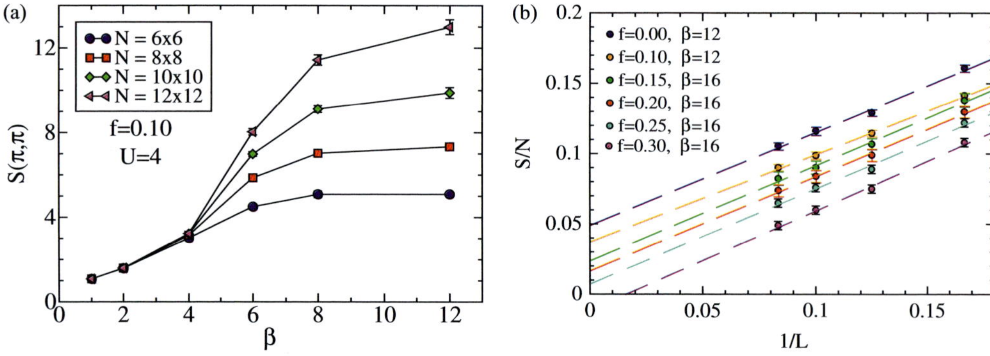

Fig. 5.3 (a) antiferromagnetic structure factor of eqn (5.6) as a function of inverse temperature β for $U = 4t$ and $f = 0.10$. Here the sites are entirely clipped from the system. By the time $\beta \approx 12$ the ground state has been reached on lattice sizes $L = 6, 8$. The system needs to be made a bit colder to saturate $S(\pi, \pi)$ for $L = 10, 12$. (b) finite size scaling analysis according to eqn (5.7). Order is lost somewhere between $f = 0.25$ and $f = 0.30$.

to sites with ϵ_i large because of the direct energy cost, and to sites with ϵ_i small (which tend to be doubly occupied), random site energies have a clear connection to site removal.

Consider site energies uniformly chosen on $-\Delta_\mu < \epsilon_{\rm i} < +\Delta_\mu$, so that Δ_μ controls the scale of the disorder. Figure 5.4 shows how increasing Δ_μ decreases the spin–spin correlations in real space. The long-distance behavior of $c(j)$ is reduced by about a factor of two when $\Delta_\mu = 1.6t$. Whether there is long-range order can only be addressed by the finite-size scaling described by eqn (5.7) and shown for $U_i = 0$ impurities in Figs. 5.1 and 5.3. Figure 5.5 exhibits the scaling for the case of site randomness. The analysis of the structure factor $S(\pi, \pi)$ and spin–spin correlation at largest separation lead to similar values of the order parameter, which is driven to zero at $\Delta_\mu^{\rm c} \sim 1.4$.

5.2.3 Interaction-driven metal-insulator transition

Having previously described the manner in which randomness affects the magnetism, this section turns to an examination of the effect of interactions on Anderson localization. A preliminary technical point is necessary to describe how the conductivity can be obtained in DQMC simulations. The difficulty is similar to that in obtaining the density of states, where the quantity directly measured in DQMC is $G_{\mathbf{p}}(\tau)$, which must be analytically continued to $A_{\mathbf{p}}(\omega)$, eqn (5.9).

The current–current correlation function in imaginary time (Scalapino *et al.*, 1993),

$$\Lambda_{xx}(\mathbf{q}, \tau) = \langle j_x(\mathbf{q}, \tau)\, j_x(-\mathbf{q}, 0)\rangle \qquad j_x(\mathbf{q}) = \sum_{\mathbf{l}} e^{i\mathbf{q}\cdot\mathbf{l}} j_x(\mathbf{l})$$

$$j_x(\mathbf{l}) = i \sum_{\sigma} t_{\mathbf{l}+\hat{x},\mathbf{l}}(c^\dagger_{\mathbf{l}+\hat{x},\sigma} c_{\mathbf{l}\sigma} - c^\dagger_{\mathbf{l}\sigma} c_{\mathbf{l}+\hat{x},\sigma}), \tag{5.11}$$

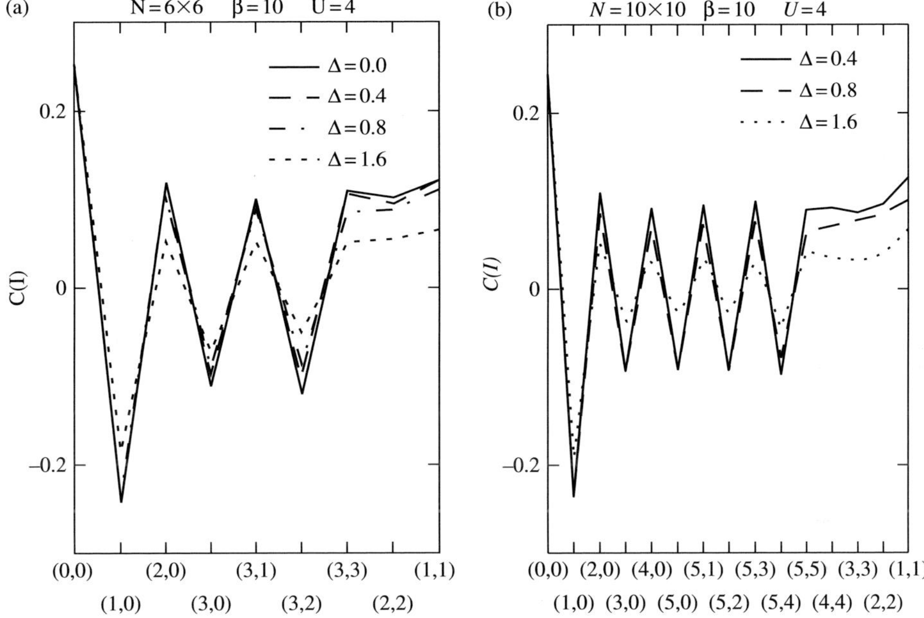

Fig. 5.4 (a) spin–spin correlations as a function of separation on 6×6 lattices at $T = t/10$ and $U = 4t$ for different amounts of chemical potential randomness. The value shown at $(0,0)$ is $C(0,0) - \frac{1}{2}$. (b) the same but for 10×10 lattices.

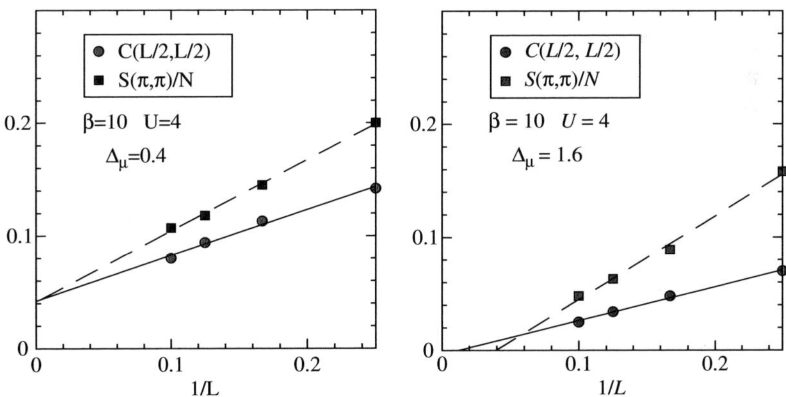

Fig. 5.5 Finite-size scaling analysis of the antiferromagnetic structure factor $S(\pi, \pi)/N$ and long-range spin correlations $c(L/2, L/2)$. These quantities, appropriately scaled, are plotted as a function of the inverse linear lattice size $1/L$. A non-zero extrapolation to $1/L = 0$ indicates antiferromagnetic long-range order. The extrapolated values should be identical.

can be directly measured in DQMC. This quantity is related, by the fluctuation–dissipation theorem, to the real frequency $\text{Im}\,\Lambda(\omega)$,

$$\Lambda_{xx}(\mathbf{q},\tau) = \int_{-\infty}^{+\infty} \frac{d\omega}{\pi} \frac{\exp(-\omega\tau)}{1-\exp(-\beta\omega)} \,\text{Im}\,\Lambda_{xx}(\mathbf{q},\omega). \tag{5.12}$$

To obtain the dc conductivity it is assumed that the temperature T is much lower than frequencies Ω at which $\text{Im}\,\Lambda$ deviates from its low-frequency behavior, $\text{Im}\,\Lambda = \omega\sigma_{\text{dc}}$. Under this assumption, for the largest accessible imaginary time $\tau = \beta/2$, the kernel in eqn (5.12) has fallen to negligible values in the region where $\text{Im}\,\Lambda$ is non-linear. It is therefore a good approximation to use the linear form over the full frequency range. In that case the integral can be performed to obtain,

$$\sigma_{\text{dc}} = \frac{\beta^2}{\pi}\Lambda_{xx}(\mathbf{q}=0,\tau=\beta/2). \tag{5.13}$$

Equation 5.13 is the central result from which σ_{dc} can be found (Trivedi *et al.*, 1996).

Figure 5.6 shows the application of this DQMC technology to the AHH at quarter filling, $\rho = 0.5$. Figure 5.6 (a) considers the case of site randomness $\Delta_\mu = 8t$. In the absence of interactions, $U = 0$, the system is expected to be an Anderson insulator and, indeed, σ_{dc} falls as the temperature is decreased. When the interactions are turned on sufficiently strongly, however, this T dependence is reversed: σ_{dc} increases as the temperature is lowered. The phenomenon is qualitatively identical in the case of bond randomness, as seen in Fig. 5.6 (b), which has $\Delta_t = 2t$. This figure reveals one of the central DQMC results for the physics of the AHH: when U is turned on in the Anderson insulator, the conductivity exhibits an insulator-to-metal transition (Chakraborty *et al.*, 2007; Denteneer *et al.*, 1999).

An obvious concern in such numerical studies is the effect of finite lattices. Clearly for weak randomness the localization length ξ_{loc} could exceed the linear lattice size, giving a false metallic signal. The choices $\Delta_\mu = 8t$ and $\Delta_t = 2t$ in Fig. 5.6 are such that $\xi_{\text{loc}} < L$, as can be demonstrated by explicitly looking at the eigenstates, or alternatively by the sign of $d\sigma_{\text{dc}}/dT$, as shown. Comparison of the data for 8×8 and 12×12 lattices in Fig. 5.6 (b) also lends assurance that the insulating signal is robust. Now that more powerful computers and algorithms are available, it would be interesting to revisit this problem and carefully study larger lattices and a bigger range of disorder and interaction values.

Further confidence in these results can be obtained by starting with the metallic phase of the site disordered case of Fig. 5.6 (a) which has $\Delta_\mu = 8t$ and $U = 4t$ and increasing Δ_μ. One expects the return of the Anderson insulator, as the delocalizing effect of U gets overwhelmed by larger randomness. Figure 5.7 (a) shows that this is indeed the case. For larger Δ_μ the temperature dependence of the conductivity becomes once again characteristic of insulating behavior. Figure 5.7 (b) shows the same data, but replotted at constant β. The curve crossing reflects the change in sign of the temperature dependence of σ_{dc} and is another way to capture the metal–insulator transition.

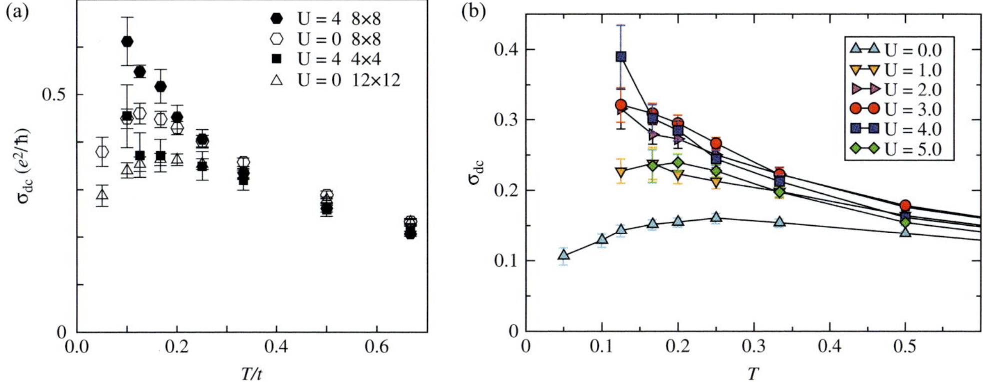

Fig. 5.6 Conductivity $\sigma_{\rm dc}$ as a function of temperature T for various values of interaction strength U for a quarter-filled Hubbard model. Data points are averages over 10–20 disorder realizations. Increasing U causes the Anderson insulating phase to convert to metallic: the temperature dependence of $\sigma_{\rm dc}$ changes sign. (a) site (chemical potential) randomness of magnitude $\Delta_\mu = 8t$, with calculations performed on an 8×8 square lattice. (b) bond (hopping) randomness of magnitude $\Delta_t = 2t$ and lattice sizes as indicated in the legend.

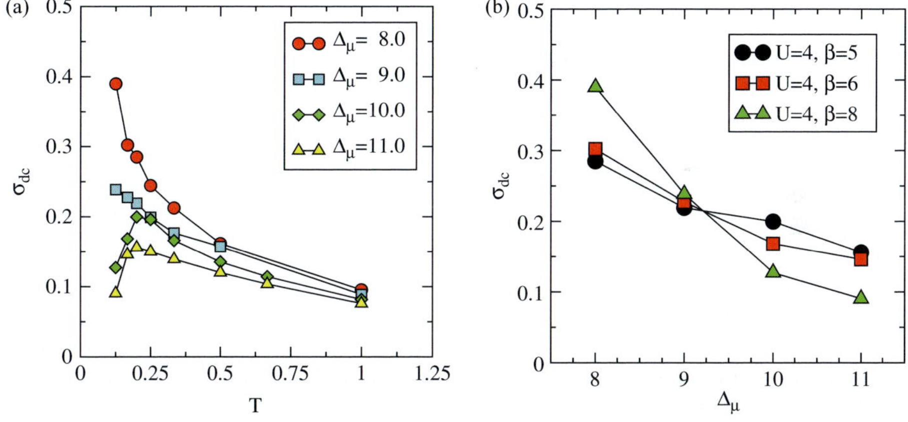

Fig. 5.7 (a) as the site randomness Δ_μ is made larger, the metallic phase induced by interactions can revert to insulating. (b) a crossing plot helps pinpoint the transition from metal to insulator. For $\Delta_\mu \leq 9t$ the conductivity $\sigma_{\rm dc}$ increases as T is lowered. For $\Delta_\mu \geq 9t$ the ordering is reversed and conductivity $\sigma_{\rm dc}$ decreases as T is lowered. Here $U = 4t$ and $\rho = 0.5$.

As noted in Section 5.1.3, there is a natural way of picturing this interaction-driven metal–insulator transition, emphasized by Dobrosavljević and coworkers (Aguiar *et al.*, 2005; Radonjic *et al.*, 2010; Tanaskovic *et al.*, 2003), in terms of screening of the random site energies by U. The most naive way to characterize this phenomenon is at the static mean-field level, where sites with large negative site energy ϵ_i accumulate higher density. This enhanced density then, through the on-site repulsion U, acts counter to the initial tendency for localization there. Farhoodfar *et al.* (2009) have also examined this screening effect using Gutzwiller wave functions for the AHH.

In Fig. 5.8 we show both the original landscape of random chemical potentials ϵ_i panel (a) and this most simple version of the "renormalized" landscape $\epsilon_i + U\langle n_i\rangle$ panel (b). While the variance is somewhat reduced, it is not dramatically altered (Chakraborty *et al.*, 2007). Further analysis of the role of site-energy renormalization in the metal–insulator transition within DQMC is being undertaken. In particular, it would be very interesting to go beyond the most naive static Hartree–Fock expression $U\langle n_i\rangle$ and consider more sensitive prescriptions in terms of the self-energy.

Greater insight into the interaction-induced metal–insulator transition can be obtained by the introduction of a Zeeman magnetic field term $B_{\|}\sum_i(n_{i\uparrow} - n_{i\downarrow})$, which couples to the spin of the fermions (but not to the hopping) (Denteneer and Scalettar, 2003). Such a term is realized experimentally through a field that is parallel to a two-dimensional sheet.

If, indeed, it is non-zero U that is causing metallic behavior, one might expect the effect to be suppressed by $B_{\|}$, since a Zeeman field spin polarizes the fermions and reduces the effect of interactions. Figure 5.9 shows precisely this effect. In Fig. 5.9 (a) a metallic phase with $U = 4t$ and disordered hopping is systematically destroyed by increasing $B_{\|}$ until, ultimately, $d\sigma_{\rm dc}/dT > 0$ at low temperatures. Figure 5.9 (b)

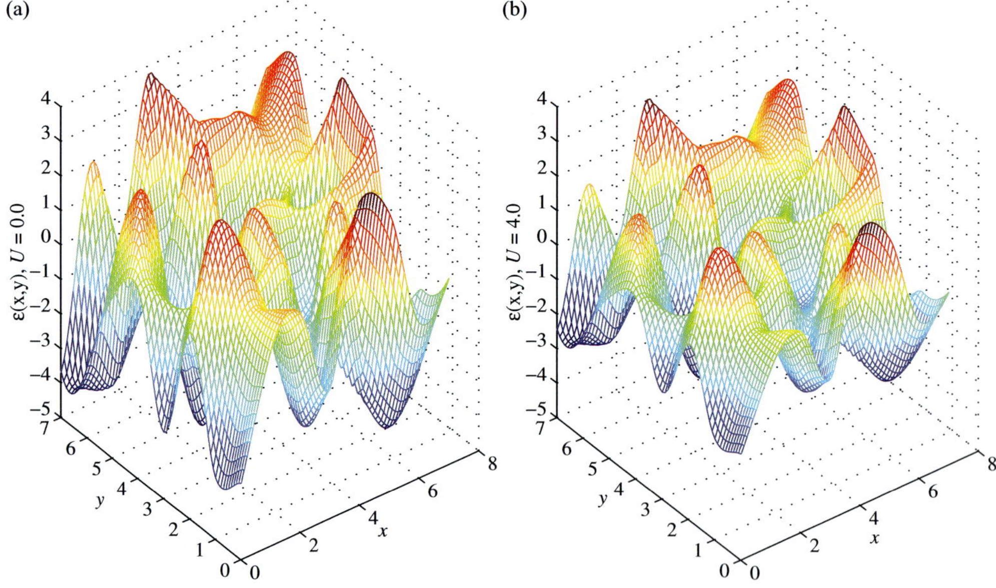

Fig. 5.8 Original (a) and renormalized (b) site energies in the AHH. The lattice size is 8×8, $U = 4.0t$, $\Delta_\mu = 8.0t$, and $\beta = 8.0/t$ with density $\rho = 0.5$. Some screening of the randomness is present, but very significant fluctuations remain.

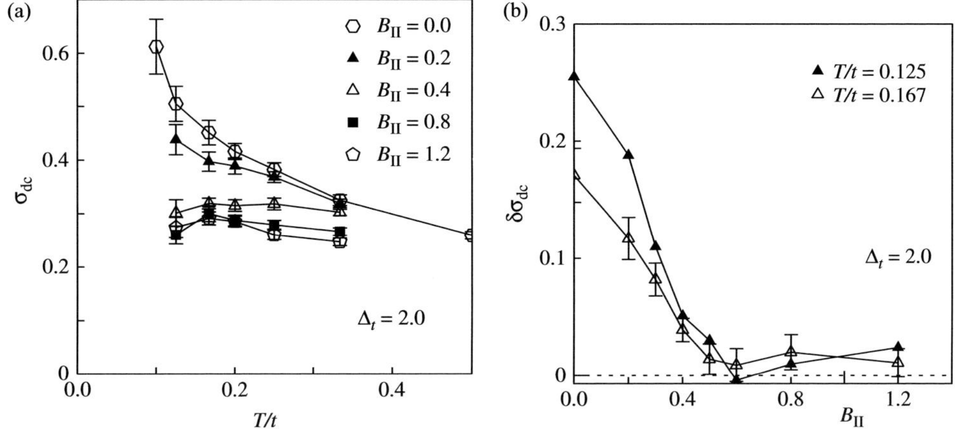

Fig. 5.9 (a) conductivity σ_{dc} as a function of temperature T for various strengths of Zeeman field $B_{\|}$. As $B_{\|}$ increases, a transition from metallic to insulating behavior is seen in σ_{dc}. Calculations are performed on $8{\times}8$ lattices for $U = 4t$ at density $\rho = 0.5$ with hopping disorder strength $\Delta_t = 2.0$; error bars result from averaging over 16 quenched disorder realizations. $B_{\|}$ and Δ_t are given in units of t. (b) conductivity difference $\delta\sigma_{\mathrm{dc}} \equiv \sigma_{\mathrm{dc}}(B_{\|}, T) -\ \sigma_{\mathrm{dc}}(B_{\|} = 4, T)$, with conductivity at very high B-field subtracted, as a function of $B_{\|}$ for low temperature T. A sharp onset of conductivity is seen at a Zeeman field at which the slope of $\sigma_{\mathrm{dc}}(T)$ changes sign in (a).

shows the difference between the conductivity at small $B_{\|}$ and its extremely high field value at $B_{\|} = 4$. A fairly sharp enhancement of the conductivity is present at low field. This signals the metallic phase.

5.2.4 The role of particle–hole symmetry

The results of the preceding section, especially Fig. 5.6, demonstrated that the particular type of disorder, chemical potential or hopping, was not especially relevant to the occurrence of an interaction-driven metal–insulator transition in the Anderson–Hubbard model. This result is in conformance with general expectations based on renormalization-group arguments: typically, one expects that as one changes length scales and follows the associated flow of Hamiltonian parameters, randomness in one term will spread to the others. As a consequence, precisely where disorder is introduced into the original Hamiltonian does not matter.

The above argument can fail when symmetries prevent the mixing of the different terms. One might ask whether the particle–hole symmetry of the half-filled AHH plays any role in the metal–insulator transition. As has been emphasized, the low temperature phase of the clean HH is an antiferromagnetic insulator. For weak interactions the ("Slater") gap can be thought of as arising predominantly from the magnetic correlations, which then evolves into a "Mott" gap for strong interactions. This gap is disturbed by site randomness, both because it harms the antiferromagnetic order (Sections 5.2.2 and 5.2.3) and because the energy penalty for double occupancy is reduced. It is plausible, therefore, that increasing site disorder might *induce* metallic behavior.

The temperature evolution of the dc conductivity at half filling is shown in Fig. 5.10. When Δ_μ (a) becomes sufficiently large, the insulating phase of the half-filled HH is destroyed, and the conductivity rises as T is lowered. In contrast, in the case of bond disorder, Fig. 5.10 (b), no such transition is observed to occur. As the strength of the hopping randomness Δ_t is tuned upwards, the conductivity is suppressed, always trending to lower values as T is decreased. The data of Fig. 5.10 in the absence of disorder also serve to validate the procedure for computing $\sigma_{\rm dc}$ described at the beginning of this section, by showing that the insulating Slater/Mott phase in the half-filled clean system is correctly captured.

Another way to emphasize this difference is through the evolution of density ρ with chemical potential μ. As previously discussed, in the clean Hubbard model, there is a plateau in $\rho(\mu)$ at $\rho = 1$, which reflects the existence of a gap at half filling: the cost μ to add a particle jumps discontinuously. QMC calculations show that this plateau is destroyed by site disorder, but not by bond disorder (Denteneer *et al.*, 2001). This is certainly consistent with the conductivity evolution presented in Fig. 5.10, which similarly suggests the destruction of an insulating gap in one case, but not the other.

That particle–hole symmetry is crucial to the interplay of randomness and interactions at half filling can be made even more compelling by considering the, somewhat artificial, case when the site randomness is reversed in sign for the up and down spin

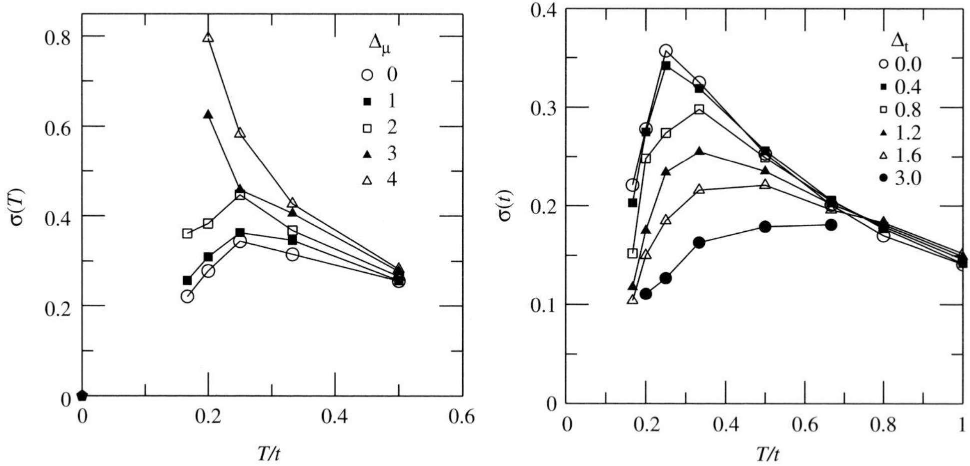

Fig. 5.10 The temperature dependence of the conductivity is shown for the half-filled AHH as the degree of randomness is increased. In the absence of disorder, the system is insulating. Chemical potential disorder (a) destroys the clean system gap and causes a transition to a metallic phase. On the other hand, the system remains insulating for all values of hopping disorder (b). Here $U = 4t$ and the lattice size is 8×8.

species. That is, the disorder in the AHH takes the form $\sum_i \epsilon_i (n_{i\uparrow} - n_{i\uparrow})$. (Such a term can be thought of as a random Zeeman field in the z direction.) Clearly, in the absence of correlations, $U = 0$, the introduction of a relative sign between spin up and down is completely irrelevant to the physics: the localization length, conductivity, and all other properties of the two species will be identical. When $U \neq 0$, however, the sign does matter. Particle–hole symmetry is preserved with this type of site disorder, similar to bond disorder, but in contrast to canonical site disorder. Studies of the conductivity at half filling with this "particle–hole-symmetric site disorder" reveal that it behaves in a fashion similar to the case of bond disorder: The conductivity bends downward as T is lowered and does not exhibit an interaction-driven metal-insulator transition (Denteneer *et al.*, 2001).

5.2.5 Summary

This section has described the results of DQMC simulations of the HH with various types of disorder. Removing sites from the lattice or adding a random chemical potential was shown to suppress the antiferromagnetism at half filling, and ultimately to preclude long-range order in the ground state. However, the magnetic order is much more robust than in the case when the half-filled model is doped with mobile carriers, a situation in which a change of just a few percent in density immediately destroys long-range order. Meanwhile, turning on interactions in the Anderson model can cause the formation of a metallic phase. Most of the data presented were for lattices of about 100 sites. Comparison of measurements on different sizes allows for a reasonable assessment of finite-size effects. Nevertheless, it would be very interesting to go back to these studies with modern computers and algorithms, which can handle lattices a factor of two larger in linear extent (500–1000 sites) and reexamine the metal–insulator transition that is suggested by the early work.

5.3 Repulsive Hubbard model: spectral function

The insulating state of the translationally invariant HH is characterized by a Mott gap in the density of states, a phenomenon that occurs only at commensurate filling. The disorder-driven Anderson insulator has no such feature at the Fermi surface for any particle density—the transition to zero conductance is characterized purely by the passage of the chemical potential through the mobility edge dividing localized and delocalized eigenstates. In this section, exact diagonalization studies are used to show that on-site repulsion and disorder combine to yield a suppression of the density of states at the Fermi energy of the AHH, where both interaction and randomness are present. Exact diagonalization works only on small clusters, but averaging over many boundary conditions (Gammel *et al.*, 1992; Gros, 1996) in addition to disorder realizations, as described in Chiesa *et al.* (2008) can mitigate some of the finite-size effects. Although completely absent in the presence of correlation effects and randomness separately, the pseudogap will be seen to occur over a remarkably wide range of fillings, repulsion, and disorder strengths in the combined system. This phenomenon can be captured at the mean-field level, at least qualitatively. As discussed in Section 5.1.3,

Efros and Shklovskii (1975) and Altshuler and Aronov (1979) have demonstrated that such density-of-states anomalies arise in a number of contexts.

5.3.1 Introduction

The preceding section focussed on the magnetic properties and the conductivity of the AHH, with only a brief discussion of the density of states $A(\omega)$, and computing with finite-temperature DQMC methods in an HH with non-magnetic (($U_i = 0$) impurity sites. The DQMC approach relies on a computation of the imaginary-time-dependent Green's function and subsequent analytic continuation. Two virtues of diagonalization in the computation of $A(\omega)$ are the absence of error bars and the broadening that results from maximum entropy constructions. This allows for the precise resolution of the features of $A(\omega)$. The formalism relies on a computation of the real-space matrix elements of the propagator and then their Fourier transform,

$$
\begin{aligned}
A(j,\omega) &= -\frac{1}{\pi}\mathrm{Im}\left\langle c_j \frac{1}{\omega - H + i\eta} c_j^\dagger \right\rangle \\
A(\mathbf{p},\omega) &= \sum_j e^{i\mathbf{p}\cdot j} A(j,\omega) \\
A(\omega) &= \frac{1}{N}\sum_{\mathbf{p}} A(\mathbf{p},\omega). \qquad (5.14)
\end{aligned}
$$

As has already been mentioned, when disorder is absent and at small U, a Slater insulating phase arises from antiferromagnetic correlations on non-frustrated (bipartite) lattices. This evolves into a Mott-insulating phase at large U. Both Mott and Slater behaviors occur at a particular commensurate density, half filling $\rho = 1$, and are characterized by a gap in $A(\omega)$. Meanwhile, when $U = 0$ and the disorder is dominant, an Anderson insulator forms, with no gap in $A(\omega)$.

There is a rich interplay between these various phenomena. Weak random chemical potentials at first aid antiferromagnetism (Janis *et al.*, 1993), a consequence of an enhanced exchange coupling $J = 2t^2/(U+\Delta) + 2t^2/(U-\Delta) > 4t^2/U$, where Δ is the difference in chemical potential on the adjacent sites. Eventually, disorder suppresses antiferromagnetism and closes the Slater gap by encouraging double occupancy and moment loss on low-energy sites. See Figs. 5.4 and 5.5. As discussed earlier, although the exact solution of the AHH is not known, numerical calculations, including those presented in the preceding section, suggest that the on-site repulsion U can destroy the Anderson insulator and cause metallic behavior (Belitz and Kirkpatrick, 1994; Castellani *et al.*, 1998; Chakravarty *et al.*, 1998; Denteneer *et al.*, 1999; Dobrosavljević *et al.*, 1997; Finkel'stein, 1983; Lee and Ramakrishnan, 1985).

The density of states is a key diagnostic of many aspects of these transitions. The Mott insulator is revealed by a gap in $A(\omega)$, while the Anderson insulator has none. The studies of Altshuler and Aronov (1979) and Efros and Shklovskii (1975) reveal, through quite different physical mechanisms, a feature in $A(\omega)$ when disorder and interactions are both present. A more refined treatment of spectral gaps, not discussed here, compares the low-energy behavior of the density of states with that of the spin and charge responses $\chi(\omega)$ and $\Pi(\omega)$ obtained from an analytic continuation of spin–spin and density–density correlation functions (Vekic *et al.*, 1995).

In Chiesa *et al.* (2008), the AHH was studied with an exact diagonalization approach. The conclusion was that, at intermediate couplings when on-site repulsion U equals the bandwidth $\Delta_\mu = 8t$ and for a wide variety of fillings, a pseudogap remains in the density of states even when the disorder becomes very large. This is a somewhat surprising conclusion, at variance with intuition that large disorder will eventually blur all structure in $A(\omega)$. Here, that work is first reviewed and is then extended to stronger and weaker couplings.

5.3.2 Exact diagonalization studies of the pseudogap

Figure 5.11 (a) shows the spectral function of the AHH at constant $U = 8t$ and $\Delta_\mu = 10t$ for decreasing fillings from $\rho = 1.0$ to $\rho = 0.6$. (Here, as before, site energies are chosen uniformly on $-\Delta_\mu < \epsilon_i < +\Delta_\mu$.) The disorder is large enough to have destroyed the Mott gap at half filling, $\rho = 1$. But it is perhaps not surprising that nevertheless a remnant pseudogap remains at this filling. What is unexpected is that the pseudogap exists for the doped case, despite the fact that there is no initial Mott gap in the clean system: for $\rho \neq 1$ the spectral function is ungapped before adding disorder. The universal nature of this pseudogap is further emphasized by changing the disorder strength at fixed filling $\rho = 0.6$ (Fig. 5.11 (b)). As in Fig. 5.11 (a), the pseudogap is pinned at the chemical potential and does not change depth or width with disorder strength.

The angle-resolved spectral function $A(\mathbf{p}, \omega)$ provides additional details concerning the pseudogap and connects directly to ARPES measurements. Figure 5.12 shows results around one-quarter filling, $\rho = 0.57$. The most important difference with a

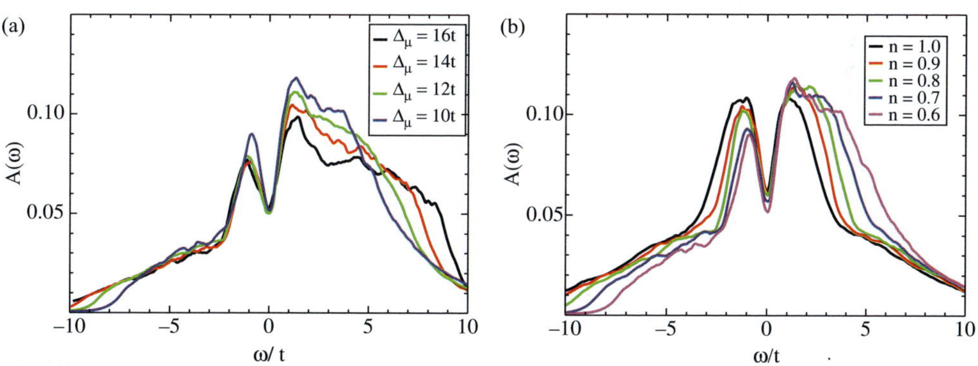

Fig. 5.11 (a) the spectral function of the AHH exhibits a filling-independent pseudogap at fixed disorder ($\Delta_\mu = 10t$) and interaction ($U = 8t$) strengths. The behavior of $A(\omega)$ in the vicinity of the Fermi energy $\omega = 0$ is surprisingly similar to that of the doped Mott insulator, even as one moves substantially away from half filling. (b) both the position and depth of the pseudogap are insensitive to the disorder strength. The spectral function is shown for $\Delta_\mu = 10t - 16t$ at $U = 8t$ and $\rho = 0.6$.

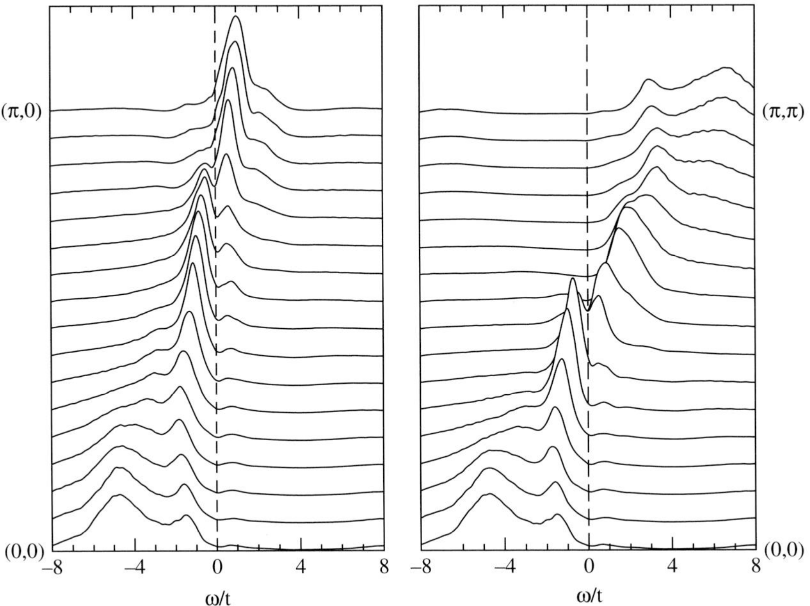

Fig. 5.12 ARPES spectrum at $\rho = 0.57$. The different traces show results for distinct cuts through the Brillouin zone. Whenever there is non-zero spectral weight at the Fermi surface, a pseudogap forms.

Fermi liquid concerns the behavior when the spectral density hits the Fermi level. In a Fermi liquid this results in the sharpening of the peak and the formation of the Fermi surface. Here the spectrum develops a pseudogap whenever the spectral density is finite at the Fermi level. Although a similar anomaly is seen in the clean $(t - t' - t'')$ Hubbard model (Sénéchal and Tremblay, 2004), in that case the anomaly shows up only around certain specific points on the Fermi surface (in the hole-doped case around $(\pi, 0)$). The results in Fig. 5.12 are for a finite eight-site cluster. A comparison with larger and smaller sizes reveals that the feature of the spectrum that is most universal is exactly the pseudogap. Similar results for filling $\rho = 0.8$ are given in Chiesa *et al.* (2008).

Does this pseudogap phenomenon persist at somewhat stronger coupling, $U = 3W/2 = 12t$? Fig. 5.13 (a) shows the expected gradual closing of the Mott gap at half filling from its clean system value $\Delta_{\rm Mott} = U/2$. This evolution continues as W passes through U, as seen in Fig. 5.13 (b). However, as for the intermediate coupled case $U = W = 8t$, the pseudogap exhibits a saturation with disorder W at larger correlation strengths.

Figure 5.14 shows results for different dopings at $U = 12t$ and fixed disorder $\Delta_\mu = 10t$. $A(\omega)$ at half filling, $\rho = 1.0$, shows markedly different structure from the other fillings, a remnant of the Mott gap, but is basically independent of doping for $\rho \neq 1$. Although the value of $A(\omega = 0)$ is largely independent of density, there is a

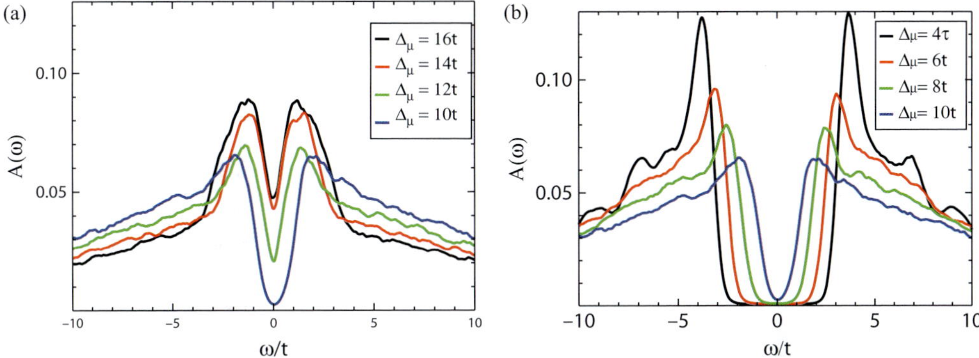

Fig. 5.13 Density of states for different disorder strengths, $\Delta_\mu = 4t - 10t$ (a) and $\Delta_\mu = 10t - 16t$ (b). The on-site $U = 12t$ is well above the bandwidth. Disorder closes the Mott gap, but an anomalous suppression of $A(\omega = 0)$ persists out to large randomness.

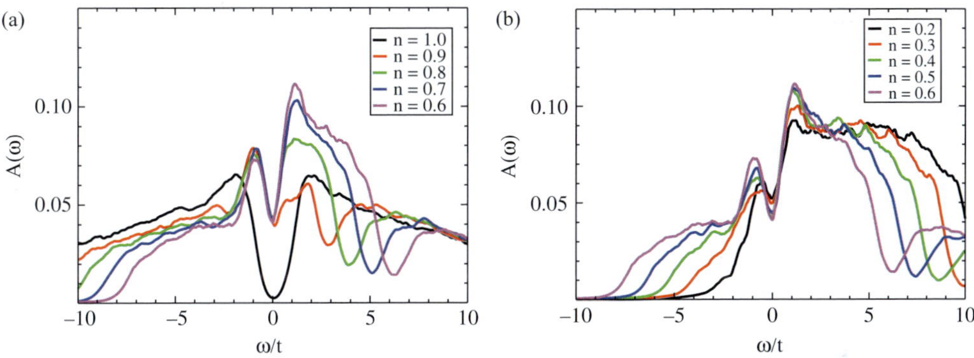

Fig. 5.14 The density of states is shown at strong coupling and disorder, $\Delta_\mu = 10t$ and $U = 12t$. (a) scan over densities between quarter- and half filling at fixed disorder and interaction strengths. (b) the density is reduced further, below quarter filling. At half filling, $\rho = 1$, the gap shows a clear remnant of Mott physics, but once the system is doped the anomaly at the Fermi surface is quite remarkably independent of density, even down to the very dilute limit.

breakaway from the universal form, which moves closer to $\omega = 0$ as $\rho \to 1$, an effect that is more apparent here than at intermediate coupling $U = 8t$ (Fig. 5.11). Figure 5.14 (b) especially emphasizes that the density-of-states suppression persists even in the most highly doped cases.

We have also explored the disorder dependence by examining two densities that bracket quarter filling, $\rho = 0.4$ and $\rho = 0.6$, and sweeping the size of the disorder up to $\Delta_\mu = 16t$ at $U = 12t$. The collapse of the data is not quite so dramatic as at intermediate coupling, shown in Fig. 5.11 (a), but the qualitative feature is similar: there is no trend to closing the pseudogap for large Δ_μ for either filling.

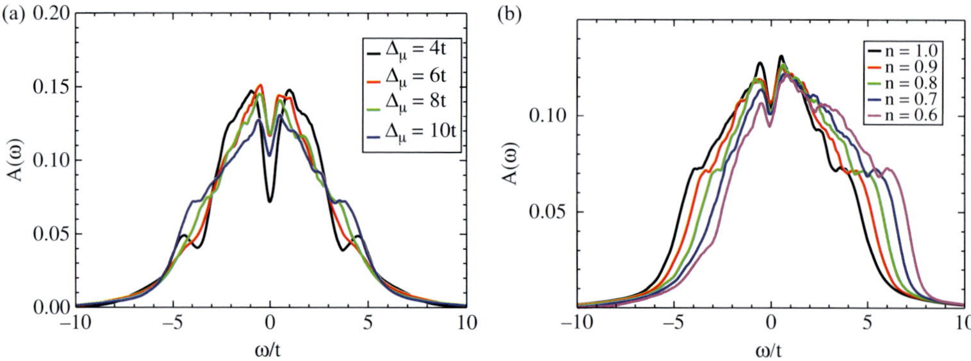

Fig. 5.15 The density of states is shown at strong coupling and disorder. (a) half filling density of states for different disorder strengths at weak coupling $U = 4t$. (b) scan over densities at fixed disorder strength $W/t = 10$ and interaction $U/t = 4$. There is a dip in $A(\omega)$, which is largely filling independent. This resembles the results at stronger coupling.

Consider now the case when U is reduced to values well below the bandwidth. The analysis parallels that of $U = 12t$. Figure 5.15 (a) exhibits $A(\omega)$ for half filling and a systematic increase in disorder strength W, while Fig. 5.15 (b) details the density dependence at fixed disorder. Once again the pseudogap persists over a wide range of fillings and disorder strengths.

5.3.3 Mean-field treatment

It is interesting to ask to what extent the preceding results can be captured already at a mean-field level. In the mean-field approach one re-writes the AHH in a way that replaces the quartic interaction terms with quadratic ones through,

$$U\,\hat{n}_{i\uparrow}\hat{n}_{i\downarrow} \Rightarrow U\big(\;\hat{n}_{i\uparrow}\langle\hat{n}_{i\downarrow}\rangle + \langle\hat{n}_{i\uparrow}\rangle\hat{n}_{i\downarrow} - \langle\hat{n}_{i\uparrow}\rangle\langle\hat{n}_{i\downarrow}\rangle\;\big). \tag{5.15}$$

The expectation values $\langle\hat{n}_{i\sigma}\rangle$ are computed self-consistently by diagonalizing the resulting non-interacting (single-particle) Hamiltonian, a process that is only cubic in the number of sites, instead of exponential as when treating the full many-body problem.

This section contains mean-field results for the AHH on lattices up to 64 sites. It is of course, possible to solve much larger lattices in the mean-field approach. However, we confine ourselves to smaller clusters that are nevertheless still an order of magnitude larger than those used in the diagonalization approach. Even on these sizes there are issues on how to construct the mean-field solution since the procedure tends to get trapped in a number of different metastable fixed points. For the smallest clusters the mean-field calculation is performed many times, starting with the density of all the degenerate ground states of the model without hopping (including different S_z sectors), retaining the one with the lowest energy. For the largest clusters, the number of permutations make this unfeasible. An alternate strategy is also employed, which solves the N particle case using a starting density based on the N eigenvectors of the $N-1$ particle calculation giving the lowest total energy for N particles. Away from half filling on small clusters, this gives results in agreement with the first method.

Figure 5.16 (b) shows results for a 64 site lattice at intermediate coupling $U = 8t$, disorder $\Delta_\mu = 8t$, and a range of fillings. The basic message is very similar to that of the diagonalization studies, namely that a robust pseudogap exists across a broad range of fillings. Indeed, the main difference between the two approaches is that the pseudogap has greater depth in the mean-field treatment. Figure 5.16 (b) shows a fixed density of quarter filling and different lattice sizes, from 8 to 64 sites. The results vary only a small amount, and the pseudogap width and depth show only minor changes. Focussing then on eight sites and doing a very careful search of the different minima yields the results of Fig. 5.17. This suggests that within mean-field the pseudogap is

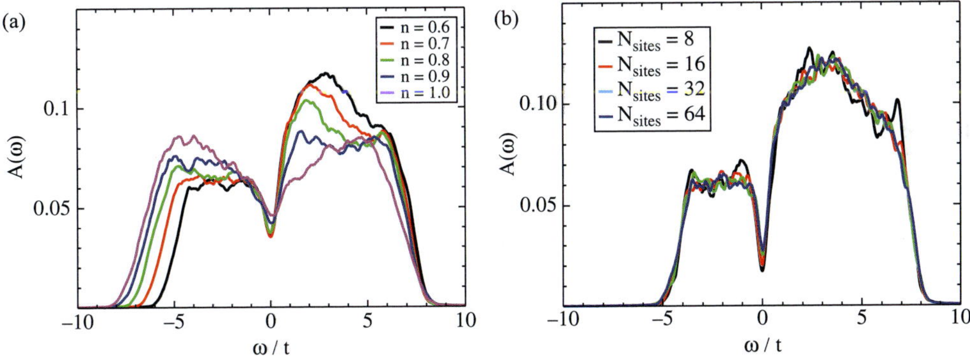

Fig. 5.16 (a) mean-field results for the density of states on 64 site lattices using starting configurations each of which is built on the solution at the preceding (lower) density (see text). (b) finite-size dependence of density of states.

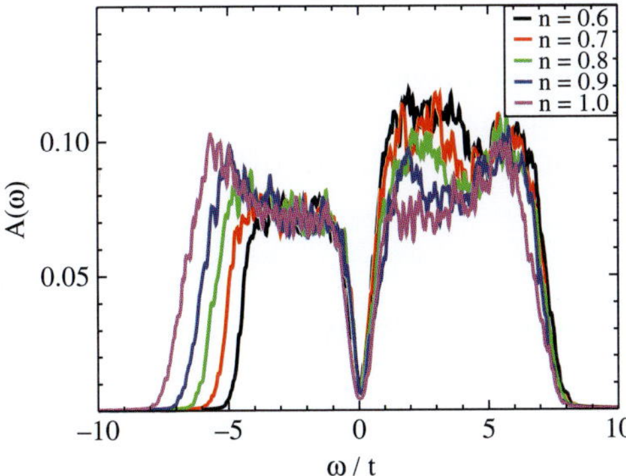

Fig. 5.17 Mean-field results for the density of states on eight site lattices. The larger depth of the pseudogap in the mean-field approach relative to diagonalization is especially evident here.

not merely deeper that exact diagonalization, but that $A(\omega = 0)$ may be in fact driven all the way to zero in this approach. Such overestimates of the effects of correlation are, of course, typical of mean-field methods, though admittedly it is not obvious that the pseudogap, which arises from interactions *and* disorder, should follow the same pattern. Besides the pseudogap itself, the structure away from the Fermi surface is broader and flatter in the mean-field case, compared to exact diagonalization.

5.3.4 Summary

This section has focussed on describing exact diagonalization results for the AHH. A very robust pseudogap exists for $U = 4t$ and $U = 12t$, being largely insensitive to density and disorder strength, as earlier reported (Chiesa *et al.*, 2008) for $U = 8t$. Comparisons of results for different cluster sizes, geometries (one and two dimensional), and connectivity (frustrated versus unfrustrated) show the pseudogap is basically unchanged in all cases, emphasizing the role of local physics (Chiesa, 2010). Although a detailed analysis in not provided here, numerics also indicate that the pseudogap energy scale is set by the kinetic energy t and does not change with U or W.

That features of the pseudogap appear to be captured even at a mean-field level is in agreement with a recent study that explored the AHH very carefully within a Hartree–Fock treatment (Fazileh *et al.*, 2006; Shinaoka and Imada, 2009). The existence of an zero-energy anomaly in the density of states was also found in one dimension using exact diagonalization on cluster with up to six sites by Shinaoka and Imada (2009). Based on an ansatz for the scaling form of the low-energy part of the spectral function, these authors have claimed that the anomaly is, in fact, a singular soft gap, i.e. the density of states is characterized by ω^α (with $\alpha = 0.075$) as ω approaches the chemical potential. Whether this is the state of affairs in two dimensions is impossible to answer conclusively by straight exact diagonalization since the system can hardly be grown beyond 12 sites. The use of mean-field theory, although supportive of the existence of the anomaly (Fazileh *et al.*, 2006; Shinaoka and Imada, 2009), cannot be used to settle such a subtle question since, as has been seen in this chapter, mean-field theory describes a soft gap even for sizes where exact diagonalization does not.

Even more recently, the pseudogap has been observed in ensembles of (two-site) Hubbard dimers (Wortis, pers. comm.). This very intriguing result indicates that the pseudogap might arise from the interplay of disorder and interactions at a very local level.

5.4 Attractive Hubbard model with site dilution

5.4.1 Introduction

This section contains a discussion of the interplay between the attractive electron–electron interactions that lead electrons to form pairs and eventually to superconduct and disorder, which tends to break the pairs and hence destroy superconductivity. The attractive HH is the simplest description where this interplay can be studied qualitatively. As we shall argue here, it is able to capture qualitatively many of the interesting features of superconducting–insulating transitions.

Specifically, this section explores the properties of the Attractive HH in the presence of site dilution. Similarly to Section 5.2, the case where the interaction is turned off in a fraction f of sites by setting $U_i = 0$ is considered. The interaction on the remaining $1 - f$ sites is attractive: $U < 0$.

In two dimensions and in the absence of dilution, there is a degeneracy between charge-density-wave (CDW) and superconducting order at half filling that leads to an effective three-component order parameter. This suppresses the critical temperature to zero, due to the Mermin–Wagner (Mermin and Wagner, 1966) theorem. Away from half filling the degeneracy is lifted and the two-component superconducting order parameter allows for finite-temperature superconductivity. Although we will not discuss the effects of finite temperature here, it is worth mentioning that impurity also breaks the CDW–superconducting order degeneracy, and the system can be superconducting at half filling at $T \neq 0$.

Here we will study the effects of dilution at $T = 0$ and the following questions will be addressed: for a given attraction strength U and a given density n, what is the fraction of non-interacting sites f that the system can possess before the breakdown of ground-state superconductivity? Can superconductivity be favoured by non-interacting impurities? What changes arise if, instead of distributing the f sites randomly, the non-interacting sites form patterns, such as checkerboard or striped?

To address the issues above we will consider results obtained with DQMC (Farias, 2009; Hurt *et al.*, 2005; Mondaini *et al.*, 2008), and within the Bogoliubov–de Gennes approximation (Aryanpour *et al.*, 2006, 2007). Comparisons with DMFT results from Shenoy (2008) and mean-field results from Litak and Györffy (2000) will also be presented.

5.4.2 Randomly distributed impurities

The attractive HH in the presence of site dilution is

$$H = -t \sum_{\langle ij \rangle \sigma} \left(c^{\dagger}_{i\sigma} c_{j\sigma} + c^{\dagger}_{j\sigma} c_{i\sigma} \right) - \mu \sum_{i\sigma} n_{i\sigma} - \sum_{i} U_i \, n_{i\uparrow} n_{i\downarrow}, \tag{5.16}$$

where $c^{\dagger}_{i\sigma}(c_{i\sigma})$ are fermion creation (destruction) operators at site i with spin σ, and $n_{i\sigma} = c^{\dagger}_{i\sigma} c_{i\sigma}$. The kinetic energy lattice sum $\langle ij \rangle$ is over nearest-neighbor sites on a two-dimensional square lattice, and μ is the chemical potential. The on-site attraction U_i is chosen to take on the two values $U_i = 0$ and $U_i = U$ with probabilities f and $1 - f$ respectively; note that $U > 0$ corresponds to attraction, according to the definition of the on-site term in eqn (5.16).

It is very important to emphasize that in eqn (5.16) the interaction term is not written in particle–hole symmetric form, as is the case in eqn (5.1). What this means is that the results of this section on site dilution are not trivial consequences of those presented for the repulsive model. In that case, setting $U_i = 0$ on some sites also meant that the chemical potential varied from site to site. Indeed, this choice causes the Mott plateau of Fig. 5.1 (b) at $n = 1 + f$ to be accompanied by a plateau at $n = 1 - f$. This convention is frequently used, e.g. in studies of the periodic Anderson model, and is essential to control the sign problem in these DQMC simulations.

The random site-diluted attractive Hubbard model was first studied by Litak and Györffy (2000), within the Hartree–Fock approximattion. The averaging over the U configurations was done with the coherent potential approximation (CPA). Later DQMC studies of this model were presented by Hurt *et al.* (2005) and Mondaini *et al.* (2008), followed by DMFT studies at (Shenoy, 2008).

Within DQMC the transition from a superconducting ground state to a non-superconducting one is located in the following way: first the configurationally averaged equal-time pairing correlation function $\Gamma(i)$ and the pairing structure factor P_s are calculated,

$$\Gamma(j) \equiv \langle \Delta(i)\Delta^\dagger(i+j) + \mathrm{H.c.}\rangle, \qquad P_\mathrm{s} = \left[\sum_j \Gamma(j)\right], \tag{5.17}$$

where $\Delta(i) = c_{i\downarrow}c_{i\uparrow}$ and $[\ldots]$ denotes the average over disorder configurations.

The $T = 0$ behavior is extracted from plots of the pair structure factor as a function of inverse temperature β as shown in Fig. 5.18, for $U = 3$ and $f = \frac{1}{16}$. The finite-size scaling behavior of P_s allows us to extract quantitative information about the superconducting transition. In analogy to the finite-size scaling of the antiferromagnetic structure factor in Section 5.2.2, the spin-wave correction (Huse, 1988) to the pair structure factor in the ground state is expected to be inversely proportional to the linear lattice size,

$$\frac{P_\mathrm{s}}{L^2} = |\Delta|^2 + \frac{a}{L}, \tag{5.18}$$

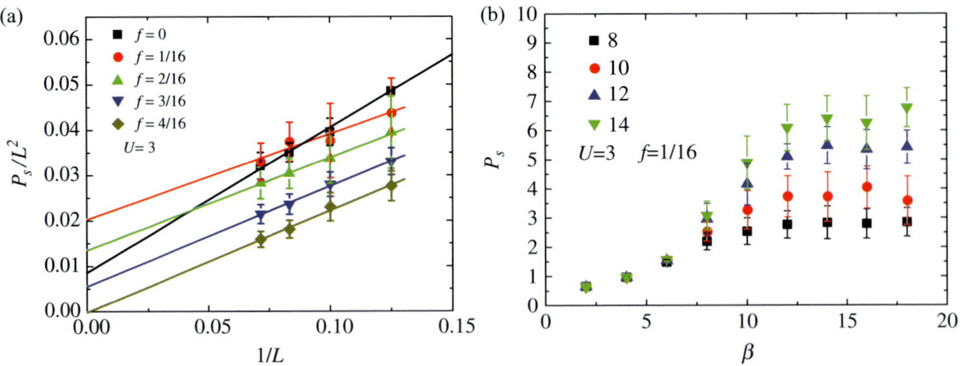

Fig. 5.18 (a) shows the pair structure factor P_s as a function of inverse temperature β for electronic density $n = 1$, attraction strength U=3, concentration of free sites $f = \frac{1}{16}$, and different system sizes. (b) shows finite size scaling analysis of the pair structure factor P_s/L^2. A non-zero extrapolation to $1/L = 0$ indicates a superconducting ground state.

where Δ is the superconducting gap function at zero temperature, and $a \equiv a(U, f)$ is independent of L.

In Fig. 5.18 (b) we see superconductivity being destroyed as the fraction of non-interacting sites f is increased at half filling for $U = 3$. For $f < \frac{4}{16}$ the extrapolation to $1/L = 0$ has a non-zero intercept with the vertical axis, indicating that Δ is non-zero and the ground state is superconducting. For $f = \frac{4}{16}$ we can see that Δ has reached zero, and the system is no longer in a superconducting ground state. Another interesting feature is the non-monotonic behavior of the zero-temperature gap with impurity concentration. This non-monotonicity is strongly dependent on interaction strength, and can be more easily seen in Fig. 5.19, where the normalized zero-temperature gap as a function of disorder strength f is shown for different values of U. The increase of the order parameter with increasing site dilution at half filling is also observed in the repulsive HH, as shown in Fig. 5.1. Once again we emphasize that the results of this section—where the Hamiltonian is not particle–hole symmetric—cannot be obtained from the particle–hole-symmetric Hamiltonian for the repulsive model presented in Section 5.2. The critical fraction f_c above which superconductivity is destroyed can also be obtained from Fig. 5.19.

Figure 5.20 presents a comparison between DQMC (Farias, 2009; Hurt *et al.*, 2005; Mondaini *et al.*, 2008), CPA (Litak and Györffy, 2000), DMFT (Shenoy, 2008), and Bogoliubov–de Gennes results for the critical fraction of non-interacting sites f_c as a function of attraction strength U for half filling and for $n = 0.875$. We can see that CPA and Bogoliubov–de Gennes agree perfectly; mean field shows f_c decreases monotonically with U. DMFT results, on the other hand, show that f_c increases with U for small U. For U larger than a critical value U_c (only considered for $n = 0.875$) $f_c \to 1$, meaning that even in the absence of attractive sites there is a non-zero pairing amplitude. This is an unexpected result and further studies are necessary to understand its source within DMFT. DQMC results are consistent with an increase

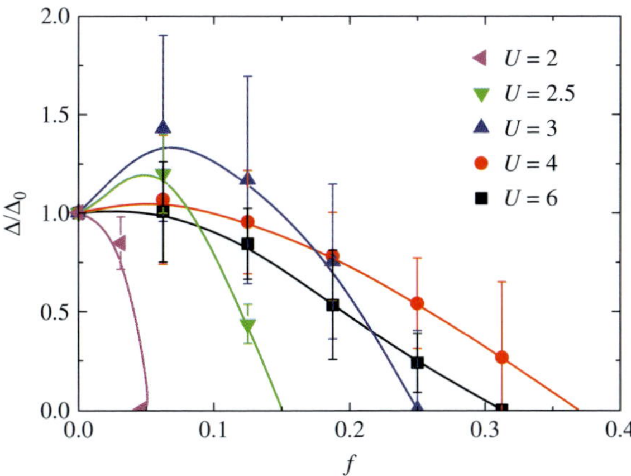

Fig. 5.19 Zero-temperature gap, normalized by the $f = 0$ gap, as a function of disorder strength for electronic density $n = 1$.

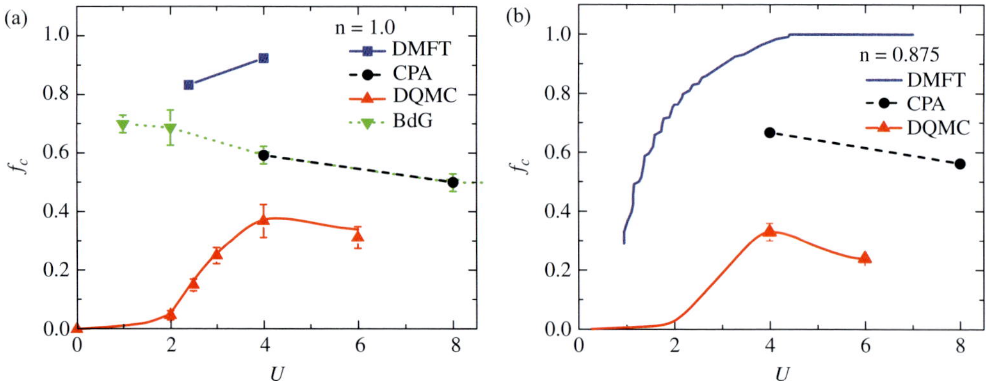

Fig. 5.20 Critical fraction f_c of non-interacting sites as a function of attraction strength U for half filling (a) and for $n = 0.875$ (b). DQMC results from Mondaini *et al.* (2008) (half filling) and from Hurt *et al.* (2005) and Farias (2009) for $n = 0.875$ are represented by red triangles; lines are guides to the eye. DMFT results from Shenoy (2008) are represented by blue lines, black circles are CPA results from Litak and Györffy (2000) and green triangles are Bogoliubov–de Gennes results (unpublished).

of f_c for small U and a decrease for larger U, with a maximum around $U = 4$ for $n = 1$. Although data for $n = 0.875$ are only available for $U \geq 4$, it is known that the non-interacting system is metallic for any filling and therefore f_c must go to zero at $U = 0$.

We next compare (Fig. 5.21) results of the critical fraction f_c for fixed interaction strength $U = 4t$ as a function of density. DMFT results for a flat band show f_c is independent of n for all densities considered. These results strongly depend on the shape of the band and, for a semicircular density of states (not shown) f_c increases with filling. DQMC results, although only available for a small range of densities, are also consistent with an independent f_c with n. CPA results, on the other hand, display a decreasing f_c for $n > 0.3$. This qualitative disagreement points to the relevant role played by fluctuations that are not being properly taken into account within the Hartree–Fock approach.

5.4.3 Site dilution with pattern formation

We now turn to the following point: given a certain amount of attractive interaction, what is best for superconductivity—distribute the interaction homogeneously through the system or concentrate the interaction on a fraction of the sites? If the interaction is concentrated on a fraction $1 - f$ of the sites, should these sites be randomly distributed throughout the lattice or should they form patterns?

Here we consider three possibilities: a random distribution of the fraction f of non-interacting sites, stripes formed with non-interacting sites along one of the axes of the square lattice, and a checkerboard pattern along both axes of the square lattice. We concentrate on three fractions of free sites f, namely $f = 0.25$, $f = 0.5$ and $f = 0.75$.

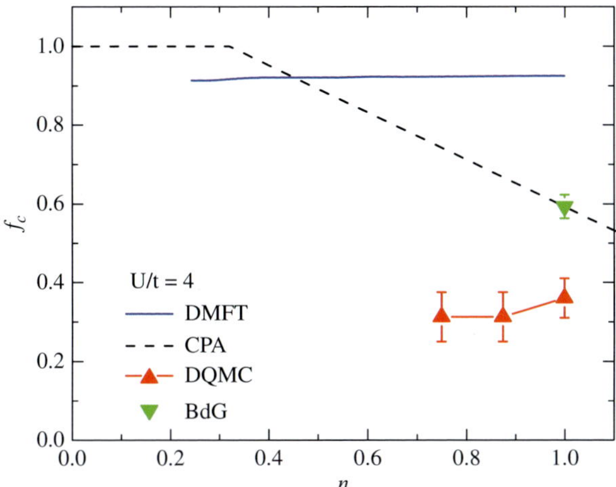

Fig. 5.21 Critical fraction of non-interacting sites f_c as a function of density for $U = 4t$. DQMC results from Hurt *et al.* (2005) (half filling) are represented by red triangles. DMFT results from Shenoy (2008) are represented by the blue line. The black dashed line represents CPA results from Litak and Györffy (2000) and the green triangle is a Bogoliubov–de Gennes result (unpublished).

In all cases, to provide a meaningful comparison with the uniform case, which focusses on the distribution of the attractive centers and not on their overall number, we fix the average interaction strength. For instance, when $f = 0.5$ the interacting sites on the inhomogeneous lattices are chosen to have interaction strength twice as large as their uniform counterpart.

This study was carried out within the Bogoliubov de Gennes approximation. With the use of symmetries, system sizes of up to 1500×1500 sites were reached for uniform, checkerboard and striped patterns. For the random site dilution, translation and rotation symmetries are broken and averages of up to 50 realizations are done for each configuration. System sizes up to 24×24 were therefore considered (Aryanpour *et al.*, 2006, 2007).

Let us discuss how the order parameter is affected by impurities. Figure 5.22 shows the zero-temperature pairing amplitude as a function of density for $\langle U \rangle = 1$ and $\langle U \rangle = 3$ for $f = 0.5$. For the uniform case, we observe that the gap is symmetrical about half filling, where it reaches a maximum. For all patterns considered, the maxima move to lower densities and for $\langle U \rangle = 3$ the gap has reached zero at half filling. Figure 5.22 (b) shows the average density on attractive (n_U) and non-interacting (n_0) sites separately and helps to explain why the pairing amplitude goes to zero at half filling. As U is increased the system reduces its energy by populating the attractive sites and emptying the non-interacting ones. For $f = 0.5$ all the electrons can be accommodated on the attractive sites, completely filling it and leaving the free sites empty. We can see in Fig. 5.22 that indeed $n_U \to 2$ while $n_0 \to 0$ as $\langle U \rangle$ increases.

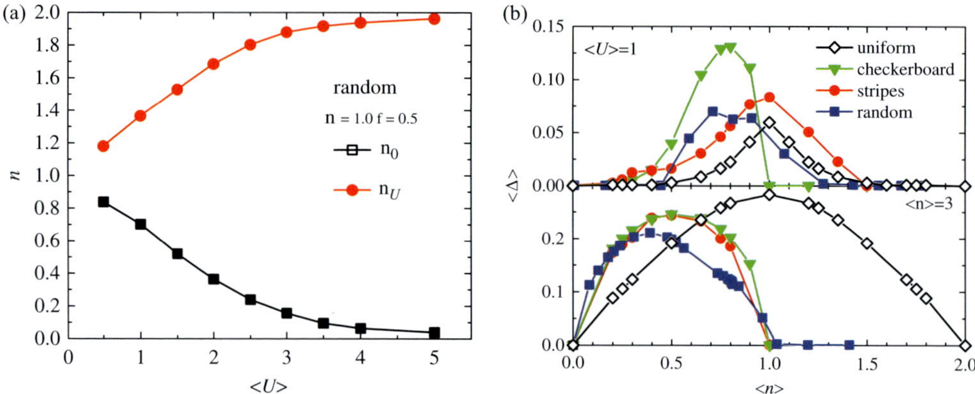

Fig. 5.22 (a) shows the average order parameter as a function of average density for the different patterns considered: uniform, randomly distributed sites, checkerboard pattern, and striped pattern, with $f = 0.5$, for both $\langle U \rangle = 1$ and $\langle U \rangle = 3$. (b) shows how electrons distribute through attractive (n_U) and non-interacting (n_0) sites as a function of attraction strength, for half filling, randomly distributed sites, and $f = 0.5$.

Therefore, for half filling and $f = 0.5$ the system becomes a charge-ordered insulator for large values of attraction.

Another interesting feature of site dilution can be seen in Fig. 5.22. Since the maxima of the pairing amplitude for the diluted systems have moved to smaller densities, we see that for small densities the order parameter for the diluted system is larger than the one for the uniform system (for instance consider $f = 0.5$, $\langle U \rangle = 3$ and $n = 0.4$). It is useful to define the ratio $r = \Delta_{\text{pattern}}/\Delta_{\text{uniform}}$. Whenever $r > 1$ superconductivity is favoured by the inhomogeneous distribution of the attractive interaction throughout the lattice. Also, when $r = 0$ the non-uniform system is not superconducting.

Figure 5.23 presents a summary of the detailed study (Aryanpour *et al.*, 2006, 2007) of the effects of inhomogeneity on superconductivity. The lines are contour plots and represent constant r for the different patterns and values of f considered. The presence of a metallic state was probed, not only by a vanishing order parameter, but also by a non-vanishing density of states at the Fermi level (not shown).

We can see that superconductivity is enhanced for small densities and U. As the density is increased, r decreases, vanishing for large enough U at $n^* = 2(1 - f)$. This density n^* corresponds, as shown in Fig. 5.22, to doubly occupied attractive sites and empty non-interacting ones, leading the system into a charge-ordered insulating state. For larger U, increasing the density drives the system into a metallic state. The situation is not so simple for small U, as the ground state can remain superconducting for $n > n^*$. One interesting point to note is that the checkerboard pattern presents reentrant superconductivity as the density is increased, both for $f = 0.25$ and $f = 0.75$.

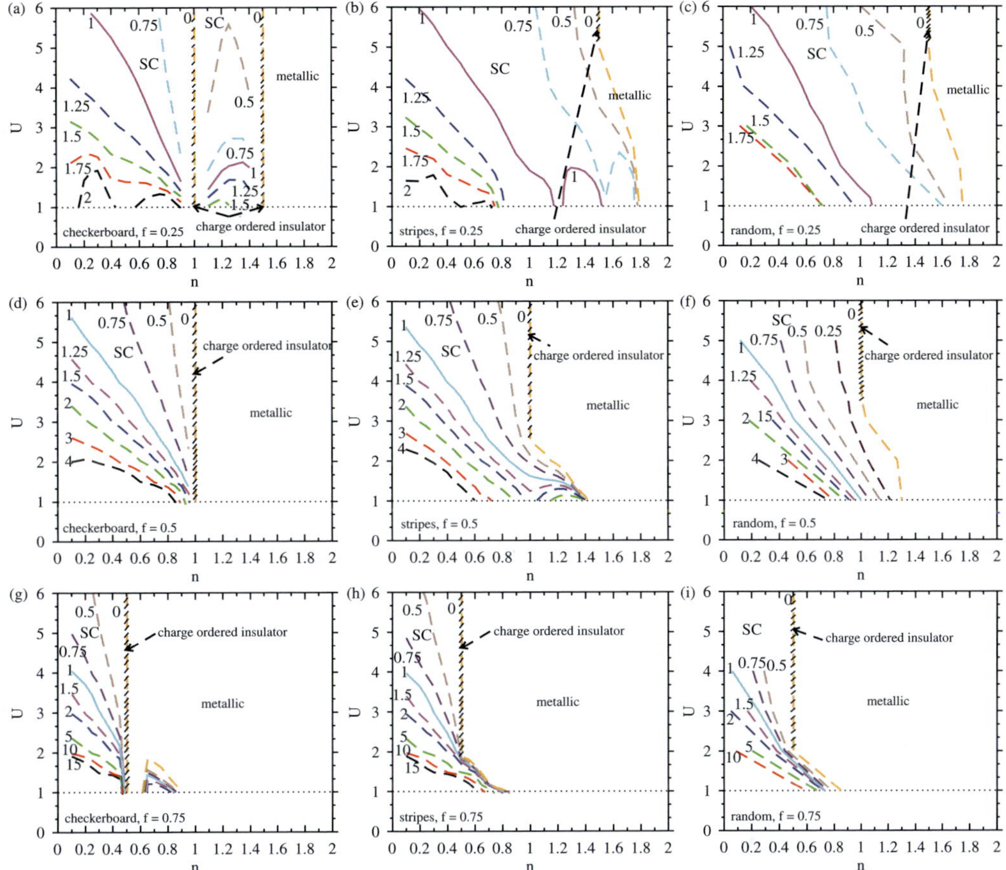

Fig. 5.23 Phase diagram for the site-diluted attractive Hubbard model within the Bogoliubov–de Gennes approximation at $T = 0$. (a) $f = 0.25$, checkerboard pattern; (b) $f = 0.25$ striped pattern; (c) $f = 0.25$, random pattern; (d) $f = 0.5$, checkerboard pattern; (e) $f = 0.5$ striped pattern; (f) $f = 0.5$, random pattern; (g) $f = 0.75$, checkerboard pattern; (h) $f = 0.75$ striped pattern; (i) $f = 0.75$, random pattern. The lines are contour plots for $r = \Delta_{\text{pattern}}/\Delta_{\text{uniform}}$.

5.4.4 Summary

This section has described the results of DQMC simulations and of Bogoliubov–de Gennes calculations for the site-diluted attractive HH. For randomly distributed non-interacting sites, DQMC shows that superconductivity in the ground state is destroyed above an inpurity concentration f_c. For $U = 4t$ $f_c \simeq 0.31$ within the density range considered. That f_c is largely independent of n is also seen within DMFT calculations. For interaction strengths smaller than $U \sim 4t$ the critical concentration increases with U, both for $n = 1.0$ and $n = 0.875$. This behavior is once again in agreement with DMFT results, but in contrast with CPA results, due to the relevant role played by fluctuations that are not included within the latter calculations. For large U, DQMC results are consistent with a decrease in f_c with U, recovering CPA results. It is worth mentioning that the transition is not mediated by purely geometrical factors, since f_c found in DQMC calculations is not in agreement with either the classical percolation value nor the quantum one. In close analogy to the site-dilution effects in antiferromagnetic transitions discussed in Section 5.2, for $2.5t \leq U \leq 6t$ the superconducting order parameter initially increases with increasing disorder f. This is likely to be a result of the break-down of the CDW–superconducting order parameter degeneracy. It would be interesting to revisit this problem and consider other densities and interaction strengths.

The study of the distribution of attractive interactions throughout a lattice forming either checkerboard, striped, or random patterns has shown a very rich phase diagram, with the presence of superconducting, metallic, insulating, and charge-ordered phases. A surprising conclusion is that, apart from a few particular features, the physics displayed in these phase diagrams is remarkably pattern independent. The density regions for which superconductivity is present is strongly dependent on the fraction f of non-interacting sites, and increases as f decreases. For large values of U, the superconducting region ends in a charge-ordered insulating state located at $n^* = 2(1 - f)$. Also surprisingly, there is a significant range of electronic densities and interaction strengths where the average superconducting order parameter is larger for a lattice with an inhomogeneous distribution of attractive sites than for the uniform lattice, provided the average interaction strength per site is kept constant. This enhancement of the order parameter is due to the proximity effect, i.e. the tunneling of pairs from interacting sites to non-interacting ones, leading to a non-zero order parameter in the latter. This conclusion is supported by the fact that this enhancement takes place in the small-U regime, where the coherence length is larger.

5.5 Conclusions

Numerical investigations of the AHH show, on the one hand, how randomness affects Mott-insulator and magnetically ordered phases and, on the other, how interactions affect localization. In the former case, the conclusion is more or less the expected one, namely that local chemical potentials and site dilution destroy magnetic moments and thereby magnetism. Somewhat less intuitive is the enhancement of conductivity by interactions in the latter case, and especially the dependence of the effect on the nature (particle–hole symmetry) of the underlying randomness. While it would be too strong to claim that computations have solved the very challenging problem of the physics of the AHH, as one can see from the brief review of Section 5.1.3, many numerical methods, from exact diagonalization to DMFT and inhomogeneous Hartree–Fock observe interaction-induced enhancement of the conductivity.

A particularly intriguing effect is the observation of a pseudogap in the density of states. This phenomenon does not appear to be a natural interpolation between the behavior of the disordered, non-interacting system and the pure Hubbard model, but instead to be entirely new physics not present in either limit.

Superconductor–insulator transitions have also a long and interesting experimental and theoretical history. The DQMC and Bogoliubov–de Gennes results reviewed here describe some of the things that numerical methods can teach us about these systems, including the counter-intuitive idea that an inhomogeneous distribution of attraction can help pairing.

Experiments and theoretical work on ultracold atomic gases have increasingly come to grips with the "smooth" inhomogeneity induced by the potential that confines the particles, and the associated presence of spatial domains exhibiting different physical phenomena. This field is now turning also to the consideration of the case of the extremely short length scale variations incorporated in the AHH. The ability to image individual "sites" in these experiments might make possible new and very direct contact between numeric and experimental work.

Acknowledgements

RTS acknowledges the support of National Nuclear Security Administration (under the Stewardship Science Academic Alliances program through DOE Research Grants DOE DE-FG52-09NA29464. S. Story was supported by the Research Experience for Undergraduates program of the National Science Foundation. T. Paiva acknowledges the support of the Brazilian agencies FAPERJ and CNPq.

We are, moreover, very grateful for the helpful discussions with colleagues which have helped shaped this work. These include V. Dobrosavljević, R.R. dos Santos, K. Held, M. Jarrell, W.E. Pickett, R.R.P. Singh, M. Ulmke, and G.T. Zimanyi. We are especially indebted to N. Trivedi, whose physical insight and numerical prowess has been invaluable.

References

Abrahams, E., Anderson, P.W., Licciardello, D.C., and Ramakrishnan, T.V. (1979). *Phys. Rev. Lett.*, **42**, 673.

Aguiar, M.C.O., Dobrosavljević, V., Abrahams, E., and Kotliar, G. (2005). *Phys. Rev. B*, **71**, 205115.

Aguiar, M.C.O., Dobrosavljević, V., Abrahams, E., and Kotliar, G. (2009). *Phys. Rev. Lett.*, **102**, 156402.

Altshuler, B.L. and Aronov, A.G. (1979). *Solid State Commun.*, **30**, 115.

Aryanpour, K., Dagotto, E.R., Mayr, M., Paiva, T., Pickett, W.E., and Scalettar, R.T. (2006). *Phys. Rev. B*, **73**, 104518.

Aryanpour, K., Paiva, T., Pickett, W.E., and Scalettar, R.T. (2007). *Phys. Rev. B*, **76**, 184521.

Bai, Z.J., Chen, W.B., Scalettar, R., and Yamazaki, I. (2009), pp. 1–110. Multi-Scale Phenomena in Complex Fluids. Higher Education Press.

Beard, B.B. and Wiese, U.J. (1996). *Phys. Rev. Lett.*, **77**, 5130.

Belitz, D. and Kirkpatrick, T.R. (1994). *Rev. Mod. Phys.*, **66**, 261.

Berkovits, R. and Kantelhardt, J.W. (2002). *Phys. Rev. B*, **65**, 125308.

Blankenbecler, R., Scalapino, D.J., and Sugar, R.L. (1981). *Phys. Rev. D*, **24**, 2278.

Byczuk, K., Hofstetter, W., Yu, U., and Vollhardt, D. (2010). *Correlated Electrons in the Presence of Disorder.* arXiv:1003.2498.

Caldara, G., Srinivasan, B., and Shepelansky, D.L. (2000). *Phys. Rev. B*, **62**, 10680.

Castellani, C., Di Castro, C., and Lee, P.A. (1998). *Phys. Rev. B*, **57**, R9381.

Chakraborty, P.B., Denteneer, P.J.H., and Scalettar, R.T. (2007). *Phys. Rev. B*, **75**, 125117.

Chakravarty, S., Yin, L., and Abrahams, E. (1998). *Phys. Rev. B*, **58**, R559.

Chiesa, S. (2010). unpublished.

Chiesa, S., Chakraborty, P.B., Pickett, W.E., and Scalettar., R.T. (2008). *Phys. Rev. Lett.*, **101**, 086401.

Chui, S.T. and Tanatar, B. (1995). *Phys. Rev. Lett.*, **74**, 458.

Denteneer, P.J.H. and Scalettar, R.T. (2003). *Phys. Rev. Lett*, **90**, 246401.

Denteneer, P.J.H., Scalettar, R.T., and Trivedi, N. (1999). *Phys. Rev. Lett.*, **83**, 4610.

Denteneer, P.J.H., Scalettar, R.T., and Trivedi, N. (2001). *Phys. Rev. Lett.*, **87**, 146401.

Dobrosavljević, V., Abrahams, E., Miranda, E., and Chakravarty, S. (1997). *Phys. Rev. Lett.*, **79**, 455.

Efros, A.L. and Shklovskii, B.I. (1975). *J. Phys. C*, **C8**, L49.

Farhoodfar, A., Chen, X., Gooding, R.J., and Atkinson, W.A. (2009). *Phys. Rev. B*, **80**, 045108.

Farias, G.J. (2009). MSc. thesis, Universidade Federal do Mato Grosso.

Fazekas, P. (1999). *Lecture Notes on Electron Correlation and Magnetism.* World Scientific, Singapore.

Fazileh, F., Gooding, R.J., Atkinson, W.A., and Johnston, D.C. (2006). *Phys. Rev. Lett.*, **96**, 046410.

Feynman, R. (1982). *Int. J. Theor. Phys.*, **21**, 467–488.

Finkel'stein, A.M. (1983). *Zh. Eksp. Teor. Fiz.*, **84**, 168. [Sov. Phys. JETP 1984, **57**, 97].

Furukawa, N. and Imada, M. (1992). *J. Phys. Soc. Japan*, **61**, 3331.

Gammel, J. Tinka, Campbell, D.K., and Loh, E.Y. (1992). *Synthetic Metals*, **57**, 6.

Gebhard, F. (1997). *The Mott Metal–Insulator Transition, Models and Methods.* Springer.

Georges, A., Kotliar, G., Krauth, W., and Rozenberg, M. (1996). *Rev. Mod. Phys.*, **68**, 13.

Gros, C. (1996). *Phys. Rev. B*, **53**, 6865.

Gubernatis, J.E., Jarrell, M., Silver, R.N., and Sivia, D.S. (1991). *Phys. Rev. B*, **44**, 6011.

Gubernatis, J.E., Loh, E.Y., Scalapino, D.J., Sugar, R.L., Scalettar, R.T., and White, S.R. (1989). In *Proceedings of the Los Alamos Conference on Quantum Simulation, Aug. 8–11, 1989, Los Alamos, NM.* World Scientific.

Herbut, I.F. (2001). *Phys. Rev. B*, **63**, 113102.

Hettler, M.H., Tahvildar-Zadeh, A.N., Jarrell, M., Pruschke, T., and Krishnamurthy, H.R. (1998). *Phys. Rev. B*, **58**, R7475.

Hirsch, J.E. (1985). *Phys. Rev. B*, **31**, 4403.

Hirsch, J.E., Scalapino, D.J., Sugar, R.L., and and, R. Blankenbecler (1982). *Phys. Rev. B*, **26**, 5033.

Hirsch, J.E. and Tang, S. (1989). *Phys. Rev. Lett.*, **62**, 591.

Hurt, D., Odabashian, E., Pickett, W.E., Scalettar, R.T., Mondaini, F., Paiva, T., and dos Santos, R.R. (2005). *Phys. Rev. B*, **72**, 144513.

Huse, D.A. (1988). *Phys. Rev B*, **37**, 2380.

Ishida, K., Nakai, Y., and Hosono, H. (2009). *To What Extent Iron-Pnictide New Superconductors Have Been Clarified: A Progress Report.* http://arxiv.org/abs/0906.2045.

Jaksch, D., Bruder, C., Cirac, J.I., Gardiner, C.W., and Zoller, P. (1998). *Phys. Rev. Lett.*, **81**, 3108.

Janis, V., Ulmke, M., and Vollhardt, D. (1993). *Europhys. Lett.*, **24**, 287.

Kawashima, N., Gubernatis, J.E., and Evertz, H.G. (1994). *Phys. Rev. B*, **50**, 136.

Kotlyar, R. and Das Sarma, S. (2001). *Phys. Rev. Lett.*, **86**, 2388.

Lee, P.A. and Ramakrishnan, T.V. (1985). *Rev. Mod. Phys.*, **57**, 287.

Lewenstein, M., Sanpera, A., Ahufinger, V., Damski, B., Sen, A., and Sen, U. (2007). *Adv. Phys.*, **56**, 24.

Litak, G. and Györffy, B.L. (2000). *Phys. Rev. B*, **62**, 6629.

Loh, E.Y., Gubernatis, J.E., Scalettar, R.T., White, S.R., Scalapino, D.J., and Sugar, R.L. (1990). *Phys. Rev. B*, **41**, 9301.

Loh Jr., E.Y., Gubernatis, J.E., Scalettar, R.T., White, S.R., and Sugar, R.L. (1989). In *Workshop on Interacting Electrons in Reduced Dimensions*, Volume 213, NATO ASI Series B; Physics, NY. Plenum.

Mermin, N.D. and Wagner, H. (1966). *Phys. Rev. Lett.*, **17**, 1133.

Mondaini, F., Paiva, T., dos Santos, R.R., and Scalettar, R.T. (2008). *Phys. Rev. B*, **78**, 174519.

Montorsi, Arianna (ed.) (1992). *The Hubbard Model.* World Scientific, Singapore.

Prokofév, N.V., Svistunov, B.V., and Tupitsyn, I.S. (1996). *J. Expt. and Theor. Phys.*, **64**, 911.

Pudalov, V.M., D'Iorio, M., Kravchenko, S.V., and Campbell, J.W. (1993). *Phys. Rev. Lett.*, **70**, 1866.

Radonjic, M.M., Tanaskovic, D., Dobrosavljević, V., and Haule, K. (2010). *Phys. Rev. B*, **81**, 075118.

Rasetti, M. (1991). *The Hubbard Model—Recent Results.* World Scientific.

Rieger, H. and Kawashima, N. (1999). *Eur. Phys. J. B*, **9**, 233.

Sandvik, A.W. (1992). *J. Phys. A*, **25**, 3667.

Sandvik, A.W. (1997). *Phys. Rev. B*, **59**, R14157.

Sandvik, A.W. and Kurkijarvi, J. (1991). *Phys. Rev. B*, **43**, 5950.

Scalapino, D.J. (1994). In *Proceedings of the International School of Physics (July 1992)* (ed. R. Broglia and J. Schrieffer), New York. North-Holland. See also references cited therein.

Scalapino, D.J., White, S.R., and Zhang, S.C. (1993). *Phys. Rev. B*, **47**, 7995.

Scalettar, R.T. (2003). Website: *Computational Quantum Magnetism.* http://research.yale.edu/boulder/Boulder-2003/index.html, July, 2003.

Sénéchal, David and Tremblay, A.-M.S. (2004). *Phys. Rev. Lett.*, **92**, 126401.

Shenoy, V.B. (2008). *Phys. Rev. B*, **78**, 134503.

Shinaoka, H. and Imada, M. (2009). *Electronic and Magnetic Properties of Metallic Phases under Coexisting Short-range Interaction and Diagonal Disorder.* arXiv: 0906.4386.

Song, Yun, Wortis, R., and Atkinson, W.A. (2008). *Phys. Rev. B*, **77**, 054202.

Sorella, S., Tosatti, E., Baroni, S., Car, R., and Parinello, M. (1988). *Int. J. Mod. Phys. B*, **1**, 993.

Srinivasan, B., Benenti, G., and Shepelansky, D.L. (2003). *Phys. Rev. B*, **7**, 205112.

Sugiyama, G. and Koonin, S.E. (1986). *Ann. Phys.*, **168**, 1.

Tanaskovic, D., Dobrosavljević, V., Abrahams, E., and Kotliar, G. (2003). *Phys. Rev. Lett.*, **91**, 066603.

Tesanovic, Z. (2009). *Physics*, **2**, 60.

Trivedi, N., Scalettar, R.T., and Randeria, M. (1996). *Phys. Rev. B*, **54**, R3756.

Ulmke, M., Denteneer, P.J. H., Scalettar, R.T., and Zimanyi, G.T. (1998). *Europhys. Lett.*, **42**, 655.

Varney, C.N., Lee, C.R., Bai, Z.J., Chiesa, S., Jarrell, M., and Scalettar, R.T. (2009). *Phys. Rev. B*, **80**, 075116.

Vekic, M., Cannon, J.W., Scalapino, D.J., Scalettar, R.T., and Sugar, R.L. (1995). *Phys. Rev. Lett.*, **74**, 2367.

Vojta, T., Epperlein, F., and Schreiber, M. (1998). *Phys. Rev. Lett.*, **81**, 4212.

Vollhardt, D. (1993). In *Correlated Electron Systems* (ed. V. Emery), Singapore, p. 57. World Scientific.

White, S.R., Scalapino, D.J., Sugar, R.L., Loh, E.Y., Gubernatis, J.E., and Scalettar, R.T. (1989). *Phys. Rev. B*, **40**, 506.

Wortis, R. private communication.

Zala, G., Narozhny, B.N., and Aleiner, I.L. (2001). *Phys. Rev. B*, **65**, 02020(R).

6

Dynamical Mean-field Theories of Correlation and Disorder

E. MIRANDA[1] and V. DOBROSAVLJEVIĆ[2]

[1] Instituto de Física Gleb Wataghin, Unicamp, R. Sérgio Buarque de Holanda, 777, Campinas, SP 13083-859, Brazil
[2] Department of Physics and National High Magnetic Field Laboratory, Florida State University, Tallahassee, FL 32306

6.1 Mott transitions in clean and disordered systems

It is this fascination with the local and with the failures, not successes, of band theory, which... contradicted the assumptions of the time...

P. W. Anderson, Nobel Prize Lecture, 1979

6.1.1 Introduction

The dynamical mean-field theory (DMFT) of interacting lattice fermions is perhaps best described as the optimal description of these systems which takes into account only *local* (on-site) correlation effects (Georges *et al.*, 1996; Pruschke *et al.*, 1995). Technically, this is implemented through the assumption of a single-electron self-energy which, in the lattice translationally invariant case, is independent of the site in real space or of wave vector in reciprocal space and therefore is a function of frequency only

$$\Sigma\left(\boldsymbol{R_i},\omega\right) \rightarrow \Sigma\left(\omega\right) \qquad \text{or} \qquad \Sigma\left(\boldsymbol{k},\omega\right) \rightarrow \Sigma\left(\omega\right).$$

This kind of approximation surely leaves out inter-site correlations. However, local approaches to strong correlations have a long history and several semi-phenomenological descriptions of classes of compounds within this framework have been shown to be consistent with their physical properties. A particularly well-studied example are heavy fermion systems (Grewe and Steglich, 1991; Stewart, 1984). Many properties of heavy fermion materials suggest that local correlations are sufficient for a good description, the large carrier effective mass (as derived from the magnetic susceptibility or the Sommerfeld specific heat linear coefficient) being the best known but by no means the only ones. Transport properties such as DC and optical conductivity and ultrasound attenuation can also be understood with the same assumptions (Varma, 1985). Other examples of systems well described by a local approach include systems close to a Mott metal–insulator transition, in particular, the vicinity of the finite-temperature critical end-point (Kagawa *et al.*, 2005; Kotliar *et al.*, 2000, 2002; Limelette *et al.*, 2003*a*,*b*; Rozenberg *et al.*, 1999). A particularly striking consequence of a local self-energy is the cancellation of many-body renormalizations in the ratio of the coefficient of the T^2-term of the resistivity to the square of the specific heat coefficient, usually called the Kadowaki–Woods ratio (Kadowaki and Woods, 1986; Miyake *et al.*, 1989). Recently, it has been shown that, when materials-specific effects (such as carrier density, density of states and Fermi velocity values) are properly taken into account, then the Kadowaki–Woods ratio appears to be universal across a much wider range of compounds, including, besides heavy fermion systems, organic charge-transfer salts, transition-metal oxides, and transition metals (Jacko *et al.*, 2009). Thus, a local approach to strong correlations seems to be much more generally valid than initially thought.

Several starting points lead to theories that ultimately predict a self-energy of this form, most notably descriptions based on the Gutzwiller wave-function (Gutzwiller, 1963, 1965; Vollhardt, 1984) or the large-N limit (Coleman, 1987; Millis and Lee, 1987). However, these theories usually end up imposing further restrictions, beyond a local self-energy, for example inelastic scattering effects are not included, higher-energy incoherent features are absent, among others. A description which incorporates *all* possible local effects in a fully self-consistent fashion is provided by DMFT.

Historically, DMFT was proposed by a recourse to the infinite-dimensional limit of lattice systems (Metzner and Vollhardt, 1989). Indeed, when appropriate rescaling of parameters is done (as is usual when considering this limit), the theory remains meaningful and non-trivial as $d \to \infty$ and the self-energy becomes completely local. The reader can find many alternative derivations of DMFT in this limit in the review by Georges *et al.* (1996). Alternatively, DMFT can also be viewed as the best local description of three-dimensional systems. The focus here will *not* be to derive the theory by resorting to the infinite-dimensional limit but rather to highlight the physical content of a local description of correlation effects. This is specially important since we will later explore other, more general local theories that do not become exact in any particular limit but which inherit the insights gained from DMFT. We will therefore focus mostly on a Bethe lattice, which is most transparent and lends itself particularly well to generalizations to the disordered case. Furthermore, we will highlight the key physical assumptions involved, which are kept in the other approximations.

6.1.2 The clean case

Consider for concreteness the Hubbard model with only nearest-neighbor hopping on a lattice with finite coordination z, in usual notation,

$$H = -\sum_{\langle ij \rangle,\sigma} t_{ij} c^{\dagger}_{i\sigma} c_{j\sigma} + U \sum_{i} n_{i\uparrow} n_{i\downarrow}. \tag{6.1}$$

We focus on a particular site, call it j, which in the clean case can be any site. The effective dynamics of this site alone can be obtained by integrating out all the other sites. This is no longer a Hamiltonian dynamics and the procedure requires an action description, which we will write in imaginary time

$$\begin{aligned} S_{\text{eff}}(j) = {} & \sum_{\sigma} \int_0^{\beta} d\tau c^{\dagger}_{j\sigma}(\tau)\left(\partial_{\tau} - \mu\right) c_{j\sigma}(\tau) \\ & + \sum_{\sigma} \int_0^{\beta} d\tau \int_0^{\beta} d\tau' c^{\dagger}_{j\sigma}(\tau)\, \Delta(\tau - \tau')\, c_{j\sigma}(\tau') \\ & + U \int_0^{\beta} d\tau n_{j\uparrow}(\tau)\, n_{j\downarrow}(\tau). \end{aligned} \tag{6.2}$$

The second term above comes from integrating out the other sites, the first and third ones being the local contributions, already present before the integration. The thing to note here is the fact that, in general, the integration over *interacting* sites generates other higher-order terms, involving four and more fermionic fields. In the high-dimensional limit or in DMFT in general, *these higher-order terms are absent or neglected.* It is clear that this means that only single-particle inter-site correlations are kept in this limit/approximation and this is precisely what is encoded in the second, retarded term in eqn (6.2). The "hybridization function" $\Delta(\tau)$ describes the "leaking" of electrons in and out of site j. It can be written as

$$\Delta(\tau) = t^2 \sum_{l,m=1}^{z} G_{lm}^{(j)}(\tau), \tag{6.3}$$

where $G_{lm}^{(j)}(\tau)$ is the Green's function for propagation from site m to site l *in a lattice from which site j has been removed* (hence the superscript (j))

$$G_{lm}^{(j)}(\tau) = -\left\langle T\left[c_{l\sigma}(\tau)\, c_{m\sigma}^{\dagger}(0)\right]\right\rangle^{(j)}, \tag{6.4}$$

and the sums extend over the z nearest-neighbors of site j. Let us not dwell on how this is calculated for now.

The action (6.2) is equivalent to the one of an Anderson single-impurity problem (Anderson, 1961), whose Hamiltonian is

$$\begin{aligned} H = & \sum_{\boldsymbol{k},\sigma} E_{\boldsymbol{k}} a_{\boldsymbol{k}\sigma}^{\dagger} a_{\boldsymbol{k}\sigma} - \sum_{\sigma} \mu c_{j\sigma}^{\dagger} c_{j\sigma} \\ & + \sum_{\boldsymbol{k},\sigma} \left(\frac{V_{\boldsymbol{k}}}{\sqrt{N_s}} a_{\boldsymbol{k}\sigma}^{\dagger} c_{j\sigma} + \text{H.c.} \right) + U c_{j\uparrow}^{\dagger} c_{j\uparrow} c_{j\downarrow}^{\dagger} c_{j\downarrow}, \end{aligned} \tag{6.5}$$

provided we choose $E_{\boldsymbol{k}}$ and $V_{\boldsymbol{k}}$ above in such a way that the Fourier transform, in Matsubara frequency space, of the hybridization function in eqn (6.3) is such that

$$\Delta(i\omega_n) = \frac{1}{N_s} \sum_{\boldsymbol{k}} \frac{|V_{\boldsymbol{k}}|^2}{i\omega_n - E_{\boldsymbol{k}}}. \tag{6.6}$$

This equivalence proves to be extremely useful since the well-studied behavior of the Anderson single-impurity problem serves as a guide to physical insight (Georges and Kotliar, 1992).

Suppose now that we can somehow find the full interacting Green's function of the system described by the action of eqn (6.2)

$$G_{jj}(\tau) = -\left\langle T\left[c_{j\sigma}(\tau)\, c_{j\sigma}^{\dagger}(0)\right]\right\rangle_{\text{eff}}, \tag{6.7}$$

where the subscript "eff" emphasizes that it is to be calculated under the dynamics dictated by (6.2). We can repackage our ignorance about this function by defining a self-energy $\Sigma(i\omega_n)$ such that

$$G_{jj}(i\omega_n) = \frac{1}{i\omega_n + \mu - \Delta(i\omega_n) - \Sigma(i\omega_n)}. \tag{6.8}$$

Having quantified the local dynamics, we now need to bring in information from the rest of the lattice. In principle, this is quite straightforward: *since only local correlations are included,* $\Sigma(i\omega_n)$ *is also the self-energy for generic lattice propagation*

$$G(\boldsymbol{k}, i\omega_n) = \frac{1}{i\omega_n - \varepsilon_{\boldsymbol{k}} + \mu - \Sigma(i\omega_n)}, \tag{6.9}$$

where $\varepsilon_{\boldsymbol{k}}$ is the non-interacting dispersion. From this expression, we can obtain the Green's function with one site removed from eqn (6.4) and from that the hybridization function (6.3), which closes the self-consistency loop. This procedure to get $\Delta(i\omega_n)$ from $G(\boldsymbol{k}, i\omega_n)$ is described, for example in the review by Georges *et al.* (1996). However, we will proceed in the simpler and more illuminating case of the Bethe lattice with coordination z, for which

$$G_{lm}^{(j)}(\tau) = \delta_{lm} G_{ll}^{(j)}, \tag{6.10}$$

since the removal of site j completely disconnects the two branches that start at the nearest-neighbors l and m, if $l \neq m$. Thus,

$$\Delta(\tau) = t^2 \sum_{l=1}^{z} G_{ll}^{(j)}(\tau) \tag{6.11}$$

$$= zt^2 G_{ll}^{(j)}(\tau), \tag{6.12}$$

since all sites are equivalent. We can now take the limit $z \to \infty$, noting that, in this limit, the removal of one nearest-neighbor site (j) is irrelevant for the local propagation at l and that an appropriate rescaling, namely $zt^2 \to \tilde{t}^2$, is necessary

$$\Delta(\tau) = \tilde{t}^2 G_{ll}(\tau) = \tilde{t}^2 G_{jj}(\tau), \tag{6.13}$$

since all sites are equivalent. This is the self-consistency condition in this case: the solution of the problem is the one for which, if we plug in the local Green's function $G_{jj}(\tau)$ in eqn (6.13), then insert it into the action (6.2) and find the expectation value in eqn (6.7), we get back $G_{jj}(\tau)$.

A more physical alternative route and one which is not restricted to the Bethe lattice is to note *that the local Green's function obtained from* eqn (6.9) *must coincide with the one defined in* eqn (6.8)

$$G_{jj}(i\omega_n) = \frac{1}{N_s} \sum_{\boldsymbol{k}} G(\boldsymbol{k}, i\omega_n) \Rightarrow$$

$$\frac{1}{i\omega_n + \mu - \Delta(i\omega_n) - \Sigma(i\omega_n)} = \int d\varepsilon \frac{\rho_0(\varepsilon)}{i\omega_n - \varepsilon + \mu - \Sigma(i\omega_n)}. \tag{6.14}$$

Here, $\rho_0(\varepsilon)$ is the bare density of states generated from $\varepsilon_{\boldsymbol{k}}$. The two procedures can be shown to be equivalent and establish the necessary self-consistency condition.

It is possible to extend the above analysis to two-particle correlation functions (Georges *et al.*, 1996) but we will not delve into this. It is, however, worthwhile to notice that the current–current correlation function, which provides the conductivity through the Kubo formula, acquires no vertex correction within DMFT and is given by a simple bubble of renormalized single-particle Green's functions.

It is perhaps wise to highlight yet again what the key assumptions of this approach are:

1. The effects of interactions that are included are on-site only, or equivalently, the self-energy is purely local.
2. Different sites "know" about each other through single-particle processes only, see Fig. 6.1 (a).
3. The local dynamics, as dictated by the local effective action and usually encoded in the local Green's function, must coincide with the local dynamics as derived from the lattice propagation.

The DMFT approach, like the original Bragg–Williams mean-field theory of magnetism (Goldenfeld, 1992), focuses on a single lattice site, but replaces its environment by a self-consistently determined "effective medium" (Georges *et al.*, 1996). Unlike the Bragg–Williams theory, the effect of the environment cannot be captured by a static external field, but must be encoded in a full complex *function* $\Delta(i\omega_n)$, which contains information about the dynamics of an electron moving in and out of the given site. The calculation then reduces to solving an appropriate quantum impurity problem, eqn (6.2), supplemented by an additional self-consistency condition, eqn (6.14), that ultimately determines this hybridization function $\Delta(i\omega_n)$.

The approach has been very successful in examining the vicinity of the Mott transition in clean systems, in which it has met spectacular success in elucidating various properties of several transition metal oxides (Georges *et al.*, 1996), heavy fermion systems, and even Kondo insulators (Rozenberg *et al.*, 1996).

6.1.3 The clean Mott transition

The Mott transition in a single-band Hubbard model can be regarded as a prototype for a interaction-driven metal–insulator transition, a phenomenon with plausible relevance to many physical systems of current interest. Its basic mechanism has been correctly understood for more than fifty years (Mott, 1949), yet the precise nature of this phase transition has long remained controversial and ill-understood. Part of the confusion stems from the fact that at low temperatures the Mott insulator is typically unstable to antiferromagnetic ordering, leading many authors (Slater, 1951) to focus on magnetism as a proposed driving force. The shortcomings of this view were most lucidly emphasized by Anderson (1978), who stressed that the Mott insulating state persists well above the Nel temperature. It is thus transmutation of conduction electrons into local magnetic moments—not the long range magnetic ordering—that should be regarded as the fundamental physical process behind the Mott transition. The two phenomena can be most clearly separated in systems where the tendency for magnetic ordering can be appreciably weakened due to frustration effects, such as often found in orbitally degenerate transition metal oxides. Here, the competition between antiferromagnetic superexchange and ferromagnetic tendencies due to Hund's

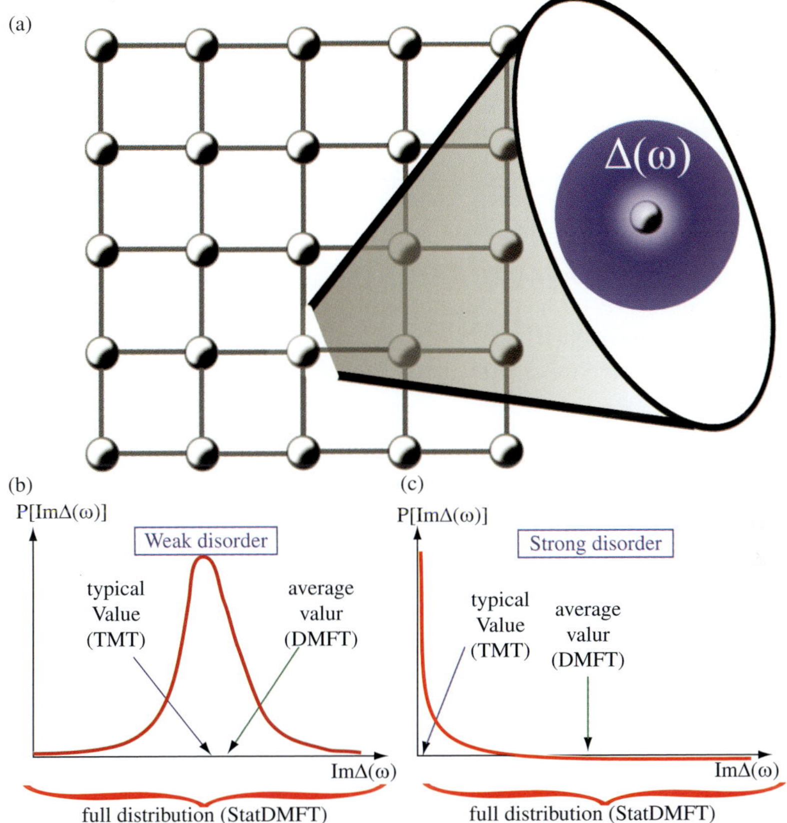

Fig. 6.1 Schematic view of all DMFT-inspired theories of disordered strongly correlated systems. (a) All these theories share the same ingredient: the local site "knows" about the other sites through the self-consistently determined hybridization function $\Delta(\omega)$ (see eqn (6.2)), which acts as an order parameter. They differ in the level of description of this order parameter, which is in general a random quantity (b and c). Both DMFT and the typical medium theory (TMT) replace the actual local realization of this random variable by a fixed, non-random function, namely, the average and the typical values of $\Delta(\omega)$, respectively. These two functions are very similar at weak disorder (b), but become increasingly different as the disorder grows (c). The statistical dynamical mean-field theory (statDMFT), by contrast, retains the full spectrum of spatial fluctuations, since each site "sees" the actual local realization $\Delta_i(\omega)$.

rule couplings typically lead to large cancellations, resulting in very weak magnetic correlations in the paramagnetic phase.

From the theoretical point of view, this situation can be most clearly formulated by focusing on the "maximally frustrated" Hubbard model, with infinite-range hopping of random sign (Georges *et al.*, 1996). In this model the magnetic frustration is so strong as to completely suppress any magnetic ordering, while the DMFT approximation becomes exact, allowing precise and detailed characterization of such an "ideal" Mott transition within the paramagnetic phase. In the following we briefly describe the main features of the resulting DMFT picture of the bandwidth-driven Mott transition.

Critical behavior and mass divergence at $T = 0$

Within DMFT, the critical regime between the Fermi liquid metal and the Mott insulator features (Georges *et al.*, 1996) a finite-temperature coexistence region and a first-order transition line ending at the critical end-point at $T = T_c$. At $T = 0$, however, the metallic solution is the stable (lower energy) one throughout the coexistence regime. It is characterized by heavy quasiparticles with an effective mass that diverges as the transition is approached.

In general, the effective mass is evaluated from the single-particle self-energy using the expression

$$\frac{m^*}{m} = \left.\frac{1 - \frac{\partial}{\partial\omega}\Sigma'(\mathbf{k},\omega)}{1 + \frac{m}{k}\frac{\partial}{\partial k}\Sigma'(\mathbf{k},\omega)}\right|_{k=k_F,\omega=0}, \tag{6.15}$$

where k_F denotes the Fermi momentum, and Σ' is the real part of the self-energy $\Sigma(\mathbf{k},\omega)$. The quasiparticle weight, on the other hand, is defined by

$$Z^{-1} \equiv 1 - \left.\frac{\partial}{\partial\omega}\Sigma'(\mathbf{k},\omega)\right|_{k=k_F,\omega=0}. \tag{6.16}$$

Within DMFT, $\Sigma(\mathbf{k},\omega) = \Sigma(\omega)$ is momentum independent, and $m^*/m = Z^{-1}$. Note that, since generally $\frac{\partial}{\partial\omega}\Sigma'(\omega)\big|_{\omega=0} < 0$, the interactions increase the effective mass. The actual divergence is obtained only if the quantity $A \equiv -\frac{\partial}{\partial\omega}\Sigma'(\omega)\big|_{\omega=0}$ itself diverges. This scenario is realized, for example, in the Brinkmann–Rice theory of the Mott transition, as well as in the more recent DMFT solution. Since the quasiparticle weight is simply $Z^{-1} = m^*/m$, it must diverge at the same place as m^* does.

We should emphasize that this result is exact within the DMFT approach and is an excellent approximation for many Mott compounds where magnetic frustration is sufficiently strong. To put this result in perspective, we contrast it with a popular but uncontrolled weak-coupling approach, based on the so-called "on-shell approximation" (Ting *et al.*, 1975) for the effective mass of the correlated electron gas. Here, an approximate expression for the effective mass is proposed

$$\frac{m^*}{m} \approx \left.\frac{1}{1 + \frac{m}{k}\frac{d}{dk}\Sigma'(\mathbf{k},\xi_\mathbf{k})}\right|_{k=k_F}, \tag{6.17}$$

where $\xi_\mathbf{k}$ is the unrenormalized band dispersion. When this approximation is applied to the low-density electron gas within the random phase approximation scheme (Zhang and Das Sarma, 2005), one finds that the effective mass diverges *before* the quasiparticle weight Z vanishes. This result seems quite pathological, since the natural interpretation of the effective mass divergence is the localization of itinerant electrons, where one also expects the breakdown of the quasiparticle picture.

To benchmark the validity of the proposed "on-shell approximation," we apply it to the maximally frustrated Hubbard model, where DMFT provides us with an exact result for both the effective mass and the full self-energy. In this case $\Sigma'(\mathbf{k}, \omega = \xi_{\mathbf{k}}) = \Sigma'(\omega{=}\xi_{\mathbf{k}})$. Noting that

$$\frac{m}{k}\frac{d}{dk}\Sigma'(\omega = \xi_{\mathbf{k}})\Big|_{k=k_{\mathrm{F}}} = \frac{\partial}{\partial\omega}\Sigma'(\omega)\Big|_{\omega=0},$$

we get the "on-shell" result

$$\frac{m^*}{m} \approx \frac{1}{1+\frac{\partial}{\partial\omega}\Sigma'(\omega)}\Big|_{\omega=0} = \frac{1}{1-A}.$$

As we can see, this expression is equivalent to the exact expression $m^*/m = 1 + A$, only to leading order, i.e. for $(1 - Z) \ll 1$. On the other hand, the positive quantity A is expected to grow with the interaction. As long as it is finite, neither the properly defined effective mass m^*/m, nor the inverse quasiparticle weight Z^{-1} will ever diverge. In contrast, if one uses the "on-shell" expression, then the effective mass will blow up as soon as $A = 1$, and this will happen at some point in any approximation where A grows with the interaction. However, as we can see, this will not lead to the divergence of the inverse quasiparticle weight Z^{-1}. What we can see from these expressions is that the essence of the "on-shell" approximation is simply to linearize the expression for $(m^*/m)^{-1}$ by expanding it in the quantity $A \equiv -\frac{\partial}{\partial\omega}\Sigma'(\omega)\big|_{\omega=0}$. Instead of appearing in the numerator of the effective mass expression, it now enters the denominator, leading to an unphysical effective mass divergence. This example provides a perfect illustration of how dangerous it is to indiscriminately apply weak-coupling results to non-perturbative phenomena near the Mott transition. It also shows how DMFT not only correctly captures the essence of strong correlations, but also provides a simple and transparent insight into their physical content.

Quantum-critical behavior at $T > T_c$

Many systems close to the metal–insulator transition often display surprisingly similar transport features in the high-temperature regime Here, the family of resistivity curves typically assumes a characteristic "fan-shaped" form, reflecting a gradual crossover from metallic to insulating transport. At the highest temperatures the resistivity depends only weakly on the control parameter (concentration of charge carriers or pressure), while as T is lowered the system seems to "make up its mind" and rapidly converges towards either a metallic or an insulating state. Since temperature acts as a natural cutoff scale for the metal–insulator transition, such behavior is precisely what one expects for quantum criticality. In some cases (Abrahams *et al.*, 2001), the entire family of curves displays beautiful scaling behavior, with a remarkable "mirror symmetry" of the relevant scaling functions (Dobrosavljević *et al.*, 1997). But under what microscopic conditions should one expect such scaling phenomenology? Should one expect similar or very different transport phenomenology in the Mott

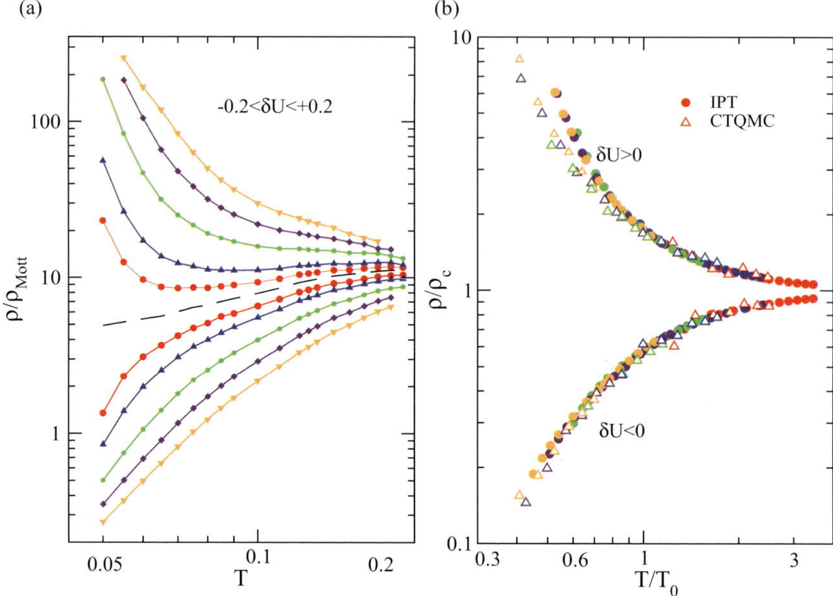

Fig. 6.2 (a) DMFT resistivity curves as functions of the temperature along different trajectories across the Mott transition in a half-filled Hubbard model (Terletska *et al.*, 2011). (b) Resistivity scaling displaying remarkable "mirror symmetry" (Dobrosavljević *et al.*, 1997).

picture? Is the paradigm of quantum criticality even a useful language to describe high-temperature transport around the Mott point?

Somewhat surprisingly, most DMFT studies of the Mott transition have focused on the lowest-temperature regime, paying little attention to the high-temperature crossover regime relevant to many experiments. On the other hand, it is well known that at very low temperatures $T < T_c \sim 0.03T_F$, this model features a first-order metal–insulator transition terminating at the critical end-point T_c (Fig. 6.2), very similar to the familiar liquid-gas transition. For $T > T_c$, however, different crossover regimes have been tentatively identified (Georges *et al.*, 1996) but they have not been studied in any appreciable detail. The fact that the first-order coexistence region is restricted to such very low temperatures provides strong motivation to examine the high-temperature crossover region from the perspective of "hidden quantum criticality." In other words, it is very plausible that the presence of a coexistence dome at $T < T_c \ll T_F$, an effect with very small energy scale, is not likely to influence the behavior at much higher temperatures $T \gg T_c$. In this high-temperature regime smooth crossover is found, which may display behavior consistent with the presence of a "hidden" quantum critical point at $T = 0$. To test this idea, very recent work (Terletska *et al.*, 2011) utilized standard scaling methods appropriate for quantum criticality and computed the resistivity curves along judiciously chosen trajectories respecting the symmetries of the problem. Characteristic scaling behavior for the entire family of resistivity curves has been identified, with the corresponding beta-function displaying striking "mirror symmetry" consistent with experiments. These findings provide compelling arguments in support of the suggestion that finite-temperature behavior in many Mott systems should be interpreted from the perspective of quantum criticality (Panagopoulos and Dobrosavljević, 2005). It should stressed, however, that the DMFT solution in question does not contain any physical processes associated with approach to magnetic or charge ordering. If quantum criticality is indeed at play here, it has a fundamentally different nature, one that is associated with the destruction of a Fermi liquid without the aid of any static symmetry breaking pattern—in dramatic contrast to most known critical phenomena.

6.1.4 The disordered case

Let us now proceed to write down the equations for the dynamical mean-field theory description of a disordered system (Dobrosavljević and Kotliar, 1993, 1994; Janis *et al.*, 1993; Janis and Vollhardt, 1992). We will focus on the case of diagonal site disorder, which for the Hubbard model reads

$$H = -\sum_{\langle ij\rangle,\sigma} t_{ij} c^{\dagger}_{i\sigma} c_{j\sigma} + \sum_{j,\sigma} \varepsilon_j c^{\dagger}_{j\sigma} c_{j\sigma} + U \sum_i n_{i\uparrow} n_{i\downarrow}, \tag{6.18}$$

where ε_j are assumed to be independent random variables drawn from a given distribution $P(\varepsilon)$ whose strength is W (say, a uniform distribution from $-W/2$ to $W/2$). Focusing once more on the local dynamics, it is clear it must now be dictated by

$$\begin{aligned} S_{\text{eff}}(j) = & \sum_{\sigma} \int_0^{\beta} d\tau c^{\dagger}_{j\sigma}(\tau)\left(\partial_{\tau} + \varepsilon_j - \mu\right) c_{j\sigma}(\tau) \\ & + \sum_{\sigma} \int_0^{\beta} d\tau \int_0^{\beta} d\tau' c^{\dagger}_{j\sigma}(\tau)\, \Delta(\tau - \tau')\, c_{j\sigma}(\tau') \\ & + U \int_0^{\beta} d\tau n_{j\uparrow}(\tau)\, n_{j\downarrow}(\tau). \end{aligned} \tag{6.19}$$

Notice that the effective action is now different for different sites because of the ε_j term. However, the hybridization function $\Delta(\tau)$ is *not* site-dependent. This is motivated again by the infinite-dimensional limit. Indeed, if the number of nearest neighbors is infinite, the sum in eqn (6.3) is effectively an *averaging procedure over all possible realizations of the Green's function.* In fact, in the infinite-dimensional Bethe lattice, eqn (6.11) becomes

$$\Delta(\tau) = zt^2 \left[\frac{1}{z} \sum_{l=1}^{z} G^{(j)}_{ll}(\tau)\right] \tag{6.20}$$

$$\xrightarrow[z\to\infty]{} \tilde{t}^2 \overline{G_{ll}(\tau)}, \tag{6.21}$$

where the overbar denotes average over quenched disorder. Again, the effect of removing one nearest neighbor is negligible in this case. Thus, the hybridization function is proportional to the *average local Green's function* (see Fig. 6.1 (b) and (c)). Solving the DMFT equations in the disordered case then entails solving an *ensemble* of single-impurity problems as in eqn (6.19), one for each value of ε_j, and finding for each of them the local Green's function and self-energy (which are now also site-dependent)

$$G_{jj}(\tau) = -\left\langle T\left[c_{j\sigma}(\tau)\, c_{j\sigma}^{\dagger}(0)\right]\right\rangle_{\text{eff}}, \tag{6.22}$$

$$G_{jj}(i\omega_n) = \frac{1}{i\omega_n - \varepsilon_j + \mu - \Delta(i\omega_n) - \Sigma_j(i\omega_n)}. \tag{6.23}$$

The self-consistency can then be written for the Bethe lattice as (cf. eqns (6.11)–(6.13))

$$\Delta(i\omega_n) = t^2\overline{G_{jj}(i\omega_n)} = t^2\int d\varepsilon \frac{P(\varepsilon)}{i\omega_n - \varepsilon + \mu - \Delta(i\omega_n) - \Sigma[\varepsilon, i\omega_n]}, \tag{6.24}$$

where we have slightly modified the notation in order to show that the denominator on the right-hand side depends on the site-energy ε both explicitly and implicitly through the self-energy $\Sigma[\varepsilon, i\omega_n]$. It is obvious that the above procedure reduces to the original DMFT in the clean case (cf. eqn (6.13)), but what does it reduce to in the non-interacting, disordered case?

It turns out that the treatment of disordered non-interacting systems obtained from these equations is equivalent to the so-called coherent potential approximations (CPAs) (Economou, 2006; Elliott *et al.*, 1974), which are known to become exact in infinite dimensions (Vlaming and Vollhardt, 1992). The CPA equations are usually obtained through a strategy that consists in replacing the effects of scattering off the exact disorder potential by an effective *average medium*. Formally, one writes the average Green's function in terms of an average-medium self-energy $\Sigma_{\text{AM}}(i\omega_n)$ (a frequency-dependent complex quantity)

$$\overline{G(\boldsymbol{k}, \boldsymbol{k}', i\omega_n)} = \frac{\delta_{\boldsymbol{k},\boldsymbol{k}'}}{i\omega_n - \varepsilon_{\boldsymbol{k}} + \mu - \Sigma_{\text{AM}}(i\omega_n)}, \tag{6.25}$$

where again $\varepsilon_{\boldsymbol{k}}$ is the clean non-interacting dispersion. $\Sigma_{\text{AM}}(i\omega_n)$ is calculated by replacing the average medium by the exact potential (as defined by the actual values of ε_j) at a *single* generic site (while keeping it at the other sites) and imposing that the difference between the exact and the average scattering t-matrices vanishes on the average (Economou, 2006; Elliott *et al.*, 1974). A similar effective-medium approach, incidentally, can be used to derive the DMFT of clean interacting systems (Georges *et al.*, 1996), so it is no surprise that one recovers CPA in this case. In the generic case of disordered interacting systems, $\Sigma_{\text{AM}}(i\omega_n)$ is obtained from the local part of the average Green's function

$$\overline{G_{jj}(i\omega_n)} = \int d\varepsilon \frac{P(\varepsilon)}{i\omega_n - \varepsilon + \mu - \Delta(i\omega_n) - \Sigma[\varepsilon, i\omega_n]}$$
$$= \frac{1}{N_s}\sum_{\boldsymbol{k}} \frac{1}{i\omega_n - \varepsilon_{\boldsymbol{k}} + \mu - \Sigma_{\mathrm{AM}}(i\omega_n)}. \tag{6.26}$$

One may wonder what is the form of the DMFT self-consistency for generic disordered interacting systems beyond the Bethe lattice case; in other words, the analogue of eqn (6.14). This is most easily done through the analogy with CPA. Once we have the local self-energy for every value of ε_j, $\Sigma_j(i\omega_n)$ for a given $\Delta(i\omega_n)$ (eqns (6.22) and (6.23)), we first find the average medium self-energy $\Sigma_{\mathrm{AM}}(i\omega_n)$ through eqn (6.26). We then note that the average local Green's function within CPA is also given by

$$\overline{G_{jj}(i\omega_n)} = \frac{1}{i\omega_n + \mu - \Delta(i\omega_n) - \Sigma_{\mathrm{AM}}(i\omega_n)}, \tag{6.27}$$

since this is what you get if you replace the actual scattering potential $\varepsilon_j + \Sigma_j(i\omega_n)$ in eqn (6.23) by the effective-medium self-energy $\Sigma_{\mathrm{AM}}(i\omega_n)$. Finally, from comparing eqns (6.26) and (6.27) we arrive at the desired self-consistency condition

$$\frac{1}{N_s}\sum_{\boldsymbol{k}} \frac{1}{i\omega_n - \varepsilon_{\boldsymbol{k}} + \mu - \Sigma_{\mathrm{AM}}(i\omega_n)} = \frac{1}{i\omega_n + \mu - \Delta(i\omega_n) - \Sigma_{\mathrm{AM}}(i\omega_n)}, \tag{6.28}$$

which would give an improved hybridization function $\Delta(i\omega_n)$ in an iterative procedure. We should note in passing that the average-medium self-energy and the average Green's function (eqn (6.25)) are the key ingredients in the calculation of the conductivity, which as mentioned before involves no vertex corrections within DMFT (Dobrosavljević and Kotliar, 1994).

It is important to know the limitations of this approach. The main one is its inability to describe the disorder-induced Anderson metal–insulator transition (Anderson, 1958). As the self-consistency condition makes quite apparent, the central order parameter of this mean-field theory is the average local Green's function, see eqns (6.21), (6.26), or (6.28). However, as explained by Anderson in the original 1958 paper (Anderson, 1958), the average local Green's function, which is non-critical and finite at the mobility edge, is unable to signal the phase transition between extended and localized states. Indeed, the spatial fluctuations of the local Green's function are so large that its *typical* value is far removed from the *average* one (see Fig. 6.1 (b) and (c)). Thus, DMFT cannot describe the Anderson localization transition and one needs to go beyond this approximation if Anderson localization effects are to be incorporated. It had long been known that CPA has no Anderson transition, so this should not come as a big surprise. We will show below, however, that one can in fact address the effects of localization while at the same time retaining the local description of all correlation effects.

6.1.5 Applications of the disordered DMFT

Early applications of the disordered DMFT scheme focused on the phase diagram of the disordered Hubbard model. In particular, the fate of the antiferromagnetic phase of the clean model when disorder is introduced was investigated (Singh *et al.*, 1998; Ulmke *et al.*, 1995). Appropriate incorporation of broken-symmetry phases, such as antiferromagnetism, requires generalizing the procedure of Section 6.1.4 through the introduction of sub-lattice structure and spin-dependent single-particle quantities, which is quite straightforward and will not be discussed here (Georges *et al.*, 1996). The main finding was a surprising enhancement of the ordering tendencies at weak disorder and strong interactions, which was attributed to a peculiar disorder-induced delocalization effect (Singh *et al.*, 1998; Ulmke *et al.*, 1995). More recently, the phase diagram of the paramagnetic Hubbard model (suitable for systems with a high degree of magnetic frustration) has been determined, showing a gradual suppression of the region of coexistence of metallic and insulating phases found in the clean case (Aguiar *et al.*, 2005).

Kondo disorder

A very attractive feature of the disordered DMFT approach is its ability to provide *full distributions of local quantities*, which in turn may have profound effects on the low-temperature behavior of physical systems. Consider for example the *ensemble* of effective actions in eqn (6.19), which, as explained before, can be viewed as an *ensemble* of Anderson single-impurity problems, with the same conduction electron bath (eqn (6.6)), but different impurity-site energies ε_j. As is well known, at sufficiently strong coupling U, a local magnetic moment can be stabilized at these impurity sites at high temperatures (Anderson, 1961). However, below an energy scale set by the Kondo temperature $T_{\mathrm{K}j}$, the moments are "quenched" by the conduction electrons and form a singlet bound state (or Kondo resonance) (Anderson, 1961; Anderson and Yuval, 1969; Hewson, 1993; Kondo, 1964; Nozières, 1974; Wilson, 1975; Yuval and Anderson, 1970). The dependence of $T_{\mathrm{K}j}$ on ε_j is given by

$$T_{\mathrm{K}j} \approx D \exp\left(-1/\rho_{\mathrm{F}} J_j\right), \tag{6.29}$$

where D and ρ_{F} are the conduction-electron half band-width and density of states at the Fermi level, respectively, and J_j is the local Kondo exchange coupling constant (Schrieffer and Wolff, 1966)

$$J_j = 2\left|V_{k_{\mathrm{F}}}\right|^2 \left[\frac{1}{|\varepsilon_j|} + \frac{1}{|\varepsilon_j + U|}\right]. \tag{6.30}$$

Therefore, because of the strong exponential dependence of the Kondo temperature on the local parameters, a distribution of site energies can give rise to a wide distribution of Kondo temperatures (Dobrosavljević *et al.*, 1992). As a consequence, depending on whether a specific site has $T_{\mathrm{K}j} < T$ or $T_{\mathrm{K}j} > T$, it will behave as a free spin in the former case or as a quenched inert impurity in the latter one, with significant effects on thermodynamic and transport properties (see Fig. 6.3).

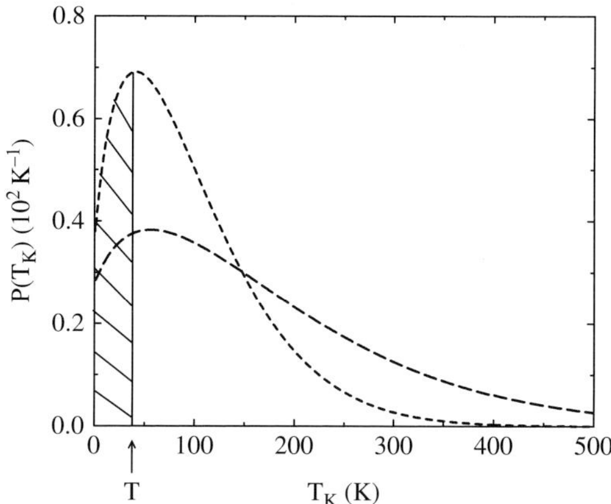

Fig. 6.3 Distribution of Kondo temperatures for the alloys UCu_4Pd (long dashed line) and $UCu_{3.5}Pd_{1.5}$ (short dashed line). Spins with T_K in the hatched area ($T > T_K$) behave as effectively free and lead to a singular thermodynamic response. From reference (Miranda *et al.*, 1996).

This scenario has received strong experimental support in the context of disordered heavy fermion systems. This was initially sparked by NMR experiments done on the Kondo alloy $UCu_{4-x}Pd_x$, whose broad temperature-dependent line-widths were analyzed in terms of a distribution of Kondo temperatures (Bernal *et al.*, 1995). The same distribution was then used to calculate the magnetic susceptibility and the specific heat, with very good agreement with the observed behavior (Bernal *et al.*, 1995). In this context, this phenomenology has been dubbed the *Kondo disorder model.* This was particularly striking because this system was among many intensively studied heavy fermion compounds (Stewart, 2001, 2006) whose properties are in apparent contradiction with Landau's theory of Fermi liquids (Landau, 1957*a*,*b*, 1959). In particular, the magnetic susceptibility showed an approximately logarithmic divergence with lowering temperatures, in contrast with the usual saturation to a constant found in weakly or even some strongly correlated Fermi liquid metals. The reason for the observed anomalous behavior was quite clear within the Kondo disorder model. Indeed, the distribution of Kondo temperatures needed to explain the NMR line-widths was so broad that $P(T_K) \approx P_0 = \text{const.}$ when $T_K < \Lambda$, where Λ is some low-energy scale of the distribution. In this case, no matter how low the temperature is, there are always a few unquenched spins left over with $T_K < T$ whose contribution to the susceptibility is Curie-like and large (Fig. 6.3)

$$\chi(T) \sim \frac{1}{T}. \tag{6.31}$$

Thus, by using a fairly accurate parametrization of the Kondo susceptibility (Wilson, 1975)

$$\chi_{\mathrm{Kondo}}(T) \sim \frac{1}{T + \alpha T_{\mathrm{K}}}, \tag{6.32}$$

one can immediately find that the bulk susceptibility obtained from an average over the local contributions calculated with the empirical $P(T_{\mathrm{K}})$ distribution is dominated by the low-T_{K} spins (with $T_{\mathrm{K}} \ll T \sim \Lambda$) and is logarithmically divergent,

$$\overline{\chi}(T) \sim \int \frac{P(T_{\mathrm{K}})}{T + \alpha T_{\mathrm{K}}} dT_{\mathrm{K}} \sim \int_0^{\Lambda} \frac{P_0}{T + \alpha T_{\mathrm{K}}} dT_{\mathrm{K}} \sim \ln\left(\frac{T_0}{T}\right), \tag{6.33}$$

where T_0 is a distribution-dependent constant.

Clearly, the Kondo disorder model found a natural setting within the DMFT approach to disordered systems, which put the phenomenology obtained from NMR on a firmer basis (Miranda *et al.*, 1996, 1997*a*,*b*). Because heavy fermion systems are characterized by a lattice of ions with incomplete f-shells, the most appropriate model Hamiltonian is a disordered Anderson lattice Hamiltonian

$$\begin{aligned} H_{\mathrm{And}} = &- \sum_{\langle ij \rangle, \sigma} t_{ij} c_{i\sigma}^{\dagger} c_{j\sigma} + \sum_{j,\sigma} (\varepsilon_j - \mu) c_{j\sigma}^{\dagger} c_{j\sigma} + \sum_{j,\sigma} (E_{fj} - \mu) f_{j\sigma}^{\dagger} f_{j\sigma} \\ &+ \sum_{j,\sigma} \left(V_j c_{j\sigma}^{\dagger} f_{j\sigma} + \mathrm{H.c.} \right) + U \sum_{j} f_{j\uparrow}^{\dagger} f_{j\uparrow} f_{j\downarrow}^{\dagger} f_{j\downarrow}, \end{aligned} \tag{6.34}$$

in usual notation, and in which we have in general assumed that both f- and c-site energies, E_{fj} and ε_j, as well as the local hybridizations V_j between them are random quantities, each with its own independent distribution. The effective local action in this case reads

$$\begin{aligned} S_{\mathrm{eff}}(j) = &\int_0^{\beta} d\tau \sum_{\sigma} \left[c_{j\sigma}^{\dagger}(\tau) (\partial_\tau + \varepsilon_j - \mu) c_{j\sigma}(\tau) + f_{j\sigma}^{\dagger}(\tau) (\partial_\tau + E_{fj} - \mu) f_{j\sigma}(\tau) \right] \\ &+ \int_0^{\beta} d\tau \int_0^{\beta} d\tau' \sum_{\sigma} \left[c_{j\sigma}^{\dagger}(\tau) \Delta(\tau - \tau') c_{j\sigma}(\tau') \right] \\ &+ \int_0^{\beta} d\tau \left[V_j \sum_{\sigma} c_{j\sigma}^{\dagger}(\tau) f_{j\sigma}(\tau) + U f_{j\uparrow}^{\dagger}(\tau) f_{j\uparrow}(\tau) f_{j\downarrow}^{\dagger}(\tau) f_{j\downarrow}(\tau) \right], \end{aligned} \tag{6.35}$$

and the self-consistency condition is analogous to the one in eqn (6.24)

$$\Delta(i\omega_n) = \tilde{t}^2 \overline{G_{jj}^{c}(i\omega_n)}, \tag{6.36}$$

where the local c-electron Green's function $G_{jj}^{c}(i\omega_n)$ is

$$\left[G_{jj}^{c}(i\omega_n) \right]^{-1} = i\omega_n - \varepsilon_j + \mu - \frac{V_j^2}{i\omega_n - E_{fj} - \Sigma_j(i\omega_n)} - \Delta(i\omega_n), \tag{6.37}$$

and the averaging procedure is performed over the random quantities ε_j, V_j and E_{fj}. Thus, T_{K} fluctuations can have several origins in general, as the local Kondo temperature is affected by ε_j, E_{fj} and V_j. Within DMFT, it was possible to better justify the *ad hoc* assumptions of the Kondo disorder model. In particular, one could quantify the validity of, and thus justify the approximation of, calculating the bulk susceptibility as an average over single-site contributions (Miranda *et al.*, 1996). In addition, good agreement was also found with the dynamic magnetic susceptibility obtained through neutron scattering experiments (Aronson *et al.*, 1995). Furthermore, going well beyond the simple Kondo disorder phenomenology, the DMFT approach is able to give direct information about transport properties. The following observations were found (Miranda *et al.*, 1996, 1997*a*,*b*).

- There is a strong interaction-induced renormalization of the disorder seen by the conduction electrons
- This, in turn, leads to a rapid suppression of the low-temperature Fermi-liquid coherence characteristic of clean heavy fermion materials, as a function of increasing disorder.
- Finally, when the quasiparticle coherence is completely destroyed and the distribution of Kondo temperatures develops a finite intercept in the limit of $T_{\mathrm{K}} \to 0$, the non-Fermi-liquid thermodynamics described above is accompanied by a non-Fermi-liquid *linear-in-T resistivity*

$$\rho(T) = \rho_0 - AT, \tag{6.38}$$

where $A > 0$. As in the case of the thermodynamic properties, the anomalous resistivity is also due to left-over low-T_{K} free spins, off which the conduction electrons scatter incoherently.

It was possible to verify that the self-consistency does not lead to a large disorder dependence of the hybridization function $\Delta(\tau)$. As a result, the distribution of Kondo temperatures is fairly sensitive to the *bare* distribution of random parameters ε_j, E_{fj} and V_j. Going beyond DMFT, as we will discuss later, one finds that this is an artifact of the approximations and, in general, self-consistency leads to a much more robust dependence on disorder.

Elastic and inelastic scattering in the disordered Hubbard model

The interplay between local correlation effects and transport is a striking feature, which is made almost obvious by the DMFT scheme. This has been demonstrated in studies of the disordered Hubbard model, eqn (6.18), (see Tanasković *et al.* (2003) and Aguiar *et al.* (2004)), as we now describe.

The study of Tanasković *et al.* (2003) was confined to $T = 0$ and thus addresses only the effects of elastic scattering. This was done using the Kotliar–Ruckenstein slave-boson mean-field theory as the impurity solver (Kotliar and Ruckenstein, 1986) (see Section 6.2.2 for further details). The relevant question is how interactions renormalize the scattering of quasiparticles by the disorder potential at the Fermi level (hence at $T = 0$). In Hartree–Fock theory, the renormalized disorder potential is determined by the self-consistently determined distribution of the electronic charge. This, in turn, is governed by the charge compressibility if the charge can be assumed to readjust itself to the disorder potential in a fashion dictated by linear response theory. For small U, the response is that of a good metal and allows for a flexible adjustment of the charge to the bare random potential, leading to a weakened renormalized disorder ("disorder screening") (Herbut, 2001). For strong interactions close to Mott localization, however, the charge compressibility is significantly reduced (the metal becomes increasingly less compressible) and Hartree–Fock theory predicts poor disorder screening.

It was shown in Tanasković *et al.* (2003) that indeed the efficient disorder screening predicted by Hartree–Fock theory is recovered by DMFT at weak interactions. However, it was found that another phenomenon intervenes and strong disorder screening does occur even as the system approaches the Mott transition. The reason why this happens is once again related to the peculiarities of the Kondo effect discussed above. Indeed, as several DMFT studies have shown (Georges *et al.*, 1996), the Mott transition is signaled by the disappearance of the metallic quasiparticles (Brinkman and Rice, 1970). The coherent nature of these quasiparticles exists only within a narrow energy range around the Fermi level, whose width is set by the Kondo temperature of the associated single-impurity problem: as $U \to U_c$, $T_K \to 0$. In the disordered case, there is a distribution of T_Ks, but all of them vanish at the transition. Now, the value assumed by the renormalized disorder potential on a given site is set by *the position of the local Kondo resonances* (within DMFT), which are known to be *strongly pinned* to the Fermi level (Hewson, 1993). Therefore, Kondo resonance pinning strongly reduces the bare-disorder fluctuations and screens the disorder rather effectively, leading to a correlation-induced suppression of the renormalized disorder, in sharp contrast to the simple Hartree–Fock prediction.

Kondo physics again comes in when one looks at inelastic scattering (Aguiar *et al.*, 2004). This was done by means of iterative perturbation theory (Georges and Kotliar, 1992; Kajueter and Kotliar, 1996; Zhang *et al.*, 1993) (see Section 6.3.2 for more details). Indeed, T_K also governs the temperature above which inelastic scattering dominates over elastic scattering. It is found that a gradual and mild temperature dependence of the resistivity is observed in the weakly correlated regime $U \approx W \ll D$, which is reasonably captured by Hartree–Fock theory. As interactions become of the order of (or larger than) the Fermi energy $U \approx W \gg D$, however, a sharper temperature dependence sets in. This is due to the strong suppression of the low-temperature scales of the associated Kondo impurity problems, which is not well described within Hartree–Fock theory.

Furthermore, varying the disorder strength W at strong interactions ($U \gg D$) leads to vastly different temperature dependences of the resistivity. Here, the distribution of Kondo temperatures defines the range over which inelastic processes become progressively more dominant. When $U \approx W$, a wide distribution of Kondo temperatures is generated and a rather slow growth of the resistivity with temperature is found. In contrast, in the cleaner case $W \ll U$, there is a much narrower distribution of T_Ks resulting in a sharp temperature dependence of the resistivity, typical of the onset of coherence in heavy fermion materials (Stewart, 1984) (see Fig. 6.4).

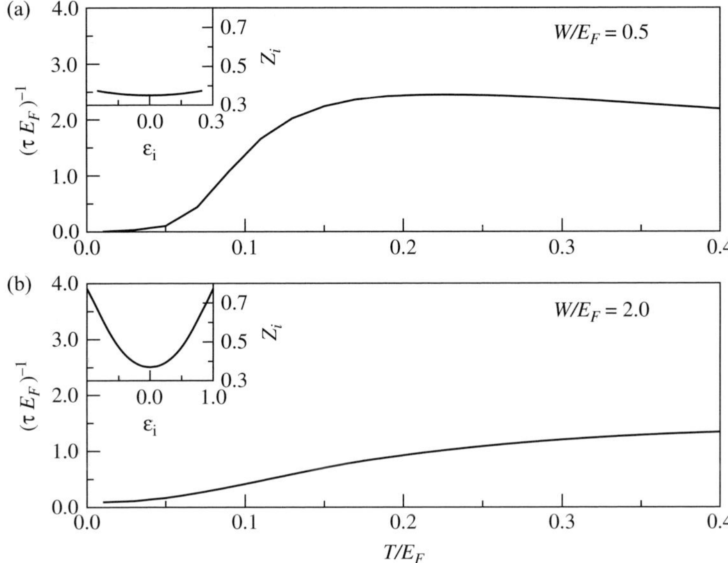

Fig. 6.4 Scattering rate $1/\tau$ as a function of temperature for the disordered Hubbard model within DMFT at $U = 2D \equiv 2E_{\mathrm{F}}$ for weak ($W = 0.5D$, (a)) and strong disorder ($W = 2D$, (b)). The insets show the dependence of the local Kondo temperature $\sim Z_i$ on the site energy ε_i. At weak disorder, a sharp distribution of Kondo temperatures gives rise to a steep rise, whereas stronger disorder leads to a broad distribution and a slow temperature dependence. From Aguiar *et al.* (2004).

6.2 Mott–Anderson transitions: typical medium theory

Although the dynamical mean-field theory described above is able to capture many features that are expected to be quite independent of its underlying assumptions (e.g. a distribution of local energy scales governing both thermodynamic and transport properties), it is clear that its limitations call for improvements at several points. In particular, it would be highly desirable to incorporate Anderson localization effects. As we have seen, these are conspicuously absent in the original DMFT, as the latter is essentially a mean-field theory whose order parameter is $\Delta(\omega)$, which in turn is determined by the average local Green's function (see eqns (6.24) or (6.27)–(6.28)). As Anderson localization originates precisely in the spatial fluctuations of this quantity, this is not enough. Physically, $\Delta(\omega)$ represents the *available electronic states* to which an electron can "jump" on its way out of a given lattice site. From Fermi's golden rule, the transition ("escape") rate to a neighboring site is proportional to the imaginary part of $\Delta(\omega)$. If this is zero at the Fermi energy, the electrons cannot hop out and are effectively localized. In a clean system, in which this quantity is the same at every site, this is a good order parameter for localization, as in the case of the clean Mott transition. In a highly disordered system, $\Delta(\omega)$ shows strong spatial fluctuations from site to site and its average value is not a good measure of the conducting properties. A "typical" site in an Anderson insulator will have a hybridization function $\Delta_i(\omega)$ with large gaps and a few isolated peaks, reflecting the nearby localized wavefunctions that have an overlap with it. The vanishing of its imaginary part signals the electron's inability to leave the site and is a good indicator of localized behavior. However, averaging over the whole sample washes out these gaps hiding the true insulating behavior. The discrepancy between the typical and average values of $\Delta_i(\omega)$ persists even on the metallic side, where $\Delta_{\text{typ}}(\omega)$ can be much smaller than $\overline{\Delta(\omega)}$.

Two alternative routes can be taken at this point. The ideal solution is to track the actual local hybridization or escape rate at each site. We will focus on this possibility in Section 6.3, where we analyze the so-called statistical dynamical mean-field theory. The other option is to focus on a simpler, yet meaningful measure of the escape rate, which, although incapable of incorporating the richness of the actual local realizations, does not "throw away the (localization) baby with the bath water." Here we should take as guidance the remark by Anderson that "no real atom is an average atom" (Anderson, 1978) and seek a more apt description of a "real atom." Indeed a good measure of the typical escape rate $\text{Im}\Delta_{\text{typ}}(\omega)$ is the *geometric average* $\exp\langle\ln[\text{Im}\Delta_i(\omega)]\rangle$. This has the great advantage of serving as an order parameter for the localization transition: indeed, in contrast to the algebraic average, the geometric average vanishes at the mobility edge (Anderson, 1958) (see also (Janssen, 1998; Mirlin, 2000; Schubert *et al.*, 2010)). We can then reason by analogy with the regular dynamical mean-field theory approach explained above and construct a self-consistent extension centered around this *typical* escape rate function. This theory has been dubbed the typical medium theory (TMT) (Dobrosavljević *et al.*, 2003*a*).

6.2.1 Formulation of the theory

We can proceed by analogy with the DMFT equations as explained in Section 6.1.4, see eqns (6.25)–6.28), to obtain the TMT. We again focus on a disordered Hubbard model, eqn (6.18), and imagine replacing the disordered medium by an effective *typical* medium described by a self-energy function $\Sigma_{\text{TMT}}(\omega)$. How do we determine this self-energy? Focusing on a generic site j, it is described by an effective action that has the same form as eqn (6.19). The hybridization function $\Delta(\omega)$ is still left unspecified at this point, but we envisage that it will reflect a "typical" site as opposed to an "average" one, so we set $\Delta(\omega) = \Delta_{\text{typ}}(\omega)$ in eqn (6.19). The local Green's function is still defined as in eqns (6.22) and (6.23). The local density of states is given by the imaginary part of the local Green's function

$$\rho_j(\omega) = \frac{1}{\pi}\text{Im}G_{jj}(\omega - i\delta). \tag{6.39}$$

The *typical* local density of states can be defined through its geometric average

$$\rho_{\text{typ}}(\omega) = \exp\left[\int d\varepsilon_j P(\varepsilon_j)\ln\rho_j(\omega)\right]. \tag{6.40}$$

Note that we have reverted to the real frequency axis because we need a positive-definite quantity in order to be able to define a geometric average. To preserve causality, the typical local Green's function is obtained through the usual Hilbert transform

$$G_{\text{typ}}(\omega) = \int_{-\infty}^{\infty} d\omega' \frac{\rho_{\text{typ}}(\omega')}{\omega - \omega'}. \tag{6.41}$$

Note the analogous *average* quantity in the second equality of eqn (6.26), which appears in DMFT. The typical medium self-energy $\Sigma_{\text{TMT}}(\omega)$ is then defined through the inversion of the following equation

$$G_{\text{typ}}(\omega) = \frac{1}{N_s} \sum_{\boldsymbol{k}} \frac{1}{\omega - \varepsilon_{\boldsymbol{k}} + \mu - \Sigma_{\text{TMT}}(\omega)}, \tag{6.42}$$

which is the analogue of the first equality in eqn (6.26). Finally, the loop is closed by setting

$$G_{\text{typ}}(\omega) = \frac{1}{\omega + \mu - \Delta_{\text{typ}}(\omega) - \Sigma_{\text{TMT}}(\omega)}, \tag{6.43}$$

which can be used in an iterative scheme to generate an updated hybridization function $\Delta_{\text{typ}}(\omega)$ and is the analogue of eqn (6.28). It becomes clear that the *crucial difference between TMT and DMFT is the replacement of the average local Green's function by the typical one* (see Fig. 6.1 (b) and (c)). This has been shown to capture even quantitative features of the Anderson localization transition, as we will discuss below. For a more comprehensive review, see Dobrosavljević (2010).

6.2.2 Applications of typical medium theory

We will now describe the most important results obtained from the TMT theory of disordered systems. We will focus first on the non-interacting case and then the disordered Hubbard model. We should also mention a study of the Falicov–Kimball model within TMT in Byczuk (2005).

Critical behavior in the non-interacting case

As a first test of its usefulness, the TMT approach was first applied to the non-interacting three-dimensional case, eqn (6.18) with $U = 0$ (Dobrosavljević *et al.*, 2003*a*). In fact, the results of applying the TMT equations to a cubic lattice were directly compared to a numerical diagonalization of the Hamiltonian. In particular, the numerically determined arithmetic and geometric averages of the local density of states at the Fermi level, $\rho_{\text{av}}(\omega = 0)$ and $\rho_{\text{geo}}(\omega = 0)$, were compared to the results of CPA (Economou, 2006; Elliott *et al.*, 1974) and of TMT (see Fig. 6.5). A remarkably accurate agreement between $\rho_{\text{av}}(\omega = 0)$ and CPA was observed. As is known, this quantity is not critical at the Anderson transition. On the other hand, the geometric average of the local density of states does vanish at a critical disorder strength W_c. A reasonably good agreement between the numerically determined $\rho_{\text{geo}}(\omega = 0)$ and

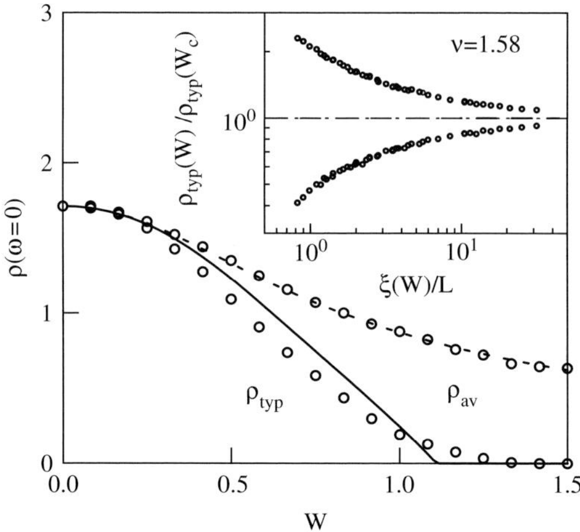

Fig. 6.5 Comparison between the TMT and CPA with numerical results (circles) on the non-interacting three-dimensional tight-binding model with diagonal disorder. The average local density of states at the band center $\rho_{\rm av}$ is remarkably well captured by the coherent potential approximation (dashed line) but is *not critical* at the Anderson localization transition at $W_{\rm c} \approx 1.38$ (in units of the clean bandwidth). The geometrically averaged local density of states $\rho_{\rm typ}$, on the other hand, vanishes at the transition and is well described by the TMT (full line), except for the detailed critical behavior. The inset shows the numerics for the critical behavior of $\rho_{\rm typ}$ as a function of the correlation length $\xi\,(W)$, where $\xi\,(W) \sim |W - W_{\rm c}|^{\nu}$, $\nu \approx 1.58$. From Dobrosavljević *et al.* (2003*a*).

TMT is found for most values of the disorder strength, even though TMT misses the correct critical behavior. This is not too surprising as TMT has the flavor of a mean-field theory. Physically, it is clear the $\rho_{\rm geo}\,(\omega)$ should be viewed as the density of extended states of system, decreasing with increasing disorder and eventually vanishing altogether for sufficiently large randomness. Thus the spectral weight described by $\rho_{\rm geo}\,(\omega)$ is not conserved.

In fact, further insight into the critical behavior of TMT can be achieved analytically (Dobrosavljević *et al.*, 2003*a*). By assuming an elliptic density of states for the clean lattice, $\rho_0\,(\omega) = \frac{2}{\pi D}\sqrt{1 - \left(\frac{\omega}{D}\right)^2}$, it can be proved that in the critical region $W \to W_{\rm c} = eD$, the typical density of states (given, as usual, by the geometric average) assumes a universal form

$$\rho_{\rm typ}\,(\omega, W) \approx \rho\,(\omega = 0, W)\, f\left[\omega/\omega_0\,(W)\right], \tag{6.44}$$

where

$$\rho\,(\omega = 0, W) = \left(\frac{4}{\pi}\right)^2 (W_{\rm c} - W)\,, \tag{6.45}$$

the frequency scale

$$\omega_0(W) = \sqrt{\frac{e}{4}(W_c - W)}, \tag{6.46}$$

and the scaling function has a simple form $f(x) = 1 - x^2$. Note that eqn (6.45) gives an order-parameter critical exponent $\beta_{\text{TMT}} = 1$, which should be compared to the accepted value in three dimensions $\beta_{3D} \approx 1.58$ (Slevin and Ohtsuki, 1999).

The disordered Hubbard model at half filling

We now direct our attention to disordered *interacting* systems. The TMT was applied to the disordered Hubbard model (Aguiar *et al.*, 2009; Byczuk *et al.*, 2005). The phase diagram of the paramagnetic half-filled case at $T = 0$ was obtained with two different impurity solvers: Wilson's numerical renormalization group (NRG) (Wilson, 1975) was used in Byczuk *et al.* (2005) and the Kotliar–Ruckenstein slave-boson mean-field theory (SB4) (Kotliar and Ruckenstein, 1986) was applied in Aguiar *et al.* (2009). The results obtained largely agree with each other but there are some small discrepancies in the details. Essentially, three different phases are observed: a disordered correlated metal phase, characterized by $\rho_c \equiv \rho_{\text{typ}}(\omega = 0) \neq 0$, a Mott-like insulating phase (which we will call simply a Mott insulator), and an Anderson-like insulating phase (for which we will use the name "Mott–Anderson insulator"), the latter two phases having $\rho_c = 0$. We will discuss each set of results and then the discrepancies. For reviews, see Dobrosavljević (2010) and Byczuk *et al.* (2009*b*).

For small values of disorder and interaction, both approaches find that the system is metallic, although this is not too surprising. There is also agreement on the fact that, for a fixed small disorder strength, the order parameter ρ_c increases with increasing interaction ($U > W$). This is a reflection of disorder screening by interactions (see Section 6.1.5). Eventually, at a critical value of the interaction strength $U_c(W)$, the order parameter ρ_c exhibits a finite jump and drops to zero, signaling a metal–Mott-insulator phase transition. Furthermore, both methods agree that, for a fixed small value of U, as one increases the disorder ($W > U$) ρ_c decreases monotonically, indicating that the spectral weight due to extended states is decreasing. At the critical value $W_c(U)$, ρ_c vanishes and the system enters a Mott–Anderson insulating phase.

The differences in the results of the two approaches are the following. The NRG-based TMT (Byczuk *et al.*, 2005) predicts the metal–Mott-insulator transition for $U > W$ to be first order in character, with typical hysteretic behavior: the metallic solution is locally stable for $U < U_{c2}(W)$ and the Mott-insulating solution is locally stable for $U > U_{c1}(W)$, where $U_{c1}(W) < U_{c2}(W)$. In the coexistence region $U_{c1}(W) < U < U_{c2}(W)$, both solutions can be stabilized (although only one is a true global energy minimum at each U). The SB4-based TMT (Aguiar *et al.*, 2009), on the other hand, predicts no hysteresis. To understand this, it should be mentioned that SB4 is a good description of the low-energy properties of the impurity spectral function, although it misses the higher-energy features that give rise to the Hubbard bands in the lattice. An important ingredient of this description is the quasiparticle weight $Z_i \equiv Z(\varepsilon_i)$, well known from Fermi liquid theory, which in the impurity problem (eqn (6.19)) determines the width of the Kondo resonance (essentially the Kondo temperature). Formally, it appears in the local Green's function at site i

$$G_{ii}(i\omega_n) = \frac{Z_i}{i\omega_n - \tilde{\varepsilon}_i + \mu - Z_i \Delta_{\text{typ}}(i\omega_n)}, \tag{6.47}$$

where $\tilde{\varepsilon}_i$ is the renormalized local site energy, which gives the position of the resonance. In the clean lattice case, Z vanishes continuously as the interaction strength is tuned to its critical value $U \to U_{\text{c}}$, $Z \sim U_{\text{c}} - U \to 0$, signaling the Mott localization of the itinerant electrons, which become localized magnetic moments. The continuous nature of the transition in the clean case, with no accompanying hysteresis, survives the introduction of disorder within the SB4-based TMT. In fact, in this approach all the Z_i vanish at a unique $U_{\text{c}}(W)$. It should be stressed that in both approaches ρ_{c} is discontinuous at the transition, a fact that can be ascribed (at least for small disorder) to the observed perfect screening of disorder ($\rho_{\text{av}}(\omega = 0) \to \rho_{\text{geo}}(\omega = 0)$ as $U \to U_{\text{c}}(W)$) and the pinning of the clean density of states at the Fermi level to its non-interacting value (see Section 6.1.5).

Moreover, the results obtained with the NRG impurity solver indicate that the transition in the region $W > U$ is such that $\rho_{\text{c}} \to 0$ continuously as $W \to W_{\text{c}}(U)$ (Byczuk *et al.*, 2005). This is in contrast to the SB4-based approach, which finds that ρ_{c} exhibits a discontinuous jump to zero at $W_{\text{c}}(U)$ (Aguiar *et al.*, 2009). In fact, the latter method brings out an important ingredient that significantly enhances the physical understanding of the TMT approach to this problem. This is achieved by tracking the behavior of the quasiparticle weights Z_i. For a given fully-converged hybridization function $\Delta_{\text{typ}}(i\omega_n)$, the ensemble of impurity problems is characterized by the function $Z(\varepsilon_i)$ for $|\varepsilon_i| < W/2$ (a uniform disorder distribution is assumed). This function has the property that, as the phase transition is approached, $Z(\varepsilon_i) \to 0$ for $|\varepsilon_i| < U/2$, whereas $Z(\varepsilon_i) \to 1$ if $|\varepsilon_i| > U/2$ (see Fig. 6.6) (Aguiar *et al.*, 2006). In the SB4 language (Kotliar and Ruckenstein, 1986), $Z \to 0$ implies a singly occupied site with a localized magnetic moment, whereas $Z \to 1$ means either a doubly occupied or a singly occupied site, either of which is essentially non-interacting. Thus, in the region $W > U$, a fraction of the sites—those with $|\varepsilon_i| < U/2$—experience Mott localization, while those sites with $|\varepsilon_i| > U/2$ undergo Anderson localization. The picture that emerges is that of a *spatially inhomogeneous system*, composed of Mott-localized droplets intermingled with Anderson-insulating regions. This situation has been dubbed a "site-selective Mott transition" (Aguiar *et al.*, 2009). Analytical insight into the SB4-results can be brought to bear in order to show that in that approach any finite U renders the vanishing of ρ_{c} *discontinuous*, in sharp contrast to the non-interacting case (Dobrosavljević *et al.*, 2003*a*).

Finally, in the intermediate region of $W \approx U$, it was suggested, based on the NRG results, that there might be a crossover from a metal to a disordered Mott insulator (Byczuk *et al.*, 2005). However, since their results clearly show regions where $\rho_{\text{c}} \neq 0$ and regions where $\rho_{\text{c}} = 0$, we believe the correct interpretation is to identify the former as metallic and the latter as insulating. Besides, this is expected from the

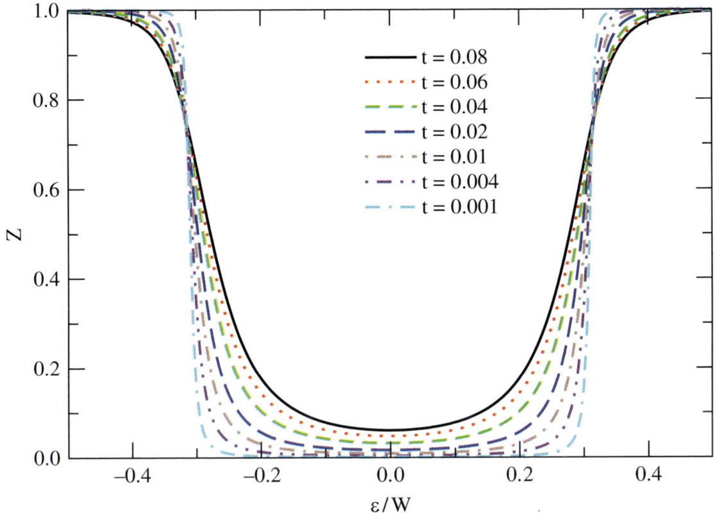

Fig. 6.6 $Z(\varepsilon_i)$ showing the two-fluid behavior of the Mott–Anderson transition within TMT. As the transition is approached ($t \to 0$) for $W = 2.8 > U = 1.75$, sites with $|\varepsilon_i| < U/2 \approx 0.31W$ represent Mott-insulating regions with $Z(\varepsilon_i) \to 0$, whereas sites with $|\varepsilon_i| > U/2$ are Anderson localized with $Z(\varepsilon_i) \to 1$. From Aguiar *et al.* (2006).

sharp distinction between an insulator and a metal at zero temperature. Having said that, however, it is clear that the nature of the transition in this problematic region certainly deserves a further more detailed investigation.

The disordered Hubbard model was also studied within TMT by allowing antiferromagnetic order on a bipartite lattice (Byczuk *et al.*, 2009*a*). The phase diagram at zero temperature as a function of disorder and interactions was determined, with the identification of both paramagnetic and antiferromagnetic metallic phases (for finite disorder only), an antiferromagnetic Mott-insulating phase, and a paramagnetic Anderson-insulating phase at large disorder strength.

6.3 Mott–Anderson transitions: statistical DMFT

As outlined above, the most natural and accurate extension of the DMFT philosophy that can incorporate Anderson localization effects is one which replaces the algebraic average of the hybridization function $\overline{\Delta(\omega)}$ (as in the original DMFT) or its typical value/geometric average $\Delta_{\text{typ}}(\omega)$ (as in the TMT), by the actual realizations of $\Delta_j(\omega)$ at each site (Dobrosavljević and Kotliar, 1997, 1998) (see Fig. 6.1 (b) and (c)). As is to be expected, the complexity of the equations increases considerably and one has to rely heavily on numerical computations. However, many insights have been obtained regarding the systems analyzed. Besides, a much larger degree of universality of the distributions is observed when compared with the much more “rigid” approaches of DMFT or TMT.

6.3.1 Formulation of the theory

Let us examine how the statDMFT works. We begin with the by now familiar disordered Hubbard model of eqn (6.18) and focus, as usual, on the dynamics of a given site j dictated by an effective action (we now revert back to imaginary time)

$$\begin{aligned} S_{\text{eff}}(j) = & \sum_{\sigma} \int_0^{\beta} d\tau c_{j\sigma}^{\dagger}(\tau)(\partial_{\tau} + \varepsilon_j - \mu) c_{j\sigma}(\tau) \\ & + \sum_{\sigma} \int_0^{\beta} d\tau \int_0^{\beta} d\tau' c_{j\sigma}^{\dagger}(\tau) \Delta_j(\tau - \tau') c_{j\sigma}(\tau') \\ & + U \int_0^{\beta} d\tau n_{j\uparrow}(\tau) n_{j\downarrow}(\tau). \end{aligned} \tag{6.48}$$

Notice the crucial difference now: the hybridization function $\Delta_j(i\omega_n)$ is now *site-dependent*. Each site, besides having a different energy ε_j, also "sees" a different local environment $\Delta_j(i\omega_n)$. The local dynamics is again encoded in a site-dependent self-energy $\Sigma_j(i\omega_n)$, obtained from the local Green's function as before, see eqns (6.22) and (6.23). It is important to note that this description assumes a *diagonal* (albeit site-dependent) self-energy function, adhering to the generic philosophy of incorporating only *local* interaction effects.

Unlike the previous DMFT or TMT approaches, statDMFT does not try to mimic this self-energy function $\Sigma_j(i\omega_n)$ through any type of effective medium. Instead, we choose to include its full spatial fluctuations. We do so by appealing to the physical picture of the self-energy as a *shift of the local site energy*, albeit a complex, frequency-dependent one: $\varepsilon_j \to \varepsilon_j + \Sigma_j(i\omega_n)$. Single-particle propagation can thus be viewed as described by the effective resolvent

$$\widehat{G}(i\omega_n) = \left[i\omega_n \widehat{1} - \widehat{t} - \widehat{\varepsilon} - \widehat{\Sigma}(i\omega_n)\right]^{-1}, \tag{6.49}$$

where $\widehat{1}$ is the identity operator, $\widehat{t}$ and $\widehat{\varepsilon}$ are, respectively, the hopping and site-energy terms (first and second terms of the Hamiltonian (6.18)), and the matrix elements of the self-energy operator $\widehat{\Sigma}(i\omega_n)$ are given in the site basis as

$$\left\langle i \left| \widehat{\Sigma}(i\omega_n) \right| j \right\rangle = \Sigma_j(i\omega_n) \delta_{ij}. \tag{6.50}$$

Although any matrix element of the resolvent (6.49), both intra- and inter-site, can in principle be calculated (which is important, for example, for a Landauer-type calculation of the conductivity), the self-consistency requires only the diagonal part, related to the local Green's function

$$\left\langle j \left| \widehat{G}(i\omega_n) \right| j \right\rangle = G_{jj}(i\omega_n) = \frac{1}{i\omega_n - \varepsilon_j - \Delta_j(i\omega_n) - \Sigma_j(i\omega_n)}. \tag{6.51}$$

This last equation closes the self-consistency loop by providing, in an iterative scheme, an updated hybridization function *for each site* $\Delta_j(i\omega_n)$. We summarize the self-consistency loop for completeness:

1. For a given realization of disorder, ε_j, start from a set of "initial trial" hybridization functions $\Delta_j(i\omega_n)$, one for each site.
2. For the set of effective actions (6.48), calculate the local Green's function $G_{jj}(i\omega_n)$ (eqn (6.22)) and the local self-energy $\Sigma_j(i\omega_n)$ (eqn (6.23)) for each site.
3. Invert the matrix resolvent (6.49) and get its diagonal elements $G_{jj}(i\omega_n)$.
4. Obtain an updated set of hybridization functions $\Delta_j(i\omega_n)$ by equating these diagonal elements to the expression of the local Green's functions, eqn (6.51).

In general, the set of eqns (6.48)–(6.51) forms the so-called statistical dynamical mean-field theory of disordered correlated electron systems. It should be noted that, besides the challenge of solving the *ensemble* of impurity problems represented by eqn (6.48), eqn (6.51) poses the numerical problem of inversion of a complex matrix for each value of the frequency, which can be very time-consuming. The pay-off is a description which incorporates all Anderson localization effects. Indeed, when interactions are turned off, the theory becomes exact since, for example, eqn (6.49) becomes the exact single-particle Green's function (from which transport properties can be obtained with the Landauer formalism). In the absence of randomness, we recover, of course, the DMFT equations. In the presence of both disorder and interactions, this is the optimal theory of disordered interacting lattice fermions that includes only *local* correlation effects.

6.3.2 Early implementations of statDMFT: the Bethe lattice

It had long been known that the Bethe lattice (or "Cayley tree") leads to considerable simplifications of the treatment of non-interacting disordered systems. For example, the so-called self-consistent theory of localization of Abou-Chacra, Anderson and Thouless (Abou-Chacra *et al.*, 1973) becomes exact on a Bethe lattice. This is so because the local Green's function with a neighboring site removed, eqn (6.4), satisfies a single compact stochastic equation in that lattice, which allows for quite an efficient analysis, both analytically and numerically. It was shown, for example, that for a coordination number $z > 2$, there is always an Anderson transition at a non-zero critical disorder strength W_c. This transition has been extensively studied (Mirlin and Fyodorov, 1991) and is regarded as a large-dimensionality limit of the Anderson transition. Although it has an anomalous critical behavior, with an exponential rather than a power-law dependence, this description has been very fruitful, especially if one is interested in the non-critical region.

Given the complexity of the full statDMFT equations, this suggested that the preliminary investigations could be carried out on a Bethe lattice. Here, it is appropriate to comment on what is, in our view, a misunderstanding of the conceptual basis of the statDMFT approach. It has been stated (Semmler *et al.*, 2010*b*) that there is somehow a *conceptual* difference between the statDMFT as applied to the Bethe lattice and the statDMFT used to analyze realistic lattices, like the square or cubic ones (for these, see Section 6.3.3). The misunderstanding comes from assuming that, whereas on a realistic lattice one has a fixed disorder realization which is then solved by statDMFT, therefore defining a *deterministic* problem, on the infinite Bethe lattice one does not deal with fixed-disorder realizations but with distributions, and the approach becomes *non-deterministic* and "statistical" (we realize the origin of the misunderstanding may have been the use of this word in the name of the method (Dobrosavljević and Kotliar, 1998)). This distinction is unfounded if one realizes that on *any* infinite lattice with a single fixed-disorder realization (with random, spatially uncorrelated, parameters), the infinite values of local quantities (such as the local density of states) on each lattice site give rise to a *statistical distribution of local quantities* (assuming the lattice-translational invariance of the distributions; see Lifshits *et al.* (1988) for a careful discussion). Thus when one solves a given Hamiltonian with statDMFT on an infinite Bethe lattice, one is actually solving, in practice as well as in principle, for a single fixed-disorder realization. Each iteration of the method outlined in Abou-Chacra *et al.* (1973) corresponds to "going outwards" on the branches of a fixed-disorder realization of an infinite Bethe lattice, while at the same time accumulating random variables and building a histogram of local quantities. By the same token, if one could solve the statDMFT equations for a single disorder realization on an *infinite* realistic (say, square) lattice, each one of its sites would contribute one random variable for a histogram of the same local quantities (which would be, obviously, different from the ones obtained from the differently connected Bethe lattice). In practice, of course, one solves many disorder realizations of *finite*, hopefully large, realistic lattices in order to generate distributions with good statistics. However, fundamentally there is no *conceptual* difference between the statDMFT solutions on the two types of lattice.

The Mott–Anderson transition

In Dobrosavljević and Kotliar (1997; 1998), the first implementation of the statDMFT theory as applied to the Mott–Anderson transition described by the disordered Hubbard model was completed. This was done by using a Fermi-liquid parametrization of the associated zero-temperature impurity problems, namely, the infinite-U slave-boson mean-field theory (Coleman, 1987; Read and Newns, 1983). Like its Kotliar–Ruckenstein finite-U counterpart (Kotliar and Ruckenstein, 1986), this theory captures the low-energy sector and is known to give a quantitatively good description of this limit. This Fermi-liquid description is encapsulated in just two parameters: the quasiparticle weight Z_j and the effective level energy (or Kondo resonance location) $\tilde{\varepsilon}_j$. The local self-energy is written as

$$\Sigma_j\left(i\omega_n\right) = \left(1 - Z_j^{-1}\right) i\omega_n + \frac{\tilde{\varepsilon}_j}{Z_j} - \varepsilon_j + \mu, \tag{6.52}$$

leading to a local Green's function

$$G_j\left(i\omega_n\right) = \frac{Z_j}{i\omega_n - \tilde{\varepsilon}_j - Z_j\Delta_j\left(i\omega_n\right)}. \tag{6.53}$$

The results of Dobrosavljević and Kotliar (1997; 1998) reveal that, as in the non-interacting case, the Mott–Anderson transition can be identified by the vanishing of the typical local density of states (as described by the geometric average) at a certain critical disorder strength. Interestingly, in contrast to non-interacting electrons, the critical behavior is conventional (power-law) and the average density of states is *divergent.* There is at present no good understanding of this divergence.

Furthermore, a great opportunity afforded by statDMFT is the ability to investigate *distributions* of local quantities. By tracking the distribution of quasiparticle weights, it was found that it broadens considerably with increasing disorder, showing a characteristic *power-law form* at large randomness

$$P(Z) \sim Z^{\alpha-1}, \tag{6.54}$$

with the exponent $\alpha = \alpha(W)$ a smooth function of disorder. It should be remembered that Z_j determines the local Kondo temperature $T_{\mathrm{K}j} \sim Z_j$ and thus governs the local contribution to thermodynamic quantities such as the magnetic susceptibility and specific heat (see Section 6.1.5). In fact, by averaging over this distribution of Zs as in eqn (6.33), one finds power-law dependences for these quantities as well

$$\chi(T) \sim C(T)/T \sim T^{\alpha-1}. \tag{6.55}$$

As disorder increases, α decreases and eventually becomes smaller than one well before the Mott–Anderson transition. When this happens, the thermodynamic response becomes singular and non-Fermi-liquid-like

$$\chi(T \to 0) \to \infty. \tag{6.56}$$

Many different correlated systems have indeed been shown to exhibit this form of anomalous behavior (Stewart, 2001), with non-universal exponents α. The presence of non-universal, smoothly varying exponents characterizing divergences in physical quantities is reminiscent of a large class of disordered systems and is usually dubbed a quantum Griffiths phase (for reviews, see Miranda and Dobrosavljević (2005) and Vojta (2006)), by analogy with a similar situation in classical systems first analyzed by Griffiths (1969). Most other known examples of quantum Griffiths phases had been found in the vicinity of magnetic phase transitions in the presence of disorder, most notably in insulating magnets (Fisher, 1992, 1995; Guo *et al.*, 1996; Motrunich *et al.*, 2001; Pich *et al.*, 1998), but also in metallic systems (Castro Neto *et al.*, 1998; Castro Neto and Jones, 2000; de Andrade *et al.*, 1998; Millis *et al.*, 2002). Here, however, the characteristic power laws are found in the vicinity of the paramagnetic Anderson metal–insulator transition and hence the name "electronic Griffiths phase" was adopted. This anomalous behavior is apparently not at all dependent on the particular details of the disordered Hubbard model. Very similar power-law distributions of Kondo temperatures were also found in Bethe lattice implementations of statDMFT for the disordered Anderson lattice Hamiltonian (6.34) (see Fig. 6.7) (Aguiar *et al.*, 2003; Miranda and Dobrosavljević, 1999, 2001).

It should be remembered that the forms of the distributions of Kondo temperatures obtained within DMFT were strongly dependent on the shape of the bare distributions

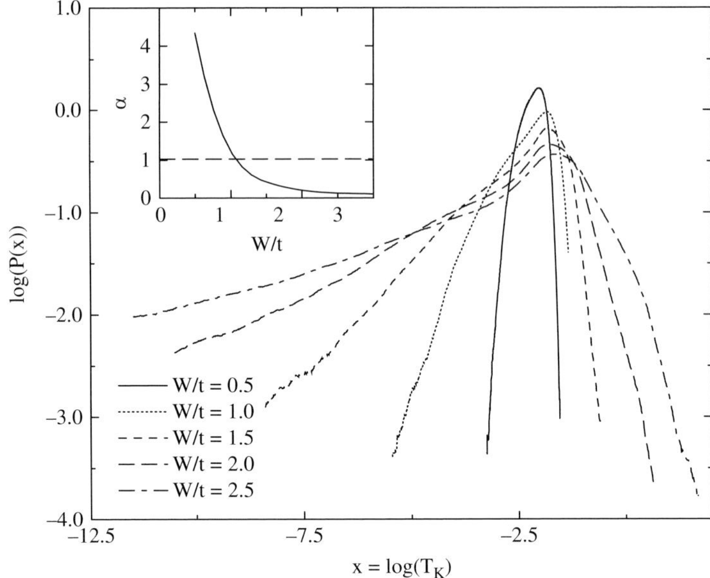

Fig. 6.7 Power-law distributions of Kondo temperatures in a disordered Anderson lattice model for different levels of disorder: the linear behavior for small values of $\log T_{\rm K}$ implies a power-law $P\left(T_{\rm K}\right) \sim T_{\rm K}^{\alpha-1}$. The inset shows the exponent α as a function of disoder W. From Miranda and Dobrosavljević (2001).

of parameters. This is in sharp contrast to the ubiquitous power-law distributions found in the statDMFT approach. Thus, although the exponent α is disorder-dependent and non-universal, the power-law *shape* is quite independent of whether the bare parameters are given by, say, uniform, Gaussian or binary distributions (Aguiar *et al.*, 2003). This is again easy to understand if we note that the local Kondo temperature depends exponentially on the local density of states at the Fermi level (and less strongly on its value at higher energies). Now, due to the extended nature of the electronic wavefunctions in metallic systems, the density of states at one site is influenced by spatial fluctuations at very distant sites and thus samples a great number of local environments. The resulting distributions of local quantities thus reflect this long-distance sampling.

In order to understand why this effect leads *specifically* to a power law, an effective model was proposed by Tanasković *et al.* (2004). The effective model consisted of a disordered Anderson lattice model with Gaussian-distributed conduction-electron disorder (ε_j in eqn (6.34))

$$P\left(\varepsilon_j\right) = \frac{1}{\sqrt{2\pi}W} \exp\left(-\varepsilon_j^2/2W^2\right), \tag{6.57}$$

treated within DMFT. In DMFT, the hybridization function $\Delta(i\omega_n)$ is *site-independent* and the Kondo temperature distribution is solely determined by fluctuations of ε_j. It can then be shown that (Tanasković *et al.*, 2004)

$$T_{Kj} = T_K^0 e^{-\lambda \varepsilon_j^2}, \tag{6.58}$$

where T_K^0 is the Kondo temperature at $\varepsilon_j = 0$ and λ is determined by other model parameters but does not depend on ε_j. It is easy to show from eqns (6.57) and (6.58) that

$$P(T_K) = P[\varepsilon_j(T_K)] \left| \frac{d\varepsilon_j}{dT_K} \right| \sim T_K^{\alpha-1}, \tag{6.59}$$

which is precisely the power-law distribution of Kondo temperatures found generically *within statDMFT treatments*. We note, in passing, that this kind of argument is generic to all known quantum Griffiths phases: the relevant energy scales are exponentially suppressed by a certain random parameter (see eqn (6.58)), whose probability is in turn also exponentially small (see eqn (6.57)) (Miranda and Dobrosavljević, 2005; Vojta, 2006).

What is the relation between these results, obtained within DMFT, and the power laws observed in the applications of statDMFT? In statDMFT, the single (average) hybridization function of DMFT gets replaced by a strongly fluctuating distribution of local hybridizations $\Delta_j(i\omega_n)$. The imaginary part of each of these functions describes the available density of states for Kondo screening at site j and enters the expression for the local Kondo temperature much as ρ_F does in eqn (6.29). From the central-limit theorem, fluctuations of the available densities of states around the mean value are generically Gaussian for weak and intermediate disorder and lead to a power-law distribution of T_Ks in a fashion quite similar to the effective model. Indeed, detailed calculations showed that the effective model is quite accurate when compared with full statDMFT results (Tanasković *et al.*, 2004). Crucially, these arguments can be used to show that the non-Fermi liquid behavior already occurs at quite moderate values of disorder and *strictly precedes* the Anderson metal–insulator transition. This elucidates then the microscopic origin of the electronic Griffiths phase.

Iterative perturbation theory as impurity solver

Most of the early statDMFT results on the Bethe lattice were obtained through the use of the infinite-U slave-boson mean-field theory (Coleman, 1987; Read and Newns, 1983) as impurity solver. These are good descriptions of the low-energy coherent Fermi-liquid part of the impurity spectrum but fail to account for inelastic scattering at low energies as well as higher-energy incoherent features such as upper and lower Hubbard bands. A technique that is able to incorporate there features is the so-called iterative perturbation theory (Georges and Kotliar, 1992; Kajueter and Kotliar, 1996; Zhang *et al.*, 1993). The iterative perturbation theory approach also lends itself more easily to an analysis of the temperature dependence of physical quantities. It has been used to analyze the disordered Anderson lattice model (Aguiar *et al.*, 2003) as well as the disordered Hubbard model (Semmler *et al.*, 2010*a*). It suffers from the disadvantage of

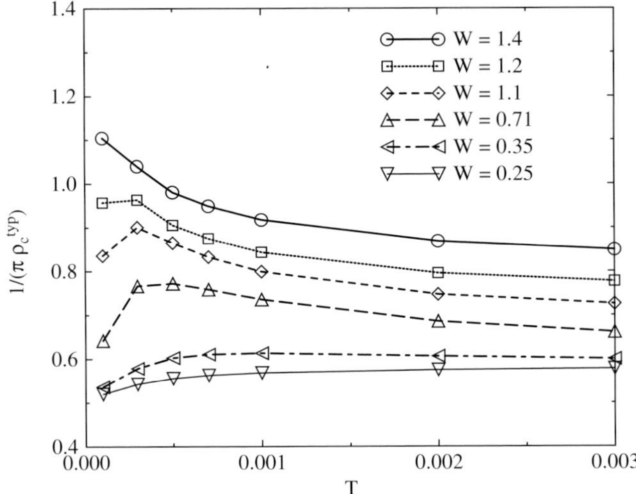

Fig. 6.8 Temperature dependence of the inverse typical conduction electron density of states of the disordered Anderson lattice model for different levels of disorder. This "fan-like" family of curves is generic to many disordered, strongly correlated systems ("Mooij correlations"). From Aguiar *et al.* (2003).

not being able to capture the correct exponential dependence of the low-energy Kondo scale, see eqns (6.29) and (6.58), leaving out, therefore, the possibility of characterizing Griffiths phase behavior (Section 6.3.2).

In the case of the disordered Anderson lattice, one of the interesting findings was the interplay between elastic scattering off the disorder potential and inelastic electron-electron scattering (Aguiar *et al.*, 2003). If one uses the inverse of the typical density of states at the Fermi level as a rough guide to the resistivity (a direct calculation of the resistivity was numerically prohibitive at the time of those studies), its *temperature dependence* is found to be quite sensitive to the amount of disorder. Indeed, low-disorder regimes are marked by an increase of the resistive properties with rising temperatures, signaling the onset of inelastic scattering processes, much as in the clean case. On the other hand, strongly disordered samples exhibit a decrease in the resistivity with increasing temperatures, because a decrease in the *effective* elastic scattering outweighs the increase in the inelastic one. This type of fan-like family of resistivity curves (see Fig. 6.8) has been seen in several disordered strongly correlated materials, being known as Mooij correlations (Mooij, 1973). They are seen to arise here within a local approach to electronic correlations and disorder.

More recently the Hubbard model with binary alloy disorder has been also studied on a Bethe lattice with iterative perturbation theory as impurity solver (Semmler *et al.*, 2010*a*). Particular attention has been paid to the dependence of the distribution of local densities of states on the small imaginary part ("broadening") that is usually added to the frequency in numerical determinations of Green's functions. The dependence of the distribution of local densities of states on this parameter can be used to characterize the localized or extended nature of the electronic states. The main result of that paper is the determination of the zero-temperature phase diagram of the model at a particular filling as a function of interaction and disorder strengths and the identification of regions with metallic, Mott–Anderson-insulating and band-insulating behaviors. In particular, the opening of the Mott gap occurs at values of the interaction and disorder strengths for which the gapless system is already Anderson localized. It is difficult to compare this phase diagram with the one obtained by the infinite-U slave-boson mean-field theory (Dobrosavljević and Kotliar, 1997, 1998) because the models were solved with different types of disorder and in different regimes.

6.3.3 StatDMFT on realistic lattices

Although the Bethe lattice implementations were very informative, the exotic connectivity involved introduces some unwanted features, especially with regard to the critical behavior of the Anderson transition (Mirlin and Fyodorov, 1991), which is believed to correspond to some infinite-dimensional limit. Therefore, it is important to check what the effects of finite dimensions are. Motivated by this, the full statDMFT equations have been implemented in realistic lattices in recent years. Given the successful application of DMFT to the description of the Mott–Hubbard transition, a natural candidate is the analysis of the disordered Mott–Hubbard transition.

The disordered Mott–Hubbard transition

As with any other phase transition, the characterization of the effects of disorder on the Mott transition poses an important yet difficult problem. In the particular case of phase transitions incontrovertibly described by an order parameter, this characterization has seen considerable advances. In insulating quantum magnets, many examples have been found in which a whole vicinity of the disordered critical point is described as a quantum Griffiths phase. This is a phase in which rare regions of nearly ordered material dominate the physics and the thermodynamic response becomes divergent (Miranda and Dobrosavljević, 2005; Vojta, 2006). The sizes and energy scales governing the rare regions span several orders of magnitude and their description requires taking account of very broad distributions. Furthermore, in many cases, as the critical point itself is approached, the relative widths of the distributions grow without limit, a situation generically described as an *infinite randomness fixed point.* This has been well established in systems with both Ising and continuous symmetry (Fisher, 1992, 1994, 1995; Fisher and Young, 1998; Guo *et al.*, 1996; Hyman and Yang, 1997; Hyman *et al.*, 1996; Motrunich *et al.*, 2001; Narayanan *et al.*, 1999*a*,*b*; Pich *et al.*, 1998; Refael *et al.*, 2002; Yang *et al.*, 1996). More recently, a symmetry-based classification scheme of these Griffiths phases has been proposed and applied to several different systems with great success (Hoyos *et al.*, 2007; Hoyos and Vojta, 2006, 2008; Vojta, 2003, 2006; Vojta *et al.*, 2009; Vojta and Schmalian, 2005*a*,*b*). The Mott transition, however, poses a problem of a different nature, as it is *not* described by an order parameter in an obvious way. It is thus not clear how to extend the above insights into its description.

The implementation of statDMFT offers a natural way out. In particular, the clean problem is aptly described by DMFT, as we mentioned at the end of Section 6.1.2. The first implementation of statDMFT for the disordered Hubbard model was performed by Song *et al.* (2008) using the so-called "Hubbard I" approximation (Hubbard, 1963) as the impurity solver. This simple impurity solver captures the physics close to the atomic limit and is therefore convenient for a study of the Mott-insulating phase. However, it suffers from the deficiency of not giving rise to a quasiparticle peak in the single-impurity spectral function, a feature known to exist for any finite U when the hybridization function is that of a good metal. The method was applied to the cases of half and quarter filling and the density of states and inverse participation ratio were obtained. Interestingly, the localization length has a non-monotonic behavior as a function of the interaction U: it increases initially with U, showing a tendency to delocalization, but eventually decreases and becomes even smaller than the non-interacting value at the Mott transition. At quarter filling, an Altshuler–Aronov (Altshuler and Aronov, 1979) density-of-states anomaly is found, which is, however, absent at half-filling due to an interaction-induced suppression of the charge susceptibility.

Shortly afterwards, the disordered Hubbard model in a two-dimensional square lattice at $T = 0$ was solved with statDMFT (Andrade *et al.*, 2009*a*,*b*) using the Kotliar–Ruckenstein slave-boson mean-field theory as the impurity solver (Kotliar and Ruckenstein, 1986). Very similar results were also obtained (Pezzoli and Becca, 2010; Pezzoli *et al.*, 2009) within an approach based on a Gutzwiller variational wavefunction (Gutzwiller, 1963, 1964, 1965). This is not too surprising, as the Kotliar–Ruckenstein theory, when applied to the Mott transition, is known to be equivalent to the Gutzwiller wavefunction approach. This kind of approach is known to be able to capture the low-energy features, such as the disappearance of the quasiparticle peak as the transition is approached, unlike the Hubbard I approximation. In the impurity-problem language, the low-energy sector is described by two parameters, as in the infinite-U case (see Section 6.3.2): the quasiparticle weight Z, well known from Fermi liquid theory, which determines the width of the Kondo resonance (essentially the Kondo temperature) and the resonance position $\tilde{\varepsilon}$, which measures its shift from the chemical potential. It is important to notice that in the clean lattice, Z also determines the effective carrier mass ($m/m^* \sim Z$), a feature unique to cases in which the self-energy only depends on the frequency. Indeed, as the interaction strength is tuned to its critical value $U \to U_c$, $Z \sim U_c - U \to 0$, signaling the transmutation of the itinerant carriers into localized magnetic moments. In the statDMFT description, these local quantities vary from site to site, $Z \to Z_i$ and $\tilde{\varepsilon} \to \tilde{\varepsilon}_i$, and their distributions were thoroughly analyzed. Surprisingly, their critical behaviors were found to be very dissimilar.

As the transition is approached, which now happens at a disorder-dependent critical interaction $U_c(W)$, all $Z_i \to 0$, just as in the conventional Gutzwiller–Brinkman–Rice scenario. However, the quasiparticle weight distribution $P(Z)$ becomes increasingly broader as $U \to U_c(W)$. In fact, the typical value $Z_{\text{typ}} = \exp(\langle \ln Z \rangle) \to 0$, whereas the mean value remains finite, indicating that although almost all sites become local moments, some remain empty or doubly occupied (Aguiar *et al.*, 2006). Besides,

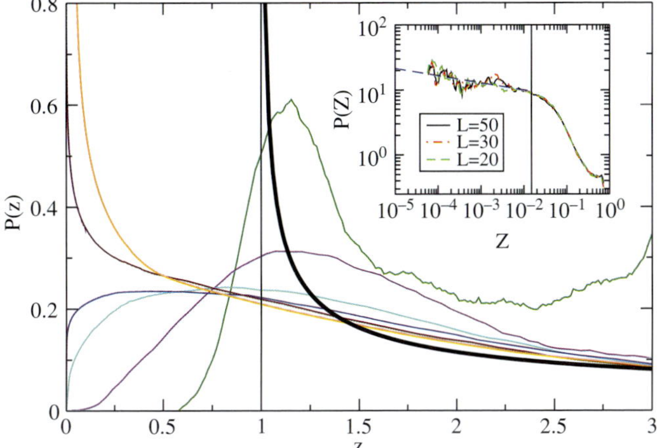

Fig. 6.9 Power-law distributions of quasiparticle weights Z in the half-filled disordered Hubbard model for different interaction strengths $(U/U_c(W) = 0.6, 0.8, 0.9, 0.92, 0.94, 0.97)$ showing characteristic power-law behavior $P(Z) \sim Z^{\alpha-1}$; as $U \to U_c$, $\alpha \to 0$ and the curves become increasingly singular at small Z. The inset shows the weak dependence on the lattice size. From Andrade *et al.* (2009*a*).

the Z distribution acquires a generic power-law shape, as in other Griffiths phases (see Fig. 6.9)

$$P(Z) \sim Z^{\alpha-1}. \tag{6.60}$$

This is very similar to the Bethe lattice studies of Section 6.3.2 and, as in those cases, these power laws generate a singular thermodynamic response, see eqn (6.55). Nevertheless, whereas before we had a disorder-driven Anderson-type transition, here this generic behavior is found in the proximity of the *interaction-driven* Mott transition *for fixed disorder strength*, clearly showing the amplifying effects of electronic correlations. The similarity to the quantum Griffiths scenario of magnets is not fortuitous. The low-Z values which dominate the thermodynamics occur in exponentially rare regions of suppressed disorder. Finally, we note that just as in other known quantum Griffiths phases, this electronic Griffiths phase seems to be tied to a phase transition characterized by an *infinite randomness fixed point*: we find that, up to the numerical uncertainty, $\alpha \to 0$ as $U \to U_c(W)$.

Interestingly, correlations have the opposite effect on the distribution of resonance positions $\tilde{\varepsilon}_i$. Indeed, the width of the $\tilde{\varepsilon}$ distribution *decreases* as $U \to U_c(W)$ (see Fig. 6.10). This is easily understood from the pinning of the Kondo resonances to the chemical potential, as already noted within DMFT (Tanasković *et al.*, 2003) (see Section 6.1.5). Just as in DMFT, this leads to a strong *disorder screening effect*, although, unlike DMFT, here the screening effect is not perfect: a small amount of disorder seems to survive as the transition is approached.

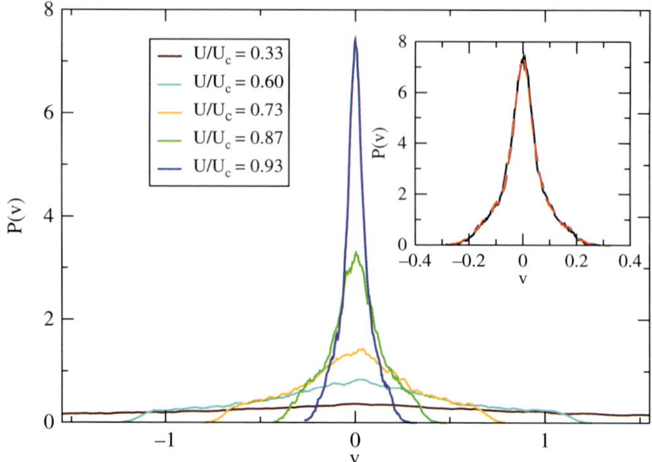

Fig. 6.10 Distributions of the ratio of the local Kondo resonance position $\tilde{\varepsilon}_i$ to the quasiparticle weight Z_i, $v_i \equiv \tilde{\varepsilon}_i/Z_i$ in the half-filled disordered Hubbard model showing a decreasing effective disorder strength, as measured by $\tilde{\varepsilon}_i$, with increasing interactions. The inset shows the weak dependence on the lattice size (full line for $L = 20$ and dashed line for $L = 50$). From Andrade *et al.* (2009*b*).

The disordered Hubbard model has also been investigated very recently in two dimensions with statDMFT and with iterative perturbation theory as impurity solver (Semmler *et al.*, 2010*a*). The disorder model used, however, was not the usual uniform diagonal disorder, but rather a combination of diagonal and off-diagonal disorder, together with disorder in the interaction term. This choice was intended to describe the so-called speckle disorder found in some set-ups of fermionic cold atoms loaded in optical lattices. The phase diagram was determined at half filling at zero as well as finite temperatures. A feature specific to this kind of disorder is the fact that it is unbounded and arbitrarily large values of on-site potentials occur. As a result, long (exponential) tails arise flanking the Hubbard bands. These contribute to filling up a possible interaction-induced Hubbard gap. The main consequence of the presence of these tails is a partial suppression of the Mott-insulating phase in favor of a disordered strongly correlated metal phase. As a result, the Mott insulator and Anderson(–Mott) insulator phases do not share a phase boundary and are separated by a metallic phase. This is in contrast to the phase diagram obtained within TMT, see Section 6.2.2.

A single impurity in a strongly correlated host

The smallest element in an analysis of the effects of inhomogeneities in a crystal is a single isolated point-like impurity in an otherwise homogeneous host. In the case of a weakly correlated host, the expected effect is the formation of a potential scattering center, which can be usually be quite simply described at low energies by a set of phase shifts with a negligible temperature dependence. If the host is a weakly correlated metal, the localized impurity generates a radially oscillatory disturbance of the host charge. The oscillation is characterized by a wave vector determined by the extremal radii of the Fermi surface and an envelope which decays as the inverse of the distance to the impurity raised to the power d in d dimensions

$$\delta n\left(\boldsymbol{x}\right) \sim \frac{\cos\left(2k_{\mathrm{F}}r\right)}{r^{d}}. \tag{6.61}$$

These are known as Friedel oscillations (Kasuya, 1956; Ruderman and Kittel, 1954; Yosida, 1957). A dilute collection of these impurities can then be treated as independent and their effects on transport properties is readily computed by conventional techniques, such as the Boltzmann equation.

The situation can be very different if the host is a strongly correlated system. The local inhomogeneity can disrupt the delicate balance of its host and lead to a spatially dependent pattern that has to be computed in a fully self-consistent manner. For example, local moments or antiferromagnetic ordering can be induced in the vicinity of the impurity (Alloul *et al.*, 2009). Superconducting correlations, specially of the unconventional type (e.g. d-wave), also lead to characteristic spatial patterns in the vicinity of impurities, which can be probed by scanning tunneling microscopy (STM) techniques (Balatsky *et al.*, 2006). Besides, strong correlations typically generate very low energy scales. Therefore, one should expect a non-negligible energy and temperature dependence with possibly observable effects. Indeed, STM studies have revealed a non-trivial interplay between spatial fluctuations and the energy dependence of the local density of states in cuprate superconductors (McElroy *et al.*, 2005). In this case, the spectral function, probed by STM, is hardly affected by disorder up to the superconducting gap energy, whereas much stronger spatial fluctuations are observed at higher energies.

In general, the localized imperfection can be viewed as a source probe coupled to all wave vectors of the host charge. Therefore, if this coupling can be treated within linear response theory, the relevant quantity to keep track of is the host wavevector-dependent charge susceptibility. The vicinity to the Mott metal–insulator transition causes the charge susceptibility to be strongly suppressed as the system becomes increasingly more localized. Therefore, we would expect considerable changes in the spatial pattern of the Friedel oscillations in a strongly correlated metal. The statDMFT is the tool of choice to study this situation. A single non-magnetic impurity in a half-filled Hubbard model has been studied using just such a tool (Andrade *et al.*, 2010). A great deal of analytical insight can then be obtained through an expansion in the *disorder strength* to first non-trivial order, using the Kotliar–Ruckenstein slave-boson impurity solver (Kotliar and Ruckenstein, 1986). Note that the procedure is fully non-perturbative in the interaction strength. In the weak coupling limit, $U \ll D$, the approach has been shown to be fully equivalent to the Hartree–Fock approximation, which predicts a static self-energy

$$\Sigma_{j}\left(i\omega_{n}\right) = U n_{j}^{(0)}, \tag{6.62}$$

where $n_j^{(0)}$ is the non-interacting charge density

$$n_i^{(0)} = 1 + 2\Pi_{i0}^{(0)} V, \tag{6.63}$$

with $\Pi_{ij}^{(0)}$ is the real-space static Lindhard polarization function and V the potential of the single impurity at site 0. This contains the usual Friedel oscillation pattern of eqn (6.61).

However, as the interaction strength is tuned to be close to the Mott-localization limit U_c, a very different situation arises. In this case, the charge density pattern decays as (Andrade *et al.*, 2010)

$$n_0 \approx -4\frac{(U_c - U)^3}{U_c^5}\left[\Pi^{(0)}\right]_{i0}^{-1} V \qquad (U \to U_c). \tag{6.64}$$

Two features of this result stand out. First, the spatial pattern is determined by the *inverse* of the static Lindhard function. Surprisingly, however, up to logarithmic corrections, the inverse static Lindhard function behaves in very much the same way as the usual Lindhard function. Second, the *amplitude* of the strongly-correlated Friedel oscillations is significantly suppressed as $U \to U_c$ down by the third power of the parametric distance to the transition. As a result, the charge perturbation dies out significantly as one recedes from the impurity, implying a short "healing length" (see Fig. 6.11). Comparison with numerical solutions of the statDMFT equations shows that the analytical expressions are quite accurate even for $|V| \lesssim D$.

The scattering T-matrix, which governs the transport properties, can also be computed with the same technique. We find its real-space expression to be

$$T_i = \left[\delta_{i,0} + U\Pi_{i0}^{(0)}\right] V \qquad (U \ll U_c), \tag{6.65}$$

at small U and

$$T_i = -\frac{U_c - U}{U_c^2}\left[\Pi^{(0)}\right]_{i0}^{-1} V \qquad (U \to U_c), \tag{6.66}$$

as one approaches the Mott transition. Here, the point to note is once again the strong suppression of scattering in the critical region, a situation that is by now familiar ("disorder screening").

The temperature dependence of the scattering is also very special. Indeed, it is known that the resistivity in the ballistic regime in two dimensions exhibits an anomalous, non-Fermi-liquid, linear-in-temperature dependence (Zala *et al.*, 2001), due to the coherent scattering off the Friedel oscillations created by the dilute impurities. Since we find these oscillations to be strongly suppressed by correlations it is important to study if the anomalous behavior survives. Besides, strong interactions also give rise to large inelastic scattering effects, and it is unclear if these are strong enough to mask the anomalous temperature dependence. The full analysis shows two things (Andrade *et al.*, 2010).

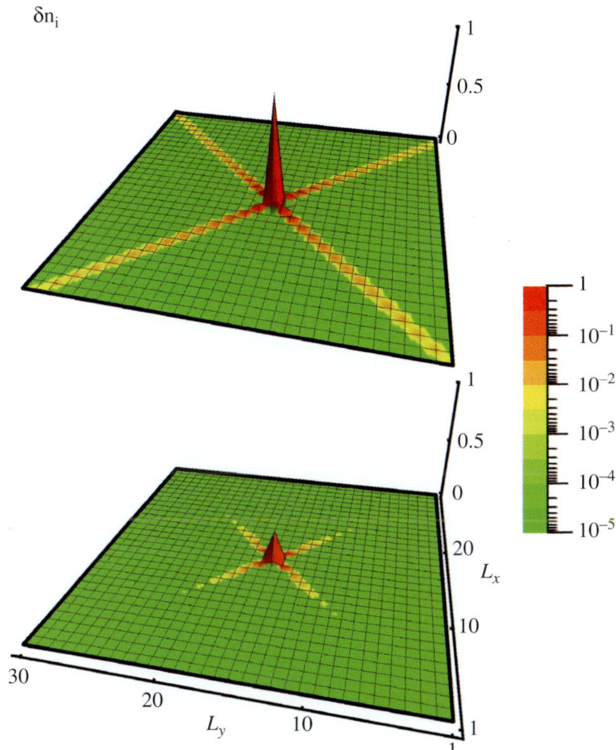

Fig. 6.11 Friedel oscillation patterns of charge oscillations generated by a localized non-magnetic impurity in a half-filled Hubbard model. The long-range oscillations of $\delta n_i = n_i - 1$ in the non-interacting case (top figure) are strongly suppressed in the strongly correlated regime (bottom figure, $m/m^* = 0.3$). From Andrade *et al.* (2010).

- The interaction suppression of scattering (disorder screening), eqn (6.66), acts to weaken both elastic and inelastic contributions from impurities, but *does not destroy* the non-Fermi-liquid, anomalous behavior.
- However, inelastic scattering limits considerably the *temperature range* in which the linear-in-T behavior of the resistivity is observed. These effects come both from the Friedel-oscillation regions, close to the impurities, as well as from the bulk, impurity-free regions of the system. It is found that, although both contributions are deleterious to the anomalous behavior, it is the bulk contribution that always dominates. In practice, even for quite modest mass renormalizations, the linear-in-T regime is probably already unobservable: for $m/m^* \sim 0.6$, for example, it is restricted to $T \lesssim 10^{-4} T_{\mathrm{F}}$, where $T_{\mathrm{F}} \approx D$.

Both effects above seem to make the observation of the anomalous resistivity in two dimensions very unlikely in the strongly correlated regime.

In more general terms, the strategy of expansion in weak disorder strength outlined above, which is quite general, seems to be a good avenue for exploration of strongly correlated spatially inhomogeneous systems.

6.3.4 Inhomogeneous systems with slab geometries

Although the local approach outlined in Sections 6.3.1–6.3.3 has been described in the context of random systems, its microscopic formulation (Dobrosavljević and Kotliar, 1997, 1998) is clearly applicable in any spatially non-uniform situation, whether random or not. Indeed, in the past several years this method has been applied to a number of non-random, yet spatially non-uniform systems. Because of this conceptual similarity, we chose to describe it here under the statDMFT "umbrella," although the systems have nothing "statistical" in them. A particular focus has been put on systems composed of different materials arranged in a slab geometry. Motivation for these investigations has come both from advances in growth techniques for epitaxial heterostructures of complex materials, particularly transition-metal oxides (Ohtomo *et al.*, 2002), and also from the use of surface-sensitive probes of strongly correlated materials, such as photoemission experiments (Mo *et al.*, 2003; Rodolakis *et al.*, 2009; Sekiyama *et al.*, 2000). A number of different spatial arrangements of complex, strongly correlated materials have been considered: semi-infinite strongly correlated systems with a free-standing planar boundary ("strongly correlated surface physics") (Borghi *et al.*, 2009, 2010; Liebsch, 2003; Potthoff and Nolting, 1999; Schwieger *et al.*, 2003), planar interfaces between two semi-infinite materials (Helmes *et al.*, 2008), heterostructures (Okamoto and Millis, 2004*a*,*b*), and semi-infinite metallic leads sandwiching barriers of strongly correlated metals or insulators (Borghi *et al.*, 2010; Chen and Freericks, 2007; Freericks, 2004; Zenia *et al.*, 2009). Studies of other types of inhomogeneity have also been performed (Snoek *et al.*, 2008).

The abrupt change of chemical properties at the planes of surfaces or interfaces can lead to novel phenomena, some of which have been identified and characterized by the local approach we are describing here. This a fast moving area of research and we will not attempt to cover the richness of behavior that has been and continues to be uncovered. We will confine ourselves to a few important findings that serve to convey the flavor of these phenomena.

The interface between a metal and a Mott insulator

Metal–insulator and metal–semiconductor interfaces have been produced and studied for many years. Phenomena such as band bending due to charge rearrangement are well known. However, the question of what happens when a metal is grown epitaxially onto a Mott insulator has received much less attention. A first attempt to elucidate this question has been made by considering an ideal interface between particle–hole symmetric metallic and Mott-insulating systems (Borghi *et al.*, 2009, 2010; Helmes *et al.*, 2008). In the absence of antiferromagnetism and within the DMFT picture, the spins of a bulk Mott insulator are not quenched because there are no conduction electrons available at the Fermi level with which they can form a singlet. When a Mott insulator is brought into contact with a metal, however, the layers that are closest to the interface can have access to the metallic carriers (through tunneling) and be quenched by means of the Kondo effect. This has been dubbed the "Kondo proximity effect." Evidently, the metallization and Kondo quenching processes are established in a spatially smooth fashion as one enters the insulator, thus creating a strongly correlated metallic surface layer. The metallic character can be characterized by means of the by now familiar quasiparticle weight $Z(x)$, which is here a spatially varying quantity that depends on the distance x from the interface. It is found that $Z(x)$ decays exponentially as a function of x

$$Z(x) \sim \exp(-x/\xi), \tag{6.67}$$

with a characteristic length scale ξ that sets the size of the metallic surface layer and is determined by the parametric distance to the Mott transition, $\xi \sim (U_c - U)^{-1/2}$. The 1/2 value of the critical exponent is expected from the mean-field character of the approach. At Mott criticality, this characteristic length diverges and

$$Z(x) = \frac{A}{x^2}. \tag{6.68}$$

The numerical value of the constant A above is found to be extremely small (~ 0.008), showing the Kondo-induced penetration of the metal into the Mott insulator to be extremely ineffective. The small values of the quasiparticle residues, which govern the local Fermi energy scale, also lead to a strong dependence on temperature, energy, or applied voltage. Very small values of these perturbations are able to completely destroy the metallic behavior, leading to the concept of a "fragile" Fermi liquid (Zenia *et al.*, 2009).

Surface dead layer

Another important phenomenon occurs at the free planar surface bounding a strongly correlated metal (Borghi *et al.*, 2009, 2010). Here again the proximity to the interface with the vacuum induces variations of the quasiparticle residue Z. Interestingly, since the outermost layer electrons cannot tunnel further out of the material, they have less of a chance to hybridize and stabilize the metallic behavior. As a consequence, there is a strong suppression of the quasiparticle weight close to the open surface. In effect, the outermost electrons are a very "fragile" Fermi liquid if not completely Mott localized, thus creating a surface layer of almost Mott-insulating character, the so-called "dead layer." Once again the characteristic width of this layer also depends on the proximity to criticality as $\xi \sim (U - U_c)^{-1/2}$.

This has important consequences for the interpretation of photoemission experiments. For a long time, a quasiparticle peak had been sought in the photoemission spectra of strongly correlated materials with little success. While there have been many attempts to explain this mystery, it is clear now that one expects on general grounds that the quasiparticle peak width (which is proportional to Z) becomes extremely narrow close to the surface. Since conventional photoemission spectra only probe the outer surface layers of the material, it is not too surprising that it has been difficult to detect a well-formed quasiparticle peak in the past. More recently, however, higher-energy photons have been used, which are able to penetrate deeper into the compound and thus probe the behavior more typical of the bulk. As expected from the theoretical results outlined above, the quasiparticle becomes much better defined with increasing incident photon energy (Mo *et al.*, 2003; Rodolakis *et al.*, 2009; Sekiyama *et al.*, 2000).

6.4 Glassy behavior of correlated electrons

We discussed so far the nature of strongly inhomogeneous metallic phases, resulting from the interplay of strong correlations and disorder. Such "electronic Griffiths phases", which can be viewed as precursors to the metal–insulator transition, reflect the formation of rare regions with anomalously slow dynamics. The resulting non-Fermi-liquid behavior is generically characterized by power-law anomalies, with non-universal, rapidly varying exponents. In contrast, many experimental data, especially in Kondo alloys, seem to show reasonably weak anomalies, close to marginal Fermi-liquid behavior (Stewart, 2001).

6.4.1 Instability of the electronic Griffiths phases to spin-glass ordering

Physically, it is clear what is missing from the theory. Similar to magnetic Griffiths phases (Miranda and Dobrosavljević, 2005; Vojta, 2006), the electronic Griffiths phase is characterized (Miranda and Dobrosavljević, 2001; Tanasković *et al.*, 2004) by a broad distribution $P(T_{\rm K}) \sim (T_{\rm K})^{\alpha-1}$ of local energy scales (Kondo temperatures), with the exponent $\alpha \sim W^{-2}$ rapidly decreasing with disorder W. At any given temperature, the local moments with $T_{\rm K}(i) < T$ remain unscreened. As disorder increases, the number of such unscreened spins rapidly proliferates. Within the existing theory (Miranda and Dobrosavljević, 2001; Miranda *et al.*, 1996, 1997*a*; Tanasković *et al.*, 2004) these unscreened spins act essentially as free local moments and provide a very large contribution to the thermodynamic response. In a more realistic description, however, even the Kondo-unscreened spins are *not* completely free, since the metallic host generates long-ranged Ruderman–Kittel–Kasuya–Yosida (RKKY) interactions even between relatively distant spins.

In a disordered metal, impurity scattering introduces random phase fluctuations in the usual periodic oscillations of the RKKY interaction, which, however, retains its power-law form (although its *average* value decays exponentially; see Jagannathan *et al.* (1988) and Narozhny *et al.* (2001)). Hence, such an interaction acquires a random amplitude J_{ij} of zero mean but finite variance (Jagannathan *et al.*, 1988; Narozhny *et al.*, 2001)

$$\langle J_{ij}^2(R)\rangle \sim \frac{1}{R^{2d}}. \tag{6.69}$$

As a result, in a disordered metallic host, a given spin is effectively coupled with random but long-range interactions to many other spins, often leading to spin-glass freezing at the lowest temperatures. How this effect is particularly important in

Griffiths phases can also be seen from the mean-field stability criterion (Bray and Moore, 1980) for spin-glass ordering, which takes the form

$$J\chi_{\mathrm{loc}}(T) = 1. \tag{6.70}$$

Here, J is a characteristic interaction scale for the RKKY interactions, and $\chi_{\mathrm{loc}}(T)$ is the disorder average of the local spin susceptibility. As we generally expect $\chi_{\mathrm{loc}}(T)$ to diverge within a Griffiths phase, this argument strongly suggests that in the presence of RKKY interactions such systems should have an inherent instability to finite- (even if very low) temperature spin-glass ordering.

Similarly as other forms of magnetic order, the spin glass ordering is typically reduced by quantum fluctuations (e.g. the Kondo effect) that are enhanced by coupling of the local moments to itinerant electrons. Sufficiently strong quantum fluctuations can completely suppress spin-glass ordering even at $T = 0$, leading to a quantum critical point separating a metallic spin glass from the conventional Fermi-liquid ground state. As in other quantum critical points, one expects the precursors to magnetic ordering to emerge even before the transition is reached and produce non-Fermi liquid behavior within the corresponding quantum critical region. Since many systems where disorder-driven non-Fermi-liquid behavior is observed are not too far from incipient spin-glass ordering, it is likely that these effects play an important role and should be theoretically examined in detail.

From the theoretical point of view, a number of recent works have examined the general role of quantum fluctuations in glassy systems and the associated quantum critical behavior. Most of the results obtained so far have concentrated on the behavior within the mean-field picture (i.e. in the limit of large coordination), where a consistent description of the quantum critical point behavior has been obtained for several models. In a few cases (Read *et al.*, 1995), corrections to mean-field theory have been examined, but the results appear inconclusive and controversial at this time. In the following, we briefly review the most important results obtained within the mean-field approaches.

Quantum critical behavior in insulating and metallic spin glasses

- *Ising spin glass in a transverse field*

The simplest framework to study the quantum critical behavior of spin glasses is provided by localized spin models such as the infinite-range Ising model in a transverse field (TFIM) with random exchange interactions J_{ij} of zero mean and variance J^2/N ($N \longrightarrow \infty$ is the number of lattice sites).

$$H_{\mathrm{TFIM}} = -\sum_{ij} J_{ij}\sigma_i^z\sigma_j^z - \Gamma\sum_i \sigma_i^x. \tag{6.71}$$

In the classical limit ($\Gamma = 0$), this model reduces to the well-studied Sherrington–Kirkpatrick model (Mézard *et al.*, 1986), where spins freeze with random orientations below a critical temperature $T_{\mathrm{SG}}(\Gamma = 0) = J$. Quantum fluctuations are introduced

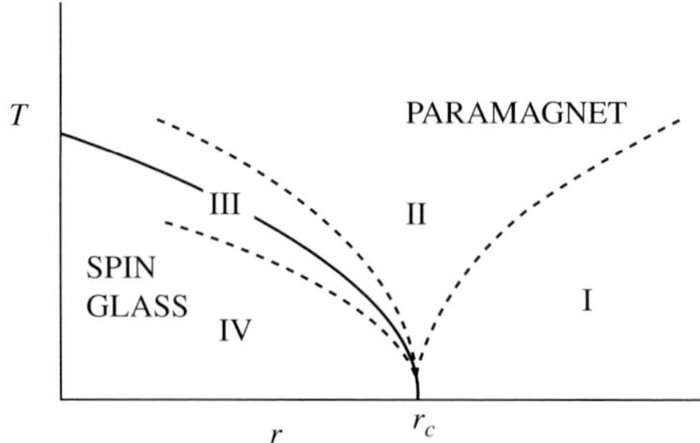

Fig. 6.12 Generic phase diagram (following Read *et al.* (1995)) of quantum critical behavior for spin glasses. The parameter r, which measures the quantum fluctuations, can represent the transverse field for localized spin models or the Fermi energy in metallic spin glasses.

by turning on the transverse field, which induces up-down spin flips with tunneling rate $\sim \Gamma$. As Γ grows, the critical temperature $T_{SG}(\Gamma)$ decreases, until the quantum critical point is reached at $\Gamma = \Gamma_c \approx 0.731J$, signaling the $T = 0$ transition from a spin glass to a quantum-disordered paramagnetic state (Fig. 6.12).

Similar to DMFT theories for electronic systems, such infinite range models can be formally reduced to a self-consistent solution of an appropriate quantum impurity problem, as first discussed in the context of quantum spin glasses by Bray and Moore (1980). Early work quickly established the phase diagram (Dobrosavljević and Stratt, 1987) of this model, but the dynamics near the quantum critical point proved more difficult to unravel, even when the critical point is approached form the quantum-disordered side. Here, the problem reduces to solving for the dynamics of a single Ising spin in a transverse field, described by an effective Hamiltonian (Miller and Huse, 1993) of the form

$$H = -\frac{1}{2}J^2 \int\int d\tau d\tau' \sigma^z(\tau)\chi(\tau - \tau')\sigma^z(\tau') + \Gamma \int d\tau \sigma^x(\tau).$$

Physically, the interaction of the considered spin with the spin fluctuations of its environment generates the retarded interaction described by the "memory kernel" $\chi(\tau - \tau')$. An appropriate self-consistency condition relates the memory kernel to the disorder-averaged local dynamical susceptibility of the quantum spin

$$\chi(\tau - \tau') = \overline{\langle T\sigma^z(\tau)\sigma^z(\tau')\rangle}.$$

A complete solution of the quantum critical behavior can be obtained, as first established in a pioneering work by Miller and Huse (1993). These authors set up a diagrammatic perturbation theory for the dynamic susceptibility, showing that the leading loop approximation already captures the exact quantum critical behavior, as the higher-order corrections provide only quantitative renormalizations. The dynamical susceptibility takes the general form

$$\chi(\omega_n) = \chi_o + \left(\omega_n^2 + \Delta^2\right)^{1/2},$$

where the local static susceptibility χ_o remains finite throughout the critical regime, and the spin excitations exist above a gap

$$\Delta \sim (r/|\ln r|)^{1/2}$$

which vanishes as the transition is approached from the paramagnetic side (here $r = (\Gamma - \Gamma_c)/\Gamma_c$ measures the distance from the critical point).

- *The quantum spin-glass phase and the replicon mode*

The validity of this solution was confirmed by a generalization (Ye *et al.*, 1993) to the M-component rotor model (the Ising model belongs to the same universality class as the $M = 2$ rotor model), which can be solved in closed form in the large-M limit. This result, which proves to be exact to all orders in the $1/M$ expansion, could be extended even to the spin-glass phase, where a full replica symmetric solution was obtained. Most remarkably, the spin excitation spectrum remains gapless ($\Delta = 0$) throughout the ordered phase. Such gapless excitations commonly occur in ordered states with broken continuous symmetry, but are generally not expected in classical or quantum models with a discrete symmetry of the order parameter. In glassy phases (at least within mean-field solutions), however, gapless excitations generically arise for both classical and quantum models. Here, they reflect the marginal stability (Mézard *et al.*, 1986) found in the presence of replica symmetry-breaking, a phenomenon which reflects the high degree of frustration in these systems. The role of the Goldstone mode in this case is played by the so-called "replicon" mode, which describes the collective low-energy excitations characterizing the glassy state.

A proper treatment of the low-energy excitations in this regime requires special attention to the role of replica symmetry-breaking (RSB) in the $T \to 0$ limit. The original work (Ye *et al.*, 1993) suggested that RSB is suppressed at $T = 0$, so that the simpler replica symmetric solution can be used at low temperatures. Later work (Georges *et al.*, 2001), however, established that the full RSB solution must be considered before taking the $T \to 0$ limit, and only then can the correct form of the leading low-temperature corrections (e.g. the linear T-dependence of the specific heat) be obtained.

- *Physical content of the mean-field solution*

In appropriate path-integral language (Dobrosavljević and Stratt, 1987; Ye *et al.*, 1993), the problem can be shown to reduce to solving a one dimensional classical Ising model with long-range interactions, the form of which must be self-consistently determined. Such classical spin chains with long range interactions in general can be highly non-trivial. Some important examples are the Kondo problem (Anderson and Yuval, 1969; Anderson *et al.*, 1970; Yuval and Anderson, 1970), and the dissipative two-level system (Leggett *et al.*, 1987), both of which map to an Ising chain with $1/\tau^2$ interactions. Quantum phase transitions in these problems correspond to the KTB transition found in the Ising chain (Kosterlitz, 1976), the description of which required a sophisticated renormalization-group analysis. Why then is the solution of the quantum Ising spin glass model so simple? The answer was provided in the paper by Ye *et al.* (1993), which emphasized that the critical state (and the RSB spin-glass state) does not correspond to the critical point, but rather to the high-temperature phase of the equivalent Ising chain, where a perturbative solution is sufficient. In Kondo language, this state corresponds to the Fermi-liquid solution characterized by a finite Kondo scale, as demonstrated by a quantum Monte-Carlo calculation of Rozenberg and Grempel (1998), which also confirmed other predictions of the analytical theory.

From a more general perspective, the possibility of obtaining a simple analytical solution for quantum critical dynamics has a simple origin. It follows from the fact that all corrections to Gaussian (i.e. Landau) theory are irrelevant above the upper critical dimension, as first established by the Hertz–Millis theory (Hertz, 1976; Millis, 1993) for conventional quantum criticality. The mean-field models become exact in the limit of infinite dimensions, hence the Gaussian solution of Miller and Huse (1993) and Ye *et al.* (1993) becomes exact. Leading corrections to mean-field theory for rotor models were examined via an ε-expansion below the upper critical dimension $d_c = 8$ for the rotor models by Read, Sachdev, and Ye, but these studies found runaway flows, presumably indicating non-perturbative effects that require more sophisticated theoretical tools. Most likely these include Griffiths phase phenomena controlled by the infinite randomness fixed point, as already discussed in Section 6.3.

- *Metallic spin glasses*

A particularly interesting role of the low-lying excitations associated with the spin-glass phase is found in metallic spin glasses. Here the quantum fluctuations are provided by the Kondo coupling between the conduction electrons and local moments, and therefore can be tuned by controlling the Fermi energy in the system. The situation is again the simplest for Ising spins, where an itinerant version of the rotor model of Sengupta and Georges (1995) can be considered. Similar results have obtained for the "spin-density glass" model of Sachdev *et al.* (1995). The essential new feature in these models is the presence of itinerant electrons, which, as in the Hertz–Millis approach (Hertz, 1976; Millis, 1993), have to be formally integrated out before an effective order-parameter theory can be obtained. This is justified *provided* that the quasiparticles remain well defined at the quantum critical point, i.e. if the quasiparticle weight $Z \sim T_K$ remains finite and the Kondo effect remains operative. The validity of these assumptions is by no means obvious, and led to considerable controversy before a detailed quantum Monte Carlo solution of the model became available (Rozenberg and Grempel, 1999), confirming the proposed scenario.

Under these assumptions, the theory can again be solved in closed form, and we only quote the principal results. Physically, the essential modification is that the presence of itinerant electrons now induces Landau damping, which creates dissipation for the collective mode. As a result, the dynamics is modified, and the local dynamic susceptibility now takes the following form

$$\chi(\omega_n) = \chi_o + \left(|\omega_n| + \omega^*\right)^{1/2}. \tag{6.72}$$

The dynamics is characterized by the crossover scale $\omega^* \sim r$ (r measures the distance to the transition), which defines a crossover temperature $T^* \sim \omega^*$ separating the Fermi-liquid regime (at $T \ll T^*$) from the quantum critical regime (at $T \gg T^*$). At the critical point $\chi(\omega_n) = \chi_o + |\omega_n|^{1/2}$, leading to non-Fermi liquid behavior of all physical quantities, which acquire a leading low-temperature correction of the $T^{3/2}$ form. This is a rather mild violation of Fermi liquid theory, since both the static spin susceptibility and the specific heat coefficient remain finite at the quantum critical point. A more interesting feature, which is specific to glassy systems, is the persistence of such quantum critical non-Fermi-liquid behavior *throughout* the metallic glass phase, reflecting the role of the replicon mode.

Spin-liquid behavior, destruction of the Kondo effect by bosonic dissipation, and fractionalization

- *Quantum Heisenberg spin glass and the spin-liquid solution*

Quantum spin glass behavior proves to be much more interesting in the case of Heisenberg spins, where the Berry phase term (Fradkin, 1991) plays a highly non-trivial role, completely changing the dynamics even within the paramagnetic phase. While the existence of a finite-temperature spin-glass transition was established even in early work (Bray and Moore, 1980), solving for the details of the dynamics proved difficult until the remarkable work of Sachdev and Ye (Sachdev and Ye, 1993). By a clever use of large-N methods, these authors identified a striking *spin-liquid* solution within the paramagnetic phase. In contrast to the nonsingular behavior of Ising or rotor quantum spin glasses, the dynamical susceptibility now displays a logarithmic singularity at low frequency. On the real axis it takes the form

$$\chi(\omega) \sim \ln(1/|\omega|) + i\frac{\pi}{2}\mathrm{sgn}(\omega).$$

A notable feature of this solution is that it is precisely of the form postulated for the "marginal Fermi liquid" phenomenology (Varma *et al.*, 1989) of doped cuprates. The specific heat is also found to assume a singular form $C \sim \sqrt{T}$, which was shown (Georges *et al.*, 2001) to reflect a non-zero extensive entropy if the spin liquid solution is extrapolated to $T = 0$. Of course, the spin-liquid solution becomes unstable at a finite-ordering temperature, and the broken-symmetry state has to be examined to discuss the low-temperature properties of the model.

Subsequent work (Georges *et al.*, 2001) demonstrated that this mean-field solution remains valid for all finite N and generalized the solution to the spin-glass (ordered) phase. A closed set of equations describing the low-temperature thermodynamics in the spin-glass phase was obtained, which was very recently re-examined in detail (Camjayi and Rozenberg, 2003), revealing fairly complicated behavior.

- *Metallic Heisenberg spin glasses and fractionalization*

Even more interesting is the fate of this spin-liquid solution in itinerant systems, where an additional Kondo coupling is added between the local moments and the conduction electrons. The mean-field approach can be extended to this interesting situation by examining a Kondo–Heisenberg spin-glass model (Burdin *et al.*, 2002) with the Hamiltonian

$$H_{\mathrm{KH}} = -t \sum_{\langle ij \rangle \sigma} (c_{i\sigma}^{\dagger} c_{j\sigma} + \mathrm{H.\ c.}) + J_{\mathrm{K}} \sum_{i} \mathbf{S}_i \cdot \mathbf{s}_i + \sum_{\langle ij \rangle} J_{ij} \mathbf{S}_i \cdot \mathbf{S}_j. \tag{6.73}$$

In the regime where the scale of the RKKY interaction $J = \langle J_{ij}^2 \rangle^{1/2}$ is small compared to the Kondo coupling J_{K}, one expects Kondo screening to result in standard Fermi-liquid behavior. In the opposite limit, however, the spin fluctuations associated with the retarded RKKY interactions may be able to adversely affect the Kondo screening, and novel metallic behavior could emerge. This intriguing possibility can be precisely investigated in the mean-field (infinite-range) limit, where the problem reduces to a single-impurity action of the form (Burdin *et al.*, 2002)

$$\begin{aligned} S_{\mathrm{eff}}^{\mathrm{KH}} = & \sum_{\sigma} \int_0^{\beta} d\tau c_{\sigma}^{\dagger}(\tau) \left(\partial_{\tau} - \mu + v_j \right) c_{\sigma}(\tau) \\ & - t^2 \sum_{\sigma} \int_0^{\beta} d\tau \int_0^{\beta} d\tau' c_{\sigma}^{\dagger}(\tau) G_{\mathrm{c}}(\tau - \tau') c_{\sigma}(\tau') \\ & + J_{\mathrm{K}} \int_0^{\beta} d\tau \mathbf{S}(\tau) \cdot \mathbf{s}(\tau) \\ & - \frac{J^2}{2} \int_0^{\beta} d\tau \int_0^{\beta} d\tau' \chi(\tau - \tau') \mathbf{S}(\tau) \cdot \mathbf{S}(\tau'). \end{aligned} \tag{6.74}$$

Such a single-impurity action (6.74) describes the so-called Bose–Fermi Kondo (BFK) impurity model (Sengupta, 2000; Si and Smith, 1996; Zaránd and Demler, 2002; Zhu and Si, 2002) where, in addition to the coupling to the fermionic bath of conduction electrons, the Kondo spin also interacts with a bosonic bath of spin fluctuations, with local spectral density $\chi(\omega_n)$. Because the same BFK model also appears in "extended" DMFT theories (Smith and Si, 2000) of quantum criticality in clean systems (Sengupta, 2000; Si *et al.*, 2001; Si and Smith, 1996; Zaránd and Demler, 2002; Zhu and Si, 2002), its properties have been studied in detail and are by now well understood.

In the absence of the RKKY coupling ($J = 0$), the ground state of the impurity is a Kondo singlet for any value of $J_{\mathrm{K}} \neq 0$. By contrast, when $J > 0$, the dissipation induced by the bosonic bath tends to destabilize the Kondo effect. For a bosonic bath of 'ohmic" form ($\chi(\omega_n) = \chi_o - C|\omega_n|$), this effect only leads to a finite decrease of the Kondo temperature, but the Fermi-liquid behavior persists. In contrast, for "subohmic" dissipation ($\chi(\omega_n) = \chi_o - C|\omega_n|^{1-\epsilon}$ with $\epsilon > 0$) two different phases exist, and for sufficiently large RKKY coupling the Kondo effect is destroyed. The two regimes are separated by a quantum phase transition (see Fig. 6.13).

Of course, in the Kondo lattice model with additional RKKY interactions considered, the form of the bosonic bath $\chi(\omega_n)$ is self-consistently determined and can take different forms as the RKKY coupling J is increased. The model was analytically

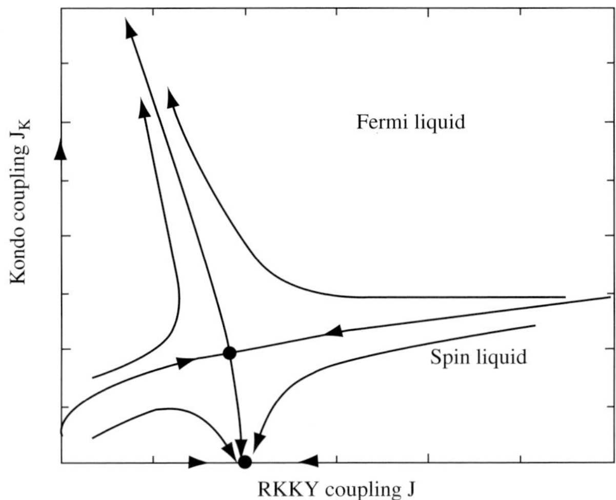

Fig. 6.13 Phase diagram of the Bose–Fermi Kondo model in the presence of a sub-ohmic bosonic bath (Sengupta, 2000; Si and Smith, 1996; Zaránd and Demler, 2002; Zhu and Si, 2002). Kondo screening is destroyed for sufficiently large dissipation (RKKY coupling to spin fluctuations).

solved within a large-N approach by Burdin *et al.* (2002), who calculated the evolution of the Fermi-liquid coherence scale T^* and the corresponding quasiparticle weight Z in the presence of RKKY interactions. Within the paramagnetic phase both $T^*(J)$ and $Z(J)$ are found to decrease with J until the Kondo effect (and thus the Fermi liquid) is destroyed at $J = J_c \approx 10T_K^o$ (here T_K^o is the $J = 0$ Kondo temperature), where both scales vanish (Burdin *et al.*, 2002; Tanaskovic *et al.*, 2005). At $T > T^*(J)$ (and of course at any temperature for $J > J_c$) the spins effectively decouple from conduction electrons and spin-liquid behavior, essentially identical to that of the insulating model, is established. Thus, sufficiently strong and frustrating RKKY interactions are able to suppress Fermi-liquid behavior, and marginal Fermi-liquid behavior emerges in a metallic system. This phenomenon, corresponding to spin–charge separation resulting from the destruction of the Kondo effect, is sometimes called "fractionalization" (Coleman and Andrei, 1987; Demler *et al.*, 2002; Kagan *et al.*, 1992; Senthil *et al.*, 2003, 2004). Such behavior has often been advocated as an appealing scenario for exotic phases of strongly correlated electrons, but with the exception of the described model, there are very few well established results and model calculations to support its validity. Finally, we should mention related work (Parcollet and Georges, 1999) on doped Mott insulators with random exchanges, which have many similarities with the above picture.

We should note, however, that this exotic solution is valid only within the paramagnetic phase, which is generally expected to become unstable to magnetic (spin-glass) ordering at sufficiently low temperatures. Since fractionalization emerges only for sufficiently large RKKY coupling (in the large-N model $J_c \approx 10T_K^o$), while in general one expects magnetic ordering to take place already at $J \sim T_{\mathrm{K}}^{o}$ (according to the famous Doniach criterion (Doniach, 1977)), one expects (Burdin *et al.*, 2002) the system to magnetically order much before the Kondo temperature vanishes. If this is true, then one expects the quantum critical behavior to be very similar to metallic Ising spin glasses, i.e. to assume the conventional Hertz–Millis form, at least for the mean-field spin-glass models we discuss here. The precise relevance of this paramagnetic spin-liquid solution thus remains unclear, at least for systems with weak or no disorder in the conduction band.

- *Fractionalized two-fluid behavior of electronic Griffiths phases*

The situation seems more promising in the presence of sufficient amounts of disorder, where the electronic Griffiths phase forms. Here the disorder generates a very broad distribution of local Kondo temperatures, making the system much more sensitive to RKKY interactions. This mechanism has recently been studied within an extended DMFT approach (Tanaskovic *et al.*, 2005), which is able to incorporate both the formation of the Griffiths phase and the effects of frustrating RKKY interactions leading to spin-glass dynamics. At the local impurity level, the problem is still reduced to the Bose–Fermi Kondo model, but the presence of conduction-electron disorder qualitatively modifies the self-consistency conditions determining the form of $\chi\left(\omega_n\right)$.

To obtain a sufficient condition for decoupling, we examine the stability of the Fermi-liquid solution, by considering the limit of infinitesimal RKKY interactions. To leading order we replace

$$\chi(\omega_n) \longrightarrow \chi_o\left(\omega_n\right) \equiv \chi(\omega_n; J=0),$$

and the calculation reduces to the "bare model" of Tanaskovic *et al.* (2005). In this case, $P\left(T_{\mathrm{K}}\right) \sim T_{\mathrm{K}}^{\alpha-1}$, where $\alpha \sim 1/W^2$, and

$$\chi_0\left(\omega_n\right) \sim \int dT_{\mathrm{K}} P\left(T_{\mathrm{K}}\right) \chi\left(\omega_n, T_{\mathrm{K}}\right) \sim \chi_0\left(0\right) - C_0\left|\omega_n\right|^{1-\epsilon}, \tag{6.75}$$

where $\epsilon = 2-\alpha$. Thus, for sufficiently strong disorder (i.e. within the electronic Griffiths phase), even the "bare" bosonic bath is sufficiently singular to generate decoupling. The critical value of W will be modified by self-consistency, but it is clear that decoupling will occur for sufficiently large disorder.

Once decoupling is present, the system is best viewed as composed of two fluids, one made up of a fraction n of decoupled spins, and the other of a fraction $(1-n)$ of Kondo screened spins. The self-consistent $\chi\left(\omega_n\right)$ acquires contributions from both fluids

$$\chi\left(\omega_n\right) = n\chi_{dc}\left(\omega_n\right) + \left(1-n\right)\chi_s\left(\omega_n\right). \tag{6.76}$$

A careful analysis (Tanaskovic *et al.*, 2005) shows that, for a bath characterized by an exponent ϵ

$$\chi_{dc}(\omega_n) \sim \chi_{dc}(0) - C\,|\omega_n|^{1-(2-\epsilon)}; \tag{6.77}$$

$$\chi_s(\omega_n) \sim \chi_s(0) - C'\,|\omega_n|^{1-(2-\epsilon-1/\nu)}, \tag{6.78}$$

where $\nu = \nu(\epsilon)$ is a critical exponent governing how the Kondo scale vanishes at the quantum critical point of the Bose–Fermi model. Since $\nu > 0$, the contribution of the decoupled fluid is more singular and dominates at lower frequencies. Self-consistency then yields $\epsilon = 1$, as in the familiar spin-liquid state of Sachdev and Ye (1993). For $\epsilon = 1$, the local susceptibility is logarithmically divergent (both in ω_n and T). This does not necessarily mean that the bulk susceptibility, which is the experimentally relevant quantity, behaves in the same manner (Parcollet and Georges, 1999). More work remains to be done to determine the precise low-temperature form of this and other physical quantities and to assert the relevance of this mechanism for specific materials.

As in the case where conduction-electron disorder is absent, the spin-liquid state is unstable towards spin-glass ordering at sufficiently low temperatures. However, numerical estimates for the Griffiths phase model (Tanaskovic *et al.*, 2005) suggest a surprisingly wide temperature window where the marginal behavior should persist above the ordering temperature. Figure 6.14 represents the predicted phase diagram of this model. For weak disorder the system is in the Fermi-liquid phase, while for $W > W_{\mathrm{c}}$ the marginal Fermi-liquid phase emerges. The crossover temperature (dashed line) delimiting this regime can be estimated from the frequency up to which the logarithmic behavior of the local dynamical susceptibility $\chi(i\omega)$ is observed. The spin-glass phase, obtained from eqn (6.70), appears only at the lowest temperatures, well below the marginal Fermi-liquid boundary. Interestingly, recent experiments (MacLaughlin *et al.*, 2001) have indeed found evidence of dynamical spin freezing in the millikelvin temperature range for the same Kondo alloys that display normal-phase non-Fermi-liquid behavior in a much broader temperature window.

The two-fluid phenomenology of the disordered Kondo lattice we have described above is very reminiscent of earlier work on the clean Kondo lattice, where the conduction electrons effectively decouple from the local moments, the latter forming a spin-liquid state (Coleman and Andrei, 1987; Demler *et al.*, 2002; Kagan *et al.*, 1992; Senthil *et al.*, 2003, 2004). The major difference between the results presented in this section and these other cases is that here local spatial disorder fluctuations lead to an *inhomogeneous* coexistence of the two fluids, as each site decouples or not from the conduction electrons depending on its local properties. The mean-field models discussed should be considered as merely the first examples of this fascinating physics. The specific features of the spin-liquid behavior that were obtained from these models may very well prove to be too restrictive and perhaps even inaccurate. For example, the specific heat enhancements may well be overestimated, reflecting the residual $T = 0$ entropy of the mean-field models. Nevertheless, the physics of Kondo screening being destroyed by the interplay of disorder and RKKY interactions will almost certainly play a central role in determining the properties of many non-Fermi-liquid systems, and clearly needs to be investigated in more detail in the future.

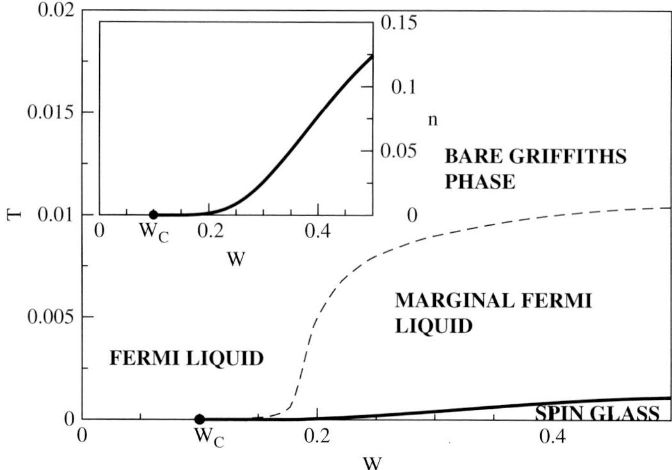

Fig. 6.14 Phase diagram of the electronic Griffiths phase model with RKKY interactions (Tanaskovic *et al.*, 2005). The inset shows the fraction of decoupled spins as a function of the disorder strength W.

6.4.2 Electron glass

Another aspect of disordered interacting electrons poses a fundamental problem. Very generally, Coulomb repulsion favors a uniform electronic density, while disorder favors local density fluctuations. When these two effects are comparable in magnitude, one can expect many different low-energy electronic configurations, i.e. the emergence of many *metastable states*. Similar to other "frustrated" systems with disorder, such as spin glasses, these processes can be expected to lead to *glassy* behavior of the electrons, and the associated anomalously slow relaxational dynamics. Indeed, both theoretical (Davies *et al.*, 1982; Pollak and Hunt, 1991) and experimental (Ben-Chorin *et al.*, 1993; Bogdanovich and Popović, 2002; Jaroszyński *et al.*, 2002; Ovadyahu and Pollak, 1997) work has found evidence of such behavior deep on the insulating side of the transition. However, at present very little is known as to the precise role of such processes in the critical region. Nevertheless, it is plausible that the glassy freezing of the electrons must be important, since the associated slow relaxation clearly will reduce the mobility of the electrons. From this point of view, the glassy freezing of electrons may be considered, in addition to the Anderson and the Mott mechanisms, as a third fundamental process associated with electron localization. Interest in understanding the glassy aspects of electron dynamics has experienced a genuine renaissance in the last few years, primarily due to experimental advances. Emergence of many metastable states, slow relaxation, and incoherent transport have been observed in a number of strongly correlated electronic systems. These included transition-metal oxides such as high-T_c materials, manganites, and ruthenates. Similar features have recently been reported in two-dimensional electron gases and even three-dimensional doped semiconductors such as Si:P.

- *Infinite-dimensional model of an electron glass*

The interplay of the electron–electron interactions and disorder is particularly evident deep on the insulating side of the metal–insulator transition (MIT). Here, both experimental (Massey and Lee, 1996) and theoretical studies (Efros and Shklovskii, 1975) have demonstrated that they can lead to the formation of a soft "Coulomb gap", a phenomenon that is believed to be related to the glassy behavior (Ben-Chorin *et al.*, 1993; Bogdanovich and Popović, 2002; Jaroszyński *et al.*, 2002; Ovadyahu and Pollak, 1997) of the electrons. Such glassy freezing has long been suspected (Belitz and Kirkpatrick, 1995) to be of importance, but more recent theoretical work (Chakravarty *et al.*, 1999; Dobrosavljević *et al.*, 1997) has suggested that it may even dominate the MIT behavior in certain low-carrier-density systems. The classic work of Efros and Shklovskii (1975) has clarified some basic aspects of this behavior, but a number of key questions have remain unanswered.

As a simplest example (Pastor and Dobrosavljević, 1999) displaying glassy behavior of electrons, we focus on a simple lattice model of spinless electrons with nearest-neighbor repulsion V in the presence of random site energies ε_i and inter-site hopping t, as given by the Hamiltonian

$$H = \sum_{<ij>} (-t_{ij} + \varepsilon_i \delta_{ij}) c_i^\dagger c_j + V \sum_{<ij>} c_i^\dagger c_i c_j^\dagger c_j. \tag{6.79}$$

This model can be solved in a properly defined limit of large coordination number (Georges *et al.*, 1996), where an extended dynamical mean-field theory (EDMFT) formulation becomes exact. We concentrate on the situation where the disorder (or more generally frustration) is large enough to suppress any uniform ordering. We then rescale both the hopping elements and the interaction amplitudes as $t_{ij} \to t_{ij}/\sqrt{z}$; $V_{ij} \to V_{ij}/\sqrt{z}$. As we will see shortly, the required fluctuations then survive even in the $z \to \infty$ limit, allowing for the existence of the glassy phase. Within this model:

1. the universal form of the Coulomb gap (Efros and Shklovskii, 1975) proves to be a direct consequence of glassy freezing
2. the glass phase is identified through the emergence of an extensive number of metastable states, which in our formulation is manifested as a replica symmetry breaking instability (Mézard *et al.*, 1986)
3. as a consequence of this ergodicity breaking (Mézard *et al.*, 1986), the zero-field cooled compressibility is found to vanish at $T = 0$, suggesting the absence of screening (Efros and Shklovskii, 1975) in disordered insulators
4. the quantum fluctuations can melt this glass even at $T = 0$, but the relevant energy scale is set by the electronic mobility and is therefore a non-trivial function of disorder.

We should stress that although this model allows us to examine the interplay of glassy ordering and quantum fluctuations due to itinerant electrons, it is too simple to describe the effects of Anderson localization. These effects require extensions to lattices with finite coordination and will be discussed in the next section.

For simplicity, we focus on a Bethe lattice at half filling and examine the $z \to \infty$ limit. This strategy automatically introduces the correct order parameters and after standard manipulations (Dobrosavljević and Kotliar, 1994) the problem reduces to a self-consistently defined single-site problem, as defined by an the effective action of the form

$$S_{\text{eff}}(i) = \sum_a \int_o^\beta \int_o^\beta d\tau d\tau' \, [c_i^{\dagger a}(\tau)(\delta(\tau-\tau')\partial_\tau + \varepsilon_i + t^2 G(\tau,\tau'))c_i^a(\tau')$$

$$+\frac{1}{2}V^2 \delta n_i^a(\tau)\chi(\tau,\tau')\delta n_i^a(\tau')] + \frac{1}{2}V^2 \sum_{a\neq b} \int_o^\beta \int_o^\beta d\tau d\tau' \, \delta n_i^a(\tau) \, q_{ab} \, \delta n_i^b(\tau'). \quad (6.80)$$

Here, we have used functional integration over replicated Grassmann fields (Dobrosavljević and Kotliar, 1994) $c_i^a(\tau)$, which represent electrons on site i and replica index a, and the random site energies ε_i are distributed according to a given probability distribution $P(\varepsilon_i)$. The operators $\delta n_i^a(\tau) = (c_i^{\dagger a}(\tau)c_i^a(\tau) - 1/2)$ represent the *density fluctuations* from half filling. The order parameters $G(\tau-\tau')$, $\chi(\tau-\tau')$ and q_{ab} satisfy the following set of self-consistency conditions

$$G(\tau-\tau') = \int d\varepsilon_i P(\varepsilon_i) < c_i^{\dagger a}(\tau)c_i^a(\tau') >_{\text{eff}}, \quad (6.81)$$

$$\chi(\tau-\tau') = \int d\varepsilon_i P(\varepsilon_i) < \delta n_i^{\dagger a}(\tau)\delta n_i^a(\tau') >_{\text{eff}}, \quad (6.82)$$

$$q_{ab} = \int d\varepsilon_i P(\varepsilon_i) < \delta n_i^{\dagger a}(\tau)\delta n_i^b(\tau') >_{\text{eff}} . \quad (6.83)$$

- *Order parameters*

In these equations, the averages are taken with respect to the effective action of eqn (6.80). Physically, the "hybridization function" $t^2 G(\tau-\tau')$ represents the single-particle electronic spectrum of the environment, as seen by an electron on site i. In particular, its imaginary part at zero frequency can be interpreted (Dobrosavljević and Kotliar, 1994) as the inverse lifetime of the local electron and as such remains finite as long as the system is metallic. We recall (Dobrosavljević and Kotliar, 1994) that for $V=0$ these equations reduce to the familiar CPA description of disordered electrons, which is exact for $z=\infty$. The second quantity $\chi(\tau-\tau')$ represents an (interaction-induced) *mode-coupling* term that reflects the *retarded* response of the density fluctuations of the environment. Note that very similar objects appear in the well-known mode-coupling theories of the glass transition in dense liquids (Cummins *et al.*, 1994). Finally the quantity q_{ab} $(a \neq b)$ is nothing but the familiar Edwards–Anderson order parameter q_{EA}. Its non-zero value indicates that the time-averaged electronic density is spatially non-uniform.

- *Equivalent infinite range model*

From a technical point of view, an RSB analysis is typically carried out by focusing on a free energy expressed as a functional of the order parameters. In our Bethe lattice approach, one directly obtains the self-consistency conditions from appropriate recursion relations (Dobrosavljević and Kotliar, 1994), without invoking a free-energy functional. However, we have found it useful to map our $z = \infty$ model to another *infinite range* model, which has *exactly* the same set of order parameters and self-consistency conditions, but for which an appropriate free-energy functional can easily be determined. The relevant model is still given eqn (6.79), but this time with *random* hopping elements t_{ij} and *random* nearest-neighbor interaction V_{ij}, having zero mean and variances t^2, and V^2, respectively. For this model, standard manipulations (Dobrosavljević and Kotliar, 1994) result in the following free-energy functional

$$F[G,\chi,q_{ab}] = -\frac{1}{2}\sum_a \int_o^\beta \int_o^\beta d\tau d\tau' \, [t^2 G^2(\tau,\tau') + V^2\chi^2(\tau,\tau')] - \frac{1}{2}\sum_{a\neq b}(\beta V)^2 q_{ab}^2$$
$$- \ln\left[\int d\varepsilon_i P(\varepsilon_i)\int Dc_i^{\dagger a} Dc_i^a \exp\{-S_{\text{eff}}(i)\}\right], \tag{6.84}$$

with $S_{\text{eff}}(i)$ given by eqn (6.80). The self-consistency conditions (eqns (6.81)–(6.83)) then follow from

$$0 = \delta F/\delta G(\tau,\tau'); \; 0 = \delta F/\delta\chi(\tau,\tau'); \; 0 = \delta F/\delta q_{ab}. \tag{6.85}$$

We stress that eqns (6.81)–(6.83) have been derived for the model with *uniform* hopping elements t_{ij} and interaction amplitudes V_{ij}, in the $z \to \infty$ limit, but the *same* equations hold for an infinite-range model where these parameters are random variables.

- *The glass transition*

In our electronic model, the random site energies ε_i play a role of static random fields. As a result, in the presence of disorder, the Edwards–Anderson parameter q_{EA} remains non-zero for any temperature, and thus cannot serve as an order parameter. To identify the glass transition, we search for an RSB instability, following standard methods (de Almeida and Thouless, 1978; Mezard and Young, 1992). We define $\delta q_{ab} = q_{ab} - q$, and expand the free-energy functional of eqn (6.84) around the replica symmetric solution. The resulting quadratic form (Hessian matrix) has the matrix elements given by

$$\frac{\partial^2 F}{\partial q_{ab}\partial q_{cd}} = (\beta V)^2\delta_{ac}\delta_{bd} + [< \delta n_a(\tau_1)\delta n_b(\tau_2) >_{\text{RS}} < \delta n_c(\tau_3)\delta n_d(\tau_4) >_{\text{RS}} \tag{6.86}$$
$$-V^4 \int_0^\beta \int_0^\beta \int_0^\beta \int_0^\beta d\tau_1 d\tau_2 d\tau_3 d\tau_4 [< \delta n_a(\tau_1)\delta n_b(\tau_2)\delta n_c(\tau_3)\delta n_d(\tau_4) >_{\text{RS}}],$$

where the expectation values are calculated in the RS solution. Using standard manipulations (de Almeida and Thouless, 1978), and after lengthy algebra, we finally arrive at the desired RSB stability criterion that takes the form

$$1 = V^2 \left[(\chi_{\text{loc}}(\varepsilon_i))^2 \right]_{\text{dis}}. \tag{6.87}$$

Here, $[...]_{\text{dis}}$ indicates the average over disorder and $\chi_{\text{loc}}(\varepsilon_i)$ is the *local compressibility*, which can be expressed as

$$\chi_{\text{loc}}(\varepsilon_i) = \frac{\partial}{\partial \varepsilon_i} \frac{1}{\beta} \int_o^\beta d\tau < \delta n_i(\tau) >, \tag{6.88}$$

and which is evaluated by carrying out quantum averages for a fixed realization of disorder. The relevant expectation values have to be carried with respect to the full local effective action $S_{\text{eff}}(i)$ of eqn (6.80), evaluated in the replica symmetric (RS) theory. In general, the required computations cannot be carried out in closed form, primarily due to the unknown "memory kernel" $\chi(\tau - \tau')$. However, as we will see, the algebra simplifies in several limits, where explicit expressions can be obtained.

- *Classical electron glass*

In the classical ($t = 0$) limit, the problem can be easily solved in closed form. We first focus on the replica symmetric (RS) solution and set $q_{ab} = q$ for all replica pairs. The corresponding equation reads

$$q = \frac{1}{4} \int_{-\infty}^{+\infty} \frac{dx}{\sqrt{\pi}} e^{-x^2/2} \tanh^2 \left[\frac{1}{2} x \left((\beta V)^2 q + (\beta W)^2 \right)^{1/2} \right], \tag{6.89}$$

where we have considered a Gaussian distribution of random site energies of variance W^2. Note that the interactions introduce an effective, *enhanced* disorder strength

$$W_{\text{eff}} = \sqrt{W^2 + V^2 q}, \tag{6.90}$$

since the frozen-in density fluctuations introduce an added component to the random potential seen by the electrons. As expected, $q \neq 0$ for any temperature when $W \neq 0$. If the interaction strength is appreciable as compared to disorder, we thus expect the resistivity to display an appreciable *increase* at low temperatures. We emphasize that this mechanism is *different* from Anderson localization, which is going to be discussed in the next section, but which also gives rise to a resistivity increase at low temperatures.

Next, we examine the instability to glassy ordering. In the classical ($t = 0$) limit eqn (6.87) reduces to

$$1 = \frac{1}{16} (\beta V)^2 \int_{-\infty}^{+\infty} \frac{dx}{\sqrt{\pi}} e^{-x^2/2} \cosh^{-4} \left[\frac{1}{2} x \beta W_{\text{eff}}(q) \right], \tag{6.91}$$

with $W_{\text{eff}}(q)$ given by eqn (6.90). The resulting RSB instability line separates a low-temperature glassy phase from a high-temperature "bad metal" phase. At large disorder these expressions simplify and we find

$$T_{\text{G}} \approx \frac{1}{6\sqrt{2\pi}} \frac{V^2}{W}, \quad W \to \infty. \tag{6.92}$$

We conclude that T_{G} decreases at large disorder. This is to be expected, since in this limit the electrons drop into the lowest potential minima of the random potential. This defines a unique ground state, suppressing the *frustration* associated with the glassy ordering and thus reducing the glassy phase. It is important to note that for the well known de Almeida–Thouless (AT) line T_{RSB} decreases *exponentially* in the strong-field limit. In contrast, we find that in our case, $T_{\text{G}} \sim 1/W$ decreases only slowly in the strong-disorder limit. This is important, since the glassy phase is expected to be most relevant for disorder strengths sufficient to suppress uniform ordering. At the same time, glassy behavior will only be observable if the associated glass transition temperature remains appreciable.

- *The glassy phase*

To understand this behavior, we investigate the structure of the low-temperature glass phase. Consider the single-particle density of states at $T = 0$, which in the classical limit can be expressed as

$$\overline{\rho}(\varepsilon, t = 0) = \frac{1}{N} \sum_i \delta(\varepsilon - \varepsilon_i^{\text{R}}), \tag{6.93}$$

where $\varepsilon_i^{\text{R}} \equiv \varepsilon_i + \sum V_{ij} n_j$ are the renormalized site energies. In the thermodynamic limit, this quantity is nothing but the probability distribution $P_R(\varepsilon_i^{\text{R}})$. It is analogous to the "local field distribution" in the spin-glass models and can be easily shown to reduce to a simple Gaussian distribution in the RS theory, establishing the *absence* of any gap for $T > T_{\text{G}}$. Obtaining explicit results from a replica calculation in the glass phase is more difficult, but useful insight can be achieved by using standard simulation methods (Palmer and Pond, 1979; Pazmandi *et al.*, 1999) on our equivalent infinite-range model; some typical results are shown in Fig. 6.15. We find that as a result of glassy freezing, a pseudo-gap emerges in the single-particle density of states, reminiscent of the Coulomb gap of Efros and Shklovskii (1975) (ES). The low-energy form of this gap appears *universal*,

$$\rho(\varepsilon) \approx C\varepsilon^{\alpha}/V^2, \text{ for } \varepsilon < E_{\text{g}}; \quad C = \alpha = 1, \tag{6.94}$$

independent of the disorder strength W, again in striking analogy with the predictions of Efros and Shklovskii. To establish this result, we have used stability arguments very similar to those developed for spin-glass models (Pazmandi *et al.*, 1999), demonstrating that the form of eqn (6.94) represents an exact *upper bound* for $\rho(\varepsilon)$. For *infinite-ranged* spin-glass models, as in our case, this bound appears to be *saturated*, leading to universal behavior. Such universality is often associated with a critical, self-organized

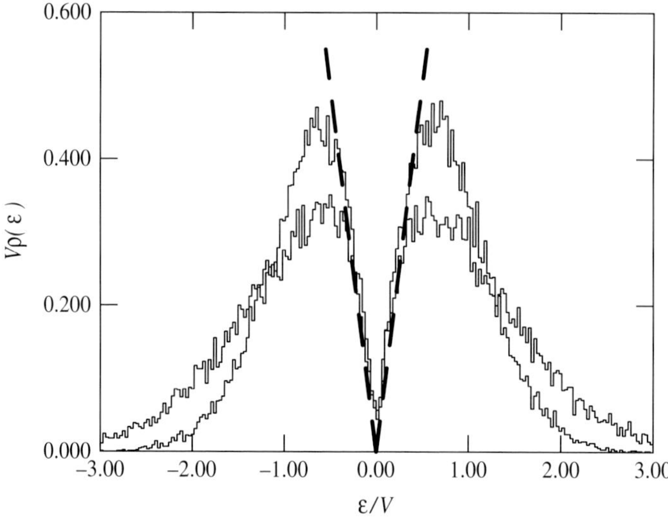

Fig. 6.15 Single particle density of states in the classical ($t = 0$) limit at $T = 0$, as a function of disorder strength. Results are shown from a simulation on $N = 200$ site system, for $W/V = 0.5$ (thin line) and $W/V = 1.0$ (full line). Note that the low-energy form of the gap takes a *universal* form, independent of the disorder strength W. The dashed line follows eqn (6.94).

state of the system. Recent work (Pazmandi *et al.*, 1999) finds strong numerical evidence of such criticality for spin-glass models; we believe that the universal gap form in our case has the same origin. Furthermore, assuming that the universal form of eqn (6.94) is obeyed immediately allows for an estimate of $T_G(W)$. Using eqn (6.87) to estimate the gap size for large disorder gives $T_G \sim E_g \sim V^2/W$, in agreement with eqn (6.92).

The ergodicity breaking associated with the glassy freezing has important consequences for our model. Again, using the close similarity of our classical infinite-range model to standard spin-glass models (Mézard *et al.*, 1986), it is not difficult to see that the *zero-field cooled* (ZFC) compressibility vanishes at $T = 0$, in contrast to the field-cooled one, which remains finite. Essentially, if the chemical potential is modified *after* the system is cooled to $T = 0$, the system immediately falls out of equilibrium and displays hysteretic behavior (Pazmandi *et al.*, 1999) with vanishing *typical* compressibility. If this behavior persists in finite dimensions and for more realistic Coulomb interactions, it could explain the absence of screening in disordered insulators.

- *Arbitrary lattices and finite coordination: mean-field glassy phase of the random-field Ising model*

The simplest theories of glassy freezing (Mézard *et al.*, 1986) are obtained by examining models with random inter-site interactions. In the case of disordered electronic systems, the interactions are not random, but glassiness still emerges due to frustration introduced by the competition of the interactions and disorder. As we have seen for the Bethe lattice (Pastor and Dobrosavljević, 1999), random interactions are generated by renormalization effects, so that standard DMFT approaches can still be used. However, one would like to develop systematic approaches for arbitrary lattices and in finite coordination. These issues already appear on the classical level, where our model reduces to the random-field Ising model (RFIM) (Nattermann, 1997). To investigate the glassy behavior of the RFIM (Pastor *et al.*, 2002), a systematic approach that can incorporate short-range fluctuation corrections to the standard Bragg–Williams theory is the method of Plefka (1982) and Georges *et al.* (1990). This work has shown that:

1. corrections to even the lowest nontrivial order immediately result in the appearance of a glassy phase for sufficiently strong randomness
2. this low-order treatment is sufficient in the joint limit of large coordination and strong disorder
3. the structure of the resulting glassy phase is characterized by universal hysteresis and avalanche behavior emerging from the self-organized criticality of the ordered state.

- *Long-range Coulomb interactions and the Efros–Shklovskii gap*

So far we have focused on the limit of large coordination, where the EDMFT approximation is essentially exact and the resulting electron glass phase shows many similarities to the standard Parisi theory of spin glasses (Mézard *et al.*, 1986). While it produced many appealing results, this mean-field approach has remained very controversial for short-range spin glasses in finite dimensions; a competing "droplet" theory (Fisher and Huse, 1986) approach suggested that many of Parisi's predictions may not survive in physical dimensions $d = 2$ or 3. The situation is more promising in the case of the physically relevant long-ranged Coulomb interaction, where the prediction of the mean-field approach may possibly persist even in low dimensions.

The most interesting test of these ideas relates to the possibility of describing the emergence of the "Coulomb gap" for $d = 2, 3$ for localized electrons interacting via long-range Coulomb interactions, as first predicted, based on heuristic arguments, a long time ago (Efros and Shklovskii, 1975). Here, the single-particle density of states is predicted to assume a power-law form $g(\varepsilon) \sim \varepsilon^{\gamma}$, notably with a dimensionality-dependent exponent $\gamma = d - 1$.

Conventional mean-field theories typically produce universal dimensionality independent exponents and thus cannot be expected to explain this unfamiliar situation. On the other hand, the EDMFT (Chitra and Kotliar, 2000) approach does include the effects of spatial correlations, as the bosonic collective modes describing the "cavity field" are treated at a Gaussian level, similar to the familiar Hertz–Millis theories of quantum criticality (Hertz, 1976; Millis, 1993). When applied to the case of long-range Coulomb interactions, the corresponding plasmon propagator does reflect (Chitra and Kotliar, 2000) both the specific form of the long-range interactions and the dimensionality of the system. When applied to disordered electrons, the replica version of this method produces self-consistency conditions which assume a very similar form as in ordinary Parisi theory or the simple infinite range electron glass model we examined in previous sections. Indeed, recent work has generalized our EDMFT method to the long-range models, where the glassy phase of the generalized Coulomb interactions (Pankov and Dobrosavljević, 2005) of the form $V(R) \sim 1/R^{\alpha}$ has been examined in finite dimensions. We will not elaborate of the details of these theories here, but we emphasize that most qualitative features of the resulting glassy phase have been found to share very similar behavior (Muller and Ioffe, 2004) to the infinite-range model of electron glasses, including the corresponding pattern of replica symmetry breaking (Müller and Pankov, 2007; Pankov, 2006). Most remarkably, however, this theory produced (Pankov and Dobrosavljević, 2005) an interaction-range- and dimensionality-dependent Coulomb-gap exponent

$$\gamma = (d - \alpha)/\alpha, \tag{6.95}$$

in precise agreement with an appropriate generalization of the Efros-Shklovskii argument (Efros and Shklovskii, 1975). This early result was later confirmed (Müller and Pankov, 2007) by a more detailed low-temperature solution of the same equations deep in the replica-symmetry broken phase, based on a new $T = 0$ solution of the Parisi equation due to Pankov (Pankov, 2006).

Another striking result of these theories should be noted. Namely, one finds (Pankov and Dobrosavljević, 2005) that the Coulomb pseudogap starts to form around crossover temperatures $T^* \sim T_{\mathrm{C}}$ (the Coulomb energy), much above the glass transition temperature $T_{\mathrm{G}} \approx 0.05T_{\mathrm{C}}$, and in perfect quantitative agreement (Fig. 6.16 with earlier numerical work (Grannan and Yu, 1993). The EDMFT theory, in fact, demonstrated the following: (1) the universal value of the Efros-Shklovskii exponent γ is a direct consequence of the marginal stability of the glassy phase; (2) in contrast, the emergence of a non-universal Coulomb pseudogap in the high-temperature regime $T_{\mathrm{G}} < T < T^*$ is not related to glassy freezing and the breakdown of screening (Muller and Ioffe, 2004; Müller and Pankov, 2007), and is a robust effect found (Efros, 1992) even in the absence of disorder. The two mechanisms for pseudogap formation have often been confused (Grannan and Yu, 1993) in previous work, leading to incorrect and even misleading interpretations of what determines its form in a given temperature range.

- *Quantum melting of the electron glass*

Next, we investigate how the glass transition temperature can be depressed by quantum fluctuations introduced by inter-site electron tunneling. For simplicity, we again focus on the simplest infinite-range model of the electron glass. As in other quantum glass problems, quantum fluctuations introduce dynamics in the problem, and the relevant self-consistency equations cannot be solved in closed form for general values of the parameters. In the following, we will see that in the limit of large randomness an exact solution is possible. The main source of difficulty in general quantum glass problems relates to the existence of a self-consistently determined

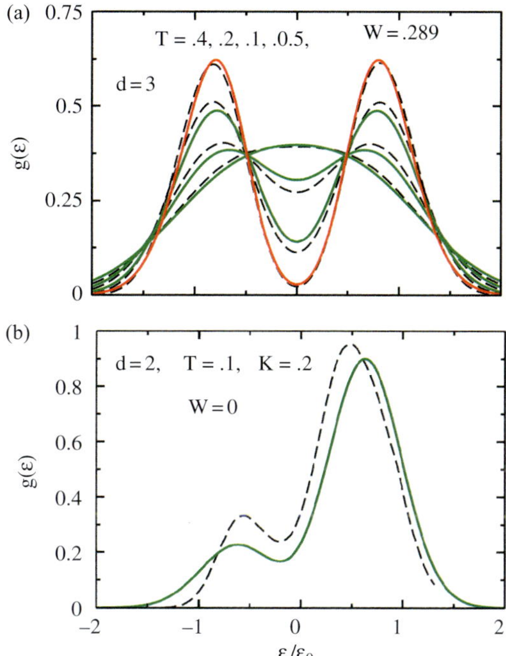

Fig. 6.16 The analytical EDMFT predictions (Pankov and Dobrosavljević, 2005) for the single-particle density of states (full lines) are found to be in excellent quantitative agreement with simulation results (dashed lines), with no adjustable parameters. Shown are results for the three-dimensional case studied in Sarvestani *et al.* (1995), corresponding to $W = 1/(2\sqrt{3})$, and temperatures $T = 0.4, 0.2, 0.1, 0.05$ (a), and the two-dimensional model of Efros (1992), corresponding to $W = 0$, $T = 0.1$, $K = 0.2$. Green lines correspond to the fluid pseudogap phase, while the red curve corresponds to $T = 0.05$, close to the glass transition temperature.

"memory kernel" $\chi(\tau - \tau')$ in the local effective action. By the same reasoning as in the classical case, one can also ignore this term since this quantity is also bounded.

The remaining action is that of *non-interacting* electrons in the presence of a strong random potential. The resulting *local* compressibility then takes the form

$$\chi_{\rm loc}(\varepsilon) = \frac{\beta}{4} \int_{-\infty}^{+\infty} d\omega \rho_\varepsilon(\omega) \cosh^{-2}(\frac{1}{2}\beta\omega). \tag{6.96}$$

Here, $\rho_\varepsilon(\omega)$ is the local density of states, which in the considered large z limit is determined by the solution of the CPA equation

$$\rho_\varepsilon(\omega) = -\frac{1}{\pi}{\rm Im}G(\omega); \;\; G(\omega) = \int \frac{d\varepsilon P(\varepsilon)}{\omega + i\eta - \varepsilon - t^2 G(\omega)}. \tag{6.97}$$

In the limit $W/t >> 1$, it reduces to a narrow resonance of width $\Delta = \pi t^2 P(0) \sim t^2/W$

$$\rho_\varepsilon(\omega) \approx \frac{1}{\pi} \frac{\Delta}{(\omega-\varepsilon)^2 + \Delta^2}. \tag{6.98}$$

The resulting expression for the quantum critical line in the large-disorder limit takes the form

$$t_G(T=0, W \to \infty) = V/\sqrt{\pi}. \tag{6.99}$$

At first glance, this result is surprising, since it means that a *finite* value of the Fermi energy is required to melt the electron glass at $T = 0$, *even* in the $W \to \infty$ limit! This is to be contrasted with the behavior of T_G in the classical limit, which according to eqn (6.92) was found to decrease as $1/W$ for strong disorder. At first puzzling, the above result in fact has a simple physical meaning: the small resonance width (or "hybridization energy") $\Delta \sim t^2/W$ can be interpreted (Anderson, 1958; Dobrosavljević and Kotliar, 1997) as the characteristic energy scale for the electronic motion. As first pointed out by Anderson (1958), according to Fermi's golden rule, the transition rate to a neighboring site is proportional to Δ and not t, and thus becomes extremely small at large disorder. Thus the "size" of quantum fluctuations, which replace the thermal fluctuations at $T = 0$, is proportional to $\Delta \sim 1/W$, and thus becomes very small in the large W limit. We can now easily understand the qualitative behavior shown in eqn (6.99) by replacing $T_G \to \Delta \sim t_G^2/W$ in eqn (6.92). The leading W-dependence *cancels out*, and we find a *finite* value for t_G in the $W \to \infty$ limit.

More generally, we can write an expression for the glass-transition critical line in the large disorder limit, as a function of $\beta = 1/T$ and t in the scaling form

$$1 = (V/t)^2 \phi(\beta t^2/W), \tag{6.100}$$

with

$$\phi(z) = \frac{1}{4} z^2 \int_{-\infty}^{+\infty} dx \left[\int_{-\infty}^{+\infty} dy \frac{1}{\pi} \frac{1}{1+(x-y)^2} \cosh^{-2}(\frac{1}{2} zy) \right]^2. \tag{6.101}$$

At finite disorder an exact solution is not possible, but we can make analytical progress motivated by our discussion of the large-W limit. Namely, one can imagine evaluating the required local compressibilities in eqn (6.88) by a "weak coupling" expansion in powers of the interaction V. To leading order, this means evaluating the compressibilities at $V = 0$, an approximation which becomes exact for W large. Such an approximation can be tested for other spin-glass problems. We have carried out the corresponding computations for the infinite range Ising spin glass model in a transverse field, where the exact critical transverse field is known from numerical studies. We can expect the leading approximation to *underestimate* the size of the glassy region, i. e. the critical field, since the omitted "memory kernel" introduces long range correlations in time, which make the system more "classical." Indeed, we find that the leading approximation underestimates the critical field by only about 30%, whereas the next order correction gives an error of less than 5%. Encouraged by these

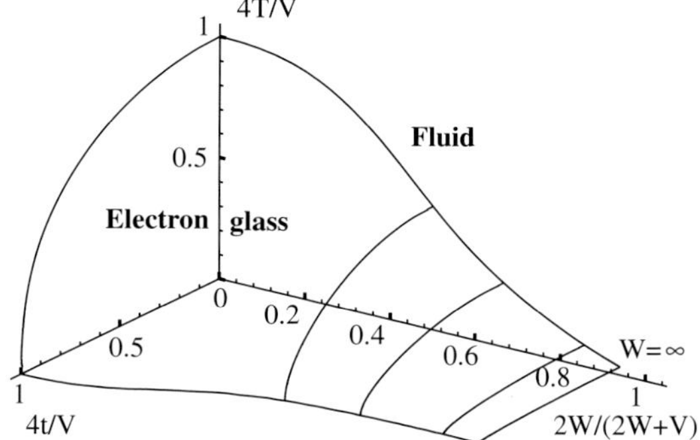

Fig. 6.17 Phase diagram as a function of quantum hopping t, temperature T, and disorder strength W. The glass transition temperature T_{G} decreases only slowly (as $\sim 1/W$) in the strong-disorder limit. In contrast, the critical value of the hoping element t_{G} remains finite as $W \to \infty$.

arguments, we use this "weak-coupling" approximation for arbitrary disorder strength W. Again, the computation of the compressibility reduces to that of noninteracting electrons in a CPA formulation; the resulting phase diagram is shown in Fig. 6.17.

- *Quantum critical behavior of the electron glass*

So far, we have seen how our extended DMFT equations can be simplified for large disorder, allowing an exact computation of the phase boundary in this limit. In our case, this quantum critical line separates a (non-glassy) Fermi liquid phase from a metallic glass phase which, as we will see, features non-Fermi liquid behavior. If one is interested in the details of the *dynamics* of the electrons near the quantum critical line, the above simplifications do not apply and one is forced to self-consistently calculate the form of the "memory kernel" (local dynamic compressibility) $\chi(\tau - \tau')$. Fortunately, this task can be carried out using methods very similar to those developed for DMFT models for metallic spin glasses (Read *et al.*, 1995). Formulating such a theory is technically possible because the exact quantum critical behavior is captured when the relevant field theory is examined at the Gaussian level (Miller and Huse, 1993), in the considered limit of large dimensions.

Because of the technical complexity of this calculation, we only report the main results, while the details can be found in (Dalidovich and Dobrosavljević, 2002). In this paper, the full RSB solution was found, both around the quantum critical line and in the glassy phase. In the Fermi-liquid phase, the memory kernel was found to take the form

$$V^2\chi(\omega_n) = D(\omega_n) + \beta q_{\mathrm{EA}}\delta_{\omega_n,0},$$

with

$$D(\omega_n) = -yq_{\mathrm{EA}}^2/V^4 - \sqrt{|\omega_n| + \Delta}.$$

Here, Δ is a characteristic energy scale that vanishes on the critical line, which also determines the crossover temperature scale separating the Fermi liquid from the quantum-critical regime. In contrast to conventional quantum critical phenomena, but similarly to metallic spin glasses, the "gap" scale Δ is equal to zero not only on the critical line, but remains zero *throughout the entire glassy phase.* As a result, the excitations in this region assume a non-Fermi-liquid form

$$D(\omega_n) = -yq_{\rm EA}^2/V^4 - \sqrt{|\omega_n|}.$$

This behavior reflects the emergence of soft "replicon" modes (Mézard *et al.*, 1986) describing in our case low-energy charge rearrangements inside the glassy phase. At finite temperatures, electrons undergo inelastic scattering from such collective excitations, leading to a temperature dependence of the resistivity that takes the following non-Fermi-liquid form

$$\rho(T) = \rho(o) + AT^{3/2}.$$

Interestingly, very recent experiments (Bogdanovich and Popović, 2002) on two-dimensional electron gases in silicon have revealed precisely such a temperature dependence of the resistivity. This behavior has been observed in what appears to be an intermediate metallic glass phase separating a conventional (Fermi liquid) metal at high carrier density from an insulator at the lowest densities.

Another interesting feature of the predicted quantum critical behavior relates to the disorder dependence of the crossover exponent ϕ describing how the gap scale $\Delta \sim \delta r^{\phi}$ vanishes as a function of the distance δr from the critical line. Calculations (Arrachea *et al.*, 2004) show that $\phi = 2$ in the presence of site energy disorder, which for our model plays the role of a random symmetry breaking field, and $\phi = 1$ in its absence. This indicates that site disorder, which is common in disordered electronic systems, produces a particularly large quantum critical region, which could be the origin of the large dephasing observed in many materials near the MIT.

- *Glassy behavior near the Mott–Anderson transition*

As we have seen, the stability of the glassy phase is crucially determined by the electronic mobility at $T = 0$. More precisely, we have shown that the relevant energy scale that determines the size of quantum fluctuations introduced by the electrons is given by the local "resonance width" Δ. It is important to recall that precisely this quantity may be considered (Anderson, 1958) as an order parameter for Anderson localization of non-interacting electrons. Recent work (Dobrosavljević and Kotliar, 1997, 1998) demonstrated that the *typical* value of this quantity plays the same role even at a Mott–Anderson transition. We thus expect Δ to generally vanish in the insulating state. As a result, we expect the stability of the glassy phase to be strongly affected by Anderson localization effects, as we will explicitly demonstrate in the next section.

On physical grounds, one expects the quantum fluctuations (Pastor and Dobrosavljević, 1999) associated with mobile electrons to suppress glassy ordering, but their precise effects remain to be elucidated. Note that even the *amplitude* of such quantum fluctuations must be a singular function of the distance to the MIT, since they are dynamically determined by processes that control the electronic mobility.

To clarify the situation, the following basic questions need to be addressed: (1) Does the MIT coincide with the onset of glassy behavior? (2) How do different physical processes that can localize electrons affect the stability of the glass phase? In the following, we provide simple and physically transparent answers to both questions. We find that: (a) glassy behavior generally emerges before the electrons localize and (b) Anderson localization (Anderson, 1958) enhances the stability of the glassy phase, while Mott localization (Mott, 1990) tends to suppress it.

In order to be able to examine both the effects of Anderson and Mott localization, we concentrate on extended Hubbard models given by the Hamiltonian

$$H = \sum_{ij\sigma}(-t_{ij} + \varepsilon_i \delta_{ij}) c^{\dagger}_{i,\sigma} c_{j,\sigma} + U \sum_i n_{i\uparrow} n_{i\downarrow} + \sum_{ij} V_{ij} \delta n_i \delta n_j .$$

Here, $\delta n_i = n_i - \langle n_i \rangle$ represent local density fluctuations ($\langle n_i \rangle$ is the site-averaged electron density), U is the on-site interaction, and ε_i are Gaussian-distributed random site energies of variance W^2. In order to allow for glassy freezing of electrons in the charge sector, we introduce weak inter-site density–density interactions V_{ij}, which we also choose to be Gaussian-distributed random variables of variance V^2/z (z is the coordination number). We emphasize that, in contrast to previous work (Dobrosavljević *et al.*, 2003*a*), we shall now keep the coordination number z finite, in order to allow for the possibility of Anderson localization. To investigate the emergence of glassy ordering, we formally average over disorder by using standard replica methods (Dobrosavljević *et al.*, 2003*b*) and introduce collective Q-fields to decouple the inter-site V-term (Dobrosavljević *et al.*, 2003*b*). A mean-field is then obtained by evaluating the Q-fields at the saddle-point level. The resulting stability criterion takes a form similar to the one previously discussed (eqn 6.87)

$$1 - V^2 \sum_j [\chi_{ij}^2]_{\text{dis}} = 0. \tag{6.102}$$

Here, the non-local static compressibilities are defined (for a fixed realization of disorder) as

$$\chi_{ij} = -\partial n_i / \partial \varepsilon_j , \tag{6.103}$$

where n_i is the local quantum expectation value of the electron density, and $[\cdots]_{\text{dis}}$ represents the average over disorder. Obviously, the stability of the glass phase is determined by the behavior of the fourth-order correlation function $\chi^{(2)} = \sum_j [\chi_{ij}^2]_{\text{dis}}$ in the vicinity of the MIT. We emphasize that this quantity is to be calculated in a disordered Hubbard model with finite range hopping, i.e. in the vicinity of the Mott–Anderson transition. The critical behavior of $\chi^{(2)}$ is very difficult to calculate in general, but we will see that simple results can be obtained in the limits of weak and strong disorder, as follows.

6.4.3 Large disorder

As the disorder grows, the system approaches the Anderson transition at $t = t_c(W) \sim W$. The first hint of singular behavior of $\chi^{(2)}$ in an Anderson insulator is seen by examining the deeply insulating, i.e. atomic limit $W \gg t$, where to leading order we set $t = 0$ and obtain $\chi_{ij} = \delta(\varepsilon_i - \mu)\delta_{ij}$, i.e. $\chi^{(2)} = [\delta^2(\varepsilon_i - \mu)]_{\text{dis}} = +\infty$! Since we expect all quantities to behave in qualitatively the same fashion throughout the insulating phase, we anticipate $\chi^{(2)}$ to diverge already at the Anderson transition. Note that, since the instability of the glassy phase occurs already at $\chi^{(2)} = V^{-2}$, the glass transition must *precede* the localization transition. Thus, for any finite inter-site interaction V, we predict the emergence of an intermediate *metallic glass phase* separating the Fermi liquid from the Anderson insulator. Assuming that near the transition

$$\chi^{(2)} \simeq \frac{A}{W^2}((t/W) - B)^{-\alpha} \tag{6.104}$$

(A and $B = t_c/W$ are constants of order unity), from eqn (6.102) we can estimate the form of the glass-transition line and we get

$$\delta t(W) = t_G(W) - t_c(W) \sim V^{2/\alpha} W^{1-2/\alpha};\; W \to \infty. \tag{6.105}$$

The glass transition and the Anderson transition lines are predicted to converge at large disorder for $\alpha < 2$ and diverge for $\alpha > 2$. Since all the known exponents characterizing the localization transition seem to grow with dimensionality, we may expect a particularly large metallic glass phase in large dimensions.

Bethe lattices. In order to confirm this scenario by explicit calculations, we compute the behavior of $\chi^{(2)}$ at the Anderson transition of a half-filled Bethe lattice of coordination $z = 3$. We use an essentially exact numerical approach (Dobrosavljević and Kotliar, 1997) based on the recursive structure of the Bethe lattice (Abou-Chacra *et al.*, 1973). In this approach, local and non-local Green's functions on a Bethe lattice can be sampled from a large ensemble and the compressibilities χ_{ij} can then be calculated by examining how a local charge density n_i is modified by an infinitesimal variation of the local site energy ε_j on another site. To do this, we have taken special care in evaluating the local charge densities n_i by numerically computing the required frequency summations over the Matsubara axis, where the numerical difficulties are minimized. Using this method, we have calculated $\chi^{(2)}$ as a function of W/t (for this lattice at half-filling $E_F = 2\sqrt{2}t$), and find that it decreases exponentially (Mirlin and Fyodorov, 1991) as the Anderson transition is approached. We emphasize that only a finite enhancement of $\chi^{(2)}$ is required to trigger the instability to glassy ordering, which therefore occurs well before the Anderson transition is reached. The resulting $T = 0$ phase diagram, valid in the limit of large disorder, is presented in Fig. 6.18. Note that the glass-transition line in this case has the form $t_G(W) \sim W$, in agreement with the fact that exponential critical behavior of $\chi^{(2)}$ corresponds to $\alpha \to \infty$ in the above general scenario. These results are strikingly different from those obtained in a theory that ignores localization (Pastor and Dobrosavljević, 1999), where $t_G(W)$ was found to be weakly dependent on disorder and remain *finite* as $W \longrightarrow \infty$. Anderson

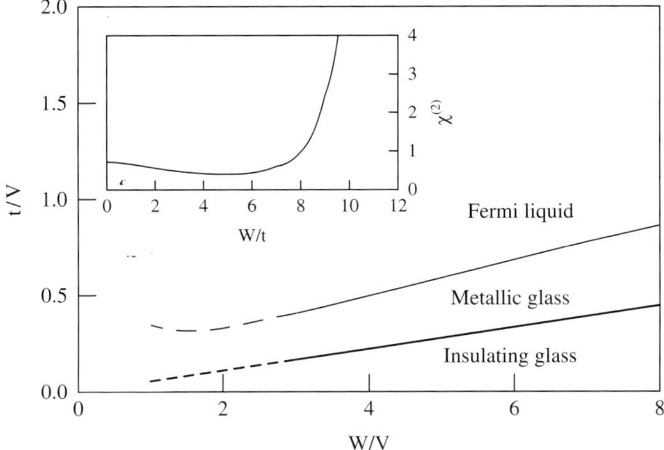

Fig. 6.18 Phase diagram for the $z = 3$ Bethe lattice, valid in the large-disorder limit. The inset shows $\chi^{(2)}$ as a function of disorder W.

localization effects thus strongly enhance the stability of the glass phase at sufficiently large disorder. Nevertheless, since the Fermi liquid to metallic glass transition occurs at a finite distance *before* the localization transition, we do not expect the leading quantum critical behavior (Dalidovich and Dobrosavljević, 2002) at the transition to be qualitatively modified by localization effects.

Typical medium treatment. As an alternative approach to the Bethe lattice calculation, in this section we introduce Anderson localization to the problem by using the formalism of TMT (Dobrosavljević *et al.*, 2003*a*), which was explained in detail in Section 6.2. We calculate the cavity field $\Delta_{\text{typ}}(\omega)$ by solving the relevant self-consistency condition (Dobrosavljević *et al.*, 2003*a*), which in turn allows us to find local compressibilities:

$$\chi_{ii} = -\frac{\partial n}{\partial \varepsilon_i} = \frac{1}{\pi}\frac{\partial}{\partial \varepsilon_i}\int_{-\infty}^{0} d\omega Im G(\varepsilon_i, \omega, W), \tag{6.106}$$

$$G(\varepsilon_i, \omega, W) = \frac{1}{\omega - \varepsilon_i - \Delta_{\text{typ}}(\omega)}, \tag{6.107}$$

needed to determine the critical line of the glass transition. These calculations were performed using a model semicircular bare density of states $\rho_0(\omega)$ and a box distribution of disorder $P(\varepsilon_i)$. The resulting phase diagram is shown in Fig. 6.19. The intermediate metallic glassy phase still exists, but shrinks as $W \to \infty$, reflecting the small value of the critical exponent $\alpha = 1$, which can be shown analytically within TMT. A more realistic value for this exponent, corresponding to $d = 3$, requires more detailed numerical calculations, which remains a challenge for future work.

Low disorder: Mott transition. In the limit of weak disorder $W \ll U, V$, interactions drive the MIT. Concentrating on the model at half-filling, the system will undergo a

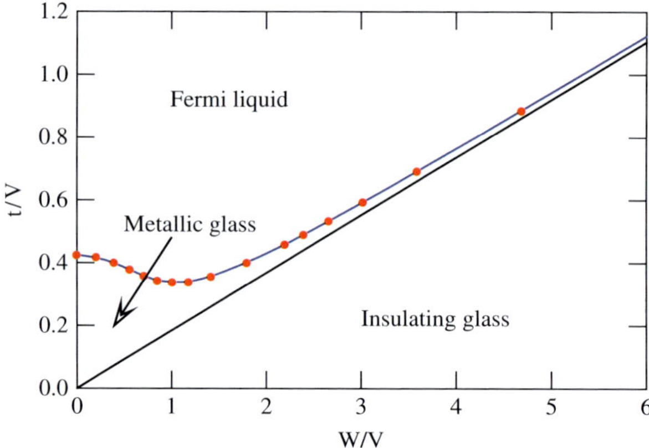

Fig. 6.19 Phase diagram from TMT of Anderson localization (Dobrosavljević *et al.*, 2003*a*), giving $\alpha = 1$. The intermediate metallic glass phase shrinks as disorder W grows. Compare this to the Bethe lattice case (Fig. 6.18), where $\alpha = \infty$.

Mott transition (Mott, 1990) as the hopping t is sufficiently reduced. Since for the Mott transition $t_{\mathrm{Mott}}(U) \sim U$, near the transition $W \ll t$ and to leading order we can ignore localization effects. In addition, we assume that $V \ll U$ and to leading order the compressibilities have to be calculated with respect to the action S_{el} of a disordered Hubbard model. The simplest formulation that can describe the effects of weak disorder on such a Mott transition is obtained from DMFT (Georges *et al.*, 1996). This formulation, which ignores localization effects, is obtained by rescaling the hopping elements $t \to t/\sqrt{z}$ and then formally taking the limit of large coordination $z \to \infty$ (see Section 6.1.4). To obtain qualitatively correct analytical results describing the vicinity of the disordered Mott transition at $T = 0$, we have solved the DMFT equations using a four-boson method (Dobrosavljević *et al.*, 2003*b*). At weak disorder, these equations can be easily solved in closed form and we simply report the relevant results. The critical value of hopping for the Mott transition is found to decrease with disorder as

$$t_{\mathrm{c}}(W) \approx t_{\mathrm{c}}^{o}\,(1 - 4(W/U)^2 + \cdots), \tag{6.108}$$

where for a simple semicircular density of states (Georges *et al.*, 1996) $t_{\mathrm{c}}^{o} = 3\pi U/64$ (in this model, the bandwidth $B = 4t$). Physically, the disorder tends to suppress the Mott insulating state, since it broadens the Hubbard bands and narrows the Mott–Hubbard gap. At sufficiently strong disorder $W \geq U$, the Mott insulator is suppressed even in the atomic limit $t \to 0$. The behavior of the compressibilities can also be calculated near the Mott transition and to leading order we find

$$\chi^{(2)} = \left[\frac{8}{3\pi t_{\mathrm{c}}^{o}}(1 - \frac{t_{\mathrm{c}}(W)}{t})\right]^2 (1 + 28(W/U)^2). \tag{6.109}$$

Therefore, as with any compressibility, $\chi^{(2)}$ is found to be very small in the vicinity of the Mott transition, even in the presence of finite disorder. As a result, the tendency to glassy ordering is strongly suppressed at weak disorder when one approaches the Mott insulating state.

Finally, having analyzed the limits of weak and strong disorder, we briefly comment on what may be expected in the intermediate region $W \sim U$. The Mott gap cannot exist for $W > U$, so in this region and for sufficiently small t (i.e. kinetic energy), one enters a gapless (compressible) Mott–Anderson insulator. For $W \sim U$, the computation of $\chi^{(2)}$ requires the full solution of the Mott–Anderson problem. The required calculations can and should be performed using the formulation of Dobrosavljević and Kotliar (1997), but that difficult task is a challenge for the future. However, based on general arguments presented above, we expect $\chi^{(2)}$ to *vanish* as one approaches the Mott insulator ($W < U$), but to *diverge* as one approaches the Mott–Anderson insulator ($W > U$). Near the tetracritical point, we may expect $\chi^{(2)} \sim \delta W^{-\beta} \delta t^{\alpha}$, where $\delta W = W - W_{\text{Mott}}(t)$ is the distance to the Mott transition line and $\delta t = t - t_{\text{c}}(W)$ is the distance to the Mott–Anderson line. Using this ansatz and eqn (6.102), we find the glass-transition line to take the form

$$\delta t = t_G(W) - t_{\text{c}}(W) \sim \delta W^{\beta/\alpha}; \; W \sim W_{\text{Mott}}. \tag{6.110}$$

We thus expect the intermediate metallic glass phase to be suppressed as the disorder is reduced, and one approaches the Mott-insulating state. Physically, glassy behavior of electrons corresponds to many low-lying rearrangements of the charge density; such rearrangements are energetically unfavorable close to the (incompressible) Mott insulator, since the on-site repulsion U opposes charge fluctuations. Interestingly, very recent experiments on low density electrons in silicon MOSFETs have revealed the existence of exactly such an intermediate metallic glass phase in low mobility (highly disordered) samples (Bogdanovich and Popović, 2002). In contrast, in high-mobility (low-disorder) samples (Jaroszyński *et al.*, 2002), no intermediate metallic glass phase is seen and glassy behavior emerges only as one enters the insulator, consistent with our theory. Similar conclusions have also been reported in studies of highly disordered InO_2 films (Ben-Chorin *et al.*, 1993; Ovadyahu and Pollak, 1997), where the glassy slowing down of the electron dynamics seems to be suppressed as the disorder is reduced and one crosses over from an Anderson-like to a Mott-like insulator. In addition, these experiments (Bogdanovich and Popović, 2002; Jaroszyński *et al.*, 2002) provide striking evidence of scale-invariant dynamical correlations inside the glass phase, consistent with the hierarchical picture of glassy dynamics, that generally emerges from mean-field approaches (Mézard *et al.*, 1986) such as the one used in this work.

6.5 Beyond DMFT: loop expansion and diffusion modes

DMFT approaches to correlation and disorder have already provided significant insight in a number of key phenomena and processes around the MIT. Most importantly, these methods have made it possible to describe strong correlation effects associated with both "Mottness" and the glassy behavior of electrons. The information provided is, in many ways, rather complementary to the conventional weak-coupling approaches which focused on the effects of long-wavelength diffusion modes within the Fermi liquid picture. Because of their focus on local correlation effects, the DMFT theories are ill-suited to describe those phenomena and regimes where long-range spatial correlations dominate. One of the biggest challenges for future work is to find ways to combine the strength of both approaches and introduce systematic methods to incorporate the non-local spatial correlations ignored by DMFT. While little has been done so far to implement this ambitious program, some approaches have already been outlined that provide guidance on how the problem should be formulated. We conclude our discussion by a brief outline of a promising approach to include both the local correlation processes and the diffusion-mode effects within a single theory.

6.5.1 Gauge-invariant models of Wegner

The task of identifying the leading non-local corrections to DMFT can be formulated in a particularly elegant and transparent fashion for certain special models of disorder. Here, the DMFT equations can be obtained (Dobrosavljević and Kotliar, 1994) from a functional-integral formulation at the saddle-point level, allowing systematic corrections using a loop expansion. In this class, the inter-site hopping elements are assumed to be random variables of the form

$$t_{ij} = y_{ij}\, g(x_i, x_j), \tag{6.111}$$

and, in addition, there can be an arbitrary distribution of site energies ε_i. Here, the y_{ij} are independent *bond* variables with a symmetric distribution, i.e. $\overline{y_{ij}^{2n+1}} = 0$, and $g(x_i, x_j)$ is an arbitrary function of local *site* variables x_i.

The special class of models that have a symmetric distribution of hopping elements has a very simple physical interpretation. As first observed by Wegner (Schaffer and Wegner, 1980; Wegner, 1979), in these "gauge-invariant" models, the phases of the electrons undergo random shifts at every lattice hop and so the mean-free path ℓ reduces to one lattice spacing. On general grounds, on length scales longer than ℓ the details of the lattice structure are washed out by disorder, so that for gauge-invariant models in the large-dimensionality limit, the details of the lattice structure become irrelevant. We contrast this with the models with arbitrary disorder discussed previously, which have a well-defined pure limit and accordingly can also have an arbitrarily large mean-free path. The presence of this intermediate length scale (ℓ can be much larger than the lattice spacing a, but much smaller than the localization length ξ) is often irrelevant to both long-wavelength phenomena such as localization and local phenomena such as the Mott transition. The gauge-invariant models avoid these unnecessary complications without disrupting any of the qualitative properties on either very short or very long length scales.

The general properties are the same for all the models in this class, but for simplicity of our presentation, we will restrict our attention to the separable case (Shiba, 1971) where $g(x_i, x_j) = x_i x_j$, with an arbitrary distribution $P_X(x_i)$ for the site variables x_i. For the trivial choice of $P_X(x_i) = \delta(x_i - 1)$, the models reduce to the gauge-invariant models of Wegner (Schaffer and Wegner, 1980; Wegner, 1979). Non-trivial distributions $P_X(x_i)$ that extend to small values of the variable x_i are useful for the study of disorder-induced local moment formation (Milovanović *et al.*, 1989). Sites with small x_i represent sites with weak hybridization. At intermediate correlations, we expect sites with small x_i to behave as local moments and give large contributions to thermodynamic quantities such as the specific heat coefficient $\gamma = C/T$, while other sites remain in the itinerant regime.

Also, we take the y_{ij} to be Gaussian random variables with zero mean and variance

$$\overline{y_{ij}^2} = \frac{1}{z} f_{ij}\, t^2. \tag{6.112}$$

Here, the (uniform) matrix f_{ij} specifies the lattice structure

$$f_{ij} = \begin{cases} 1, & \text{for connected sites,} \\ 0, & \text{for disconnected sites,} \end{cases} \tag{6.113}$$

and we have scaled the (square of the) hopping elements by the coordination number $z = \sum_j f_{ij}$, in order to obtain finite results in the $z \to \infty$ limit.

6.5.2 Functional integral formulation

At this point, it is convenient to explicitly perform the averaging over the Gaussian random (bond) variables y_{ij}, using the standard replica formulation (Schaffer and Wegner, 1980; Wegner, 1979). The hopping part of the action then assumes the form (Dobrosavljević and Kotliar, 1994)

$$S_{\text{hop}} = \frac{1}{2} t^2 \sum_{ij} \frac{1}{z} f_{ij}\, x_i^2\, x_j^2 \left[\sum_{\alpha,s} \int_o^\beta d\tau \left[c_{s,i}^{\dagger\,\alpha}(\tau) c_{s,j}^{\alpha}(\tau) + h.c. \right] \right]^2 . \tag{6.114}$$

As we can see from this expression, the averaging over disorder has generated a *quartic* term in the action that is non-local in (imaginary) time, spin and replica indices. We are now in a position to introduce collective Q-fields (Finkel'stein, 1983, 1984; Schaffer and Wegner, 1980; Wegner, 1979) of the form (in terms of Matsubara frequencies $\omega = 2n\pi T$; the indices "n" are omitted for brevity)

$$Q_{\omega_1 \omega_2}^{\alpha_1 \alpha_2, s_1 s_2}(i) = \frac{1}{z} \sum_j f_{ij}\, x_j^2\, c_{j,s_1}^{\dagger\,\alpha_1}(\omega_1) c_{j,s_2}^{\alpha_2}(\omega_2), \tag{6.115}$$

by decoupling the (quartic) hopping term using a Hubbard–Stratonovich transformation. For simplicity, as before, we will ignore the superconducting phases, as well as the fluctuations in the particle–particle (Cooper) channel, so that the Q-field does not have anomalous components. The procedure can be straightforwardly generalized to include the omitted terms (Efetov *et al.*, 1980).

It is now possible to formally integrate out the electron (Grassmann) fields and the resulting action for the Q-fields can be written as

$$S[Q] = S_{\rm hop}[Q] \; + \; S_{\rm loc}[Q]. \tag{6.116}$$

The non-local part of the action $S_{\rm hop}[Q]$ takes a simple quadratic form in terms of the Q fields

$$S_{\rm hop}[Q] = -\frac{1}{2}\, t^2 \sum_{ij} \sum_{\alpha_1\alpha_2} \sum_{s_1 s_2} \sum_{\omega_1\omega_2} K_{ij}\, Q^{\alpha_1\alpha_2, s_1 s_2}_{\omega_1\omega_2}(i) Q^{\alpha_2\alpha_1, s_2 s_1}_{\omega_2\omega_1}(j), \tag{6.117}$$

where, $K_{ij} = \frac{1}{z} f^{-1}_{ij}$ is the inverse lattice matrix, scaled by coordination number z. In contrast, all the non-linearities are contained in the *local* part of the action

$$S_{\rm loc}[Q] = -\sum_i \ln \int dx_i P_X(x_i) \int d\varepsilon_i P_S(\varepsilon_i) \int Dc_i^\dagger Dc_i \exp\left\{ -S_{\rm eff}[c_i^\dagger, c_i, Q_i, x_i, \varepsilon_i] \right\}, \tag{6.118}$$

where the local effective action takes the form

$$\begin{aligned} S_{\rm eff}[c_i^\dagger, c_i, Q_i, x_i, \varepsilon_i] = \\ -\sum_{\alpha_1\alpha_2} \sum_{s_1 s_2} \sum_{\omega_1\omega_2} c^{\dagger\,\alpha_1}_{i,s_1}(\omega_1)[(i\omega_1 + \mu - \varepsilon_i)\,\delta_{\alpha_1\alpha_2}\delta_{s_1 s_2}\delta_{\omega_1\omega_2} - x_i^2\, t^2\, Q^{\alpha_1\alpha_2, s_1 s_2}_{\omega_1\omega_2}(i)] c^{\alpha_2}_{i,s_2}(\omega_2) \\ +U \sum_\alpha \sum_{\omega_1+\omega_3=\omega_2+\omega_4} c^{\dagger\,\alpha}_{i,\uparrow}(\omega_1) c^{\alpha}_{i,\uparrow}(\omega_2) c^{\dagger\,\alpha}_{i,\downarrow}(\omega_3) c^{\alpha}_{i,\downarrow}(\omega_4). \end{aligned} \tag{6.119}$$

The local effective action $S_{\rm eff}[c_i^\dagger, c_i, Q_i, x_i, \varepsilon_i]$ is identical to the action of a (generalized) Anderson impurity model embedded in an electronic bath characterized by a hybridization function $x_i^2\, t^2\, Q^{\alpha_1\alpha_2, s_1 s_2}_{\omega_1\omega_2}(i)$. We can thus interpret our system as a *collection* of Anderson impurity models (Anderson, 1961) that are "connected" through the existence of collective Q-fields. Here we note that, in contrast to an ordinary Anderson model, the hybridization function is now *non-diagonal* in frequency, spin, and replica indices. Physically, this reflects the fact that in general dimensions a given site can be regarded as an Anderson impurity model in a *fluctuating* bath that breaks translational invariance in time, space, and spin.

6.5.3 Saddle-point solution

The action has a general form that is very similar to standard lattice models investigated in statistical mechanics (Goldenfeld, 1992). As usual, the problem simplifies considerably in the limit of large coordination number, when the spatial fluctuations of the Hubbard–Stratonovich field (Q in our case) are suppressed and the mean-field theory becomes exact. It is worth pointing out that there are two classes of lattices that can have large coordination.

1. Lattices with short-range bonds but living in a space of *large dimensionality.* For example, on a hypercubic lattice with nearest neighbor hopping in d dimensions, $z = 2d$.
2. Lattices embedded in a finite dimensional space but having *long hopping range.* In this case, the lattice matrix f_{ij} takes the form

$$f_{ij} = \begin{cases} 1, |i-j| < L, \\ 0, \text{ otherwise,} \end{cases} \tag{6.120}$$

and the coordination number $z \sim L^d$.

In either case, when $z \to \infty$, the functional integral over Q-fields, representing the partition function, can be evaluated (exactly) by a saddle-point method and we obtain a mean-field theory. In order to derive the mean-field equations in our case, we look for extrema of the action $S[Q]$ with respect to variations of the Q-fields, i.e.

$$\frac{\delta\, S[Q]}{\delta\, Q^{\alpha_1\alpha_2,s_1s_2}_{\omega_1\omega_2}(i)} = 0. \tag{6.121}$$

Since the saddle-point solution is translationally invariant in time and space and conserves spin, it is *diagonal* in all indices

$$\left[Q^{\alpha_1\alpha_2,s_1s_2}_{\omega_1\omega_2}(i)\right]|_{\mathrm{SP}} = \delta_{\alpha_1\alpha_2}\delta_{s_1s_2}\delta_{\omega_1\omega_2}Q^{\mathrm{SP}}_s(\omega), \tag{6.122}$$

and the saddle-point equations assume the form

$$Q^{\mathrm{SP}}_s(\omega) = \int d\varepsilon_i P_S(\varepsilon_i) \int dx_i P_X(x_i)\, x_i^2\; G_{i,s}(\omega), \tag{6.123}$$

where

$$G_{i,s}(\omega) =< c^{\dagger}_s(\omega)c_s(\omega) >_{S_{\mathrm{eff}}[c^{\dagger},c,Q^{\mathrm{SP}},x_i,\varepsilon_i]}\,. \tag{6.124}$$

If we identify

$$W_{s,i}(\omega) \equiv x_i^2\, t^2 Q^{\mathrm{SP}}_s(\omega), \tag{6.125}$$

we see that our saddle-point equations become *identical* to the standard DMFT ($d \to \infty$) equations, when applied to the appropriate model of hopping disorder. We emphasize that the present equations are exact at $z \to \infty$ for an *arbitrary* lattice, due to the presence of the "gauge-invariant" form of the hopping disorder. Since the saddle-point equations determine the local effective action, this means that all the *local* correlation functions will be insensitive to the lattice structure in this mean-field limit. However, other properties, such as the tendency to the formation of spin and charge density waves, *are* very sensitive to the details of the lattice structure.

As an example, we can compare the case of simple hopping disorder, $t_{ij} = y_{ij}$, in a bipartite lattice such as the Bethe lattice and the case of a lattice with infinite range hopping (the limit $L \to \infty$ of the model (2) above). The self-consistency (mean-field) equations are *identical* in the two cases and in fact reduce to those of a *pure* Hubbard model on a Bethe lattice with hopping t. On the other hand, it is well established (Rozenberg *et al.*, 1992) that in the first case the system is unstable towards the formation of an antiferromagnetic ground state, even for arbitrarily small U/t, while in the second, the system remains paramagnetic for any U/t, due to the large frustration.

In many physical systems, such as doped semiconductors (Paalanen and Bhatt, 1991), disorder introduces large amounts of frustration and magnetic ordering does not occur, even though the system is strongly correlated. In order to study such situations, it is useful to have at hand microscopic models that have a non-magnetic ground state and allow one to study the approach to the MIT that occurs at $T = 0$.

6.5.4 Loop expansion

This present approach is particularly convenient for the study of the effects of strong correlations on *disorder-driven* transitions and the interplay of Anderson localization and strong correlations in general. This is especially true since Anderson localization is not present in $d = \infty$ (or infinite-range) models and so one has to extend the approach to include spatial fluctuations missing from the mean-field description. In order to systematically study the fluctuation effects, we proceed to carry out an expansion in terms of the deviations of the collective Q-fields from their saddle-point value, i.e. in powers of $\delta Q(i) = Q(i) - Q^{\mathrm{SP}}$. This procedure, also known as a *loop expansion* (Goldenfeld, 1992) has been used in other disordered problems, such as spin-glasses (Dominicis *et al.*, 1991), to generate systematic corrections to the mean-field theory. The method is particularly convenient when applied to long-range models (Dominicis *et al.*, 1991) (class (2) above), since in that case the loop corrections are ordered by a small parameter $1/z$. The loop expansion can be applied also to large dimensionality models, but in that case a given order in a loop expansion can be considered to be an infinite resummation of the simple $1/d$ expansion (Georges *et al.*, 1990), since each term contains all powers of $1/d$.

When the expansion of the effective action in terms of the δQs is carried to lowest, quadratic order, we obtain a theory describing gaussian fluctuations around the saddle point, which represent weakly interacting collective modes (Finkel'stein, 1983, 1984; Schaffer and Wegner, 1980; Wegner, 1979). Higher-order terms in the expansion then generate effective interactions of these modes, which under appropriate conditions can lead to fluctuation-driven phase transitions. In practice, if all the components of the collective Q-fields are retained in this analysis, the theory becomes prohibitively complicated and cumbersome. However, in order to analyze the *critical behavior*, it is not necessary to keep track of all the degrees of freedom. It suffices to limit the analysis to *soft modes*, i.e. those that represent low-energy excitations. In disordered metallic phases, charge and spin conservation laws lead to the existence of *diffusion modes*, which are the hydrodynamic modes describing charge- and spin-density relaxation. In the Fermi liquid regime (Castellani *et al.*, 1987), all the other collective excitations require higher energy and can be neglected in a hydrodynamic description of the system. One is then led to construct a theory of interacting diffusion modes, as a theory of critical phenomena for disordered interacting systems.

This line of reasoning was used in field-theoretical approaches to the localization problem of noninteracting electrons, as first developed by Wegner (Schaffer and Wegner, 1980; Wegner, 1979). In this theory, collective Q-fields, similar to the ones presented in this paper, are introduced. At the saddle-point level, no phase transition occurs and all the states are extended. An analysis of the fluctuations of the Q-fields is then performed and a *subset* of those fluctuations identified, which represent the hydrodynamic (diffusion) modes. Only the fluctuations of these modes are retained and an effective hydrodynamic theory is constructed—the *non-linear sigma model* (Schaffer and Wegner, 1980; Wegner, 1979) . The interactions of these modes lead to the metal–insulator (localization) transition, which was analyzed using renormalization-group techniques and $2+\varepsilon$ expansions. In subsequent work, Finkelshtein (Finkel'stein, 1983, 1984) was able to apply a similar procedure to *interacting* disordered electrons. However, this theory is based on a number of implicit assumptions that restrict its validity to *Fermi-liquid* regimes. In the language of Q-fields, this theory again expands around a noninteracting saddle-point and the interaction effects appear only at the level of the fluctuation corrections.

When strongly correlated electronic systems are considered, much of the physics relates precisely to the *destruction of coherent quasiparticles* by inelastic processes, so that one needs a description that is not limited to Fermi-liquid regimes. In the language of hydrodynamics, new soft modes appear, which indicate the tendency towards interaction-driven instabilities. In particular, strong correlations can lead to local moment formation and the Mott transition. In both cases, charge fluctuations are suppressed and low-energy spin fluctuations dominate the physics.

In the present approach, in contrast to the work of Wegner and Finkelshtein, the strong correlations are treated in a non-perturbative fashion already at the saddle-point (mean-field) level, so that *all* the soft modes can be systematically included. In particular, even at the one-loop level, we can address the question of how the disorder-induced local moment formation and the approach to the Mott transition affect the weak localization (diffusion) corrections. Ultimately, our approach indicates how a more general low energy theory can be constructed that extends the σ-model description so as to include strong correlation effects.

In the present discussion, we will limit our attention to the form of the Gaussian fluctuations of the Q-fields, which allow one in principle to compute the leading corrections to mean-field theory. The Gaussian part of the action takes the form

$$S^{(2)}[Q] = -\frac{1}{2}t^2 \sum_{l_1\cdots l_4} \int \frac{dk}{(2\pi)^d}\,\delta Q_{l_1 l_2}(k)\left[\left(L^2k^2+1\right)\delta_{l_1 l_4}\delta_{l_2 l_3}\right.$$
$$\left. -t^2 W(l_1)W(l_3)\delta_{l_1 l_2}\delta_{l_3 l_4} + t^2\Gamma(l_1\cdots l_4)\right]\,\delta Q_{l_3 l_4}(-k). \qquad (6.126)$$

This expression is appropriate for the long-ranged model (b) above, in which case the inverse lattice matrix in momentum space takes the form $K(k)\approx 1+L^2k^2$ and we cut off the momentum integrals at $\Lambda = 2\pi/L$. Note that the coefficient of k^2, which can be interpreted as the *stiffness* of the δQ modes, is $\sim L^2$, so we see that indeed the fluctuations are suppressed at $L\to\infty$. In the above formula, the index l_m is used to represent the frequency, spin and replica indices. The local vertex function $\Gamma(l_1 \cdots l_4)$ is given by

$$\Gamma(l_1 \cdots l_4) = \int d\varepsilon_i P_S(\varepsilon_i) \int dx_i P_X(x_i)\, x_i^4 < c_i^\dagger(l_1) c_i(l_2) c_i^\dagger(l_3) c_i(l_4) >_{S_{\mathrm{eff}}[Q^{\mathrm{SP}}]} . \tag{6.127}$$

At this level, the dynamics of the collective fluctuations δQ is governed by the form of $S^{(2)}[\delta Q]$, which is expressed in terms of the *local correlation functions* of the saddle-point theory, i. e., of the $d = \infty$ disordered Hubbard model. Accordingly, a detailed study of the $d = \infty$ limit does not provide only a mean-field description of the problem, but also determines the form of the leading corrections resulting from fluctuations. Extensions of the theory to include the effects of these fluctuation corrections remain to be addressed in more detail in future work.

Acknowledgements

This work was supported by FAPESP through grant 07/57630-5 (EM), CNPq through grant 304311/2010-3 (EM), and by NSF through grants DMR-0542026 and DMR-1005751. (VD).

References

Abou-Chacra, R., Anderson, P.W., and Thouless, D.J. (1973). *J. Phys. C*, **6**, 1734.

Abrahams, E., Kravchenko, S.V., and Sarachik, M.P. (2001). *Rev. Mod. Phys.*, **73**, 251–266.

Aguiar, M.C.O., Dobrosavljević, V., Abrahams, E., and Kotliar, G. (2005). *Phys. Rev. B*, **71**, 205115.

Aguiar, M.C.O., Dobrosavljević, V., Abrahams, E., and Kotliar, G. (2006, Mar). *Phys. Rev. B*, **73**(11), 115117.

Aguiar, M.C.O., Dobrosavljević, V., Abrahams, E., and Kotliar, G. (2009, Apr). *Phys. Rev. Lett.*, **102**(15), 156402.

Aguiar, M.C.O., Miranda, E., and Dobrosavljević, V. (2003). *Phys. Rev. B*, **68**, 125104.

Aguiar, M.C.O., Miranda, E., Dobrosavljević, V., Abrahams, E., and Kotliar, G. (2004). *Europhys. Lett.*, **67**, 226.

Alloul, H., Bobroff, J., Gabay, M., and Hirschfeld, P.J. (2009, Jan). *Rev. Mod. Phys.*, **81**(1), 45.

Altshuler, B.L. and Aronov, A.G. (1979). *Solid State Commun.*, **30**, 115.

Anderson, P.W. (1958). *Phys. Rev.*, **109**, 1492–1505.

Anderson, P.W. (1961). *Phys. Rev.*, **124**, 41.

Anderson, P.W. (1978). *Rev. Mod. Phys.*, **50**(2), 191.

Anderson, P.W. and Yuval, G. (1969). *Phys. Rev. Lett.*, **23**, 89.

Anderson, P.W., Yuval, G., and Hamman, D.R. (1970). *Phys. Rev. B*, **1**, 4464.

Andrade, E.C., Miranda, E., and Dobrosavljević, V. (2009*a*). *Phys. Rev. Lett.*, **102**, 206403.

Andrade, E.C., Miranda, E., and Dobrosavljević, V. (2009*b*). *Physica B: Condensed Matter*, **404**(19), 3167–3171. Proceedings of the International Conference on Strongly Correlated Electron Systems.

Andrade, E.C., Miranda, E, and Dobrosavljević, V. (2010). *Phys. Rev. Lett.*, **104**, 236401.

Aronson, M.C., Obsorn, R., Robinson, R.A., Lynn, J.W., Chau, R., Seaman, C.L., and Maple, M.B. (1995). *Phys. Rev. Lett.*, **75**, 725.

Arrachea, L., Dalidovich, D., Dobrosavljević, V., and Rozenberg, M.J. (2004). *Phys. Rev. B*, **69**, 064419.

Balatsky, A.V., Vekhter, I., and Zhu, Jian-Xin (2006, May). *Rev. Mod. Phys.*, **78**(2), 373–433.

Belitz, D. and Kirkpatrick, T.R. (1995). *Phys. Rev. B*, **52**, 13922.

Ben-Chorin, M., Ovadyahu, Z., and Pollak, M. (1993). *Phys. Rev. B*, **48**, 15025–15034.

Bernal, O.O., MacLaughlin, D.E., Lukefahr, H.G., and Andraka, B. (1995). *Phys. Rev. Lett.*, **75**, 2023.

Bogdanovich, S. and Popović, D. (2002). *Phys. Rev. Lett.*, **88**, 236401.

Borghi, G., Fabrizio, M., and Tosatti, E. (2009, Feb). *Phys. Rev. Lett.*, **102**(6), 066806.

Borghi, G., Fabrizio, M., and Tosatti, E. (2010, Mar). *Phys. Rev. B*, **81**(11), 115134.

Bray, A.J. and Moore, M.A. (1980). *J. Phys. C*, **13**, L655.

Brinkman, W.F. and Rice, T.M. (1970). *Phys. Rev. B*, **2**, 4302.

Burdin, S., Grempel, D.R., and Georges, A. (2002). *Phys. Rev. B*, **66**, 045111.

Byczuk, K. (2005). *Phys. Rev. B*, **71**, 205105.

Byczuk, K., Hofstetter, W., and Vollhardt, D. (2005, Feb). *Phys. Rev. Lett.*, **94**(5), 056404.

Byczuk, Krzysztof, Hofstetter, Walter, and Vollhardt, Dieter (2009*a*, Apr). *Phys. Rev. Lett.*, **102**(14), 146403.

Byczuk, K., Hofstetter, W., Yu, U., and Vollhardt, D. (2009*b*). *EPJ Special Topics*, **180**, 135.

Camjayi, A. and Rozenberg, M.J. (2003). *Phys. Rev. Lett.*, **90**, 217202.

Castellani, C., Kotliar, B.G., and Lee, P.A. (1987). *Phys. Rev. Lett.*, **56**, 1179.

Castro Neto, A.H., Castilla, G., and Jones, B.A. (1998). *Phys. Rev. Lett.*, **81**, 3531.

Castro Neto, A.H. and Jones, B.A. (2000, December). *Phys. Rev. B*, **62**, 14975–15011.

Chakravarty, S., Kivelson, S., Nayak, C., and Voelker, K. (1999). *Phil. Mag. B*, **79**, 859.

Chen, Ling and Freericks, J.K. (2007, Mar). *Phys. Rev. B*, **75**(12), 125114.

Chitra, R. and Kotliar, G. (2000). *Phys. Rev. Lett.*, **84**, 3678.

Coleman, P. (1987). *Phys. Rev. B*, **35**, 5072.

Coleman, P. and Andrei, N. (1987). *J. Phys.: Condens. Matter*, **1**, 4057.

Cummins, H.Z., Li, G., Du, W.M., and Hernandez, J. (1994). *Physica A*, **204**, 169.

Dalidovich, D. and Dobrosavljević, V. (2002). *Phys. Rev. B*, **66**, 081107.

Davies, J.H., Lee, P.A., and Rice, T.M. (1982). *Phys. Rev. Lett.*, **49**, 758–761.

de Almeida, J.R.L. and Thouless, D.J. (1978). *J. Phys. A*, **11**, 983.

de Andrade, M.C., Chau, R., Dickey, R.P., Dilley, N.R., Freeman, E.J., Gajewski, D.A., Maple, M.B., Movshovich, R., Neto, A.H. Castro, Castilla, G., and Jones, B.A. (1998). *Phys. Rev. Lett.*, **81**, 5620.

Demler, E., Nayak, C., Kee, H.-Y., Kim, Y.-B., and Senthil, T. (2002). *Phys. Rev. B*, **65**, 155103.

Dobrosavljević, V. (2010). *Int. Jour. Mod. Phys. B*, **24**(12-13), 1680–1726.

Dobrosavljević, V., Abrahams, E., Miranda, E., and Chakravarty, S. (1997). *Phys. Rev. Lett.*, **79**, 455.

Dobrosavljević, V., Kirkpatrick, T.R., and Kotliar, B.G. (1992). *Phys. Rev. Lett.*, **69**, 1113.

Dobrosavljević, V. and Kotliar, G. (1993). *Phys. Rev. Lett.*, **71**, 3218.

Dobrosavljević, V. and Kotliar, G. (1994). *Phys. Rev. B*, **50**, 1430.

Dobrosavljević, V. and Kotliar, G. (1997). *Phys. Rev. Lett.*, **78**, 3943.

Dobrosavljević, V. and Kotliar, G. (1998). *Phil. Trans. R. Soc. Lond. A*, **356**, 1.

Dobrosavljević, V., Pastor, A.A., and Nikolić, B.K. (2003*a*). *Europhys. Lett.*, **62**, 76–82.

Dobrosavljević, V. and Stratt, R.M. (1987). *Phys. Rev. B*, **36**, 8484.

Dobrosavljević, V., Tanasković, D., and Pastor, A.A. (2003*b*). *Phys. Rev. Lett.*, **90**, 016402.

Dominicis, C. De, Kondor, I., and Temisvari, T. (1991). *J. de Physique I*.

Doniach, S. (1977). *Physica B*, **91**, 231.

Economou, E.N. (2006). *Green's Functions in Quantum Physics*. Springer, Berlin.

Efetov, K.B., Larkin, A.I., and Khmel'nitskii, D.E. (1980). *Zh. Eksp. Teor. Fiz.*, **79**, 1120.

Efros, A.L. (1992, Apr). *Phys. Rev. Lett.*, **68**(14), 2208–2211.

Efros, A.L. and Shklovskii, B.I. (1975). *J. Phys. C*, **8**, L49.

Elliott, R.J., Krumhansl, J.A., and Leath, P.L. (1974). *Rev. Mod. Phys.*, **46**, 465.

Finkel'stein, A.M. (1983). *Zh. Eksp. Teor. Fiz.*, **84**, 168. [Sov. Phys. JETP **57**, 97 (1983)].

Finkel'stein, A.M. (1984). *Zh. Eksp. Teor. Fiz.*, **86**, 367. [Sov. Phys. JETP **59**, 212 (1983)].

Fisher, D.S. (1992). *Phys. Rev. Lett.*, **69**, 534.

Fisher, D.S. (1994). *Phys. Rev. B*, **50**, 3799.

Fisher, D.S. (1995). *Phys. Rev. B*, **51**, 6411.

Fisher, D.S. and Huse, D.A. (1986). *Phys. Rev. Lett.*, **56**, 1601.

Fisher, D.S. and Young, A.P. (1998, 14 October). *Phys. Rev. B*, **58**, 9131.

Fradkin, E. (1991). *Field Theories of Condensed Matter Systems*. Adisson-Wesley, Redwood City, California.

Freericks, J.K. (2004, Nov). *Phys. Rev. B*, **70**(19), 195342.

Georges, A. and Kotliar, G. (1992). *Phys. Rev. B*, **45**, 6479.

Georges, A., Kotliar, G., Krauth, W., and Rozenberg, M.J. (1996). *Rev. Mod. Phys.*, **68**, 13.

Georges, A., Mezard, M., and Yedida, J.S. (1990). *Phys. Rev. Lett.*, **64**, 2937.

Georges, A., Parcollet, O., and Sachdev, S. (2001). *Phys. Rev. B*, **63**, 134406.

Goldenfeld, N. (1992). *Lectures on Phase Transitions and the Renormalization Group.* Addison-Wesley, Reading.

Grannan, Eric R. and Yu, Clare C. (1993, Nov). *Phys. Rev. Lett.*, **71**(20), 3335–3338.

Grewe, N. and Steglich, F. (1991). In *Handbook on the Physics and Chemistry of Rare Earths* (ed. K.A. Geschneider Jr. and L. Eyring), Volume 14, p. 343. Elsevier, Amsterdam.

Griffiths, R.B. (1969). *Phys. Rev. Lett.*, **23**, 17.

Guo, Muyu, Bhatt, R.N., and Huse, David A. (1996, August). *Phys. Rev. B*, **54**, 3336–3342.

Gutzwiller, Martin C. (1963, Mar). *Phys. Rev. Lett.*, **10**(5), 159–162.

Gutzwiller, Martin C. (1964, May). *Phys. Rev.*, **134**(4A), A923–A941.

Gutzwiller, Martin C. (1965, Mar). *Phys. Rev.*, **137**(6A), A1726–A1735.

Helmes, R.W., Costi, T.A., and Rosch, A. (2008, Aug). *Phys. Rev. Lett.*, **101**(6), 066802.

Herbut, I.F. (2001). *Phys. Rev. B*, **63**, 113102.

Hertz, J.A. (1976). *Phys. Rev. B*, **14**, 1165.

Hewson, A.C. (1993). *The Kondo Problem to Heavy Fermions.* Cambridge University Press, Cambridge.

Hoyos, J.A., Kotabage, C., and Vojta, T. (2007, 4 December). *Phys. Rev. Lett.*, **99**, 230601.

Hoyos, J.A. and Vojta, T. (2006, 11 October). *Phys. Rev. B*, **74**, 140401(R).

Hoyos, J.A. and Vojta, T. (2008, 17 June). *Phys. Rev. Lett.*, **100**(24), 240601.

Hubbard, J. (1963). *Proc. R. Soc. (London) A*, **276**, 238.

Hyman, R.A. and Yang, K. (1997). *Phys. Rev. Lett.*, **78**, 1783.

Hyman, R.A., Yang, K., Bhatt, R.N., and Girvin, S.M. (1996). *Phys. Rev. Lett.*, **76**, 839.

Jacko, A.C., Fjrestad, J.O., and Powell, B.J. (2009). *Nature Physics*, **5**, 422–425.

Jagannathan, A., Abrahams, E., and Stephen, M.J. (1988). *Phys. Rev. B*, **37**, 436.

Janis, V., Ulmke, M., and Vollhardt, D. (1993). *EPL (Europhysics Letters)*, **24**(4), 287.

Janis, V. and Vollhardt, D. (1992, Dec). *Phys. Rev. B*, **46**(24), 15712–15715.

Janssen, M. (1998). *Phys. Rep.*, **295**, 1.

Jaroszyński, J., Popović, D., and Klapwijk, T.M. (2002). *Phys. Rev. Lett.*, **89**, 276401.

Kadowaki, K. and Woods, S.B. (1986). *Solid State Commun.*, **58**(8), 507–509.

Kagan, Yu., Kikoin, K.A., and Prokof'ev, N.V. (1992). *Physica B*, **182**, 201.

Kagawa, F., Miyagawa, K., and Kanoda, K. (2005). *Nature*, **436**, 534–537.

Kajueter, H. and Kotliar, G. (1996). *Phys. Rev. Lett.*, **77**, 131.

Kasuya, T. (1956). *Progr. Theoret. Phys.*, **16**, 45.

Kondo, J. (1964). *Prog. Theor. Phys.*, **32**, 37.

Kosterlitz, J.M. (1976). *Phys. Rev. Lett.*, **37**, 1577.

Kotliar, G., Lange, E., and Rozenberg, M.J. (2000). *Phys. Rev. Lett.*, **84**, 5180.

Kotliar, G., Murthy, S., and Rozenberg, M.J. (2002). *Phys. Rev. Lett.*, **89**, 046401.

Kotliar, G. and Ruckenstein, A.E. (1986). *Phys. Rev. Lett.*, **57**(11), 1362.

Landau, L.D. (1957*a*). *Sov. Phys. JETP*, **5**, 101.

Landau, L.D. (1957*b*). *Sov. Phys. JETP*, **3**, 920.
Landau, L.D. (1959). *Sov. Phys. JETP*, **8**, 70.
Leggett, A.J., Chakravarty, S., Dorsey, A.T., Fisher, M.P.A., Garg, A., and Zwerger, W. (1987). *Rev. Mod. Phys.*, **59**, 1.
Liebsch, A. (2003, Mar). *Phys. Rev. Lett.*, **90**(9), 096401.
Lifshits, I.M., Gredeskul, S.A., and Pastur, L.A. (1988). *Introduction to the Theory of Disordered Systems.* Wiley, New York.
Limelette, P., Georges, A., Jerome, D., Wzietek, P., Metcalf, P., and Honig, J.M. (2003*a*). *Science*, **302**(5642), 89–92.
Limelette, P., Wzietek, P., Florens, S., Georges, A., Costi, T.A., Pasquier, C., Jerome, D., Meziere, C., and Batail, P. (2003*b*). *Phys. Rev. Lett.*, **91**, 016401.
MacLaughlin, D.E., Bernal, O.O., Heffner, R.H., Nieuwenhuys, G.J., Rose, M.S., Sonier, J.E., Andraka, B., Chau, R., and Maple, M.B. (2001). *Phys. Rev. Lett.*, **87**, 066402.
Massey, J.G. and Lee, M. (1996). *Phys. Rev. Lett.*, **77**, 3399.
McElroy, K., Lee, J., Slezak, J.A., Lee, D.-H., Eisaki, H., Uchida, S., and Davis, J.C. (2005). *Science*, **309**(5737), 1048.
Metzner, W. and Vollhardt, D. (1989, Jan). *Phys. Rev. Lett.*, **62**(3), 324–327.
Mézard, M., Parisi, G., and Virasoro, M.A. (1986). *Spin Glass Theory and Beyond.* World Scientific, Singapore.
Mezard, M. and Young, A.P. (1992). *Europhys. Lett.*, **18**, 653–659.
Miller, J. and Huse, D.A. (1993). *Phys. Rev. Lett.*, **70**, 3147.
Millis, A.J. (1993). *Phys. Rev. B*, **48**(10), 7183.
Millis, A.J. and Lee, P.A. (1987, Mar). *Phys. Rev. B*, **35**(7), 3394.
Millis, A.J., Morr, D.K., and Schmalian, J. (2002). *Phys. Rev. B*, **66**, 174433.
Milovanović, M., Sachdev, S., and Bhatt, R.N. (1989). *Phys. Rev. Lett.*, **63**, 82.
Miranda, E. and Dobrosavljević, V. (1999). *Physica B*, **259-261**, 359.
Miranda, E. and Dobrosavljević, V. (2001). *Phys. Rev. Lett.*, **86**, 264.
Miranda, E. and Dobrosavljević, V. (2005). *Rep. Prog. Phys.*, **68**, 2337.
Miranda, E., Dobrosavljević, V., and Kotliar, G. (1996). *J. Phys.: Condens. Matter*, **8**, 9871.
Miranda, E., Dobrosavljević, V., and Kotliar, G. (1997*a*). *Phys. Rev. Lett.*, **78**, 290.
Miranda, E., Dobrosavljević, V., and Kotliar, G. (1997*b*). *Physica B*, **230**, 569.
Mirlin, A.D. (2000). *Phys. Rep.*, **326**, 259.
Mirlin, A.D. and Fyodorov, Y.V. (1991). *Nucl. Phys. B*, **366**, 507.
Miyake, K., Matsuura, T., and Varma, C.M. (1989). *Solid State Commun.*, **71**(12), 1149–1153.
Mo, S.-K., Denlinger, J.D., Kim, H.-D., Park, J.-H., Allen, J.W., Sekiyama, A., Yamasaki, A., Kadono, K., Suga, S., Saitoh, Y., Muro, T., Metcalf, P., Keller, G., Held, K., Eyert, V., Anisimov, V.I., and Vollhardt, D. (2003, May). *Phys. Rev. Lett.*, **90**(18), 186403.
Mooij, J.H. (1973). *Phys. Status Solidi A*, **17**, 521–530.
Motrunich, O., Mau, S.-C., Huse, D.A., and Fisher, D.S. (2001). *Phys. Rev. B*, **61**, 1160.
Mott, N.F. (1949). *Proc. Roy. Soc. (London)*, **A197**, 269.

Mott, N.F. (1990). *Metal-Insulator Transition*. Taylor & Francis, London.

Muller, M. and Ioffe, L.B. (2004). *Phys. Rev. Lett.*, **93**, 256403.

Müller, M. and Pankov, S. (2007, Apr). *Phys. Rev. B*, **75**(14), 144201.

Narayanan, R., Vojta, T., Belitz, D., and Kirkpatrick, T.R. (1999*a*, October). *Phys. Rev. B*, **60**, 10150–10163.

Narayanan, R., Vojta, T., Belitz, D., and Kirkpatrick, T.R. (1999*b*, June). *Phys. Rev. Lett.*, **82**, 5132–5135.

Narozhny, B.N., Aleiner, I.L., and Larkin, A.I. (2001). *Phys. Rev. B*, **62**, 14898.

Nattermann, T. (1997). *Theory of the Random Field Ising Model*. World Scientific, Singapore.

Nozières, P. (1974). *J. Low Temp. Phys.*, **17**, 31.

Ohtomo, A., Muller, D.A., Grazul, J.L., and Hwang, H.Y. (2002). *Nature*, **419**, 378.

Okamoto, Satoshi and Millis, Andrew J. (2004*a*). *Nature*, **428**, 630.

Okamoto, Satoshi and Millis, Andrew J. (2004*b*, Dec). *Phys. Rev. B*, **70**(24), 241104.

Ovadyahu, Z. and Pollak, M. (1997). *Phys. Rev. Lett.*, **79**, 459–462.

Paalanen, M.A. and Bhatt, R.N. (1991). *Physica B*, **169**, 231.

Palmer, R.G. and Pond, C.M. (1979). *J. Phys. F*, **9**, 1451.

Panagopoulos, C. and Dobrosavljević, V. (2005). *Phys. Rev. B*. cond-mat/0410111, in press.

Pankov, S. (2006, May). *Phys. Rev. Lett.*, **96**(19), 197204.

Pankov, S. and Dobrosavljević, V. (2005). *Phys. Rev. Lett.*, **94**, 046402.

Parcollet, O. and Georges, A. (1999). *Phys. Rev. B*, **59**(8), 5341.

Pastor, A.A. and Dobrosavljević, V. (1999). *Phys. Rev. Lett.*, **83**, 4642.

Pastor, A.A., Dobrosavreljević, V., and Horbach, M.L. (2002). *Phys. Rev. B*, **66**, 014413.

Pazmandi, F., Zarand, G., and Zimanyi, G.T. (1999). *Phys. Rev. Lett.*, **83**, 1034–1037.

Pezzoli, M.E. and Becca, F. (2010, Feb). *Phys. Rev. B*, **81**(7), 075106.

Pezzoli, M.E., Becca, F., Fabrizio, M., and Santoro, G. (2009, Jan). *Phys. Rev. B*, **79**(3), 033111.

Pich, C., Young, A.P., Rieger, H., and Kawashima, N. (1998). *Phys. Rev. Lett.*, **81**, 5916.

Plefka, T. (1982). *J. Phys. A*, **15**, 1971–1978.

Pollak, M. and Hunt, A. (1991). In *Hopping Transport in Solids* (ed. M. Pollak and B. Shklovskii), pp. 175–206. Elsevier, Amsterdam.

Potthoff, M. and Nolting, W. (1999, Sep). *Phys. Rev. B*, **60**(11), 7834–7849.

Pruschke, T., Jarrell, M., and Freericks, J.K. (1995). *Adv. Phys.*, **44**, 187.

Read, N. and Newns, D.M. (1983). *J. Phys. C*, **16**, L1055.

Read, N., Sachdev, S., and Ye, J. (1995). *Phys. Rev. B*, **52**, 384.

Refael, G., Kehrein, S., and Fisher, D.S. (2002). *Phys. Rev. B*, **66**, 060402(R).

Rodolakis, F., Mansart, B., Papalazarou, E., Gorovikov, S., Vilmercati, P., Petaccia, L., Goldoni, A., Rueff, J.P., Lupi, S., Metcalf, P., and Marsi, M. (2009, Feb). *Phys. Rev. Lett.*, **102**(6), 066805.

Rozenberg, M.J., Chitra, R., and Kotliar, G. (1999). *Phys. Rev. Lett.*, **83**, 3498.

Rozenberg, M.J. and Grempel, D. (1998). *Phys. Rev. Lett.*, **81**, 2550.

Rozenberg, M.J. and Grempel, D. (1999). *Phys. Rev. B*, **60**, 4702.

Rozenberg, M.J., Kotliar, G., and Kajueter, H. (1996). *Phys. Rev. B*, **54**, 8542.
Rozenberg, M.J., Zhang, X.Y., and Kotliar, G. (1992). *Phys. Rev. Lett.*, **69**, 1236.
Ruderman, M.A. and Kittel, C. (1954). *Phys. Rev.*, **96**, 99.
Sachdev, S., Read, N., and Oppermann, R. (1995). *Phys. Rev. B*, **52**, 10286.
Sachdev, S. and Ye, J. (1993). *Phys. Rev. Lett.*, **70**, 3339.
Sarvestani, M., M.Schreiber, and Vojta, T. (1995). *Phys. Rev. B*, **52**, R3820.
Schaffer, L. and Wegner, F. (1980). *Phys. Rev. B*, **38**, 113.
Schrieffer, J.R. and Wolff, P.A. (1966). *Phys. Rev.*, **149**, 491.
Schubert, G, Schleede, J., Byczuk, K., Fehske, H., and Vollhardt, D. (2010, Apr). *Phys. Rev. B*, **81**(15), 155106.
Schwieger, S., Potthoff, M., and Nolting, W. (2003, Apr). *Phys. Rev. B*, **67**(16), 165408.
Sekiyama, A., Iwasaki, T., Matsuda, K., Saitoh, Y., nuki, Y, and Suga, S. (2000). *Nature*, **403**, 396.
Semmler, D., Byczuk, K., and Hofstetter, W. (2010*a*, Mar). *Phys. Rev. B*, **81**(11), 115111.
Semmler, D., Wernsdorfer, J., Bissbort, U., Byczuk, K., and Hofstetter, W. (2010*b*, Dec). *Phys. Rev. B*, **82**(23), 235115.
Sengupta, A.M. (2000). *Phys. Rev. B*, **61**, 4041.
Sengupta, A.M. and Georges, A. (1995). *Phys. Rev. B*, **52**, 10295.
Senthil, T., Sachdev, S., and Vojta, M. (2003). *Phys. Rev. Lett.*, **90**, 216403.
Senthil, T., Vojta, M., and Sachdev, S. (2004). *Phys. Rev. B*, **69**, 035111.
Shiba, H. (1971). *Prog. Teor. Phys.*, **46**, 77.
Si, Q., Rabello, S., Ingersent, K., and Smith, J.L. (2001). *Nature*, **413**, 804.
Si, Q. and Smith, J.L. (1996). *Phys. Rev. Lett.*, **77**, 3391.
Singh, A., Ulmke, M., and Vollhardt, D. (1998, Oct). *Phys. Rev. B*, **58**(13), 8683–8693.
Slater, J.C. (1951). *Phys. Rev.*, **82**, 538.
Slevin, K. and Ohtsuki, T. (1999, Jan). *Phys. Rev. Lett.*, **82**(2), 382.
Smith, J.L. and Si, Q. (2000). *Phys. Rev. B*, **61**, 5184.
Snoek, M, Titvinidze, I, Toke, C, Byczuk, K, and Hofstetter, W (2008). *New J. Phys.*, **10**(9), 093008.
Song, Y., Wortis, R., and Atkinson, W.A. (2008, Feb). *Phys. Rev. B*, **77**(5), 054202.
Stewart, G.R. (1984). *Rev. Mod. Phys.*, **56**, 755.
Stewart, G.R. (2001). *Rev. Mod. Phys.*, **73**, 797.
Stewart, G.R. (2006, 28 July). *Rev. Mod. Phys.*, **78**, 743.
Tanasković, D., Dobrosavljević, V., Abrahams, E., and Kotliar, G. (2003). *Phys. Rev. Lett.*, **91**, 066603.
Tanaskovic, D., Dobrosavljević, V., and Miranda, E. (2005). *Phys. Rev. Lett.*, **95**, 167204.
Tanasković, D., Miranda, E., and Dobrosavljević, V. (2004). *Phys. Rev. B*, **70**, 205108.
Terletska, H., Vučičević, J., Tanasković, D., and Dobrosavljević, V. (2011, Jul). *Phys. Rev. Lett.*, **107**(2), 026401.
Ting, C.S., Lee, T.K., and Quinn, J.J. (1975, Apr). *Phys. Rev. Lett.*, **34**(14), 870–874.
Ulmke, M., Janiš, V., and Vollhardt, D. (1995, Apr). *Phys. Rev. B*, **51**(16), 10411–10426.

Varma, C.M. (1985). *Comments Solid State Phys.*, **11**, 221.
Varma, C.M., Littlewood, P.B., Schmitt-Rink, S., Abrahams, E., and Ruckenstein, A.E. (1989). *Phys. Rev. Lett.*, **63**, 1996.
Vlaming, R. and Vollhardt, D. (1992, Mar). *Phys. Rev. B*, **45**(9), 4637–4649.
Vojta, T. (2003). *Phys. Rev. Lett.*, **90**, 107202.
Vojta, T. (2006, 16 May). *J. Phys. A: Math. Gen.*, **39**, R143–R205.
Vojta, T., Kotabage, C., and Hoyos, J. A. (2009, 5 January). *Phys. Rev. B*, **79**(2), 024401.
Vojta, T. and Schmalian, J. (2005*a*, 29 November). *Phys. Rev. Lett.*, **95**, 237206.
Vojta, T. and Schmalian, J. (2005*b*, 15 July). *Phys. Rev. B*, **72**, 045438.
Vollhardt, D. (1984). *Rev. Mod. Phys.*, **56**, 99.
Wegner, F. (1979). *Phys. Rev. B*, **35**, 783.
Wilson, K.G. (1975). *Rev. Mod. Phys.*, **47**, 773.
Yang, K., Hyman, R.A., Bhatt, R.N., and Girvin, S.M. (1996). *J. Appl. Phys.*, **79**, 5096.
Ye, J., Sachdev, S., and Read, N. (1993). *Phys. Rev. Lett.*, **70**, 4011.
Yosida, K. (1957). *Phys. Rev.*, **106**, 893.
Yuval, G. and Anderson, P.W. (1970). *Phys. Rev. B*, **1**, 1522.
Zala, G., Narozhny, B.N., and Aleiner, I.L. (2001). *Phys. Rev. B*, **64**, 214204(31).
Zaránd, G. and Demler, E. (2002). *Phys. Rev. B*, **66**, 024427.
Zenia, H., Freericks, J.K., Krishnamurthy, H.R., and Pruschke, Th. (2009, Sep). *Phys. Rev. Lett.*, **103**(11), 116402.
Zhang, X.Y., Rozenberg, M.J., and Kotliar, G. (1993). *Phys. Rev. Lett.*, **70**, 1666.
Zhang, Y. and Das Sarma, S. (2005, Jan). *Phys. Rev. B*, **71**(4), 045322.
Zhu, L. and Si, Q. (2002). *Phys. Rev. B*, **66**, 024426.

7
Visualizing Critical Correlations Near the Metal–Insulator Transition in $Ga_{1-x}Mn_xAs$

Ali YAZDANI

Joseph Henry Laboratories and Department of Physics, Princeton University, Princeton, New Jersey

7.1 Introduction

Since Anderson first proposed that disorder could localize electrons in solids fifty years ago (Anderson, 1958), studies of the transition between extended and localized quantum states have been at the forefront of physics (Lagendijk *et al.*, 2009). Realizations of Anderson localization occur in a wide range of physical systems, from seismic waves to ultracold atomic gases, in which localization has recently been achieved using random optical lattices (Aspect and Inguscio, 2009). In electronic systems, the signatures of localization have long been examined through electrical transport measurements (Belitz and Kirkpatrick, 1994; Lee and Ramakrishnan, 1985), and more recently using local scanning probe techniques that have imaged localized electronic states (Hashimoto *et al.*, 2008; Ilani *et al.*, 2004). For non-interacting systems, the electronic states at the mobility edge are predicted to have a diverging localization length with scale-independent power-law characteristics, which are described as being multifractal (Evers and Mirlin, 2008). Given the poorly understood nature of the metal–insulator transition in the presence of disorder and electron–electron interactions, direct imaging of electronic states can provide insights into the interplay between localization and interactions. Recently, we carried out a series of scanning tunneling microscopy (STM) and spectroscopy studies of electronic states in the dilute magnetic semiconductor $\mathrm{Ga}_{1-x}\mathrm{Mn}_x\mathrm{As}$, over a range of Mn concentrations near the metal–insulator transition ($x = 1.5$–5%).(Richardella *et al.*, 2010) Over the last decade, $\mathrm{Ga}_{1-x}\mathrm{Mn}_x\mathrm{As}$ has emerged as a promising material for spintronic applications, having a high ferromagnetic transition temperature (Chiba *et al.*, 2008; Ohno *et al.*, 2000). Mn atoms substituted at Ga sites act both as acceptors that drive the metal–insulator transition and as localized spins that align at low temperatures to give rise to magnetism. The nature of the electronic states underlying magnetism in these heavily doped semiconductors is still debated. It is often assumed that the carriers that mediate magnetism in $\mathrm{Ga}_{1-x}\mathrm{Mn}_x\mathrm{As}$ are either Bloch states associated with the valence bands or extended states originating from an impurity band (Burch *et al.*, 2006; Dietl *et al.*, 2000; Jungwirth *et al.*, 2007). However, the validity of these assumptions has been questioned (Dietl, 2008; Moca *et al.*, 2009). Moreover, many recent low-temperature transport studies show evidence of electron–electron interaction and weak localization of carriers, even for samples with high doping levels (Chun *et al.*, 2002; Neumaier *et al.*, 2007; Rokhinson *et al.*, 2007). We use atomic-scale imaging and high-resolution spectroscopy with the STM to visualize electronic states in $\mathrm{Ga}_{1-x}\mathrm{Mn}_x\mathrm{As}$ and to examine the spatial structure of electron–electron correlations in this system. Our results indicate that spatial heterogeneity and electronic correlations must be considered in understanding the mechanism of magnetism in highly doped semiconductors. More broadly, these experiments also provide the first atomic-scale perspective on the structure of electronic states near a metal–insulator transition and evidence for critical phenomena in the spatial structure of electronic states in real space.

7.2 Experimental results

Figure 7.1 shows STM topographs of cleaved $Ga_{1-x}Mn_xAs$ samples (2000 Å thick) grown using molecular-beam epitaxy (MBE) on a p-type Be-doped GaAs buffer layer (Myers *et al.*, 2006). Prior to measurements at a temperature of 4.2 K, the degenerately doped substrates are cleaved *in situ* to expose a (110) or equivalent surface of the heterostructure in cross section (Fig. 7.1 (a) inset). The topographs show in-gap states, dominated by individual Mn acceptor wavefunctions, although other defects such as As antisites are observed as well. Using previous STM studies of samples with more dilute Mn concentrations (Garleff *et al.*, 2008; Kitchen *et al.*, 2006) and the results of tight-binding model calculations (Strandberg *et al.*, 2009; Tang and Flatté, 2004), we can identify the topographic signatures of individual Mn acceptors in layers from the surface to the third subsurface layer (Fig. 7.1 (b)). The size of an individual Mn acceptor state wavefunction is about 20 Å, due to its deep binding energy, which results in a metal–insulator transition at a relatively high level of doping (between 1 and 2%) in $Ga_{1-x}Mn_xAs$. Increasing the Mn concentration from weakly insulating samples with variable-range hopping resistivity at 1.5% ($T_c = 30$ K) (Sheu *et al.*, 2007) to relatively conducting samples at 5% ($T_c = 86$ K, annealed), we find higher concentrations of dopants appear in STM topographs on top of the atomically ordered GaAs lattice (Fig. 7.1 (c)). All characteristic lengths, such as dopant separation or mean-free path (∼10 Å), are much shorter for $Ga_{1-x}Mn_xAs$ in comparison to other semiconductors doped with shallow dopants (Moca *et al.*, 2009).

Spectroscopic mapping with the STM can be used to show that Mn acceptors and other defects give rise to atomic-scale fluctuations in the local electronic density of states (LDOS) over a wide range of energies. Figure 7.2 (a) shows tunneling spectra of states from within the valence band ($V < 0$) to the conduction band edge ($V > 1.5$), measured along a line perpendicular to the buffer-layer–film interface. Contrasting the spatial dependence of electronic states in the buffer layer (with 2×10^{18} per cm^3 Be acceptors) to those of $Ga_{1-x}Mn_xAs$ (with $x = 0.015$) in Fig. 7.2 (a), we find the Mn-doped region to have strong spatial variations in the electronic states at the valence and conduction band edges and a broad distribution of states within the GaAs band gap. Increasing the Mn concentration gives rise to a larger number of in-gap states (compare Fig. 7.2 (b) and (c)). These features of the local electronic structure of $Ga_{1-x}Mn_xAs$ are difficult to reconcile with a weakly disordered valence or impurity band picture, and show the importance of compensation and disorder in this compound. The Fermi energy E_F lies within the range of electronic states that are spatially inhomogeneous.

In addition to strong spatial variations, electronic states of $Ga_{1-x}Mn_xAs$ are influenced by electron–electron interactions (Fig. 7.2 (d).). STM spectra, spatially averaged across large areas for several samples with increasing doping levels, illustrate a strong suppression of the tunneling density of states near E_F. The evolution from weakly insulating (1.5%) to relatively conducting samples (5%) is well correlated with the increase in the density of states at the Fermi level, yet a suppression centered at E_F is observed at all doping levels. This feature is indicative of an Altshuler–Aronov

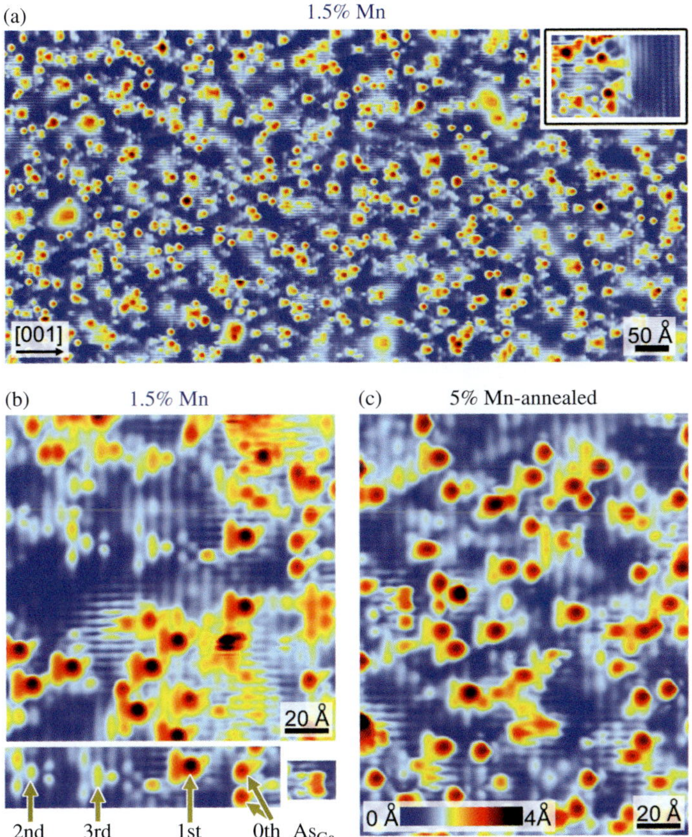

Fig. 7.1 STM topography of the in-gap states of $Ga_{1-x}Mn_x$. (a) STM topograph (+1.5 V, 20 pA) of $Ga_{0.985}Mn_{0.015}As$ over a 1000×500 Å area. The inset shows a topograph (+1.8 V, 20 pA) of the heterostructure junction between the Mn-doped and Be-doped layers of GaAs. The size of this area is 80×125 Å. (b) A topograph (+1.5 V, 30 pA) over a smaller area of size 150×150 Å of $Ga_{0.985}Mn_{0.015}As$. Several substitutional Mn atoms in various layers are identified in the lower part of the panel. They are marked by which layer they are in, with the surface labeled as zero. An As antisite is also shown. (c) STM topograph (+1.5 V, 20 pA) of the $Ga_{0.95}Mn_{0.05}As$ sample after annealing for 6 h at 200°C. The size of the area is 150×195 Å. (b) and (c) have the same scale. From Richardella *et al.* (2010) with permission.

correlation gap, which is expected to occur in the tunneling density of states of a disordered material on the metallic side of the phase transition (Altshuler and Aronov, 1979), appearing as a square-root singularity in the conductance near E_F (Fig. 7.2 (d)). Previous spectroscopic measurements of $Ga_{1-x}Mn_xAs$ using macroscopic tunneling junctions have reported similar correlation gaps (Chun *et al.*, 2002).

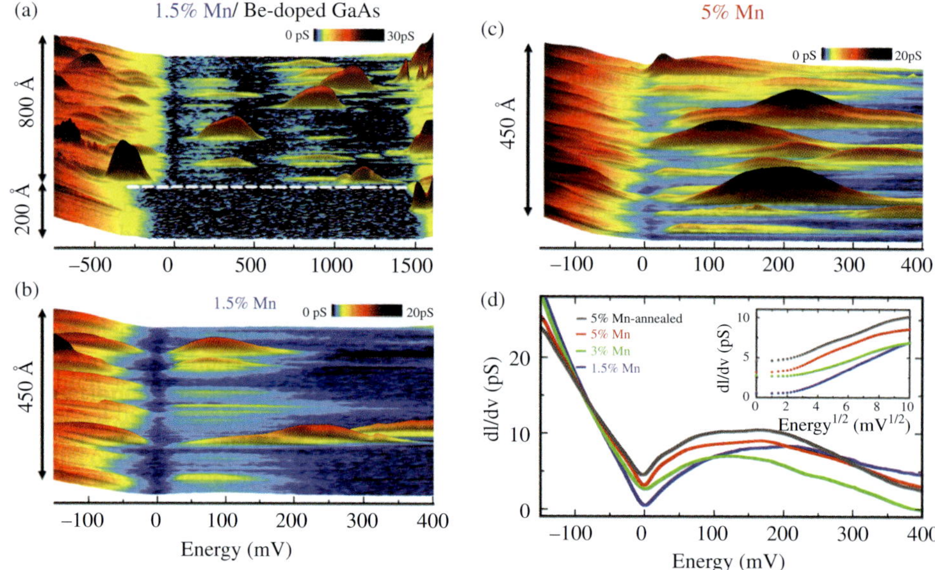

Fig. 7.2 Local dI/dV spectroscopy for various dopings. (a) The differential conductance (dI/dV) ($\Delta V = 10$ mV) across a line normal to the growth surface of length 1000 Å. The first 200 Å is the Be-doped GaAs buffer layer, and 800 Å of 1.5% Mn-doped GaAs follows. The junction between the two is marked by a dashed line. (b), (c) The dI/dV spectra (ΔV=5 mV) for energies close to Fermi level, across a line of length 450 Å over the 1.5% sample (b), and the 5% sample (c). Top of the valence band and the in-gap states are shown. (d) The spatially averaged differential conductance for several samples. The inset shows the same data as the main panel, with the square root of the voltage on the horizontal axis.From Richardella *et al.* (2010) with permission.

7.3 Analysis

To determine whether there are any specific length scales associated with the spatial variation of the LDOS in $Ga_{1-x}Mn_xAs$, we now examine energy-resolved STM conductance maps. In Fig. 7.3, we show examples of such maps at different energies relative to E_F for the 1.5% doped sample. These maps show that, in addition to modulations on the length scales of individual acceptors, there are spatial structures in the local density of states with longer length scales. To characterize these variations, for each conductance map, we compute the angle-averaged autocorrelation function between two points separated by r,

$$C(E,r) = 1/(2\pi)\int d\theta \int d^2r'[g(E,\mathbf{r}') - g_0(E)] \times [g(E,\mathbf{r}'+\mathbf{r}) - g_0(E)],$$

in which $g(E,r)$ is the local value of the differential tunneling conductance that is proportional to the LDOS, $g_0(E)$ is the average value of the conductance at each energy E. Displaying $C(E,r)$ in Fig. 7.4 (a), we find a dramatic increase of the long-distance

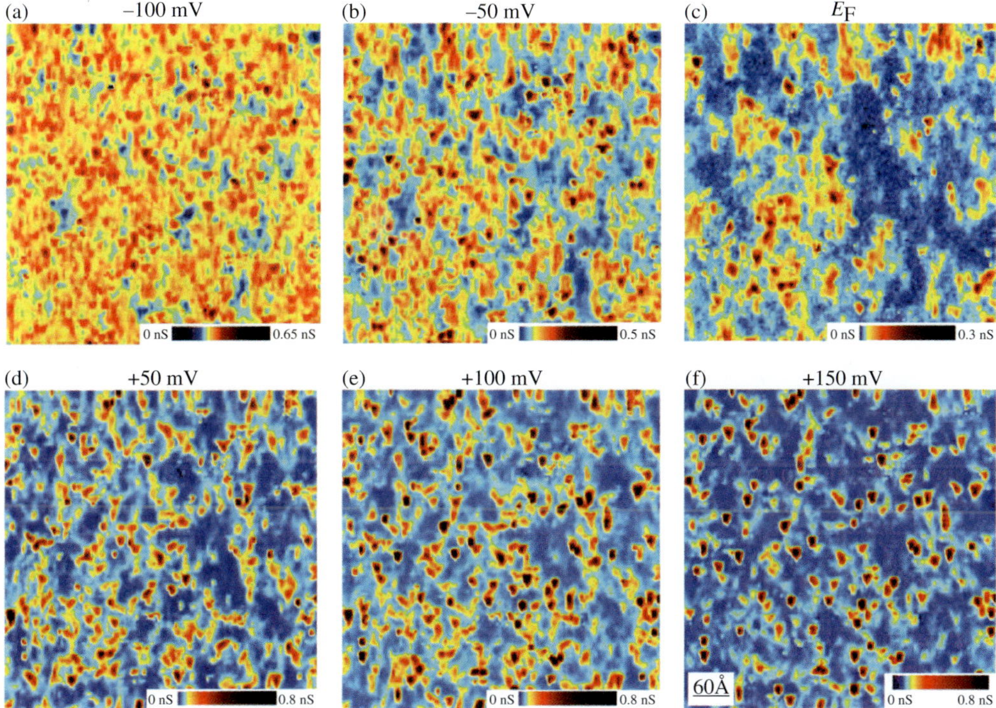

Fig. 7.3 Spectroscopic mapping in $Ga_{0.985}Mn_{0.015}As$. The differential conductance (dI/dV) measurements over an area of 500×500 Å show the spatial variations in the LDOS for the states in the valence band as well as states deep inside the semiconductor gap: (a), −100 mV; (b), −50 mV; (c), 0 mV; (d), +50 mV; (e), +100 mV; and (f), +150 mV.From Richardella *et al.* (2010) with permission.

correlations near E_F. At this energy, the correlations remain measurable to length scales well beyond that of single Mn acceptor states, which dominate the behavior on short length scales at all energies. The increased correlation length can also be seen directly in the size of the patches of high and low conductance (Fig. 7.3 (c)). We have observed the enhanced spatial correlations near E_F for all doping levels examined in this study (up to 5%, Fig. 7.4 (b)). However, this effect is most dramatic for the least doped samples (1.5%) closest to the metal–insulator transition. Control experiments on Zn- or Be-doped GaAs samples show no evidence of any special length scale or of a sharp peak near the E_F in the autocorrelation function.

Continuous phase transitions, such as the metal–insulator transition, are typically characterized by a correlation length, which describes the exponential decay of spatial fluctuations when a system is tuned near the phase transition. At the critical point, this correlation length diverges and spatial fluctuations and other physical properties display power-law spatial characteristics. In the non-interacting limit, the transition between a metal and an insulator occurs by tuning the chemical potential relative to

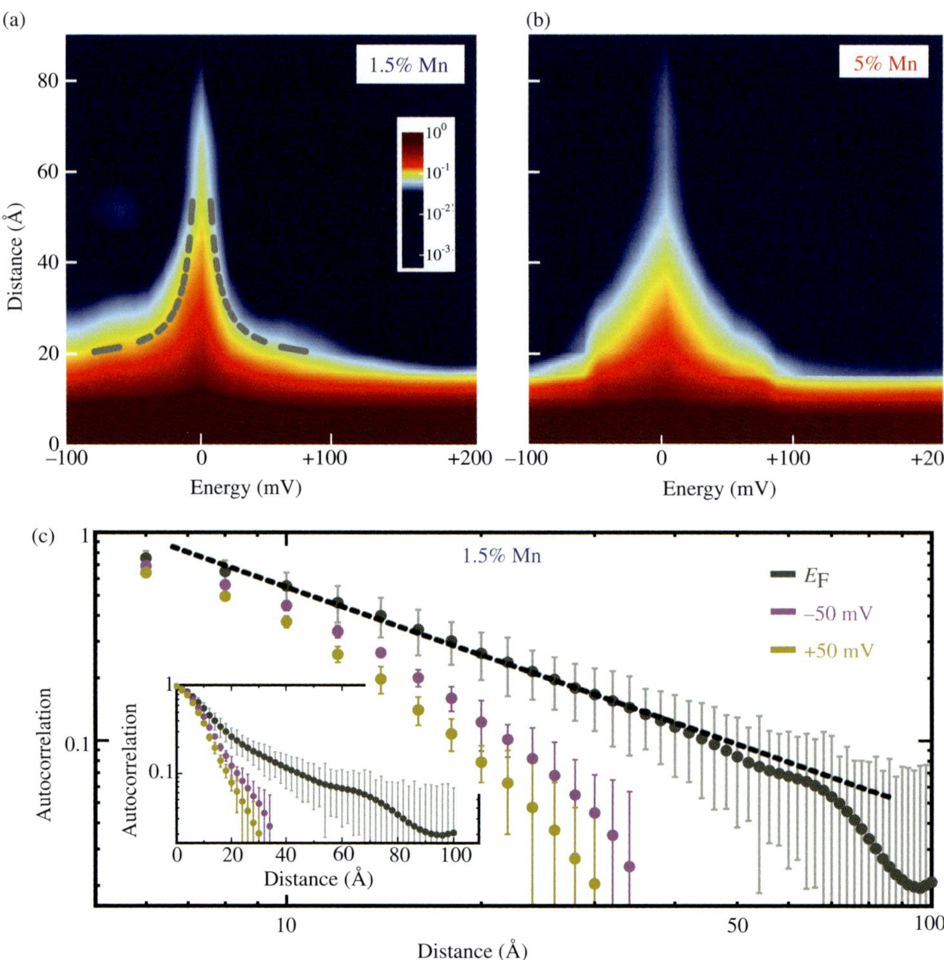

Fig. 7.4 Correlation length for Mn dopings close to the metal–insulator transition. The autocorrelation was calculated from LDOS maps and is presented on a logarithmic scale for (a) $Ga_{0.985}Mn_{0.015}As$ and (b) $Ga_{0.95}Mn_{0.05}As$. The dashed line is $\sim(E - E_F)^{-1}$ and is a guide to the eye. (c) The autocorrelation for the dI/dV map at the Fermi level of $Ga_{0.985}Mn_{0.015}As$, as well as one valence band (–50 mV) and one in-gap energy (+50 mV) are plotted. The inset shows the autocorrelation for the same energies on a semi-logarithmic scale.From Richardella *et al.* (2010) with permission.

the mobility edge. Mapping the spatial structure of the electronic states as function of energy can be used to determine the correlation length and to probe the critical properties for such a transition between extended and localized states (Belitz and Kirkpatrick, 1994; Evers and Mirlin, 2008; Lee and Ramakrishnan, 1985). In our experiments, the distance dependence of the energy-resolved autocorrelation function

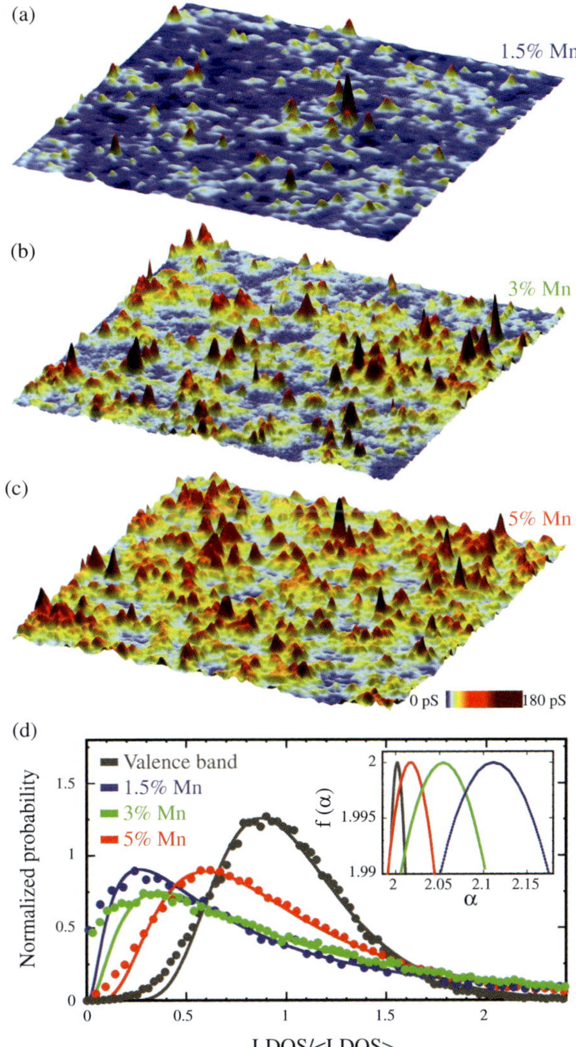

Fig. 7.5 The spatial variations of the LDOS at the Fermi level, their histogram, and multifractal spectrum. The LDOS mapping of a 700 × 700 Å area of (a) $Ga_{0.985}Mn_{0.015}As$, (b) $Ga_{0.97}Mn_{0.03}As$, and (c) $Ga_{0.95}Mn_{0.05}As$. (d) The normalized histogram of the maps presented in parts (a) to (c). The local values of the dI/dV are normalized by the average value of each map. The inset shows the multifractal spectrum, f(α), near the value α_0 where the maximum value occurs. For comparison, the results of a similar analysis over an LDOS map at –100 mV (valence band states) for the 1.5% doped sample are also shown.From Richardella *et al.* (2010) with permission.

$C(E, r)$ for the 1.5% sample (Fig. 7.4 (c)) appears to follow a power law at E_F, while at nearby energies it falls off exponentially (see inset). These observations, together with the apparent divergence of the correlation at a specific energy, are indeed signatures of the critical phenomena associated with a metal–insulator transition. However, our observation that the longest-ranged correlations are centered at E_F, as opposed to some other energy, which could be identified as a mobility edge, signifies the importance of electron–electron interactions in the observed correlations.

Given the importance of electron–electron interactions, the conductance maps are perhaps more precisely identified as probing the spatial nature of quasiparticle excitations of a many-body system rather than simply imaging single-electron states in the non-interacting limit. Currently there are no theoretical models of the real space structure of these excitations near the metal–insulator transition in a strongly interacting and disordered system, although there is continued effort in understanding the nature of such transitions in the presence of interactions (Basko *et al.*, 2006; Belitz and Kirkpatrick, 1994). Nevertheless, we suspect that the correlation length associated with these excitations will become shorter due to multi-particle processes and inelastic effects at energies away from E_F. Our experimental results for the least conducting sample (1.5%) indicate that the correlation length ξ is indeed suppressed away from E_F, roughly following $(E - E_F)^{-1}$ (dashed line in Fig. 7.4 (a)). At E_F, for this sample, these correlations decay in space following a power law $r^{-\eta}$, with $\eta = 1.2 \pm 0.3$.

Despite the importance of strong interactions, many of the predictions for the non-interacting limit still appear to apply. Weakly disordered extended states are expected to show Gaussian distributions of the LDOS indicating that these states have a finite probability to be present over the entire system. In contrast, near the metal–insulator transition wide distributions are expected, especially in local quantities such as the LDOS, which begin to cross over from Gaussian to log-normal distributions even in the limit of weak localization (Altshuler *et al.*, 1989; Lerner, 1988). Spectroscopic maps of the density of states at E_F for three different dopings (Fig. 7.5 (a–c)) show different degrees of spatial variations; however, their histograms (Fig. 7.5 (d)) are similar in being skewed log-normal distributions where the mean is not representative of the distribution due to rare large values. Decreasing the doping skews the distribution further, in a systematic fashion, away from Gaussian and toward a log-normal distribution. For comparison, a histogram of the LDOS for states deep in the valence band for the least doped sample (dark gray circles in Fig. 7.5 (d)) shows a Gaussian distribution.

7.4 Multifractal Wavefunctions

Based on the predictions for the non-interacting limit, we expect critical states to have a spatial structure that is multifractal in nature. This property is directly related to the scale-invariant nature of critical wavefunctions and has been examined in great detail by numerical simulations of the single-particle quantum states near an Anderson transition (Evers and Mirlin, 2008). Multifractal patterns, which are ubiquitous in nature, are usually described by analysis of their self-similarity at different length scales through their singularity spectrum f(α). Physically, f(α) describes all the fractal dimensions embedded in a spatial pattern, such as those associated with a quantum wavefunction and its probability distribution. It is calculated by splitting the probability distribution into sets of locations $\{r_i\}$ that share a common exponent α, where the distribution scales locally with distance like $|\psi(r_i)|^2 \sim L^{-\alpha}$, and measuring the fractal dimension of each set (Evers and Mirlin, 2008). A variety of techniques have been developed to compute f(α), which has been used to distinguish between various models of the Anderson transition (Chhabra and Jensen, 1989). Application of such an analysis to our conductance maps (Fig. 7.5 (d), inset), shows an f(α) spectrum that is peaked at a value away from two, which is indicative of anomalous scaling in a two-dimensional map. The f(α) spectrum also shows a systematic shift with decreasing doping, indicating a trend from weak toward strong multifractality with decreasing doping. In contrast, these signatures of multifractal behavior are absent for states deep in the valence band (gray curve) that, despite the strong disorder, show scaling consistent with those expected for extended states.

7.5 Conclusion

These experiments represent the first time spatial structure of a semiconductor have been examined in proximity to the metal-insulator transition. The spectroscopic mapping with STM reveal strong fluctuation of electronic states in real space but they contain spatial variations indicating their proximity to a quantum phase transition. Analysis of these fluctuations reveal features that have been long predicted for spatial structure of electronic states near the metal-insulator transition. However, our findings are not from a system that is weakly interacting, so it is rather remarkable that the ideas from non-intreating picture of a metal-insulator transition appear to be directly relevant. Our findings also suggest that proximity to the metal–insulator transition and electronic correlations may play a more significant role in the underlying mechanism of magnetism of $Ga_{1-x}Mn_xAs$ than previously anticipated. Beyond its application to understand the nature of states of $Ga_{1-x}Mn_xAs$, our experimental approach provides a direct method to examine critical correlations for other material systems near a quantum phase transition. In principle, experiments at the lowest temperatures for samples closest to the metal–insulator transition should provide accurate measurements of power-law characteristics, which can be directly compared to theoretically predicted critical exponents.

Acknowledgements

This work was done in collaboration with A. Richardella, P. Roushan, S. Mack, B. Zhou, D. Huse, and D. Awschalom. This work was supported by grants from ONR, ARO, NSF, and the NSF-MRSEC program through the Princeton Center for Complex Material.

References

Altshuler, B.L. and Aronov, A.G. (1979). *Solid State Commun.*, **30**, 115–117.

Altshuler, B.L., Kravtsov, V.E., and Lerner, I.V. (1989). *Phys. Lett. A*, **134**, 488–492.

Anderson, P.W. (1958). *Phys. Rev.*, **109**, 1492–1505.

Aspect, A. and Inguscio, M. (2009). *Phys. Today*, **62**, 30–35.

Basko, D.M., Aleiner, I.L., and Altshuler, B.L. (2006). *Ann. Phys.*, **321**, 1126–1205.

Belitz, D. and Kirkpatrick, T.R. (1994). *Rev. Mod. Phys.*, **66**, 261–390.

Burch, K.S., Shrekenhamer, D.B., Singley, E.J., Stephens, J., Sheu, B.L., Kawakami, R.K., Schiffer, P., Samarth, N., Awschalom, D.D., and Basov, D.N. (2006). *Phys. Rev. Lett.*, **97**, 087208.

Chhabra, A. and Jensen, R.V. (1989, Mar). *Phys. Rev. Lett.*, **62**, 1327–1330.

Chiba, D., Sawicki, M., Nishitani, Y., Nakatani, Y., Matsukura, F., and Ohno, H. (2008). *Nature*, **455**, 515–518.

Chun, S.H., Potashnik, S.J., Ku, K.C., Schiffer, P., and Samarth, N. (2002). *Phys. Rev. B*, **66**, 100408(R).

Dietl, T. (2008). *J. Phys. Soc. Jpn.*, **77**, 031005–031015.

Dietl, T., Ohno, H., Matsukura, F., Cibert, J., and Ferrand, D. (2000). *Science*, **287**, 1019–1022.

Evers, F. and Mirlin, A.D. (2008). *Rev. Mod. Phys.*, **80**, 1355–1417.

Garleff, J.K., Çelebi, C., Roy, W. Van, Tang, J. M., Flatté, M.E., and Koenraad, P.M. (2008). *Phys. Rev. B*, **78**, 075313.

Hashimoto, K., Sohrmann, C., Wiebe, J., Inaoka, T., Meier, F., Hirayama, Y., Römer, R.A., Wiesendanger, R., and Morgenstern, M. (2008). *Phys. Rev. Lett.*, **101**, 256802.

Ilani, S., Martin, J., Teitelbaum, E., Smet, J.H., Mahalu, D., Umansky, V., and Yacoby, A. (2004). *Nature*, **427**, 328–332.

Jungwirth, T., Sinova, J., MacDonald, A.H., Gallagher, B.L., Novák, V., Edmonds, K.W., Rushforth, A.W., Campion, R.P., Foxon, C.T., Eaves, L., Olejnik, E., Mašek, J., Eric Yang, S.-R., Wunderlich, J., Gould, C., Molenkamp, L.W., Dietl, T., and Ohno, H. (2007). *Phys. Rev. B*, **76**, 125206.

Kitchen, D., Richardella, A., Tang, J. M., Flatté, M.E., and Yazdani, A. (2006). *Nature*, **442**, 436–439.

Lagendijk, A., van Tiggelen, B., and Wiersma, D.S. (2009). *Phys. Today*, **62**, 24–29.

Lee, P.A. and Ramakrishnan, T.V. (1985). *Rev. Mod. Phys.*, **57**, 287–337.

Lerner, I.V. (1988). *Phys. Lett. A*, **133**, 253–256.

Moca, C.P., Sheu, B.L., Samarth, N., Schiffer, P., Janko, B., and Zarand, G. (2009). *Phys. Rev. Lett.*, **102**, 137203.

Myers, R.C., Sheu, B.L., Jackson, A.W., Gossard, A.C., Schiffer, P., Samarth, N., and Awschalom, D.D. (2006). *Phys. Rev. B*, **74**, 155203.

Neumaier, D., Wagner, K., Geißler, S., Wurstbauer, U., Sadowski, J., Wegscheider, W., and Weiss, D. (2007). *Phys. Rev. Lett.*, **99**, 116803.

Ohno, H., Chiba, D., Matsukura, F., Omiya, T., Dietl, E. Abe T., Ohno, Y., and Ohtani, K. (2000). *Nature*, **408**, 944–946.

Richardella, A., Roushan, P., Mack, S., Zhou, B., Huse, D., Awschalom, D., and Yazdani, A. (2010). *Science*, **327**, 665.

Rokhinson, L.P., Lyanda-Geller, Y., Ge, Z., Shen, S., Liu, X., Dobrowolska, M., and Furdyna, J.K. (2007). *Phys. Rev. B*, **76**, 161201.

Sheu, B. L, Myers, R.C., Tang, J.-M., Samarth, N., Schiffer, D.D. Awschalom P., and Flatté, M.E. (2007). *Phys. Rev. Lett.*, **99**, 227205.

Strandberg, T.O., Canali, C.M., and MacDonald, A.H. (2009). *Phys. Rev. B*, **80**, 024425.

Tang, J. M. and Flatté, M.E. (2004). *Phys. Rev. Lett.*, **92**, 047201.

8
Glassy Dynamics of Electrons Near the Metal–Insulator Transition

Dragana Popović

National High Magnetic Field Laboratory, Florida State University
Tallahassee, Florida 32310, USA

8.1 Introduction

The discovery of many novel materials over the past couple of decades has revived interest in the metal–insulator transition (MIT), one of the longstanding, fundamental problems of condensed-matter physics. Many such materials, including manganites, diluted magnetic semiconductors and high-temperature superconductors, are created by doping an insulating host and thus are close to a conductor–insulator transition. There is substantial evidence that, in many cases, strong electronic correlations (Mott localization) and disorder (Anderson localization) both play an important role in this regime (see Miranda and Dobrosavljević (2005) for a review). In general, the competition between Coulomb repulsion, which favors a uniform distribution of electrons (Wigner crystallization), and disorder, which favors a random one, leads to frustration in the system. This means that electrons are unable to satisfy these requirements simultaneously, resulting in the absence of a unique ground state and the emergence of a large number of metastable states or minima ("valleys") in the free-energy landscape. Metastable states, which correspond to different charge configurations, are separated by high barriers, leading to slow dynamics, divergence of the equilibration time, and breaking of ergodicity. Such a system is called a Coulomb glass. However, there are many other types of glassy systems. A common denominator in all of them is their complex or "rugged" energy landscape (Fig. 8.1). The most extensively studied and best known glasses in condensed-matter physics are probably conventional spin glasses (Binder and Young, 1986) such as Cu:Mn. On the other hand, Coulomb glasses were first anticipated theoretically almost three decades ago (Davies *et al.*, 1982, 1984; Grünewald *et al.*, 1982; Pollak, 1984; Pollak and Ortuño, 1982) in situations where electrons are strongly localized due to disorder, but experimental studies have been scarce. Some recent work on electron glasses in such a strongly localized regime, away from the MIT, has been reviewed by Amir *et al.* (2011). In the opposite limit of well-delocalized electrons, one expects a single, well-defined ground state and the absence of glassiness. Not surprisingly, it is the behavior in the intermediate region, near the MIT, that has been most difficult to understand.

In the presence of disorder, the local electron density undergoes strong spatial fluctuations. Therefore, it is plausible to expect that in the vicinity of the MIT the

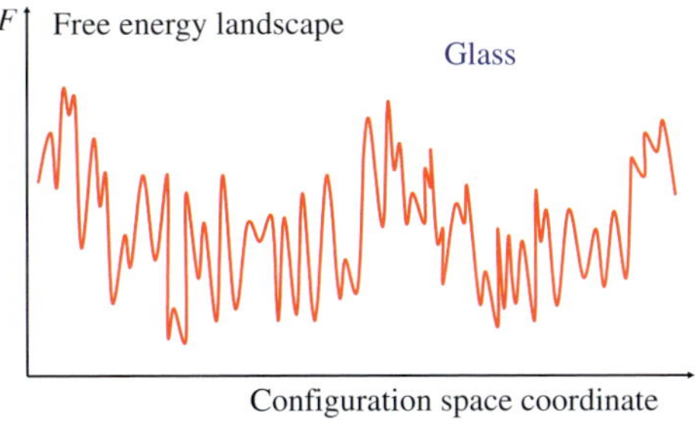

Fig. 8.1 The free-energy (F) landscape of a glassy system. The horizontal axis represents the one-dimensional projection of the configurational coordinates of the degrees of freedom.

local density in some areas may be higher than the average and correspond to the metallic state, whereas in the remaining regions the local density may be lower than the average and the electrons are localized. In fact, it has been proposed that in weakly doped Mott insulators near the MIT, the system will settle for a nanoscale phase separation[1] between a conductor and an insulator (Dagotto, 2002; Gor'kov and Sokol, 1987; Kivelson *et al.*, 2003). New and powerful experimental techniques, including imaging by scanning tunneling microscopy (Kohsaka *et al.*, 2007), have indeed firmly established the existence of charge inhomogeneities in strongly correlated electron systems, such as cuprates (Dagotto, 2005; Orenstein and Millis, 2000). This obviously leads to the possibility of a myriad of competing charge configurations and the emergence of the associated glassy dynamics. Recent studies (Jelbert *et al.*, 2008; Raičević *et al.*, 2007, 2008, 2011) of a lightly doped, insulating $La_{2-x}Sr_xCuO_4$, $x = 0.03$, have found several clear signatures of glassy charge dynamics as temperature $T \rightarrow 0$, consistent with an underlying glassy ground state that results from Coulomb interactions. Further work is needed, however, to see how this glassy state evolves as the system approaches the transition from an insulator to a conductor.

It is also desirable to extend studies of charge dynamics to other types of material to explore a possible generality of glassy freezing in strongly correlated systems near the MIT (Dobrosavljević *et al.*, 2003; Miranda and Dobrosavljević, 2005; Pankov and Dobrosavljević, 2005). Many materials do exhibit complex behavior due to the existence of several competing ground states (Dagotto, 2005). In fact, the frustration caused by the competition of interactions on different length scales may give rise to glassy dynamics even in the absence of disorder (Schmalian and Wolynes, 2000), while even a small amount of disorder may favor glassiness over various static charge-ordered states (Pankov and Dobrosavljević, 2005). Even though the emergence of glassiness thus appears to be ubiquitous at low temperatures, Coulomb glasses and out-of-equilibrium systems in general remain poorly understood. Experimentally, studies of charge dynamics near the MIT in many materials are often complicated by the accompanying changes in magnetic or structural symmetry, as well as by the glassy freezing of spins. Doped semiconductors, such as Si:P and Si:B, are free from such complications. They have been used extensively to study the critical behavior near the MIT (Miranda and Dobrosavljević, 2005; Sarachik, 1995) but, in spite of some early hints of glassiness in the insulating regime (Monroe, 1990; Monroe *et al.*, 1987; Paalanen *et al.*, 1983), charge dynamics was not studied further until recently (Kar *et al.*, 2003; Thorsmølle and Armitage, 2010), as discussed below.

Two-dimensional electron systems (2DESs) in semiconductors (Ando *et al.*, 1982) such as Si provide another relatively simple system for exploring the interplay of electronic correlations and disorder (for a review of disordered electronic systems, see Lee and Ramakrishnan (1985)). They have an additional advantage in that all the relevant parameters—carrier concentration, disorder and interactions—can be varied relatively easily. This review will describe some of the work that has demonstrated that the 2DES in Si is an excellent model system not only for studying the MIT in two

[1] Global phase separation is not possible as it would violate charge neutrality.

dimensions, but also for investigating the nearly universal non-equilibrium behavior exhibited by a large class of both two- and three-dimensional (2D, 3D) systems (e.g. spin glasses, supercooled liquids, granular films). In fact, the work on the 2DES in Si provides additional strong evidence that many such universal features are robust manifestations of glassiness, regardless of the dimensionality of the system.

The dimensionality plays an important role in the MIT. In two dimensions, for example, the very existence of the metal and the MIT has been questioned for many years. Recently, considerable experimental evidence has become available in favor of such a transition. Some of that work has been described in several review papers (e.g. Kravchenko and Sarachik (2004)) with a focus on very clean (low-disorder) samples. This review will first extend that discussion to the cases of higher disorder in order to (a) demonstrate that the 2D MIT is observed in all 2DESs in Si regardless of the amount of disorder, (b) point out important similarities to the MIT in 3D systems, and (c) set the stage for the discussion of glassy dynamics near the MIT. As summarized below, the 2DES in Si exhibits all the main manifestations of glassiness: slow, correlated dynamics, non-exponential relaxations, diverging equilibration time as $T \to 0$, aging and memory. The results provide strong support to theoretical proposals describing the 2D MIT as the melting of a Coulomb glass (Chakravarty *et al.*, 1999; Dalidovich and Dobrosavljević, 2002; Dobrosavljević *et al.*, 2003; Pastor and Dobrosavljević, 1999; Thakur and Neilson, 1996, 1999). The review concludes with further comparison to other complex materials, both 2D and 3D, and by identifying some open directions for future research.

8.2 Metal–insulator transitions in two dimensions

In 2D systems in semiconductor heterostructures, the electron density n_s can be varied easily over two orders of magnitude simply by applying voltage V_g to the so-called gate electrode. At low n_s, the 2DES is strongly correlated: $r_s \gg 1$, where $r_s = E_C/E_F \propto n_s^{-1/2}$ (E_C is the average Coulomb energy per electron and E_F is the Fermi energy; see also Kravchenko and Sarachik (2004) for more details). At the same time, the electrons “feel” a random potential (disorder) caused by charged impurities that are located away from the 2DES. Some of the striking experimental evidence for the MIT that occurs in a variety of “clean”, i.e. low-disordered 2D systems has been presented also in the chapter by S. V. Kravchenko (this volume).

This section reviews work that demonstrates the existence of the 2D MIT regardless of the amount or type of disorder. The disorder, however, does affect the values of n_c, the critical density for the MIT, and the precise form of the temperature dependence of conductivity $\sigma(T)$. For the sake of clarity, it is important to recall that a qualitative distinction between a metal and an insulator exists only at $T = 0$: $\sigma(T = 0) \neq 0$ in the metal and $\sigma(T = 0) = 0$ in the insulator. The experiments were performed on a 2DES in Si metal-oxide-semiconductor field-effect transistors (MOSFETs), where the low-density regime can be reached more easily than in other semiconductors (Kravchenko and Sarachik, 2004). In Si MOSFETs, the (Drude) mobility $\mu = \sigma/(n_s e)$ peaks as a function of n_s because, at very high n_s that are not of interest here, the scattering due to the roughness of the Si–SiO_2 interface becomes increasingly important (Ando *et al.*, 1982). The peak mobility at 4.2 K is commonly used as a rough measure of the amount of disorder.

8.2.1 Effects of disorder

High-mobility (low-disorder) samples

In high-mobility Si MOSFETs where, roughly speaking, the 4.2 K peak $\mu > 2\ \mathrm{m}^2/(\mathrm{Vs})$, the low-temperature resistivity ρ exhibits a strong, metallic drop ($d\rho/dT > 0$) with decreasing T for $n_\mathrm{s} > n_\mathrm{s}^*$ (Fig. 8.2) (Kravchenko and Sarachik, 2004). $d\rho/dT$ changes sign at a density n_s^* and becomes insulating-like ($d\rho/dT < 0$) for $n_\mathrm{s} < n_\mathrm{s}^*$. Even though $d\rho/dT < 0$ does not necessarily imply that ρ diverges at zero temperature (i.e. that $\sigma(T = 0) = 0$), the density n_s^* has often been assumed to represent the critical density for the MIT. However, other, more appropriate methods have also been used to identify n_c. For example, n_c was determined based on both a vanishing activation energy and a vanishing non-linearity of current–voltage characteristics when extrapolated from the insulating phase (Pudalov *et al.*, 1993; Shashkin *et al.*, 2001). It was established that $n_\mathrm{c} \approx n_\mathrm{s}^*$, although a small but systematic difference of a few percent has been reported such that $n_\mathrm{c} < n_\mathrm{s}^*$ (Altshuler *et al.*, 2001; Pudalov *et al.*, 1998*a*). The inset to Fig. 8.2 illustrates the use of activation energies, determined from the fits of the data at the lowest n_s and T to the simply activated form,[2] $\rho \propto \exp(E_A/k_BT)$.

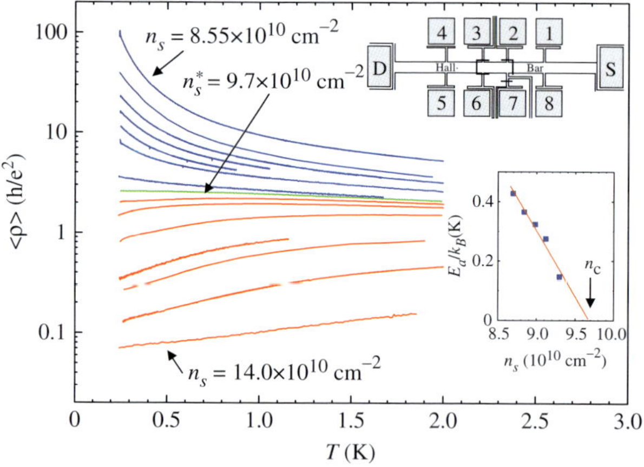

Fig. 8.2 High-mobility sample ($\mu \approx 2.5\ \mathrm{m}^2/\mathrm{Vs}$): resistivity $\rho = 1/\sigma$ vs T for $n_\mathrm{s} = 8.55$, 8.70, 8.84, 8.99, 9.13, 9.27, 9.56, 9.71, 9.85, 9.99, 10.4, 11.2, 11.6, 12.9, and $14.0 \times 10^{10}\mathrm{cm}^{-2}$ (from the top) (Jaroszyński *et al.*, 2002*b*). Insets: a schematic of the sample (top view) and activation energies vs n_s; $n_\mathrm{c} \approx n_\mathrm{s}^*$ (here $r_\mathrm{s} \approx 17$).

[2]Close enough to n_c, the data can be fitted almost equally well to the variable-range hopping law. This gives the same values of n_c within the error of the fit.

In the vicinity of n_c (or n_s^*), the resistivity curves can be collapsed onto the same scaling function of a single parameter T_0, i.e. $\rho(T, n_s) = \rho_c f(T/T_0)$ (Kravchenko *et al.*, 1994). T_0 is the same function of $\delta_n \equiv (n_s - n_c)/n_c$ on both the metallic and the insulating sides of the transition, $T_0 \propto |\delta_n|^{z\nu}$, where z is the dynamical exponent, ν the correlation-length exponent, and $z\nu = 1.6$ (Kravchenko *et al.*, 1995). Here $\rho_c(n_c)$ is the so-called "separatrix," the temperature-independent critical resistivity curve. On the metallic side, the scaling does not work at higher T, in the regime where $\rho(T)$ goes through a maximum and then decreases as T is raised further (Kravchenko and Sarachik, 2004). On general grounds, the scaling of $\rho(n_s, T)$ at low enough T represents a strong indication of the existence of the MIT at $T = 0$ (Sachdev, 1999).

At somewhat higher n_s in the metallic phase, it becomes apparent that after the initial rapid drop $\rho(T)$ becomes a very weak function as $T \to 0$ (Pudalov *et al.*, 1998*a*,*b*). In fact, it is so weak that, on a logarithmic scale, it appears to saturate. As a result of this "saturation" of $\rho(T)$, single-parameter scaling fails at the lowest temperatures (Pudalov *et al.*, 1998*a*). However, a careful analysis of the data shows (Kim *et al.*, 1998; Washburn *et al.*, 1999*a*,*b*) that all of the $\sigma(n_s, T)$ (or $\rho(n_s, T)$) curves can be scaled according to the more general form (Belitz and Kirkpatrick, 1994) $\sigma(n_s, T) = \sigma_c(T) f(T/T_0)$, where the critical conductivity $\sigma_c = \sigma(n_c, T) \propto T^x$, i.e. it vanishes as $T \to 0$. In other words, the critical conductivity is not the separatrix, but instead it belongs to the insulating family of curves: $n_c < n_s^*$, consistent with other studies (Altshuler *et al.*, 2001; Pudalov *et al.*, 1998*a*). The exponent x, however, is very small, with values between 0.1 and 0.2 (Kim *et al.*, 1998; Washburn *et al.*, 1999*a*,*b*), so that a direct observation of $\sigma_c(T)$ is very difficult at experimentally accessible temperatures. A small value of x is also the reason for the apparent success of the single-parameter scaling in the limited range of T and n_s. It is important to note that scaling with $x \neq 0$, also observed in 3D materials near the MIT (Sarachik, 1995), does not contradict (Belitz and Kirkpatrick, 1994) any fundamental principle for 2D systems. Indeed, violations of "Wegner scaling" (Belitz and Kirkpatrick, 1994), where $x = (D - 2)/z$ (where D is the dimensionality), were predicted for certain microscopic models (Belitz and Kirkpatrick, 1995; Castellani *et al.*, 1987; Kirkpatrick and Belitz, 1994) with strong spin-dependent components of the Coulomb interactions.

While the general scaling form describes the data satisfactorily as $T \to 0$, it would be desirable to explore scaling and the precise form of $\sigma_c(T)$ over a wider range of T. It turns out that the situation becomes simpler in samples with more disorder, where the onset of "saturation" of σ (or ρ) in the metallic phase is pushed to higher T.

Intermediate-mobility samples and effects of local magnetic moments

Although the metallic temperature dependence of conductivity, $d\sigma/dT < 0$ (i.e. $d\rho/dT > 0$), is less pronounced in samples with moderate mobility (4.2 K peak $\mu \sim 1$ m^2/Vs), it was demonstrated early on (Popović *et al.*, 1997) that scaling $\sigma(n_s, T) = \sigma_c f(T/T_0)$ is obeyed with exactly the same value of the exponent ($z\nu = 1.6$) as in high-mobility samples. While the value of n_c, defined as n_s^*, was higher than in

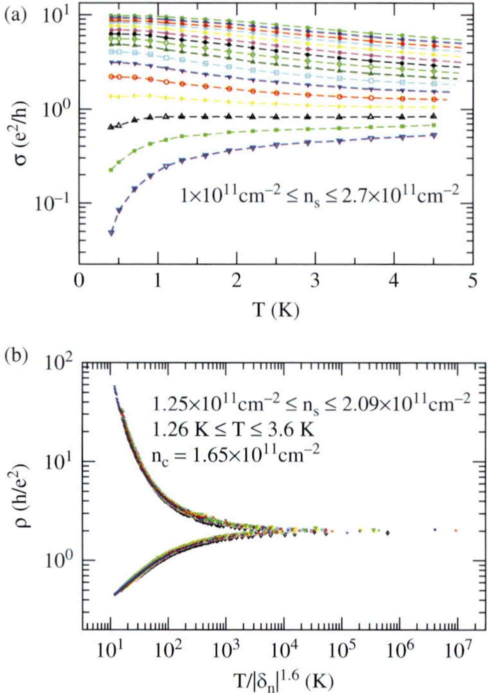

Fig. 8.3 Intermediate-mobility sample ($\mu \approx 1$ m^2/Vs; sample 12). (a) Conductivity σ vs T for different $n_{\rm s}$. (b) Scaling of σ with temperature for the same data (and other $n_{\rm s}$ not shown) in the $n_{\rm s}$ range given on the plot. $\delta_n \equiv (n_{\rm s} - n_{\rm c})/n_{\rm c}$ is the reduced density; $n_{\rm c} \approx n_{\rm s}^*$ (the corresponding $r_{\rm s} \approx 15$). For $n_{\rm s} \geq n_{\rm s}^*$, the scaling works only for $T \geq 1.1$ K. The substrate (back-gate) bias on the sample was $V_{\rm sub} = -40$ V (Feng *et al.*, 1999).

high-mobility devices,[3] the corresponding $r_{\rm s}$, at approximately 13, was still comparably large.

By extending the measurements to lower T (Feng *et al.*, 1999), however, the "saturation" of $\sigma(T)$ was observed to take place in the metallic phase at T as high as ~ 1 K (Fig. 8.3 (a)). Moreover, the separatrix acquired an insulating-like ($d\sigma/dT > 0$) temperature dependence. As a result, the simple, single-parameter scaling around the separatrix works only at $T > 1.1$ K on the metallic side (Fig. 8.3 (b)). Thus the data are qualitatively similar to those on high-mobility samples, indicating that $n_{\rm c} < n_{\rm s}^*$. However, since $\sigma(T)$ in the metallic phase ($n_{\rm s} > n_{\rm c}$) does not have a simple enough form over a sufficiently wide range of T, it is still difficult to make reliable extrapolations to $T = 0$. Fortunately, in Si MOSFETs it is also possible to change the type of disorder in the same sample. As shown below, this results in a surprisingly

[3] A study of Si MOSFETs with different peak mobilities indicated a systematic increase of $n_{\rm c}$ with disorder; $n_{\rm c}$ was defined as the "separatrix" $n_{\rm s}^*$ (Pudalov *et al.*, 1998*a*).

simple, precise form of $\sigma(T)$ over a very wide range of T, allowing reliable zero-temperature extrapolations and striking scaling behavior.

For a 2DES in Si, the disorder is due to oxide charge scattering (scattering by ionized impurities randomly distributed in SiO_2 within a few Å of the interface) and to the roughness of the Si–SiO_2 interface (Ando *et al.*, 1982). For fixed n_s, it is possible to change the disorder by applying bias V_{sub} to the Si substrate (back gate). In particular, the reverse (negative) V_{sub} moves the electrons closer to the interface, which increases the disorder. It also increases the splitting between the subbands since the width of the triangular potential well at the interface is reduced by applying negative V_{sub}. Usually, only the lowest subband is occupied at low T, giving rise to the 2D behavior. In sufficiently disordered samples, however, the band tails associated with the upper subbands can be so long that some of their strongly localized states may be populated even at low n_s, and act as additional scattering centers for 2D electrons. In particular, since at least some of them must be singly populated due to a large on-site Coulomb repulsion (tens of meV), they may act as local magnetic moments. Clearly, the negative V_{sub} reduces this type of scattering by depopulating the upper subband.

Therefore, by varying V_{sub}, it is possible to study the effect of local magnetic moments on the transport properties of the conduction electrons in a systematic and controlled way. It has been established (Feng *et al.*, 1999) that scattering by local magnetic moments suppresses the metallic behavior with $d\sigma/dT < 0$. Indeed, the data suggest that in the $T \to 0$ limit, the $d\sigma/dT < 0$ (*i. e.* $d\rho/dT > 0$) behavior is suppressed by an arbitrarily small amount of scattering of the conduction electrons by disorder-induced local moments (Feng *et al.*, 1999). Figures 8.3 (a) and 8.4 (a) illustrate the striking effect of local moments on $\sigma(T)$ in the same sample: while the "usual', metallic behavior with $d\sigma/dT < 0$ is observed in the absence of local moments (Fig. 8.3 (a)), $\sigma(T)$ curves become insulating-like ($d\sigma/dT > 0$) for all n_s after many local moments are introduced with V_{sub} (Fig. 8.4 (a)). However, this does not necessarily indicate the destruction of the metallic phase. In disordered 3D metals, for example, it is well known that the derivative $d\sigma/dT$ can be either negative or positive near the MIT (Lee and Ramakrishnan, 1985). On the other hand, the metallic behavior where σ decreases but does not go to zero (as expected for an insulator) as $T \to 0$ is new and unexpected in two dimensions.

The analysis of the insulating-like $\sigma(T)$ curves in Fig. 8.4 (a) shows (Feng *et al.*, 2001) that they follow a very simple and precise form over a broad (two orders of magnitude) range of T and n_s: $\sigma(n_s, T) = \sigma(n_s, T = 0) + A(n_s)T^2$ [Fig. 8.4 (b)]. The high quality of the fits allows a reliable extrapolation of $\sigma(n_s, T = 0)$, whose finite (i.e. non-zero) values mean that, in spite of the decrease of $\sigma(n_s, T)$ with decreasing T, the 2D system is in the metallic state. In particular, the zero-temperature conductivity $\sigma(n_s, T = 0)$ is a power-law function of δ_n (Fig. 8.4 (c)): $\sigma(n_s, T = 0) \propto \delta_n^{\mu}$ ($\mu \approx 3$), as expected in the vicinity of a quantum critical point (Goldenfeld, 1992) such as the MIT. The power law holds over a very wide range of δ_n (up to 5) similar to what has been observed (Rosenbaum *et al.*, 1980) in Si:P near the MIT. In addition, even though the MIT occurs at different n_c in different samples, the critical exponents μ are the same (Fig. 8.4 (c)), as expected from general arguments (Goldenfeld, 1992). It has also been demonstrated (Feng *et al.*, 2001) that near the MIT the data obey dynamical scaling

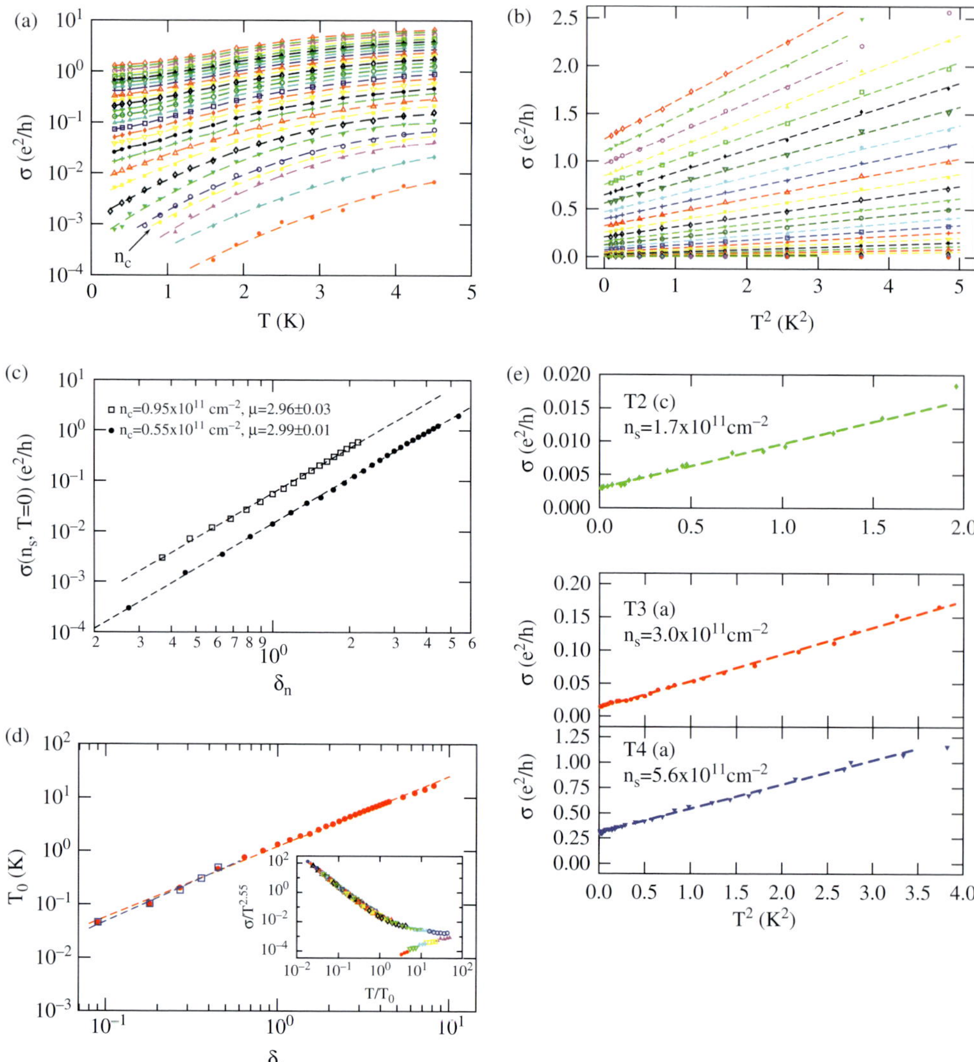

Fig. 8.4 (a)–(d) The same intermediate-mobility sample (sample 12) as in Fig. 8.3, but here $V_{\rm sub} = +1$ V. Adapted from Feng *et al.* (2001). (a) $\sigma(T)$ for $0.3 \leq n_{\rm s}(10^{11}{\rm cm}^{-2}) \leq 3.0$ (bottom to top) in steps of $0.1 \times 10^{11}{\rm cm}^{-2}$. (b) The same $\sigma(T)$ data plotted vs T^2; the lowest $n_{\rm s} = 0.7 \times 10^{11}{\rm cm}^{-2}$ and $0.3 \leq T(K) \leq 2.2$ K. Other measurements show that this $\sigma(T)$ holds at least down to 0.020 K (Eng *et al.*, 2002; Feng *et al.*, 2001). (c) $\sigma(n_{\rm s}, T = 0)$ vs δ_n for samples 12 (dots) and 9 (squares). The dashed lines are fits with the slopes equal to the critical exponent μ. At the MIT, the corresponding $r_{\rm s} \approx 22$ and 17 for the two samples, respectively. (d) Scaling parameter T_0 as a function of $|\delta_n|$ for sample 12. Open symbols: $n_{\rm s} < n_{\rm c}$, closed symbols: $n_{\rm s} > n_{\rm c}$. The dashed lines are fits with slopes 1.4 ± 0.1 and 1.32 ± 0.01, respectively. Inset: scaling of raw data $\sigma/\sigma_{\rm c} \sim \sigma/T^x$ in units of $e^2/h{\rm K}^{2.55}$ for all $n_{\rm s}$ shown in (a) and $T < 2$ K. (e) : $\sigma(T)$ vs T^2 for three different samples and densities. The 4.2 K peak mobility was between 0.15 and 0.33 m^2/Vs. Adapted from Sjöstrand and Stiles (1975) and Sjöstrand *et al.* (1976).

$\sigma(n_s, T) = \sigma_c(T) f(T/\delta_n^{z\nu})$ (Fig. 8.4 (d)) with a temperature dependent critical conductivity $\sigma_c = \sigma(n_s = n_c, T) \propto T^x$ ($z\nu = 1.3 \pm 0.1$, $x \approx 2.6$, $\mu = x(z\nu) = 3.4 \pm 0.4$), both in agreement with theoretical expectations near a quantum phase transition (Belitz and Kirkpatrick, 1994) and consistent with the extrapolations of $\sigma(T)$ to $T = 0$.

The T^2 form of $\sigma(T)$ is well established for metals containing local magnetic moments, and it is believed to result from the Kondo effect (Hewson, 1993). In fact, there is no other known mechanism that results in an *increase* of σ as T^2. Here this feature provides the most direct evidence of the presence of local magnetic moments. It should be noted that, in general, one expects T^2 behavior for a quantum impurity embedded in a Fermi liquid in any dimension. While the nature of this novel metallic state in two dimensions may require further study, its simple $\sigma(T)$ allows for an unambiguous extrapolation to $T = 0$. The zero-temperature conductivity $\sigma(n_s, T = 0)$ decreases continuously, and follows a distinct power-law behavior as the MIT is approached. In particular, metallic σ as small as $10^{-3}e^2/h$ has been observed (Feng *et al.*, 2001), in striking contrast to anything that has been reported in other 2D systems when $d\sigma/dT < 0$. A similar observation in 3D systems (Rosenbaum *et al.*, 1980) has demonstrated the absence of minimum metallic conductivity, and has had a profound impact on shaping theoretical ideas about the MIT.

2DESs in Si MOSFETs have been studied extensively for more more than four decades, so it is interesting that the T^2 behavior was identified only relatively recently (Feng *et al.*, 2001). A thorough examination of the early literature reveals, however, that the samples discussed here are representative of a broad class of Si MOSFETs known historically (and somewhat unfairly) as "non-ideal" samples (Ando *et al.*, 1982). Non-ideal samples could be made more "ideal" by applying V_{sub} and vice versa (Ando *et al.*, 1982), consistent with the recent studies (Feng *et al.*, 1999, 2001) discussed above. Figure 8.4 (e) shows an example of $\sigma(T)$ measured on samples with a modest peak mobility between 0.15 and 0.33 m^2/Vs (Sjöstrand *et al.*, 1976; Sjöstrand and Stiles, 1975). The "saturation" (on a log scale) of $\sigma(T)$ observed for $T < 1$ K was puzzling at the time, but the replotted data show clear T^2 behavior.

Low-mobility (highly disordered) samples

Samples with very low 4.2 K peak mobility ($\mu < 0.1$ m^2/Vs) have attracted less attention because they do not exhibit a pronounced, if any, $d\sigma/dT < 0$ metallic behavior. However, they not only exhibit the 2D MIT, but also other similarities to the behavior of "clean" 2DESs, such as the onset of glassy charge dynamics (Section 8.3). In some ways, it is even advantageous to investigate samples with a lot of disorder. For example, the relevant electron densities, such as n_c, are pushed to higher values (higher E_F), so that it is possible to reach much lower effective temperatures T/T_F (where T_F is the Fermi temperature) than in high-mobility samples. Therefore, comparative studies of samples with varying amounts of disorder should provide valuable insights into the physics of systems near the MIT.

Detailed studies of transport and electron dynamics near the MIT have been performed on a set of Si MOSFETs with a 4.2 K peak mobility of only 0.06 m^2/Vs with the applied $V_{\text{sub}} = -2$ V (Bogdanovich and Popović, 2002*b*). This value of V_{sub}

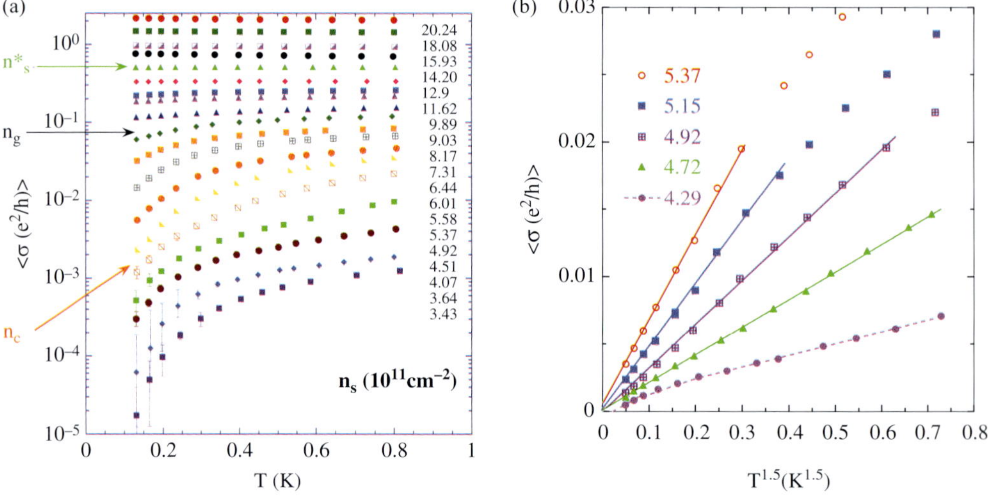

Fig. 8.5 Low-mobility sample ($\mu \approx 0.06$ m^2/Vs). (a) time-averaged $\langle\sigma\rangle$ vs T for different n_s. The error bars show the size of the fluctuations with time. n_s^*, n_g, and n_c are marked by arrows; n_g is the glass-transition density. (b) $\langle\sigma\rangle$ vs $T^{1.5}$ for a few $n_s(10^{11}\mathrm{cm}^{-2})$ near n_c. The solid lines are fits, the dashed line is a guide to the eye, clearly showing insulating behavior: $\langle\sigma(T \to 0)\rangle = 0$. From Bogdanovich and Popović (2002*b*)

maximizes the peak mobility by removing the contribution of scattering by local magnetic moments (see Section 8.2.1 above), at least in the experimental-T range. Because of the glassy fluctuations of σ with time t at low n_s and T (see Section 8.3), the carrier density had to be changed at relatively high T (here ~ 0.8 K) in small steps in order to obtain reproducible values of the time-averaged conductivity $\langle\sigma\rangle$. The behavior of $\langle\sigma(n_s, T)\rangle$ (Fig. 8.5 (a)) turns out to be quite similar to that of high-mobility Si MOSFETs. At the highest n_s, for example, the devices exhibit metallic behavior with $d\langle\sigma\rangle/dT < 0$. The change of $\langle\sigma\rangle$ in a given T range, however, is small (only 6% for the highest $n_s = 20.2 \times 10^{11}\mathrm{cm}^{-2}$), as observed in other Si MOSFETs with a large amount of disorder (Pudalov *et al.*, 1998*a*). Even though the density $n_s^* = 12.9 \times 10^{11}\mathrm{cm}^{-2}$ at the separatrix (Bogdanovich and Popović, 2002*a*,*b*) is much higher than in "clean" samples, the value of $\langle\sigma(n_s^*)\rangle = 0.5\ e^2/h$ is similar, according to the Drude formula corresponding to $k_F l \lesssim 1$, where k_F is the Fermi wave vector and l the mean-free path. Likewise, at the lowest n_s the data are best described by the simply activated form $\langle\sigma\rangle \propto \exp(-E_A/k_B T)$.

Perhaps the most striking difference between high- and low-mobility samples first becomes apparent when the vanishing of activation energy E_A is used to determine n_c. While in "clean" samples this gives $n_c \lesssim n_s^*$ (Section 8.2.1), here E_A vanishes at $n_c \approx 5 \times 10^{11}\mathrm{cm}^{-2}$, which is more than a factor of two smaller than n_s^* (Fig. 8.5 (a)). This suggests that there is a wide range of n_s on the metallic side of the MIT where $\sigma(T)$ is insulating-like. Indeed, close to n_c, the best phenomenological fit to the data is metallic power-law behavior $\langle\sigma(n_s, T)\rangle = a(n_s) + b(n_s)T^x$ with $x \approx 1.5$ (Fig. 8.5 (b))

(Bogdanovich and Popović, 2002*b*). The fitting parameter $a(n_s)$ is relatively small and in fact vanishes for $n_s(10^{11}\mathrm{cm}^{-2}) = 4.72$ and 4.92. Therefore, $n_c = (5.0 \pm 0.3) \times 10^{11}\mathrm{cm}^{-2}$ ($r_s \sim 7$), based on the data on both metallic and insulating sides of the MIT. Of course, a simple power law $\langle\sigma(n_c, T)\rangle \propto T^x$ is consistent not only with general expectations near the MIT and the behavior observed in 3D systems (Belitz and Kirkpatrick, 1994), but also with a careful analysis of high-mobility 2DES (Kim *et al.*, 1998; Washburn *et al.*, 1999*a*,*b*) (Section 8.2.1) and those with local magnetic moments (Feng *et al.*, 2001) (Section 8.2.1). Here the exponent x takes a distinctly different value, presumably reflecting the different universality classes of those situations.

The surprising non-Fermi liquid $T^{3/2}$ behavior is consistent with theory (Dalidovich and Dobrosavljević, 2002) for the transition region between a Fermi liquid and an (insulating) electron glass. Indeed (see Section 8.3), the transition into a charge (Coulomb) glass in low-mobility samples takes place as $T \to 0$ at a density n_g, such that $n_c < n_g < n_s^*$ (Fig. 8.5 (a)). The $T^{3/2}$ correction is characteristic of transport in the intermediate $n_c < n_s < n_g$ region, where the dynamics are glassy, but where σ is still metallic—$\sigma(T \to 0) \neq 0$—albeit so small that $k_F l < 1$. Such "bad metals" include a variety of strongly correlated materials with unusual properties (Emery and Kivelson, 1995). Interestingly, the $T^{3/2}$ behavior can also be revealed in high-mobility 2DESs by applying a parallel magnetic field.

8.2.2 Two-dimensional metal–insulator transition in a parallel magnetic field

Since magnetic fields B applied parallel to the 2DES plane couple only to electrons' spins, they are often used to probe the importance of spin, as opposed to charge, degrees of freedom. Some of the intriguing results that have been obtained in parallel B in the vicinity of the zero-field 2D MIT have been described in the review by Kravchenko and Sarachik (2004). One of the main issues in such studies has been the fate of the metallic phase in a parallel B. From the insulating side, the critical density $n_c(B)$ can be determined by extrapolating to zero the activation energy and non-linearity of the current–voltage characteristics (Section 8.2.1). For $n_s > n_c(B)$, however, the metallic $d\sigma/dT < 0$ behavior observed in high-mobility samples is suppressed by B, making it even more difficult to determine the critical density from the metallic side. Nevertheless, a careful analysis reveals exactly the same metallic $T^{3/2}$ correction in high-mobility samples in parallel magnetic fields as in highly disordered samples in zero magnetic field, confirming the existence of the MIT in these two cases.

Figure 8.6 (a) shows the $(n_s, B, T = 0)$ phase diagram obtained for a high-mobility sample (Jaroszyński *et al.*, 2004*b*). The critical densities $n_c(B)$ were first determined using the activation energy method. Good agreement was found with the $n_c(B)$ dependence obtained from very similar samples using both activation energies and non-linear current–voltage characteristics (Shashkin *et al.*, 2001).[4] At low fields, n_c increases with B, and then it saturates for $B \gtrsim 4$ T, consistent with other studies that

[4]Because of the small sample-to-sample differences in the amount of disorder, the data from Shashkin *et al.* (2001) have been shifted up by $0.85 \times 10^{10}\mathrm{cm}^{-2}$ to make the $n_c(B = 0)$ values coincide.

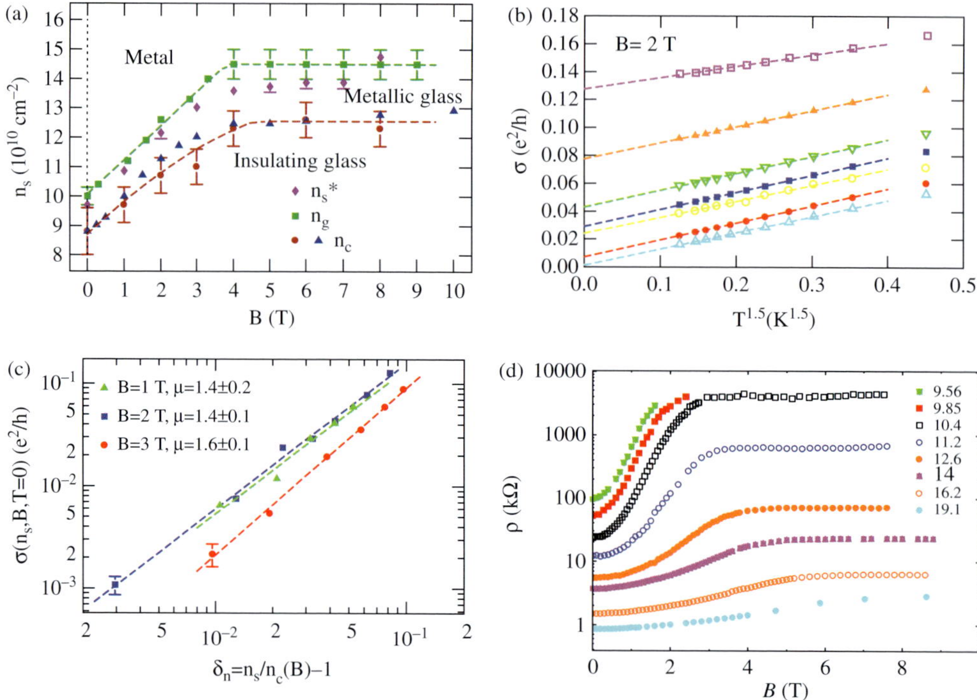

Fig. 8.6 High-mobility Si MOSFET. (a) $T = 0$ phase diagram in a parallel magnetic field (Jaroszyński *et al.*, 2004*b*). The dashed lines guide the eye. The n_c values are from Jaroszyński *et al.* (2004*b*) (dots) and Shashkin *et al.* (2001) (triangles). The glass transition takes place at $n_g(B) > n_c(B)$, giving rise to an intermediate, metallic glass phase. The density at the separatrix $n_s^* \approx n_g$ within the error for all B. (b) σ vs $T^{1.5}$ for several n_s in the metallic glass phase for $B = 2$ T (from top n_s = 11.9, 11.6, 11.3, 11.2, 11.0, 10.9, 10.7 $\times$ 10^{10}cm^{-2}; $n_c(B = 2\text{ T}) =$ 10.67 $\times$ 10^{10}cm^{-2}) for the same sample. Dashed lines are fits. (c) Zero-temperature conductivity $\sigma(T - 0) \propto \delta_n^\mu$ obtained from the fits shown in (b). (d) The positive magnetoresistance for the same sample (Jaroszyński *et al.*, 2004*b*) and different densities n_s(10^{10}cm^{-2}), as shown. A strong increase in magnetoresistance reflects a magnetic-field-driven MIT at $n_c(B)$.

show that the 2DES is fully spin polarized here (Okamoto *et al.*, 1999; Tutuc *et al.*, 2001; Vitkalov *et al.*, 2000). Even though $\sigma(T)$ is very weak at higher n_s, it is interesting to attempt to determine the separatrix, where $d\sigma/dT = 0$. The corresponding density $n_s^*(B) > n_c(B)$ and, most intriguingly, within the measurement error it coincides with $n_g(B)$, the glass-transition density (Jaroszyński *et al.*, 2004*a*). $n_g(B)$ was determined independently, based on the measurements of the fluctuations in $\sigma(t)$ (Jaroszyński *et al.*, 2004*b*) (see Section 8.3.1).

The important question to address is the form of $\sigma(T)$ in the narrow $n_c(B) < n_s < n_g(B)$ region. Here the data are best described by the metallic ($\langle\sigma(T = 0)\rangle \neq 0$) power-law form $\langle\sigma(n_s, B, T)\rangle = \langle\sigma(n_s, B, T = 0)\rangle + b(n_s, B)T^{1.5}$ (see Fig. 8.6 (b)). This is similar to what was observed in the metallic glassy phase of highly disordered samples at $B = 0$ (Section 8.2.1). The extrapolated $T = 0$ conductivities go to zero precisely at $n_c(B)$, in a power-law fashion $\langle\sigma(n_s, B, T = 0)\rangle \propto \delta_n^{\mu}$, with $\mu \approx 1.5$ as shown in Fig. 8.6 (c) where $\delta_n = n_s/n_c(B) - 1$ (Jaroszyński *et al.*, 2004*b*). This is in agreement with theoretical expectations near a quantum phase transition (Belitz and Kirkpatrick, 1994). Interestingly, there is some evidence (Fletcher *et al.*, 2001) of similar behavior at $B = 0$ with $\mu \sim 1$–1.5, obtained by extrapolating to $T = 0$ the "saturation" of $\sigma(T)$ in the $d\sigma/dT < 0$ regime of different high-mobility Si MOSFETs. The striking power-law behavior shown in Fig. 8.6 (c) and the remarkable agreement between $n_c(B)$ obtained from $\sigma(T)$ on both insulating and metallic sides of the MIT are strong evidence for the survival of the MIT and the metallic phase in parallel magnetic fields B.

In low-disorder ("clean") samples at $B = 0$, the critical density $n_c \lesssim n_s^* \approx n_g$ (Section 8.2.1, Fig. 8.6 (a)) and the intermediate, metallic glass phase is practically absent. A parallel magnetic field B, however, increases its width, allowing the emergence of the characteristic $T^{3/2}$ correction to σ. The increase of both n_g and n_c, and the broadening of the metallic glass phase with B can be understood to result from the suppression of screening by a parallel B (Dolgopolov and Gold, 2000; Herbut, 2001), which increases the effective disorder. This, in turn, favors glassiness, consistent with theoretical expectations (Dobrosavljević *et al.*, 2003), and makes the behavior of "clean" samples more similar to that of highly disordered ones (Section 8.2.1). The existence of the glass transition in high parallel B, where the 2DES is spin polarized, provides evidence that charge, not spin, degrees of freedom are responsible for glassy ordering. This result clearly imposes a strong constraint on the types of theory that can be formulated to describe this phenomenon. Likewise, the broadening of the metallic glass phase with B also indicates that its existence is not due to spin.

For n_s near $n_c(B = 0)$, 2DESs in various semiconductors exhibit a strong, positive magnetoresistance (Kravchenko and Sarachik, 2004), which has been the subject of great interest. The magnetoresistance saturates at the field that corresponds to the full spin polarization (Okamoto *et al.*, 1999; Tutuc *et al.*, 2001; Vitkalov *et al.*, 2001, 2000). As shown in Fig. 8.6 (d) for the same high-mobility Si MOSFET discussed above, the magnetoresistance jump is strong only for n_s not too far from $n_c(B = 0)$ since it takes place as the system undergoes a magnetic-field driven MIT at $n_c(B)$ (see Fig. 8.6 (a)). In the metallic phase where $n_s \gg n_c(B > 4\text{ T})$, the magnetoresistance is weak.

In low-mobility Si MOSFETs, where the metallic glass phase is already clearly observable at $B = 0$, the $(n_s, B, T = 0)$ phase diagram has not been studied. However, it is plausible that the parallel B will broaden the metallic glass phase even further.

Finally, in intermediate-mobility 2DESs with local magnetic moments, it is relatively easy to map out the $(n_s, B, T = 0)$ phase diagram because of the simple form of $\sigma(T)$ that holds over a wide range of T. In parallel B, $\sigma(n_s, T)$ data (Fig. 8.7 (a)) are qualitatively similar to the $B = 0$ case (Section 8.2.1) (Eng *et al.*, 2002). At the lowest $n_s < n_c$, for example, σ decreases exponentially with decreasing T, indicating an insulating state at $T = 0$. For $n_s > n_c$, $\sigma(T)$ is weaker and its *curvature* is the opposite from that expected for an insulating state. It clearly extrapolates to a finite value as $T \to 0$, indicating a metallic phase (Section 8.2.1). n_c is identified as the

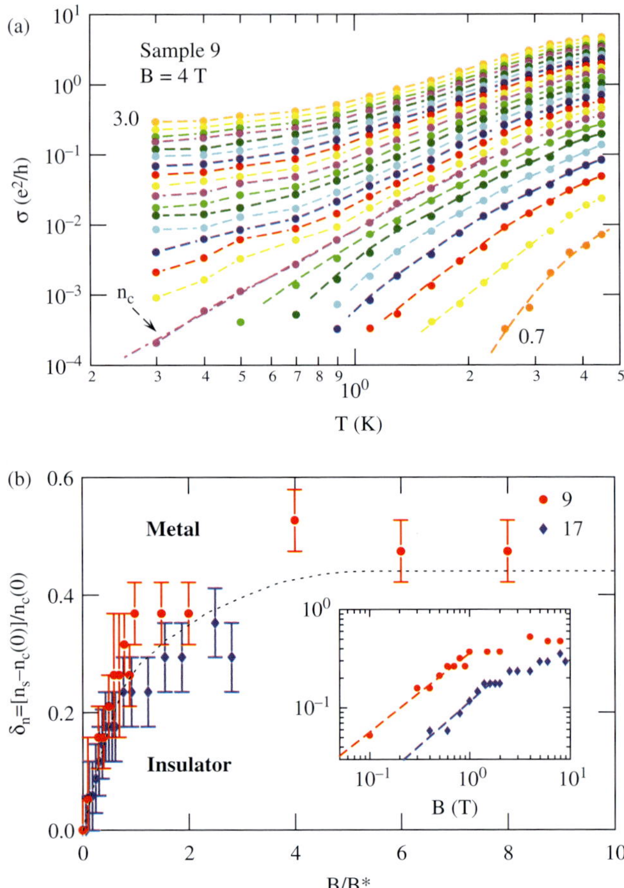

Fig. 8.7 Intermediate-mobility sample ($\mu \approx 1$ m^2/Vs) with local magnetic moments ($V_{\mathrm{sub}} = +1$ V) (Eng *et al.*, 2002). (a): $\sigma(T)$ for sample 9 at $B = 4$ T. n_{s} varies from $3.0 \times 10^{11}\mathrm{cm}^{-2}$ (a) to $0.7 \times 10^{11}\mathrm{cm}^{-2}$ (b) in steps of $0.1 \times 10^{11}\mathrm{cm}^{-2}$. $n_{\mathrm{c}}(B = 4\ \mathrm{T}) = 1.4 \times 10^{-11}\mathrm{cm}^{-2}$ and is marked by the arrow. The dashed lines guide the eye. σ_{c} clearly follows a simple power-law dependence on $T : \sigma_{\mathrm{c}} \propto T^x$. (b): $T = 0$ phase diagram for two samples. The dashed line guides the eye. The boundary between metallic and insulating phases is described by a power-law relation (see inset) $(n_{\mathrm{c}}(B) - n_{\mathrm{c}}(0))/n_{\mathrm{c}}(0) \propto (B/B^*)^\beta$ at low fields, with the same crossover exponent $\beta \approx 0.9$ for both samples ($B^* = 1$ T for sample 9). Inset: the same data vs B on a log-log scale. The dashed lines are fits with the slopes equal to β. At $B = 0$, $n_{\mathrm{c}}(10^{11}\mathrm{cm}^{-2}) = 0.95 \pm 0.05$ ($r_{\mathrm{s}} \approx 17$) and 0.85 ± 0.05 ($r_{\mathrm{s}} \approx 18$) for samples 9 and 17, respectively.

density where $\sigma_{\mathrm{c}} = \sigma(n_{\mathrm{s}} = n_{\mathrm{c}}, T) \propto T^x$ (Fig. 8.7 (a)), consistent with the $B = 0$ case and in agreement with general arguments (Belitz and Kirkpatrick, 1994). The exponent $x = 2.7 \pm 0.4$ remains constant as a function of B. The critical density n_{c} determined in this way at each given B allows the mapping of the $(n_{\mathrm{s}}, B, T = 0)$ phase diagram

(Fig. 8.7 (b)). It should be noted that at low fields ($B \lesssim 2$ T), $\sigma(n_s, B, T)$ curves exhibit beautiful scaling (Eng *et al.*, 2002) in agreement with $T \to 0$ extrapolations and general considerations (Belitz and Kirkpatrick, 1994), providing additional strong evidence for a quantum phase transition in this system in parallel B. Analogous to high-mobility samples, the magnetoresistance exhibits a strong increase in the region of the magnetic-field-driven MIT near $n_c(B)$ (Eng *et al.*, 2001). For $B > 2$ T, where the 2DES is spin polarized, $n_c(B)$ saturates, indicating that the MIT occurs between a spin-polarized metal and a spin-polarized insulator. The existence of the metallic phase at fields up to 18 T has been confirmed by measurements of $\sigma(T)$, which retains a simple power-law form, albeit with a different exponent (Eng *et al.*, 2002). The charge dynamics have not been studied in these samples yet.

It is interesting that the $n_c(B)$ dependence in samples with local moments (Fig. 8.7 (b)) is quite similar to that obtained on high-mobility samples (Fig. 8.6 (a)). In both cases, at low fields $n_c(B)$ increases with B in a power-law fashion: $(n_c(B) - n_c(0))/n_c(0) \propto B^{\beta}$. The crossover exponent $\beta \approx 0.9$–1 in Fig. 8.7 (b) and $\beta = 1.1 \pm 0.1$ in Fig. 8.6 (a). In other words, n_c increases approximately linearly with B at low fields. It is remarkable that this dependence is essentially the same in both cases, even though the $d\sigma/dT$ behaviors in the metallic phase at $B = 0$ are strikingly different. The key features of the $n_c(B)$ phase diagram have been reproduced theoretically, based on a scenario of quantum melting of a Wigner crystal as the mechanism of the MIT in sufficiently clean samples (Camjayi *et al.*, 2008). In general, a power-law shift of n_c with B is expected to occur in the case of a true MIT (Belitz and Kirkpatrick, 1994), and has been observed in several 3D systems (Bogdanovich *et al.*, 1997; Rosenbaum *et al.*, 1989; Sarachik *et al.*, 1998; Watanabe *et al.*, 1999).

8.3 Glassy freezing of electrons in two dimensions

Understanding the dynamics of glasses and other systems out of equilibrium is one of the most challenging and rapidly evolving topics in condensed-matter research (see Barrat *et al.* (2003)). Since very different types of system exhibit similar behavior, it is tempting to search for common organizing principles and unified theoretical approaches. However, despite some progress, there are still no well established theoretical frameworks for treating non-equilibrium behavior, which "remains largely uncharted territory" (Committee on CMMP 2010, 2007). For example, although glassy behavior may dominate the low-temperature properties of many complex materials near quantum phase transitions such as the MIT (Miranda and Dobrosavljević, 2005) (Section 8.1), quantum glasses are even less understood than their classical counterparts. Experimental studies of charge or Coulomb glasses (Davies *et al.*, 1982, 1984; Grünewald *et al.*, 1982; Pollak, 1984; Pollak and Ortuño, 1982), which are of particular relevance to the MIT, have been relatively scarce (Ben-Chorin *et al.*, 1993; Bielejec and Wu, 2001; Grenet, 2003; Grenet *et al.*, 2007; Hernandez *et al.*, 2003; Kar *et al.*, 2003; Lee *et al.*, 2005; Martinez-Arizala *et al.*, 1998, 1997; Monroe *et al.*, 1987; Orlyanchik and Ovadyahu, 2004; Ovadyahu, 2006*a*,*b*; Ovadyahu and Pollak, 1997; Thorsmølle and Armitage, 2010; Vaknin *et al.*, 1998, 2000, 2002), and mostly limited to insulating systems far from the MIT. Recent observations (Bogdanovich and

Popović, 2002*a*,*b*; Jaroszyński and Popović, 2006, 2007*a*,*b*; Jaroszyński *et al.*, 2002*a*,*b*, 2004*a*,*b*; Popović *et al.*, 2003) of glassiness in a 2DES in Si MOSFETs near the MIT open up opportunities for exploring glassy phenomena in this important regime over a wide range of all the relevant parameters.

There are two basic ways to explore the dynamics of glassy systems. The first one is to measure the response of the system to some kind of a perturbation. In a spin glass, for example, this would typically involve a study of the relaxation of the magnetization following some combination of rapid cooling and a change in the applied magnetic field (see Vincent (2007) for a pedagogical review). The second one is to measure the fluctuations of an observable with time (i.e. noise), which provides information on correlations.[5] In spin glasses, transport-noise measurements were required in order to provide definitive information on the details of glassy ordering and dynamics (Weissman, 1993). Both approaches have been used to probe the dynamics of the 2DES in Si, focussing on the conductivity σ as the variable most relevant to the MIT. Measurements were performed on both high- and low-mobility Si MOSFETs, which also differ substantially in their geometry, size, and many other fabrication details, spanning essentially the entire range of Si technology. Thus the emergence of glassy dynamics proves to be a universal phenomenon in Si inversion layers, at least in the absence of disorder-induced local magnetic moments. The effect of local moments on charge dynamics still remains to be investigated.

8.3.1 Fluctuations of conductivity

The experimental protocol for measuring the $\sigma(t)$ fluctuations is simple, in principle: the measurement of σ at a given V_g or electron density n_s is set up at high temperatures, the sample is then cooled to the measurement T, and σ is measured as a function of time. At the end of the measurement, the sample is warmed up to a temperature that is so high that a subsequent cool-down to the same measurement T would result in a reproducible value of the time-averaged $\langle\sigma(t)\rangle$ within the experimental uncertainty. The carrier density n_s is changed at a high temperature and the protocol is repeated for different n_s and T. In practice, the measurement set-up itself is often somewhat complicated (Bogdanovich and Popović, 2002*b*; Jaroszyński *et al.*, 2002*b*; Popović *et al.*, 2003) because it is important to rule out various extraneous effects, such as the fluctuations of T, V_g, or contacts, as possible sources of the measured noise. However, there are now a number of standard methods to accomplish this (see, e.g. Scofield (1987) and Verbruggen *et al.* (1989)).

In both low- and high-mobility samples, σ exhibits strong fluctuations with time at low n_s and T. Figure 8.8 (a) shows the fluctuations of $(\sigma - \langle\sigma\rangle)/\langle\sigma\rangle$ in a low-mobility sample[6] for several n_s at $T = 0.13$ K. It is striking that, for the lowest n_s, the fluctuation amplitude is of the order of 100%. In addition to rapid, high-

[5] In equilibrium systems, the connection between spontaneous fluctuations of a variable and the response of such a variable to a small perturbation in its conjugated field is given by the fluctuation–dissipation relation. See Leuzzi and Nieuwenhuizen (2008) for the review and discussion of thermodynamics of out-of-equilibrium systems.

[6] $\langle \ldots \rangle$ represents averaging over time intervals of, typically, several hours.

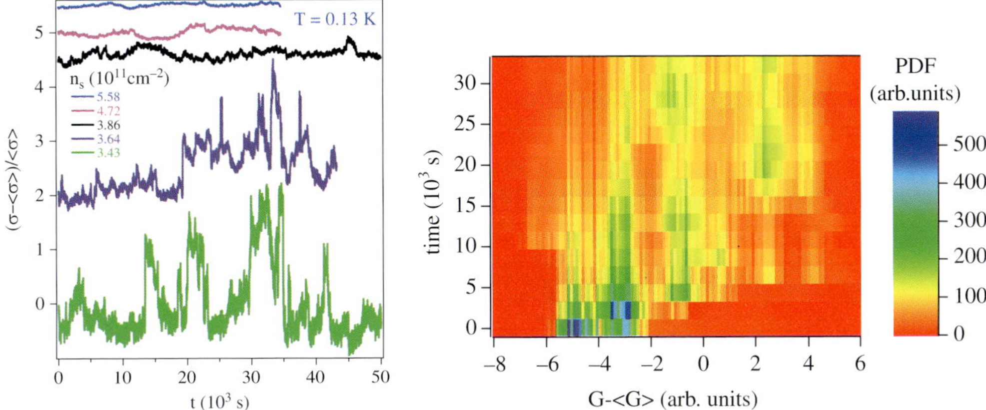

Fig. 8.8 Low-mobility sample (Bogdanovich and Popović (2002*b*) and Fig. 8.5). (b) Relative fluctuations of σ vs time for different n_s at $T = 0.13$ K. Different traces have been shifted for clarity; the lowest n_s is at the bottom and the highest n_s at the top. (a) The color map of the probability density function (PDF) of the conductance (G) fluctuations as a function of sampling time t for the $n_c < n_s(10^{11}\text{cm}^{-2}) = 5.58 < n_g$ data shown on the left.

frequency fluctuations, slow changes over periods of several hours are also evident. The probability density function of the fluctuations illustrates how the sample explores the free-energy landscape. For $n_s < n_g$, the probability density function is always not only non-Gaussian but also has a very complex structure that changes with time (Fig. 8.8 (a)). It broadens with increasing sampling time t, as the system has more time to explore the free-energy landscape. However, it explores it so slowly that it remains non-ergodic on experimental time scales. For $n_s > n_g$, on the other hand, the probability density functions become perfectly Gaussian on much shorter time scales, suggesting that the system reaches equilibrium. The amplitude of the fluctuations decreases dramatically from $\sim$100% to less than 1% with increasing n_s (Fig. 8.8 (b)) or T.

An even more dramatic density dependence of the noise at low T is revealed by the study of the power spectra $S(f)$ (where f is frequency) of the relative changes in the conductivity $(\sigma(t) - \langle\sigma\rangle)/\langle\sigma\rangle$. Most of the spectra were obtained in the $f = (10^{-4}\text{–}10^{-1})$ Hz bandwidth, where they follow the well-known empirical law $S \propto 1/f^{\alpha}$ (Hooge, 1976; Weissman, 1988). At the highest n_s, $S(f)$ does not depend on n_s (Fig. 8.9 (a)), but as n_s is reduced below n_g, S increases enormously, by up to six orders of magnitude at low f. Moreover, for a given $n_s < n_g$, $S(f)$ increases exponentially with decreasing T (Fig. 8.9 (a), inset). The observed $dS/dT < 0$ makes it possible to rule out various simple models of noise (see, e.g. Bogdanovich and Popović (2002*b*) and Popović *et al.* (2003) for discussion). The most striking feature of the data, however, is the sharp jump of the exponent α at $n_s \approx n_g$ (Fig. 8.9 (b)). While $\alpha \approx 1$ ("pure" $1/f$ noise) for $n_s > n_g$, $\alpha \approx 1.8$ below n_g, reflecting a sudden shift of the spectral weight towards lower frequencies. Similarly large values of α have been observed in

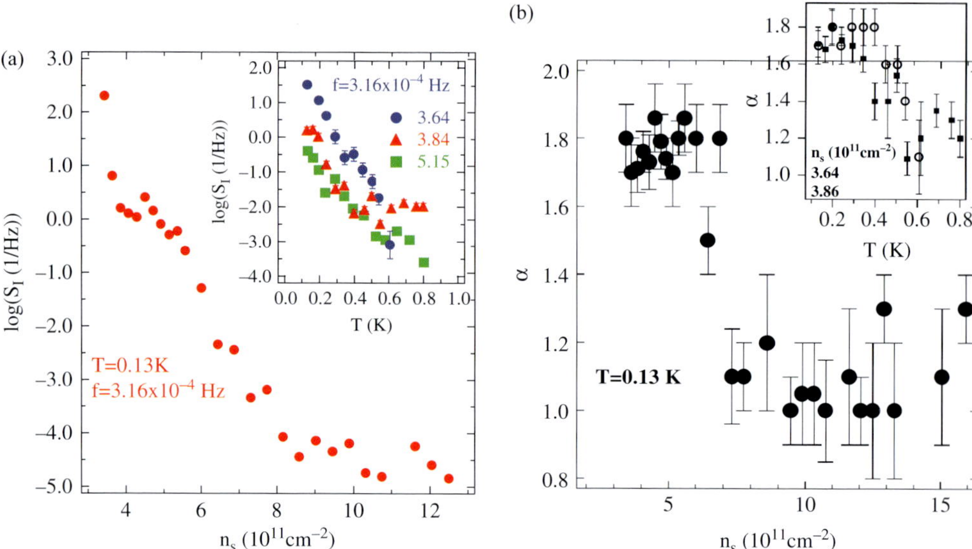

Fig. 8.9 Low-mobility sample (Bogdanovich and Popović (2002*b*) and Fig. 8.5). (a) The normalized noise power $S(f = 3.16 \times 10^{-4}\text{Hz})$ vs n_s at $T = 0.13$ K (Bogdanovich and Popović, 2002*a*). Below $n_g \approx 7.5 \times 10^{11}\text{cm}^{-2}$, the noise increases exponentially with decreasing n_s. Inset: S vs T for three different n_s (10^{11}cm^{-2}) given on the plot. (b) At $n_s \approx n_g$, the exponent α exhibits a sharp jump from ≈ 1 at high n_s ("pure" $1/f$ noise) to ≈ 1.8 at low n_s (Bogdanovich and Popović, 2002*b*). Inset: α vs T for two different n_s (10^{11}cm^{-2}) (3.64 – open symbols, 3.86 – solid symbols) in the glassy phase (Popović *et al.*, 2003).

some spin glasses above the MIT (Jaroszyński *et al.*, 1998; Neuttiens *et al.*, 2000), and in submicron wires (Wróbel *et al.*, 1998). Both the increase in the magnitude of the noise at low f and the jump in α reflect a sudden and dramatic slowing down of the electron dynamics at n_g, indicating glassy freezing. The onset of glassy dynamics on the metallic side of the MIT, i.e. at $n_g > n_c \approx 5 \times 10^{11}\text{cm}^{-2}$, implies the existence of the metallic-glass phase for $n_c < n_s < n_g$, which is consistent with the predictions of the model of interacting electrons near a disorder-driven MIT (Dobrosavljević *et al.*, 2003). Since, in the glassy phase, α decreases with increasing T (Fig. 8.9 (b), inset), the large values of α and the jump in $\alpha(n_s)$ are observable only at relatively low T.

Qualitatively, the same behavior has been observed in the resistance (or conductance) noise of high-mobility samples, except that the glassy freezing takes place at $n_g \approx n_c$ (Jaroszyński *et al.*, 2002*b*). Importantly, in both types of sample, the character of the noise also changes with density: while at low n_s both the "shape" and the variance of the noise exhibit random, nonmonotonic changes with time, at high enough n_s the noise always "looks" the same. This is illustrated in Fig. 8.10 (a) for a high-mobility sample. The figure shows $(\rho - \langle\rho\rangle)/\delta\rho$, where $\delta\rho = \langle(\rho - \langle\rho\rangle)^2\rangle^{1/2}$, in order to make it easier to compare the signals. A quantitative measure of the spectral wandering with time, such as that observed at low n_s, is the so-called second

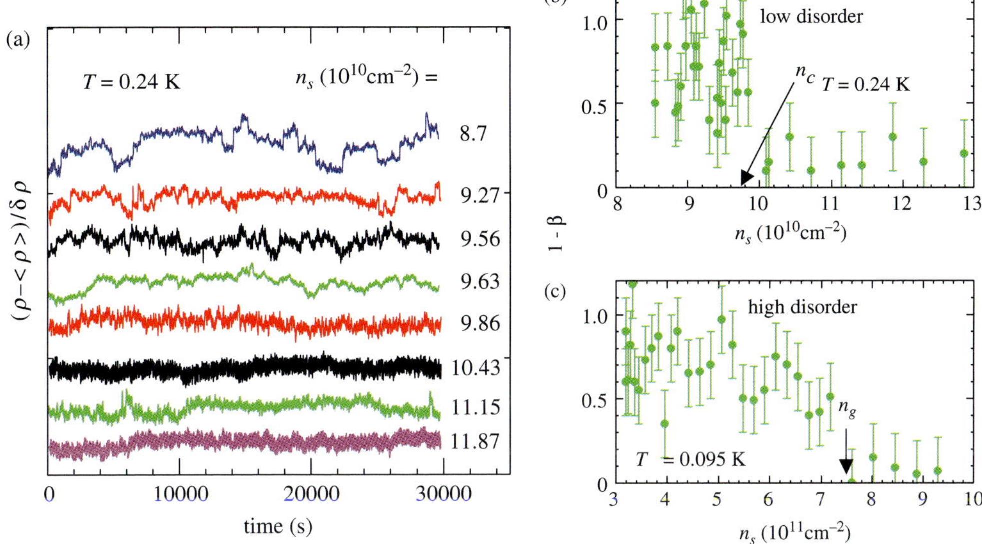

Fig. 8.10 (a) Resistance noise in a high-mobility (low-disordered) sample for several n_{s} shown on the plot (adapted from Jaroszyński *et al.* (2002*b*)). $(\rho - \langle\rho\rangle)/\delta\rho$ is plotted ($\delta\rho^2$ = variance, ρ = resistivity) in order to make the change in the character of the noise with n_{s} more apparent. Different traces have been shifted vertically for clarity. Exponent $1-\beta$, a measure of correlations, vs n_{s} for (b) high-mobility ($n_{\mathrm{c}} \approx 9.7 \times 10^{10}\mathrm{cm}^{-2}$) (Jaroszyński *et al.*, 2002*b*) and (c) low-mobility samples ($n_{\mathrm{c}} \approx 5.0 \times 10^{11}\mathrm{cm}^{-2}$) (Bogdanovich and Popović, 2002*b*).

spectrum $S_2(f_2, f)$, which is a fourth-order noise statistic. S_2 is the power spectrum of the fluctuations of $S(f)$ with time (Weissman, 1993; Weissman *et al.*, 1992), i.e. the Fourier transform of the autocorrelation function of the time series of $S(f)$. In general, if the fluctuators (e.g. two-level systems) are not correlated, $S_2(f_2, f)$ is white (independent of f_2) (Weissman, 1988, 1993; Weissman *et al.*, 1992) and equal to the square of the first spectrum. Such noise is called Gaussian. On the other hand, S_2 has a nonwhite character, $S_2 \propto 1/f_2^{1-\beta}$, for interacting fluctuators (Weissman, 1988, 1993; Weissman *et al.*, 1992). Therefore, the deviations from Gaussianity, or the exponent $(1-\beta)$, provide a direct probe of correlations between fluctuators. Indeed, S_2 has been an important tool in studies of other glasses.

A detailed dependence of the exponent $(1-\beta)$ on n_{s} has been determined for both high- and low-mobility samples (Figs. 8.10 (b) and (c), respectively) (Jaroszyński *et al.*, 2002*b*). In both cases, S_2 is white ($(1-\beta) = 0$) for $n_{\mathrm{s}} > n_{\mathrm{g}}$, indicating that the observed $1/f$ noise results from uncorrelated fluctuators. It is quite remarkable that S_2 changes its character in a dramatic way exactly at n_{g} in both types of sample. For $n_{\mathrm{s}} < n_{\mathrm{g}}$, S_2 is strongly non-Gaussian, which demonstrates that the fluctuators are strongly correlated and provides unambiguous evidence for the onset of glassy dynamics in a 2DES at n_{g}.

In the studies of spin glasses, the scaling of S_2 with respect to f and f_2 has been used (Weissman, 1993; Weissman *et al.*, 1992) to unravel the glassy dynamics and, in particular, to distinguish generalized models of interacting droplets or clusters from hierarchical pictures. In the former case, the low-f noise comes from a smaller number of large elements because they are slower, while the higher-f noise comes from a larger number of smaller elements, which are faster. In this picture, which assumes compact droplets and short-range interactions between them, big elements are more likely to interact than small ones and hence non-Gaussian effects and S_2 will be stronger for lower f. $S_2(f_2, f)$, however, need to be compared for fixed f_2/f, i.e. on timescales determined by the timescales of the fluctuations being measured, since spectra taken over a fixed time interval average the high-frequency data more than the low-frequency data. Therefore, in the interacting "droplet" model $S_2(f_2, f)$ should be a decreasing function of f at constant f_2/f. In the hierarchical picture, on the other hand, $S_2(f_2, f)$ should be scale invariant: it should depend only on f_2/f, not on the scale f (Weissman, 1993; Weissman *et al.*, 1992). Figures 8.11 (a) and (b) show that no systematic dependence of S_2 on f is seen in either high- or low-mobility 2DESs (Jaroszyński *et al.*, 2002*b*). The observed scale invariance of $S_2(f_2, f)$ signals that the system wanders collectively between many metastable states related by a kinetic hierarchy. Metastable states correspond to the local minima or "valleys" in the free-energy landscape (Fig. 8.1), separated by barriers with a wide, hierarchical distribution of heights and thus relaxation times. Intervalley transitions, which are reconfigurations of a large number of electrons, thus lead to the observed strong, correlated, $1/f$-type noise, remarkably similar to what was observed in spin glasses with a long-range correlation of spin configuration (Weissman, 1993; Weissman *et al.*, 1992). On the other hand, in systems where both long-range and short-range interactions are present, such as lightly doped $La_{2-x}Sr_xCuO_4$, $S_2(f_2, f)$ reveals clearly (Raičević *et al.*, 2008, 2011) (Fig. 8.11 (c)) the presence of some characteristic energy scale, indicating that hierarchical models are not applicable and suggesting instead the formation of a cluster glass state. This is exactly what is expected in Coulomb systems with competing interactions (Section 8.1).

In the resistance noise measurements, the glass transition in a 2DES is thus manifested by (a) a sudden and dramatic increase of $S(f)$ and a jump of α from ≈ 1 to ≈ 1.8, indicating the slowing down of the dynamics, and (b) a change of the exponent $(1-\beta)$ from a white (zero) to a non-white (non-zero) value, indicating an abrupt change to correlated statistics, consistent with the hierarchical pictures of glassy dynamics (Binder and Young, 1986). For high-mobility 2DESs in Si MOSFETs, low-T noise measurements were also performed in parallel B (Jaroszyński *et al.*, 2004*b*). By adopting the same criteria for the glass transition in B, it was possible to determine the $n_g(B)$ shown in Fig. 8.6 (a) and to establish that charge, not spin, degrees of freedom are responsible for glassy ordering (see Section 8.2.2). Measurements in both $B = 0$ and $B \neq 0$ show that glassy behavior generally emerges before the electrons localize (i.e. $n_g > n_c$), consistent with theory (Dobrosavljević *et al.*, 2003). The glassy signatures in the noise become gradually stronger as T decreases (e.g. Fig. 8.9 insets), suggesting that the glass transition takes place as $T \rightarrow 0$. Strong evidence for $T_g = 0$

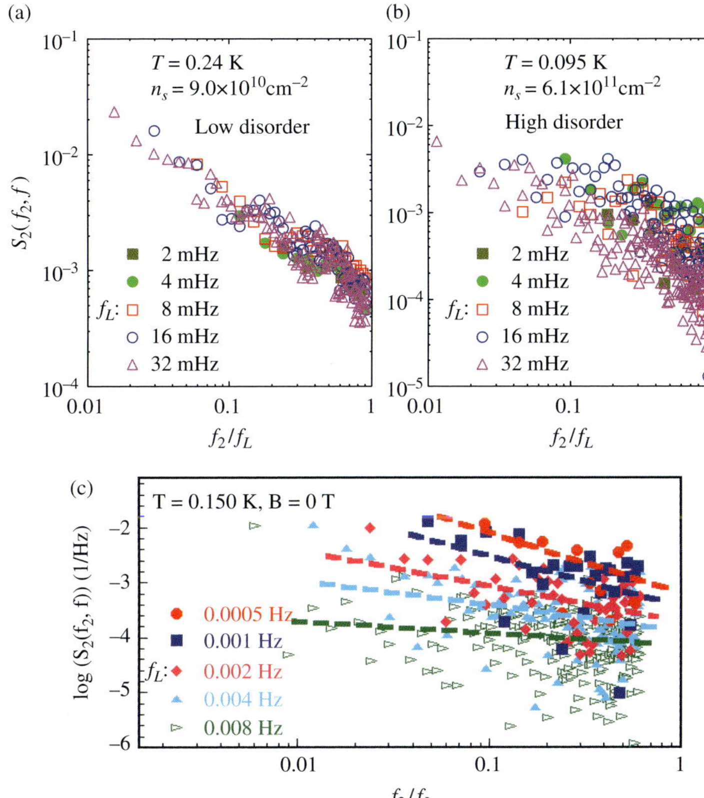

Fig. 8.11 Scaling of the second spectra S_2 measured at frequency octaves $f = (f_L, 2f_L)$ for a given n_s in (a) high-mobility (for $n_s < n_c$) and (b) low-mobility (for $n_c < n_s < n_g$) 2DES (Jaroszyński *et al.*, 2002*b*). There is no characteristic time (or energy) scale observed, in agreement with hierarchical models. This is in contrast to (c), the insulating $La_{1.97}Sr_{0.03}CuO_4$, where S_2 clearly shows the presence of some characteristic energy scale in that system (Raičević *et al.*, 2008, 2011), suggesting that charge carriers form a cluster glass state. Dashed lines are linear fits to guide the eye.

and further support for n_g as the glass transition density is provided by measurements of the response of the 2DES to a strong perturbation.

8.3.2 Glassy response

In MOSFET structures, the easiest and the most obvious way to excite the system is a sudden change of V_g. This method has been applied to several electron glasses (Ben-Chorin *et al.*, 1993; Grenet, 2003; Grenet *et al.*, 2007; Martinez-Arizala *et al.*, 1998, 1997; Orlyanchik and Ovadyahu, 2004; Ovadyahu, 2006*a*,*b*; Ovadyahu and Pollak, 1997; Vaknin *et al.*, 1998, 2000, 2002), including a 2DES in Si (Jaroszyński and

Popović, 2006, 2007*a*,*b*, 2009). So far, only the low-mobility samples discussed above (Sections 8.2.1 and 8.3.1) have been probed in this way. In particular, three different experiments have been carried out, as described below.

Relaxations of conductivity after a rapid change of carrier density

A systematic study of the relaxations (i.e. time dependence) of the conductivity $\sigma(t)$ has been performed at different n_s and T after a large, rapid change of n_s (controlled by V_g) (Jaroszyński and Popović, 2006). The sample was first cooled from high temperature (10 K) to the measurement temperature T with an initial gate voltage V_g^i. Then, at $t = 0$, the gate voltage was switched rapidly (within 1 s) to a final value V_g^f, and $\sigma(t, V_g^f, T)$ was measured. In general, the results did not depend on any of the following:

- initial temperature, so long as it was ≥ 10 K
- V_g^i
- the cooling time, which was varied between 30 min and 10 h
- the time the sample was kept at 10 K (from 5 min to 8 h)
- the time the sample spent at the measurement T before V_g was changed (from 5 min to 8 h).

Figures 8.12 (a) and (b) show a typical experimental run with $\sigma(t > 0)$ exhibiting a rapid (< 10 s) initial drop followed by a slower relaxation.[7]

After going through a minimum, $\sigma(t)$ increases and approaches a value $\sigma_0(V_g^f, T)$. A subsequent warm-up to 10 K and a cool down to the same measurement $T = 3.3$ K, while keeping the gate voltage fixed at V_g^f, shows that $\sigma_0(V_g^f, T)$ represents the equilibrium conductivity corresponding to the given V_g^f and T. It is interesting that even though initially it drops to a value close to σ_0, σ first goes away from equilibrium before it starts approaching σ_0 again. At the end of the run, the sample was warmed up to 10 K and the gate voltage was changed back to the same V_g^i, and the experiment was repeated at a different T for the same V_g^f. The whole process was then repeated for different values of V_g^f, in order to map out the density dependence of various relaxation parameters. It was established (Jaroszyński and Popović, 2006) that, for $n_s < n_g$, the relaxations have the following properties.

After a sufficiently long t, σ relaxes exponentially to its equilibrium value σ_0. The corresponding equilibration time obeys the simply activated form $\tau_{eq} \propto \exp(E_{act}/T)$, $E_{act} \approx 57$ K. While the microscopic origin of the activation energy E_{act} is not known yet, the activation to an upper subband in Si MOSFETs or to Si–SiO_2 interface traps has been ruled out (Jaroszyński and Popović, 2007*b*). However, regardless of the equilibration mechanism, the important result is that $\tau_{eq} \to \infty$ as $T \to 0$, so that, strictly speaking, the system cannot reach equilibrium only at $T = 0$. In other words, the glass transition takes place at $T_g = 0$. Remarkably, a Monte Carlo study of the 2D Coulomb glass model has also found (Grempel, 2004) an exponential divergence of

[7] The RC (R, resistance; C, capacitance) charging times of the device and the circuit were at most 10 ms.

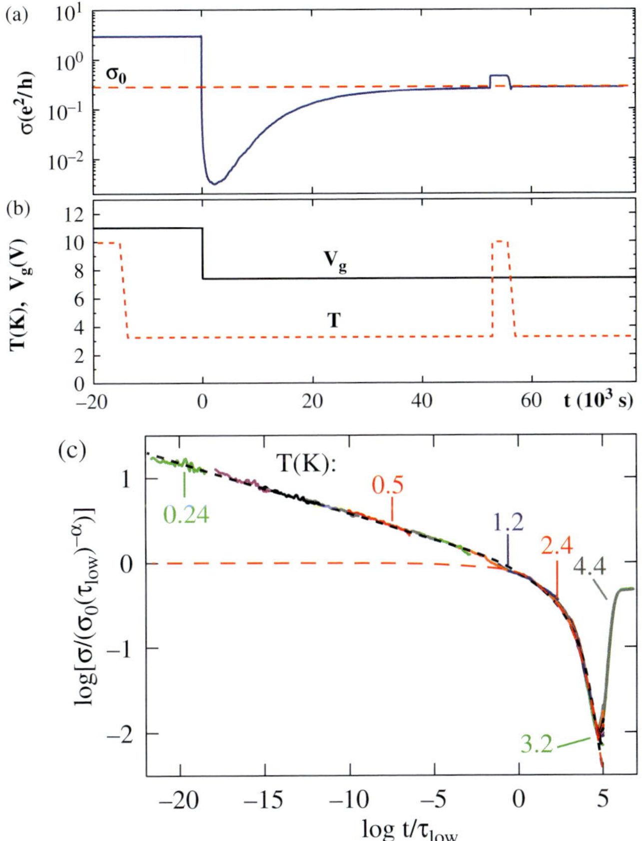

Fig. 8.12 Low-mobility sample with $n_g \approx 7.5 \times 10^{11}\text{cm}^{-2}$ and $n_c \approx 4.5 \times 10^{11}\text{cm}^{-2}$. (a) $\sigma(t)$ for $V_g^i = 11$ V $[n_s(10^{11}\text{cm}^{-2}) = 20.26]$, $V_g^f = 7.4$ V $[n_s(10^{11}\text{cm}^{-2}) = 4.74]$, and $T = 3.3$ K. (b) Experimental protocol: $V_g(t)$ and $T(t)$. (c) All the data at short times, (i.e. before the minimum) are consistent with the Ogielski relaxation, $\sigma(t,T)/\sigma_0(T) \propto (\tau_{\text{low}})^{-\alpha}(t/\tau_{\text{low}})^{-\alpha}\exp[-(t/\tau_{\text{low}})^{\beta}]$ (dotted line). The lowest-T data clearly deviate from the stretched exponential dependence (dashed line). The data have been collapsed with respect to the 2.4 K curve. Adapted from Jaroszyński and Popović (2006).

τ_{eq}, signaling a glass transition at $T_g = 0$. There are, however, some differences in the detail between the model and the experiment, indicating a need for further refinement of theory.

For short enough t, $\sigma(t)$ obeys a non-exponential, Ogielski form (Ogielski, 1985) $\sigma(t,T)/\sigma_0(T) \propto t^{-\alpha}\exp[-(t/\tau_{\text{low}})^{\beta}]$ $(0 < \alpha(n_s) < 0.4,\ 0.2 < \beta(n_s) < 0.45)$ (see Fig. 8.12 (c)), which is a product of a power law and a stretched exponential function. Both types of relaxation are considered to be typical signatures of glassy behavior and reflect the existence of a broad distribution of relaxation times. The scaling parameter $\tau_{\mathrm{low}} \propto \exp(E_a/kT)$ ($E_a \approx 20$ K) so that, as $T \to 0$, $\tau_{\mathrm{low}} \to \infty$ and the relaxations attain a pure power-law form $\propto t^{-\alpha}$. The Ogielski form, the divergence of τ_{low}, and the resulting power-law relaxation at T_{g} are consistent with the general scaling arguments (Hohenberg and Halperin, 1977) near a continuous phase transition occurring at $T_{\mathrm{g}} = 0$. These results are very similar to scaling observed in spin glasses above T_{g} (Pappas *et al.*, 2003).[8] In a 2DES, $\alpha \to 0$ as $n_{\mathrm{s}} \to n_{\mathrm{g}}$, providing further evidence for n_{g} as the glass-transition density.

τ_{low} exhibits a very pronounced and precise dependence on the density: $\tau_{\mathrm{low}} \propto \exp(\gamma n_{\mathrm{s}}^{1/2})$, where γ is a proportionality constant. Since $1/r_{\mathrm{s}} = E_{\mathrm{F}}/U \propto n_{\mathrm{s}}^{1/2}$ in two dimensions (Section 8.2), this particular form of $\tau_{\mathrm{low}}(n_{\mathrm{s}})$ provides strong evidence that the observed out-of-equilibrium behavior is dominated by the Coulomb interactions between 2D electrons.

Perhaps the most peculiar finding is that the 2DES equilibrates only after it first goes *farther away* from equilibrium. While the precise mechanism for non-monotonic relaxation remains controversial, studies of other materials—spin glasses (Jönsson and Takayama, 2005; Jonsson *et al.*, 1999), manganites (Levy *et al.*, 2002), insulating granular metals (Kurzweil and Frydman, 2007), biological systems (Nelson, 2003) and some theoretical models (Miyashita *et al.*, 2007; Morita and Kaneko, 2005)—suggest that it may be a general feature of systems that are far from equilibrium.

It should be noted that it is quite remarkable that this one experiment has provided so many important results. To put things into a perspective, for example, it has not been possible to determine T_{g} in other electron glasses so far, and scaling in spin glasses (Pappas *et al.*, 2003) has been observed only relatively recently, in spite of many more years of study.

Relaxations of conductivity after a waiting time protocol: aging and memory loss

A key characteristic of relaxing glassy systems is the loss of time translation invariance, reflected in aging effects (Hodge, 1995; Rubi and Perez-Vicente, 1997; Struik, 1978). The system is said to exhibit aging if its response to an external excitation depends on the system history in addition to the time t. In a systematic study of the history dependence in a 2DES (Jaroszyński and Popović, 2007*b*), $\sigma(t)$ was measured after the temporary change of n_{s} during the waiting time t_{w} (Figs. 8.13 (a) and (b)). The history was varied by changing t_{w} and T for several initial (final) n_{s}.

Two types of response have been observed:

1. monotonic, for relatively "small" excitations, where $\sigma(t)$ depend on t_{w}, i.e. the 2DES shows aging;
2. non-monotonic, for sufficiently "large" excitations, where $\sigma(t)$ "overshoots" σ_0 (i.e. it first goes farther away from equilibrium) and relaxations no longer depend on t_{w} (memory loss).

[8] T_{g} is finite in spin glasses.

The monotonic relaxations (Fig. 8.13 (a)) are consistent with a power-law form at the shortest times (or lowest T) and, at the longest t, with a simple exponential approach to equilibrium.

The aging is observed when $t_{\rm w} \ll \tau_{\rm eq}(T)$, i.e. if the system is unable to equilibrate under the new conditions during $t_{\rm w}$ (Fig. 8.13 (c)). In that case, $\sigma(t)$ depends also on $t_{\rm w}$: the system has a *memory* of the time $t_{\rm w}$. This is very similar to spin glasses, where T or B play a role analogous to that of $n_{\rm s}$. In the opposite case, when $\tau_{\rm eq}(T) \ll t_{\rm w}$ and the 2DES equilibrates at a new state, the relaxations do not depend on $t_{\rm w}$ since the system, naturally, has no memory of the waiting time. Therefore, the overshooting is observed when the system is excited out of thermal equilibrium. This is analogous to the experiment described above (Section 8.3.2) and thus sheds some light on that intriguing phenomenon.

The gate voltage changes $\Delta V_{\rm g}$ employed in the relaxation experiments in a 2DES have been relatively large. For example, $n_{\rm s}$ was changed up to a factor of 7, and thus the 2DES could go from the conducting to the insulating regime. Such large $\Delta V_{\rm g}$ are expected to trigger major rearrangements of the electron configuration since the corresponding shifts of the Fermi energy $\Delta E_{\rm F} \gg k_{\rm B}T$ (Müller and Lebanon, 2005).[9] It is possible to speculate that such large perturbations might be somehow responsible for the peculiar overshooting effect. Considerable charge rearrangements, coupled with possibly substantial changes in the screening of the 2DES across the MIT (Müller and Ioffe, 2004; Pankov and Dobrosavljević, 2005; Pastor and Dobrosavljević, 1999), present a fundamentally different situation to the cooling of the 2DES at a fixed $n_{\rm s}$ when $k_{\rm B}\Delta T \ll E_{\rm F}$, where no relaxations have been observed. On the other hand, smaller perturbations of a Coulomb glass are expected to lead to memory effects (Lebanon and Müller, 2005), in agreement with the observations for $t_{\rm w} \ll \tau_{\rm eq}(T)$. In that case, the final state has a large configurational similarity with the original state due to the shortness of $t_{\rm w}$. In InO_x films, another well-studied electron glass, the overshooting of equilibrium has not been seen, but aging and memory effects have been observed (Ben-Chorin *et al.*, 1993; Ovadyahu and Pollak, 1997; Vaknin *et al.*, 1998, 2000, 2002). Those experiments were done in the regime of small perturbations because the typical change in the carrier density due to $\Delta V_{\rm g}$ was $\sim 1\%$, the system always remained deep in the insulating state and $t_{\rm w} \ll \tau_{\rm eq}$ was satisfied.

Aging effects across the metal–insulator transition in two dimensions

Aging effects have been instrumental as a probe of complex non-equilibrium dynamics in many types of material. In a 2DES, where the onset of glassy dynamics takes place on the metallic side, a study of aging, especially across the MIT, is of great interest. Aging is observed if $t_{\rm w} \ll \tau_{\rm eq}$ (Fig. 8.13 (c)). Since the equilibration time $\tau_{\rm eq}$ diverges exponentially as $T \to 0$ (Section 8.3.2), strictly speaking, the system can reach equilibrium at all $T > 0$. However, even at T that are not too low (e.g. ~ 1 K), $\tau_{\rm eq}$ easily exceeds not only the experimental time window but also the age of the Universe (Jaroszyński and Popović, 2006). This makes it relatively easy to study the out-of-

[9] $E_{\rm F}$ [K]$= 7.31 n_{\rm s}$ ([10^{11}]cm^{-2}) for electrons in Si (Ando *et al.*, 1982).

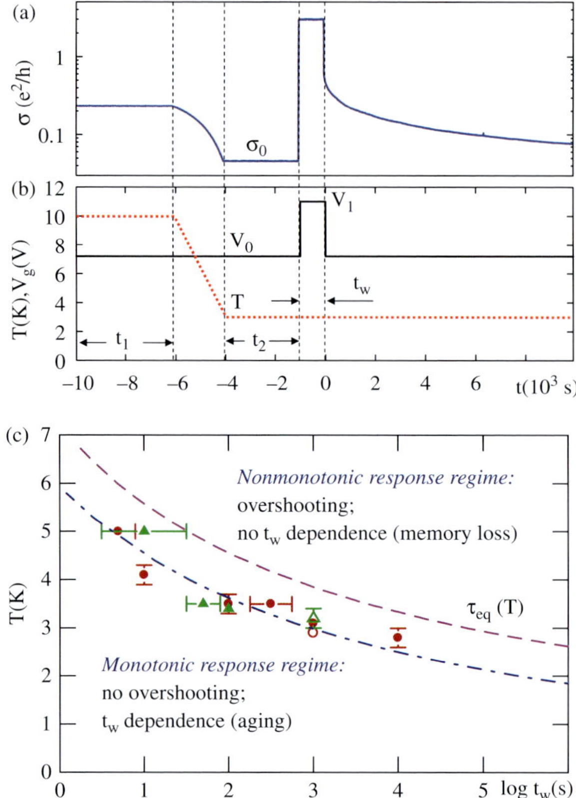

Fig. 8.13 (a) $\sigma(t)$ for $V_0 = 7.2$ V, (n_s $(10^{11}\text{cm}^{-2}) = 3.88 < n_c$), $V_1 = 11$ V (n_s $(10^{11}\text{cm}^{-2}) = 20.26$), $t_w = 1000$ s, and $T = 3.0$ K. (b) The corresponding experimental protocol: $V_g(t)$ and $T(t)$. The results do not depend on the cooling time (varied from 30 min to 10 h), nor on t_1 and t_2 (varied from 5 min to 8 h each). (c) The symbols show the values of (T, t_w) where the overshooting vanishes, i.e. the conditions that separate the two regimes. Open and solid symbols correspond to different samples; the blue dot-dashed line guides the eye. The purple dashed line represents $\tau_{eq}(T)$. Adapted from Jaroszyński and Popović (2007*b*).

equilibrium relaxation of σ at times $t \ll \tau_{eq}$, where one expects to find properties common to other types of glass. In strongly localized systems such as InO_x films the aging function $\sigma(t, t_w)$ is just a function of t/t_w (Ben-Chorin *et al.*, 1993; Ovadyahu and Pollak, 1997; Vaknin *et al.*, 1998, 2000, 2002). This is known as simple, or full aging. It is interesting that, in spin glasses, full aging has been demonstrated only relatively recently (Rodriguez *et al.*, 2003). In general, however, the existence of a characteristic time scale t_w does not necessarily imply simple t/t_w scaling (Barrat *et al.*, 2003). In the mean-field models of glasses, for example, two different cases are distinguished: one where full aging is expected, and the other where no t/t_w scaling is expected (Bouchaud *et al.*, 1997). Experimentally, departures from full aging are common (Rubi and Perez-Vicente, 1997; Struik, 1978).

Aging was investigated in detail in a 2DES (Jaroszyński and Popović, 2007*a*) using the waiting time protocol (Section 8.3.2; Figs. 8.14 (a), (b)), but T was kept fixed at 1 K such that $\tau_{\rm eq}$ was astronomical and hence the 2DES was always deep in the $t_{\rm w} \ll \tau_{\rm eq}$ limit (Fig. 8.13 (c)). $\sigma(t, t_{\rm w})$ were then explored systematically both as a function of final $n_{\rm s}$ and of the difference in densities during and after $t_{\rm w}$. Figure 8.14 (c) illustrates the significant effect of $t_{\rm w}$ on $\sigma(t)$. In fact, all the $\sigma(t, t_{\rm w})$ data can be collapsed onto a single curve simply by rescaling the time axis by $t_{\rm w}$ (Fig. 8.14 (d)). Therefore, in this case, the system exhibits full aging at least up to $t \approx (10^2\text{–}10^3)t_w$. The relaxations can be described by a power law $\sigma(t)/\sigma_0 \propto (t/t_{\rm w})^{-\alpha}$ for times up to about $t_{\rm w}$, followed by a slower relaxation at longer t. This means that the memory of $t_{\rm w}$ is imprinted on the form of each $\sigma(t)$.

For all $n_{\rm s} < n_{\rm c}$, $\sigma(t, t_{\rm w})$ exhibit simple or full aging. However, as $n_{\rm s}$ increases above $n_{\rm c}$, there is an increasingly strong departure from full aging. In some other glasses (Alba *et al.*, 1987, 1986; Struik, 1978), it was found that the data could be scaled with a modified waiting time $(t_{\rm w})^{\mu}$, where μ is a fitting parameter ($\mu = 1$ for full aging). Even though μ may not have a clear physical meaning, the μ-scaling approach has proved to be a useful tool for studying departures from full aging (Barrat *et al.*, 2003; Struik, 1978). By adopting a similar method in the study of aging in a 2DES, it was possible to achieve an approximate collapse of the data (Jaroszyński and Popović, 2007*a*). The plot of μ vs n_0 (Fig. 8.14 (e)) shows a clear distinction between the full aging regime for $n_{\rm s} < n_{\rm c}$, and the aging regime where significant departures from full scaling are seen. It is striking that the largest departure occurs at $n_{\rm s} \approx n_{\rm g}$. For $n_{\rm s} > n_{\rm g}$, some small relaxations are observed (only for $k_{\rm F}l < 1$) resulting from finite-temperature effects, but they vanish in the $T \to 0$ limit (Section 8.3.2). It was also determined that μ does not depend on T (Jaroszyński and Popović, 2009).

These results are striking because they reveal an abrupt change in the nature of the glassy phase exactly at the 2D MIT itself, before glassiness disappears completely at a higher density $n_{\rm g}$. In other words, this is strong evidence that the insulating glassy phase and the metallic glassy phase are different. Therefore, the difference in the aging properties below and above $n_{\rm c}$ puts constraints on the theories of glassy freezing and its role in the physics of the 2D MIT.

Furthermore, for a given $t_{\rm w}$, the amplitude of $\sigma(t)/\sigma_0$ has a peak at $n_{\rm s} \lesssim n_{\rm c}$ (Fig. 8.14 (f)), reflecting an interesting and surprising suppression of the relaxations on the insulating side of the 2D MIT. While a clear understanding of this effect is lacking, it is plausible that collective charge rearrangements that are responsible for the slow dynamics will be suppressed as the 2DES becomes strongly localized. It would be also of interest to study aging deeper in the insulator, in the variable-range hopping regime, but that is not possible because of the small relaxation amplitudes and the large intrinsic sample noise. Of course, the effects of disorder could be explored further by extending the relaxation studies to cleaner (high-mobility) 2DESs, where $n_{\rm g} \gtrsim n_{\rm c}$ (Jaroszyński *et al.*, 2002*b*, 2004*b*). Regardless of the origin of the non-monotonic behavior in Fig. 8.14 (f), however, it is important to note that it reflects another change in the aging properties that occurs at the MIT.

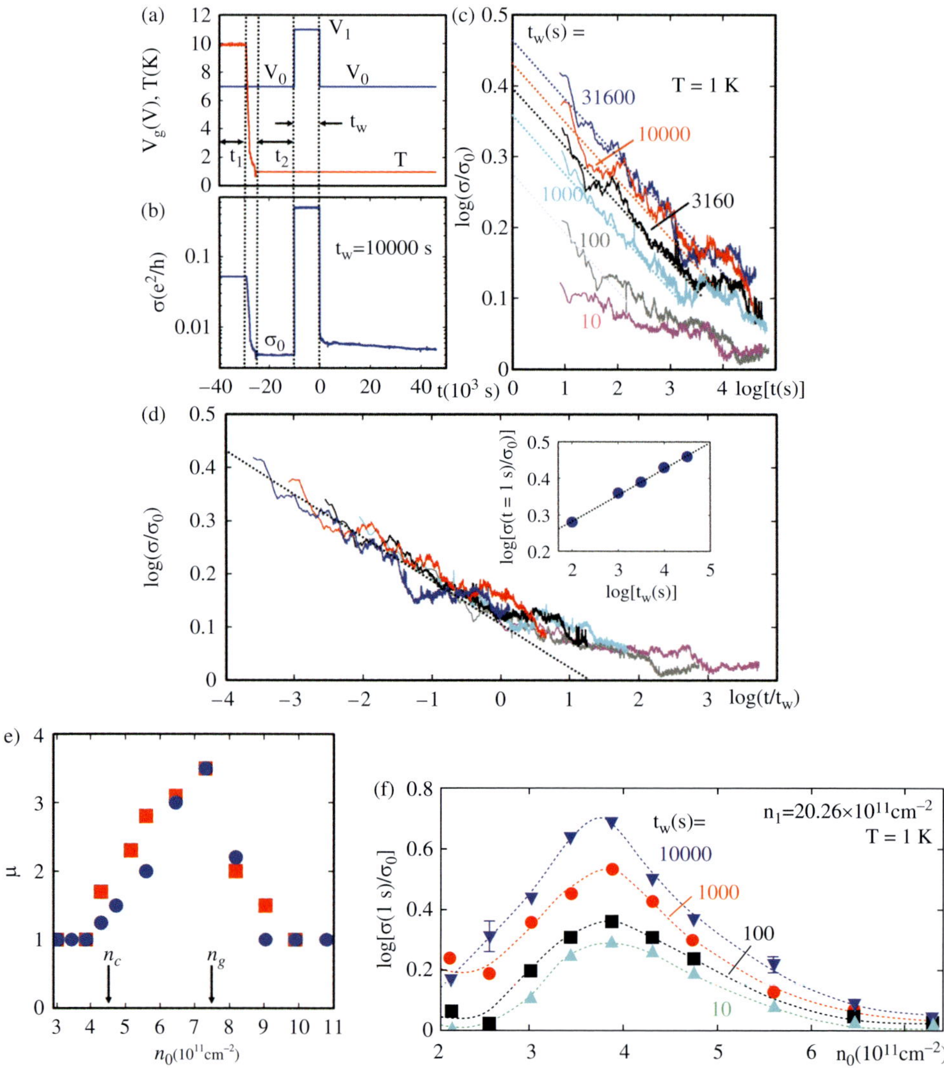

Fig. 8.14 (a) $V_g(t)$ and $T(t)$ in a typical experimental protocol, which always starts with the 2DES in equilibrium at 10 K (Jaroszyński and Popović, 2007*b*). (b) The corresponding $\sigma(t)$. The relaxation of σ during t_w is too small to be seen on this scale. (c) $\sigma(t > 0)$ for several t_w, as shown; $V_0 = 7.0$ V ($n_0(10^{11}\text{cm}^{-2}) = 3.02 < n_c$), $V_1 = 11$ V ($n_1(10^{11}\text{cm}^{-2}) = 20.26$). The dotted lines are linear fits for $t \leq t_w$. (d) The same data as in (c) but plotted vs t/t_w. The dotted line is a fit for $t \leq t_w$ with slope $-\alpha = -0.081 \pm 0.005$. Inset: $\sigma(t = 1\text{s})/\sigma_0$ vs t_w. The dotted line is a fit with slope $\alpha = 0.076 \pm 0.005$. (e) μ vs n_0 for two samples. μ does not depend on n_1. (f) Relaxation amplitudes vs n_0 for several t_w. Dotted lines guide the eye. Adapted from Jaroszyński and Popović (2007*a*), and Jaroszyński and Popović (2009).

8.4 Summary

Careful studies have provided strong evidence that all 2DESs in Si MOSFETs exhibit an MIT regardless of the amount or type of disorder (Fig. 8.15). This is confirmed by extrapolating $\sigma(T)$ on both metallic and insulating sides of the MIT, and by dynamical scaling of $\sigma(n_s, T)$ over a wide range of parameters in both $B = 0$ and $B \neq 0$. In $B = 0$, the scaling form $\sigma(n_s, T) = \sigma_c(T) f(T/\delta_n^{z\nu})$ is obeyed with a temperature dependent critical conductivity $\sigma_c = \sigma(n_s = n_c, T) \propto T^x$, where the value of x depends on the disorder. Thus $\sigma_c(T)$ belongs to the insulating family of curves. The results are consistent with general arguments near the MIT (Belitz and Kirkpatrick, 1994) and they are similar to the behavior in doped semiconductors near the 3D MIT (Sarachik, 1995). The 2D metallic phase survives in very high parallel magnetic fields, long after the 2DES becomes fully spin polarized.

Measurements of the charge dynamics in a 2DES have established that the glass transition takes place at $T_g = 0$ for all $n_s < n_g$ (Fig. 8.15). In general, the glassiness sets in on the metallic side of the MIT, i.e. at a density $n_g > n_c$, thus giving rise to an intermediate, metallic glass phase between the "normal" metal (i.e. in the $k_F l > 1$ regime) and the glassy insulator. The glass transition and various glassy effects are

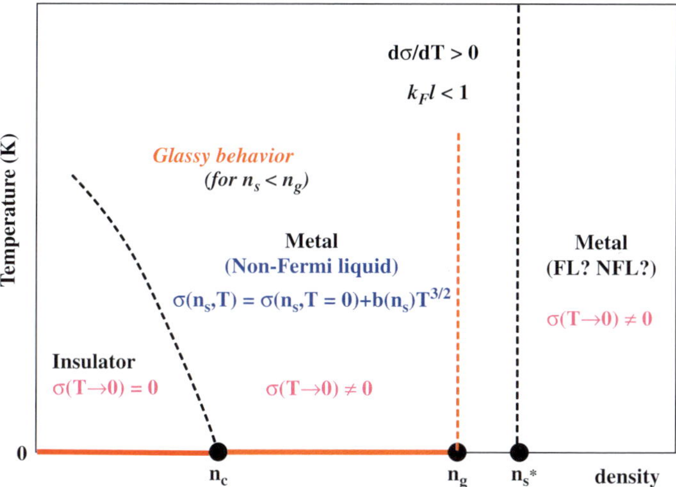

Fig. 8.15 Experimental phase diagram of the 2DES in Si MOSFETs. The MIT takes place at the density n_c and at $T = 0$ in all samples, regardless of the amount of disorder. The nature of the metallic phase at high n_s, such that $k_F l > 1$, is still under debate (see Kravchenko and Sarachik (2004)). The glass transition takes place at $T_g = 0$ for all $n_s < n_g$. In general, the glass transition is a precursor to the MIT, i.e. $n_c < n_g$, giving rise to an intermediate metallic phase with a particular form of the non-Fermi-liquid temperature dependence of conductivity. The aging properties of the glass change abruptly at n_c, indicating different natures of the insulating and metallic glass phases. For sufficiently low disorder, the intermediate phase vanishes: $n_c \lesssim n_s^* \approx n_g$.

observable only for $k_F l < 1$, so that the intermediate phase is a very poor metal ($\sigma(T = 0) \neq 0 \ll e^2/h$). It is characterized by a very specific, non-Fermi-liquid $T^{3/2}$ correction to σ in both low-mobility samples in $B = 0$ and high-mobility samples in a magnetic field. In high-mobility samples in $B = 0$, the intermediate phase practically vanishes and the glass transition coincides with the MIT.

The manifestations of the glass transition in a 2DES for $n_s < n_g$, as demonstrated by resistance noise measurements, include a dramatic slowing down of the electron dynamics and correlated statistics consistent with the hierarchical picture of glasses. The results have been further confirmed and supported by studies of relaxations, which provide evidence for the diverging equilibration time as $T \to 0$, broad distribution of relaxation times in the system, and scaling (in the time domain), consistent with a continuous phase transition at $T_g = 0$. The aging, one of the hallmarks of glassy dynamics, shows an abrupt change in its properties at the MIT, indicating that the natures of the insulating glass and metallic glass phases are different. The 2DES in Si thus exhibits all the characteristics common to other out-of-equilibrium systems, regardless of their dimensionality. Numerous similarities to spin glasses with long-range RKKY interaction are particularly remarkable. Here, however, only long-range Coulomb interactions and potential scattering are present, and measurements in a parallel magnetic field confirm that charge degrees of freedom, not spin, are responsible for glassy freezing. Therefore, the experiments show that the 2D MIT is closely related to the melting of this Coulomb glass.

8.5 Discussion

Experiments on a 2DES in Si clearly provide strong support to theoretical proposals describing the 2D MIT as the melting of a Coulomb glass (Chakravarty *et al.*, 1999; Dalidovich and Dobrosavljević, 2002; Dobrosavljević *et al.*, 2003; Pastor and Dobrosavljević, 1999; Thakur and Neilson, 1996, 1999). In particular, the model of the MIT as a Mott transition with disorder (Dobrosavljević *et al.*, 2003) predicts the emergence of an intermediate metallic glass phase in sufficiently disordered systems (Fig. 8.16). Changing the carrier density in highly disordered samples, for example, corresponds to trajectory 2 in the theoretical phase diagram in Fig. 8.16,[10] where the intermediate phase is predicted to be relatively broad. On the other hand, for low-disordered samples (trajectory 1), the metallic glass phase is very narrow or vanishes, in agreement with experimental observations. The same theory predicts (Dalidovich and Dobrosavljević, 2002) the $T^{3/2}$ correction to conductivity in the metallic glass phase, which is precisely what is found in the experiment.

The emergence of the metallic glass phase, its dependence on the disorder, and the specific form of $\sigma(T)$ in this regime were obtained using a mean-field theory approach, which is known to produce hierarchical dynamics and aging in models of other glasses. Therefore, this approach also seems promising in describing the glassy charge dynamics

[10] Since $E_F \sim n_s$ and $U \sim n_s^{1/2}$ in two dimensions and the disorder $W \approx$ const, the change of n_s in the experiment is described by $(E_F/U) \sim (W/U)^{-1}$ in the phase diagram.

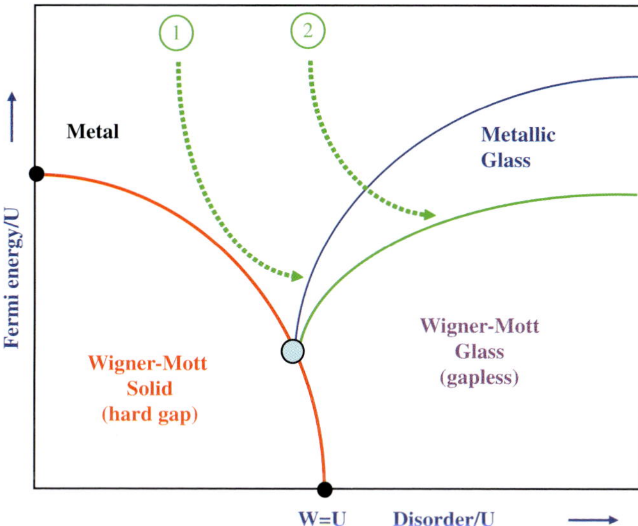

Fig. 8.16 Theoretical phase diagram (adapted from Dobrosavljević *et al.* (2003)). The Fermi energy and disorder on the two axes are expressed in units of the on-site interaction U. For large enough disorder, the MIT is preceded by the glass transition, giving rise to an intermediate, metallic glass phase. Experimental trajectories, where the MIT is approached by varying n_s, are represented schematically by the green curved arrows for the cases of low (trajectory 1) and high (trajectory 2) disorder.

in a 2DES in Si. There is currently no other theory available that predicts any of these experimental features.

A molecular dynamics simulation of the crossover from a Wigner liquid to a Wigner glass in a 2D system of interacting electrons with disorder (Reichhardt and Reichhardt, 2004) has reproduced the main noise results (Section 8.3.1). In particular, a strong, orders-of-magnitude increase in the noise power and a jump in the exponent α was found at low T and n_s, as well as the emergence of non-Gaussianity. By looking at the electron trajectories for a fixed period of time (Fig. 8.17), it was found that, at the highest n_s, electrons can flow freely throughout the sample. As n_s is reduced, electron motion consists of a mixture of 2D and 1D regions (Fig. 8.17 (b)), but over longer times the motion still occurs throughout the whole sample. At even lower n_s (Fig. 8.17 (c)), motion occurs mostly through 1D channels that percolate through the sample. Importantly, the channel structures change very slowly with time: some channels close and others open up. Since transport is dominated by a small number of channels, this gives rise to large fluctuations in the conductivity and strong correlations. Therefore, in this model, the large noise is due to dynamical inhomogeneities, similar to observations in other glasses (Glotzer, 2000; Richert, 2002). At the lowest n_s, the channels disappear and transport occurs via jumps between localized regions. This corresponds to the deep insulating regime.

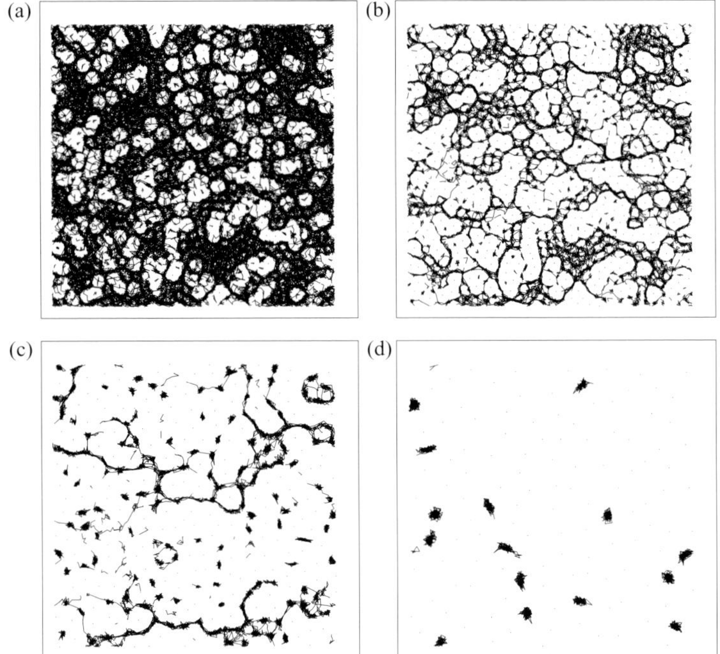

Fig. 8.17 Electron trajectories for a fixed period of time for a fixed T and four electron densities, with the highest n_1 at (a), $n_2 < n_1$ (b), $n_3 < n_2$ (c), and the lowest $n_4 < n_3$ (d). Adapted from Ref. (Reichhardt and Reichhardt, 2004).

In addition, a completely different technique, namely, a Monte Carlo simulation of a 3D Coulomb glass out of equilibrium, has also found evidence for similar dynamical heterogeneities (Kolton *et al.*, 2005).

There have been few experimental attempts so far to probe the charge dynamics near the MIT in other materials. The most exciting and important has been the study of the resistance noise in bulk doped Si (Kar *et al.*, 2003), which has been a prototypical system for investigating the critical behavior near the MIT (Miranda and Dobrosavljević, 2005; Sarachik, 1995) for several decades. The results are strikingly similar to those on a 2DES in Si, namely:

- the noise power increases by several orders of magnitude at n_c (Fig. 8.18 (a))
- as T decreases, the noise magnitude increases essentially exponentially for dopings $n/n_c \leq 1$ (Fig. 8.18 (a) right inset),
- near the MIT, the exponent α rapidly increases to a value much larger than 1 (Fig. 8.18 (b))
- the exponent $(1 - \beta)$ of the second spectrum becomes strongly non-white ($\neq 0$) near n_c (Fig. 8.18 (c)), indicating the onset of correlated dynamics.

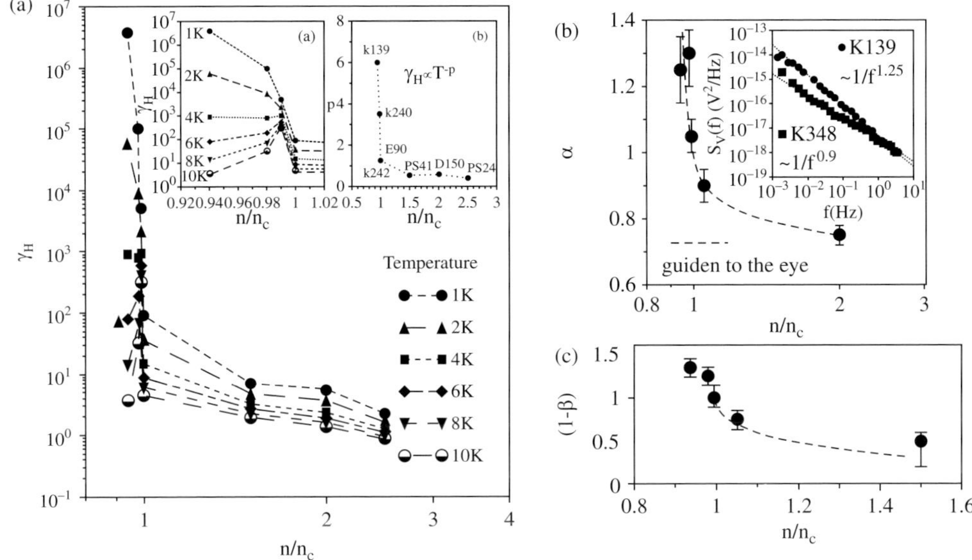

Fig. 8.18 Resistance noise in P-doped Si. Adapted from Kar *et al.* (2003). (a) Noise magnitude as a function of doping at different T. Left inset: the same graph, magnified around n_c. Right inset: the temperature dependence of the noise magnitude vs doping. (b) Exponent α vs doping. Inset: the power spectrum $S(f) \propto 1/f^{\alpha}$. (c) Exponent $(1-\beta)$, a measure of correlations, vs doping.

The similarity of the non-Gaussian spectra observed in both 2DES and in 3D doped Si near the MIT supports the intriguing possibility that such correlated dynamics may indeed be a universal feature of the MIT regardless of the dimensionality.

Resistance noise was also measured in very clean 2D hole systems (2DHS) in GaAs (Deville *et al.*, 2005, 2006; Leturcq *et al.*, 2003). Even though an orders-of-magnitude enhancement of the noise was observed with decreasing T and hole density p_s, similar to the results for 2DES in Si and bulk doped Si, no evidence for strong correlations or glassiness was found. It is likely that the (lack of) disorder here plays a crucial role, since glassiness is not expected theoretically in sufficiently clean systems (Dobrosavljević *et al.*, 2003) (Fig. 8.16). It is interesting, however, that the behavior of transport and noise in an intermediate regime of p_s was attributed to the coexistence of nanoscale conducting and insulating regions, consistent with other studies that favor the percolation picture in 2DHSs and 2DESs in GaAs (Allison *et al.*, 2006; Gao *et al.*, 2002; Ilani *et al.*, 2000, 2001). The local probes provide evidence (Ilani *et al.*, 2000, 2001) that, as the density approaches the critical hole density p_c from the metallic side, the 2DHS fragments into localized charge configurations that are distributed in space, so that the insulating phase is spatially inhomogeneous. The complicated structure emerges already on the metallic side of the MIT, reminiscent of the onset of glassiness in disordered 2DESs in Si. Clearly, more work is needed to examine the charge dynamics in these systems, especially in the presence of a higher amount of disorder.

Resistance noise techniques, long used in studies of various glassy systems, are starting to be recognized as powerful tools in the investigation of the dynamics of cuprates as well (Bonetti *et al.*, 2004; Caplan *et al.*, 2010; Fruchter *et al.*, 2007; Raičević *et al.*, 2007, 2008, 2011). The currently available data, however, are scarce and difficult to compare because they probe very different materials and regions of phase space. This warrants further systematic studies of the charge dynamics in various cuprates as a function of all the relevant parameters, such as doping and disorder, as well as different growth techniques. Comparative studies of charge dynamics in a 2DES in Si, as an excellent model system for the MIT and out-of-equilibrium behavior, and in more complex materials, such as cuprates, should provide important information on these two fundamental problems of condensed-matter physics.

Acknowledgements

This work was supported by NSF Grant DMR-0905843 and the National High Magnetic Field Laboratory via NSF Grant DMR-0654118.

References

Alba, M., Hammann, J., Ocio, M., Refregier, Ph., and Bouchiat, H. (1987). *J. Appl. Phys.*, **61**, 3683.

Alba, M., Ocio, M., and Hammann, J. (1986). *Europhys. Lett.*, **2**, 45.

Allison, G.D., Galaktionov, E.A., Savchenko, A.K., Safonov, S.S., Fogler, M.M., Simmons, M.Y., and Ritchie, D.A. (2006). *Phys. Rev. Lett.*, **96**, 216407.

Altshuler, B.L., Maslov, D.L., and Pudalov, V.M. (2001). *Physica (Amsterdam)*, **9E**, 209.

Amir, A., Oreg, Y., and Imry, Y. (2011). *Annu. Rev. Condens. Matter Phys.*, **2**, 235.

Ando, T., Fowler, A.B., and Stern, F. (1982). *Rev. Mod. Phys.*, **54**, 437.

Barrat, J.-L., Feigelman, M., Kurchan, J., and Dalibard, J. (ed.) (2003). *Slow Relaxations and Nonequilibrium Dynamics in Condensed Matter*. Springer, Berlin.

Belitz, D. and Kirkpatrick, T.R. (1994). *Rev. Mod. Phys.*, **66**, 261.

Belitz, D. and Kirkpatrick, T.R. (1995). *Z. Phys. B*, **98**, 513.

Ben-Chorin, M., Ovadyahu, Z., and Pollak, M. (1993). *Phys. Rev. B*, **48**, 15025.

Bielejec, E. and Wu, W. (2001). *Phys. Rev. Lett.*, **87**, 256601.

Binder, K. and Young, A.P. (1986). *Rev. Mod. Phys.*, **58**, 801.

Bogdanovich, S., Dai, P., Sarachik, M.P., Dobrosavljević, V., and Kotliar, G. (1997). *Phys. Rev. B*, **55**, 4215.

Bogdanovich, S. and Popović, D. (2002*a*). *Physica E*, **12**, 604.

Bogdanovich, S. and Popović, D. (2002*b*). *Phys. Rev. Lett.*, **88**, 236401. Erratum, *Phys. Rev. Lett.* **89**, 289904 (2002).

Bonetti, J.A., Caplan, D.S., Harlingen, D.J. Van, and Weissman, M.B. (2004). *Phys. Rev. Lett.*, **93**, 087002.

Bouchaud, J.-P., Cugliandolo, L.F., Kurchan, J., and Mezard, M. (1997). In *Spin Glasses and Random Fields* (ed. A. Young). World Scientific, Singapore.

Camjayi, A., Haule, K., Dobrosavljević, V., and Kotliar, G. (2008). *Nature Phys.*, **4**, 932.

Caplan, D.S., Orlyanchik, V., Weissman, M.B., Harlingen, D.J. Van, Fradkin, E.H., Hinton, M.J., and Lemberger, T.R. (2010). *Phys. Rev. Lett.*, **104**, 177001.

Castellani, C., Kotliar, G., and Lee, P.A. (1987). *Phys. Rev. Lett.*, **59**, 323.

Chakravarty, S., Kivelson, S., Nayak, C., and Voelker, K. (1999). *Philos. Mag. B*, **79**, 859.

Committee on CMMP 2010, National Research Council (2007). *Condensed-Matter and Materials Physics: The Science of the World Around Us.* The National Academies Press, Washington, D.C.

Dagotto, E. (2002). *Nanoscale Phase Separation and Colossal Magnetoresistance.* Springer-Verlag, Berlin.

Dagotto, E. (2005). *Science*, **309**, 257–262.

Dalidovich, D. and Dobrosavljević, V. (2002). *Phys. Rev. B*, **66**, 081107.

Davies, J.H., Lee, P.A., and Rice, T.M. (1982). *Phys. Rev. Lett.*, **49**, 758.

Davies, J.H., Lee, P.A., and Rice, T.M. (1984). *Phys. Rev. B*, **29**, 4260.

Deville, G., Leturcq, R., L'Hôte, D., Tourbot, R., Mellor, C.J., and Henini, M. (2005). *AIP Conf. Proc.*, **780**, 139.

Deville, G., Leturcq, R., L'Hôte, D., Tourbot, R., Mellor, C.J., and Henini, M. (2006). *Physica E*, **34**, 252.

Dobrosavljević, V., Tanasković, D., and Pastor, A.A. (2003). *Phys. Rev. Lett.*, **90**, 016402.

Dolgopolov, V.T. and Gold, A. (2000). *JETP Lett.*, **71**, 27.

Emery, V.J. and Kivelson, S.A. (1995). *Phys. Rev. Lett.*, **74**, 3253.

Eng, K., Feng, X.G., Popović, D., and Washburn, S. (2001). *Springer Proceedings in Physics*, **87**, 741.

Eng, K., Feng, X.G., Popović, D., and Washburn, S. (2002). *Phys. Rev. Lett.*, **88**, 136402.

Feng, X.G., Popović, D., and Washburn, S. (1999). *Phys. Rev. Lett.*, **83**, 368.

Feng, X.G., Popović, D., Washburn, S., and Dobrosavljević, V. (2001). *Phys. Rev. Lett.*, **86**, 2625.

Fletcher, R., Pudalov, V.M., Radcliffe, A.D.B., and Possanzini, C. (2001). *Semicond. Sci. Tech.*, **16**, 386.

Fruchter, L., Raffy, H., and Li, Z.Z. (2007). *Phys. Rev. B*, **76**, 212503.

Gao, X.P.A., Mills, A.P., Ramirez, A.P., Pfeiffer, L.N., and West, K.W. (2002). *Phys. Rev. Lett.*, **89**, 16801.

Glotzer, S.C. (2000). *J. Non-Cryst. Solids*, **274**, 342.

Goldenfeld, N. (1992). *Lectures on Phase Transitions and the Renormalization Group.* Addison-Wesley.

Gor'kov, L.P. and Sokol, A.V. (1987). *JETP Lett.*, **46**, 420.

Grempel, D.R. (2004). *Europhys. Lett.*, **66**, 854.

Grenet, T. (2003). *Eur. Phys. J. B*, **32**, 275.

Grenet, T., Delahaye, J., Sabra, M., and Gay, F. (2007). *Eur. Phys. J. B*, **56**, 183.

Grünewald, M., Pohlman, B., Schweitzer, L., and Würtz, D. (1982). *J. Phys. C*, **15**, L1153.

Herbut, I.F. (2001). *Phys. Rev. B*, **63**, 113102.

Hernandez, L.M., Bhattacharya, A., Parendo, K.A., and Goldman, A.M. (2003). *Phys. Rev. Lett.*, **91**, 126801.

Hewson, A.C. (1993). *The Kondo Problem to Heavy Fermions*. Cambridge Univ. Press, Cambridge.

Hodge, I.M. (1995). *Science*, **267**, 1945.

Hohenberg, P.C. and Halperin, B.I. (1977). *Rev. Mod. Phys.*, **49**, 435.

Hooge, F.N. (1976). *Physica (Amsterdam)*, **83B**, 14.

Ilani, S., Yacoby, A., Mahalu, D., and Shtrikman, H. (2000). *Phys. Rev. Lett.*, **84**, 3133.

Ilani, S., Yacoby, A., Mahalu, D., and Shtrikman, H. (2001). *Science*, **292**, 1354.

Jaroszyński, J. and Popović, D. (2006). *Phys. Rev. Lett.*, **96**, 037403.

Jaroszyński, J. and Popović, D. (2007*a*). *Phys. Rev. Lett.*, **99**, 216401.

Jaroszyński, J. and Popović, D. (2007*b*). *Phys. Rev. Lett.*, **99**, 046405.

Jaroszyński, J. and Popović, D. (2009). *Physica B*, **404**, 466.

Jaroszyński, J., Popović, D., and Klapwijk, T.M. (2002*a*). *Physica E*, **12**, 612.

Jaroszyński, J., Popović, D., and Klapwijk, T.M. (2002*b*). *Phys. Rev. Lett.*, **89**, 276401.

Jaroszyński, J., Popović, D., and Klapwijk, T.M. (2004*a*). *Proceedings of SPIE*, **5469**, 95.

Jaroszyński, J., Popović, D., and Klapwijk, T.M. (2004*b*). *Phys. Rev. Lett.*, **92**, 226403.

Jaroszyński, J., Wróbel, J., G.Karczewski, Wojtowicz, T., and Dietl, T. (1998). *Phys. Rev. Lett.*, **80**, 5635.

Jelbert, G.R., Sasagawa, T., Fletcher, J.D., Park, T., Thompson, J.D., and Panagopoulos, C. (2008). *Phys. Rev. B*, **78**, 132513.

Jönsson, P.E. and Takayama, H. (2005). *J. Phys. Soc. Jpn*, **74**, 1131.

Jonsson, T., Jonason, K., and Nordblad, P. (1999). *Phys. Rev. B*, **59**, 9402.

Kar, S., Raychaudhuri, A.K., Ghosh, A., v. Löhneysen, H., and Weiss, G. (2003). *Phys. Rev. Lett.*, **91**, 216603.

Kim, N.-J., Popović, D., and Washburn, S. (1998). cond-mat/9809357.

Kirkpatrick, T.R. and Belitz, D. (1994). *Phys. Rev. Lett.*, **74**, 1178.

Kivelson, S.A., Bindloss, I.P., Fradkin, E., Oganesyan, V., Tranquada, J.M., Kapitulnik, A., and Howald, C. (2003). *Rev. Mod. Phys.*, **75**, 1201.

Kohsaka, Y., Taylor, C., Fujita, K., Schmidt, A., Lupien, C., Hanaguri, T., Azuma, M., Takano, M., Eisaki, H., Takagi, H., Uchida, S., and Davis, J.C. (2007). *Science*, **315**, 1380.

Kolton, A.B., Grempel, D.R., and Domínguez, D. (2005). *Phys. Rev. B*, **71**, 024206.

Kravchenko, S.V., Kravchenko, G.V., Furneaux, J.E., Pudalov, V.M., and D'Iorio, M. (1994). *Phys. Rev. B*, **50**, 8039.

Kravchenko, S.V., Mason, W.E., Bowker, G.E., Furneaux, J.E., Pudalov, V.M., and D'Iorio, M. (1995). *Phys. Rev. B*, **51**, 7038.

Kravchenko, S.V. and Sarachik, M.P. (2004). *Rep. Prog. Phys.*, **67**, 1.

Kurzweil, N. and Frydman, A. (2007). *Phys. Rev. B*, **75**, 020202(R).

Lebanon, E. and Müller, M. (2005). *Phys. Rev. B*, **72**, 174202.

Lee, M., Oikonomou, P., Segalova, P., Rosenbaum, T.F., Hoekstra, A.F.Th., and Littlewood, P.B. (2005). *J. Phys. Condens. Matter*, **17**, L439.

Lee, P.A. and Ramakrishnan, T.V. (1985). *Rev. Mod. Phys.*, **57**, 287.

Leturcq, R., L'Hôte, D., Tourbot, R., Mellor, C.J., and Henini, M. (2003). *Phys. Rev. Lett.*, **90**, 076402.

Leuzzi, L. and Nieuwenhuizen, T.M. (2008). *Thermodynamics of the Glassy State*. Taylor & Francis, New York.

Levy, P., Parisi, F., Granja, L., Indelicato, E., and Polla, G. (2002). *Phys. Rev. Lett.*, **89**, 137001.

Martinez-Arizala, G., Christiansen, C., Grupp, D.E., Markovic, N., Mack, A.M., and Goldman, A.M. (1998). *Phys. Rev. B*, **57**, R670.

Martinez-Arizala, G., Grupp, D.E., Christiansen, C., Mack, A.M., Markovic, N., Seguchi, Y., and Goldman, A.M. (1997). *Phys. Rev. Lett.*, **78**, 1130.

Miranda, E. and Dobrosavljević, V. (2005). *Rep. Prog. Phys.*, **68**, 2337.

Miyashita, S., Tanaka, S., and Hirano, M. (2007). *J. Phys. Soc. Jpn*, **76**, 083001.

Monroe, D. (1990). In *Hopping and Related Phenomena 5* (ed. H. Fritzsche and M. Pollak). World Scientific, Singapore.

Monroe, D., Gossard, A.C., English, J.H., Golding, B., Haemmerle, W.H., and Kastner, M.A. (1987). *Phys. Rev. Lett.*, **59**, 1148.

Morita, H. and Kaneko, K. (2005). *Phys. Rev. Lett.*, **94**, 087203.

Müller, M. and Ioffe, L.B. (2004). *Phys. Rev. Lett.*, **93**, 256403.

Müller, M. and Lebanon, E. (2005). *J. Phys. IV France*, **131**, 167.

Nelson, P. (2003). Chapter 12. W.H. Freeman & Co., New York.

Neuttiens, G., Strunk, C., Haesendonck, C.V., and Bruynseraede, Y. (2000). *Phys. Rev. B*, **62**, 3905.

Ogielski, A.T. (1985). *Phys. Rev. B*, **32**, 7384.

Okamoto, T., Hosoya, K., Kawaji, S., and Yagi, A. (1999). *Phys. Rev. Lett.*, **82**, 3875.

Orenstein, J. and Millis, A.J. (2000). *Science*, **288**, 468.

Orlyanchik, V. and Ovadyahu, Z. (2004). *Phys. Rev. Lett.*, **92**, 066801.

Ovadyahu, Z. (2006*a*). *Phys. Rev. B*, **73**, 214204.

Ovadyahu, Z. (2006*b*). *Phys. Rev. B*, **73**, 214208.

Ovadyahu, Z. and Pollak, M. (1997). *Phys. Rev. Lett.*, **79**, 459.

Paalanen, M.A., Rosenbaum, T.F., Thomas, G.A., and Bhatt, R.N. (1983). *Phys. Rev. Lett.*, **51**, 1896.

Pankov, S. and Dobrosavljević, V. (2005). *Phys. Rev. Lett.*, **94**, 046402.

Pappas, C., Mezei, F., Ehlers, G., Manuel, P., and Campbell, I.A. (2003). *Phys. Rev. B*, **68**, 054431.

Pastor, A.A. and Dobrosavljević, V. (1999). *Phys. Rev. Lett.*, **83**, 4642.

Pollak, M. (1984). *Philos. Mag. B*, **50**, 265.

Pollak, M. and Ortuño, M. (1982). *Sol. Energy Mater.*, **8**, 81.

Popović, D., Bogdanovich, S., Jaroszyński, J., and Klapwijk, T.M. (2003). *Proceedings of SPIE*, **5112**, 99.

Popović, D., Fowler, A.B., and Washburn, S. (1997). *Phys. Rev. Lett.*, **79**, 1543.

Pudalov, V.M., Brunthaler, G., Prinz, A., and Bauer, G. (1998*a*). *JETP Lett.*, **68**, 442.

Pudalov, V.M., Brunthaler, G., Prinz, A., and Bauer, G. (1998*b*). *JETP Lett.*, **68**, 534.

Pudalov, V.M., D'Iorio, M., Kravchenko, S.V., and Campbell, J.W. (1993). *Phys. Rev. Lett.*, **70**, 1866.

Raičević, I., Jaroszyński, J., Popović, D., Jelbert, G., Panagopoulos, C., and Sasagawa, T. (2007). *Proceedings of SPIE*, **6600**, 660020.

Raičević, I., Jaroszyński, J., Popović, D., Panagopoulos, C., and Sasagawa, T. (2008). *Phys. Rev. Lett.*, **101**, 177004.

Raičević, I., Popović, D., Panagopoulos, C., and Sasagawa, T. (2011). *Phys. Rev. B*, **83**, 195133. arXiv:1012.0292.

Reichhardt, C. and Reichhardt, C.J. Olson (2004). *Phys. Rev. Lett.*, **93**, 176405.

Richert, R. (2002). *J. Phys.: Condens. Matter*, **14**, R703.

Rodriguez, G.F., Kenning, G.G., and Orbach, R. (2003). *Phys. Rev. Lett.*, **91**, 037203.

Rosenbaum, T.F., Andres, K., Thomas, G.A., and Bhatt, R.N. (1980). *Phys. Rev. Lett.*, **45**, 1723.

Rosenbaum, T.F., Field, S.B., and Bhatt, R.N. (1989). *Europhys. Lett.*, **10**, 269.

Rubi, M. and Perez-Vicente, C. (ed.) (1997). *Complex Behavior of Glassy Systems.* Volume 492, Lecture Notes in Physics. Springer, Berlin.

Sachdev, S. (1999). *Quantum Phase Transitions.* Cambridge University Press, Cambridge.

Sarachik, M.P. (1995). In *The Metal-Nonmetal Transition Revisited: A Tribute to Sir Nevill Mott* (ed. P. Edwards and C. Rao). Francis and Taylor Ltd., London.

Sarachik, M.P., Simonian, D., Kravchenko, S.V., Bogdanovich, S., Dobrosavljević, V., and Kotliar, G. (1998). *Phys. Rev. B*, **58**, 6692.

Schmalian, J. and Wolynes, P.G. (2000). *Phys. Rev. Lett.*, **85**, 836.

Scofield, J.H. (1987). *Rev. Sci. Instrum.*, **58**, 985–993.

Shashkin, A.A., Kravchenko, S.V., and Klapwijk, T.M. (2001). *Phys. Rev. Lett.*, **87**, 266402.

Sjöstrand, M.E., Cole, T., and Stiles, P.J. (1976). *Surf. Sci.*, **58**, 72.

Sjöstrand, M.E. and Stiles, P.J. (1975). *Solid State Commun.*, **16**, 903.

Struik, L.C.E. (1978). *Physical Aging in Amorphous Polymers and Other Materials.* Elsevier, Amsterdam.

Thakur, J.S. and Neilson, D. (1996). *Phys. Rev. B*, **54**, 7674.

Thakur, J.S. and Neilson, D. (1999). *Phys. Rev. B*, **59**, R5280.

Thorsmølle, V.K. and Armitage, N.P. (2010). *Phys. Rev. Lett.*, **105**, 086601.

Tutuc, E., De Poortere, E.P., Papadakis, S.J., and Shayegan, M. (2001). *Phys. Rev. Lett.*, **86**, 2858.

Vaknin, A., Ovadyahu, Z., and Pollak, M. (1998). *Phys. Rev. Lett.*, **81**, 669.

Vaknin, A., Ovadyahu, Z., and Pollak, M. (2000). *Phys. Rev. Lett.*, **84**, 3402.

Vaknin, A., Ovadyahu, Z., and Pollak, M. (2002). *Phys. Rev. B*, **65**, 134208.

Verbruggen, A.H., Stoll, H., Heeck, K., and Koch, R.H. (1989). *Appl. Phys. A*, **48**, 233–236.

Vincent, E. (2007). In *Ageing and the Glass Transition* (ed. M. Henkel, M. Pleimling, and R. Sanctuary), Volume 716, Lecture Notes in Physics, pp. 7–60. Springer, Heidelberg.

Vitkalov, S.A., Sarachik, M.P., and Klapwijk, T.M. (2001). *Phys. Rev. B*, **64**, 073101.

Vitkalov, S.A., Zheng, H., Mertes, K.M., and Sarachik, M.P. (2000). *Phys. Rev. Lett.*, **85**, 2164.

Washburn, S., Kim, N.J., Feng, X.G., and Popović, D. (1999*a*). *Ann. Phys. (Leipzig)*, **8**, 569.

Washburn, S., Kim, N.-J., Li, K.P., and Popović, D. (1999*b*). *Mol. Phys. Rept.*, **24**, 150.

Watanabe, M., Itoh, K.M., Ootuka, Y., and Haller, E.E. (1999). *Phys. Rev. B*, **60**, 15817.

Weissman, M.B. (1988). *Rev. Mod. Phys.*, **60**, 537.

Weissman, M.B. (1993). *Rev. Mod. Phys.*, **65**, 829.

Weissman, M.B., Israeloff, N.E., and Alers, G.B. (1992). *J. Magn. Magn. Mater.*, **114**, 87.

Wróbel, J., Jaroszyński, J., Dietl, T., Regiński, K., and Bugajski, M. (1998). *Physica (Amsterdam)*, **256B-258B**, 69.

9

Phase Competition and Inhomogeneous States as a New Paradigm for Complex Materials

Elbio DAGOTTO[1], Shuai DONG[2], Rong YU[3], Cengiz ŞEN[4], Gonzalo ALVAREZ[5], and Adriana MOREO[6]

[1] Department of Physics and Astronomy, University of Tennessee, Knoxville, TN 37996, USA, and Materials Science and Technology Division, Oak Ridge National Laboratory, Oak Ridge, TN 37831, USA

[2] Department of Physics and Astronomy, University of Tennessee, Knoxville, TN 37996, USA, and Materials Science and Technology Division, Oak Ridge National Laboratory, Oak Ridge, TN 37831, USA

[3] Department of Physics and Astronomy, University of Tennessee, Knoxville, TN 37996, USA, and Materials Science and Technology Division, Oak Ridge National Laboratory, Oak Ridge, TN 37831, USA

[4] Department of Physics, Louisiana State University, Baton Rouge, LA 70803, USA, and Center for Nanophase Materials Science, Oak Ridge National Laboratory, Oak Ridge, TN 37831, USA

[5] Computer Science and Mathematics Division, and Center for Nanophase Materials Science, Oak Ridge National Laboratory, Oak Ridge, TN 37831, USA

[6] Department of Physics and Astronomy, University of Tennessee, Knoxville, TN 37996, USA, and Materials Science and Technology Division, Oak Ridge National Laboratory, Oak Ridge, TN 37831, USA

9.1 Introduction: general aspects of phase competition

The understanding of strongly correlated electrons is among the most important challenges to the condensed-matter community at present (Dagotto, 2005). Electrons in materials belonging to this class behave in manners very different from those described by the "standard" paradigms. For example, a good metal follows qualitatively, and often quantitatively, the predictions of one-particle approximations, where the effect of interactions among the carriers is neglected. Fermi-liquid ideas are behind the justification for this successful story. However, a large variety of other materials of much importance simply cannot be fitted into the one-particle paradigm. These materials exhibit very interesting properties. For instance, some of them, such as the Cu oxide layered materials, become superconducting at "high" temperatures, i.e. halfway between room temperature and absolute zero. Moreover, the normal state of the Cu-oxide high-T_c superconductors is not a standard metal. In fact, at the optimal critical temperature, the resistivity is linear with temperature, as opposed to quadratic for a Fermi liquid. In addition, in the underdoped regime, i.e. closer to the parent antiferromagnetic compound when the hole-doping concentration is varied, the properties of these compounds are simply bizarre. A pseudogap exists in the density of states, and almost all measurements lead to results that contradict established paradigms. Other unconventional materials, such as the manganites, present huge variations in their transport properties under the influence of relatively small magnetic fields, thus generating the label "colossal" magnetoresistance (CMR) for these properties. While magnetic fields on the scale of Teslas are not small for technological applications, they are indeed small from the energy perspective: a field of a few Teslas is not supposed to change the resistivity of a standard metal by several orders of magnitude. Thus, there is no doubt that theories well beyond those of the one-particle approximation must be used to understand these complicated systems.

In recent years, a new line of thinking has been established in this context. Its basis is the notion that many of the strange properties of these complex materials arise from competition of phases. In simple terms, while in some region of the phase diagram you may have a good metal as the ground state at zero temperature, a "competing" state is not too far away in energy and, as a consequence, when increasing the temperature or when testing excitations the behavior of these materials can be drastically different from the expectations for a standard metal. In the area of manganites, a body of literature has shown that competition between the ferromagnetic metallic state—induced by the so-called double-exchange mechanism—and the charge-ordered orbital-ordered spin antiferromagnetic insulators that are very prominent in phase diagrams—induced by lattice distortions or superexchange interactions—are at the heart of the CMR phenomenon (Dagotto, 2005; Dagotto *et al.*, 2001).

Figure 9.1 illustrates the essence of a scenario proposed by Dagotto and collaborators to understand some of the features of these materials. In the clean limit, i.e. without sources of disorder, the two states competing are very different and the phase diagram has a first-order transition separating them. This regime is in principle already sufficient to observe exotic properties, but one must tune the parameters available to

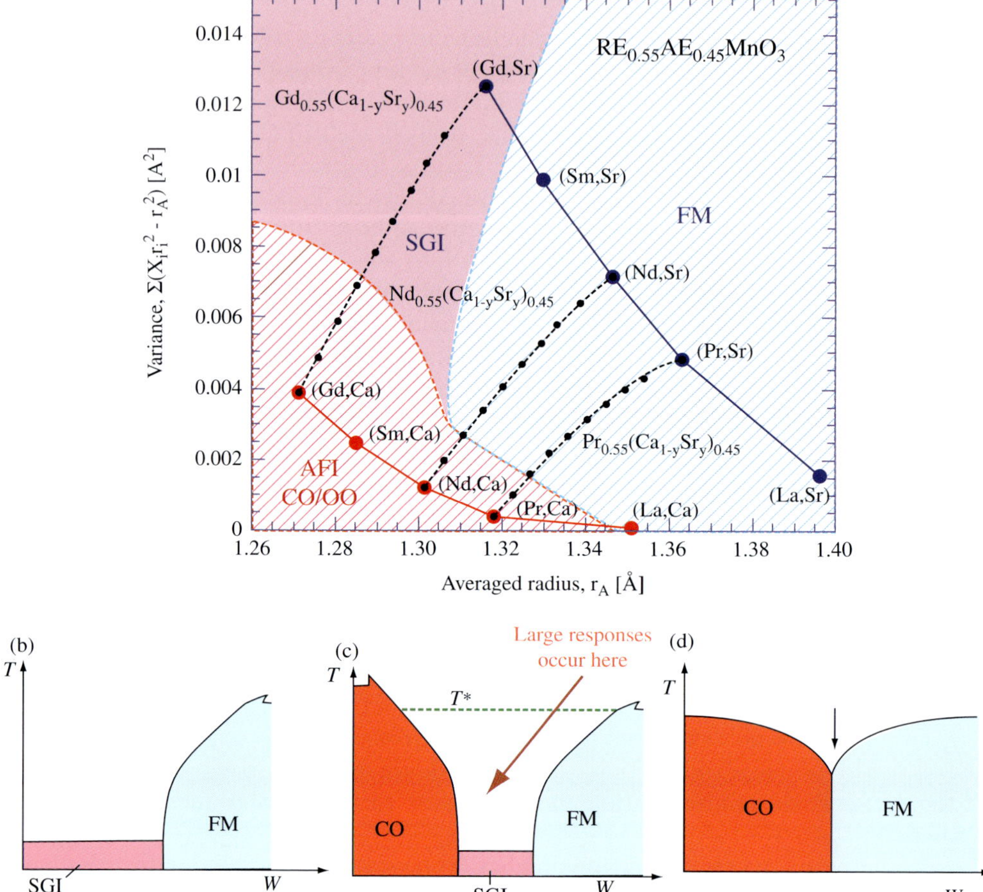

Fig. 9.1 (a) is a generic experimental phase diagram of manganites with varying bandwidth and quenched disorder, reproduced from Dagotto and Tokura (2008). At a low degree of quenched disorder, a first-order transition separates the competing ferromagnetic (FM) metallic and charge-ordered (CO) states (see (d)). As this disorder strength increases, a window opens between the competing states, and a glassy region with nanoscale inhomogeneities is formed (see (c)). The large response to magnetic fields (colossal magnetoresistance, or CMR) appears to occur in this regime, near the (now-reduced) Curie temperature. A new crossover temperature $T^* > T_C$ is observed in this region, where the nanoscale clusters start forming. Finally, at large values of the disorder strength, even the CO state is no longer observed (see (b)). SGI denotes a spin-glass insulator.

be very close to the first-order transition. A procedure to amplify this region of exotic properties is via the addition of quenched disorder (or incorporating strain or other means to locally unbalance one state over the other). The figure shows that for disorder of moderate strength a window opens between the competing states, and in this region is where one state dominates locally over the other. However, since the states are very close in space, the situation can be the opposite, in which case globally none of the two states dominates. This regime is the exotic one where inhomogeneities can be observed. Moreover, investigations show that this is a fragile state and "small" perturbations can cause important non-linearities. Thus, clearly this regime is not describable by the one-particle approximation, particularly at intermediate temperatures.

Figure 9.2 is similar to the previous figure, but adapted to the case of the competition between an antiferromagnetic insulator and a superconducting state, as occurs in high-T_c Cu-oxide superconductors. Theoretical studies using very simple

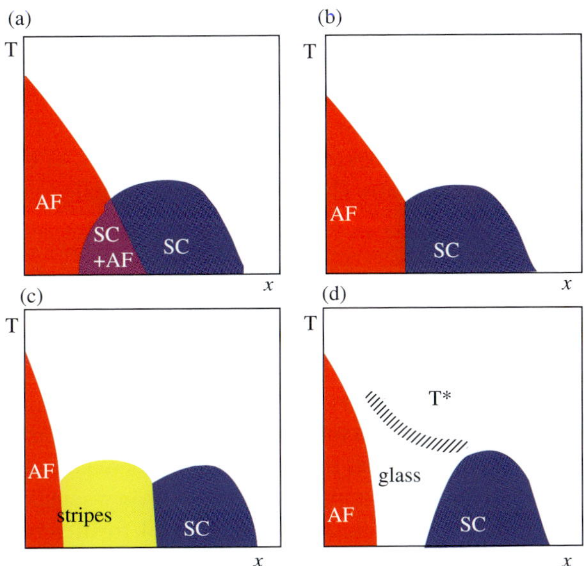

Fig. 9.2 Schematic representation of the phase diagrams that the models studied in Alvarez *et al.* (2005) present, in the clean (a), (b), (c), and dirty (d) limits. In the clean limit, regions with local coexistence of AF and SC (a), or a first-order transition separating AF from SC (b), or an intermediate striped regime (c) are possible. However, none of these reproduce the phase diagram of cuprates with an intermediate glassy state. The experimental phase diagram is achieved only by adding quenched disorder (d). In this regime, a T^* characteristic temperature is found, below which a pseudogap exists in the density of states. This phase diagram has similarities with those proposed before for manganites (see Dagotto *et al.* (2001) and Fig. 9.1), and certainly it is in excellent agreement with the experimental phase diagram of LSCO. For more details, see Alvarez *et al.* (2005). AF, antiferromagnetic; SC, superconducting.

phenomenological models have shown (Alvarez *et al.*, 2005; Mayr *et al.*, 2006) that in the clean limit there are several possible ways to transition from one competing state to the other:

(a) via regions of local coexistence of the two order parameters
(b) via first-order transitions
(c) via stripes of the two competing states.

However, none of these states matches experiments. Moreover, adding quenched disorder creates a glassy separation between the two dominant states, as shown in (d), that better resembles the phase diagram of $La_{2-x}Sr_xCuO_4$ (LSCO). From this perspective, the underdoped regime of the cuprates arises from phase competition of the two dominant states, and has nothing to do with new exotic third states. Moreover, the T^* is a remnant of the original critical temperatures in the clean limit.

Regarding the organization of this chapter, the language and presentation will be colloquial to facilitate the understanding of the main ideas. Moreover, it is important that the reader realizes that this is *not* a comprehensive review, and the focus is on the work of our group. But there is a large number of other experimental and theoretical groups that have provided crucial results on the subject of inhomogeneities in strongly correlated systems over many years. Doing justice to all of them here would be in practice impossible due to space restrictions. Thus, for a full grasp of the vast literature, readers should consult the references provided here and follow the citations in those references. By this procedure the sources of the detailed explanations of many of the items discussed in this chapter will be found.

9.2 Colossal magnetoresistance trapped in a box

9.2.1 Introduction and models for manganites

As explained in Section 9.1, the subject of CMR has attracted considerable attention in recent years (Dagotto, 2005; Dagotto *et al.*, 2001). One reason for this attention is the possibility of using CMR materials in the read sensors of computers. The shrinking size of bits in recording media needs materials that are very sensitive to small changes in magnetic fields. However, CMR in its truly colossal form tends to occur only at relatively low temperatures that may not be of immediate relevance to applications. Nevertheless, the problem of CMR as a basic-science issue remains as one of the most outstanding challenges to condensed-matter theory at present. How is it possible that changes by orders of magnitude occur in the resistivity of CMR materials in the presence of small (in typical electronic scale terms) magnetic fields?

As also reviewed in Section 9.1, considerable progress has been made in developing an understanding of CMR materials based on the notion of phase competition (Dagotto, 2005). The main idea is that for CMR to occur, there has to be an insulating state close to the ferromagnetic metallic state. Here, the word “close” is used in the energy sense, namely an insulating state has to be close in energy to the true ferromagnetic metallic state for the CMR to happen. This usually translates into having the insulating state being the true ground state in nearby regions of the phase diagram, varying for instance the bandwidth or the electronic density via chemical substitutions. Theory says that the CMR cannot occur in regions of the phase diagram where the ferromagnetic state is far from other insulating competitors (Dagotto *et al.*, 2001). In this case, a metallic state will be found even above the Curie temperature and the magnetoresistance will be very small.

These ideas about CMR were originally formulated in terms of phenomenological models and scenarios, as reviewed in Section 9.1 above. However, in recent years improvements in computer simulations and accessibility to large computer clusters have allowed the CMR effect to be studied directly using model Hamiltonians that the community at large considers to be realistic. These model Hamiltonians contain the double-exchange kinetic energy term, with a hopping modulated by the orientations of the localized spins. It also contains the Heisenberg interaction between the localized t_{2g} spins, and in addition the electron–phonon interaction between the mobile carriers and the lattice distortions, mainly of the Jahn–Teller variety (Dagotto *et al.*, 2001). Studies at the mean-field level have suggested that Hubbard interactions among the mobile electrons can be considered via renormalizations of couplings (Hotta *et al.*, 2000). Note also that by itself a large Hund coupling linking the mobile and localized degrees of freedom is sufficient to prevent double occupancy, at least at each orbital. Regarding the number of orbitals at each Mn site, clearly having two e_g orbitals is the most realistic case, but its study comes at a heavy cost, since many algorithms have a computer time growing as the fourth power of the number of degrees of freedom. This is because of the need to diagonalize fermionic matrices during Monte Carlo evolutions. Thus changing from one to two orbitals implies a 16-fold increase in the computer time required. For this reason, many computational studies have been carried out using one orbital only. Fortunately, these studies have shown that one orbital is often enough to capture the essence of the CMR effect. The one-orbital model used in the studies to be reviewed in this chapter is given by:

$$
\begin{aligned}
H_{1o} = & -t \sum_{\langle ij \rangle, \alpha} (c^{\dagger}_{i,\alpha} c_{j,\alpha} + \text{h.c.}) - J_{\mathrm{H}} \sum_{i,\alpha,\beta} c^{\dagger}_{i,\alpha} \vec{\sigma}_{\alpha,\beta} c_{i,\beta} \cdot \vec{S}_i \\
& - \lambda t \sum_{i,\gamma,\alpha} (u_{i,-\gamma} - u_{i,\gamma}) c^{\dagger}_{i,\alpha} c_{i,\alpha} + t \sum_{i,\gamma} (u_{i,\gamma})^2 + \sum_{i,\alpha} (\Delta_i - \mu) n_{i,\alpha}, \qquad (9.1)
\end{aligned}
$$

where $c^{\dagger}_{i,\alpha}$ creates an electron at site i with spin α, $\sigma_{\alpha,\beta}$ are the Pauli spin matrices, $\langle ij \rangle$ indicates the nearest-neighbor sites, and t is the nearest-neighbor hopping amplitude ($t = 1$ also sets the energy unit). Classical localized spins $\vec{S}_i$ represent the t_{2g} degrees of freedom. The third term in the Hamiltonian accounts for the energy corresponding to the lattice–carrier interaction, with λ being the strength of the electron–phonon coupling. $u_{i,\gamma}$ are the distortions of the oxygen atoms surrounding an Mn ion at site i. The index γ in three dimensions (two dimensions) runs over three (two) directions x, y and z (x and y). The tendency toward increasing the magnitude of the lattice distortions is balanced by the fourth term in the Hamiltonian, which represents the stiffness of the Mn–O bonds. Since the study of quantum phonons in this context is not possible with currently available algorithms, the oxygen displacements are considered classical, an approximation widely used in studies of manganites (Dagotto *et al.*, 2001). Δ_i represents the strength of the quenched disorder at a given site, caused by chemical substitution, and are chosen from a bimodal distribution of width 2Δ with mean zero. The overall electronic density n is controlled by the chemical potential μ.

A much-used additional approximation is to work in the limit of an infinite Hund coupling. This is reasonable since the Hund coupling has the typical scale of a few electronvolts, like any coupling in atomic physics, while hopping amplitudes are typically in the range of just a fraction of an electronvolt. The most important advantage of using the infinite Hund coupling limit is the elimination of the spin degree of freedom for the mobile carriers, since this spin is now always pointing in the same direction as the classical spins. The computer time is much reduced by taking this limit into account. The one-orbital Hamiltonian at infinite Hund coupling becomes:

$$\begin{aligned} H_{1o,J_H=\infty} &= -t\sum_{\langle ij\rangle}\{[\cos\frac{\theta_i}{2}\cos\frac{\theta_j}{2}+\sin\frac{\theta_i}{2}\sin\frac{\theta_j}{2}e^{i(\phi_i-\phi_j)}]d_i^\dagger d_j+\text{h.c.}\} \\ &\quad -\lambda t\sum_{i,\gamma}(u_{i,-\gamma}-u_{i,\gamma})d_i^\dagger d_i+t\sum_{i,\gamma}(u_{i,\gamma})^2+\sum_i(\Delta_i-\mu)n_i, \end{aligned} \tag{9.2}$$

where θ_i and ϕ_i are the spherical coordinates of the core spin at site i (assumed classical). The operators $d_i^\dagger$ now create an electron at site i with spin parallel to the core spin at i and $n_i = d_i^\dagger d_i$. Note that for an infinite Hund coupling, the system can be shown to be particle–hole symmetric with respect to density $n = 0.5$. Thus, results at densities n and $1-n$ are equivalent.

In the more realistic case of two orbitals, and still using the infinite Hund coupling limit, the model (without explicit disorder and chemical potential) becomes:

$$\begin{aligned} H_{2o} &= \sum_{\gamma,\gamma',i,\alpha} t^{\alpha}_{\gamma\gamma'}\mathcal{S}(\theta_i,\phi_i,\theta_{i+\alpha},\phi_{i+\alpha})c^\dagger_{i,\gamma}c_{i+\alpha,\gamma'} \\ &\quad +\lambda\sum_i(Q_{1i}\rho_i+Q_{2i}\tau_{xi}+Q_{3i}\tau_{zi})+\sum_i\sum_{\alpha=1}^{\alpha=3}D_\alpha Q^2_{\alpha i}, \end{aligned} \tag{9.3}$$

where the index γ now denotes the two orbitals in the e_g sector, α denotes the spatial directions, and the factor that renormalizes the hopping in the $J_H{=}\infty$ limit is the same one that appears in the case of one orbital, namely:

$$\mathcal{S}(\theta_i,\phi_i,\theta_j,\phi_j)=\cos(\frac{\theta_i}{2})\cos(\frac{\theta_j}{2})+\sin(\frac{\theta_i}{2})\sin(\frac{\theta_j}{2})e^{-i(\phi_i-\phi_j)}. \tag{9.4}$$

The parameters $t^{\alpha}_{\gamma\gamma'}$ are the hopping amplitudes between the orbitals γ and γ' in the direction α. Restricting to just two dimensions, the hoppings are $t^x_{aa} = -\sqrt{3}t^x_{ab} = -\sqrt{3}t^x_{ba} = 3t^x_{bb} = 1$ and $t^y_{aa} = \sqrt{3}t^y_{ab} = \sqrt{3}t^y_{ba} = 3t^y_{bb} = 1$. Q_{1i}, Q_{2i}, and Q_{3i} are normal modes of vibration (Q_1 is the breathing mode, while Q_2 and Q_3 are the Jahn–Teller modes) that can be expressed in terms of the oxygen coordinate $u_{i,\alpha}$ as:

$$Q_{1i} = \frac{1}{\sqrt{3}}[(u_{i,z} - u_{i-z,z}) + (u_{i,x} - u_{i-x,x}) + (u_{i,y} - u_{i-y,y})],$$

$$Q_{2i} = \frac{1}{\sqrt{2}}[(u_{i,x} - u_{i-x,x}) - (u_{i,y} - u_{i-y,y})],$$

$$Q_{3i} = \frac{1}{\sqrt{6}}[2(u_{i,z} - u_{i-z,z}) - (u_{i,x} - u_{i-x,x}) - (u_{i,y} - u_{i-y,y})]. \quad (9.5)$$

Also, $\tau_{xi} = c_{ia}^{\dagger}c_{ib} + c_{ib}^{\dagger}c_{ia}$, $\tau_{zi} = c_{ia}^{\dagger}c_{ia} - c_{ib}^{\dagger}c_{ib}$, and $\rho_i = c_{ia}^{\dagger}c_{ia} + c_{ib}^{\dagger}c_{ib}$. The constant λ is the electron–phonon coupling related to the Jann–Teller distortion of the MnO_6 octahedron (Dagotto *et al.*, 2001). Regarding the phononic stiffness, and in units of $t_{aa}^{x} = 1$, the D_α parameters are $D_1 = 1$ and $D_2 = D_3 = 0.5$, as discussed in previous literature (Hotta *et al.*, 1999).

9.2.2 Colossal magnetoresistance in Monte Carlo simulations of models for manganites

The long history of studies of CMR models will not be reviewed here. Instead, we will directly jump to the most interesting recent developments. The important fact to highlight is that the CMR effect has been clearly observed in Monte Carlo simulations of the models reviewed in the previous sections. This is quite remarkable: in a relatively tiny cluster of just 100 or 200 sites, the transport properties observed in numerical simulations of fairly simplified model Hamiltonians lead to a phenomenology quite similar to that of complicated real bulk oxides. This is a very successful example of how theory can reach a high level of sophistication once physical intuition is combined with powerful tools to carry out the calculations needed to obtain the results. It also shows how, from Hamiltonians where all the interactions are "reasonable" and easy to understand at the atomic level, a fairly non-trivial complex behavior involving unexpected nonlinearities emerges.

To show the CMR phenomenon discovered in Monte Carlo simulations, here some representative figures will be reproduced from the existing literature. The description of the figures provided here will be qualitative, with emphasis on the most important aspects of the problems. The readers are encouraged to consult the original references for details. In Fig. 9.3, the results of Monte Carlo simulations are shown for the case of a small lattice with just 64 sites. Shown using the left axis are results for the spin correlations: when these correlations develop from zero the system goes magnetic. Finding ferromagnetism in these models deserves a long discussion by itself. But this is not the main objective of this chapter, and here it is sufficient to say that the "old" ideas involving double-exchange mechanisms are sufficient to understand the tendencies toward ferromagnetism. Much more interesting for our purposes is to focus on the behavior of the resistivity vs temperature. This is shown in the same figure, on the right-hand axis. The case presented there is for an electron–phonon coupling λ that is not so large as to induce an insulating ground state, but close. In the case studied, the system is metallic at all temperatures in the case of zero disorder (the clean limit, shown in the figure as $\Delta = 0$). However, as the strength of the quenched disorder grows, the resistivity vs temperature curve develops the interesting profile expected

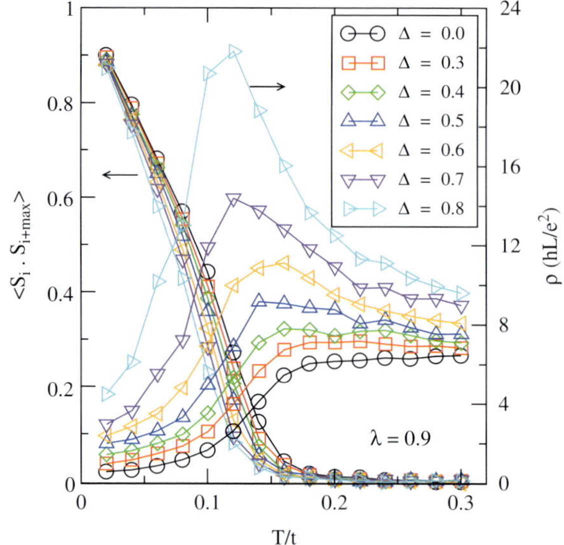

Fig. 9.3 Monte Carlo results obtained using a $4\times4\times4$ lattice, reproduced from Şen *et al.* (2006). The resistivity and spin–spin correlations are shown against temperature, the latter at the maximum allowed distance $(2\sqrt{3})$, working at $\lambda = 0.9$, electronic density $n = 0.3$, and for the disorder strengths Δ indicated. The results shown are mainly for one configuration of quenched disorder, but as many as ten configurations were used in particular cases of temperatures and Δs, and no substantial deviations were observed between disorder configurations.

from the CMR phenomenon: insulating behavior at high temperature, a peak near the location of the ferromagnetic transition, followed by metallic behavior upon further cooling. This is a great theoretical result, illustrating how disorder can generate CMR if the system is already close to a transition to an insulating state.

Figure 9.4 shows other results for the same model used in the previous figure. Figure 9.4 (a) shows results in the clean limit, i.e. without disorder. In this case, the metallic behavior at $\lambda = 0.9$ becomes CMR-like as λ increases. In a relatively narrow range of λ, a resistivity curve resembling those of CMR materials is found. The narrow range shows that fine tuning is needed in the clean limit, but this problem is avoided by increasing the strength of the quenched disorder. The increase of resistivity at very low temperatures for all λs is an effect observed in some, but not necessarily all CMR compounds, and thus we will not focus on that regime. In Fig. 9.4 (b) the influence of magnetic fields is shown for a particular case. Note that magnetic fields that are small in units of the lattice spacing $t = 1$, are sufficient to suppress the peak in the resistivity, making it broader and moving it to higher temperatures, as in experiments.

Figure 9.5 shows results at electronic density $n = 0.1$, and using a small two-dimensional lattice. Figure 9.5 (a) is in spirit similar to Fig. 9.4 (a), namely λ is varied in the clean limit and CMR emerges. However, an interesting difference is the scale of the effect. In Fig. 9.5 (a), a logarithmic scale is used, while in Fig. 9.4 (a),

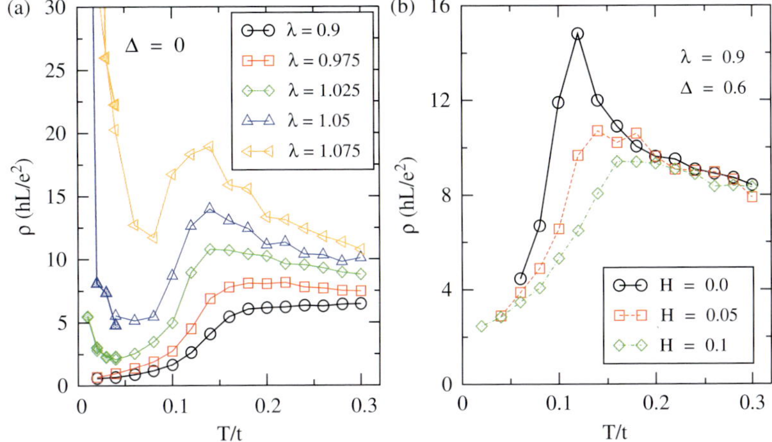

Fig. 9.4 (a) Influence of the electron–phonon coupling λ in the clean limit $\Delta = 0$, and (b) effect of magnetic fields (values indicated) for $\lambda = 0.9$ and $\Delta = 0.6$, on the resistivity curves, using a 4×4×4 lattice at density $n = 0.3$. The Hamiltonian is a one orbital model for manganites. Results reproduced from Şen *et al.* (2006).

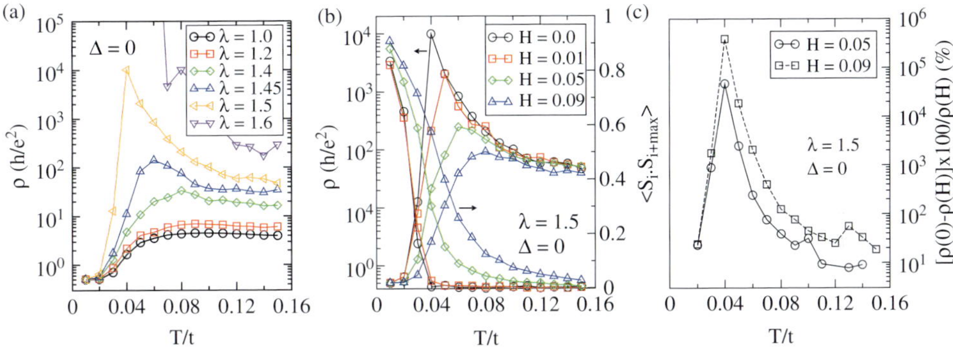

Fig. 9.5 (a) Influence of the electron–phonon coupling λ on the resistivity vs temperature curves, (b) influence of magnetic fields on the resistivity curve and on the spin–spin correlation at the maximum allowed distance $(4\sqrt{2})$ on an 8×8 lattice, and (c) magnetoresistance ratios vs temperature, calculated for two representative magnetic fields in the clean limit $\Delta = 0$, at $n = 0.1$, and using an 8×8 lattice. Results reproduced from Şen *et al.* (2006).

the scale is linear. Thus decreasing the number of carriers dramatically increases the CMR effect. Note also that the low-temperature side of the resistivity peak appears to be almost discontinuous, as in a first-order transition, and as observed in many CMR manganites (although not in all). In Fig. 9.5 (b) the influence of magnetic fields is shown. In excellent agreement with experiments, the peak now moves towards larger temperatures and becomes broader. These are remarkable results, closely matching

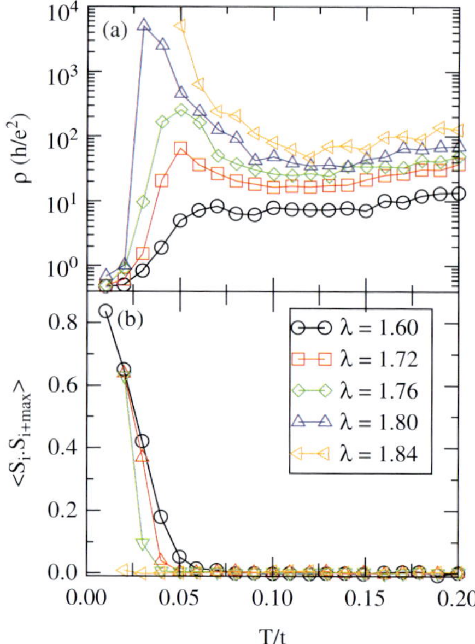

Fig. 9.6 Results for the two-orbital model at $n = 0.1$ and the λs indicated. (a) Resistivity vs temperature using an 8×8 lattice. Note how sharp the low temperature transition from low to high resistance is. (b) Spin–spin correlation at the maximum distance. Results reproduced from Şen *et al.* (2006).

CMR experiments, at least qualitatively. Finally, in Fig. 9.5 (c) the magnitude of the CMR is shown: there is a huge peak at the ferromagnetic critical temperature where the metal–insulator transition also occurs. What is really the most remarkable is that all these results are obtained on very small lattices, yet the resistivity curves are already qualitatively very similar to those in experiments!

The results in most of the investigations reviewed here have focused on the one-orbital model. However, some simulations were performed for two-orbital models. These are the results shown in Fig. 9.6. As in the case of one orbital, for two orbitals the resistivity also clearly exhibits the CMR peak, and it is the most prominent and similar to many experiments right before the transition to the insulating state as ground state with further increasing the electron–phonon coupling. Also shown are the spin correlations that denote the presence of ferromagnetism (Fig. 9.6 (b)).

9.2.3 Fragility of some states of manganites: insulator-to-metal transition induced by adding quenched disorder to an insulator

The results shown in Fig. 9.7 (a,b) illustrate some of the exotic effects observed in theoretical studies of manganites. Let us start with Fig. 9.7 (b). Here, the conductance of a ferromagnetic state is shown vs temperature for three different values of the strength of the quenched disorder. As might be intuitively expected, the conductance decreases

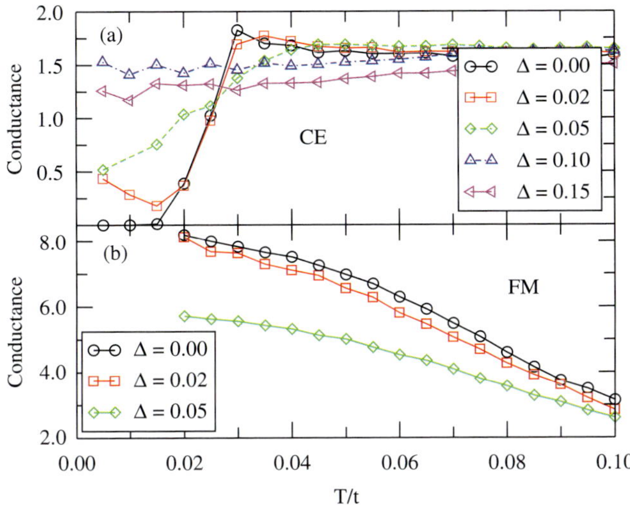

Fig. 9.7 (a) Results illustrating an exotic insulator-to-metal transition induced by increasing disorder, focusing on the regime of the CE phase in a three-dimensional lattice. Conductance vs T of the CE phase, is shown, with and without quenched disorder of strength Δ. The calculation was done on a 4^3 lattice, at $J_{\mathrm{AF}} = 0.20$, $\lambda = 0$, and using 5,000 for thermalization and 15,000 for measurement Monte Carlo steps. The disorder is bimodal, $J_{\mathrm{AF}} \pm \Delta$, and averages over five disorder configurations are shown. (b) Conductance vs T for the ferromagnetic phase, with and without disorder. In this case, where a metal exists at $\Delta = 0$, the disorder simply decreases the conductance, as intuitively expected. The lattice, Monte Carlo steps, number of disorder configurations, and type of link disorder are as in (a). The superexchange coupling is $J_{\mathrm{AF}} = 0.10$. Results reproduced from Alvarez *et al.* (2006).

with increasing disorder. This is clear for a metal: the best conductance is obtained without the sources of extra scattering caused by quenched disorder. However, the results shown in Fig. 9.7 (a) are far from obvious. These show the conductance of the CE phase vs temperature. Here CE is an old notation that denotes a state with charge, orbital, and spin order. Consider first the clean limit: in this case, the ground state is insulating at low temperatures, and thus the conductance vanishes, which is fine. However, as the strength of the disorder grows, one would naïvely expect that the insulating properties of the CE state would become even more clear. However, paradoxically, the conductance *increases* (the system is more metallic) with increasing disorder strength. This is a remarkable result, which also shows the fragility of the CE phase: the amount of disorder needed to generate a metallic (poor) state is very small.

The unexpected transition from the CE insulator to a metallic state with increasing disorder requires a careful discussion. Figure 9.8 is an illustration of qualitative ideas that may be applicable to understand this exotic behavior. Figure 9.8 (a) shows the characteristic zigzag chains that form the CE state. Each chain behaves like a band insulator, due to their particular geometry. If these chains could be stretched into a

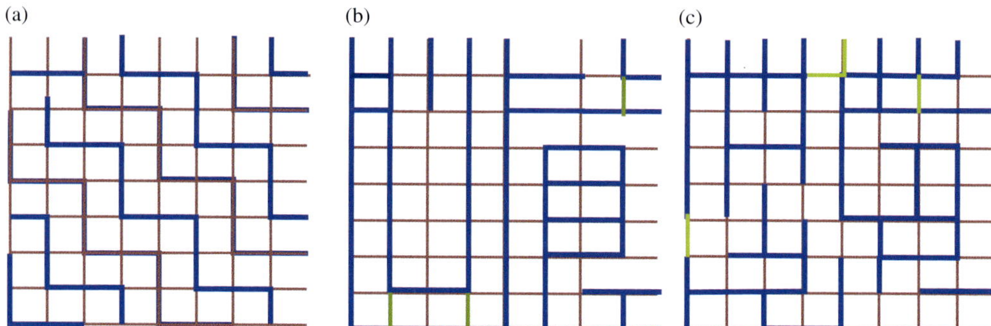

Fig. 9.8 Results of Monte Carlo simulations that summarize some of the main conclusions of this section. The figure shows the signs of the classical spin correlation $\vec{S}_i \cdot \vec{S}_j$ considering nearest-neighbor sites i, j, indicated as red/thin (negative; antiferromagnetic) and blue/thick (positive; ferromagnetic) connections. The lattice is 8×8, $J_{\mathrm{AF}} = 0.17$, and the disorder strength is (a) $\Delta = 0.01$, (b) $\Delta = 0.02$, (c) $\Delta = 0.08$. In (a), the disorder is too weak to change the CE spin pattern, which is clear in the result. However, in (b) the CE phase is replaced by long ferromagnetic (metallic) chains, which in this case are vertical. These chains provide conductance channels in the vertical direction. An equal number of horizontally conducting systems are found when analyzing many quenched disorder realizations, thus restoring rotational invariance. In (c), the state is more reminiscent of the high-disorder regime, with shorter chains. Of the three configurations, the largest conductance is obtained in (b). Results from Alvarez *et al.* (2006).

one-dimensional chain without altering any of the couplings, the resulting material would be a metal since there would be no reason for interference effects that could open a gap as in a band insulator. Figure 9.8 (b) illustrates the effect of disorder: now the chains are very distorted from their original zigzag shape. The figure shows that in this case there are chains that allow the transport of charge from one end of the small cluster to the other, thus rendering the system metallic. This geometry-altering effect of the quenched disorder may be at the heart of the paradoxical transition from an insulator to a metal that is found in Monte Carlo simulations. Figure 9.8 (c) shows the case of large disorder strength. In this situation, even the straight metallic chains get affected by the disorder and eventually, for large enough Δ, an insulator should be recovered again.. Thus, the prediction by this picture, confirmed numerically, is that the conductance should reach a maximum at intermediate Δ.

This interesting effect is not restricted to the CE state. Other states, such as the A-type antiferromagnetic state exhibit a similar phenomenology, as shown in Fig. 9.9, where again the maximum conductance is obtained at an intermediate value of Δ.

9.2.4 Recent results and the colossal magnetoresistance effect at realistic electronic densities

Some of the results presented in the previous section were obtained at electronic densities $n = 0.1$ or 0.9, and they show a very prominent CMR effect. The results at other densities have a CMR but it is not as large. This is not fully satisfying since

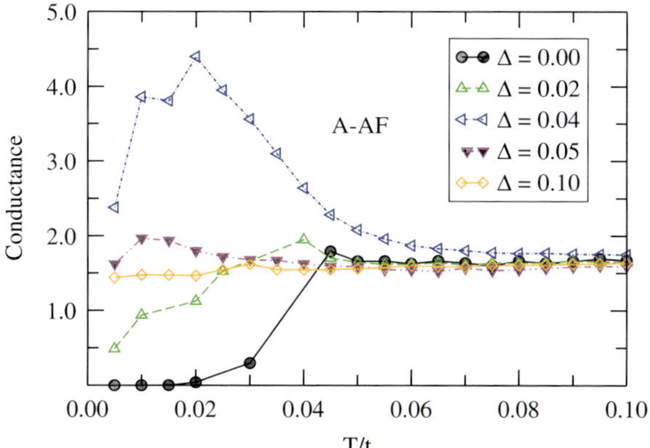

Fig. 9.9 Results illustrating the appearance of a metal upon adding quenched disorder to an insulator, in this case an A-type antiferromagnet. The figure shows numerically calculated conductance vs temperature T for the A-type antiferromagnetic phase in three dimensions, with the antiferromagnetic spin arrangement along the z-axis, with and without disorder. The calculation was done on a 4^3 lattice, working at $J_{\mathrm{AF}} = 0.15$ and $\lambda = 0$. For each temperature and Δ, 5,000 thermalization and 15,000 measurement Monte Carlo steps through the whole lattice were carried out. The link disorder is bimodal, $J_{\mathrm{AF}} \pm \Delta$, and the average shown is over five quenched disorder configurations. Results reproduced from Alvarez *et al.* (2006).

the CMR in the famous intermediate bandwidth manganites, such as those based on Ca doping, have CMR at densities close to $x = 0.3$ (i.e. $n = 0.7$). Fortunately, recent intense investigation has shown that large CMR (namely, one that needs a logarithmic scale for its appreciation) can be found at $n = 0.75$. This is not the end of the story, since these recent investigations only used the one-orbital model, so as to simplify the computational demands, and for this reason work using two orbitals at the same electronic density is still badly needed to complete the CMR picture. But this effort has been postponed for the near future, and therefore let us concentrate here on the description of the (very good) results obtained for one orbital.

In Fig. 9.10, results corresponding to the one-orbital model at electronic density $n = 0.75$ (i.e. hole density $x = 0.25$) are shown. They correspond to the phase diagram varying the superexchange interaction J_{AF} at a fixed electron–phonon coupling constant. Since in this case the Hund coupling is infinite and t is the unit of energy, there is no other parameter to change in the model when working at the clean limit (with quenched disorder we have the extra parameter Δ that regulates the strength of this disorder). At small superexchange coupling, a ferromagnetic metallic phase is found in the phase diagram. This is in good agreement with the well-known double-exchange-based argumentations that predict a ferromagnetic phase in this regime. However, increasing J_{AF} a first-order transition is found, and a new insulating phase is stabilized. This phase is shown schematically in the figure, and it contains spin and charge order (not orbital order, since this is just the one-orbital model) in a highly complex pattern. In fact, this phase is the one-orbital analog of CE-like phases that

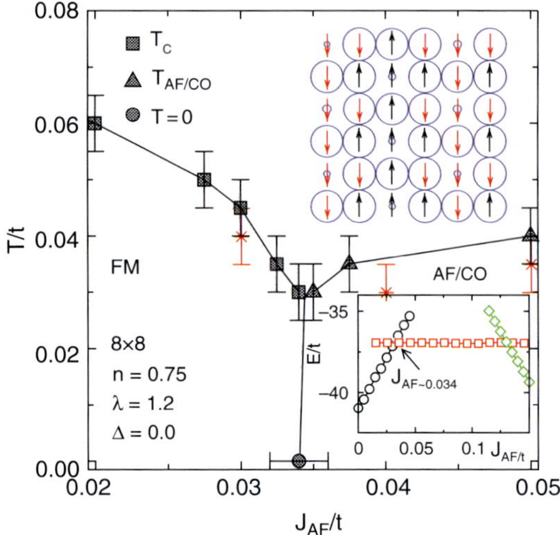

Fig. 9.10 Clean limit Monte Carlo phase diagram of the one-orbital model, using 8×8 and 12×12 lattices, at electronic density $n = 0.75$, and $\lambda = 1.2$. The AF/CO state is schematically shown, with the radius of the circles proportional to the electronic density, and arrows representing the t_{2g} spins. Details regarding the number of iterations in the Monte Carlo simulation and other issues can be found in the original (Şen *et al.*, 2007). *Inset*: energy vs. J_{AF} at very low $T{\sim}0$, with the FM (CO/AF) phase in black (red). Green dots indicate a G-ordered AF regime. This energy plot is typical: in between the extreme cases of FM and G-AFM states, there are complex states with highly nontrivial distributions of spins. FM, ferromagnetic; AF, antiferromagnetic; CO, charge-ordered.

are found in the two-orbital model. Readers should not focus on the fine details of this insulating state, since a two-orbital version will likely alter those details. The important point is that for each electronic density that was investigated, there is an insulating state with complex behavior that is stabilized for sufficiently large couplings. This is generic: in intermediate- or narrow-bandwidth manganites, insulating competitors to the ferromagnetic metallic state exist and can be observed both in theoretical studies and experiments.

Figure 9.11 is one of our best results in the area of CMR investigations. Figure 9.11 (a) shows the resistivity vs temperature, parametric with couplings such as J_{AF} and λ. The two lower curves with black open circles are results in the middle of the ferromagnetic phase. Here the resistivity shows metallic behavior as expected. In the other extreme, of a large J_{AF} or large λ, the opposite case is found: deep into the competing charge-ordered state, the resistivity is, obviously, insulating. The non-trivial message of Fig. 9.11 (a) is found at intermediate couplings, in the region of competition of the two orders. Starting in the ferromagnetic regime, as this competition becomes stronger by increasing J_{AF}, a portion of the resistivity vs temperature curve becomes insulating, right above the ferromagnetic ordering temperature. And as J_{AF} increases further, the resistivity rapidly approaches that of the competing insulating state, even

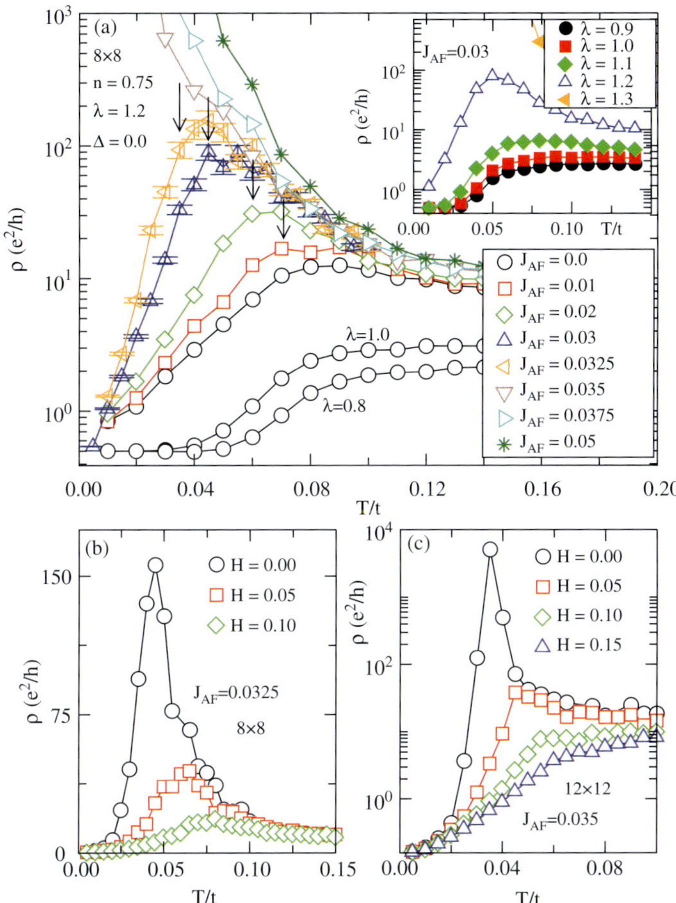

Fig. 9.11 Resistivity ρ vs T curves illustrating the remarkable similarity between theoretical calculations in small clusters and real CMR curves. The results shown were obtained with Monte Carlo simulations and in the clean limit. (a) Results fixing $\lambda = 1.2$ and varying J_{AF}. Arrows indicate T_{c}s. Results at $\lambda = 0.8$ and $\lambda = 1.0$, with $J_{\mathrm{AF}} = 0.0$, are also shown. Inset: results fixing $J_{\mathrm{AF}} = 0.03$ and varying λ. (b) Effect of magnetic fields (indicated, in t units) on ρ using $J_{\mathrm{AF}} = 0.0325$, on an 8×8 lattice. (c) Same as (b) but for $J_{\mathrm{AF}} = 0.035$, on a 12×12 lattice. More details can be found in Şen *et al.* (2007).

though at low temperatures (below the Curie temperature) a transition to a metallic ferromagnetic state is still observed (see arrows in the figure). Note also the logarithm scale used, showing that these effects are substantial.

The results shown in Fig. 9.11 are remarkably similar to those of typical CMR experiments: insulating behavior above T_{c}, followed by a sharp peak in the resistivity at approximately T_{c} upon cooling, this being followed in turn by a rapid decrease in resistivity in the ferromagnetic metallic regime. Moreover, introducing magnetic fields, the sharp peak is suppressed fast (see Fig. 9.11 (b)), with the peak becoming broader and moving to higher temperatures. Figure 9.11 (c) shows that similar results are obtained by using a slightly larger lattice than in (a) and (b).

Figure 9.12 (a) illustrates an important result. The addition of quenched disorder enhances the height of the peak of the resistivity vs temperature curves. For the parameters used, clearly increasing the strength of the disorder transforms a curve with only a small peak into a much larger one. Not only is this an interesting effect because it increases resistivity values but, more importantly, it makes the CMR effect more universal. In the case of Fig. 9.11 (a), it was important to tune J_{AF} in a narrow range close to the first-order transition of the phase diagram in order to obtain the desired CMR effect. But the quenched disorder increases the range of couplings where CMR occurs, making the effect more generic. With quenched disorder incorporated, there is no longer a need to fine tune the couplings.

Figure 9.12 (b) shows that the results using a small three-dimensional cluster are similar to those obtained in two dimensions. Figure 9.12 (c) shows that using the

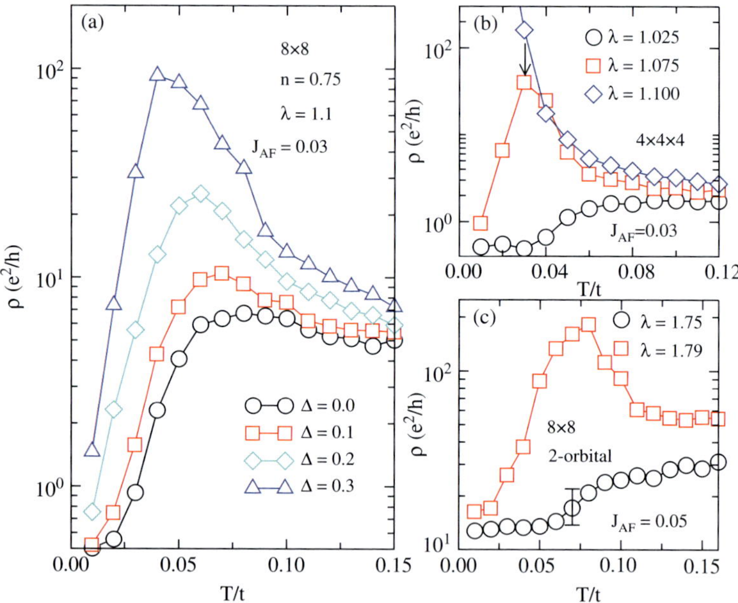

Fig. 9.12 (a) ρ vs T in the presence of quenched disorder Δ. These results show that disorder enhances the values of the resistivity and, in addition, including disorder avoids fine tuning of couplings. Up to ten different disorder realizations were used in these Monte Carlo calculations. Only small changes between configurations were observed. (b) ρ vs T using a 4×4×4 lattice, parametrized with λ, at $J_{\mathrm{AF}} = 0.03$, showing that the results in three dimensions are similar to those in two dimensions. (c) Two-orbital ρ vs T results using an 8×8 lattice for $J_{\mathrm{AF}} = 0.05$. (c) shows that the results are similar using one or two orbitals. Details about the three panels can be found in Şen *et al.* (2007).

two-orbital model similar large peaks in the resistivity are found. Thus, the effect appears to be robust: it happens with and without disorder, in two and three dimensions, and using one and two orbitals.

What is the intuitive origin of the CMR peak in the model simulations? Having the CMR "trapped" in a small box (the cluster we are simulating) allow us to formulate a variety of questions to try to find an intuitive understanding of the effect. In Fig. 9.13, the resistivity is shown on a linear scale together with two other quantities. One of them is the inverse of the density of states at the Fermi level. This quantity has a peak at approximately the same value of temperature at which the resistivity has its peak. This implies that the resistivity is correlated with a depletion in the density of states: as the system is cooled down, a gap appears to open in the regime where the resistivity grows. However, at the Curie temperature a metallic state is restored, the gap (or more accurately, the pseudogap) disappears, and a metallic state is found.

An even more important quantity that correlates with the resistivity is shown in black in Fig. 9.13 (a). These are the charge correlations at short distances, in particular $\sqrt{5}$. This distance appears very prominently in the competing charge-ordered state. The fact that with decreasing temperature this correlation follows the resistivity, first increasing upon cooling and then decreasing in the ferromagnetic state, shows that the CMR peak is correlated with tendencies, at short distance, toward the formation of the competing insulating state near the Curie temperature. For reasons still to be investigated (perhaps entropy), the system tries to form a charge-ordered state upon cooling but then, on reducing the temperature further the true ferromagnetic

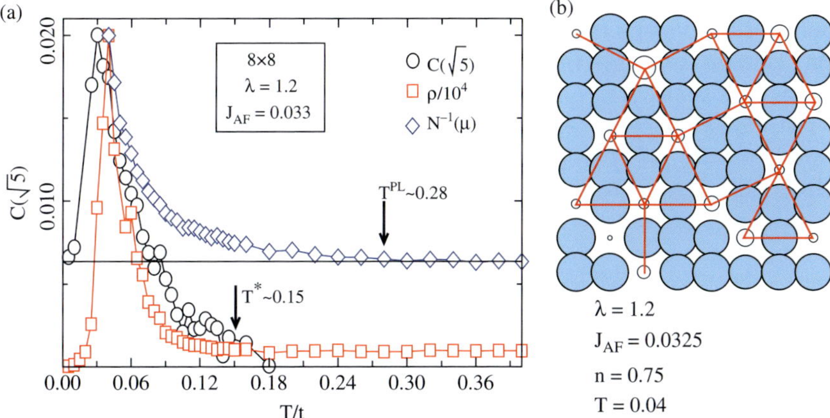

Fig. 9.13 (a) Monte Carlo averaged charge-charge correlation at distance $\sqrt{5}$, $\mathrm{C}(\sqrt{5})$, vs temperature T, showing a qualitative similarity with the rescaled resistivity (also shown). This agreement occurs below the T^* indicated. At higher T, ρ is flat and $\mathrm{C}(\sqrt{5})$ nearly vanishes. Also shown is the inverse of $N(\omega = \mu)$, to indicate the pseudogap formation at T^{pl}, where polarons form. (b) Typical Monte Carlo snapshot with the radius of the circles proportional to the local charge density. Also shown are the hole–hole distances of relevance: $\sqrt{5}$ and 2 (see text). For more details see Şen *et al.* (2007).

metallic ground state prevails. Further evidence of this arises from the Monte Carlo snapshots. In Fig. 9.13 (b), a typical Monte Carlo snapshot is shown, near the peak in the resistivity. Very clear short-range 2 and $\sqrt{5}$ charge correlations exist in this configuration. Moreover, in a more recent study (Yu *et al.*, 2008) it was revealed that this configuration also shows interesting short-range antiferromagnetic spin–spin correlations, which are coupled to the $\sqrt{5}$ charge correlations. It is clear that, at least at short distances, this configuration resembles the competing insulating state. This then proves that phase competition is crucial to achieve CMR, although more work is needed to entirely clarify the fine details of the effect.

9.3 Superconducting clustered state for underdoped cuprates

9.3.1 Competing antiferromagnetism and superconductivity in a phenomenological model

In the same spirit as in the previous sections, here a qualitative description of recent results in the context of high-temperature superconductors will be given. The main references that we will follow are the publications by Alvarez *et al.* (2005), Alvarez and Dagotto (2008) and also Mayr *et al.* (2006). In these references, readers can find a much more extended set of references, properly citing the efforts of many other groups in this context.

The main idea presented in this section builds on the results for manganites described before, but now adapted to cuprates. In manganites, competition between ferro- and antiferromagnetic states was shown to be at the heart of the CMR phenomenon. But phase competition occurs in many other compounds, and there is no reason to expect that such competition will bring interesting non-linear consequences only in the manganites. In fact, the main motivation for the work by Alvarez *et al.* (2005) was precisely to establish a "dictionary" between the cases of CMR manganites and high-T_c superconductors. If in manganites the influence of quenched disorder (or strain) near first-order transitions creates an inhomogeneous state at the nanoscale, a similar result could occur in cuprates. After all, the phase diagram of "clean" cuprates is unknown, since hole and electron doping is achieved via chemical doping, where an element in the chemical formula, typically a trivalent ion, is substituted by another, typically divalent, that has a different valence and also a different atomic size. Thus, disorder is introduced in the system together with carrier doping. It may be, although it is not yet proven, that the true phase diagram of cuprates, namely the phase diagram in the clean limit, contains either a first-order transition separating antiferromagnetic and superconducting phases, or a region of homogeneous coexistence (Alvarez *et al.*, 2005).

The Hamiltonian used in the first steps of the investigations reviewed here involves a complex number $\Delta_{\mathbf{i}}=|\Delta_{\mathbf{i}}|e^{i\phi_{\mathbf{i}}}$ and a real vector $\mathbf{S}_{\mathbf{i}}$, representing the superconducting and antiferromagnetic order parameters, respectively, at site $\mathbf{i}$ of a two-dimensional square lattice. These two order parameters are dominant in the cuprates, and any phenomenological theory must incorporate them. The interactions considered by

Alvarez *et al.* (Alvarez and Dagotto, 2008; Alvarez *et al.*, 2005; Mayr *et al.*, 2006) are the standard:

(1) terms with up to fourth powers of the order parameters, locally favoring superconducting or antiferromagnetism
(2) nearest-neighbor couplings that spread the range of the order (i.e. the gradient terms in the continuum formulation)
(3) an interaction between the order parameters, with strength $u_{\mathrm{SC|AF}}$.

Quenched disorder is also included to represent chemically doped cuprates. More specifically, the Landau–Ginzburg Hamiltonian is,

$$H = r_{\mathrm{SC}} \sum_{\mathrm{i}} |\Delta_{\mathrm{i}}|^2 + \frac{u_{\mathrm{SC}}}{2} \sum_{\mathrm{i}} |\Delta_{\mathrm{i}}|^4 + \sum_{\mathrm{i},\alpha} \rho_{\mathrm{AF}}(\mathbf{i},\alpha)\mathbf{S}_{\mathrm{i}} \cdot \mathbf{S}_{\mathrm{i}+\alpha}$$
$$+ \sum_{\mathrm{i},\alpha} \rho_{\mathrm{SC}}(\mathbf{i},\alpha)|\Delta_{\mathrm{i}}||\Delta_{\mathrm{i}+\alpha}| \cos(\phi_{\mathrm{i}} - \phi_{\mathrm{i}+\alpha}) + r_{\mathrm{AF}} \sum_{\mathrm{i}} |\mathbf{S}_{\mathrm{i}}|^2$$
$$+ \frac{u_{\mathrm{AF}}}{2} \sum_{\mathrm{i}} |\mathbf{S}_{\mathrm{i}}|^4 + u_{\mathrm{SC|AF}} \sum_{\mathrm{i}} |\Delta_{\mathrm{i}}|^2 |\mathbf{S}_{\mathrm{i}}|^2. \tag{9.6}$$

This model can be studied numerically using standard Monte Carlo techniques. Some of the main results are shown in Fig. 9.14, reproduced from Alvarez and Dagotto (2008). In Fig. 9.13 (a), a rough sketch of the phase diagram is presented, showing results for the cases of the clean limit—with an intermediate region where the expectation values for the antiferromagnetic and superconducting order parameters are non-zero simultaneously and homogeneously—and for the dirty case—where disorder opens a “gap” between the antiferromagnetic and superconducting phases similarly to the case for the manganites. In Fig. 9.13 (b), the existence of two temperature regimes is shown: at low temperature, below a T_c, superconductivity develops as clearly shown by the long-distance behavior of the pair correlations. However, above this temperature and below a T^*, short-range pairing correlations are still robust. In Fig. 9.13 (c), typical configurations of the order parameters are shown, using a representation of the superconducting order parameter via a color (for the phase) and an intensity for the amplitude. At low temperature, the near-uniform tone indicates long-range order. At high temperatures, the amplitudes are very weak after the results shown are averaged over a few Monte Carlo steps. The most interesting result is at intermediate temperatures, where the amplitudes are strong but the configuration has many colors and thus the phase of the complex order parameter is changing substantially from site to site. In this regime, there is short-range order but not long-range. This is the regime where interesting results that can be compared with experiments, to be reviewed below, were discovered in recent studies.

The main message of Fig. 9.14 is that the influence of quenched disorder on a region where antiferromagnetism and superconductivity compete leads to the formation of superconducting clusters. These clusters appear to have a robust amplitude in the order parameter, but the superconducting phases change from cluster to cluster in the temperature regime between T_c and T^*, preventing the development of long-range

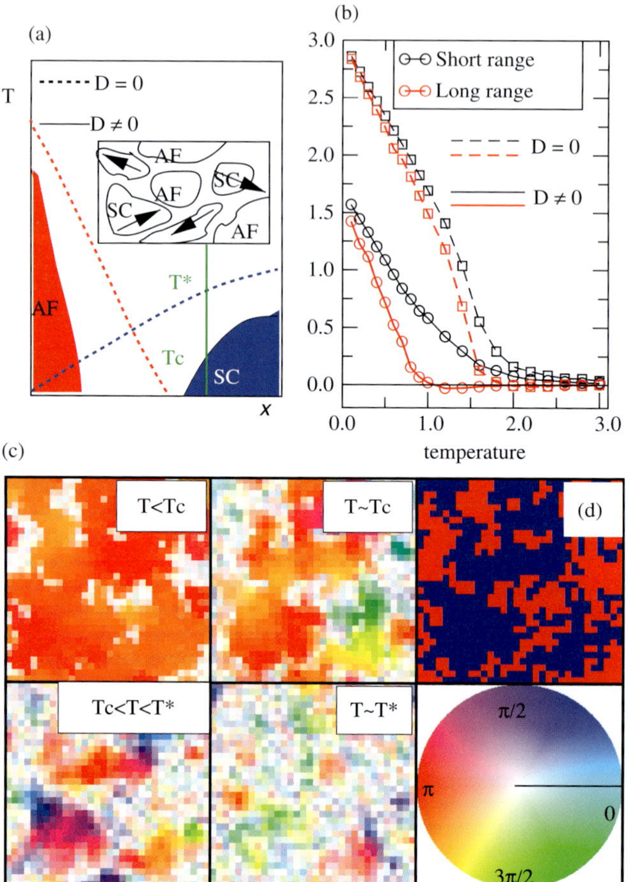

Fig. 9.14 (a) Schematic T vs doping (x) phase diagram for the competition of antiferromagnetism vs superconducting: $D = 0$ ($D \neq 0$) is the clean (dirty) limit. The vertical (green) line is the temperature range emphasized here: at T_c the superconducting long-range order develops, while at T^* the superconducting clusters (short-range order) are formed. Details can be found in the original publication (Alvarez and Dagotto, 2008). (b) Superconducting correlation (SS) vs T at short and long distances, in the clean and dirty limits. Results are from Monte Carlo simulations on a 32×32 cluster with periodic boundary conditions. Other details can be found in Alvarez and Dagotto (2008). (c) Classical superconducting order parameter Δ for a typical Monte Carlo equilibrated configuration at several temperatures (0.2, 1.0$\sim T_c$, 1.5 and 2.0$\sim T^*$). Color intensity represents $< \Delta >$ averaged over short Monte Carlo times, and the actual colors represent the superconducting angle ϕ at each site (see color wheel). For more details, see Alvarez and Dagotto (2008).

order. This state is different from other proposals in the literature to account for the properties in the $[T_c, T^*]$ range:

(a) some of these proposals use new states, unrelated to superconductivity, to explain the pseudogap regime;
(b) the other proposals invoke the large-attraction limit of BCS-like models, where upon cooling first Cooper pairs form and then, at a lower temperature, they condense.

The state described by Alvarez *et al.* has more similarities with (b), in the sense that it is the phase fluctuations that destroy superconductivity, but this clustered state can be formed even in a regime of weak pairing attraction where the Cooper pairs are not small (the standard BCS regime). It is the phase fluctuations between the clusters that cause the suppression of long-range order. This state is intrinsically inhomogeneous, as opposed to the states in the large attraction limit, which are homogeneous.

9.3.2 The Fermi arcs

The phase diagrams and results of the previous section illustrate how superconducting clustered states may emerge from phase competition and quenched disorder. To further investigate this state, it is now necessary to obtain results for the behavior of fermions (carriers) in the presence of these inhomogeneous states. Thus, somehow we must couple fermions to the classical order parameters. In Alvarez *et al.* (2005), Mayr *et al.* (2006), and Alvarez and Dagotto (2008) the approach adopted was to use a model for fermions and for those classical order parameters. The fermions move via hopping terms on a two-dimensional square lattice, they do not interact with each other (i.e. there is no Hubbard repulsion), but they locally couple to the classical order parameters. It is expected, although it was not done explicitly, that "integrating out" these fermions (carriers) would lead to an effective Hamiltonian that would resemble the Landau–Ginzburg Hamiltonian for the superconducting–antiferromagnetic competition described in the previous section.

More explicitly, the Hamiltonian for coupled fermions and classical variables is:

$$\begin{aligned} H_{\mathrm{F}} = {} & -t \sum_{<\mathrm{ij}>,\sigma} (c^{\dagger}_{\mathrm{i}\sigma} c_{\mathrm{j}\sigma} + H.c.) + 2\sum_{\mathrm{i}} J_{\mathrm{i}} S^z_{\mathrm{i}} s^z_{\mathrm{i}} \\ & + \frac{1}{2}\sum_{\mathrm{i},\alpha} V_{\mathrm{i}} |\Delta_{\mathrm{i}\alpha}|^2 - \sum_{\mathrm{i},\alpha} V_{\mathrm{i}} (\Delta_{\mathrm{i}\alpha} c_{\mathrm{i}\uparrow} c_{\mathrm{i}+\alpha\downarrow} + H.c.), \end{aligned} \tag{9.7}$$

where $c_{\mathrm{i}\sigma}$ are fermionic operators, $s^z_{\mathrm{i}}=(n_{\mathrm{i}\uparrow} - n_{\mathrm{i}\downarrow})/2$, $n_{\mathrm{i}\sigma}$ is the number operator, and $\Delta_{\mathrm{i}\alpha}=\beta|\Delta_{\mathrm{i}}|e^{i\phi_{\mathrm{i}}}$ are complex numbers for the superconducting order parameter now defined at the links $(\mathbf{i},\mathbf{i}+\alpha)$($\alpha$ = unit vector along the x or y direction; to account for the d-wave nature of the superconducting state we use $\beta = 1(-1)$ for α along x (y)).

We will not describe in detail how this model was studied numerically, but note that the Hamiltonian belongs to the same class as those analyzed in the area of manganites: both contain fermionic operators that appear only quadratically (since there is no Hubbard term). Therefore, via a library subroutine, using a finite cluster, and for a fixed configuration of the classical order parameters, the fermionic sector can be exactly diagonalized. So in practice, we use the Landau–Ginzburg Hamiltonian and Monte Carlo methods to generate configurations of antiferromagnetic and superconducting order parameters, and then for each configuration the fermions are exactly handled and a variety of observables can be calculated.

In Fig. 9.15, the main results of recent investigations (Alvarez and Dagotto, 2008) are reproduced, showing the one-particle spectral functions that are of relevance to compare theory with angle-resolved photoemission (ARPES) experiments. The results of Fig. 9.15 (a) show the spectral function at several representative temperatures,

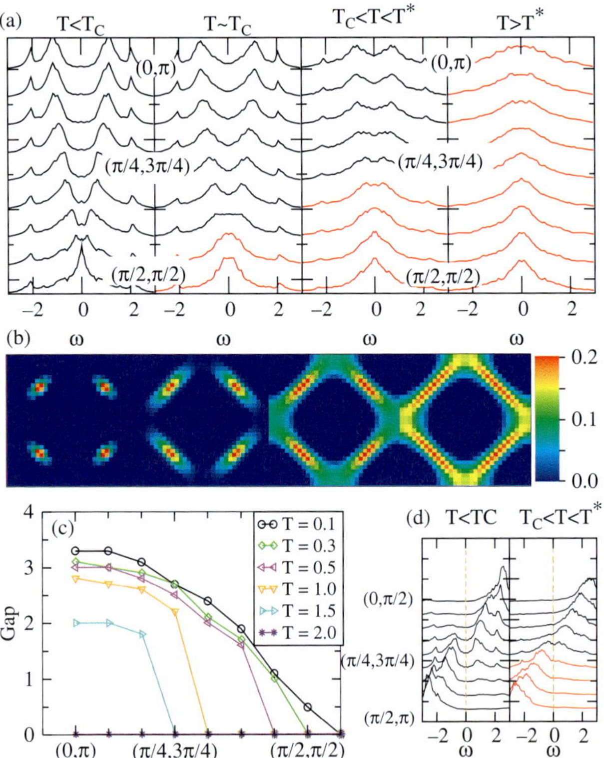

Fig. 9.15 (a) Spectral function $A(\mathbf{k},\omega)$ for equally-spaced momenta along the direction from the nodal $(\frac{\pi}{2},\frac{\pi}{2})$ point (bottom) to the antinodal $(0,\pi)$ point (top). The results show the response of fermions in the presence of the superconducting/antiferromagnetic background generated by Monte Carlo simulations using the Landau–Ginzburg Hamiltonian. The qualitative regimes of temperatures are indicated. (b) $A(\mathbf{k},\omega{=}\,0)$ in the k_x–k_y plane for the same parameters and temperatures as in (a). The results shown are those within a small energy window from the Fermi level, namely they represent the Fermi surface. (c) Superconducting gap (distance between peaks) vs momentum for the same parameters as in (a) where $T_c{\sim}0.4$ and $T^*{\sim}1.5$. (d) $A(\mathbf{k},\omega{=}\,0)$ from $(0,\frac{\pi}{2})$ to $(\frac{\pi}{2},\pi)$, at $T{=}1.0$ and 0.1, and parameters as in (a). For more details the reader should consult Alvarez and Dagotto (2008).

moving from the nodal to the antinodal points in momentum space. The results at low temperature are clear: the nodal point has no gap since it is the location of the node of the d-wave superconducting order parameter, while the rest of the momenta have a gap. The results at very high temperatures are also clear: there is no gap at any momentum, and superconductivity is absent both at long and short distances. But the results at intermediate temperatures are highly non-trivial. In this case, the gap starts closing at the nodal region and, as the temperature increases, the region without a gap extends towards the antinodal point. In Fig. 9.15 (b), the consequences of this exotic behavior are shown for the Fermi surface: in the intermediate temperature range, Fermi arcs are formed, quite similarly to experiment (see references quoted in Alvarez and Dagotto (2008)). Figure 9.15 (c) shows explicitly the gap vs momentum between antinodal and nodal points, parametric with temperature. Note the broad temperature range where the antinodal gap is almost unchanged, while the gapless nodal regions increases. Finally, in (d) the behavior along a direction perpendicular to the nodal–antinodal line is shown: at low temperature typical BCS behavior is found, exhibiting Bogoliubov quasiparticles, while in the intermediate temperature range there is no vestige of the Bogoliubov dispersions. The point of Fig. 9.15 (d) is that the closing of the gaps is not merely caused by temperature-induced broadening but by more subtle effects, involving a reconstruction of the bands.

9.3.3 Intuitive explanation for the ARPES results

The results reviewed in the previous sections are very interesting in the sense that they reproduce the ARPES experiments via the simple picture of invoking a superconducting clustered state. However, there is still considerable work to be done regarding the intuitive understanding of the numerical results. Why do the Fermi arcs form? This important question is not fully answered, but Alvarez and Dagotto (2008) have provided some first steps towards its clarification. For instance, to gain intuition, suppose that instead of using randomly located superconducting clusters, that these clusters are artificially spaced regularly, as sketched in Fig. 9.16 (a). If all the phases of the regularly-spaced clusters are made the same, of course the uniform d-wave superconducting state is recovered, concomitant with the band dispersion from the nodal to the antinodal points shown in Fig. 9.16 (b). However, if theses phases are made random then the results in Fig. 9.16 (c) are obtained: here the gap closes near the nodal point but it remains open at the antinodal, and its associated Fermi surface (see Fig. 9.16 (d)) has a Fermi arc.

From the results in Fig. 9.16 (d), clearly the mere presence of randomly oriented superconducting phases in regularly spaced clusters appears to be sufficient to generate the Fermi arcs. The intuitive understanding of these results can be further improved by considering just two of those clusters, as sketched in Fig. 9.16 (e), for the extreme case of a π phase shift. In this figure, it is shown that for two nearest-neighbor clusters where the order parameters differ by their sign (i.e. phase difference π), the superconducting order parameter must vanish in between. Figure 9.16 (f) shows some of the consequences of having the vanishing order parameter between clusters. In this case, the local density of states has the d-wave gap both in one cluster and the other, of course, but in between this it resembles that of the non-interacting

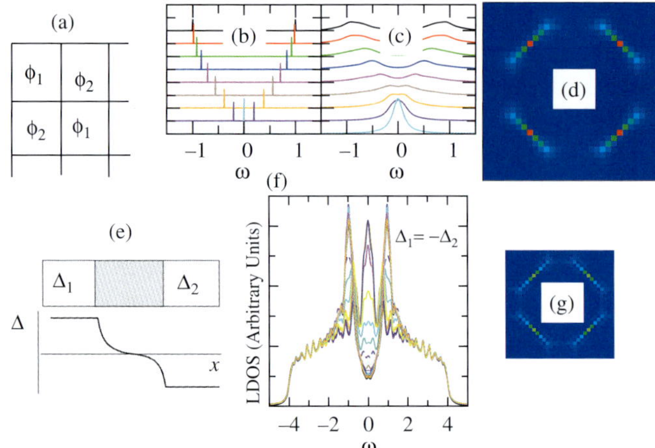

Fig. 9.16 (a) Schematic representation of a toy model configuration to understand the numerical results of the superconducting clustered state (Alvarez and Dagotto, 2008; Alvarez *et al.*, 2005). ϕ_1 and ϕ_2 refer to the SC order parameter phases in 4×4 squares. (b) $A(\mathbf{k},\omega)$ for $\mathbf{k}$ along the direction $(\pi/2,\pi/2)$ to $(0,\pi)$ for the case of regularly spaced clusters, as in (a), with ϕ_1=ϕ_2=0 (i.e. perfect *d*-wave superconductor). (c) Same as (b), but with ϕ_1 and ϕ_2 randomly chosen between 0 and π. (d) $A(\mathbf{k},\omega=0)$ in the k_x-k_y plane for case (c). (e) Schematic representation of a Josephson-junction-like structure. *Top*: two SC regions with order parameters Δ_1 and Δ_2, separated by a Δ=0 area of width w (grey). *Bottom*: expected gap interpolation for Δ_1=-Δ_2. (f) Local density of states on a 32×32 lattice containing 4 equally-spaced 12×12 SC clusters separated by 4 lattice sites. The order parameters are staggered between clusters using Δ_1=1 and Δ_2=-1. Shown are results from the center of one SC region to a nearest-neighbor. The point is to illustrate the existence of a region where the local density of states resembles that of a non-superconducting metallic state. The sharp peak is simply a van Hove singularity in two dimensions in the non-interacting limit. (g) $A(\mathbf{k},\omega=0)$ in the k_x-k_y plane for $\Delta_1=-\Delta_2=1$, with a setup as described in (f). For more details, the reader should consult the original publication (Alvarez and Dagotto, 2008).

system (which has a van Hove singularity in two dimensions). Thus, at the location where the order parameter vanishes, a small metallic region is created that does not have superconductivity. With this simple arrangement of equal-size, equally spaced, superconducting clusters with phase factors differing by π, a Fermi arc is also observed as shown in Fig. 9.16 (g). Thus, the π-shifted clusters and the metallic region in between appear to be correlated with the existence of the arcs.

9.3.4 Summary for underdoped cuprates

Let us summarize this section. Here, phenomenological models for the competition between antiferromagnetism and superconductivity were used (Alvarez and Dagotto, 2008; Alvarez *et al.*, 2005; Mayr *et al.*, 2006). The models were studied with a combination of Monte Carlo and fermionic exact-diagonalization techniques. These models were not derived: they were applied based on previous experience of Landau–Ginzburg calculations and intuitive arguments. In the presence of quenched disorder, the phase diagram resembles that of the cuprates, in the sense that a region without long-range order in either the antiferromagnetic or superconducting channels is present. However, by analyzing this glassy state microscopically, it is found that it contains superconducting clusters where the amplitude is developed but the phases are basically random. This occurs in a range of temperature between T^*, where the superconducting amplitude develops in the clusters, and T_c, where long-range order develops. These results are in excellent agreement with the scanning tunneling microscopy experiments (Gomes *et al.*, 2007), that unveiled superconducting clusters at temperatures much higher than T_c in real cuprates. However, note that in the very underdoped regime, the scanning tunneling microscopy results show an almost uniform d-wave state in a wide range of temperatures above T_c, and it is only later, upon further increasing the temperature, that the clusters become clearly visible. A full understanding of the cuprate results thus requires not only the superconducting clustered state introduced by our group but also physics that arises from the large attraction limit of the BCS models, where the non-superconducting state above T_c is uniform in space. More work is needed to address this matter.

As described in this section, the superconducting clustered state leads to interesting results for ARPES experiments. The existence of clusters that are close to one another in space, but differ in π regarding the superconducting phase, creates metallic regions that appear to be correlated with the presence of Fermi arcs. However, more work is also certainly needed in this context to fully understand the role of the metallic areas between the superconducting clusters.

Another interesting consequence of having a superconducting clustered state is the prediction of large non-linearities, similar to those that occur in manganites when ferromagnetic and antiferromagnetic states compete. In other words, via proximity to a well-developed superconducting state, the phases of the superconducting clusters may orient themselves, similar to when the moments of ferromagnetic clusters rapidly orient themselves when a small magnetic field is introduced. This "giant proximity effect" is worth exploring in more detail. The reader can find more information regarding the theoretical considerations of this problem in Alvarez *et al.* (2005).

In short, phase competition and quenched disorder can lead to nanoscopic scale clusters in Cu-oxide superconductors. These may induce interesting and unusual behavior that can be probed with a variety of experimental techniques. More work should be devoted to the analysis of the consequences of these states and to establishing the existence of these superconducting clusters in high-T_c materials.

9.4 Other cases of inhomogeneities in model Hamiltonians for oxides

In this brief review, the many cases of inhomogeneities found in simulations of models for transition metal oxides cannot be all described in detail. However, at least for some completeness, we will mention here a few extra cases of inhomogeneities beyond those that have been described thus far. Since the description is here terse, readers are encouraged to consult the literature cited and references therein to get deeper into the details of the states mentioned here.

9.4.1 Stripes in manganites

In Fig. 9.17, ground-state results for the case of ferromagnetic manganites in the presence of a robust Jann–Teller coupling are shown for three different electronic densities. The results, obtained via unbiased computer simulations of model Hamiltonians (Hotta *et al.*, 2001), show the presence of diagonal stripes that are spontaneously generated to minimize the energy. In the original reference (Hotta *et al.*, 2001), even more complex patterns were reported; in fact, including the spin degree of freedom leads to fairly exotic arrangements. The stripes described here are a form of local phase separation (Kivelson *et al.*, 1998; Tranquada *et al.*, 1995), but one that is different from the focus of the previous chapters where first-order transitions and electronic phase separation were emphasized.

9.4.2 Stripes in spin fermion models for cuprates

In Fig. 9.18, results of Monte Carlo simulations for a spin–fermion model are reproduced. The presence of vertical stripes (degenerate in energy with the horizontal ones, not shown) are clear. This spin–fermion model is a phenomenological model

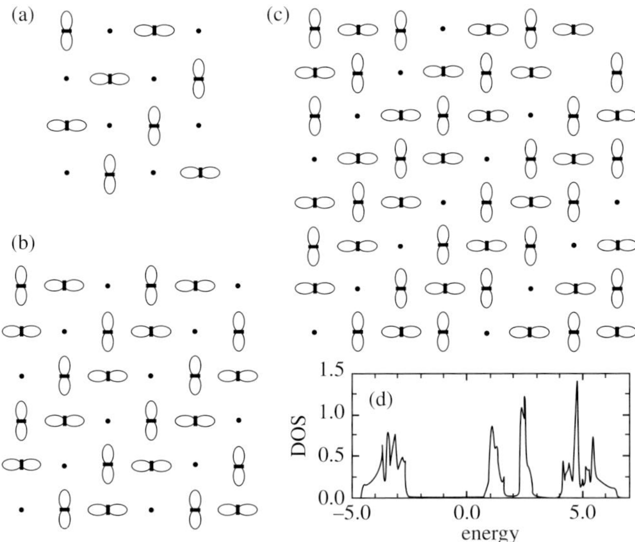

Fig. 9.17 Orbital arrangement in the charge-ordered ferromagnetic phase of a two-orbital model for manganites. Results are shown at (a) $x = \frac{1}{2}$, (b) $x = \frac{1}{3}$, and (c) $x = \frac{1}{4}$, with a large electron–phonon coupling $\lambda = 2.0$. The charge density in the lower-energy orbital is shown, with the size of the orbital being in proportion to this density. (d) The total density of states vs energy for the case $x = \frac{1}{3}$ and $\lambda = 2.0$. Results are reproduced from Hotta *et al.* (2001).

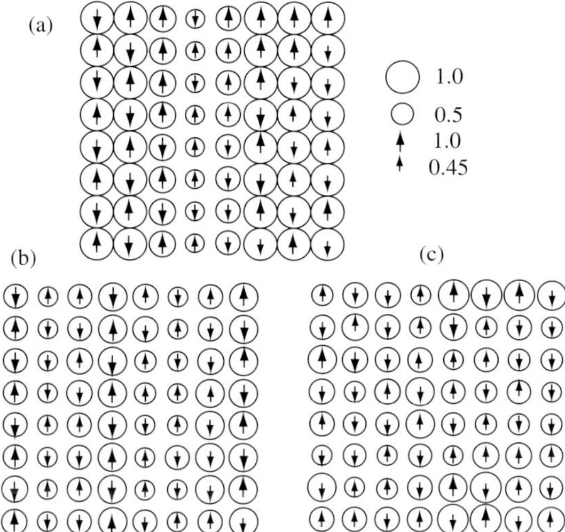

Fig. 9.18 Spin and charge distribution obtained using the spin-fermion model for the cuprates. (a) Results obtained with Monte Carlo simulations on an 8×8 cluster and electronic density per site $n = 0.85$. (b) Spin and charge distribution for a Monte Carlo snapshot at $n = 0.75$. (c) is same as (b) but at a higher temperature, illustrating the melting of the stripes. Periodic boundary conditions are used. Only the z component of the spin is shown because it is the most representative in this case. The circles are proportional to the electronic density. The reader can find the values of the temperature and parameters in the Hamiltonian used here, and other details, in the original references (Buhler *et al.*, 2000; Moraghebi *et al.*, 2001, 2002*a*,*b*).

for cuprates, similar to those described in the section on underdoped cuprates and Fermi arcs. The model contains both fermionic and classical degrees of freedom, the latter representing the antiferromagnetic order parameter. Details can be found in Buhler *et al.* (2000). As in the case of the stripes in manganites, the stripes here are yet another manifestation of local phase-separation tendencies, which in principle are unrelated to the physics of a disorder-smeared first-order transition when a metal and an insulator compete, which was described in the section devoted to manganites and the CMR effect.

9.4.3 Stripes in multiferroics

We end this section with a brief description of some remarkable very recent results. Variational studies and Monte Carlo simulations of models for manganites have recently unveiled a new phase at electronic density $n = 0.75$ (Dong *et al.*, 2009). The arrangements of spins is shown in Fig. 9.19 (a). There, the presence of spin domains that are pointing horizontally or vertically is clear. The state is called SOS, which stands for "spin orthogonal stripe." This form of self-organization is novel, since while stripes have certainly be found before, the spin domains were not orthogonal but usually the spins are collinear, perhaps with a π shift in the antiferromagnetic

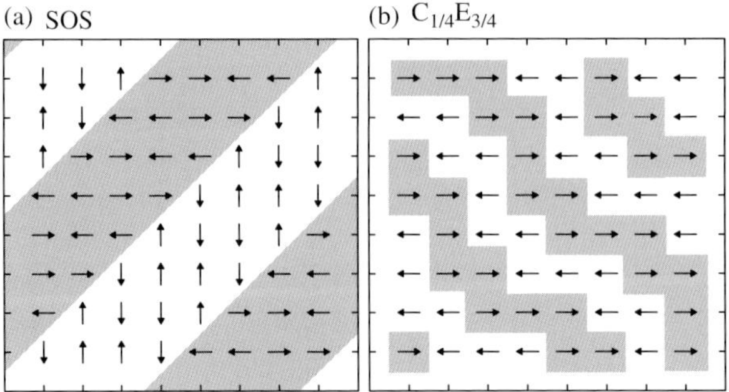

Fig. 9.19 Spin patterns of the (a) spin orthogonal stripe (SOS) state, and (b) $C_{1/4}E_{3/4}$. These states are described in the main text. Readers are referred to the original reference (Dong *et al.*, 2009) for details. The spin stripes with different (relative) directions are highlighted by color blocks.

order across the stripe (Kivelson *et al.*, 1998; Tranquada *et al.*, 1995). The holes of this doped multiferroic system tend to accumulate in the regions where the domains touch, forming diagonal charge stripes. What is even more exciting is that this spin arrangement produces a non-zero cross product between the nearest-neighbor spins at the stripes. This non-zero value triggers the spin-orbit coupling and the associated lattice distortions, favoring a ferroelectric state as explained in the original reference (Dong *et al.*, 2009). Thus, this form of inhomogeneity leads to implications in the broad field of multiferroics, where apparently non-uniform charge states may also be of relevance, as suggested by the study described here.

Figure 9.19 (b) shows another interesting form of local self-organization. This is a state called $C_{1/4}E_{3/4}$. The state is an interpolation between the full E-state of undoped manganites and the CE state of half-doped manganites. It contains interesting zigzag chains with the same orientation of the spins within the chain, and antiferromagnetic between chains. Note that since all spins are collinear, the cross products are zero and no ferroelectricity is predicted.

The three interesting cases shown in this section provide just a glimpse of the many exotic inhomogeneous states known to exist in models for transition metal oxides. Having charge distributions that are not uniform at the nanoscale appears to be a generic feature of realistic models and real materials where there are competing tendencies: as a compromise of this frustration the spin, charge, and orbital patterns become very complex and highly non-trivial ground states are found.

9.5 Conclusions: a new paradigm has developed

This chapter described very briefly some of the most recent results obtained by our group in the study of inhomogeneous states in strongly correlated electrons. Readers should consult the cited literature, particularly the references therein, and the other chapters in this book to get a full view of the enormous effort and remarkable results that the community has been gathering in recent years in this field. Until the end of the 1990s, it was very unusual to use the language of inhomogeneities in the context of transition metal oxides. In those days the experimental evidence of such inhomogeneities was considered evidence of a poor sample, and as a consequence those features were not emphasized. However, in the mid and late 1990s theoretical work started showing evidence that both in clean systems or in the presence of weak quenched disorder, inhomogeneous states were formed. Moreover, evidence started gathering that non-linearities under the influence of weak external fields were expected and found in those inhomogeneous states. Now, after more than ten years of effort, it is very common to read literature on this subject where competing tendencies and inhomogeneous states are considered the reason for the exotic properties of a variety of materials. This clearly shows that ideas based on the electronic degree of freedom forming inhomogeneous states with complex charge, spin, and orbital patterns now form a new paradigm for the description of electronic properties of a variety of compounds. This by itself is a remarkable achievement, but the effort must continue in this area since the work is far from done. Ever more sophisticated microscopic experimental techniques are needed to visualize the nanometer-scale inhomogeneities of many of these materials, while theoretical techniques, particularly computational ones, need to vastly improve to handle the real-space complexity of these states. In addition, possible new functionalities of these inhomogeneous states should be explored. The future will be very bright if we continue our focus on this very interesting and challenging area of research.

Acknowledgements

This work was supported in part by the US Department of Energy Office of Basic Energy Sciences , Materials Science and Engineering Division and by the CNMS, sponsored by the Scientific User Facilities Division, BES–DOE.

References

Alvarez, G., Aliaga, H., Şen, C., and Dagotto, E. (2006). *Phys. Rev. B*, **73**, 224426.

Alvarez, G. and Dagotto, E. (2008). *Phys. Rev. Lett.*, **101**, 177001.

Alvarez, G., Mayr, M., Moreo, A., and Dagotto, E. (2005). *Phys. Rev. B*, **71**, 014514.

Buhler, C., Yunoki, S., and Moreo, A. (2000). *Phys. Rev. Lett.*, **84**, 2690–2693.

Dagotto, E. (2005). *Science*, **309**, 257–262.

Dagotto, E., Hotta, T., and Moreo, A. (2001). *Phys. Rep.*, **344**, 1–153.

Dagotto, E. and Tokura, Y. (2008). *MRS Bulletin*, **33**, 1037–1045.

Dong, S., Yu, R., Liu, J.-M., and Dagotto, E. (2009). *Phys. Rev. Lett.*, **103**, 107204.

Gomes, K. K., Pasupathy, A. N., Pushp, A., Ono, S., Ando, Y., and Yazdani, A. (2007). *Nature*, **447**, 569–572.

Hotta, T., Feiguin, A., and Dagotto, E. (2001). *Phys. Rev. Lett.*, **86**, 4922–4925.

Hotta, T., Malvezzi, A. L., and Dagotto, E. (2000). *Phys. Rev. B*, **62**, 9432–9452.

Hotta, T., Yunoki, S., Mayr, M., and Dagotto, Elbio (1999). *Phys. Rev. B*, **60**, R15009–R15012.

Kivelson, S. A., Fradkin, E., and Emery, V. J. (1998). *Nature*, **393**, 550–553.

Mayr, M., Alvarez, G., Moreo, A., and Dagotto, E. (2006). *Phys. Rev. B*, **73**, 014509.
Moraghebi, M., Buhler, C., Yunoki, S., and Moreo, A. (2001). *Phys. Rev. B*, **63**, 214513.
Moraghebi, M., Yunoki, S., and Moreo, A. (2002*a*). *Phys. Rev. B*, **66**, 214522.
Moraghebi, M., Yunoki, S., and Moreo, A. (2002*b*). *Phys. Rev. Lett.*, **88**, 187001.
Şen, C., Alvarez, G., and Dagotto, E. (2007). *Phys. Rev. Lett.*, **98**, 127202.
Şen, C., Alvarez, G., Aliaga, H., and Dagotto, E. (2006). *Phys. Rev. B*, **73**, 224441.
Tranquada, J. M., Sternlieb, B. J., Axe, J. D., Nakamura, Y., and Uchida, S. (1995). *Nature*, **375**, 561–563.
Yu, R., Dong, S., Şen, C., Alvarez, G., and Dagotto, E. (2008). *Phys. Rev. B*, **77**, 214434.

Part II

Superconductor-Insulator Transitions: Present Status and Open Questions

10
Superconductor-Insulator Transitions: Present Status and Open Questions

Nandini TRIVEDI

Department of Physics, The Ohio State University, 191 W Woodruff Avenue, Columbus, OH 43210

10.1 Why study superconductor–insulator transitions?

Superconductivity, as a loss of resistance at a transition temperature T_c, was first discovered in 1911 by Kamerlingh Onnes in liquid metallic mercury and understood almost half a century later by Bardeen, Cooper and Schrieffer in 1957 through their celebrated BCS theory. What emerged was an understanding of this new state of matter in which electrons are held together in a many-body phase-coherent state of pairs. A direct consequence of this pairing is a finite excitation gap above the ground state, although with a magnitude that is exponentially suppressed compared to the bare pairing scale. Furthermore the macroscopic phase coherence results in electron flow without resistance. Superconductivity is fragile—usually disappearing a few degrees above absolute zero—and yet it is ubiquitous, occurring in almost half of the elements of the periodic table. The BCS theory, is without doubt, one of the most significant and towering developments of the twentieth century.

After this initial understanding, a natural direction was to seek ways of enhancing T_c by simple alloying. A breakthrough occurred with the discovery of superconductivity in complex layered copper-oxide materials (Zaanen *et al.*, 2006) with a transition temperature T_c that was initially ~ 40 K and has by now skyrocketed to about 160 K. However, what is even more exciting than a high value of the transition temperature is that a new paradigm, completely different from the BCS theory, is necessary to understand superconductivity in the cuprates. Starting with a parent compound that is an antiferromagnetic Mott insulator, superconductivity emerges upon doping with a pairing scale that is completely divorced from the coherence scale, both in magnitude and in doping dependence (Anderson, 2007; Anderson *et al.*, 2004; Paramekanti *et al.*, 2001, 2004). For the first time, magnetism and superconductivity, which had sat in two different parts of the periodic table as two competing phases, were showing a curious interplay. Such an interplay between antiferromagnetism and superconductivity as well as ferromagnetism and superconductivity is ubiquitous in heavy fermion systems (Stewart, 1984). Another milestone was the discovery of paramagnon-mediated superfluidity in ^{3}He. This made it clear that the pairing "glue" did not have to be the retarded electron–phonon interaction, but could, in fact, have a magnetic origin.

It is evident that 100 years after its discovery in mercury, superconductivity continues to be an extremely active field, with the discoveries in magnesium diboride (Canfield and Crabtree, 2003), pnictides (Johannes, 2008), alkali-doped buckyballs (Rosseinsky *et al.*, 1991; Varma *et al.*, 1991), and most recently the development of topological superconductors (Hasan and Kane, 2010). Outside of condensed-matter physics, the concepts of spontaneously broken gauge symmetry and the resultant Higgs mechanism, and pairing in nuclei between quarks and in neutron stars, all continue to play important roles in high-energy, nuclear, and astro-physics.

In 1958 Anderson wrote his seminal paper on localization of (non-interacting) electron waves in random media because of quantum interference effects (Anderson, 1958). How is a superconductor affected by disorder? It was argued by Anderson that three-dimensional superconductivity is quite robust, persisting even in polycrystalline or amorphous materials. Two dimensions turns out to be particularly intriguing because it is the marginal dimension for localization and superconductivity. In other words, the interplay of localization and superconductivity brought forward a sharp question: does a two-dimensional (2D) system develop superconducting behavior when all its eigenstates are localized due to a random potential and, if so, what is the mechanism? It is seen from experiments that superconductivity in two dimensions does exist but can be destroyed by a large variety of tuning parameters including temperature, inverse thickness (characterized by sheet resistance) disorder, gate voltage, Coulomb blockade, perpendicular magnetic field, and parallel magnetic field. In some cases this results in a metallic state. But, more often, one obtains an insulating state where the conductivity exhibits weak localization, strong localization, or activated behavior. This is the superconductor–insulator transition (SIT), the topic of the latter half of this book.

The SIT is a quantum phase transition (Sachdev, 1999) of strongly correlated fermions, which is an important topic spanning condensed matter, ultracold atoms, and high-energy physics. As the coupling between the fermions increases, the BCS regime of many-body pairing crosses over to a bosonic regime of preformed pairs. Such regimes are accessible in strongly correlated superconductors such as the high-temperature superconductors and with possibly even greater tunability of interactions in cold-atom systems. Within the extreme limit of tightly bound Cooper pairs, the SIT has a bosonic analog: the superfluid–insulator transition. This problem has been investigated by studying superfluidity in helium in confined geometries, granular media, and porous media such as Vycor, aerogel, and xerogel (Chan *et al.*, 1996; Reppy, 1992). The destruction of superfluidity by rotation is analogous to the destruction of superconductivity by a magnetic field.

The earliest studies of the SIT were DC transport measurements (Haviland *et al.*, 1989; Hebard and Paalanen, 1990; Shahar and Ovadyahu, 1992) as a function of temperature and field. However, in recent years, a host of new experimental probes have become available, such as scanning tunneling spectroscopy (STS) (Sacépé *et al.*, 2008), penetration depth measurements at low frequencies (Hetel *et al.*, 2007), and AC conductivity measurements at microwave (Liu *et al.*, 2010) and terahertz frequencies (Valdés Aguilar *et al.*, 2010). At the time of writing, the subject is very much alive and there are many unanswered questions.

For a comprehensive description of the field, the reader may consult the excellent review article by Gantmakher and Dolgopolov (2010). In the following sections we introduce some basic concepts, and we then give a brief overview of the different types of SIT, with attention to the most recent developments. Due to constraints of time and space, it has not been possible to include references to every single work in this exciting field, and we sincerely regret any omissions.

10.2 Energy scales in a superconductor: the amplitude–phase dichotomy

A singlet s-wave superconductor is described by a complex order parameter $\Delta(\mathbf{R}) = |\Delta(\mathbf{R})|\, e^{i\theta(\mathbf{R})}$. At zero temperature the pairing amplitude $|\Delta(\mathbf{R})|$ takes a uniform value,

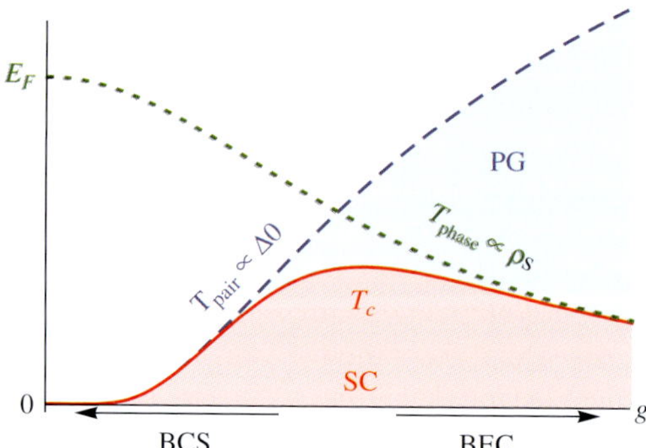

Fig. 10.1 Energy and temperature scales of a generic 2D superconductor as a function of attractive interaction g.

Δ_0. This is the energy scale associated with pairing. It typically manifests itself as an energy gap $E_g = \Delta_0$, and it also sets the maximum temperature, $T_{pair} = 0.57\Delta_0$, for the formation of Cooper pairs. On the other hand, the fluctuations of the phase $\theta(\mathbf{R})$ are controlled by the superfluid density (or phase stiffness) ρ_s. For 2D superconductors ρ_s has the dimensions of energy, and it can be directly interpreted as the energy scale for phase fluctuations.[1] ρ_s can be measured using mutual inductance techniques. It also sets the maximum temperature T_{phase} for long-range phase coherence.

Figure 10.1 shows the typical dependence of these scales on the attractive interelectron coupling g.[2] With increasing attraction, the pairing scale $T_{pair} \propto \Delta_0$ increases, whereas the phase-coherence scale $T_{phase} \propto \rho_s$ decreases (because tightly bound pairs have a larger effective mass in a lattice). The transition temperature T_c is determined by the lower of the two scales T_{pair} and T_{phase}.

Weak-coupling superconductors such as Al are described by the BCS limit of large overlapping Cooper pairs, where $T_{pair} \ll T_{phase}$. The phase-coherence scale is of the order of the Fermi energy, $T_{phase} \sim E_F \sim 10^4$ K, whereas the pairing scale is exponentially suppressed, $T_{pair} \sim 1$ K. Thus the critical temperature is determined by pairing, and we have a mean-field-like, amplitude-driven transition. Conversely, in the Bose–Einstein condensation (BEC) limit where $T_{pair} \gg T_{phase}$, the Cooper pairs are small and tightly bound, and the transition is determined by the temperature T_{phase}, above which phase coherence is lost. However, a pseudogap—a suppression in the density of states around the Fermi energy—persists up to a higher temperature T_{pair} (Randeria, 1995, 2010).

Although the crossover from BCS to BEC regimes is not directly relevant to superconductivity in the weakly coupled traditional superconductors, we will see that the concepts of the amplitude scale T_{pair} and phase scale T_{phase} are extremely useful as the SIT is approached. For example, transport measurements are sensitive to

[1] In this article we shall focus on the main ideas, glossing over factors of $\hbar$, e, k_B, and μ_B.

[2] A concrete model in which these scales can be defined would be the attractive 2D Hubbard model away from half-filling. The 2D continuum and half-filled 2D lattices are pathological.

global phase coherence and hence to $T_{\rm phase}$, whereas tunneling densities of states are sensitive to local pairing correlations. Moreover, in cold atoms and strongly coupled superconductors these concepts will be seen to play a central role. We will rely heavily on these ideas in Section 10.4.

10.3 Quantum phase transitions

As mentioned earlier, the SIT is an excellent example of a continuous quantum phase transition (QPT)—a zero-temperature phase transition driven by quantum zero-point fluctuations (Sachdev, 1999) controlled by a parameter g. For $g < g_{\rm c}$ the system is a superconductor with well-defined Bogoliubov quasiparticles and collective modes of a superconductor, whereas for $g > g_{\rm c}$ the system is an insulator, also with well-defined excitations. On the superconducting side, there is an energy scale—the 2D superfluid density or phase stiffness $\rho_{\rm s}$—that vanishes at $g_{\rm c}$. Similarly, on the insulating side, one generally expects an energy scale that vanishes at $g_{\rm c}$ (although it is not always clear what this scale is).

The QPT also profoundly affects behavior of the system at finite temperature. The superconducting transition temperature $T_{\rm c}$ decreases with increasing g, finally vanishing at the quantum critical point (QCP) at $g = g_{\rm c}$, as illustrated in Fig. 10.2. The fan emanating from the QCP is a quantum critical region that is neither superconducting nor insulating, where the spectrum is broad, and where quasiparticles are ill-defined. The primary aim of experimental and theoretical research in this field is to clearly elucidate the properties of the phases and excitations, and then go on to understand

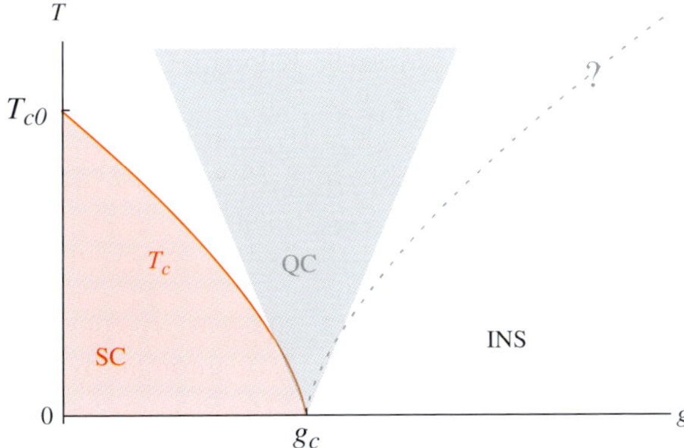

Fig. 10.2 Generic phase diagram of a SIT driven by a parameter g. For $g = 0$, thermal fluctuations produce an ordinary superconductor-to-normal transition at $T_{\rm c0}$. The superconducting transition temperature $T_{\rm c}$ decreases with increasing g and vanishes at the quantum critical point $g_{\rm c}$. Beyond $g_{\rm c}$ the system enters an insulating phase. The dashed line represents a (possibly unidentified) energy scale in the insulator that also vanishes at $g_{\rm c}$. The shaded gray region indicates the quantum critical (QC) region, where fluctuations are enhanced and QC scaling is obeyed.

the quantum critical region. For a more detailed commentary, see Section 11.2 on a scaling analysis of the SIT.

10.4 The disorder-tuned superconductor–insulator transition

10.4.1 Theories

We will begin with a discussion of the destruction of superconductivity by disorder, where the interplay of localization and superconductivity opens up a rich field of physics. A more detailed description is given in Chapter 17 and in Gantmakher and Dolgopolov (2010). Here we provide a brief summary, restricted to the case of singlet s-wave superconductivity and uncorrelated non-magnetic disorder in two dimensions, which nevertheless provides for rich physics.

Anderson's theorem

Let us begin with a clean s-wave superconductor and imagine turning on disorder gradually. The first effect of disorder is that scattering changes the "ballistic" plane-wave electronic eigenstates $|\mathbf{k}\rangle$ into "diffusive" eigenstates $|\alpha\rangle$. In the perturbative regime, when $k_F\ell \gg 1$ (where k_F is the Fermi wavevector and ℓ is the elastic mean-free path), the transition temperature T_c is basically unaffected by disorder![3] This surprising result (often referred to as "Anderson's theorem") was predicted by Anderson (1959) by considering pairing of electrons in time-reversed exact eigenstates $|\alpha\rangle$ and $|\bar{\alpha}\rangle$, and by Abrikosov and Gor'kov (1959) using diagram techniques.[4]

Superconductivity from localized eigenstates

As the elastic mean-free path ℓ becomes shorter than the clean coherence length, one passes from the clean limit to the dirty limit, and the true coherence length decreases (de Gennes, 1966; Tinkham, 1996). Using a Ginzburg–Landau approach, Kotliar and Kapitulnik demonstrated the growing importance of thermal fluctuations as the metal–insulator transition (MIT) is approached (Kapitulnik and Kotliar, 1985; Kotliar and Kapitulnik, 1986). As the disorder is further increased towards the Ioffe–Regel limit ($k_F\ell \sim 1$), all the single-particle eigenstates become localized. This corresponds to the MIT in the absence of interactions.[5] However, Ma and Lee (1985) showed that even in this regime superconductivity still persists, with a uniform pairing amplitude, as long as the condition that the number of states in a coherence volume $\rho_0\Delta_0\xi^d > 1$ is satisfied, where ρ_0 is the density of states at the Fermi energy, Δ_0 is the pairing

[3] In reality disorder can affect T_c by changing the density of states or electron–phonon coupling. There have in fact been reports of an *increase* of T_c in granular Al compared to bulk Al. We will not discuss such material specifics here.

[4] Abrikosov and Gor'kov further showed that *magnetic* impurities *do* reduce the T_c of an s-wave superconductor. In a d-wave superconductor, because of the changing sign of the order parameter on the Fermi surface, even non-magnetic impurities can dramatically degrade T_c. These effects are beyond the scope of this overview.

[5] In fact, the MIT in a 2D system occurs at zero disorder: an infinitesimal disorder strength is sufficient to produce weak localization.

amplitude of the clean superconductor, ξ is the localization length, and d is the dimensionality. Thus, the MIT does *not* imply a concomitant SIT!

Fermionic mechanism

Obviously, as one goes toward the limit of infinite disorder, where all electrons are localized on single sites, superconductivity must be destroyed. How does this happen? Two main paradigms have been proposed to explain this SIT. In the work of Finkel'stein (1994), disorder and Coulomb repulsion are treated using a diagrammatic renormalization approach within the assumption that the amplitude remains uniform across the system. This suggests an amplitude-driven "fermionic mechanism" for the SIT, in which disorder and Coulomb repulsion renormalize each other to higher values, reducing the pairing amplitude and ultimately driving T_c to zero.

The fermionic mechanism has had reasonable success in describing the general decrease of T_c. Later theoretical and experimental work, however, has suggested that in other systems the pairing amplitude becomes strongly inhomogeneous in the vicinity of the SIT, invalidating the uniform-amplitude assumption and pointing to the importance of phase fluctuations.

Bosonic mechanism

A complementary point of view was taken by Fisher *et al.* (1990), who modelled the SIT using charge-$2e$ bosons in a random potential. In this model, increasing disorder enhances phase fluctuations that eventually destroy long-range superconducting order to produce a localized Bose glass. This bosonic theory is expected to give the correct universal low-energy physics. Furthermore, it predicts a universal sheet resistance at the SIT of the order of the quantum of resistance $R_Q = h/(2e)^2$. However, it is a literal description only in the strong-coupling limit of tightly bound fermion pairs, whereas condensed matter systems are typically at weak coupling. The bosonic theory is also unable to make predictions for fermionic quantities, which are important experimental probes.

Attractive Hubbard model

With this context in mind, considerable progress has been made by considering a "minimal model for the SIT"—the disordered attractive Fermi–Hubbard model—in which Coulomb repulsion is omitted, but the attraction and disorder potential are treated as fully as possible. This model is discussed extensively in Chapter 17; here we give an overview. Ghosal, Randeria, and Trivedi (Ghosal *et al.*, 1998, 2001) developed Anderson's idea of "pairing of exact eigenstates" (PoEE) to extend all the way from weak disorder to strong disorder. They further performed Bogoliubov-de Gennes (BdG) calculations that self-consistently include spatial variations of the order parameter amplitude and electron density. The PoEE and BdG results were in qualitative agreement for single-particle properties across the whole disorder range. However, these methods are both mean-field treatments that neglect phase fluctuations. As the disorder increases, they predict that the superfluid density falls to very small values, but never vanishes; furthermore, the temperature scale for the development of

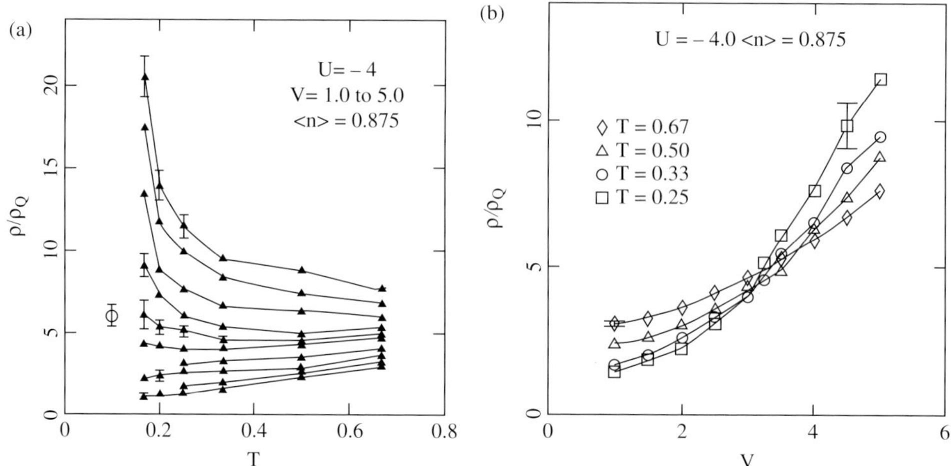

Fig. 10.3 Sheet resistance ρ as a function of temperature T and disorder strength V for the 2D attractive Hubbard model with a random potential, estimated from quantum Monte Carlo simulations (Trivedi *et al.*, 1996). There is (a) a separatrix in $\rho(T;V)$ and (b) a crossing point in $\rho(V;T)$. For $T < T_c(V)$ quantum Monte Carlo gives a finite superfluid stiffness, so an actual experiment would measure zero resistance, but this has not been incorporated in the plots.

spontaneous pairing amplitude remains roughly constant at large disorder (due to the existence of rare superconducting puddles). The phase fluctuations can be included, in a somewhat ad hoc manner, using the self-consistent harmonic approximation (SCHA), which renormalizes the small superfluid density to zero beyond a certain critical disorder, thus predicting a SIT. The BdG+SCHA results are qualitatively consistent with the phase diagram obtained from determinant quantum Monte Carlo (QMC)simulations (Scalettar *et al.*, 1999; Trivedi *et al.*, 1996). In principle, QMC provides the exact solution to the "minimal model" within statistical error bars. QMC calculations of the DC conductivity $\sigma(V;T)$ (Fig. 10.3) give a crossing point, suggesting that there is indeed a transition from a superconducting state ($\sigma = \infty$ for $T < T_c$) to an insulator ($\sigma \to 0$ when $T \to 0$). More recently, Bouadim *et al.* (2011) calculated one- and two-particle spectral functions using QMC and the maximum entropy method (MEM) for analytic continuation, which gave much additional insight into the nature of the phases and of the transition.

Unified view of the V-tuned SIT

We now summarize the present understanding of the 2D disorder-tuned SIT in Fig. 10.4, in an attempt to marry the various theories described above.

It is instructive to begin by recalling the *finite-temperature superconducting-normal transition* in a clean superconductor, illustrated in Fig. 10.4 (a). Within BCS theory, the energy gap E_g and the pairing amplitude (order parameter) Δ are equal, and $\Delta(T=0)/k_B$ is typically a few Kelvin. The superfluid stiffness ρ_s

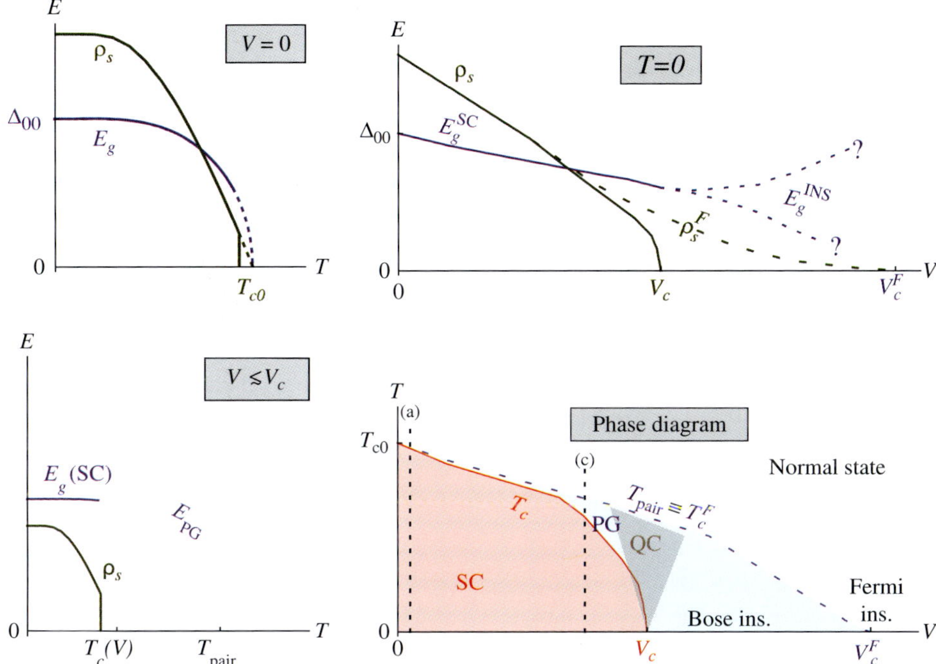

Fig. 10.4 A plausible scenario incorporating both the fermionic mechanism (Coulomb amplitude suppression) and the bosonic mechanism (phase fluctuations) for the disorder-tuned SIT. (a) Temperature dependence of gap E_g and superfluid stiffness ρ_s for a clean weak-coupling 2D superconductor. (b) With increasing disorder, the pairing amplitude develops a granular structure and ρ_s eventually falls below E_g. (c) In this regime, phase fluctuations are important. The gap below T_c becomes a pseudogap above T_c. (d) Phase diagram.

is of the order of the Fermi energy E_F, which is of the order of 10^4 Kelvin. Thus, $\rho_s \gg \Delta_0$. Phase fluctuations only become relevant in a tiny temperature range near T_c given by the Ginzburg criterion; because of phase fluctuations ρ_s drops to zero at $T_{BKT} \lesssim T_c$ according to the universal jump criterion for a Berezinskii–Kosterlitz–Thouless (BKT) transition. The quantities E_g, Δ, and ρ_s vanish almost simultaneously at T_c.

Now turn to the *zero-temperature superconductor-insulator transition* as a function of disorder V, illustrated in Fig. 10.4 (b). At weak disorder, where $\rho_s \gg \Delta_0$, the dominant effect is the Coulomb suppression of the uniform pairing amplitude predicted by Finkel'stein's theory. As the SIT is approached, the pairing amplitude $\Delta(\mathbf{R})$ becomes strongly inhomogeneous, so it is no longer meaningful to plot Δ versus V. However, the lowest energy gap E_g in the disordered inhomogeneous system is still a meaningful quantity. In this regime, invoking the results of the "minimal model" predicts that the stiffness ρ_s should fall to zero at a critical disorder strength V_c (which is *less* than the value V_c^F from Finkel'stein's theory), but the gap E_g remains finite across the SIT. This is consistent with the bosonic mechanism for the transition, where one passes from a superconductor of mobile pairs to a "Bose insulator" of localized pairs. In the disordered attractive Hubbard model (Bouadim *et al.*, 2011), increasing disorder V decreases the localization length ξ_{loc}, which confines the Cooper pairs to smaller volumes and causes the gap to *increase* as $U/2\xi_{\mathrm{loc}}^2$. On the other hand, disorder also enhances Coulomb repulsion, which may cause the gap to *decrease* towards zero as $V \to V_{\mathrm{F}}$. The actual behavior of the gap is a topic of further research.

Near the SIT, on the superconducting side, the pairing amplitude is highly inhomogeneous and the superfluid stiffness is small. Thus, *even in a weak-coupling superconductor*, disorder can drive the system into a regime where $\rho_{\mathrm{s}} \ll \Delta_0$ and phase fluctuations become important. The situation is illustrated in Fig. 10.4 (c). The critical temperature is governed by ρ_{s}, and BKT physics is pronounced. The gap, E_{g}, does not close at T_{c}; rather, it turns into a *pseudogap*, which is gradually filled in above T_{c} (Bouadim *et al.*, 2011).

Finally, consider the *disorder-temperature phase diagram* in Fig. 10.4 (d). Finkel'stein's result $T_{\mathrm{c}}^{\mathrm{F}}$ is accurate at weak disorder. However, as the SIT is approached, phase fluctuations reduce T_{c} below $T_{\mathrm{c}}^{\mathrm{F}}$. Eventually, T_{c} falls to zero at a critical disorder V_{c} that is less than V_{c}^{F}. It is plausible that $T_{\mathrm{c}}^{\mathrm{F}}$ now plays the role of the temperature scale T_{pair} for *precursor pairing*. Then, one expects that for $T_{\mathrm{c}}(V) < T < T_{\mathrm{pair}}(V)$ there is a *pseudogap phase*, in which the density of states is suppressed near the Fermi level even though $\rho_{\mathrm{s}} = 0$. A pseudogap was indeed found in the QMC+MEM calculations of Bouadim *et al.* (2011).

In the above we have presented a plausible picture, with a crossover from amplitude suppression far from the SIT to phase-fluctuation-dominated behavior near the SIT described by PoEE+SCHA, BdG+SCHA, QMC, and the bosonic model. A "grand unified theory" of the SIT, in which phonon-mediated attraction, disorder, and Coulomb repulsion are treated on an equal footing, is yet to come. Such a theory may be necessary to obtain quantitative fits to experimental data.

Emergent inhomogeneity

A notable feature of the BdG calculations is the prediction of "emergent electronic granularity" in a "structurally amorphous" material. Even for homogeneous disorder (an uncorrelated random potential at every site), the pairing amplitude develops inhomogeneity on the scale of the coherence length ξ_0, forming "superconducting islands" or "puddles" weakly connected by the Josephson effect. This is illustrated in Fig. 10.5 (a). The SIT is associated with the loss of phase coherence between these islands.

Pseudogap

As mentioned earlier, QMC+MEM give a pseudogap in the density of states, as illustrated in Fig. 10.6.

One should bear in mind that adding ingredients to the minimal model may change the physics. For example, if the on-site attraction U is allowed to vary from site to site, there is a possibility of a superconductor-to-metal transition where the single-particle

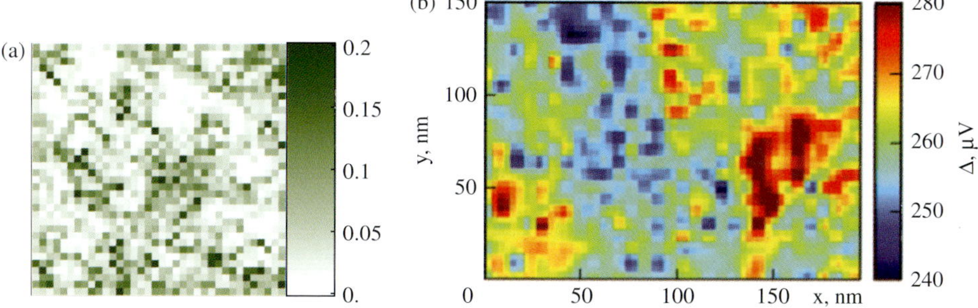

Fig. 10.5 (a) BdG results for the pairing amplitude $\Delta(\mathbf{R})$ for a 36×36 model near the SIT, showing puddles on the scale of the coherence length. (b) Spatial variation of the gap $E_g(\mathbf{R})$ measured by scanning tunneling spectroscopy on a TiN film. From Sacépé *et al.* (2011).

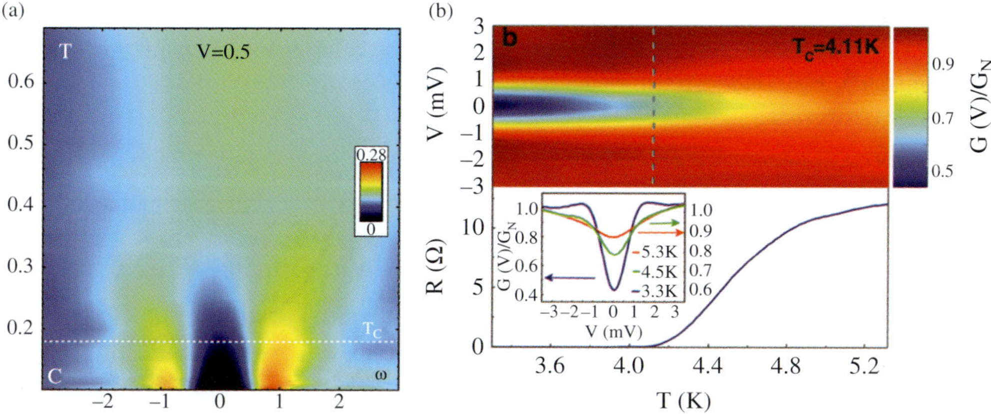

Fig. 10.6 (a) average density of states $N(\omega; T)$ as a function of energy $E = \hbar\omega$ and temperature T, calculated using QMC+MEM for a value of disorder V on the superconducting side of the SIT (Bouadim *et al.* (2011)). A pseudogap (suppression in the density of states) is clearly visible up to 2–3 times the critical temperature T_c (white line). Top (b) a similar pseudogap in the tunnel conductance $G(V; T)$ of a NbN film as a function of energy $E = eV$ and temperature T, measured using STS and averaged over 32 points on the film (Mondal *et al.* (2011)).

gap closes (see Chapter 5) The nature of the randomness in the potential and in the attractive interaction ultimately depends on material specifics.

The SIT has also been studied (Feigel'man *et al.*, 2007) in terms of the fractality of the non-interacting electron eigenstates near the MIT (Evers and Mirlin, 2008). This is a useful approach provided that the SIT and MIT are in proximity.

10.4.2 Experiments

Thin-film superconductors

The study of the SIT in thin films has a long history (Haviland *et al.*, 1989; Imry and Strongin, 1981; Strongin *et al.*, 1970). Much of the experimental work has involved quench-condensed films, which have an amorphous or granular structure depending on deposition parameters. For 2D films (thinner than the coherence length), the disorder strength is usually characterized by the sheet resistance $R_\square$. Increasing the deposition thickness or annealing time decreases $R_\square$, and hence decreases the effective disorder. Chemical doping can also be used.

The nature of the observed SIT depends on material and microstructure, as illustrated in Fig. 10.7 (and also in Section 1.3 of Gantmakher and Dolgopolov (2010)). In amorphous films with "homogeneous disorder" on an atomic scale (of materials such as Bi, Al, Pb, and MoGe), the superconductor typically exhibits a BCS-like transition where $R_\square(T)$ is constant above T_c and suddenly drops to zero below T_c, and the insulator is only weakly localized. Tunneling studies in such films are consistent with the *fermionic* mechanism, in which the gap closes at the SIT (Valles *et al.*, 1992) (remembering that the latter is defined according to transport behavior). It is possible that in these materials the critical region where phase fluctuations become important is quite small, and very careful measurements are necessary to access it.

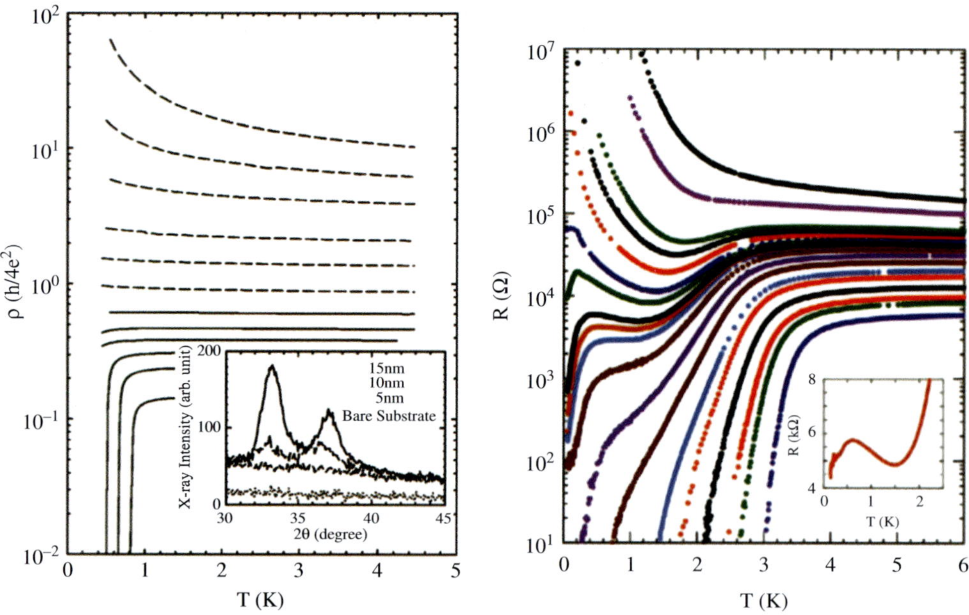

Fig. 10.7 Sheet resistance as a function of temperature and thickness, for amorphous Ta films (Qin *et al.*, 2006) and granular Bi films (Parendo *et al.*, 2007). The shapes of the curves are typical for amorphous and granular films respectively.

Conversely, in granular films (Bi, Ga, Al) (Lin and Goldman, 2010), nanostructured Bi (Nguyen *et al.*, 2009), amorphous TiN (Sacépé *et al.*, 2010), InO_x (Sacépé *et al.*, 2008, 2011; Shahar and Ovadyahu, 1992), and NbN films (Mondal *et al.*, 2011), the superconducting transition in $R_\square(T)$ is much broader, and the insulator exhibits activated transport (Lin and Goldman, 2010; Shahar and Ovadyahu, 1992). Furthermore, it appears that the gap, measured using a tunnel junction or scanning tunneling spectroscopy, *remains finite* as the SIT is approached, consistent with the *bosonic* mechanism. Scanning tunneling spectroscopy data, showing significant inhomogeneity of the gap, further supports the aforementioned picture of emergent granularity; see Fig. 10.5. Although coherence peaks in the density of states disappear at about T_c, a pseudogap persists up to much higher temperatures (Mondal *et al.*, 2011; Sacépé *et al.*, 2010), in agreement with theory (Bouadim *et al.*, 2011); see Fig. 10.6.[6]

Of course, this is a gross oversimplification of a complicated experimental situation. Nevertheless, it is encouraging that some general trends can be identified, and the dichotomy between amplitude-/phase-driven transitions and fermionic/bosonic mechanisms seems to be a useful paradigm (see Chapter 13).

So far we have focused on bulk transport and tunneling experiments. Much useful insight has also been obtained from other probes, such as the superfluid density (Hetel *et al.*, 2007), microwave conductivity (Liu *et al.*, 2010), and terahertz conductivity (Valdés Aguilar *et al.*, 2010). For example, the finite-temperature and finite-frequency conductivity of InO_x films is consistent with BKT scaling (Liu *et al.*, 2010). It is worth noting that some of these experiments (Hetel *et al.*, 2007; Valdés Aguilar *et al.*, 2010) have been performed on high-temperature cuprate superconductors, which are very different from traditional superconductors. Somewhat ironically, the obvious quantum critical scaling between T_c and $\rho_s(0)$ at the SIT in d-wave superconductors (Hetel *et al.*, 2007) remains to be tested experimentally in s-wave superconducting films.

Ultracold atoms in random potentials

Cold atomic gases offer a high degree of tunability and controllability, and they are becoming a promising option for emulating condensed-matter systems. The realization of attractive fermions in a disordered potential is still some way off. However, disordered boson systems can, and have, been produced. Two groups have already observed Anderson localization of *non-interacting* bosons in random potentials produced by laser speckle (Aspect and Inguscio, 2009). More relevantly, the *repulsive* Bose–Hubbard model in a random potential has been realized in an optical lattice. The superfluid–Bose glass transition has been studied (Pasienski *et al.*, 2010) and compared with theory (Fisher *et al.*, 1989; Gurarie *et al.*, 2009; Krauth *et al.*, 1991; Pollet *et al.*, 2009). This may certainly lead to a deeper understanding of the disorder-tuned SIT.

[6] A practical limitation of tunneling experiments is the inability to measure the tunnel resistance into an insulator, this being swamped by the sheet resistance of the insulator itself. The density of states of an insulator is nevertheless a well-defined theoretical quantity, and it is conceivable that experimental ingenuity may one day find a way to measure it.

10.5 Superconductor–insulator transitions in clean systems

While the history of the SIT has its roots in amorphous and granular films, developments of new materials and new areas in physics have made it possible to study SITs in clean systems.

Josephson junction arrays

Regular Josephson junction arrays (Rimberg *et al.*, 1997) exhibit a SIT driven by the competition between charging energy (the Coulomb blockade effect, which localizes Cooper pairs) and Josephson coupling (which tends to delocalize Cooper pairs). This is probably the most well-understood SIT, and we will refer the reader to Section 5.1 of Gantmakher and Dolgopolov (2010) for a detailed discussion and further references. Nano-honeycomb Bi films (Stewart Jr *et al.*, 2007) can be thought of as a type of Josephson junction array. Their transport exhibits a broadened superconducting transition and activated behavior on the insulating side, in line with the "granular" classification mentioned earlier (see Chapter 13).

Cold atoms

The superfluid–Mott-insulator transition in the clean Bose–Hubbard model has been studied extensively (Capogrosso-Sansone *et al.*, 2007; Kato *et al.*, 2008) and realized in cold atom experiments (Greiner *et al.*, 2002; Spielman *et al.*, 2007). See Section 5.2 of Gantmakher and Dolgopolov (2010).

Gate-tuned SITs

By applying a gate voltage to a thin film, one can perform electrostatic doping to control the carrier density. This can be done in a clean system (so it is not merely equivalent to tuning the disorder indirectly via Coulomb screening).

In the past year there have been extremely noteworthy developments in this area. One of these is the use of ionic liquids as gate materials in an electric double layer

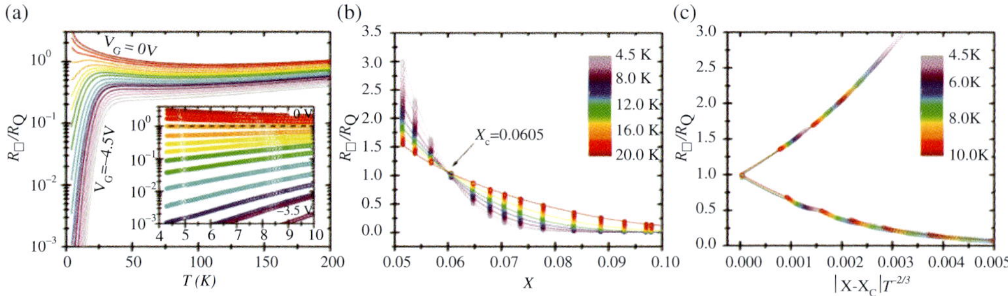

Fig. 10.8 Gate-tuned SIT in $La_{2-x}Sr_xCuO_4$ (from Bollinger et al.), as a function of temperature T and carrier concentration x, showing (a) a separatrix in $R(T,x)$, (b) a crossing point in $R(x,T)$, and (c) quantum critical scaling collapse. The universal critical sheet resistance is equal to the pair quantum resistance, $R_Q = h/(2e)^2 = 6.45\text{k}\Omega$.

transistor configuration. By using an ionic liquid in place of a solid dielectric, one can achieve very large gate capacitances while avoiding dielectric breakdown, thus achieving charge densities two orders of magnitude higher. In this way, Bollinger *et al.* (2011) succeeded in doping $La_{2-x}Sr_xCuO_4$ monolayer films through a SIT. They observed a 2D SIT consistent with the bosonic theory, and they were even able to confirm quantum critical scaling and identify the critical resistance as $R_\square = (6.45 \pm 0.10)$ kΩ (see Fig. 10.8). In another beautiful study, Lee *et al.* (2011) studied single crystals of $SrTiO_3$ over a large range of doping and carrier concentrations, obtaining insulating, conducting, and superconducting states.

10.6 The parallel-field-tuned superconductor–insulator transition

We now turn to SITs driven by magnetic fields. In a singlet s-wave superconductor, Pauli paramagnetism is absent, and there is Meissner diamagnetism. An external magnetic field combats both of these phenomena: it tends to align the spins parallel (spin effect), and to induce screening currents (orbital effect). It is of interest to study these two effects separately.

A magnetic field that couples only to the spin—a "Zeeman field"—can be engineered by applying a field $h_\parallel$ parallel to a film to minimize orbital depairing. At low fields there is a fully paired superconducting state with zero spin susceptibility. At high fields, the ground state is a polarized Fermi liquid with unequal Fermi surfaces for majority and minority spin species; this is an insulator due to weak localization in two dimensions. Hence, in two dimensions the $h_\parallel$-tuned transition is indeed an SIT.

It has usually been assumed that there is a first-order hysteretic spin paramagnetic transition at a field $h_{CC} \approx \Delta_0/\sqrt{2}$, as predicted by Chandrasekhar and Clogston (Chandrasekhar, 1962; Clogston, 1962). See Chapter 15 on spin effects for details. However, there have been many theoretical predictions that the first-order transition should be pre-empted by second-order transitions to an unusual Fulde-Ferrell-Larkin-Ovchinnikov (FFLO) state, in which pairing and magnetization coexist in a microscale self-organized pattern (Burkhardt and Rainer, 1994; Fulde and Ferrell, 1964; Larkin and Ovchinnikov, 1964; Loh and Trivedi, 2010; Machida and Nakanishi, 1984; Yoshida and Yip, 2007). Interest in FFLO physics crosses traditional boundaries between condensed matter, cold atomic gases (Radzihovsky and Sheehy, 2010), quantum chromodynamics (Casalbuoni and Nardulli, 2004), nuclear physics, and astrophysics (Alford *et al.*, 2001), and there is currently an intense effort to search for FFLO phases in superconductors as well as in cold atoms (Liao *et al.*, 2010). See Chapter 17 for further details.

In condensed-matter systems the situation is complicated by the presence of disorder. Nonetheless, there is evidence that—for some systems and parameters—the $h_\parallel$-tuned SIT occurs via a disordered FFLO phase (dLO) in which the gap is *filled in* by Andreev bound states (Cui and Yang, 2008; Loh *et al.*, 2011; Yanase, 2009) (see Fig. 10.9). This stands in contrast to both the fermionic mechanism for the disorder-tuned SIT, where the gap *closes*, and the bosonic mechanism, where the gap *persists*.

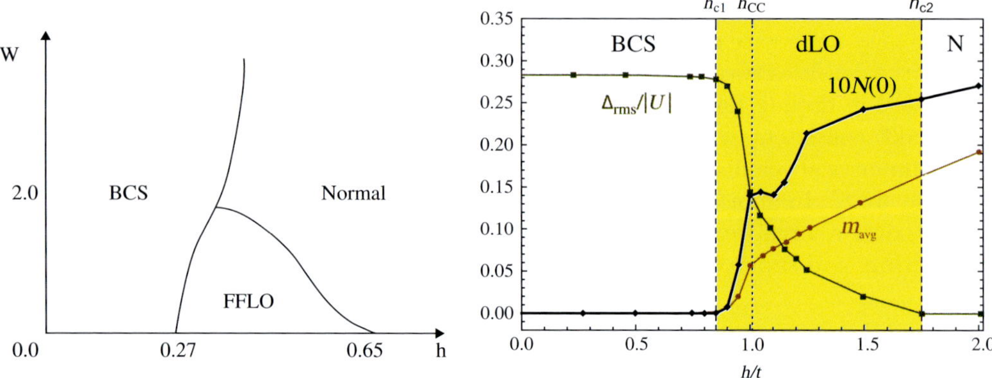

Fig. 10.9 Left: approximate BdG phase diagram as a function of Zeeman field h and disorder strength W, showing a sizeable disordered LO (dLO) regime Cui and Yang (2008). Right: root-mean-square BdG pairing amplitude Δ_{rms}, average magnetization m_{avg}, and Fermi-level density of states $N(0)$ as functions of Zeeman field h. For $h_{c1} < h < h_{c2}$ there is a disordered FFLO state with coexistent pairing and magnetization, in which the gap is partially filled in (Loh *et al.*, 2011).

Figure 10.10 is a schematic showing the evolution of the density of states across the $h_{\|}$-tuned SIT, according to our present understanding. At weak fields the density of states has BCS coherence peaks at $\pm\Delta \pm h$ and a hard gap. For intermediate fields there may be a dLO state with Andreev bound states in the gap. At high fields there is an intriguing and non-trivial feature at $E^* = h + \sqrt{h^2 - \Delta^2/4}$ called the pairing resonance, which is a signature of superconducting fluctuations in the normal state. In addition, there is an Altshuler–Aronov zero-bias anomaly due to a combination of disorder and electron-electron interaction. This zero-bias anomaly is closely related to the weak localization corrections that suppress the conductivity $\sigma(\omega)$ to zero at $\omega = 0$, leading to insulating behavior in the normal state. See Chapter 15 for details.

Figure 10.11 illustrates a dLO state, which consists of domains of positive and negative pairing amplitude separated by magnetized domain walls. At the time of writing we are not aware of any STS experiments on superconducting films in pure Zeeman fields.

The $h_{\|}$-tuned transition has been studied from other points of view (Dubi *et al.*, 2007, 2008; Lopatin *et al.*, 2005; Zhou and Spivak, 1998). These are not necessarily inconsistent with the FFLO paradigm. For example, considering the free-energy landscape of the superconducting-phase field leads to an XY spin glass (Zhou and Spivak, 1998), whose ground state is very similar to the dLO states obtained in BdG simulations. The islands of large $\Delta(\mathbf{R})$ reported in Dubi *et al.* (2007; 2008) are possibly related to the domains of positive and negative Δ in a dLO state.

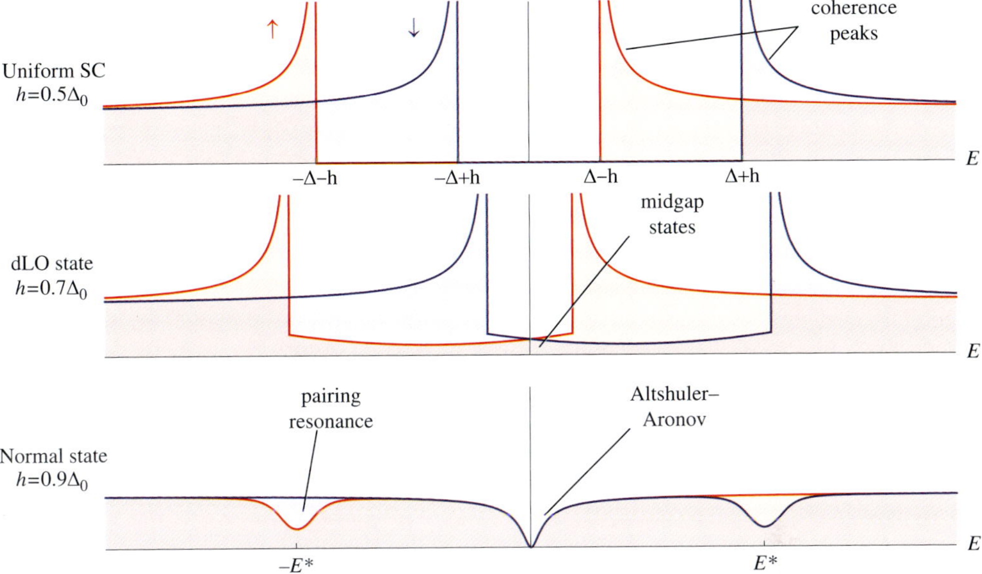

Fig. 10.10 Schematic of the density of states $N(E)$ of a disordered superconducting film with increasing parallel field. SC, superconductor.

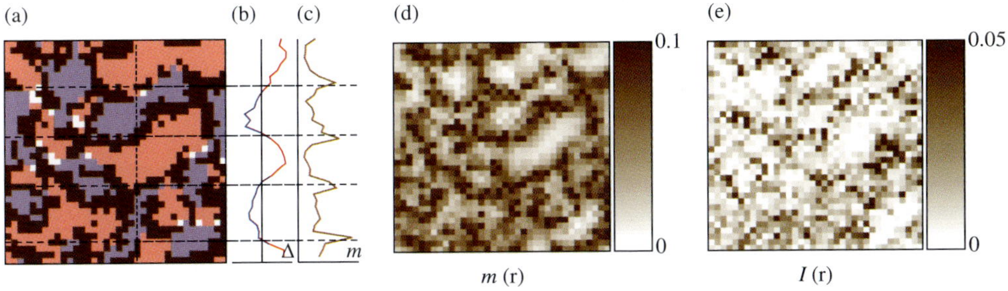

Fig. 10.11 (a) Combined plot of magnetization $m(\mathbf{r})$ and pairing amplitude $\Delta(\mathbf{r})$ in a dLO state obtained from BdG on a 36×36 lattice. Red (blue) indicates regions where $\Delta(\mathbf{r})$ is large and positive (negative). Brown regions, where the magnetization $m(\mathbf{r})$ is large, occur at domain walls where Δ changes sign. White regions are hills or valleys of the disorder potential corresponding to empty sites or localized pairs that participate in neither superconductivity nor magnetism. (b) and (c) show oscillations of Δ and m along the vertical dashed line in (a). (d) and (e) show the correspondence between magnetization $m(\mathbf{r})$ and low-energy spectral weight $I(\mathbf{r}) = \int_{-0.1t}^{0.1t} dE\ N_{\mathbf{r}}(E)$.

10.7 The perpendicular-field-tuned superconductor–insulator transition

The basic effect of a perpendicular magnetic field $h_\perp$ on a superconductor is to induce screening currents via the orbital effect. A sufficiently large $h_\perp$ generates vortices—topological defects—such that the phase winds by 2π around a normal vortex core. With increasing magnetic field, the normal cores overlap and superconductivity disappears. Whether the system is superconducting, metallic, or insulating depends on whether the vortices are pinned or mobile. Vortex matter is the source of much complexity; for example, see Chapter 18 on vortices in superconductors near a metallic gate. Theoretical work on the $h_\perp$-tuned SIT includes (Dubi *et al.*, 2007, 2008; Galitski *et al.*, 2005; Shah and Lopatin, 2007; Spivak and Zhou, 1995).

In some systems, it appears that the magnetic field first produces a transition to an insulator of localized pairs (with activated transport), and at high fields the

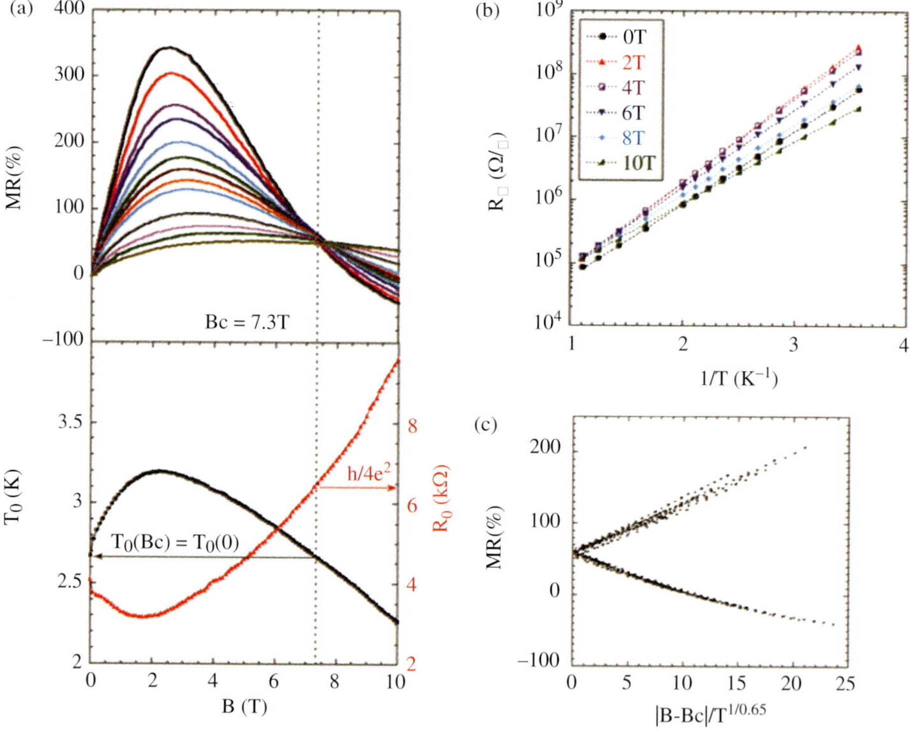

Fig. 10.12 Perpendicular-field-tuned SIT in amorphous Bi films [Lin et al. (2011)]. The magnetoresistance (MR), as a function of field B for various temperatures T from 300 mK to 900 mK, exhibits a crossing point (a) It obeys quantum critical scaling (c) The resistances can be fit by an Arrhenius form $R = R_0 \exp(T_0/T)$ (b). It obeys quantum critical scaling (c). At the SIT ($B = B_c$), the prefactor R_0 is equal to the pair quantum resistance $R_Q = h/(2e)^2 = 6.45$ kΩ.

pairs break up to reveal a weakly localized insulator (or even a normal metal in which $h_\perp$ has suppressed weak localization). Thus there can be a superconductor–insulator–metal transition. This unusual sequence of states shows up in experiments as a magnetoresistance peak as tall as nine orders of magnitude. Figure 10.12 shows a detailed characterization of the $h_\perp$-tuned SIT in amorphous Bi (Lin and Goldman, 2011). For a more in-depth discussion, see Chapter 12 on the novel insulating phase induced by a perpendicular field.

Interestingly, signs of Cooper pair localization also appear in $h_\perp$-tuned transitions in homogeneous films, whose disorder-tuned transitions are dominated by the fermionic mechanism. Clear pairing fluctuations emerge in the transport (see Chapter 14) and a pseudogap persists in the density of states (see Chapter 13).

In regular Josephson junction arrays and specially patterned amorphous Bi films, commensurability effects lead to yet more types of vortex matter. The period of well-defined magnetoresistance oscillations is strong evidence for charge-$2e$ pairs on the insulating side of the transition (Stewart Jr *et al.*, 2007); see Chapter 13.

In neutral systems such as ^{4}He (Bewley *et al.*, 2006; Yarmchuk *et al.*, 1979), ^{3}He, dilute Bose superfluids (Madison *et al.*, 2000; Matthews *et al.*, 1999), and dilute Fermi superfluids (Lin *et al.*, 2009; Zwierlein *et al.*, 2005), superfluidity can be destroyed by stirring to cause rotation, or by an artificial gauge field cleverly constructed using lasers. In particular, experiments on superfluid helium in porous media (Chan *et al.*, 1996; Reppy, 1992) are closely analogous to SITs in electronic systems.

10.8 Conclusions and open questions

To summarize, much progress has been made in the field of SITs since its inception. However, there are still some open questions (each question applies separately to each type of SIT):

- What is the nature of the normal state? Is it a Fermi liquid, or a quantum critical phase with a pseudogap?
- What is the nature of the insulator? Is it a localized Anderson insulator, a Mott insulator, a Fermi glass, a Bose glass, or something else?
- What are the energy scale(s) in the insulator that vanish at the SIT?
- What is the universality class (critical exponents, amplitude ratios, and scaling functions)?
- Is the critical resistance universal and equal to the quantum of resistance $R_Q = h/(2e)^2$? If so (as in Bollinger *et al.* (2011)), does this indicate self-duality?
- How do the single-particle spectral functions and dynamical conductivity behave in the superconductor, the insulator, and near the SIT?
- Can one develop a theory that captures both Coulomb amplitude suppression and phase fluctuations, and thus unify the fermionic and bosonic pictures of the SIT?
- Does the SIT happen concomitantly with the MIT or not?
- What is the origin of the gigantic peak in the magnetoresistance in the insulating state?

A large number of experiments exploring SITs in conventional s-wave materials have focused on transport measurements. In order to get deeper insights into the phases, it is useful to bring to bear a larger variety of experiments on SIT, similar to the activity on the high-T_c cuprates, such as a detailed and systematic analysis of STS and dynamical conductivity, some of which have already begun. In the cuprates, considerable insight about the pseudogap region was obtained by measuring the entropy carried by vortices from the Nernst effect (Wang *et al.*, 2003). Similar experiments for the regular SIT should also be extremely useful (Pourret *et al.*, 2006). It is evident that there should be large diamagnetic effects in the insulator because of puddles of superconducting regions embedded in an insulating matrix. Once again direct probes of diamagnetism in the insulator and its evolution toward the SIT, similar to the explorations in the cuprates (Li *et al.*, 2010; Wang *et al.*, 2005), should be extremely instructive. Unlike the disorder-driven transition, the effect of a perpendicular magnetic field is not as well understood. The experiments show that the sharp normal-to-superconductor transition at T_c for zero field develops long tails in the presence of a magnetic field. In other words there appears to be some source of dissipation in the superconducting state. Is this due to unpinned vortices? Why is the disorder, even when it is as high as the quantum of resistance, unable to pin the vortices?

And finally, once the phases are understood, the next frontier will be investigations of the quantum phase transition. In this context the ability to tune across the transition by gating as discussed above is a breakthrough. Gating allows one to separately tune the effects of screening and disorder and will hopefully provide a much better understanding of quantum criticality.

References

Abrikosov, A.A. and Gor'kov, L.P. (1959). *Soviet Physics JETP*, **9**, 220–221. (Zhurnal Eksperimental'noi i Teoreticheskoi Fiziki **36**, 319–320).

Alford, M., Bowers, J.A., and Rajagopal, K. (2001). *Phys. Rev. D*, **63**, 074016.

Anderson, P.W. (1958). *Phys. Rev.*, **109**, 1492–1505.

Anderson, P.W. (1959). *J. Phys. Chem. Solids*, **11**, 26–30.

Anderson, P.W. (2007). *Science*, **316**, 1705–1707.

Anderson, P.W., Lee, P.A., Randeria, M., Rice, T.M., Trivedi, N., and Zhang, F.C. (2004). *J. Physics: Cond. Matter*, **16**, R755.

Aspect, A. and Inguscio, M. (2009). *Phys. Today*, **62**, 30.

Bewley, G.P., Lathrop, D.P., and Sreenivasan, K.R. (2006). *Nature*, **441**, 588.

Bollinger, A.T., Dubuis, G., Yoon, J., Pavuna, D., Misewich, J., and Bozovic, I. (2011). *Nature*, **472**, 458–460.

Bouadim, K., Loh, Y.L., Randeria, M., and Trivedi, N. (2011). arxiv: 1011.3275. Accepted in *Nature Physics*.

Burkhardt, H. and Rainer, D. (1994). *Ann. Physik*, **3**, 181.

Canfield, P.C. and Crabtree, G.W. (2003). *Phys. Today*, **56**, 34–40.

Capogrosso-Sansone, B., Prokof'ev, N.V., and Svistunov, B.V. (2007). *Phys. Rev. B*, **75**, 134302.

Casalbuoni, R. and Nardulli, G. (2004). *Rev. Mod. Phys.*, **76**, 263–320.
Chan, M., Mulders, N., and Reppy, J. (1996). *Phys. Today*, **49**, 30–37.
Chandrasekhar, B.S. (1962). *Appl. Phys. Lett.*, **1**, 7–8.
Clogston, A.M. (1962, Sep). *Phys. Rev. Lett.*, **9**, 266–267.
Cui, Q.H. and Yang, K. (2008). *Phys. Rev. B*, **78**, 054501.
de Gennes, P.G. (1966). *Superconductivity in Metals and Alloys*. Benjamin, New York.
Dubi, Y., Meir, Y., and Avishai, Y. (2007). *Nature*, **449**, 876–880.
Dubi, Y., Meir, Y., and Avishai, Y. (2008). *Phys. Rev. B*, **78**, 024502.
Evers, F. and Mirlin, A.D. (2008). *Rev. Mod. Phys.*, **80**, 1355–1417.
Feigel'man, M.V., Ioffe, L.B., Kravtsov, V.E., and Yuzbashyan, E.A. (2007). *Phys. Rev. Lett.*, **98**, 027001.
Finkel'stein, A.M. (1994). *Physica B*, **197**, 636–648.
Fisher, M.P.A., Grinstein, G., and Girvin, S.M. (1990). *Phys. Rev. Lett.*, **64**, 587.
Fisher, M.P.A., Weichman, P.B., Grinstein, G., and Fisher, D.S. (1989). *Phys. Rev. B*, **40**, 546–570.
Fulde, P. and Ferrell, R.A. (1964). *Phys. Rev.*, **135**, A550.
Galitski, V.M., Refael, G., Fisher, M.P.A., and Senthil, T. (2005). *Phys. Rev. Lett.*, **95**, 077002.
Gantmakher, V.F. and Dolgopolov, V.T. (2010). *Physics-Uspekhi*, **53**, 3.
Ghosal, A., Randeria, M., and Trivedi, N. (1998). *Phys. Rev. Lett.*, **81**, 3940.
Ghosal, A., Randeria, M., and Trivedi, N. (2001). *Phys. Rev. B*, **65**, 014501.
Greiner, M., Mandel, O., Esslinger, T., Hansch, T.W., and Bloch, I. (2002). *Nature*, **415**, 39–44.
Gurarie, V., Pollet, L., Prokof'ev, N.V., Svistunov, B.V., and Troyer, M. (2009). *Phys. Rev. B*, **80**, 214519.
Hasan, M.Z. and Kane, C.L. (2010). *Rev. Mod. Phys.*, **82**, 3045–3067.
Haviland, D.B., Liu, Y., and Goldman, A.M. (1989). *Phys. Rev. Lett.*, **62**, 2180.
Hebard, A.F. and Paalanen, M.A. (1990). *Phys. Rev. Lett.*, **65**, 927–930.
Hetel, I., Lemberger, T.R., and Randeria, M. (2007). *Nat. Phys.*, **3**, 700–702.
Imry, Y. and Strongin, M. (1981). *Phys. Rev. B*, **24**, 6353–6360.
Johannes, M. (2008). *Physics*, **1**, 28.
Kapitulnik, A. and Kotliar, G. (1985). *Phys. Rev. Lett.*, **54**, 473–476.
Kato, Y., Zhou, Q., Kawashima, N., and Trivedi, N. (2008). *Nat. Phys.*, **4**, 617–621.
Kotliar, G. and Kapitulnik, A. (1986). *Phys. Rev. B*, **33**, 3146–3157.
Krauth, W., Trivedi, N., and Ceperley, D. (1991). *Phys. Rev. Lett.*, **67**, 2307–2310.
Larkin, A.I. and Ovchinnikov, Y.N. (1964). *Zh. Eksp. Teor. Fiz.*, **47**, 1136. (Sov. Phys. JETP **20**, 762 (1965)).
Lee, Y., Clement, C., Hellerstedt, J., Kinney, J., Kinnischtzke, L., Leng, X., Snyder, S.D., and Goldman, A.M. (2011). *Phys. Rev. Lett.*, **106**, 136809.
Li, L., Wang, Y., Komiya, S., Ono, S., Ando, Y., Gu, G.D., and Ong, N.P. (2010). *Phys. Rev. B*, **81**, 054510.
Liao, Y.A, Rittner, A.S.C., Paprotta, T., Li, W.H., Partridge, G.B., Hulet, R.G., Baur, S.K., and Mueller, E.J. (2010). *Nature*, **467**, 567–569.
Lin, Y.-J., Compton, R.L., Jimenez-Garcia, K., Porto, J.V., and Spielman, I.B. (2009). *Nature*, **462**, 628–632.

Lin, Y.H. and Goldman, A.M. (2010). *Phys. Rev. B*, **82**, 214511.

Lin, Y.H. and Goldman, A.M. (2011). *Phys. Rev. Lett.*, **106**, 127003.

Liu, W., Kim, M., Sambandamurthy, G., and Armitage, N.P. (2010). arXiv: 1010.2996.

Loh, Y.L., Trivedi, N., Xiong, Y.M., Adams, P.W., and Catelani, G. (2011). arXiv: 1102.3889. Accepted in *Phys. Rev. Lett.*

Loh, Yen Lee and Trivedi, N. (2010). *Phys. Rev. Lett.*, **104**, 165302.

Lopatin, A.V., Shah, N., and Vinokur, V.M. (2005). *Phys. Rev. Lett.*, **94**, 037003.

Ma, M. and Lee, P.A. (1985). *Phys. Rev. B*, **32**, 5658–5667.

Machida, K. and Nakanishi, H. (1984). *Phys. Rev. B*, **30**, 122–133.

Madison, K.W., Chevy, F., Wohlleben, W., and Dalibard, J. (2000). *Phys. Rev. Lett.*, **84**, 806–809.

Matthews, M.R., Anderson, B.P., Haljan, P.C., Hall, D.S., Wieman, C.E., and Cornell, E.A. (1999). *Phys. Rev. Lett.*, **83**, 2498–2501.

Mondal, M., Kamlapure, A., Chand, M., Saraswat, G., Kumar, S., Jesudasan, J., Benfatto, L., Tripathi, V., and Raychaudhuri, P. (2011). *Phys. Rev. Lett.*, **106**, 047001.

Nguyen, H.Q., Hollen, S.M., Stewart, M.D., Shainline, J., Yin, Aijun, Xu, J.M., and Valles, J.M. (2009). *Phys. Rev. Lett.*, **103**, 157001.

Paramekanti, A., Randeria, M., and Trivedi, N. (2001). *Phys. Rev. Lett.*, **87**, 217002.

Paramekanti, A., Randeria, M., and Trivedi, N. (2004). *Phys. Rev. B*, **70**, 054504.

Parendo, K.A., Tan, K.H. Sarwa B., and Goldman, A.M. (2007). *Phys. Rev. B*, **76**, 100508.

Pasienski, M., McKay, D., White, M., and DeMarco, B. (2010). *Nat. Phys.*. Advance online publication, http://dx.doi.org/10.1038/nphys1726.

Pollet, L., Prokof'ev, N.V., Svistunov, B.V., and Troyer, M. (2009). *Phys. Rev. Lett.*, **103**, 140402.

Pourret, A., Aubin, H., Lesueur, J., Marrache-Kikuchi, C.A., Berge, L., Dumoulin, L., and Behnia, K. (2006). *Nat. Phys.*, **2**, 683–686.

Qin, Y., Vicente, C.L., and Yoon, J. (2006). *Phys. Rev. B*, **73**, 100505.

Radzihovsky, L. and Sheehy, D. E (2010). *Rep. Prog. Phys.*, **73**, 076501.

Randeria, M. (1995). In *Bose-Einstein Condensation* (ed. A. Griffin, D. Snoke, and S. Stringari), pp. 355–392. Cambridge University Press, Cambridge.

Randeria, M. (2010). *Nat. Phys.*, **6**, 561–562.

Reppy, J.D. (1992). *J. Low Temp. Phys.*, **87**, 205–245. 10.1007/BF00114905.

Rimberg, A.J., Ho, T.R., Kurdak, Ç., Clarke, J., Campman, K.L., and Gossard, A.C. (1997). *Phys. Rev. Lett.*, **78**, 2632–2635.

Rosseinsky, M.J., Ramirez, A.P., Glarum, S.H., Murphy, D.W., Haddon, R.C., Hebard, A.F., Palstra, T.T.M., Kortan, A.R., Zahurak, S.M., and Makhija, A.V. (1991). *Phys. Rev. Lett.*, **66**, 2830–2832.

Sacépé, B., Chapelier, C., Baturina, T.I., Vinokur, V.M., Baklanov, M.R., and Sanquer, M. (2008). *Phys. Rev. Lett.*, **101**, 157006.

Sacépé, B., Chapelier, C., Baturina, T.I., Vinokur, V.M., Baklanov, M.R., and Sanquer, M. (2010). *Nat. Comms.*, **1**, 140.

Sacépé, B., Dubouchet, T., Chapelier, C., Sanquer, M., Ovadia, M., Shahar, D., Feigel'man, M., and Ioffe, L. (2011). *Nat. Phys.*, **7**, 239–244.

Sachdev, S. (1999). *Quantum Phase Transitions.* Cambridge University Press, London.

Scalettar, R.T., Trivedi, N., and Huscroft, C. (1999). *Phys. Rev. B*, **59**, 4364.

Shah, N. and Lopatin, A. (2007). *Phys. Rev. B*, **76**, 094511.

Shahar, D. and Ovadyahu, Z. (1992). *Phys. Rev. B*, **46**, 10917–10922.

Spielman, I.B., Phillips, W.D., and Porto, J.V. (2007). *Phys. Rev. Lett.*, **98**, 080404.

Spivak, B. and Zhou, F. (1995). *Phys. Rev. Lett.*, **74**, 2800–2803.

Stewart, G.R. (1984). *Rev. Mod. Phys.*, **56**, 755–787.

Stewart Jr, M.D., Yin, A., Xu, J.M., and Valles, J.M. (2007). *Science*, **318**, 1273.

Strongin, M., Thompson, R.S., Kammerer, O.F., and Crow, J.E. (1970). *Phys. Rev. B*, **1**, 1078–1091.

Tinkham, M. (1996). *Introduction to Superconductivity.* McGraw-Hill, NY.

Trivedi, N., Scalettar, R.T., and Randeria, M. (1996). *Phys. Rev. B*, **54**, R3756.

Valdés Aguilar, R., Bilbro, L.S., Lee, S., Bark, C.W., Jiang, J., Weiss, J.D., Hellstrom, E.E., Larbalestier, D.C., Eom, C.B., and Armitage, N.P. (2010). *Phys. Rev. B*, **82**, 180514.

Valles, J.M., Dynes, R.C., and Garno, J.P. (1992). *Phys. Rev. Lett.*, **69**, 3567.

Varma, C.M., Zaanen, J., and Raghavachari, K. (1991). *Science*, **254**, 989–992.

Wang, Y., Li, L., Naughton, M.J., Gu, G.D., Uchida, S., and Ong, N.P. (2005). *Phys. Rev. Lett.*, **95**, 247002.

Wang, Y., Ono, S., Onose, Y., Gu, G., Ando, Y., Tokura, Y., Uchida, S., and Ong, N.P. (2003). *Science*, **299**, 86–89.

Yanase, Y. (2009). *New Journal of Physics*, **11**, 055056.

Yarmchuk, E.J., Gordon, M.J.V., and Packard, R.E. (1979). *Phys. Rev. Lett.*, **43**, 214–217.

Yoshida, Nobukatsu and Yip, S.-K. (2007). *Phys. Rev. A*, **75**, 063601.

Zaanen, J., Chakravarty, S., Senthil, T., Anderson, P., Lee, P., Schmalian, J., Imada, M., Pines, D., Randeria, M., Varma, C., Vojta, M., and Rice, M. (2006). *Nature Physics*, **2**, 138–143.

Zhou, F. and Spivak, B. (1998). *Phys. Rev. Lett.*, **80**, 5647–5650.

Zwierlein, M.W., Abo-Shaeer, J.R., Schirotzek, A., Schunck, C.H., and Ketterle, W. (2005). *Nature*, **435**, 1047–1051.

11 Scaling Analysis of Direct Superconductor-Insulator Transitions in Disordered Ultrathin Films of Metals

A.M. Goldman

School of Physics and Astronomy
University of Minnesota

11.1 Introduction

Interest in two-dimensional (2D) superconductivity began with the proposals of (Ginzburg and Kirzhnitz, 1964) to the effect that superconductivity might occur at elevated temperatures in two dimensions (2D). The specific configurations considered were electrons in surface energy states. Recent studies of interfacial phenomena in oxides seem to represent a realization of this phenomenon, albeit at very low temperatures (Ueno *et al.*, 2008). The net consequence of the above proposal was that the study of very thin films was thought to be a route to high temperature superconductivity. Thus far, this has turned out not to be the case, but it nevertheless has led to a substantial amount of new physics. The study of 2D systems has been a mainstay of the investigation of fluctuation phenomena in superconductors. This occurred in two stages. The first, involved the investigation of precursive superconductivity, the highlight of which was the establishment of the validity of the Aslamazov-Larkin theory of the fluctuation contribution to the conductivity in the normal state (For a review see (Larkin and Varlamov, 2009)). Sometime later it was established that equilibrium long-range order in 2D could be topological, or quasi-long range, especially in the case of the XY model universality class (Kosterlitz and Thouless, 1973). It was then realized that the effective penetration length of 2D superconductors could be a macroscopic length. This meant that Abrikosov vortices in the 2D limit could interact logarithmically, and the transition to the superconducting state in 2D could be a topological transition as well (Beasley *et al.*, 1979) This led to the second stage of investigation of 2D superconductivity, an activity that continues to this day in part because the cuprate high temperature superconductors are layered compounds that behave as coupled or weakly coupled 2D layers (Steiner *et al.*, 2008).

An important stream in the evolution of this subject was the realization that disorder was an important element in determining the properties of 2D superconductors. Early in the development it was realized that magnetic impurities were pair breaking, whereas nonmagnetic impurities should not have a deleterious impact on superconducting pairing. This came to be known as Anderson's theorem (Anderson, 1959). On the other hand, with a high enough level of disorder Anderson localization occurs (Abrahams *et al.*, 1979). Under strong conditions of electron localization, superconductivity should disappear, even with an attractive interaction (Ebisawa *et al.*, 1985). This led to another stage of the development of the subject in which transition temperatures vs. sheet resistances were investigated and compared with theory. The fundamental idea was that disorder reduces the screening of the electron-electron repulsion and the electronic density of states, leading to a net weakening of the net electron-electron attractive interaction and the superconductivity. Figure 11.1 shows the data of (Graybeal and Beasley, 1984) of the reduction of the transition temperature of Mo_xGe_y films with increasing sheet resistance, or disorder. The solid line is generated by a theory due to (Finkelstein, 1994). This model, which effectively describes the reduction of the transition temperature, should not work in the limit of really strong disorder. The reason for this is that as the transition temperature is driven down towards zero temperature, fluctuations increase, and eventually they should be quantum mechanical. This raises the issue as to whether the endpoint of

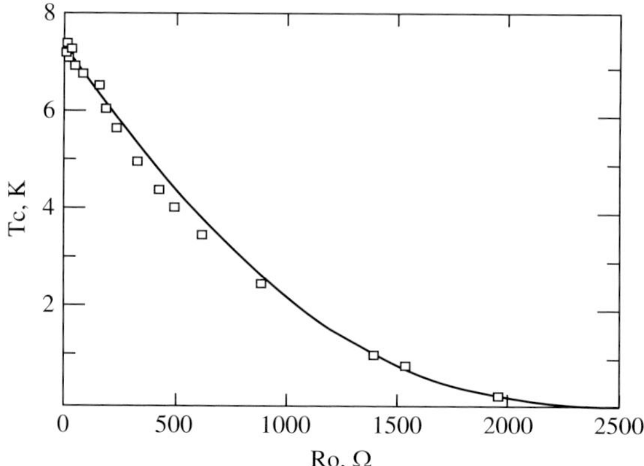

Fig. 11.1 Suppression of superconductivity in amorphous Mo_xGe_y with increasing disorder

the reduction of the transition temperature is a quantum critical point, with disorder controlling the ground state of the system.

The first real indication that disorder might drive the transition temperature down to a low enough value that quantum fluctuations could be important was the work of (Haviland *et al.*, 1989) in which the thickness variation of resistance vs. temperature, R(T) was studied in amorphous Bismuth (a -Bi) films. This investigation was carried out in an apparatus in which the films could be grown in situ, at low temperatures and measurements carried out in alternation with increments of growth. Bismuth is a semimetal in its usual form, but is metallic and superconducting under high pressure, or in thin film amorphous form, which results when it is grown on substrates held at liquid helium temperatures (Buckel and Hilsch, 1954). Curves of $R(T)$ for an amorphous Bi film are presented in Fig. 11.2. These resemble renormalization flows to an unstable fixed point at zero temperature, which for this particular set of films correspond to the quantum resistance for pairs, $h/4e^2$, or 6450 Ω. This result set the stage for consideration of the SI transition as a continuous quantum phase transition (Sachdev, 1999). The experiments described in this chapter, involve tuning or controlling the transition in ultrathin films of metals with disorder, parallel and perpendicular magnetic field, and charge density. They are focused on scaling analyses of low temperature resistance data. As will be demonstrated, such work can in principle reveal the universality class of the continuous quantum phase transitions, leading to an understanding of the possible underlying physical models at work. Unfortunately the field is a work in progress as other experiments have yielded different results dependent upon the tuning or control parameter and the material. In some instances, for certain materials, there is no direct superconductor-insulator transition. The features of materials that lead to these different outcomes are not known. Many complex systems, in particular, strongly correlated electron systems, exhibit quantum phase transitions as a function of presssure, composition, or magnetic field (Sachdev, 1999). Indeed the quantum phase transition paradigm is central to some theories of high temperature superconductivity (Broun, 2008)(Sebastian *et al.*, 2010). The advantage of studying

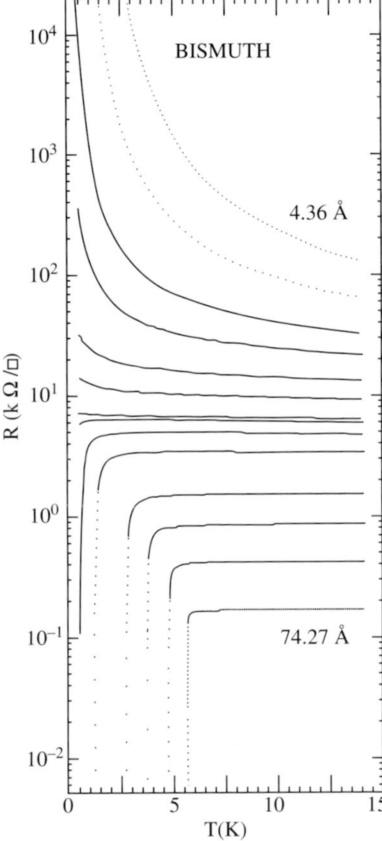

Fig. 11.2 Evolution of the temperature dependence of the sheet resistance R(T) with thickness for an *a*-Bi film deposited onto Ge. Fewer than half of the traces actually acquired are shown. Film thicknesses that are shown range from 4.36 to 74.27Å. Film thicknesses are nominal being measured by changes in a crystal oscillator monitor.

ultrathin films of metals is that in some sense they are the simplest systems exhibiting such transitions and as a consequence can provide significant insight into the nature of the phenomenon,

11.2 Quantum Phase Transitions

Quantum phase transitions (QPTs) are transitions that take place at absolute zero, where the crossing of the phase boundary changes the quantum mechanical ground state. The control parameter, rather than being temperature, is some parameter of the Hamiltonian of the system. For 2D superconducting films the parameter could be a measure of disorder, perpendicular or parallel magnetic field, or charge density (For an early review see: (Goldman and Markovic, 1998)). The understanding of quantum phase transitions requires that quantum effects need to be considered. This is to be contrasted with non-zero temperature transitions, which in some sense are classical even though the underlying microscopic physics is quantum mechanical. The reason for this is that although quantum fluctuations for such transitions may be important at microscopic length scales, they are not relevant at the longer length scales that are associated with critical behavior. The critical fluctuations can be understood in the context of classical statistical mechanics with an effective Hamiltonian for the order parameter given by the Landau free-energy functional. Continuous phase transitions at nonzero temperature are accompanied by divergent correlation lengths and correlation times (See: (Sondhi *et al.*, 1997) and (Sachdev, 1999)). The frequency associated with critical fluctuations vanishes as the transition temperature. The system, and the critical fluctuations, will behave classically whenever the energy associated with them is less than k_BT. Quantum, or zero-temperature phase transitions, are another story entirely as the quantum mechanical ground state changes when the critical point is crossed. In contrast with phase transitions at nonzero temperature, quantum effects are central to QPTs. Fluctuations behave quantum mechanically.

The existence of divergent length scales in both classical and quantum phase transitions and often divergent behavior of measurable quantities near the phase transition lead to the use of scaling analyses which reveal critical exponents that characterize the transition. These critical exponents correspond to specific universality classes which can be represented by specific statistical mechanical models. Phase transitions governed by different microscopic physics can actually belong to the same universality class.

Usually the correlation length of a thermal phase transition is written as

$$\xi = \xi_0 |t|^{-\nu} \tag{11.1}$$

where ν is the correlation length exponent and t is the reduced temperature $t = |T - T_c|/T_c$. Similarly the relaxation time τ of the order parameter in the critical region is given by

$$\tau \sim \xi^z \tag{11.2}$$

Here z is the dynamical critical exponent. Frequently τ is denoted as ξ_τ which is the dynamical correlation length.

In the general case, and specifically for zero-temperature quantum phase transitions, t is replaced by δ, which is a dimensionless distance of a control parameter from its critical value. The parameter for ultrathin films of metals, could be perpendicular or parallel magnetic field strength, disorder, or charge density in the systems we are considering. With these definitions we can write

$$\xi_\tau \sim |\delta|^{-\nu z} \tag{11.3}$$

The key feature of QPTs is the interplay of dynamics and thermodynamics. As a consequence of this, a d-dimensional quantum system at finite temperature is described in the $T \Longrightarrow 0$ limit as a classical system of $d + z$ dimensions, with the finite extent of the system in the extra dimensions being given by $-i\,\hbar\,\beta$ in units of time, where

$\beta = 1/k_BT$. The extent of these extra dimensions are divergent only in $T \Longrightarrow 0$ limit. What is remarkable is that the universality class of the quantum transition may be one studied extensively in some classical context. This also allows for the possibility of a computational treatment of the quantum mechanical problem using simulations of the $d + z$ dimensional classical problem. Disorder on the other hand can change the universality class of the equivalent classical problem. It is also not true in general that space and time enter in the same fashion in the equivalent classical problem. For this to happen the dynamical exponent, z, must be unity, and whether or not this happens depends upon the detailed quantum dynamics of the system. As mentioned above, the effect of considering $T \neq 0$ in the statistical mechanics is to force the "temporal" dimension of the problem to be finite. The formal model used to analyze data at nonzero temperatures is finite-size scaling. The success of finite size scaling analyses of the various superconductor-insulator transitions is part of the evidence for there being QPTs. Scaling can be used to characterize properties measured at nonzero temperatures in the regime of critical fluctuations and thus to determine the critical exponents and universality class of the transition.

When $z = 1$, the quantum partition function looks like a classical partition function in $d + 1$ dimensions, except for the fact that the extra dimension is finite in extent given by $-i\hbar\beta$ in units of imaginary time. In the $T \Longrightarrow 0$ limit this length diverges, and the effective dimensionality of the system is $d + 1$. For systems with long-range interactions, z is believed to be unity. This means that as far as quantum phase transitions are concerned, a 2D system at finite temperatures such as the 2D XY model would have an effective dimension of 3 and would have the critical exponents of the 3D XY model. This will play a role in the discussion of the results of experiments later in this chapter.

Fisher (Fisher, 1990) presented a scaling theory that can be used to characterize measured resistances at nonzero temperatures in the regime of critical fluctuations and to thus determine the critical exponents and universality class of the transition. For the resistance near the transition, it is of the form:

$$R = R_c F(\delta/T^{1/\upsilon z}, \delta/E^{1/\upsilon(z+1)}) \tag{11.4}$$

Here R is the sheet resistance, R_c and F are an arbitrary constant and an arbitrary function. The energy scale for the fluctuations $\Omega \sim \xi_\tau^{-z}$, is cut off at nonzero temperature by k_BT, defining a cut off length L_T given by $k_BT \sim L_T^{-z}$. This gives rise to the first term in the argument of the arbitrary function F. The second term comes from a characteristic length associated with the electric field as compared with the correlation length.

The applicability of these ideas assumes that there is a continuous and direct SI transition controlled by a parameter of the Hamiltonian such as perpendicular or parallel magnetic field, disorder, or charge transfer. The first scaling analysis of a perpendicular field driven transition was carried out by (Hebard and Paalanen, 1990) on InO_x films. They found that the exponent product $\upsilon z \sim 1.2$ from scaling with only the first argument of Eq. 11.4. Later Yazdani and Kapitulnik(Yazdani and Kapitulnik, 1995) analysed the properties of Mo_xGe_y films, carrying out both temperature and electric field scaling. They were able to determine both υz and $\upsilon(z+1)$ and thus ascertain that $z=1$ in agreement with expectations (Herbut, 2001). It should also be noted that Gantmakher and collaborators carried out scaling analyses of the field-driven SI transition of InO_x films (Gantmakher *et al.*, 2000).

The relevance of Eq. 11.4 to the actual behavior of films depends upon whether there is a direct SI transition controlled by varying some control parameter, independent of the underlying mechanism for the transition. For direct transitions there are two distinctive schools of thought. One approach has the insulating behavior resulting from the localization of charged Bosons, *i. e.*, Cooper pairs (Fisher, 1990; Fisher *et al.*, 1990). The second has the insulator resulting from the localization of fermions that result from the enhancement of Coulomb repulsion that suppresses pairing (Finkelstein, 1994). There is also substantial evidence to support this latter view (Valles *et al.*, 1989, 1992).

There are also ideas ((Dubi *et al.*, 2007; Y. Dubi and Avishai, 2005) and (Spivak *et al.*, 2008)) and some experimental evidence (Qin *et al.*, 2006) for an indirect transition in which there is an intermediate inhomogeneous phase in which a film contains superconducting and insulating puddles. There is some evidence for an intermediate metallic regime (Qin *et al.*, 2006). It is clear that there may be several types of SI transitions, depending upon the material, its level of disorder and the control parameter. In the following we will consider a series of investigations of SI transitions in films of *a*-Bi, all of which are very close to the disorder-tuned SI transition of Fig. 11.2. In all of the cases considered, the transition appears to be direct, and it was possible to carry out a scaling analysis using the first argument of Eq. 11.4 with the in-plane electric field fixed in a linear regime.

An important caveat is that the success of scaling by itself and the identification of the universality class of the transition from the values of the critical exponents may not elucidate the microscopic physics of the transition or the nature of the insulating state.

11.3 Experimental Approach

The films that were investigated were grown on substrates held at or near liquid helium temperatures. Single-crystals of (100) $SrTiO_3$ served as substrates. Platinum electrodes, 100Å thick, configured for four-terminal electrical measurements were pre-evaporated on the epi-polished front surfaces of the substrates. Experiments were initiated with the placement of the substrate in a dilution refrigerator/UHV deposition apparatus. In the early work this was an Oxford minifridge (Orr and Goldman, 1985). In later work an Oxford Kelvinox 400 dilution refrigerator was employed (Hernandez and Goldman, 2002). Films were grown by first depositing a 10 Å thick underlayer of *a*-Ge or *a*-Sb, followed by subsequent layers of *a*-Bi. All depositions were carried out under UHV conditions with the substrates held at or near liquid-helium temperatures during deposition. The geometries of the positions of the source and substrate in the two different setups was identical. In the apparatus that employed the minfridge, the cryostat was moved in and out of the growth chamber, whereas in that employing a Kelvinox 400 the substrates were held on a helium-cooled transfer rod which was used to move samples on and off the refrigerator platform. Films grown at liquid helium temperatures with the indicated underlayers are believed to be homogeneously disordered (Strongin et al., 1970). This approach allows for repeated cycles of in situ evaporation and measurement permitting the generation of curves of R(T) at different thicknesses as shown in Fig. 11.2.

11.4 Thickness- and Perpendicular Magnetic Field-Tuned SI Transitions

We now consider data of $R(T, \delta)$ for a set of films different from those shown in Fig. 11.2. As is typical of such experiments, at some critical thickness, d_c, R becomes temperature independent, while for even thicker films it decreases rapidly with decreasing temperature, indicating the onset of superconductivity. The critical thickness is found by plotting R vs. d at different temperatures (inset of Fig. 11.3) and identifying the crossing point for which the resistance is temperature independent, or by plotting dR/dT vs. d at the lowest temperatures and finding the thickness for which $dR/dT \Longrightarrow 0$. In the quantum critical regime the resistance of a two-dimensional system is expected to obey the scaling law of Eq. 11.4, with the in-plane electric field taken to be a constant and from the current-voltage characteristics taken to be in the linear regime.

To analyze the data, we obtain curves of R vs d at various temperatures. Having determined the critical thickness in the manner described above, we then rewrite Eq. (11.4) as $R(t, \delta) = R_c F(\delta t)$, where $t = T^{-1/vz}$, and treat the parameter $t(T)$ as an unknown variable which is determined at each temperature to obtain the best collapse

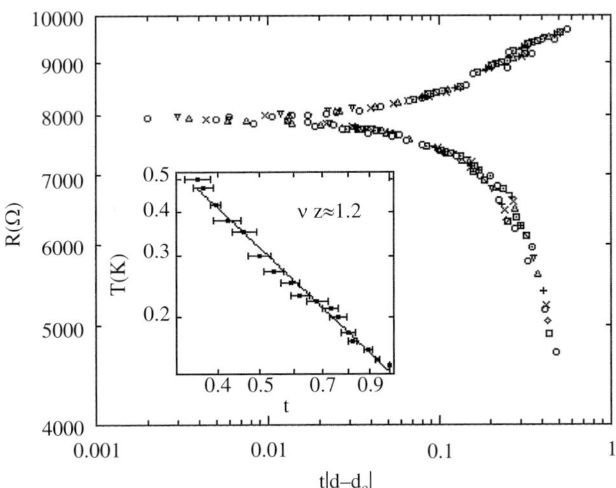

Fig. 11.3 Resistance per square as a function of the scaling variable, $t|d - d_c|$, for seventeen different temperatures, ranging from 0.14 to 0.5 K. Different symbols represent different temperatures. Inset: temperature dependence of t.

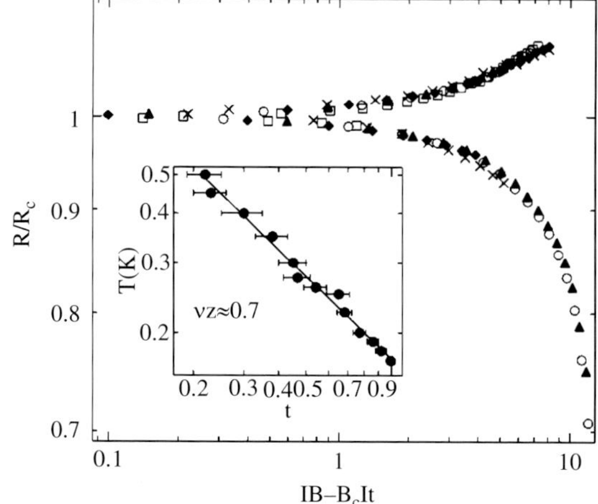

Fig. 11.4 Normalized resistance per square as a function of the scaling variable $t|B - Bc|$. Each symbol represents one film at different temperatures. Only a small portion of the data is shown for clarity. Inset: The fit of a power law to the temperature dependence of the parameter t determines the value of vz.

of the data. Specifically, t is determined by performing a numerical minimization between a curve at a particular temperature and the lowest temperature curve. The exponent ηz is then found from the temperature dependence of t, which must be a power law in temperature for the procedure to make physical sense. This procedure does not require detailed knowledge of the functional form of the temperature or thickness dependence of R, or prior knowledge of the critical exponents. A different method of obtaining critical exponents was also used to check the consistency of this procedure. A log-log plot of $(\partial R/\partial d)|_{d_c}$ vs. T^{-1} was constructed. Its slope is equal to $1/vz$ if Eq. (11.4) is obeyed. Exponents obtained this way were essentially identical, within the quoted experimental uncertainty, to those obtained using the first procedure.

The collapse of the data of $R(T, \delta)$ is shown in Fig. 11.3 (Markovic *et al.*, 1999). The exponent product νz is found to be 1.2 ± 0.2, with the error determined from the power law fit. This agrees with theoretical predictions from which $z = 1$ would be expected for a bosonic system with long range Coulomb interactions independent of dimensionality, and with $v \geq 1$ in 2D.

In addition, the magnetoresistance as a function of temperature and magnetic field with perpendicular field orientation was studied. By sorting this data the perpendicular field tuned transition in a nonzero magnetic field can be investigated. A scaling analysis resulted in a critical exponent product of 1.4 ± 0.2, apparently independent of magnetic field.

Values of $R(T)$ for five films of different thickness were studied in magnetic fields of up to 1.2 T, applied perpendicular to the plane. The exponent product was determined with magnetic field rather than thickness as the tuning parameter. The collapse of the data in this instance is shown in Fig. 11.4. The exponent product determined in a similar manner was found to be $vz = 0.7 \pm\ 0.2$ for each of the five films. The same value was obtained from log-log plots of the derivative at the critical field vs. T^{-1}. The fact that the exponent product differs from that of the thickness tuned transition suggests a universality class different from that of the thickness-tuned transition.

It is possible to combine the data obtained from the thickness-tuned transitions in fixed magnetic field and field-tuned transitions at fixed thickness to construct a phase diagram with thickness and magnetic field as independent variables. This is shown in Fig. 11.5. The films characterized by parameters which lie above the boundary are superconducting and those below are insulating, distinguished by the sign of dR/dT.

The critical resistance, in contrast with the predictions of the dirty Boson model is not universal. Figure 11.6 shows that it decreases as the field increases, roughly in a linear fashion. Since thicker films have lower normal-state resistances and higher critical fields, this means that T_c decreases with increasing thickness and decreasing normal state resistance. Variants of the dirty boson model predict values of the critical resistance close to those found here.

It is important to note several features of these two different SI transitions. The transitions are direct in that there appear to be no intermediate metallic phases and there is no resistance saturation at the lowest temperatures. This is in contrast with results reported for other types of systems (Mason and Kapitulnik, 2002) and (Seo

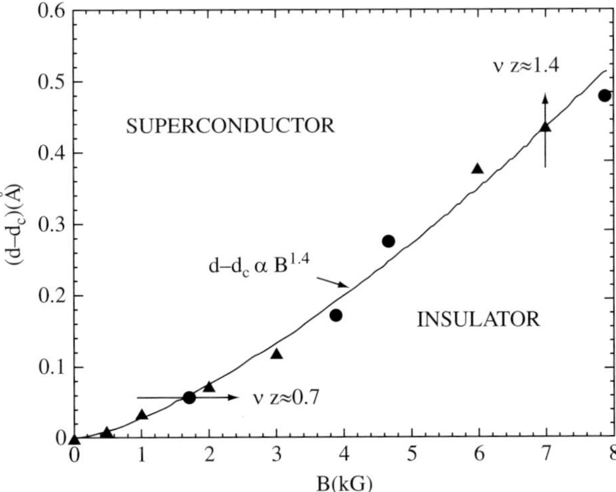

Fig. 11.5 Phase diagram in the $d - B$ plane in the $T \Longrightarrow 0$ limit. The points on the phase boundary were obtained from disorder-driven transitions (triangles) and magnetic-field-driven transitions (circles). Values of the critical exponent products are shown next to arrows giving the direction in which the boundary was crossed. Here d_c is the zero-field critical thickness.

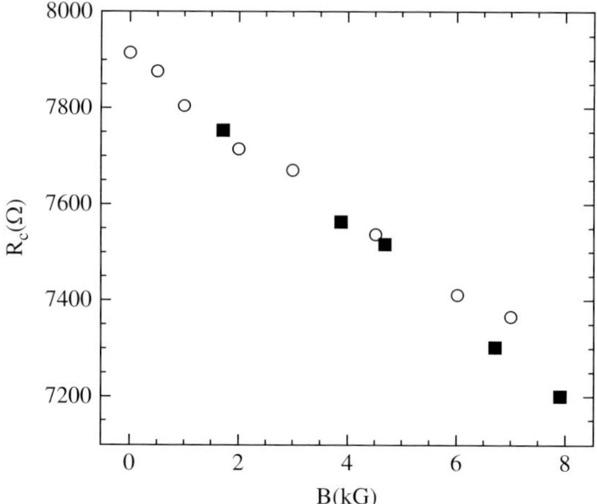

Fig. 11.6 The critical resistance as a function of the critical field for a series of bismuth films. Here R_c decreases with increasing thickness, as thicker films have lower normal-state resistances and higher critical fields.

et al., 2006), or what is found in granular films (Jaeger *et al.*, 1989). It should also be noted that the films which have been studied in the case of the perpendicular field tuned SI transition are very close in their properties to the insulating regime. Subsequent works, in our laboratory on *a*- Bi films (Lin and Goldman, unpublished), and by others on *a*-Pb films (Parker *et al.*, 2006), which were less disordered have revealed the presence of an intermediate regime that may have two phases. The precise nature of the phase diagram is not known at this time.

11.5 Parallel Magnetic Field SI Transition

The superconductor-insulator transition of *a*-Bi films was studied in parallel magnetic fields in two types of samples, those in which superconductivity was induced by electrostatic doping, which will be discussed below, and those which were intrinsic superconductors with thicknesses close to the critical thickness for the appearance of superconductivity (Parendo et al., 2006a). In the case of a 10.22 Å thick film induced into the superconducting state with a charge (electron) transfer of 3.35 x $10^{13}/\mathrm{cm}^2$, the best collapse of the data was for $\upsilon z = 0.65 \pm\ 0.1$. The results of the scaling analysis are shown in Fig. 11.7. The results of a similar study of an intrinsically superconducting film 10.25 Å thick were nearly identical. A distinct crossing point was found with $R_c = 10460\ \Omega$ and $B_c = 9.18$T. The exponent product that produced the best collapse of data was 0.75 ± 0.1, slightly higher than that found for the field-tuned transition of an electrostatically induced superconducting film, but agreeing within the uncertainty of the analysis.

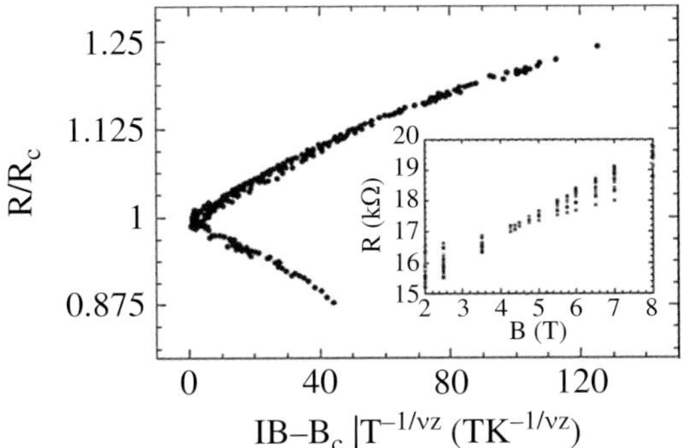

Fig. 11.7 Finite size scaling plot for the 10.22 Å thick film with Δn =3.35x10^{13} cm^{-2} with B as the tuning parameter for 150 mK $< T <$ 340 mK. The best collapse of the data was for $vz = 0.65$ with an uncertainty of ± 0.1. Inset: Isotherms of $R(B)$ at temperatures between 150 and 340 mK.

11.6 Electrostatic Control of the SI Transition

Since there may be issues of sample inhomogeneity and variable strength vortex pinning in thickness and perpendicular magnetic field tuned transitions, it is of interest to study SI transitions in which the level of physical disorder is fixed, and in which the outcome of the study was not dependent upon the degree of vortex pinning. This can be accomplished by inducing superconductivity in an insulator using the electric field effect (Ahn *et al.*, 2006). The presumption is that the same level of disorder is present in the insulating and superconducting states.

The field effect geometry was one in which $SrTiO_3$ (STO) crystal served as both a substrate and gate insulator. The preparation of this device involved several steps. First a small section of the unpolished back surface of a 500μm thick single crystal of (100) STO was mechanically thinned *ex situ*, resulting in this surface and the epi-polished front surface being parallel and separated by 45$\pm$ 5μm (Bhattacharya *et al.*, 2004). A 0.5 mm by 0.5 mm 1000 Å thick Pt gate electrode was deposited *ex situ* onto the thinned section of the back surface directly opposite the eventual location of the measured square of film. Platinum electrodes, 100Å thick were deposited *ex situ* onto the substrates polished front surface to form a four-probe measurement geometry. After the substrate was placed in a Kelvinox-400 dilution refrigerator a 10Å thick underlayer of a-Sb and successive layers of a-Bi were deposited under UHV conditions through a shadow mask onto the substrate's front surface.

A sequence of a-Bi films was studied (Parendo *et al.*, 2005, 2006*a*). The temperature dependence of insulating films was governed by the Mott variable range hopping form:

$$R(T) = R_0 \exp\left[(T_0/T)^{1/3}\right] \tag{11.5}$$

In the case of an insulating film, 10.22Å in thickness, the addition of electrons to the film, using the field effect, induced superconductivity. The evolution of this electrostatically tuned SI transition with charge transfer is shown in Fig. 11.8. An important feature of the data was the crossover from Mott hopping in the insulating regime to a $ln(T)$ dependence of the conductance on temperature in the normal state for films which underwent a transition to superconductivity. What was also found is that the slope of the $ln(T)$ term in the conductance was a linear function of the charge transfer, as shown in the inset to Fig. 11.8.

This SI transition, as shown in Fig. 11.9 was successfully analyzed using finite size scaling(Sondhi *et al.*, 1997), employing the charge transfer Δn as the control parameter. This strongly suggests that the electrostatically tuned transition is also a continuous quantum phase transition. In this instance, the exponent product υz was $0.7 \pm\ 0.1$, all the way up to 1K if the $ln(T)$ dependence of the conductance is first removed. This was done by assuming that there were two parallel conductance channels, one involving weak localization and electron-electron interactions, and the other the critical fluctuations.

The electrostatically tuned SI transition appears to involve little change in physical disorder as determined by noting that the resistance above 1K changes very little as superconductivity develops. This is in contrast with what is found for the disorder or thickness tuned transition. In Fig. 11.10 we show the resistances at 120 mK as a function of the resistances at 1K for the electrostatic tuned transition and a thickness-tuned transition. The thickness-tuned transition takes place with a much larger change in the high temperature resistance than does the electrostatically tuned transition. The apparent coincidence of an insulator-metal transition with the insulator superconductor transition suggests that the charge-tuned SI transition is an electronic phenomenon and is not associated with the localization of bosons.

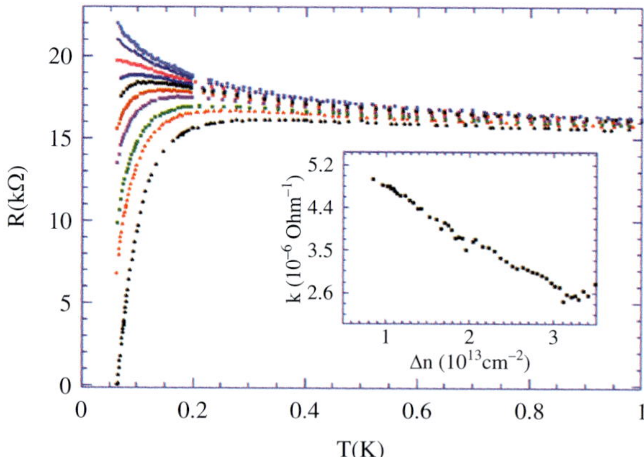

Fig. 11.8 Resistance versus temperature at various values of Δn for the 10.22 Å thick film with B =0. Data are shown from 60 mK to 1 K. The values of Δn that are shown, from top to bottom, are 0, 0.62, 1.13, 1.43, 1.61, 1.83, 2.04, 2.37, 2.63, and 3.35 x 10^{13} cm-2. Forty four curves of R(T) for other values of Δn are omitted from the plot for clarity. Inset: slope of $ln(T)$ vs. Δn.

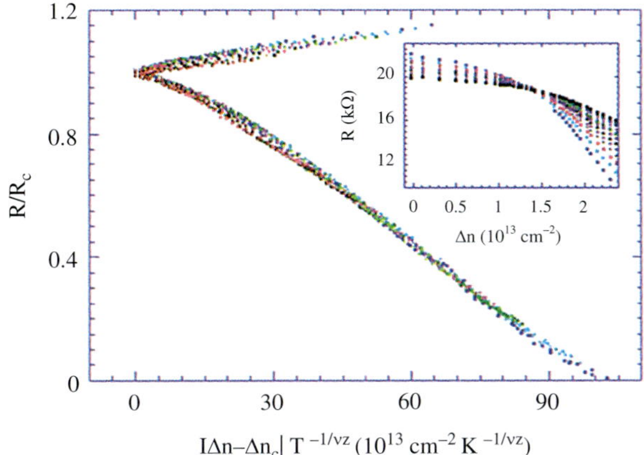

Fig. 11.9 Finite size scaling plot for the 10.22 Å thick film with $B = 0$, including data from 60 mK to 140 mK with Δn as the tuning parameter. Fifty four values of Δn between 0 and 3.35x10[13] cm^{-2} were included. The best collapse of the data wasfor υz =0.7 with an uncertainty of $\pm$ 0.1. Inset: $R(\Delta n)$ for isotherms between 60 mK and 140 mK.

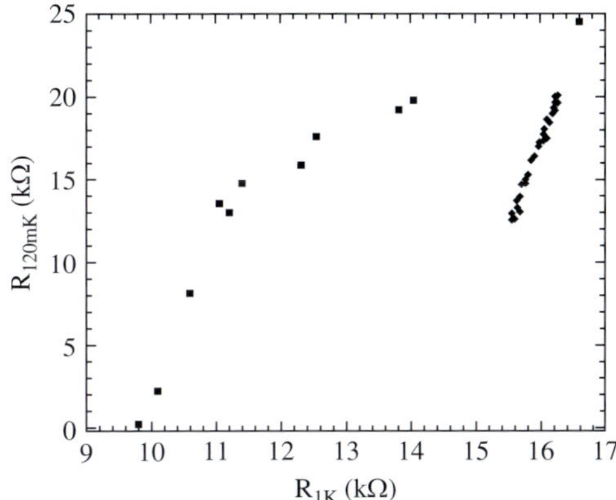

Fig. 11.10 Resistance at 1 K as a function of resistance at 120 mK for electrostatic (circles) and thickness (squares) tuned transitions. As superconductivity develops, the resistance at 1 K changes much more for thickness tuning than for electrostatic tuning.

11.7 Nonlinear and Hot Electron Effects

The nature of a quantum phase transition is determined by the values of the correlation length exponent and the dynamical critical exponent respectively. Nonlinear conductivity near the quantum critical point can be important because the results of a finite size scaling analysis under changes of in-plane electric field can be combined with scaling under changes of temperature to determine separate values of the two exponents. In this analysis, it is important to determine the extent to which intrinsic nonlinear response is more important than Joule heating (Sondhi *et al.*, 1997). It has been possible to demonstrate that the dominant effect in *a*-Bi films is actually electron heating. An electric field scaling analysis that successfully collapses data has been shown to be a direct consequence of heating and not a quantum critical nonlinear electric response. This is a consequence of being able to map electric field scaling onto temperature scaling by relating in-plane electric fields to elevated electron temperatures.

Figure 11.11 shows a differential resistance vs. measuring current for a parallel field tuned SI transition of an *a*–Bi film. The value of R_D/R_{D_c} is plotted as a function of $|B - B_c|V^{-1/v(z+1)}$ where V is the in-plane voltage across the film. The result of the temperature scaling which is not shown is $vz = 0.68 \pm 0.05$, whereas for the electric field scaling, $v(z+1) \sim 2.0$. This then yields $z \sim 0.5$ and $v \sim 1.3$ (Parendo *et al.*, 2006*b*). The value of z is believed to be unphysical as it should be 1 or 2 for charged Bose systems, depending upon their interactions (Herbut, 2001).

After detailed scrutiny of the variation of the differential resistance with temperature and magnetic field, it was seen that values of $R_D(T)$ with decreasing temperature become temperature independent at particular values of current. This suggests that these currents heat the electrons to the particular temperature for all lower refrigerator temperatures. Upon quantitative analysis of this hypothesis it is found that $T_{electron} \sim P^{0.18}$ where P is the power dissipated in the film by the current.

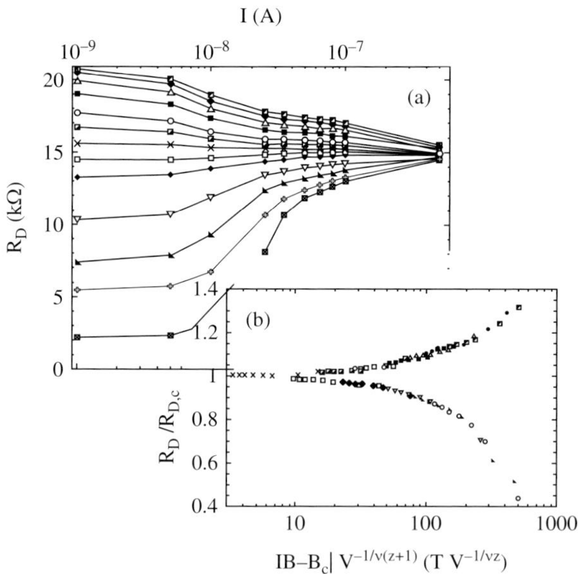

Fig. 11.11 (a) Differential resistance vs. measurement current for a parallel magnetic field tuned transition. From bottom to top, the magnetic fields are 2, 3, 4, 5, 6, 6.5, 7, 7.5, 8, 9, 10, 11, and 12 T. (b) Electric field scaling for the transition in (a).

This is very close to the relationship proposed by Wellstood et al.(Wellstood *et al.*, 1994) to describe the relationship between a disordered metal film's minimum electron temperature as measured using Johnson noise, and measurement power (Parendo *et al.*, 2006*b*).

The result is that electric field scaling can be mapped onto temperature scaling and for the particular a Bi film in question leads to $vz = 0.72$, which agrees within experimental uncertainty with the value obtained by temperature scaling. Electric field scaling then appears to work because it is really temperature scaling and the increased measurement current to enter the nonlinear regime simply heats the electrons.

An important question is whether this is a general result. In the zero temperature limit, nonlinear transport effects are expected to compete with Joule heating with material specific properties being important. The electron temperature varies with power as $T_{electron} \sim P^{1/\theta}$ with $\vartheta = p + 2$ where p is the power of the temperature in the expression for the temperature dependence of the inelastic electron-phonon scattering rate, $\tau_{in}^{-1} \sim P^{1/\vartheta}$. There is a criterion that for $2/\vartheta < z/(z+1)$, Joule heating rather than intrinsic nonlinear effects will dominate (Sondhi *et al.*, 1997). This criterion arises from a dimensional analysis that may ignore important multiplicative factors. Assuming z to be 1 or 2, the value of θ for a-Bi meets this criterion. However, even in the earlier work of Yazdani and Kapitulnik (Yazdani and Kapitulnik, 1995) on MoGe thin films, in which scaling analyses revealed apparent values of $vz \sim 1.35$ and $v(z+1) \sim 2.65$, leading to $v \sim 1.3$ and $z \sim 1$, heating may actually be responsible. It is known that $p = 2$ in MoGe, so that $\theta = 4$. This puts MoGe in the "marginally dangerous" category in which both heating and intrinsic effects may be important. With the particular value of p for this material, one can also map electric field scaling onto temperature scaling. The resultant conclusion from all of this is that measuring currents can heat electrons out of equilibrium with their environment. This prevents electric field scaling from yielding information allowing the separate determination of v and z. Because this analysis depends upon material parameters it is possible that there exist materials for which this analysis would work.

11.8 Discussion

In the above we have reviewed the results of a series of investigations of SI transitions tuned by disorder (thickness), parallel and perpendicular magnetic fields, and electrostatic charging. All of these examples have exhibited direct SI transitions. In all but the case of thickness tuning the exponent products have been ~ 0.7, within experimental error. If the dynamical critical exponent $z = 1$, $\nu \sim 0.7$ is very close to what might be expected for a 3D XY model. This would be expected if the nonzero temperature behavior was governed by the 2D XY model and if $z = 1$, which would increase the dimensionality of the zero-temperature quantum phase transition by one. The value of ν with $z = 1$ also corresponds to the 2D Bose Hubbard model without disorder. The films studied in each instance were either superconducting, with resistances very close to the critical resistance for the thickness tuned transition, or insulating, in the case of the electrostatically induced transition, with resistances again very close to the critical resistance.

There was no evidence of metallic regimes at low temperatures near criticality. The Bose Hubbard model of the SI transition in 2D predicts a metallic ground state only at the critical value of the tuning parameter. In experiments reported by others (Mason and Kapitulnik, 2001; Qin *et al.*, 2006), temperature independent resistances have been found on both sides of the SI transition over an extended range of tuning parameters. The nature of this resistance saturation is far from certain because of specific difficulties with electrical measurements at mK temperatures. The electronic heat capacities and coupling of the electronic degrees of freedom to the heat bath are so weak that even modest levels of Joule heating may prevent cooling of films' electrons. A refrigerator could thus cool phonons more effectively than electrons resulting in a temperature independent resistance since the film would be measured at its minimum achievable electron temperature over an extended range of lower phonon or lattice temperatures.

Scaling of the thickness-tuned transition led to an exponent product ~ 1.3 which is very close to that found for the perpendicular-field tuned transition of MoGe. Recently Steiner, Breznay and Kapitulnik (Steiner *et al.*, 2008) reported the results of a study of perpendicular field tuned SI transitions, which combined data from Ta, MoGe, and In_xO_y films. It is claimed that the In_xO_y are frequently more disordered than MoGe films and that there are two branches of properties, a low disorder branch in which a scaling analysis leads to $z \sim 1$ and $\upsilon \sim 1.35$, and a high disorder branch with $\upsilon \sim 7/3$ and near the intersection of the branches, $\upsilon \sim 4/3$, but with less accurate scaling, again taking $z = 1$. The scaling with $\upsilon \sim 1.35$ they associate with classical percolation and the scaling with $\upsilon \sim 7/3$ with quantum percolation. The phase diagram with the two branches is in the space of critical conductance and critical field normalized to $H_{c2}(0)$. The less disordered branch they assert exhibits an intermediate metallic regime, whereas the disordered branch has a direct SI transition. The disordered branch it is claimed exhibits a critical conductance equal to that expected for the quantum conductance of pairs. Also the films in the high disorder branch exhibit a large peak in magnetoresistance in the insulating regime, which has been described by others for In_xO_y or TiN films (Baturina *et al.*, 2007; Crane *et al.*, 2007; G. Sambandamurthy and Shahar, 2004; Paalanen *et al.*, 1992; Sambandamurthy *et al.*, 2005; Steiner and Kapitulnik, 2005). Similar results have not been found for any metallic films grown by quench condensation.

One may ask, why the difference in behavior. It should be noted that in situ evaporated films of metals (a-Bi, a-Pb, and a-Be) are significantly thinner ($\sim$10 to 20Å) than the In_xO_y, MoGe, or even TiN films (30 to 300Å) that have been studied by various groups. Although they may have similar sheet resistances in the normal state, and short mean free paths, their carrier concentrations may be significantly higher. Also, although there have been studies of structure of the compound films that classifies them as amorphous and homogeneous, the spatial variation of the chemistry is not known and this can affect the homogeneity of the pairing potential. Indeed, recent STM studies at 0.050K of TiN films, which were macroscopically superconducting revealed a spatially varying nonzero gap (Sacepe *et al.*, 2008). Whether this is a result of chemical disorder was not established, but the data are reminiscent of similar studies of cuprates where local inhomogeneity plays a role.

In the case of the perpendicular field driven transition of quench evaporated films, when the sheet resistance falls below about 4000 to 5000 Ω the transition is no longer direct, and there appears to be an extended region where at least a nonzero temperatures there appears to be a two-phase regime, possibly superconducting droplets in a nonsuperconducting matrix (See for example refs. (Ghosal *et al.*, 1998; Spivak *et al.*, 2008; Y. Dubi and Avishai, 2005). This has been found in both *a*-Bi (Lin and Goldman 2009) and *a*-Pb films (Parker *et al.*, 2006). Thicker Bi films ∼ 24 Å, behave as high κ Type-II superconductors and undergo a superconductor to metal transtion.

11.9 Conclusions

The causes of the differences in the data between low carrier density compounds and quench-evaporated metals exhibiting SI transitions are not understood. There are distinct differences in the behavior of these two different types of systems which likely have their origins in differences in carrier density, and perhaps chemical or physical disorder on a mesoscopic scale. With the exception of the thickness-tuned SI transition, which returns exponent products $vz \sim 1.3$, all of the scaling analyses of SI transitions of elemental metals, tuned by charge transfer, and perpendicular and parallel magnetic field, which describe direct transitions are consistent with the 3D XY model. The exponent product $vz \sim 0.7$, and if $z = 1$ as is anticipated for systems with long range interactions, the exponent $v \sim 1$. It is anticipated that systems exhibiting nonzero temperature continuous phase transitions, in the limit in which the transition is driven to zero temperature will acquire extra dimensions. Thus systems belonging to the 2D XY model universality class, such as 2D superconductors would be expected to undergo a continuous quantum phase transitions belonging to the 3D XY model if $z = 1$. Knowledge of the universality class of the phase transition does not imply knowledge of the microscopic model. Whereas the electrostatically tuned SI transition appears to involve a simultaneous transition between an insulator exhibiting Mott variable range hopping and a 2D quantum corrected metal, which would appear to be electronic, the field induced transitions are not susceptible to such simple interpretation, and their microscopic nature appears to be an open question.

References

Abrahams, E., Anderson, P.W., D.C.Licciardello, and T.V.Ramakrishnan (1979). Scaling theory of localization: Absence of quantum diffusion in two dimensions. *Phys. Rev. Lett.*, **42**, 673–676.

Ahn, C. H., Bhattacharya, A, DiVentra, M., Eckstein, J. N., Frisbie, C. Daniel, Gershenson, M. E., Goldman, A. M., Inoue, I. H., Mannhart, J., Millis, Andrew J., Morpurgo, Alberto F., Natelson, Douglas, and Triscone, Jean-Marc (2006). Electrostatic modification of novel materials. *Reviews of Modern Physics*, **78**, 1185–1212.

Anderson, P.W. (1959). Theory of dirty superconductors. *Journal of Physics and Chemistry of Solids*, **11**, 26–30.

Baturina, T. I., Strunk, C., Baklanov, M. R., and Satta, A. (2007). Quantum metallicity on the high-field side of the superconductor-insulator transition. *Phys Rev Lett*, **98**, 127003.

Beasley, M.R., Mooij, J.E., and Orlando, T.P. (1979). Possibility of vortex-antivortex pair dissociation in 2-dimensional superconductors. *Phys Rev Lett*, **42**, 1165–1168.

Bhattacharya, A., Eblen-Zayas, M., Staley, N.E., Huber, W.H., and Goldman, A.M. (2004). Micromachined srtio3 single crystals as dielectrics for electrostatic doping of thin films. *Appl Phys Lett*, **85**, 997–999.

Broun, D. M. (2008). What lies beneath the dome? *Nat Phys*, **4**, 170–172.

Buckel, W. and Hilsch, R. (1954). Influence of condensation at low temperatures on the electrical resistance and superconductivity of various metals. *Z. Physik*, **138**, 109–120.

Crane, R., Armitage, N. P., Johansson, A., Sambandamurthy, G., Shahar, D., and Gruner, G. (2007). Survival of superconducting correlations across the two-dimensional superconductor-insulator transition: A finite-frequency study. *Phys Rev B*, **75**, 184530.

Dubi, Yonatan, Meir, Yigal, and Avishai, Yshai (2007). Nature of the superconductor-insulator transition in disordered superconductors. *Nature*, **449**, 876–880.

Ebisawa, Hiromichi, Fukuyama, Hidetoshi, and Maekawa, Sadamichi (1985). Superconducting transition temperature of dirty thin films in weakly localized regime. *Journal of the Physical Society of Japan*, **54**, 2257–2268.

Finkelstein, A. (1994). Suppression of superconductivity in homogeneously disordered-systems. *Physica B*, **197**, 636–648.

Fisher, M.P.A. (1990). Quantum phase-transitions in disordered two dimensional superconductors. *Phys. Rev. Lett.*, **65**, 923–926.

Fisher, M.P.A., Grinstein, G., and Girvin, S.M. (1990). Presence of quantum diffusion in 2 dimensions - universal resistance at the superconductor-insulator transition. *Phys Rev Lett*, **64**, 587–590.

G. Sambandamurthy, L.W. Engel, A. Johansson and Shahar, D. (2004). Superconductivity-related insulating behavior. *Phys Rev Lett*, **92**, 107005.

Gantmakher, V., Golubkov, M., Dolgopolov, V., Tsydynzhapov, G., and Shashkin, A. (2000). Scaling analysis of the magnetic field-tuned quantum transition in superconducting amorphous in-o films. *JETP Letters*, **71**, 160–164.

Ghosal, A., Randeria, M., and Trivedi, N. (1998). Role of spatial amplitude fluctuations in highly disordered s-wave superconductors. *Phys Rev Lett*, **81**, 3940–3943.

Ginzburg, V. L. and Kirzhnitz, D.A. (1964). Superconductivity of electrons occupying levels on the surface (of solids). *Sov. Phys. JETP*, **19**, 269.

Goldman, A.M. and Markovic, N. (1998). Superconductor-insulator transitions in the two-dimensional limit. *Phys Today*, **51**, 39–44.

Graybeal, J. M. and Beasley, M. R. (1984). Localization and interaction effects in ultrathin amorphous superconducting films. *Phys. Rev. B*, **29**, 4167–4169.

Haviland, D.B., Liu, Y., and Goldman, A.M. (1989). Onset of superconductivity in the two-dimensional limit. *Phys Rev Lett*, **62**, 2180–2183.

Hebard, A.F. and Paalanen, M.A. (1990). Magnetic-field-tuned superconductor-insulator transition in 2-dimensional films. *Phys Rev Lett*, **65**, 927–930.

Herbut, I.F. (2001). Quantum critical points with the coulomb interaction and the dynamical exponent: When and why z=1. *Phys Rev Lett*, **87**, 137004.

Hernandez, L.M. and Goldman, A.M. (2002). Bottom-loading dilution refrigerator with ultrahigh vacuum deposition capability. *Rev Sci Instrum*, **73**, 162–164.

Jaeger, H., D.Haviland, Orr, B.G., and Goldman, A.M. (1989). Onset of superconductivity in ultrathin granular metal-films. *Phys Rev B*, **40**, 182–196.

Kosterlitz, J.M. and Thouless, D.J. (1973). Ordering, metastability and phase-transitions in 2 dimensional systems. *J Phys C Solid State*, **6**, 1181–1203.

Larkin, Anatoli and Varlamov, Andre (2009). *Theory of fluctuations in superconductors* (Revised edn). Oxford University Press, Oxford.

Markovic, N., Christiansen, C., Mack, A., Huber, W., and Goldman, A.M. (1999). Superconductor-insulator transition in two dimensions. *Phys Rev B*, **60**, 4320–4328.

Mason, N. and Kapitulnik, A. (2001). True superconductivity in a two-dimensional superconducting-insulating system. *Phys Rev B*, **64**, 060504.

Mason, N. and Kapitulnik, A. (2002). Superconductor-insulator transition in a capacitively coupled dissipative environment. *Phys Rev B*, **65**, 220505.

Orr, B.G. and Goldman, A.M. (1985). Ultrahigh-vacuum evaporation system with low temperature measurement capability. *Rev Sci Instrum*, **56**, 1288–1290.

Paalanen, M., Hebard, A., and Ruel, R. (1992). Low-temperature insulating phases of uniformly disordered 2-dimensional superconductors. *Phys Rev Lett*, **69**, 1604–1607.

Parendo, K.A., Tan, K.H.S.B., Bhattacharya, A., Eblen-Zayas, M., Staley, N.E., and Goldman, A.M. (2005). Electrostatic tuning of the superconductor-insulator transition in two dimensions. *Phys Rev Lett*, **95**, 049902.

Parendo, Kevin A., Tan, K. H. Sarwa B, and Goldman, A. M. (2006*a*). Electrostatic and parallel-magnetic-field tuned two-dimensional superconductor-insulator transitions. *Phys Rev B*, **73**, 174527.

Parendo, Kevin A, Tan, K. H. Sarwa B., and Goldman, A. M. (2006*b*). Hot-electron effects in the two-dimensional superconductor-insulator transition. *Phys Rev B*, **74**, 134517.

Parker, J., Read, D., Kumar, A., and Xiong, P. (2006). Superconducting quantum phase transitions tuned by magnetic impurity and magnetic field in ultrathin a-pb films. *Europhysics Letters*, **75**, 950–956.

Qin, Y.G., Vicente, C.L., and Yoon, J. (2006). Magnetically induced metallic phase in superconducting tantalum films. *Phys Rev B*, **73**, 100505.

Sacepe, B., Chapelier, C., Baturina, T. I., Vinokur, V. M., Baklanov, M. R., and Sanquer, M. (2008). Disorder-induced inhomogeneities of the superconducting state close to the superconductor-insulator transition. *Phys Rev Lett*, **101**, 157006.

Sachdev, Subir (1999). *Quantum phase transitions.* Cambridge University Press, Cambridge and New York.

Sambandamurthy, G., Engel, L.W., Johansson, A., Peled, E., and Shahar, D. (2005). Experimental evidence for a collective insulating state in two-dimensional superconductors. *Phys Rev Lett*, **94**, 017003.

Sebastian, Suchitra E., Harrison, N., Altarawneh, M. M., Mielke, C. H., Liang, Ruixing, Bonn, D. A., and Lonzarich, G. G. (2010). Metal-insulator quantum critical point beneath the high tc superconducting dome. *Proceedings of the National Academy of Sciences*, **107**, 6175–6179.

Seo, Y., Qin, Y., Vicente, C. L., Choi, K. S., and Yoon, Jongsoo (2006). Origin of nonlinear transport across the magnetically induced superconductor-metal-insulator transition in two dimensions. *Phys Rev Lett*, **97**, 057005.

Sondhi, S.L., Girvin, S.M., Carini, J.P., and Shahar, D. (1997). Continuous quantum phase transitions. *Rev. Mod. Phys*, **69**, 315–333.

Spivak, B., Oreto, P., and Kivelson, S. (2008). Theory of quantum metal to superconductor transitions in highly conducting systems. *Phys Rev B*, **77**, 18.

Steiner, M. and Kapitulnik, A. (2005). Superconductivity in the insulating phase above the field-tuned superconductor-insulator transition in disordered indium oxide films. *Physica C-Superconductivity and Its Applications*, **422**, 16–26.

Steiner, Myles A, Breznay, Nicholas P, and Kapitulnik, Aharon (2008). Approach to a superconductor-to-bose-insulator transition in disordered films. *Phys Rev B*, **77**, 212501.

Ueno, K., Nakamura, S., Shimotani, H., Ohtomo, A., Kimura, N., Nojima, T., Aoki, H., Iwasa, Y., and Kawasaki, M. (2008). Electric-field-induced superconductivity in an insulator. *Nat Mater*, **7**, 855–858.

Valles, J.M., Dynes, R.C., and Garno, J.P. (1989). Superconductivity and the electronic density of states in disordered two dimensional metals. *Phys Rev B*, **40**, 6680–6683.

Valles, J.M., Dynes, R.C., and Garno, J.P. (1992). Electron-tunneling determination of the order-parameter amplitude at the superconductor-insulator transition in 2d. *Phys Rev Lett*, **69**, 3567–3570.

Wellstood, F.C., Urbina, C., and Clarke, J. (1994). Hot-electron effects in metals. *Phys Rev B*, **49**, 5942–5955.

Y. Dubi, Y. Meir and Avishai, Y. (2005). Quantum hall criticality, superconductor-insulator transition, and quantum percolation. *Phys Rev B*, **71**, 125311.

Yazdani, A. and Kapitulnik, A. (1995). Superconducting-insulating transition in 2-dimensional alpha-moge thin films. *Phys Rev Lett*, **74**, 3037–3040.

12

Magnetic Field-Induced Novel Insulating Phase in 2D Superconductors

G. SAMBANDAMURTHY[1] and N. Peter ARMITAGE[2]

[1] Department of Physics, University at Buffalo, The State University of New York, Buffalo, NY 14260, USA.
[2] Department of Physics and Astronomy, The Johns Hopkins University, Baltimore, MD 21218, USA.

In a superconductor, a zero-resistance quantum many-body state emerges at low temperatures (T) inhibiting all scattering of electrons that can produce resistance. On the other hand, in an insulator the motion of electrons is greatly hindered rendering them unable to carry current at the $T = 0$ limit. At low T and with a zero magnetic field (B), superconducting thin-films have an immeasurably low resistivity (ρ). An application of B perpendicular to the film's surface weakens superconductivity and it is believed that at a well-defined critical field (B_c), superconductivity disappears and an insulating behavior sets in. This B-induced superconductor–insulator transition (SIT) in two dimensions, first observed by Hebard and Paalanen (1990), has been the subject of a large number of both theoretical (Dubi *et al.*, 2007; Feigel'man *et al.*, 2007; Finkel'stein, 1994; Fisher, 1990; Fisher *et al.*, 1990; Ghosal *et al.*, 1998) and experimental (Baturina *et al.*, 2007; Bielejec and Wu, 2002; Butko and Adams, 2001; Crane *et al.*, 2007*b*; Gantmakher *et al.*, 1998; Goldman and Markovic, 1998; Sacepe *et al.*, 2008; Sambandamurthy *et al.*, 2004; Steiner and Kapitulnik, 2005; Stewart *et al.*, 2007; White *et al.*, 1986; Yazdani and Kapitulnik, 1995) studies. The SIT in two-dimensional systems is theoretically considered within the framework of continuous quantum phase transitions (Sachdev, 1999; Sondhi *et al.*, 1997). Fisher and co-workers (Fisher, 1990; Fisher *et al.*, 1990) postulated a dual description of the SIT in which the superconducting and the insulating phases are caused by the condensation of Cooper pairs and vortices, respectively into a superfluid state. In such a picture, local superconductivity must persist beyond the transition. According to Fisher's theory (Fisher, 1990), the transition is manifested by a change in the macroscopic vortex state, and Cooper pairs will exist, albeit localized, in the insulating phase to support the formation of vortices. While several experimental (Christiansen *et al.*, 2002; Gantmakher *et al.*, 1998; Goldman and Markovic, 1998; Hebard and Paalanen, 1990; Sambandamurthy *et al.*, 2004; Yazdani and Kapitulnik, 1995) and theoretical (Fisher, 1990; Fisher *et al.*, 1990) studies are supportive of this conjecture, others (Valles, Jr. *et al.*, 1992) correlate the SIT with the vanishing of the superconducting gap and amplitude of the order parameter, indicating that Cooper pairs do not survive the transition to the insulator. In this chapter, we present data obtained from our DC and AC conductivity studies of disordered thin films of amorphous indium oxide (a:InO), which point to a possible relationship between the conduction mechanisms in the superconducting and insulating phases in the presence of a magnetic field.

12.1 Magnetic field as the tuning parameter

A typical magnetic field tuned superconductor–insulator transition in an a:InO film is shown in Fig. 12.1. In Fig. 12.1 (a) ρ vs B isotherms, taken at several T between 0.014 and 0.8 K, for a typical superconducting film are shown. Here, we focus on the lower B-range of the data, which includes the critical point of the B-driven SIT. The transition point is identified with the crossing point of the different isotherms at $B_c = 7.23$ T. For $B < B_c$ the superconducting phase prevails, with ρ increasing with T, while for $B > B_c$ an insulating behavior takes over and ρ is a strongly decreasing function of T. For this sample, the value of ρ at the transition is 4.3 kΩ. The behavior in this low-B range is in accordance with previous observations of the B-driven SIT (Bielejec

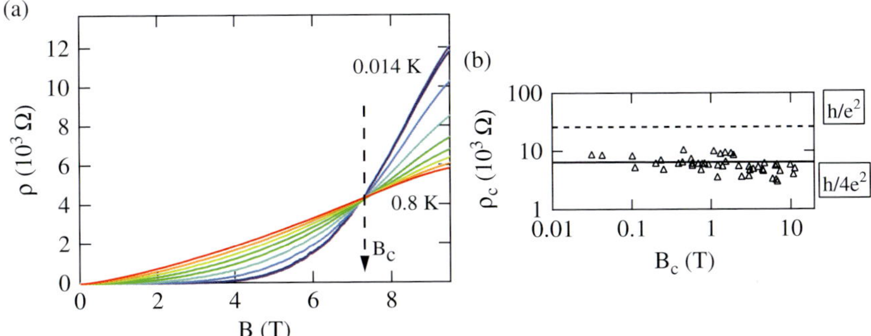

Fig. 12.1 (a) ρ versus B isotherms at a low B range for an a:InO film at $Ts = 0.014$ K, and 0.1–0.8 K in 0.1 K intervals. The critical point of the B-driven SIT, B_c, is indicated by the vertical arrow. (b) Critical resistance at the crossing point for our superconducting samples are plotted against the corresponding B_c values. The solid horizontal line marks $h/4e^2$ and the dashed horizontal line marks h/e^2.

and Wu, 2002; Gantmakher *et al.*, 1998; Goldman and Markovic, 1998; Hebard and Paalanen, 1990; Sambandamurthy *et al.*, 2004; Yazdani and Kapitulnik, 1995). In the next section, the novel behavior in the insulating phase when the magnetic field is increased well beyond B_c is discussed.

Hebard and Paalanen's results (1990) were in accord with Fisher's scaling theory argument (Fisher, 1990), which identified a universal sheet resistance separating a superconducting phase of localized vortices and Bose-condensed electron pairs from an insulating phase of Bose-condensed vortices and localized electron pairs . It is therefore interesting to note the resistance values at the critical point of the phase transition in our samples. In Fig. 12.1 (b) the ρ value at B_c, ρ_c from several superconducting a:InO samples that exhibited a well-defined crossing point are plotted. The data are scattered around 5.8 kΩ with a standard deviation of 1.8 kΩ. A scatter of about a factor of 3 in ρ_c cannot usually be taken as an indication of a universal number. However, if we consider the fact that the measured ρs themselves span more than ten orders of magnitude, a threefold variation is not that large, indicating that the Cooper-pair quantum resistance $h/4e^2$ ($\approx$ 6.45 kΩ) is of special significance for the transition. This value is in rough agreement with the theoretical prediction for this transition (Fisher, 1990) and with other experimental studies in the literature (Hebard and Paalanen, 1990; Liu *et al.*, 2068), with the notable exception of experiments done on MoGe films (Yazdani and Kapitulnik, 1995).

12.2 Novel transport properties in the insulating phase

The transport behavior of a superconducting film changes dramatically when we consider a broader range of B. In Fig. 12.2 isothermal curves from another sample, taken over a larger B range, at several T between 0.07 and 1 K are presented. As expected, when B increases beyond B_c (0.45 T for this film), the insulating behavior

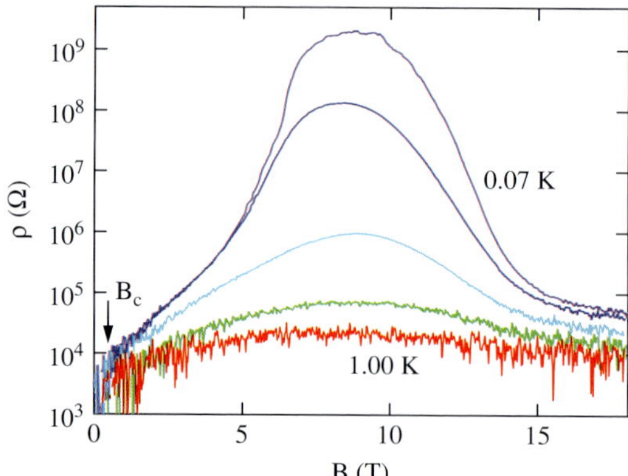

Fig. 12.2 ρ versus B isotherms at a large B range for a sample at T of 0.07, 0.16 , 0.35, 0.62 and 1.00 K. The critical point of the B-driven SIT, B_c, is indicated by the vertical arrow.

initially becomes more pronounced. The surprising feature in our data is that, even though ρ increases by more than five orders of magnitude from its value at B_c (the ordinate in Fig. 12.2 is plotted using a logarithmic scale) (Butko and Adams, 2001), beyond approximately 9 T this trend reverses, the insulator becomes increasingly weaker, and eventually ρ approaches a value of 70 kΩ (at 0.07 K). This represents a drop of more than four orders of magnitude from the peak value of 2 GΩ. The demise of the insulating behavior seems to take place in two steps. An initial, rapid, drop beyond 10 T followed by a transition at 14 T to a slower decrease that continues all the way to, and beyond, our highest B of 18 T.

It is unclear at which B value this trend will revert to the positive magnetoresistance expected at very high B. First, we consider the details of the transport just above B_c and closer to the resistance peak. We present three pieces of evidence that suggest a possible relationship between the transport mechanism in the insulating and superconducting phases in a:InO films.

12.2.1 Activation energy scales in the insulating phase

Inspecting the details of the T dependence of ρ in the B-induced insulating phase reveals that the insulator follows an Arrhenius T dependence,

$$\rho = \rho_0 \; exp(T_I/T) \tag{12.1}$$

indicating the existence of a mobility gap at the Fermi energy, with $k_B T_I$ the magnitude of the gap (k_B is the Boltzmann constant). Fitting our data to an Arrhenius form as in eqn (12.1) yields $T_I(B)$, the activation temperature, which depends on B. In Fig. 12.3 (a) we plot T_I versus B for the sample of Fig. 12.2. At B_c, the onset of the insulating behavior, T_I is zero. It then increases, with B reaching a maximum at 9 T and, similarly to ρ itself, decreases at higher B. A rough extrapolation of our data indicates that T_I will vanish at 20 T, although at B values larger than 14 T the estimates of T_I suffer from increasingly large errors, thus making this extrapolation inaccurate.

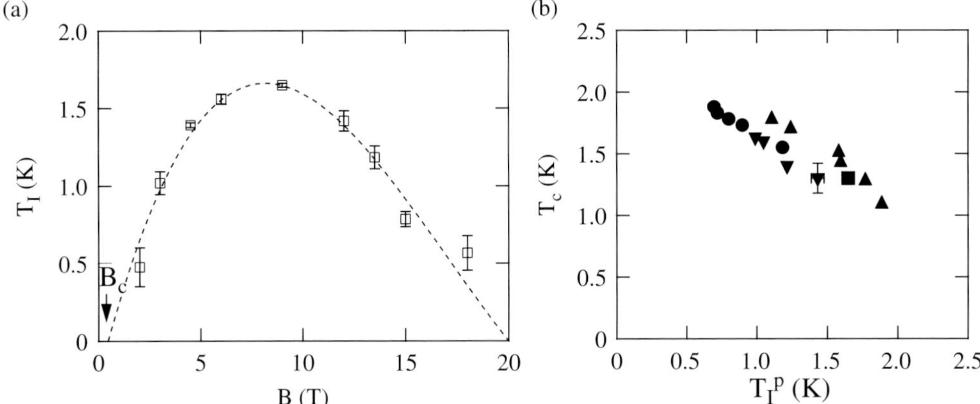

Fig. 12.3 (a) T_I, calculated from fits to eqn (12.1), as a function of B for the sample in Fig. 12.2. T_I has a peak at 9 T. The vertical arrow marks B_c ($= 0.45$ T) where $T_I = 0$. The dashed line is a guide to the eye. (b) T_c at $B = 0$ for several superconducting samples is plotted against T_I at the high-B insulating peaks. Each physical sample is marked with a different symbol and samples have been annealed to vary T_c. Error bars indicated are typical of most data points.

An interesting observation can be made when considering the values of T_I. At the insulating peak $T_I = 1.65$ K, conspicuously close to the $B = 0$ value of the superconducting transition temperature (T_c) for this film of 1.27 K. To test this observation we plot in Fig. 12.3 (b), T_c at $B = 0$ against T_I at the insulating peak for several of our superconducting samples. Overall, T_I is close to T_c, supporting the observation made above. However, in the cases where we are able to follow one physical sample through several anneal stages, we see that as T_c increases upon lowering of the disorder, T_I decreases, indicating that a more complete theory is required to account for the details of the T_c–T_I dependence. The closeness of T_I to T_c suggests the existence of a relationship between the conduction mechanism of the superconductor and that of the insulator, more details of which are given below. Although the non-monotonic behavior of the B-induced insulator has been observed before in a:InO films (Gantmakher *et al.*, 1998; Paalanen *et al.*, 1992) and in other materials (Baturina *et al.*, 2007; Butko and Adams, 2001), two new observations of this insulator require elucidation. They are:

(a) the insulating peak in a:InO samples has an unusually large ρ value
(b) the activation temperature of the insulator at the peak is close to the superconducting transition temperature of the film at $B = 0$.

In the remainder of this chapter, we present more evidence that suggests that it may be possible to account for both observations by adopting the point of view in which the properties of a:InO films are not uniform over the entire sample (Ghosal *et al.*, 1998; Kowal and Ovadyahu, 1994; Meyer and Simons, 2001).

12.2.2 Current-voltage characteristics

Several recent experiments, such as the ones presented above, provide strong evidence that at least some of the superconducting correlations remain in the insulating state of a:InO and other two-dimensional superconducting films (Baturina *et al.*, 2007; Gantmakher and Golubkov, 1995; Gantmakher *et al.*, 2000; Paalanen *et al.*, 1992; Sambandamurthy *et al.*, 2004; Stewart *et al.*, 2007). We now turn to the current–voltage characteristics of the insulator, looking for more clues in support of the above observations. However, first it is instructive to examine the evolution of the non-linear current–voltage characteristics from the superconducting state through the transition and into the insulator.

In Fig. 12.4 (a) we plot the four-terminal differential resistance (dV/dI), as a function of dc current ($I_{\rm dc}$), taken from an a:InO film and measured at $T = 0.01$ K. We show three representative traces that are measured in the $B = 0$–0.65 T range. The bottom trace in Fig. 12.4 (a), taken at $B = 0$ T, is typical of a superconductor: dV/dI is immeasurably low as long as $I_{\rm dc}$ is below a well-defined value, $I^c_{\rm dc}$, which is the critical current of the superconductor at that particular B value. When $I_{\rm dc} > I^c_{\rm dc}$, superconductivity is destroyed and a dissipative state emerges. The situation is different for the middle trace of Fig. 12.4 (a) taken at $B = 0.35$ T. Even at $I_{\rm dc} = 0$ a zero-resistance state is not observed down to the lowest T. However, the current–voltage characteristic still maintains a superconducting flavor: it is non-ohmic and dV/dI increases with increasing $I_{\rm dc}$. We therefore consider the $B = 0.35$ T trace to represent a transitional state in the path the system takes from being a superconductor to an insulator. It is a matter of some debate whether this transitional state will develop into a full superconductor in the $T = 0$ limit (Chervenak and Valles, Jr.,

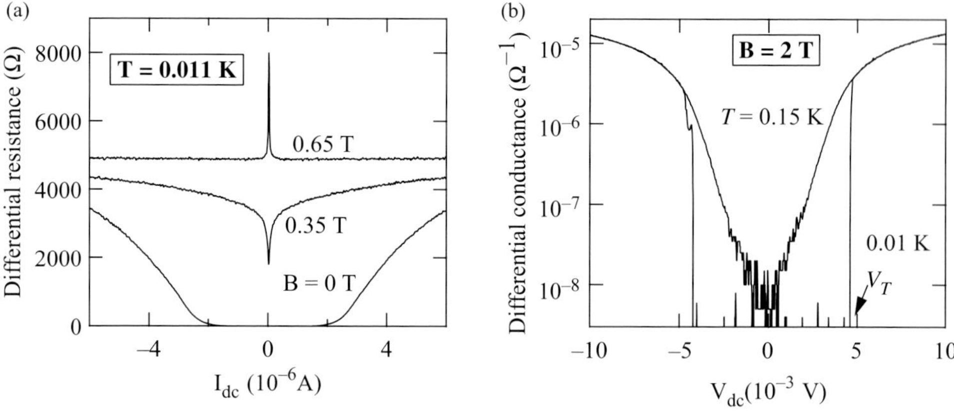

Fig. 12.4 Comparison of the current–voltage characteristics of the B-driven insulating phase at two different T (0.01 K (a) and 0.15 K (b)). The traces show the two-terminal differential conductance measured at $B = 2$ T as a function of DC voltage. The ac excitation voltage applied is 10 μV. The sample used is Ja5 with $B_c = 0.4$ T. V_T marks the threshold voltage for conduction at $T = 0.01$ K.

2000; Ephron *et al.*, 1996; Phillips and Dalidovich, 2003; Yazdani and Kapitulnik, 1995). While the bottom two traces in Fig. 12.4 (a) exhibit superconducting traits, the top trace taken at $B = 0.65$ T clearly does not. Instead, it has the opposite low-I_{dc} dependence indicative of an insulating state: it is again non-ohmic, but this time an increase in I_{dc} results in a decrease of dV/dI. This is consistent with the ρ–B data of these samples, where the transition to the insulating phase occurs at $B_c = 0.4$ T. While the non-linear current–voltage characteristic deep in the superconducting phase is usually attributed to current-induced vortex depinning (Rzchowski *et al.*, 1990), the origin of non-linearity in the insulating phase is less clear.

Let us now focus on the insulating phase well above B_c. Since we are now dealing with an insulator, it is natural to consider the differential conductance (dI/dV) rather than dV/dI. In Fig. 12.4 (b) we plot two-terminal dI/dV traces against the applied dc voltage (V_{dc}) measured at $B = 2$ T, well above the B_c ($= 0.4$ T) of this sample. In the figure, we contrast two traces that are measured at $T = 0.15$ K and 0.01 K. The data taken at $T = 0.15$ K are typical of an insulator, i.e. they are strongly non-ohmic, having a low but measurable value at $V_{dc} = 0$, which increases smoothly with increasing V_{dc}. No clear conduction threshold can be identified at this T. The response drastically changes when the film is cooled to 0.01 K: as long as V_{dc} is below a well-defined threshold value (V_T), the value of dI/dV is immeasurably low. At that V_{dc} ($= 4.65$ mV for this sample) dI/dV increases abruptly, by several orders of magnitude, and remains finite for higher values of V_{dc}. This clearly shows the emergence, at low T, of finite V_T for conduction in the insulating phase.

We have so far described the emergence, at a well-defined temperature, of sharp thresholds for conduction on the insulating side of the SIT in a:InO films. Before the implications of this observation are discussed, it is illuminating to construct a map of the B-driven insulating phase in the B–V_{dc} parameter space available in our

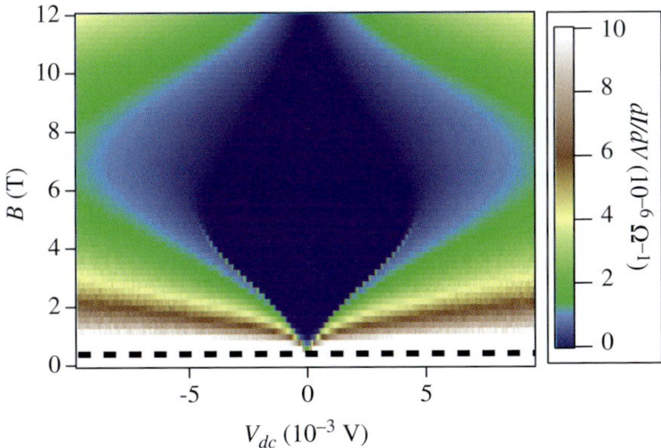

Fig. 12.5 Two-dimensional map of the dI/dV values in the B–V_{dc} plane. For the sample of Fig. 12.4, dI/dV traces as a function of V_{dc} at B intervals of 0.2 T and at $T = 0.01$ K are measured. The color scale legend on the right hand side shows the various colors used to represent the values of dI/dV. The horizontal dashed line denotes B_c ($= 0.4$ T) of this sample.

experiments. In Fig. 12.5 the behavior of a:InO films is summarized in the form of a two-dimensional color map of the dI/dV values in the B–$V_{\rm dc}$ plane for the sample of Fig. 12.4. This map was constructed by measuring dI/dV as a function of $V_{\rm dc}$ at B intervals of 0.2 T at 0.01 K. The colors in the map represent the values of dI/dV, changing from $dI/dV = 0$ (dark blue) to $dI/dV = 10^{-5}\ \Omega^{-1}$ (white). The horizontal dashed line marks $B_{\rm c}$ ($= 0.4$ T) of this sample. Sharp conduction thresholds in the insulating phase can be seen in the map as a sudden change in color in the insulating regime between $B_{\rm c} = 0.4$ T and ~ 5 T. As B is increased the threshold behaviour appears at higher values of $V_{\rm dc}$. This trend continues until B is close to the ρ-peak, which is at $B = 6$ T for this sample. Near the ρ-peak and beyond it, the dI/dV traces no longer exhibit the sharp thresholds for conduction and dI/dV increases smoothly with $V_{\rm dc}$. This manifests itself as a gradual change in colours as $V_{\rm dc}$ is changed for $B \geq 6$ T.

The sharp drop in dI/dV values at low T may be evidence for the condensation of individual charge carriers into a collective state (Christiansen *et al.*, 2002; Fisher, 1990). This brings about interesting analogies with a diverse class of physical systems showing similar thresholds for conduction, which are considered as signatures of collective phenomena. We recall two such examples here: the first is in one-dimensional organic and inorganic solids, where thresholds to conduction have been treated as the signature of the depinning of charge density waves (Gorkov and Grüner, 1989). Similarly, in two-dimensional electron systems confined in semiconductor heterostructures, thresholds for conduction have been attributed to the depinning of a magnetically induced electron solid akin to the Wigner crystal (Goldman *et al.*, 1990; Jiang *et al.*, 1991; Williams *et al.*, 1991). While the appearance in disordered superconducting samples of a:InO of a clear threshold voltage for conduction in the insulating phase can be considered as evidence for vortices condensing into a collective state at a well-defined temperature (Fisher, 1990), more recent studies in a:InO and TiN films (Altshuler *et al.*, 2009; Ovadia *et al.*, 2009; Petkovic *et al.*, 2010) have shown that the non-linear jumps in current–voltage characteristics result from a thermal decoupling of the electrons from the phonon bath in the B-induced insulating phase. This leads to interesting open questions about how such a phonon-decoupling mechanism may arise. Clearly more research is needed to understand this novel transport behavior in two-dimensional superconductors.

12.2.3 Power-law behavior in resistance

The purpose of this section is to show that the resistivity of the superconducting a:InO films can be described by a single function covering a wide range of our measurements, which includes the B-driven SIT. This function can be written as follows:

$$\rho(B,T) = \rho_{\rm c}\left(\frac{B}{B_{\rm c}}\right)^{T_0/2T} \tag{12.2}$$

where $\rho_{\rm c}$, $B_{\rm c}$, and T_0 are sample-specific parameters. The phenomenological form introduced above is consistent with the collective-pinning model of transport in thin superconducting films, which predicts a vortex-pinning energy proportional to $\ln(B)$

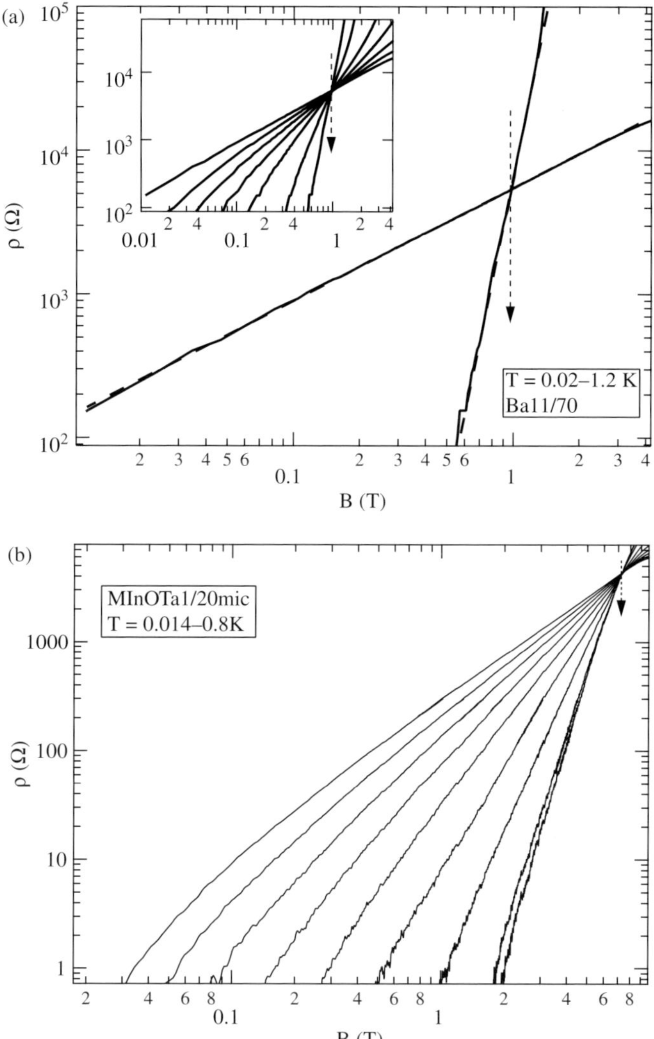

Fig. 12.6 ρ vs B isotherms for two of our samples presented in log–log graphs. (a) Isotherms at $T = 0.02$ and 1.2 K are only shown for clarity. The inset shows the full set taken at $T = 0.02$ K, 0.2 K, 0.4 K, 0.6 K, 0.8 K, 1.0 K and 1.2 K. (b) Data from another sample taken at the same T values as in Fig. 12.1. Vertical arrows mark the crossing point of different ρ isotherms. The ρ isotherms display a power-law dependence on B, shown by the dotted lines in the main figure in (a) which holds on either side of the crossing point.

(Blatter *et al.*, 1994). This form has been observed before in high-T_c layered systems (Inui *et al.*, 1989; Palstra *et al.*, 1988), as well as in amorphous superconductors (Chervenak and Valles Jnr, 1996; Ephron *et al.*, 1996; White *et al.*, 1993). Here we show that this behavior is not restricted to the superconducting phase but continues, uninterrupted, well into the insulating state(Shahar *et al.*, 1998).

We perform a quantitative analysis of the B and T dependence of the resistivity data, in the low-B superconducting regime and closer to B_c. In Fig. 12.6, ρ vs B traces taken at various T for two a:InO samples are plotted using log–log graphs. For the sample in Fig. 12.6 (a), special care is taken to extend the measurements over a large range of ρ. Each curve is well described by a power-law dependence that holds over more than two orders of magnitude in B and more than three in ρ, with non-random deviations that are only seen at high T and as the B values approach the insulating peak. The different curves are distinguished by their power, which is a function of T. A power-law B dependence of ρ in two-dimensional superconductors is associated with the collective-pinning flux-creep transport model, predicting an activation (pinning) energy U_0 that depends logarithmically on B (Blatter *et al.*, 1994):

$$U_0 \propto ln(B_c/B). \tag{12.3}$$

This, in association with activated transport that can be described by

$$\rho(B,T) = \rho_c e^{-T_0(B)/T} \tag{12.4}$$

leads to the power-law dependence of eqn (12.2). Here we use T_0 as the sample-specific activation energy extracted from our data and U_0 as the activation energy as described in the collective pinning theory, though they imply the same physical meaning. Similar power-law behavior in ρ has been observed in disordered thin films (Chervenak and Valles Jnr, 1996; Ephron *et al.*, 1996; White *et al.*, 1993) and in layered high-T_c compounds (Inui *et al.*, 1989; Palstra *et al.*, 1988), and may be indicative of the central role played by vortices in our system. Reiterating the intriguing feature in our results we note that the power-law behavior described by eqn (12.2) continues uninterrupted through B_c and into the insulating state (see inset of Fig. 12.6 (a)). Since, in the insulating phase, $B > B_c$, the pinning energy U_0 in eqn (12.3) becomes negative. This is a natural outcome of the physical picture described by Fisher (Fisher, 1990; Fisher *et al.*, 1990), according to which the vortices, collectively pinned at $B < B_c$, condense at $B > B_c$ rendering the response of the system insulating. Consequently, each vortex that is thermally activated away from the condensate will cause a reduction in the resistance leading to the apparent negative U_0.

12.3 Finite-frequency behavior: superconducting correlations on the insulating side

Complementary to the dc studies, finite-frequency microwave measurements also reveal aspects of the anomalous insulating state. Using a custom low-T multimode cavity resonator, we studied the complex ac conductivity of thin disordered InO_x films as a function of B through the nominal SIT (Crane *et al.*, 2007a,b). We resolved a significant finite-frequency superfluid stiffness well into the insulating regime. These results can also be interpreted as evidence for collective effects and localized Cooper pairs on the superconducting side of the transition.

In general, ac measurements may provide a more complete characterization of low-energy properties of systems than dc measurements alone. The use of finite frequencies allows various time and frequency scales to be extracted. Importantly in an ac measurement both the real and imaginary parts of various response functions can be quantified also. These can be, for example, the complex conductivity $\sigma_1 + i\sigma_2$, the complex dielectric constant $\varepsilon_1 + i\varepsilon_2$, or the complex sheet impedance of a thin film $1/\sigma d = R + i\omega L$. The ability to measure both the real and imaginary components brings complementary information; the real part of the conductivity (the in-phase response) holds information about the dissipative response of the system, while the imaginary part (the out-of-phase response) tells something about the polarizability of charge carriers and/or their ability to move dissipationlessly.

AC measurements as such are particularly useful in the characterization of superconductors. For frequencies much less than the superconducting gap, the imaginary conductivity is proportional to the superfluid density, which is related to the 'phase stiffness' as $\sigma_2 = \sigma_Q \frac{k_B T_\theta}{\hbar\omega}$, where T_θ is the superfluid stiffness in Kelvin and σ_Q is the quantum of conductance for Cooper pairs divided by the sample thickness $\frac{4e^2}{hd}$. The superfluid stiffness is the energy scale for introducing phase slips in the superconducting order parameter $\Psi = \Delta e^{i\phi}$ and is a fundamental quantity for understanding fluctuation phenomena in superfluids and superconductors (Ambegaokar *et al.*, 1980; Berezinskii, 1971; Bishop and Reppy, 1978; Halperin and Nelson, 1979; Hebard and Paalanen, 1985; Kosterlitz and Thouless, 1973). Note that a number of different representations have been used in the literature to characterize the size of the out-of-phase ac response of superconductors (penetration depth, penetration depth squared, superfluid density, etc.). In principle all of these are equivalent. We prefer to use the phase stiffness as it is the most intrinsic quantity.

σ_2 in a true long-range-ordered superconductor has a $1/\omega$ dependence that results from the Kramers–Kronig compatibility with the delta function at $\omega = 0$ in σ_1. The Kramers–Kronig relations are integral relations between real and imaginary parts of response functions guaranteed by causality. The coefficient of the $1/\omega$ is proportional to the superfluid stiffness, and because the quantity $\omega\sigma_2$ is frequency independent in a long-range-ordered superconductor so is the superfluid stiffness. When one has fluctuating superconductivity the delta function at $\omega = 0$ broadens and, due to spectral weight conservation, its moment must move to slightly higher frequencies.

The fact that the Kramers–Kronig relation is an integral relation between σ_1 and σ_2 weighted by $1/\omega$ means that if this spectral weight in σ_1 is still concentrated narrowly enough at reasonably low frequencies, then one can still have a large σ_2. In general, if σ_1 is low and distributed over a broad spectral range as in a normal metal then σ_2 will be small. Indeed with their low normal-state conductivity and strong scattering, we cannot detect any significant out-of-phase component in our highly disordered thin films well above T_c. Thereby we can assign all of the enhancement of σ_2 that we detect to superconductivity.

In the case of a fluctuating superconductor where the delta function in σ_1 is removed σ_2 no longer has a strict $1/\omega$ dependence. However, if the spectral weight in σ_1 is concentrated at low frequency then σ_2 can still be appreciable. If this enhanced σ_1 is due to fluctuating superconductivity, one can still define a generalized superfluid stiffness, but now it has an explicit frequency dependence. Our use of a frequency-dependent stiffness is very similar to the treatment of the superfluid stiffness in the context of the finite-frequency measurements of the Kosterlitz–Thouless–Berezinskii (KTB) transition in liquid He4 films, where one sees that the region of the universal discontinuous drop of the superfluid density broadens and shifts to higher temperatures as the probing frequency is increased. In the context of a KTB transition one considers that as the frequency increases, one probes on shorter and shorter length scales and thereby is less sensitive to the renormalizing effects of intervening unbound vortex–antivortex pairs. The expectation is that in the high-frequency limit the superfluid stiffness will behave in an approximately mean-field-like manner and reflect the underlying background superconductivity. Aspects of this behavior has been observed in thin granular Al layers (Hebard and Fiory, 1980), thin indium/indium-oxide composites (Fiory *et al.*, 1983), He4 thin films (Adams and Glaberson, 1987), as well as other related systems (Corson *et al.*, 1999).

Despite their potential impact there have been very few experiments using finite frequencies through the SIT. This is largely due to the general difficulties and experimental constraints (the need for high, but not too high frequencies, low temperature, high magnetic fields) in performing such measurements. In order to reveal the ground-state fluctuation behavior, one generally needs $\hbar\omega \gg K_B T$, with $\hbar\omega < 2\Delta$, where Δ is the superconducting gap. This generally means that measurements must be performed in the microwave range (20 GHz corresponds approximately to 1 K) in devices such as microwave cavities. We performed such measurements in a novel cryomagnetic resonant microwave cavity system using the cavity perturbation technique. In our case, the cavity diameter was optimized for performance in the 22 GHz ($\hbar\omega/k_B = 1.06$ K) TE011 mode. However, a number of other discrete frequencies from 9 to 106 GHz were accessible by insertion of an additional sapphire puck (for the low frequencies) or use of a very short "pan"-shaped cavity (above 100 GHz).

Our highest operating frequency of 106 GHz corresponds to an energy that is below the threshold for above-gap excitation (158 GHz) in the sample studies here. Relations between the cavity's resonance frequency shift $\Delta\omega$ and changes in quality factor $\Delta(1/Q)$ upon sample introduction to the complex conductivity are obtained by a cavity perturbation technique (Brandt, 1993; Klein *et al.*, 1993; Kötzler *et al.*, 1994; Peligrad *et al.*, 2001) (see Waldron (1967) for a very thorough treatment). This standard experimental technique is based on the adiabatic modification of the electromagnetic fields in a cavity that arises from the introduction of a small sample to the interior of the cavity. Due to their thin-film geometry, our samples fall in the depolarization regime, in which the field penetrates the entire volume of the sample. It has been shown that for extremely thin films, only in-plane ac electric fields or out-of-plane ac magnetic fields at the sample position can affect an appreciable change in a cavity's resonance characteristics (Peligrad *et al.*, 1998). Samples were placed along the cavity's central axis, where due to symmetry consideration and depending on the particular TE mode being used, if the electric field is in-plane then there is a zero out-of-plane magnetic field and vice versa. Modes with both these field configurations were exploited in our setup and their analysis differed (see Crane *et al.* (2007*a*) for details).

We begin by discussing the zero-field behavior along the thermal axis of the field–temperature phase diagram. Here we are interested in understanding the fluctuation behavior of the complex order parameter $\Psi = \Delta e^{i\phi}$ as the temperature is lowered towards the superconducting state. The expectation is that there will be a temperature scale below which the amplitude of the order parameter will become defined, but the phase will still fluctuate. At some lower temperature, the phase will order and a true superconducting state will occur. In the dc data, as shown in Fig. 12.7 (a), we observe a broad region over which the superconducting transition occurs. The contribution of Gaussian amplitude fluctuations can be obtained by fitting to the Aslamazov–Larkin dc form. Using the procedure of Gantmakher (2001), a lower bound on T_{c0} can be estimated as the lowest temperature that does not cause an inflection point in the extracted effective normal-state resistance $R_N(T)$, as defined by the full expression for the Aslamazov–Larkin fluctuation resistivity. Within this analysis, 2.28 K is the best lower bound on T_{c0} and represents the temperature scale below which the superconducting amplitude is relatively well defined. Lower values of T_{c0} produce a kink in the extracted resistivity, where for instance a T_{c0} of 2.2 K is clearly too low, as shown in Fig. 12.7 (a).

In Fig. 12.7 (b), we plot the superfluid stiffness vs temperature as defined above. We see a broad temperature region characterized by a gentle roll-off of the superfluid stiffness with increasing temperature. Moreover, we observe a distinct temperature where the superfluid stiffness acquires a frequency dependence. Within the standard theory, such an occurrence is indicative of the approach to a KTB transition. In this model, the *zero-frequency* superfluid stiffness is renormalized discontinuously to zero at a temperature T_{KTB} set by the superfluid stiffness itself at this temperature. Above T_{KTB} the system still appears superconducting on short length scales set by the separation between thermally generated free vortices. Therefore at finite frequencies we expect the superfluid stiffness to approach zero *continuously*. Since T_{KTB} is the temperature where vortices proliferate, in at least moderate-fugacity superconductors (fugacity being a quantity related to the vortex core energy), the temperature where the superfluid stiffness curves measured at different frequencies deviate from each other can be identified as the approximate location of T_{KTB}. With our units of superfluid stiffness, the predicted transition temperature is $T_{\mathrm{KTB}} = T_\theta/4$, which is shown as a solid line in the figure. We point out here that although a frequency dependence as such is seen in Fig. 12.7, the stiffness where T_θ acquires its frequency dependence is well above the stiffness predicted to be the critical value. This point is expanded on in Crane *et al.* (2007*a*).

Note that any residual normal electron contribution should give an insignificant contribution to σ_2 (Crane *et al.*, 2007*a*,*b*). In principle there can also be a small contribution to σ_2 from "normal" electrons excited above the gap. However, as these

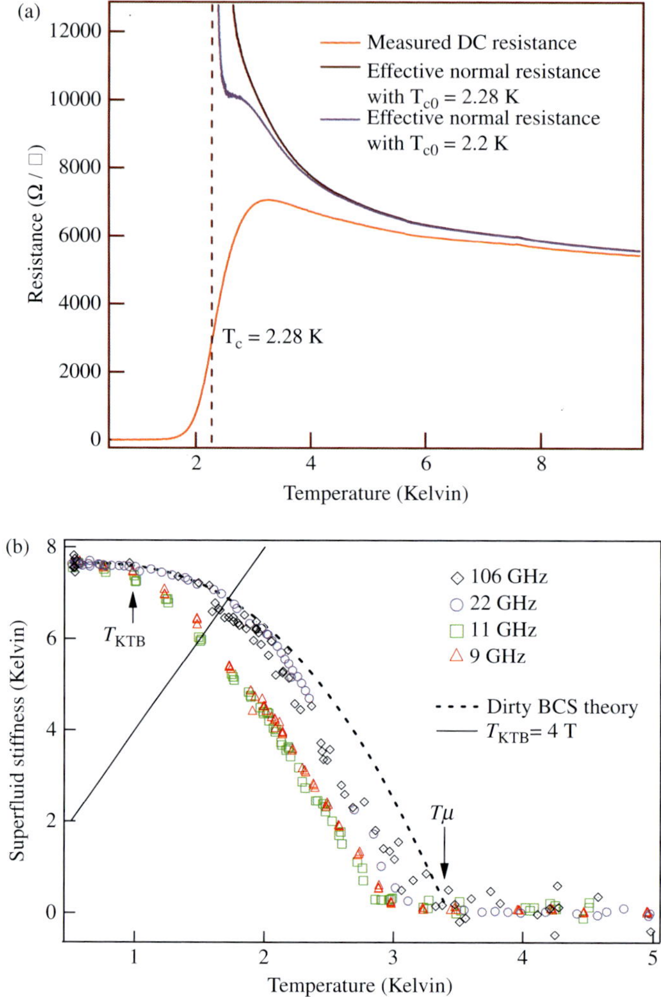

Fig. 12.7 (a) dc sheet resistance. The normal-state resistance curves are generated using the procedure described in the text. T_{c0} is found to be 2.28 K. (b) Temperature dependence of the superfluid stiffness T_θ. The prediction of the dirty Bardeen–Cooper–Schrieffer (BCS) model is shown as a dashed line, which goes to zero at T_μ. The lower-frequency data show a significant deviation from mean-field behavior. The temperature where T_θ acquires a frequency dependence can be identified with T_{KTB}.

electrons are subject to essentially normal-state dissipative processes their contribution to σ_2 is very small because the normal-state scattering is so strong (the material is amorphous with an extremely short mean-free path). These considerations should be particularly true at our, relatively low, frequencies (low as compared to the normal-state scattering rate) in a highly disordered material with a very small normal state

σ_1 and large normal-state scattering rate ($1/\tau$). One can see this via the general theory of metallic conduction, in for instance the Drude model, where the magnitude of σ_2 becomes very small at measurement frequencies well below the normal-state scattering rate, i.e. $\sigma_2(\omega)$ is generally very small when $\omega\tau \ll 1$. From an estimate of the mean-free path in this highly disordered system (6 nm) (from Steiner and Kapitulnik (2005), who used materials very similar to ours) and a reasonable order-of-magnitude guess of the Fermi velocity (1/200th of the speed of light), one gets a very large normal-state scattering rate of 250 THz. At even our largest experimental frequencies (106 GHz) $\omega\tau \ll 1$. Using these numbers, we can make a simple estimate of the order of magnitude of possible "contamination" of our measured superfluid stiffness from above-gap electrons. If *all* the charge carriers that participate in pairing became subject to normal-state dissipative processes they would contribute a contamination signal $T_{\rm cont}$ to the superfluid stiffness, which can be estimated using the relation from the Drude model $T_{\rm cont} = T_\theta[T = 0](\omega\tau)^2$. This is a worst-case scenario of the contribution to σ_2 from these carriers and follows from the Drude formula and our definition of T_θ. We find that even at our maximum frequency of 106 GHz, the largest possible contamination to our superfluid stiffness is of the order of 2×10^{-6} Kelvin, which can be compared to our low-temperature value of 7.66 K. This is far below our sensitivity level. It is a consequence of the large normal-state scattering and is one of the added benefits of using highly disordered samples. In clean materials, in principle it is possible to have a contribution from normal or above-gap electrons. Note that even if we have overestimated the scattering rate by a factor of a hundred, this contamination signal becomes only of order 0.02 K, which is still well below our sensitivity.

Along the "quantum" axis, we can follow the superfluid stiffness signal across the critical point and into the insulating state. In Fig. 12.8 we display the field and temperature dependence of the generalized finite-frequency superfluid stiffness measured at 22 GHz, T_θ. Similar plots can be made at our other measurement frequencies. The finite-frequency superfluid stiffness falls quickly with increasing field, but remains finite above $H_{\rm SIT}$ and well into the insulating regime, to fields almost three times the critical field $H_{\rm SIT} = 3.68$ T. Here the critical field has been defined by the iso-resistance point from dc measurements performed concurrently (Crane *et al.*, 2007*a*). This is one of the first direct model-free measures of superconducting correlations on the insulating side of the two-dimensional superconductor–insulator transition in an amorphous film. It is the use of relatively high probing frequencies that allows us to resolve superconducting fluctuations into the insulating part of the phase diagram.

Our observation of a finite-frequency superfluid stiffness at $H > H_{\rm SIT}$ is not inconsistent with an insulating $T = 0$ ground state. As alluded to above, our experiments are sensitive to superfluid *fluctuations* because we probe the system on short timescales via an experimental frequency $\omega_{\rm Exp}$ that is presumably high compared to an intrinsic order parameter fluctuation rate $\omega_{\rm QC}$ close to the transition. We note that at low temperatures and well into the insulating side of the phase diagram, the superfluid stiffness becomes temperature independent as $T \to 0$. This shows that the observed effects are not thermally driven and is indicative of their intrinsic quantum-mechanical

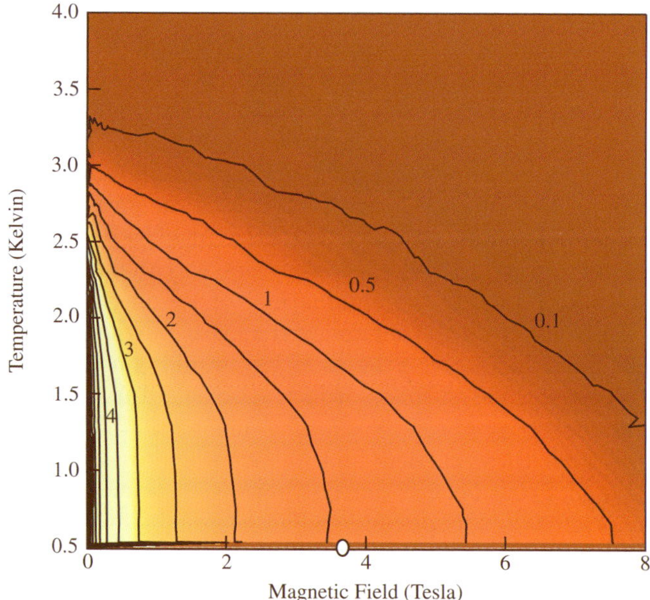

Fig. 12.8 Superfluid stiffness ($\propto \omega\sigma_2$) at 22 GHz given in units of degrees Kelvin. The critical field, H_{SIT}, defined as the iso-resistance point in dc measurements, is shown as a white dot at $H = 3.68$ T. Yellow indicates maximum. Contours with markers are the detection limit for T_θ at specified frequencies. Black dots on the temperature axis denote T_{amp} ($= T_{c0}$) and T_{phase} ($= T_{\mathrm{KTB}}$), which are temperatures signifying the freezing of amplitude and phase fluctuations, respectively. H_{SIT} appears as a black dot on the horizontal axis. Open symbols represent data obtained from a small linear extrapolation beyond our maximum field of 8 Tesla.

nature. Although we cannot rule out inhomogeneous superconducting patches (Ghosal *et al.*, 1998), we consider the large ($10^6\Omega/\square$) and strongly diverging resistance of our samples at the highest fields and low temperatures to at least favor an interpretation of a material that is globally insulating.

These measurements give the first direct evidence for quantum superconducting fluctuations around an insulating ground state and, moreover, the unambiguous evidence for a state of matter with localized Cooper pairs. The finite-frequency superfluid stiffness was an increasing function of ω, showing that superconductivity, although fluctuating on longer length scales, was increasingly well defined on shorter length scales. These measurements give further evidence for a novel insulating state that has collective properties in accord with the the dc measurements discussed in detail above.

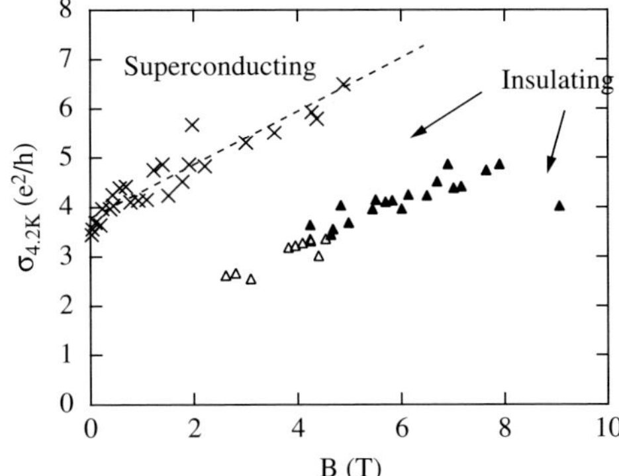

Fig. 12.9 Experimental phase diagram for a:InO superconducting thin films. The crosses mark the B_c of each sample at the SIT. The dashed line best fits the B_c data. Empty and filled triangles mark the positions of the B-induced insulating peak for samples that, at $B = 0$, are insulating and superconducting respectively.

12.4 Summary

To summarize, intriguing results from dc and ac transport measurements of thin films of a:InO that were driven through a SIT by the application of perpendicular B are presented. The observations are discussed from the point of view that the results point to a possible relationship between the conduction mechanisms in the superconducting and insulating phases in these disordered films. We summarize our findings in the form of an experimental phase diagram (Fisher, 1990), shown in Fig. 12.9, plotted in the B-disorder plane. It is constructed based on our data from 11 different a:InO films at 33 different annealing stages. The normal-state conductivity of the films at $T = 4.2$ K ($\sigma_{4.2\mathrm{K}}$) is used as a measure of disorder. Conductivity at higher T could have equally been used with no essential change to the phase diagram. Two kinds of data points are plotted. The crosses mark the B_c of each film and thus define the boundary between the superconducting and insulating phases. The dashed line is a best fit to the B_c data. Extrapolation of this fit to $B = 0$ gives a $\sigma_{4.2K}$ value of $3.8{\pm}0.1\ e^2/h$ (e is the electronic charge and h is Planck's constant). Triangles are used to identify the B-position of the insulating peak for samples that at $B = 0$ were insulating (empty triangles) or superconducting (filled triangles). Our phase diagram is similar to the schematic phase diagram for a two-dimensional superconductor suggested in (Paalanen *et al.*, 1992).

Acknowledgements

GS acknowledges support from NSF DMR-0847324 and University at Buffalo's IRDF grant. NPA is supported by NSF DMR-0847652. The majority of the dc measurements were performed at the Weizmann Institute of Science under the guidance of Prof. Dan Shahar. Some measurements were performed at the National High Magnetic Field Laboratory, which is supported by the NSF Cooperative Agreement No. DMR-0084173 and by the State of Florida.

References

Adams, P.W. and Glaberson, W.I. (1987). *Phys Rev. B*, **35**, 4633–4652.

Altshuler, B.L., Kravtsov, V.E., Lerner, I.V., and Aleiner, I.L. (2009). *Phys. Rev. Lett.*, **102**, 176803.

Ambegaokar, V., Halperin, B.I., Nelson, D.R., and Siggia, E.D. (1980). *Phys. Rev. B*, **21**, 1806–1826.

Baturina, T.I., Mironov, A.Y., Vinokur, V.M., Baklanov, M.R., and Strunk, C. (2007). *Phys. Rev. Lett.*, **99**, 257003.

Berezinskii, V.L. (1971). *Sov Phys JETP*, **32**, 493.

Bielejec, E. and Wu, W. (2002). *Phys. Rev. Lett.*, **88**, 206802.

Bishop, D.J. and Reppy, J.D. (1978). *Phys. Rev. Lett.*, **40**, 1727–1730.

Blatter, G., Feigel'man, M.V., Geshkenbein, V.B., Larkin, A.I., and Vinokur, V.M. (1994). *Rev. Mod. Phys.*, **66**, 1125.

Brandt, E.H. (1993). *Phys. Rev. Lett.*, **71**, 2821.

Butko, V.Y. and Adams, P.W. (2001). *Nature*, **409**, 161.

Chervenak, J.A. and Valles Jnr, J.M. (1996). *Phys. Rev. B*, **54**, R15649.

Chervenak, J.A. and Valles, Jr., J.M. (2000). *Phys. Rev. B*, **61**, R9245.

Christiansen, C., Hernandez, L.M., and Goldman, A.M. (2002). *Phys. Rev. Lett.*, **88**, 037004.

Corson, J., Mallozzi, R., Orenstein, J., Eckstein, J.N., and Bozovic, I. (1999). *Nature*, **398**, 221.

Crane, R.W., Armitage, N.P., Johansson, A., Sambandamurthy, G., Shahar, D., and Gruner, G. (2007*a*). *Phys. Rev. B*, **75**, 094506.

Crane, R.W., Armitage, N.P., Johansson, A., Sambandamurthy, G., Shahar, D., and Gruner, G. (2007*b*). *Phys. Rev. B*, **75**, 184530.

Dubi, Y., Meir, Y., and Avishai, Y. (2007). *Nature*, **449**, 876.

Ephron, D., Yazdani, A., Kapitulnik, A., and Beaseley, M.R. (1996). *Phys. Rev. Lett.*, **76**, 1529.

Feigel'man, M.V., Ioffe, L.B., Kravtsov, V.E., and Yuzbashyan, E.A. (2007). *Phys. Rev. Lett.*, **98**, 027001.

Finkel'stein, A.M. (1994). *Physica B*, **197**, 636.

Fiory, A.T., Hebard, A.F., and Glaberson, W.I. (1983). *Phys Rev. B*, **28**, 5075–5087.

Fisher, M.P.A. (1990). *Phys. Rev. Lett.*, **65**, 923.

Fisher, M.P.A., Grinstein, G., and Girvin, S.M. (1990). *Phys. Rev. Lett.*, **64**, 587.

Gantmakher, V.F. and Golubkov, M.V. (1995). *JETP Letters*, **61**, 606.

Gantmakher, V.F. and Golubkov, M.V. (2001). *JETP Letters*, **73**, 131.

Gantmakher, V.F., Golubkov, M.V., Dolgopolov, V.T., Tsydynzhapov, G.E., and Shashkin, A.A. (1998). *JETP Lett*, **68**, 363.

Gantmakher, V.F., Golubkov, M.V., Dolgopolov, V.T., Tsydynzhapov, G.E., and Shashkin, A.A. (2000). *JETP Letters*, **71**, 160.

Ghosal, A., Randeria, M., and Trivedi, N. (1998). *Phys. Rev. Lett.*, **81**, 3940.

Goldman, A.M. and Markovic, N. (1998). *Phys. Today*, **51**, 39.

Goldman, V.J., Santos, M., Shayegan, M., and Cunningham, J.E. (1990). *Phys. Rev. Lett.*, **65**, 2189.

Gorkov, L.P. and Grüner, G. (ed.) (1989). *Charge Density Waves in Solids.* North-Holland, Amsterdam.

Halperin, B.I. and Nelson, D.R. (1979). *J Low Temp Phys*, **36**, 599.

Hebard, A.F. and Fiory, A.T. (1980). *Phys. Rev. Lett.*, **44**, 291–294.

Hebard, A.F. and Paalanen, M.A. (1985). *Phys. Rev. Lett.*, **54**, 2155.

Hebard, A.F. and Paalanen, M.A. (1990). *Phys. Rev. Lett.*, **65**, 927–930.

Inui, M., Littlewood, P.B., and Coppersmith, S.N. (1989). *Phys. Rev. Lett.*, **63**, 2421.

Jiang, H.W., Stormer, H.L., Tsui, D.C., Pfeiffer, L.N., and West, K.W. (1991). *Phys. Rev. B*, **44**, 8107.

Klein, O., Donovan, S., Dressel, M., and Grüner, G. (1993). *Int J Infrared Millim Waves*, **14**, 2423.

Kosterlitz, M. and Thouless, D. (1973). *J. Phys. C*, **6**, 1181.

Kötzler, J., Nakielski, G., Baumann, M., Behr, R., Goerke, F., and Brandt, E.H. (1994). *Phys. Rev. B*, **50**, 3384.

Kowal, D. and Ovadyahu, Z. (1994). *Solid State Comm.*, **90**, 783.

Liu, Y., McGreer, K.A., Nease, B., Haviland, D.B., Martinez, G., Halley, J.W., and Goldman, A.M. (2068). *Phys. Rev. Lett.*, **67**, 1991.

Meyer, J.S. and Simons, B.D. (2001). *Phys. Rev. B*, **64**, 134516.

Ovadia, M., Sacépé, B., and Shahar, D. (2009). *Phys. Rev. Lett.*, **102**, 176802.

Paalanen, M.A., Hebard, A.F., and Ruel, R.R. (1992). *Phys. Rev. Lett.*, **69**, 1604.

Palstra, T.T.M., Batlogg, B., Schneemeyer, L.F., and Waszczak, J.V. (1988). *Phys. Rev. Lett.*, **61**, 1662.

Peligrad, D.-N., Nebendahl, B., Kessler, C., Mehring, M., Dulcic, A., Pozek, M., and Paar, D. (1998). *Phys. Rev. B*, **58**, 11652.

Peligrad, D.-N., Nebendahl, B., Mehring, M., Dulcic, A., Pozek, M., and Paar, D. (2001). *Phys Rev. B*, **64**, 224504.

Petkovic, A., Chtchelkatchev, N.M., Baturina, T.I., and Vinokur, V.M. (2010). *Phys. Rev. Lett.*, **105**, 187003.

Phillips, P. and Dalidovich, D. (2003). *Science*, **302**, 243.

Rzchowski, M.S., Benz, S.B., Tinkham, M., and Lobb, C. J. (1990). *Phys. Rev. B*, **42**, 2041.

Sacepe, B., Chapelier, C., Baturina, T.I., Vinokur, V.M., Baklanov, M.R., and Sanquer, M. (2008). *Phys. Rev. Lett.*, **101**, 157006.

Sachdev, S. (1999). *Quantum Phase Transitions.* Cambridge University Press, Cambridge.

Sambandamurthy, G., Engel, L.W., Johansson, A., and Shahar, D. (2004). *Phys. Rev. Lett.*, **92**, 107005.

Shahar, D., Hilke, M., Li, C.C., Tsui, D.C., Sondhi, S.L., Cunningham, J.E., and Razeghi, M. (1998). *Solid State Comm.*, **107**, 19.

Sondhi, S.L., Girvin, S.M., Carini, J.P., and Shahar, D. (1997). *Rev. Mod. Phys.*, **69**, 315.

Steiner, M. and Kapitulnik, A. (2005). *Physica C*, **422**, 16.

Stewart, M.D., Yin, A., Xu, J.M., and Valles, Jr., J.M. (2007). *Science*, **318**, 1273.

Valles, Jr., J.M., Dynes, R.C., and Garno, J.P. (1992). *Phys. Rev. Lett.*, **69**, 3567.

Waldron, R.A. (1967). *The Theory of Waveguides and Cavities.* Maclaren & Sons Ltd, London.

White, A.E., Dynes, R.C., and Garno, J.P. (1986). *Phys. Rev. B*, **33**, 3549.

White, W.R., Kapitulnik, A., and Beasley, M.R. (1993). *Phys. Rev. Lett.*, **70**, 670.

Williams, F.I.B., Wright, P.A., Clark, R.G., Andrei, E.Y., Deville, G., Glattli, D.C., Probst, O., Etienne, B., Dorin, C., Foxon, C.T., and Harris, J.J. (1991). *Phys. Rev. Lett.*, **66**, 3285.

Yazdani, A. and Kapitulnik, A. (1995). *Phys. Rev. Lett.*, **74**, 3037.

13
Evidence of Cooper Pairs on the Insulating Side of the Superconductor–Insulator Transition

---

M.D. Stewart, Jr. and James M. Valles, Jr.

Department of Physics, Brown University, Providence, RI 02912

13.1 Introduction: the superconducting order parameter and possible insulating phases

Amorphous film systems have posed something of a puzzle for workers in the field of the superconductor–insulator transition (SIT). Some films, such as InO_x (Gantmakher *et al.*, 1998; Hebard and Paalanen, 1990; Sambandamurthy *et al.*, 2004; Steiner and Kapitulnik, 2005), TiN (Baturina *et al.*, 2007*a*) and Be (Bielejec *et al.*, 2001) exhibit behaviors that qualitatively agree with boson-dominated models of the transition (Fisher, 1990). Their magnetic-field-tuned SITs appear to be Cooper-pair delocalization-to-localization transitions. On the other hand, the behavior of amorphous, ultra-thin elemental films conforms more with fermion dominated models of the transition (Phillips and Dalidovich, 2001; Valles, Jr. *et al.*, 1994). In particular, insulating films that are too thin to superconduct exhibit transport properties dominated by single electrons (Chervenak and Valles, Jr., 1999; Haviland *et al.*, 1989) and a large density of quasiparticle states at the Fermi energy (Hsu *et al.*, 1995; Valles, Jr. *et al.*, 1992). There are signs, however, that some Cooper pairs remain when thicker superconducting films are driven into the insulating phase with a magnetic field, suggesting a "mixed" insulating state composed of quasiparticles and Cooper pairs (Hsu *et al.*, 1995; Parker *et al.*, 2006).

Experimental techniques that detect quasi-particles or Cooper pairs directly can be very useful for determining the insulating ground state (Barber *et al.*, 1994; Valles, Jr. *et al.*, 1992). Electron tunneling, which directly measures the quasiparticle density of states has provided some of the strongest evidence that fermions dominate the insulating phase of amorphous elemental film systems. Similarly direct probes sensitive to Cooper pairs have not been available. Recently, we developed an amorphous film system that can be driven through SITs and probed directly for Cooper pairs (Stewart Jr *et al.*, 2007). It is composed of amorphous elemental films deposited on substrates patterned with a regular array of holes that can be subjected to a perpendicular magnetic field (see Fig. 13.1; Stewart Jr *et al.* (2007)). The coupling of the magnetic vector potential to the phase of the wavefunction of charged carriers and the multiply connected geometry created by the hole array combine to make the transport properties vary periodically with the magnetic field. For charge-$2e$ carriers of Cooper pairs, the period is dictated by the superconducting flux quantum.

In recent experiments we have shown that these insulating amorphous nano-honeycomb (NHC) films contain Cooper pairs. Their magnetoresistances oscillate strongly with a period coinciding with the superconducting flux quantum, belying the presence of charge-$2e$ carriers. In addition, the patterning qualitatively transforms the thickness-tuned and magnetic-field-tuned SITs to render them more similar to Cooper-pair localization-to-delocalization transitions (Nguyen *et al.*, 2009; Stewart Jr *et al.*, 2007).

In this chapter, we first review the evidence that fermions dominate the SITs of unpatterned amorphous elemental films, with the goal of providing a contrast for the SITs of the NHC film systems. Subsequently, we present the SITs of the NHC film system, providing the evidence of Cooper pairing in its insulating phases and the characteristics of their transitions that we can ascribe to bosonic behavior.

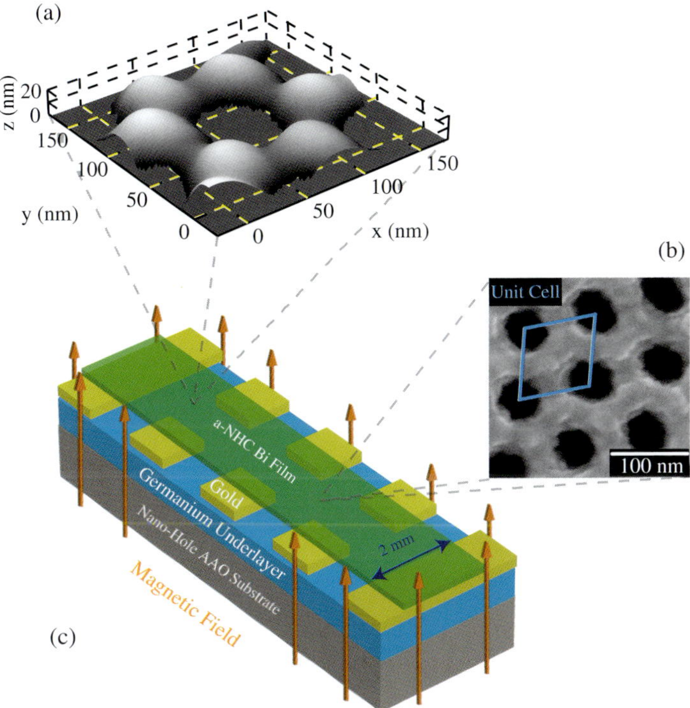

Fig. 13.1 Schematic representation of the nano-honeycomb film system discussed in the chapter (a) illustrating the nano-honeycomb structure (b) of a nano-honeycomb film taken postexperiment and an AFM topograph (c) accompanied by an SEM micrograph. The holes have a center- to-center spacing, a = 100 nm and a radius r_{hole} of 27 nm. Adapted from Stewart Jr *et al.* (2007).

We finish the chapter with comparisons of the behavior in the SITs in NHC-films and indium-oxide-like film systems, which have insulating phases that have been considered to be Cooper-pair- (i.e. boson-) dominated.

13.2 Evidence of fermionic insulators: thin amorphous films

13.2.1 Resistive transitions near the superconductor–insulator transition

A number of amorphous-film systems (Dynes *et al.*, 1986; Graybeal and Beasley, 1984; Haviland *et al.*, 1989; Lee and Ketterson, 1990; Okuma *et al.*, 1998; Strongin *et al.*, 1970) undergo a transition from a non-superconducting state to a superconducting state with increasing film thickness or decreasing sheet resistance, which appears like the one shown in Fig. 13.2. These systems, which include elemental superconductors (e.g. Bi, Pb, and Sn) quench-condensed onto amorphous semiconductor layers (Sb, Ge), alloys such as MoGe (Graybeal and Beasley, 1984), MoC (Lee and Ketterson,

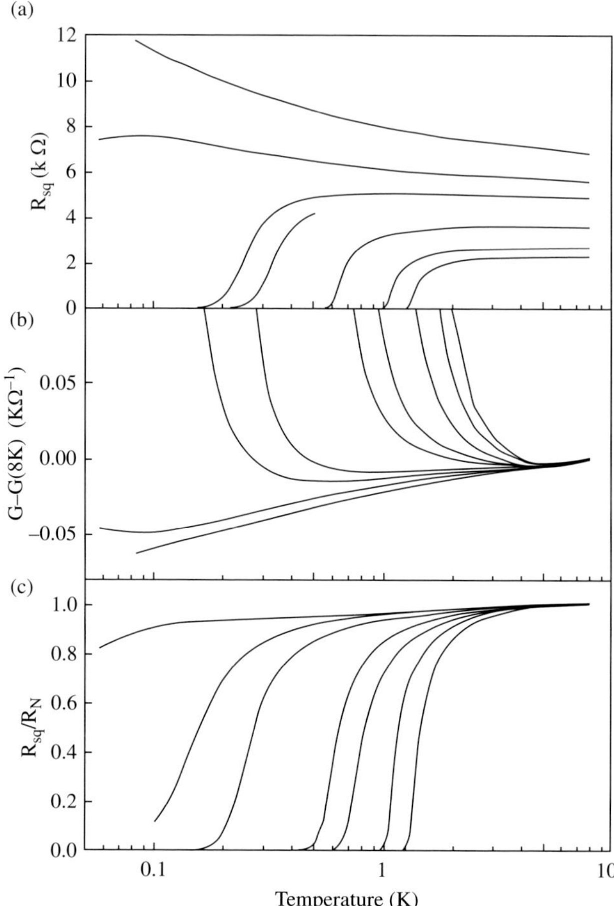

Fig. 13.2 (a) Sheet resistance as a function of temperature of a series of Bi/Sb films near the SIT. The temperature is shown on a logarithmic scale to illustrate the relative broadening of the transitions with proximity to the SIT. (b) Change in the conductance as a function of temperature of the same films as in (a) measured relative to the conductance at 8 K. (c) Sheet resistance as a function of temperature with the logarithmic conductance change removed.

1990) and NbSi (Okuma *et al.*, 1998), possess short elastic mean-free paths l so that $k_F l \simeq 1$, where k_F is the Fermi wavevector become, electrically continuous at near monolayer thicknesses and appear structurally smooth. Figure 13.2 shows the sheet resistance as a function of temperature, $R(T)$, of a series of Bi films formed by repeated quench-condensation of Bi onto a thin layer of quench-condensed Sb (adapted from Chervenak and Valles, Jr. (1999)). $R(T)$ rises exponentially with decreasing temperature for the thinnest films, consistent with the exponential localization of the charge carriers and a zero-temperature insulator phase. This rise becomes more gradual and difficult to discern with increasing thickness as the normal-state sheet resistance, $R_N \equiv R(8\,\text{K})$, becomes comparable to the quantum of resistance for pairs $R_Q = h/4e^2 = 6\ k\Omega$. A transition to a zero-resistance state emerges in slightly thicker films. The transition temperature, as measured at the midpoint of the resistance drop, T_{c0}, is substantially below the 6.2 K bulk transition temperature. Subsequent evaporations produce films with sharper transitions and higher T_{c0}.

Multiple characteristics of these transitions imply that the non-superconducting state is composed of unpaired, weakly localized electrons and the appearance of superconductivity coincides with the appearance of pairs. To begin, all of the non-superconducting films exhibit a monotonically rising resistance with decreasing T and thus do not exhibit signs of the positive paraconductance fluctuations that accompany pairing fluctuations (Chervenak and Valles, Jr., 1999; Haviland *et al.*, 1989; Lee and Ketterson, 1990). The $R(T)$ of all films that do exhibit such paraconductance fluctuations drop monotonically below the temperature at which the fluctuations appear (i.e. they superconduct). Correspondingly, none of the $R(T)$ exhibit the reentrant dip displayed by films composed of superconducting grains that are close to but on the insulating side of the SIT (Barber Jr *et al.*, 2006; Jaeger *et al.*, 1986). These behaviors are emphasized by the plot in Fig. 13.2 (b), which shows the change in the conductance, $G(T)$, below 8 K for the films in Fig. 13.2 (a). All of the $G(T)$ appear to rise steadily at low temperatures from the nearly logarithmic dependence of the lowest-conductance film. The logarithmic dependence of the conductance of the non-superconducting films is expected for an insulating phase consisting of weakly localized electrons. Both the weak-localization and disorder-enhanced electron–electron-interaction corrections to the conductance go like $AG_{00}l\ln(T)$, where $G_{00} = e^2/(2\pi^2\hbar)$ and $A \simeq 1$. The slope of the log dependence in Fig. 13.2 (b) gives $A = 1.25$.

The behavior of the superconducting $R(T)$ also indicate that pairs or the amplitude of the superconducting order parameter disappear on approaching the critical point from the superconducting side. The primary indicator is the continuous reduction of the mean-field transition temperature, estimated as $R(T_{c0}) = R_N/2$, with increasing R_N or decreasing film thickness (see Figs 13.3 and 13.6 for example). This temperature marks the formation of pairs and thus the pair-formation temperature decreases continuously to zero. As shown in Fig. 13.3, this reduction, which has been observed in many thin-film systems, fits well to theories of the detrimental effects of disorder-enhanced electron–electron interaction effects on the pairing interaction and thus T_{c0} (Belitz and Kirkpatrick, 1994; Finkelstein, 1994). As discussed in the next section, electron tunneling experiments provide strong evidence for this scenario.

13.2.2 Tunneling density of states near the superconductor–insulator transition

Electron tunneling measurements have provided some of the strongest evidence that pairs appear at the SIT. This pair-detection method relies on the fact that the density of pairs in a superconductor is proportional to the superconducting energy gap Δ, $n_{CP} \propto N(E_F)\Delta$ and the normalized, superconducting tunneling density of states depends on Δ:

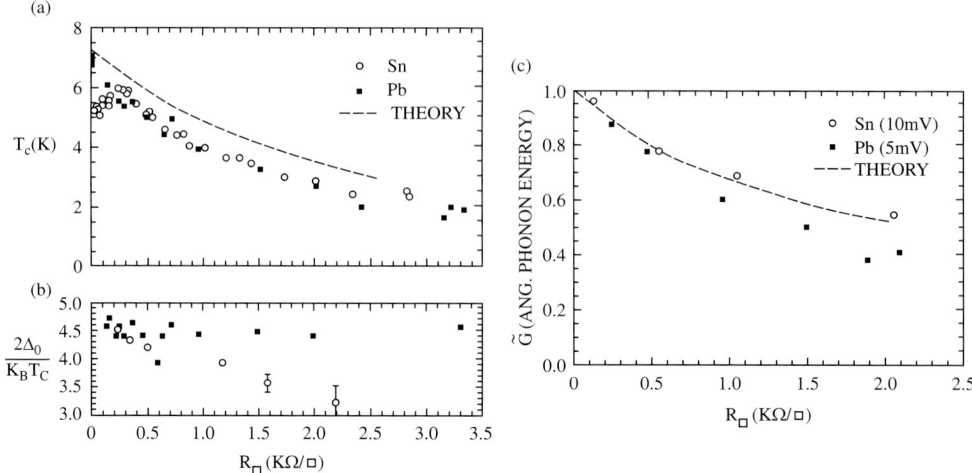

Fig. 13.3 (a) Transition temperature as a function of normal-state sheet resistance for Pb/Ge and Sn/Ge films obtained in multiple experimental runs. The dashed line gives a theoretical prediction based on the measurements of the reduction in the density of states shown in the (c) (b) Gap ratio as a function of sheet resistance. (c) Tunneling density of states measured at the average phonon frequency. The dashed line is the predicted decline of this density of states due to disorder-enhanced electron–electron interaction effects. Adapted from Valles, Jr. *et al.* (1989)

$$N_S(\epsilon) = \frac{|\epsilon|}{\sqrt{\epsilon^2 - \Delta^2}} \tag{13.1}$$

Here, $N(E_F)$ is the density of states at the Fermi energy E_F and ϵ is the quasiparticle energy measured relative to E_F (see for example, Tinkham (2004)).

Tunneling measurements of $N_S(\epsilon)$ for films near the SIT have been made for quench-condensed films using normal metal-film (Al)-oxide-superconducting film tunnel junction structures. The voltage dependence of the differential conductance of these junctions is given by:

$$G_j(V) \propto \int_{-\infty}^{\infty} N_N(\epsilon) N_S(\epsilon) \frac{df(eV + \epsilon)}{dE} d\epsilon \tag{13.2}$$

where $N_N(\epsilon)$ is the normal-state density of states of the superconducting film and f is the Fermi distribution function. For films near the SIT, which have normal-state sheet resistances of more than hundreds of ohms, $N_N(\epsilon)$ exhibits a logarithmic singularity due to disorder-enhanced electron–electron interaction effects. This contribution to $G_j(V)$ can be normalized away to high accuracy by dividing the conductance of the junction below T_{c0} by the conductance obtained at high magnetic fields or at a temperature just above T_{c0}. At temperatures $T \ll T_{c0}$, this normalized conductance

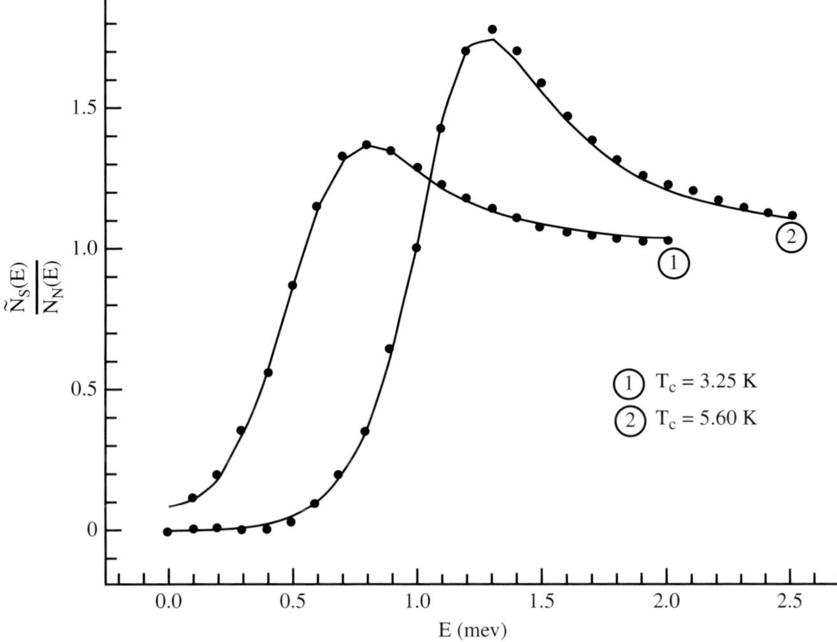

Fig. 13.4 Normalized tunnel junction conductance of two Pb/Ge films with the indicated values of T_{c0}. The lines are fits to the thermally smeared BCS density of states.

is directly proportional to the superconducting density of states, $\tilde{G}_j(\mathrm{V}) \propto \mathrm{N_S(eV)}$ (Dynes *et al.*, 1986; Valles, Jr. *et al.*, 1994).

Dynes and coworkers measured the tunneling density of states (TDOS) of a series of homogeneous Pb/Ge films using a single tunnel junction through successive evaporations of Pb (see Fig. 13.4, which is adapted from Dynes *et al.* (1986)). They showed that the gap ratio $\frac{2\Delta_0}{k_B T_{c0}}$ remained near its strongly coupled value and the TDOS maintained its BCS form for films with T_{c0} as low as 3 K. Thus Δ_0 and T_{c0} decreased at the same rate as the SIT was approached. Subsequent work indicated similar behavior for homogeneous Sn/Ge films and, moreover, correlated the suppression of Δ_0 with a suppression in the normal-state tunneling density of states. Figure 13.3 (a) shows the reduction of T_{c0} for series of Pb/Ge and Sn/Ge films that were simultaneously probed with electron tunneling (Valles, Jr. *et al.*, 1989). The dashed line in Fig. 13.3 (a) is a theoretical prediction for the T_{c0} reduction with R_N (Belitz and Kirkpatrick, 1994). The theory attributes the T_{c0} depression mostly to the reduction in the density of states involved in pairing (see Fig. 13.3 (c)) due to disorder-enhanced electron–electron interactions. Other than the offset, which can be corrected for if the bulk T_{c0} of the films is lower than for crystalline Pb, the agreement between theory and experiment suggests that the growth of disorder-enhanced electron–electron interaction effects drives down the order-parameter amplitude to suppress T_{c0} (Valles, Jr. *et al.*, 1989).

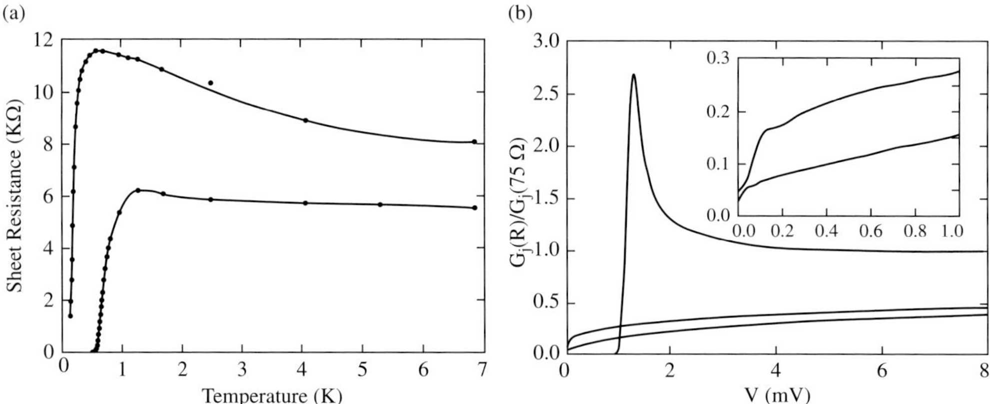

Fig. 13.5 (a) Resistive transitions of two Bi/Sb films close to the SIT. The estimated values of T_{c0}, defined by $R(T_{c0}) = R(7\text{ K})/2$ are 0.19 and 0.7 K. (b) Tunnel junction conductance measured at 0.13 K for the two films on the left and at 0.3 K for a third, much thicker film, with sheet resistance of 75 Ω and $T_{c0} = 6.2$ K. All of the tunnel junction conductances are normalized by the conductance of the 75 Ω film measured at $T = 7\text{ K} > T_{c0}$. Inset: expanded view of the low-voltage data.

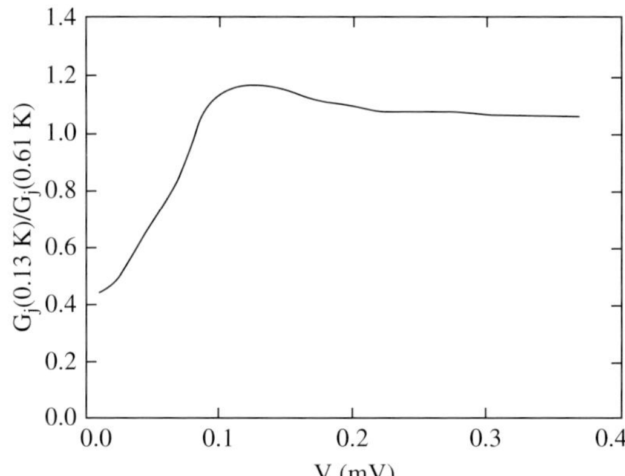

Fig. 13.6 Tunneling density of states of the $T_{c0} = 0.7$ K film as measured at 0.13 K. These data correspond to $\frac{G_j(\text{V},0.13\text{ K})}{G_j(\text{V},0.6\text{ K})}$. The normalizing curve was obtained at a temperature just below T_{c0} where the film sheet resistance was low enough to obtain an accurate measure of the tunnel junction conductance.

Later experiments pushed very close to the SIT, investigating Bi films with T_{c0} as much as a factor of thirty below the bulk T_{c0}. As shown in Fig. 13.5, the superconducting density of states became progressively broader, states appeared in the gap and the energy gap, as measured by the position of the gap edge, continued to decrease toward zero with T_{c0}. The estimated gap ratio, $2\Delta_0/k_BT_{c0}$ appeared to fall below the BCS value as $T_{c0} \rightarrow 0$ (Valles, Jr. *et al.*, 1992).

13.3 Evidence of bosons in insulating thin amorphous films

Superconducting ultra-thin amorphous films that are close to the SIT can be driven to insulating behaviour by applying a magnetic field perpendicular to the film plane. There are transport and tunneling features that suggest that a small fraction of Cooper pairs persist into this insulating phase. Thus magnetic-field and thickness tuning through the SIT appear to produce slightly different insulating ground states. First, the quality of scaling analyses of the $R(T)$, inspired by the dirty-boson model of Fisher (1990), of films including MoGe (Yazdani and Kapitulnik, 1995), Bi (Markovic *et al.*, 1999), and Be (Bielejec and Wu, 2002) strongly suggest that Cooper-pair degrees of freedom survive and play an important role through the magnetic-field-tuned SIT. Such analyses are the subject of Chapter 11 and will not be discussed further here. More qualitatively, closeups of the $R(T)$ in the critical regime reveal non-monotonic temperature dependencies that can be ascribed to pairing fluctuations. Figure 13.7 shows this as a small dip in the $R(T)$ of films that insulate in the low-temperature limit. This slight reentrance is consistent with pairing correlations developing with decreasing temperature, which give rise to local, but not global superconductivity. Xiong and coworkers emphasized this point in a comparison of magnetic-field- and magnetic-impurity- tuned SITs in a similar film system (see Chapter 14). In the case of tunneling, the tunnel junction conductance in the transition region also shows a depleted density of states at the Fermi energy, which is a sign of pairing. For the data in Fig. 13.7, the depletion, which is measured by the suppression of the $G_{\rm N}$ below 1, is more than 10% at the estimated critical field, $H_{\rm c} = 2.5$ T, for the transition. Taken at face value,

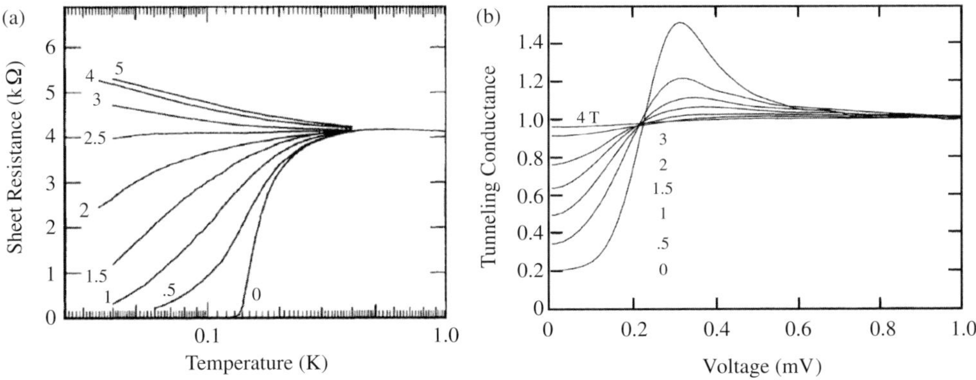

Fig. 13.7 (a) Magnetic-field-tuned SIT for PbBi/Ge film for perpendicular magnetic field values indicated in Tesla. Notice that the $R(T)$ becomes nearly temperature-independent at the "critical field", $\mu_0 H_{\rm c} = 2.5$ T. (b) Tunneling density of states measured at 0.3 K for the same PbBi/Ge film in film perpendicular magnetic fields. The data have been normalized to the tunnel junction conductance above $T_{\rm c0}$. Notice that the density of states at zero bias is suppressed from the normal-state value, even for $H > H_{\rm c}$. This suppression suggests that pairing fluctuations persist into the insulating phase.

these data suggest that about ten percent of the pairs existing at zero field persist to H_c.

13.4 Evidence of bosons in other homogeneous films

One of the vexing issues in the SIT field has been that other disordered films with homogeneous morphologies exhibit qualitatively different behavior from the quench-condensed films considered above. Signatures associated with bosons on the insulating side of the magnetic-field-tuned SIT are much stronger for films of indium oxide and TiN. These films exhibit a larger dip or reentrance in their $R(T)$ and a giant positive magnetoresistance. The latter feature has been attributed to magnetic-field-enhanced localization of Cooper pairs or enhanced mobility of vortices. In addition, the temperature dependence of the $R(T)$ is activated and thus much stronger than the logarithmic dependence exhibited by the quench-condensed amorphous films. These features are described in Chapter 12. Altogether, the behavior has been presumed to indicate a boson-dominated SIT. What the salient differences are between these films and the quench-condensed films are not clear. Like the quench-condensed films, the indium oxide films are also amorphous. On the other hand they are substantially thicker and thus have a lower carrier density. What could drive stronger boson localization in these systems is receiving much attention.

13.5 Amorphous nano-honeycomb films: direct evidence of Cooper pairs

The litany of behaviors discussed above seem to imply that elemental amorphous film systems contain different insulating states than their compound counterparts. To address this question, we have investigated the behavior of ultra-thin elemental films patterned with an NHC array of holes. This patterning technique allows a simple, effective, and direct probe of phase coherence and thus Cooper pairing in the films patterned. For the duration of this chapter, we show how the NHC probe works, describe the features of data taken on NHC films, and compare NHC characteristics to those observed in other film systems.

13.5.1 Phase coherence and nano-honeycomb films

To understand why studying NHC films near the SIT should address the onset of Cooper pairing and the nature of the insulating state, we outline how phase coherence in a film with a multiply connected geometry affects magneto-transport in the film. In the presence of an applied vector potential $\vec{A}$, the phase of the superconducting order parameter ϕ is modified as

$$\nabla\phi = \frac{2e}{h}\left(\vec{A} + \mu_0\lambda^2\vec{J}_\mathrm{s}\right). \tag{13.3}$$

Here, $\vec{J}_\mathrm{s}$ is the supercurrent density and λ is the penetration depth. This form is often invoked to derive fluxoid quantization in superconductors by integrating $\nabla\phi$,

$$\oint \nabla\phi \cdot dl = 2\pi n, \tag{13.4}$$

as demanded by single-valuedness of the wave function. Usually the integral is simplified by choosing a path that encloses Meissner screening currents but where $\vec{J}_\mathrm{s} = 0$ (i.e. a path many λ away from the edge of the hole piercing the superconductor). This condition cannot generally be met in quench-condensed films since disorder pushes λ toward the limit of the film dimension. Regardless of the form of $\nabla\phi$ and the difficulty of the integration, eqn (13.4) must still hold. The above therefore results in a periodic modulation of the kinetic energy of the Cooper pairs in a multiply connected superconductor as well as the free energy. This modulates the (Ginzburg–Landau) solutions for the critical temperature, T_c, which directly implies an oscillation in the resistance of the film. It is these resistance oscillations that NHC films were designed to measure.

As implied by eqn (13.3), the period of the oscillations observed in the resistance depends on the charge of the carriers, given as $2e$ above. This period, referred to hereafter as the matching field,

$$H_\mathrm{M} = h/2eS, \tag{13.5}$$

or one superconducting flux quantum per area, S. Naturally, the area must be specified in order to distinguish between single electron oscillations and those of Cooper pairs. In a typical fluxoid quantization experiment, this area corresponds to the area enclosed by the line integral of eqn (13.3). If the superconductor were a film rolled on its edge into a hollow cylinder, as in the classic experiment of Little and Parks, the area would correspond to the area of the hole in the cylinder. NHC films represent a networked set of Little–Parks cylinders in which resistance oscillations with a number of periods depending on the geometrical arrangement of holes can be observed. Linearized Ginzburg–Landau theory implies the dominant period should be that implied by a unit cell of holes (Bezryadin and Pannetier, 1995) as shown in Fig. 13.1.

While the above is certain to lead to the observation of resistance oscillations in a superconducting NHC film, one may doubt whether resistance oscillations can be observed in a film exhibiting insulating behavior. The key to the observation of oscillations on the insulating side of the SIT lies in the size of the phase-coherence length for Cooper pairs. So long as pairs remain phase-coherent on the length scale of $\sim\sqrt{S}$, resistance oscillations should be observable even if the system has not attained the global phase-coherence necessary for the zero-resistance state. The importance of the small feature size in NHC films is apparent as smaller features increase the sensitivity to phase coherence on the shortest possible length scale.

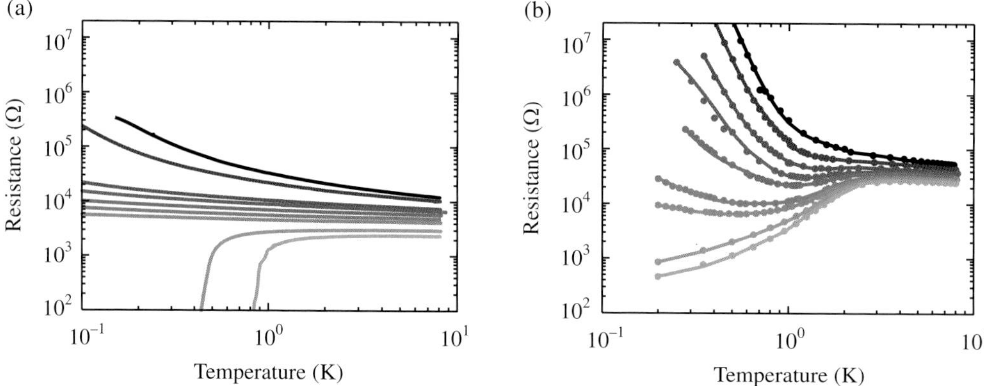

Fig. 13.8 Thickness-tuned SIT transition in Bi/Sb films produced simultaneously (a) without and (b) with the NHC structure. In (a), thicknesses range from 0.57 nm to 0.95 nm. In (b) the thickness ranges from 1.09 nm to 1.32 nm; lines are guides for the eye.

13.5.2 Exponential insulator

We now move on to discuss the behavior of NHC films and their relation to films prepared in the same way but without patterning. Figure 13.8 shows a series of Bi/Sb films of increasing thickness, quench-condensed onto a fire-polished glass substrate and onto an NHC substrate with $r_{\text{hole}} = 27$ nm and hole spacing $a = 100$ nm. Both sets of films pass from an insulating state, where $dR/dT < 0$ at the lowest temperatures, to a superconducting-like state, where $dR/dT > 0$ at the lowest temperatures. Our unpatterned films exhibit the same characteristics as those discussed earlier in this chapter. The $R(T)$ of the thinnest NHC films increases monotonically. However, as the thickness of the film is increased the $R(T)$ nearest the critical point develop a minimum as the $R(T)$ first drop and then begin to increase at a lower temperature. The insulating trend of each of the seven insulators in Fig. 13.8 becomes very dramatic at the lowest temperatures. Each exhibits an activated or nearest-neighbor hopping temperature dependence,

$$R(T) = R_0 e^{(T_0/T)}, \tag{13.6}$$

which can be seen in the Arrhenius plot of Fig. 13.9 (a). Here, the resistance for all seven insulating films shown in Fig. 13.8 can be well described by fits with a single value of $R_0 = 6.2$ kΩ and variable T_0.

The feasibility of other temperature dependencies (i.e $R(T) = R_0 e^{(T_0/T)^{\chi}}$ with different values of χ) while simultanously relaxing the requirement that each film be fit with the same R_0 were investigated as well. In particular, $\chi = \frac{1}{2}, \frac{1}{3}, \frac{1}{4}$, corresponding to variable-range hopping conduction in dimensions $d = 1$, 2, and 3 respectively, were fit and the residuals analyzed. Consistently, $\chi = 1$ minimized these residuals. While the fits shown restrict the value of R_0 for each film to be the same, emphasizing its relative constancy through the transition, this restriction does not significantly affect the value

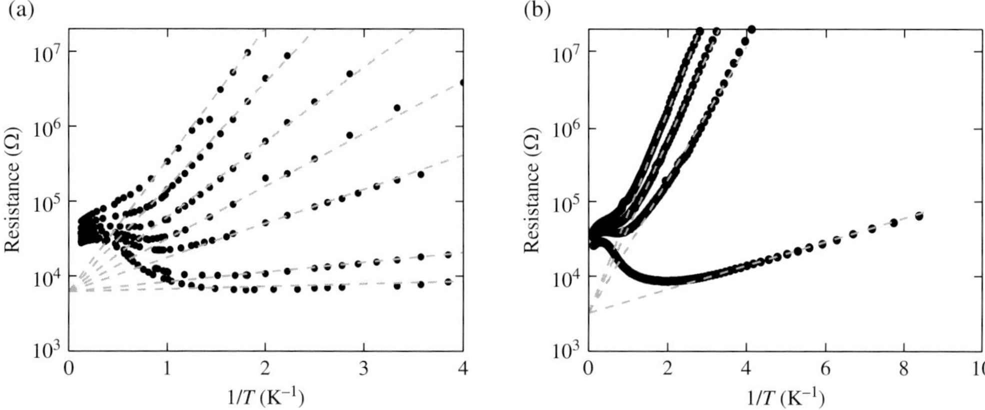

Fig. 13.9 Activated insulating behavior for two NHC film series with (a) $r_{\text{hole}} = 27$ nm and (b) $r_{\text{hole}} = 23$ nm. Dashed lines are fits to an activated form, $R = R_0 \exp^{(T_0/T)}$.

of T_0. The activated form for the temperature dependence is further supported by the data on a second series of films of a different hole size ($r_{\text{hole}} = 23$ nm), as shown in Fig. 13.9 (b). Here, an insulating film, measured over twice the range of $\frac{1}{T}$ as those films shown in Fig. 13.9 (a), does not deviate from the activated form. With this form for the low-temperature resistance, the activation energies T_0 decrease from 4 K to nearly zero at the transition. Thus the critical point of the insulator-to-superconductor in NHC films is associated with $T_0 = 0$, $R_C = R_0$, and metallic, or temperature-independent, behavior at the transition.

It is clear from Fig. 13.8 that the insulating states of films with and without patterning exhibit different characteristics. The activated form for the resistance is reminiscent of the disorder-tuned SIT of TiN (Baturina *et al.*, 2007*a*) and indium oxide (Sambandamurthy *et al.*, 2004) where workers have used this behavior as support for a Cooper-pair localized state. Additionally, as has already been pointed out, the dip in the $R(T)$ shown prominently in Figs 13.8 and 13.9, is also present in indium oxide films. Already, NHC films appear to be more closely related to compound thin films than they do to the ultra-thin, elemental films that comprise them. This is a sign that patterning the NHC films has not only allowed detection of Cooper pairs but also altered their behavior.

13.5.3 Magnetoresistance oscillations

Several of the NHC insulating films depicted above show giant magneto-oscillations in the $R(H)$ at lower temperatures. A sample of these oscillations is plotted in Fig. 13.10 (a) for the sixth film in the $r_{\text{hole}} = 27$ nm series of Fig. 13.9 (a). The period of the oscillations, $H_M = 0.23$ T that normalizes the magnetic-field axis in the figure, is very well defined. Fourier transforming the data yields the same period (to within 1%) and strong suppression of all other periods save higher harmonics of H_M. The oscillations begin at relatively high temperatures, along the portion of the

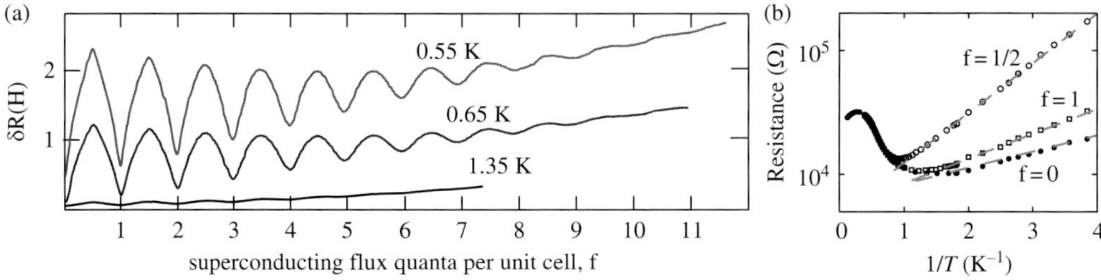

Fig. 13.10 Magnetoresistance oscillations in an insulating NHC film. (a) The fractional change in resistance $\delta R(H)$ against normalized magnetic field $f = H/H_{\mathrm{M}}$. The oscillations have a period $H = H_{\mathrm{M}} = h/2eS$. (b) Arrhenius plot of the resistance in static applied fields corresponding to $f = H/H_{\mathrm{M}} = 0$, $\frac{1}{2}$, and 1, which shows the resistance oscillation in (a) derives from oscillations in the activation energy scale T_0. Adapted from Stewart Jr *et al.* (2007).

trajectory where the $R(T)$ begins to decrease (though oscillations can be observed without a minimum) and extends to the lowest temperatures measured. In the data of Fig. 13.10 (a) the magnetoresistance executes about ten oscillations on a rising background resistance before their amplitude decays to zero. These data also illustrate the alacrity with which the oscillations grow as the temperature decreases, reaching over 200% by $T = 0.55$ K.

The oscillations shown in Fig. 13.10 (a) originate from oscillations in the activation energy, T_0. This can be seen in Fig. 13.10 (b), where the $R(T)$ are plotted for $f = H/H_{\mathrm{M}} = 0$, $\frac{1}{2}$, and 1 for the same film measured in Fig. 13.10 (a). The resistances in field remain activated but the activation energy for $f = \frac{1}{2}$ is much larger than the activation for $f = 0$ or $f = 1$. By contrast the Arrhenius prefactor R_0 varies by only a few percent. Thus the change in the resistance at low fields is due primarily to the periodic dependence of $T_0(f)$. Since the resistance remains activated in field, the rapid growth of the oscillations shown in Fig. 13.10 (a) is exponential.

The period of the magnetoresistance oscillations of Fig. 13.10, $H_{\mathrm{M}} = 0.23$ T, closely corresponds to one superconducting flux quantum, $h/2e$, threading a unit cell of holes of area S (Little and Parks, 1962). This implies that the insulators in Fig. 13.9 are composed of localized Cooper pairs at the lowest temperatures (Deaver and Fairbank, 1961; Doll and Näbauer, 1961). It is worth noting that it is possible to observe magneto-oscillations with the same $2e$ period in multiply connected normal metal insulating films as well (Aronov and Sharvin, 1987). However, when in the hopping regime, as is the case here, these oscillations have the opposite phase to those observed in Fig. 13.10 (Meir *et al.*, 1991). In other words, in normal metals the resistance first decreases with increasing field. Here, the initial resistance increase with field is consistent with an increase in screening currents and a reduction in the phase coherence of Cooper pairs. In addition, single-electron oscillations do not produce such large changes in the resistance. Typically, single-electron oscillations are $\leq 1\%$ of the resistance measured with $H = 0$, while oscillations in NHC films exceed 100% very quickly. The large difference in the magnitude of the two effects is presumably due to the the substantial difference in the phase-coherence length between single electrons in a disordered film and Cooper pairs in the same.

13.5.4 Periodic magnetic-field-tuned SIT

With clear evidence of bosonic behavior in zero magnetic field established through the observation of $h/2e$ magnetoresistance oscillations, we move on to an investigation of the perpendicular magnetic-field-tuned SIT. Many claims of a bosonic insulating state are made for films driven to insulating behavior in this context (Baturina *et al.*, 2007*b*; Hebard and Paalanen, 1990; Sambandamurthy *et al.*, 2004, 2006). As has been discussed previously, even ultra-thin elemental films exhibit some signs of pairing in the magnetic-field-induced insulating state. Importantly, however, the insulating state induced in this way need not be the same as that induced through decreasing thickness or increasing disorder. However, given the bosonic nature of $H = 0$ NHC insulators in the previous section, one might expect bosonic behavior in the magnetic-field-induced insulators as well.

Figure 13.11 (a) shows the magnetic-field-driven SIT transition that takes place between $H = 0$ and $H = H_{\rm M}/2$. Here, the resistance is plotted versus $\log(T)$ to emphasize the lowest-temperature behavior and $H_{\rm M} = h/2eS$ again, where S is the area of a unit cell of holes. As the field is increased, the low-temperature dR/dT changes from superconducting-like, $dR/dT > 0$ at $H = 0$ to insulating-like, $dR/dT < 0$. Between these fields the data have a very weak temperature dependence at the lowest temperatures and are nearly metallic. Even so, Fig. 13.11 (b) shows the resistance can successfully be parametrized by

$$R(T, H) = R_0(H)e^{T_0(H)/T} \tag{13.7}$$

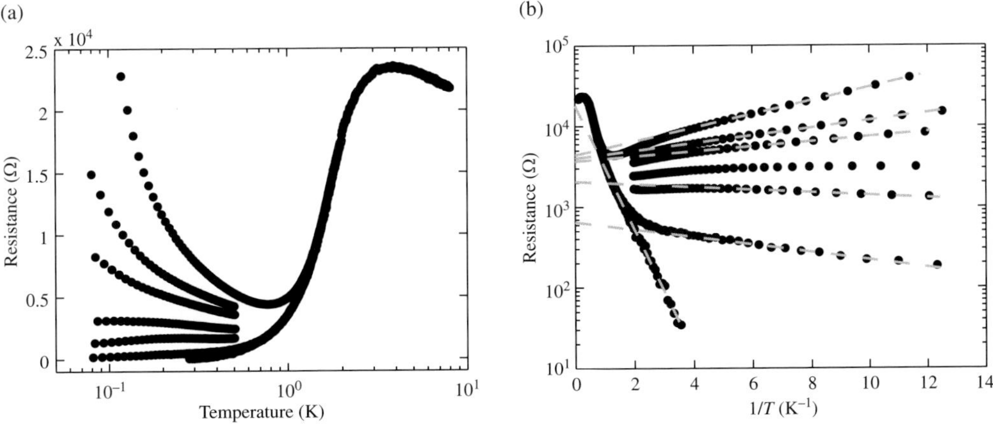

Fig. 13.11 Field-driven SIT in nano-honeycomb films plotted as (a) resistance against temperature and (b) in an Arrhenius plot. The dashed lines in (b) are fits to the activated form used to parameterize the data. Applied magnetic fields from bottom to top are $H = 0$, $0.04H_{\rm M}$, $0.1H_{\rm M}$, $0.15H_{\rm M}$, $0.25H_{\rm M}$, $0.33H_{\rm M}$, and $H = 0.5H_{\rm M}$. Adapted from Stewart *et al.* (2008).

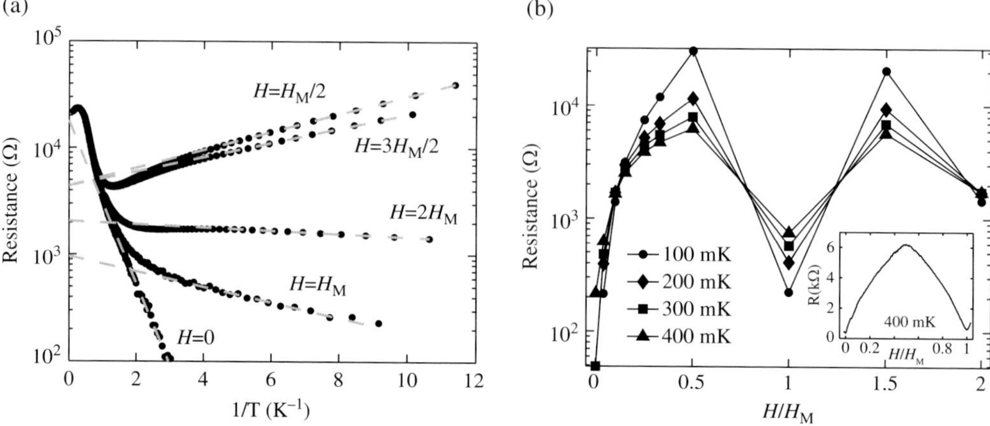

Fig. 13.12 (a) Resistance as a function of temperature for matching and half-matching fields showing the multiplicity of the SIT. Those data at $H = nH_{\rm M}$ show a tendency to superconduct ($dR/dT > 0$) while those at $H = H_{\rm M}(n+1)/2$ insulate. Adapted from Stewart *et al.* (2008)

at the lowest temperatures in all fields except for the data for $H = 0.15H_{\rm M}$. The energy scale T_0 extracted from some of the fits shown is small and so cannot be associated with a simple activated process. Nonetheless, the superconducting and most insulating ($H_{\rm M}/2$) data are well described by this form, and it seems the most reasonable procedure for interpolating between those curves. The value of T_0 characterizes dR/dT in the lowest temperature data and allows identification of the critical field as $H_{\rm c} \approx 0.15H_{\rm M}$ where $dR/dT \approx 0$.

Figure 13.12 shows another feature of these NHC films: dR/dT changes sign multiple times with increasing field. As the field is increased above $H = H_{\rm M}/2$ the resistance begins to decrease again so that by $H = H_{\rm M}$, $dR/dT > 0$ and $T_0 < 0$ at the lowest temperatures. The transitions that take place with $H > H_{\rm M}/2$ differ from the initial SIT shown in Fig. 13.11 in their behavior on the "superconducting" side of the transition. First, the low-resistance phase can emerge from the insulating phase as the field is *increased* (e.g. $H_{\rm M}/2 \leq H \leq H_{\rm M}$). Moreover, the superconducting-like states for these transitions in $H = nH_{\rm M}$ are much weaker than their $H = 0$ counterpart. The data in $H_{\rm M}$ resemble the data in $0.04H_{\rm M}$ as the data in $2H_{\rm M}$ resemble those in $0.1H_{\rm M}$, resulting in similar values of T_0 (–0.14 and –0.03 K for $H_{\rm M}$ and $2H_{\rm M}$ respectively). A slightly weaker temperature dependence is also observed in $H = 3H_{\rm M}/2$ as compared with $H = H_{\rm M}/2$. This weakening temperature dependence on both sides of the higher-n SITs combine to decrease the amplitude of the magnetoresistance oscillations as H increases. In a field of a few Tesla, these oscillations damp out completely.

13.5.5 High-field magnetoresistance peak

Increasing magnetic field initially leads to greater Cooper-pair localization. Beyond the first few magnetic-field-tuned SITs, superconducting $R(T)$ disappear altogether

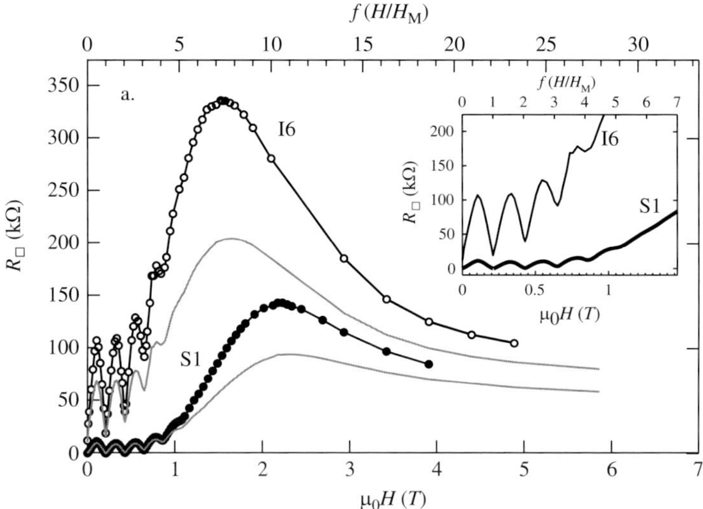

Fig. 13.13 Magnetoresistance at two temperatures for a film, I6, that insulates at $H = 0$ and a film, S1, that superconducts at $H = 0$. The curves with symbols were obtained at $T = 100$ mK and the solid curves at $T = 120$ mK. Note the similarities between I6 and S1: both exhibit oscillations and a peak. Inset: expanded view of the low-field magnetoresistance. The period of these oscillations are consistent with charge-$2e$ carriers.

and the $R(T)$ are insulating even at the matching fields, as shown in Fig. 13.13. In addition, the magnetoresistance oscillations dampen and appear atop a rising resistance "background." Thus, the Cooper-pair insulator phase exhibits a giant positive magnetoresistance. Analysis of the $R(T)$ in this regime show them to remain activated: the rising resistance and oscillations reflect a rising and oscillating activation energy (Nguyen *et al.*, 2009).

At even higher fields, the magnetoresistance peaks (Nguyen *et al.*, 2009). The peak appears beyond the field where the superimposed oscillations dissipate completely and very near the expected upper critical magnetic field for breaking pairs. Around the peak field the $R(T)$ continue to fit well to an activated form. Moving beyond the peak, the conductance, $G(T) = 1/R(T)$, transforms from activated to logarithmic, $G(T) \sim \log(T)$. This logarithmic behavior and its magnitude is consistent with weakly localized electrons dominating the high-field transport (Altshuler *et al.*, 1980). These observations strongly suggest that the peak magnetic field is sufficient to dissociate the majority of the Cooper pairs and single-electron transport starts to turn on. This dissociation is complete at the highest fields, where signatures of single-electron localization appear. Qualitatively similar peak behavior occurs in InO_x and TiN films, supporting the contention that localized Cooper pairs dominate their insulating phases.

13.6 Signatures of Cooper-pair localization and comparison to other film systems

The thickness tuned and magnetic-field-tuned transitions shown above possess very similar characteristics. First, the insulating R(T) exhibit a strong minimum in the $R(T)$ at higher temperatures. Second, the activated temperature dependence for the $R(T)$ at the lowest temperatures, indicative of a hard gap in the density of states in the insulating state, exists in both transitions. Third, the transition between superconducting and insulating states is due to the diminution of the hard-gap energy scale T_0 (see Fig. 13.14). Lastly, and perhaps most importantly, both transitions are boson-dominated (i.e. boson localization to delocalization) as indicated by the $h/2e$ periodicity in the magnetoresistance of the thickness-tuned transition and the $h/2e$ periodic transitions in the field-driven case. Taken together these main characteristics of NHC film SITs suggest that decreasing film thickness affects the superconducting state in a way very similar to the application of an increasing perpendicular magnetic field and further suggests a commonality in the mechanism of boson localization.

As a preliminary assessment of the mechanism(s) for Cooper-pair localization in NHC films it is helpful to make a comparison to Josephson junction arrays (JJAs). The activated form for the resistance and the presence of the magnetoresistance oscillations are reminiscent of JJA behavior (Delsing *et al.*, 1994; Geerligs *et al.*, 1989). JJAs are explicitly islanded systems produced with lithographic techniques to have characteristic dimensions on the micron or fraction-of-a-micron scale. They consist of well-defined superconducting islands with some form of weak link connecting them, and are characterized by the inter-island resistance and capacitance (Delsing *et al.*, 1994). Their lithographically well-defined geometries allow calculation of the Josephson coupling, E_{J}, and the charging energy, E_{C}, respectively. It is these parameters, along with the superconducting energy gap, which provide an intuitive way to characterize the charge-transport properties of the array. To calculate the Josephson coupling one usually uses the Ambegaokar–Baratoff equation (Tinkham, 2004),

$$E_J = \left(\frac{R}{R_{\mathrm{Q}}}\right)\Delta/2, \tag{13.8}$$

where $R_{\mathrm{Q}} = h/4e^2$ and Δ is the superconducting energy gap. The charging energy is (Tinkham, 2004),

$$E_{\mathrm{C}} = \frac{q^2}{2C} = \frac{q^2}{\kappa L} \tag{13.9}$$

where q is the charge of the current carriers, C is the inter-island capacitance, $\kappa = \epsilon/\epsilon_0$, and L is the length scale characterizing the capacitance. In JJAs, these two energy scales compete with E_{J}, favoring the delocalization of pairs, and E_{C}, favoring their localization. This competition drives JJAs through their own insulator-to-superconductor transitions (Geerligs *et al.*, 1989). If $E_{\mathrm{C}} \gg E_{\mathrm{J}}$, moving current through the array becomes energetically prohibitive and the $R(T)$ diverges as $T \to 0$

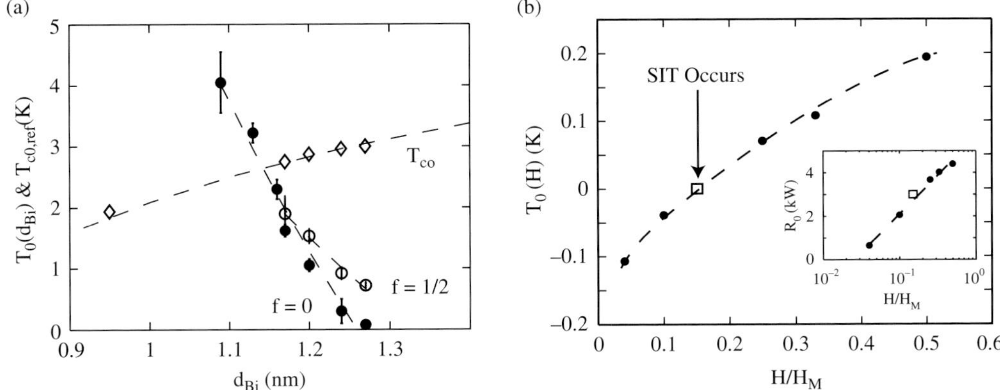

Fig. 13.14 Comparison of the behavior of T_0 through the (a) thickness-driven and (b) magnetic-field-driven SITs as defined by the fits to the temperature dependence of the data. Adapted from Stewart Jr *et al.* (2007).

due to Coulomb blockade, yielding an insulating ground state. In JJAs this is observed through activated resistances (Delsing *et al.*, 1994) similar to those in Figs 13.11 and 13.12.

Following the descriptions of JJAs, we can estimate the charging energy in NHC insulators as well as the Josephson coupling. If we inspect the NHC structure it is natural to assign islands of radius $r_{\text{island}} = L \simeq 50$ nm at the "nodes" in the structure. It is not clear what value of $\kappa = \epsilon\epsilon_0$ is appropriate for films near the SIT, but $\kappa \approx 10$ is probably reasonable (Beloborodov, pers. comm.). Somewhat sidestepping the problem of ϵ, we can calculate the charging energy per unit ϵ to get an idea of the scale. This calculation yields an enormous energy $E_C \sim 1 \times 10^4$ K/ϵ (with $q = 2e$ for Cooper pairs) when compared with T_0 (see Fig. 13.14). We can also take the opposite approach and use the activation energy as an approximate charging energy to estimate the length scale of the islands. Using $T_0 = 2$ K as a typical activation energy, this estimate yields $L = 400\ \mu$m/ϵ. This value of L encompasses thousands of holes and does not correspond to any structure in the NHC films. Thus treating the NHC Cooper-pair localization in the same way as if the film were a JJA requires some considerations beyond the lumped-element analysis typically applied to JJAs.

The Josephson coupling energy can be estimated using eqn (13.8), with $\Delta \simeq 1.7k_BT_{c0}$, and where T_{c0} is the *unpatterned* film mean-field critical temperature. With a $T_{c0} = 2$ K, and $R = R_N \approx 20$ kΩ, $E_J = 5$ K. When comparing this estimate to the data, one must use $\delta T_0 = T_0(\frac{1}{2}) - T_0(0)$. This is because eqn (13.8) applies to the $H = 0$ case and the presence of a magnetic field induces a phase difference $\Delta\phi$ between either side of a junction. This contributes to the free energy as $E_J(H) = E_J \cos(\Delta\phi)$. Since $\Delta\phi \propto H$, $E_J \simeq E_{J0}\cos(2\pi H/H_M)$ (Tinkham, 2004) so that in this context $E_J \approx \delta T_0$ if JJA physics applies. The estimate compares fairly well with the maximal $\delta T_0 \approx 1$ K. This is fairly good evidence that some Josephson coupling is involved in tuning T_0 and also that it plays some role in driving the SIT.

The qualitative similarities in the data and the estimate of E_J make a compelling argument for Josephson physics and one cannot help but think some of it applies. The extent to which JJA physics is relevant is limited by the charging energy scale estimation and the absence of a mechanism to form islands in an amorphous film. These two difficulties are related and two scenarios to deal with them can be imagined. First, if T_0 is an accurate reflection of the island size then these islands must spontaneously form. One possibility is that disorder induces fluctuations in the amplitude of the superconducting order parameter, which give rise to fluctuating islands (Dubi *et al.*, 2006; Ghosal *et al.*, 2001). This is discussed further in Chapter 17, however it is not clear whether the length scales for the phenomena in NHC films matches that which occurs theoretically or how crucial this match is to the viability of an explanation in this context.

Secondly, it is possible that a source of low-energy dissipation in the system could renormalize the charging energy to agree with T_0 while keeping the physical size of islands small (Takahide *et al.*, 2000). This possibility is supported by a comparison of the NHC data to that of unpatterned films. Such a comparison forces one to regard NHC films as a new ultra-thin film system that lies closer in its behavior to that of indium oxide or TiN than to unpatterned films. NHC films possess three distinct types of inhomogeneity, the effects of which, at present, remain entangled. They are the introduction of holes, disorder in the hole array, and thickness fluctuations that occur as a result of slight undulations in the surface of the NHC substrate (Stewart *et al.*, 2009). Much more experimental progress is necessary to properly quantify the extent of these inhomogeneities, however, thickness fluctuations on the $\sim$ 10-nm length scale are a compelling candidate for the dominant effect. This picture is also physically satisfying, motivated as it is by quantifiable surface roughness.

Regardless of the mechanism(s) of CPL in NHC films, some form of inhomongeneity appears necessary to understand the data presented in this chapter. The transformation of unpatterned films dominated by fermions at the thickness-tuned transition to films dominated by bosons through patterning is a fairly surprising result. NHC data strongly suggest that the inhomogeneities are present in other film systems that purport to be Cooper-pair insulators. The extent of any such inhomogeneity in other film systems and the degree to which its effect on the SIT may be mitigated or amplified based on carrier density, natural variations in film composition, or other film preparation considerations remains to be elucidated. The observed transformation of Bi/Sb films through patterning may light a path toward understanding the mechanisms of CPL in a more general context.

References

Altshuler, B., Aronov, A., and Lee, P.A. (1980). *Phys. Rev. Lett.*, **44**, 1288.

Aronov, A.G. and Sharvin, Yu. V. (1987). *Rev. Mod. Phys.*, **59**, 755.

Barber, R.P., Merchant, L.M., LaPorta, A., and Dynes, R.C. (1994). *Phys. Rev. B*, **49**, 3409–3412.

Barber Jr, R.P., Hsu, S.Y., Valles, Jr., J.M., Dynes, R.C., and Glover, R.E. (2006). *Phys. Rev. B*, **73**, 134516.

Baturina, T.I., Mironov, A.Y., Vinokur, V.M., Baklanov, M.R., and Strunk, C. (2007*a*). *Phys. Rev. Lett.*, **99**, 257003.

Baturina, T.I., Strunk, C., Baklanov, M.R., and Satta, A. (2007*b*). *Phys. Rev. Lett.*, **98**, 127003.

Belitz, D. and Kirkpatrick, T.R. (1994). *Rev. Mod. Phys.*, **66**, 261.

Bezryadin, A. and Pannetier, B. (1995). *J. Low Temp. Phys.*, **98**, 251.

Bielejec, E., Ruan, J., and Wu, W. (2001). *Phys. Rev. Lett.*, **87**, 036801.

Bielejec, E. and Wu, W. (2002). *Phys. Rev. Lett.*, **88**, 206802.

Chervenak, J.A. and Valles, Jr., J.M. (1999). *Phys. Rev. B*, **59**, 11209–11212.

Chervenak, J.A. and Valles, Jr., J.M. (1999). *Phys. Rev. B*, **59**, 11209.

Deaver, B.S. and Fairbank, W.M. (1961). *Phys. Rev. Lett.*, **7**, 43–46.

Delsing, P., Chen, C.D., Haviland, D.B., Harada, Y., and Claeson, T. (1994). *Phys. Rev. B*, **50**, 3959–3971.

Doll, R. and Näbauer, M. (1961). *Phys. Rev. Lett.*, **7**, 51–52.

Dubi, Y., Meir, Y., and Avishai, Y. (2006). *Phys. Rev. B*, **73**, 054509.

Dynes, R.C., White, A.E., Graybeal, J.M., and Garno, J.P. (1986). *Phys. Rev. Lett.*, **57**, 2195–2198.

Finkelstein, A.M. (1994). *Physica B*, **197**, 636.

Fisher, M.P.A. (1990). *Phys. Rev. Lett.*, **65**, 923.

Gantmakher, V.F., Golubkov, M.V., Dolgopolov, V.T., Tsydynzhapov, G.E., and Shashkin, A.A. (1998). *JETP Lett.*, **68**, 363–369.

Geerligs, L.J., Peters, M., de Groot, L.E.M., Verbruggen, A., and Mooij, J.E. (1989). *Phys. Rev. Lett.*, **63**, 326–329.

Ghosal, A., Randeria, M., and Trivedi, N. (2001). *Phys. Rev. B*, **65**, 014501.

Graybeal, J and Beasley, M (1984). *Phys. Rev. B*, **29**, 4167–4169.

Haviland, D.B., Liu, Y., and Goldman, A.M. (1989). *Phys. Rev. Lett.*, **62**, 2180–2183.

Hebard, A.F. and Paalanen, M.A. (1990). *Phys. Rev. Lett.*, **65**, 927–930.

Hsu, S.-Y., Chervenak, J.A., and Valles, Jr., J.M. (1995). *Phys. Rev. Lett.*, **75**, 132–135.

Jaeger, H.M., Haviland, D.B., Goldman, A.M., and Orr, B.G. (1986). *Phys. Rev. B*, **34**, 4920–4923.

Lee, S.J. and Ketterson, J.B. (1990). *Phys. Rev. Lett.*, **64**, 3078–3081.

Little, W.A. and Parks, R.D. (1962). *Phys. Rev. Lett.*, **9**, 9.

Markovic, N., Christiansen, C., Mack, A., Huber, W., and Goldman, A. (1999). *Phys. Rev. B*, **60**, 4320–4328.

Meir, Y., Wingreen, N.S., Entin-Wohlman, O., and Altshuler, B.L. (1991). *Phys. Rev. Lett.*, **66**, 1517–1520.

Nguyen, H.Q., Hollen, S.M., Stewart, M.D., Shainline, J., Yin, A., Xu, J.M., and Valles, Jr., J.M. (2009). *Phys. Rev. Lett.*, **103**, 157001.

Okuma, S., Terashima, T., and Kokubo, N. (1998). *Phys. Rev. B*, **58**, 2816–2819.

Parker, J, Read, D, Kumar, A, and Xiong, P (2006). *Europhys. Lett.*, **75**, 950–956.

Phillips, P and Dalidovich, D (2001). *Phil. Mag. B*, **81**, 847–854.

Sambandamurthy, G., Engel, L.W., Johansson, A., and Shahar, D. (2004). *Phys. Rev. Lett.*, **92**, 107005.

Sambandamurthy, G., Johansson, A., Peled, E., Shahar, D., Bjornsson, P.G., and Moler, K.A. (2006). *Europhys Lett.*, **75**, 611–617.

Steiner, M.A. and Kapitulnik, A (2005). *Physica C*, **422**(1-2), 16–26.

Stewart, M.D., Nguyen, H.Q., Hollen, S.M., Yin, A., Xu, J.M., and Valles, Jr., J.M. (2009). *Physica C*, **469**, 774–777.

Stewart, M.D., Yin, A., Xu, J.M, and Valles, Jr., J.M. (2008). *Phys. Rev. B*, **77**, 140501(R).

Stewart Jr, M.D., Yin, A., Xu, J.M., and Valles, Jr., J.M. (2007). *Science*, **318**, 1273–1275.

Strongin, M, Thompson, R.S., Kammerer, O.F., and Crow, J.E. (1970). *Phys. Rev. B*, **1**, 1078–1091.

Takahide, Y., Yagi, R., Kanda, A., Ootuka, Y., and Kobayashi, S. (2000). *Phys. Rev. Lett.*, **85**, 1974–1977.

Tinkham, M. (2004). *Introduction to Superconductivity* (2nd edn). Dover, Mineola NY.

Valles, Jr., J.M., Dynes, R.C., and Garno, J.P. (1989). *Phys. Rev. B*, **40**, 6680–6683.

Valles, Jr., J.M., Dynes, R.C., and Garno, J.P. (1992). *Phys. Rev. Lett.*, **69**, 3567–3570.

Valles, Jr., J.M., Hsu, S.Y., Dynes, R.C., and Garno, J.P. (1994). *Physica B*, **197**(1-4), 522–529.

Yazdani, A. and Kapitulnik, A. (1995). *Phys. Rev. Lett.*, **74**, 3037–3040.

14
Superconductor–Insulator Transitions in Ultra-thin a-Pb Films: Disorder, Magnetic-field, and Magnetic-impurity Tuning

---

Peng Xiong, Ashwani Kumar, H. Jeffrey Gardner and Liuqi Yu

Department of Physics, Florida State University

14.1 Introduction

Superconductor–insulator transitions (SITs) in two dimensions (Goldman and Markovic, 1998) are prime examples of quantum phase transitions (Sachdev, 1999). The understanding of two-dimensional (2D) SITs has broad implications for quantum phase transitions in a variety of systems (Sondhi, 1997), and studies in systems of relatively simple structures may provide useful insights into the physics of much more complex materials (Merchant *et al.*, 2001). Phenomenologically, SITs in 2D systems can be divided into two distinct groups depending on the morphologies of the thin films. The film morphology is known to have a significant impact on the manifestations of superconductivity and other characteristics of electron transport.

Granular films grown by common vapor-deposition techniques typically consist of relatively large grains with bulk-like electronic properties, and the overall transport properties of the films are determined primarily by the degree of electronic coupling between the grains. For granular films of superconducting metals, an insulator-to-superconductor transition results from a continuous increase of the Josephson coupling between the grains as the average film thickness increases. SITs in disordered 2D granular films (Jaeger *et al.*, 1989) bear much resemblance to SITs in Josephson junction arrays (van der Zant *et al.*, 1996). More importantly, tunneling experiments (Barber *et al.*, 1994; Merchant *et al.*, 2001) have produced direct evidence that the individual grains are fully superconducting even deep in the (transport) insulating state. Therefore, there is general agreement that SITs in 2D granular films are driven by increasing phase fluctuations of the pair wavefunctions.

In contrast, the picture of SITs in 2D homogeneous films is far murkier, despite the fact that the overall features of these SITs appear to be less complex. There are apparently conflicting experimental results, which have led to a number of different theoretical models, primarily in two distinct frameworks: disordered bosonic (Fisher, 1990; Fisher *et al.*, 1990) and fermionic (Finkelstein, 1994; Fukuyama, 1984) transitions. In the dirty-boson model (Fisher, 1990; Fisher *et al.*, 1990), the fluctuations in the phase of the superconducting order parameter (SOP) play the dominant role in the SIT. The model requires robust coexistence of Cooper pairs and vortices on both sides of the SIT and implies an electronically inhomogeneous insulating state. The inherent charge–flux duality in the model predicts specific finite-size scaling behavior as well as a universal critical resistance, $h/4e^2$, for the SITs (Fisher *et al.*, 1990). The fermionic picture (Finkelstein, 1994; Fukuyama, 1984), based on enhancement of the Coulomb interaction and suppression of the electronic density of states, points to an electronically uniform phase transition driven by progressive suppression of the SOP amplitude with increasing disorder or magnetic field, and is signified by a vanishing SOP at the phase boundary.

Clearly, an issue central to the controversy is the electronic homogeneity of the phases (the existence of a well-defined phase boundary) across the SITs and the nature of the insulating state in a structurally homogeneous film. Here we describe and discuss the results of a set of experiments aimed at directly comparing the characteristics of SITs in ultra-thin homogeneous *a*-Pb films driven by three different tuning parameters: disorder (d), magnetic field (B), and paramagnetic impurity (MI). Among them, the

SIT by continuous tuning of MI pair-breaking is experimentally implemented for the first time. We consider the MI-tuned transition a useful reference, in which the microscopic mechanism for the destruction of superconductivity is well understood, the phase boundary is well-defined, and the insulating state is purely fermionic. Similarly, we regard the d-tuned SIT in a granular film as an unambiguous example of a bosonic transition. We will make straightforward comparisons of the phenomenology of d-tuned and B-tuned SITs in homogeneous films with that of the two model systems in order to gain new insight into the mechanisms of these two SITs.

14.2 Experiments

Our experiments were carried out in a modified dilution refrigerator. A substrate with pre-deposited thin gold electrodes was mounted on a stage thermally linked to the mixing chamber and located at the center of a superconducting magnet. Two tantalum coils and a NiCr wire were mounted on an electrical feedthrough at the bottom of the vacuum can, providing sources for the evaporation of Sb, Pb, and the MI, respectively. The film thickness was monitored with a quartz crystal monitor. Before the deposition of Pb, an insulating layer of a-Sb (1 nm) was evaporated. Such a wetting layer ensures electrical (Strongin, 1970) and possibly structural uniformity down to a single atomic layer for Pb. Electrical transport measurements were performed *in situ* at increasing Pb film thicknesses, which resulted in an insulator-to-superconductor transition. Electrical continuity is reached at monolayer Pb thickness ($\sim 3\mathring{A}$) with measurable conductivity at high temperatures. At the SIT (sheet resistance $R \sim 7$ kΩ) the nominal Pb thickness is approximately 9 $\mathring{A}$. A set of $R_{\square}(T)$ data for a film at thicknesses from 7.00 to 11.49 $\mathring{A}$ is shown in Fig. 14.1. This is commonly referred to as the disorder (as measured by the normal-state sheet resistance, R_{N}, of the 2D film) tuned SIT. At selected R_{N} or T_{c} on the superconducting side, perpendicular magnetic fields were applied, which drove the film into an insulating state (B-tuned SIT). Finally, MI was deposited in precisely controlled increments on the Pb film, by heating a NiCr wire with a fixed current and controlling the deposition time with a shutter, which resulted in the MI-tuned transition. Since the thickness of the Pb films (< 2 nm) in our experiments was always much smaller than the superconducting coherence length, the effect of the MI should be uniform throughout the entire thickness despite their residing on the surface of the films.

14.3 Experimental approach

14.3.1 Overall features of the superconductor–insulator transitions

14.3.1.1 d-tuned SIT

Figure 14.1 shows a typical set of $R_{\square}(T)$ data for a Pb film at thicknesses from 7.00 to 11.49 $\mathring{A}$. Here the tuning parameter is disorder, as measured by R_{N} of the 2D films. The overall characteristics of the d-tuned SIT are similar to previously published results (Haviland *et al.*, 1989; Parker *et al.*, 2006). At high R_{N} (small thickness) the film is a strongly localized insulator characterized by $R_{\square}(T)$ consistent with hopping transport.

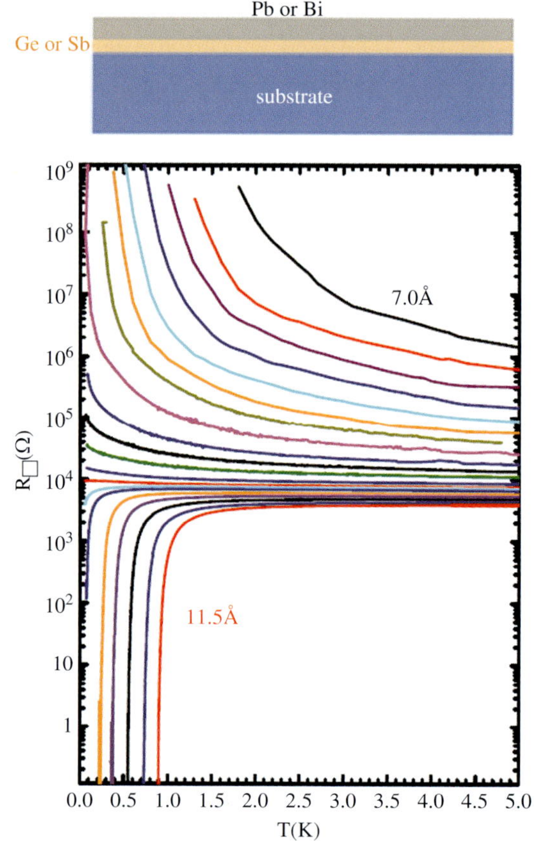

Fig. 14.1 Top: schematic showing the morphology of a homogeneous film grown on a substrate with a pre-deposited thin layer of Ge or Sb. Bottom: sheet resistance as a function of temperature with increasing film thickness for a homogeneous Pb film.

With decreasing disorder (decreasing $R_{\rm N}$ or increasing thickness), the temperature dependence of the insulating behavior becomes progressively weaker and crosses over to a logarithmic dependence, i.e. the weak-localization regime. Further decrease of disorder below $R_{\rm N} \sim 7$ kΩ (nominal Pb thickness of approximately 9 $\mathring{A}$) leads to the emergence of superconductivity. On the superconducting side, the resistive transitions are sharp, with well-defined critical temperatures, signifying long-range phase coherence; the $T_{\rm c}$ increases continuously with increasing film thickness (decreasing disorder or $R_{\rm N}$). The SIT is well-defined and there is a clear phase boundary at a critical $R_{\rm N}$.

14.3.1.2 ***B-tuned SIT***

At any particular $R_{\rm N}$ or $T_{\rm c}$ on the superconducting side in Fig. 14.1, we can apply a magnetic field on the sample and induce a B-tuned SIT. Figure 14.2 shows the $R_{\Box}(T)$ data for a representative B-tuned SIT in a Pb film with a zero-field $T_{\rm c}$

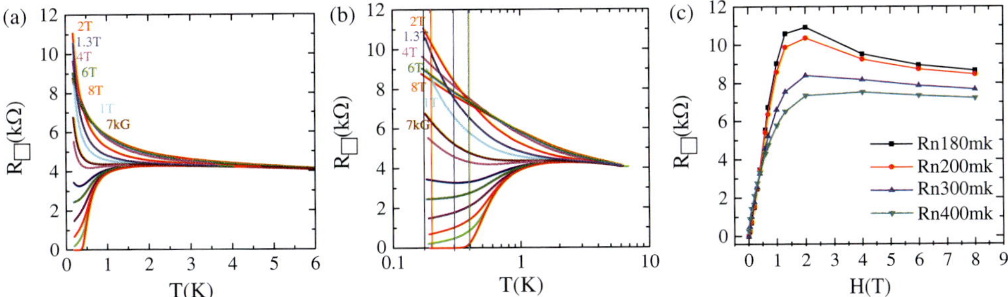

Fig. 14.2 Magnetic-field-tuned SIT for a homogeneous Pb film with initial T_c of 0.55 K and R_N of 4.1 kΩ. (a) $R_\square(T)$ for the film in applied magnetic fields of (from bottom to top) 0.0, 0.04, 0.1, 0.2, 0.3, 0.4, 0.5, 0.7, 1.0, 1.3, 2.0, 4.0, 6.0, and 8.0 T. (b) The same set of data plotted on a logarithmic temperature scale. (c) $R(H)$ for the same film at 180, 200, 300, and 400 mK (vertical lines in (b)).

of 0.55 K (R_N = 4.1 kΩ). The application of an increasingly large perpendicular magnetic field progressively broadens the resistive transition, and eventually destroys the superconducting state ($R(0) = 0$) and results in an insulating state ($R(0) = \infty$). The B-tuned SIT has several significant differences with the d-tuned SIT. Perhaps the most pronounced feature is the lack of a well-defined phase boundary between the superconducting and insulating states. In fact, in a range of R_N near the transition, the $R_\square(T)$ data exhibit reentrance or even double-reentrance. On the insulating side, increasing magnetic field actually weakens the temperature dependence of $R_\square(T)$, as illustrated by the crossing of the $R_\square(T)$ curves (Fig. 14.2 (b)) and a magnetoresistance (MR) peak (Fig. 14.2 (c)) at low temperatures.

14.3.1.3 **MI-*tuned SIT***

The biggest challenge in the experimental realization of an MI-tuned SIT is the necessity for controllable and precise variation of the magnetic impurity concentration in the samples. In conventional setups, each MI concentration requires a different specimen, hence the results invariably suffer from sample-to-sample variations and lack of fine control in the MI density, especially in the ultra-thin films required for SIT studies. In our experimental setup, we are capable of preparing an ultra-thin film and then incrementally varying the magnetic impurity density on the same film *in situ*, hence the same sample can be measured as superconductivity is suppressed successively and an SIT is effected. MI was deposited in precisely controlled increments on the Pb film, by heating a NiCr wire with a fixed current and controlling the deposition time with a shutter. In order to find out the identity of the MI, we performed *ex-situ* X-ray photoelectron spectroscopy (XPS) measurements on several samples at the conclusion of the respective experimental runs. The samples were transferred to an XPS system from the dilution fridge with exposure to air. Figure 14.3 (a) shows a broad-range survey for all common elements that may be present on the surface of the sample. The presence of gold is due primarily to the gold-coated sample holder, while the presence

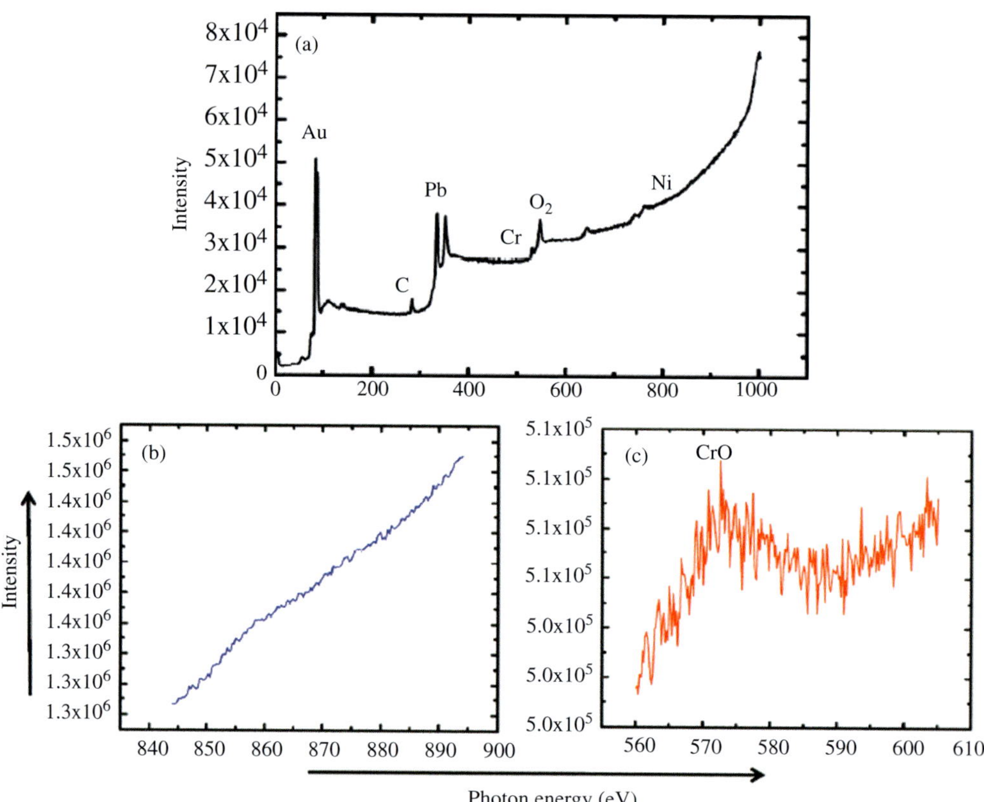

Fig. 14.3 XPS spectra for a quench-condensed Pb film with magnetic impurities deposited from a heated NiCr wire. (a) Broad-range survey reveals the presence of gold (Au), carbon (C), oxygen (O), and lead in both elemental (Pb) and oxide (PbO) forms. There is no visible peak corresponding to chromium (Cr) or nickel (Ni) in the spectrum due to the minute amount of impurities. (b) Absence of any peak at 852 eV indicates no measurable Ni. (c) Peak at 577 eV indicates the presence of Cr in an oxidized form.

of carbon and oxygen is clearly an environmental effect. Peaks corresponding to both pure and oxidized Pb are readily identified. The broad-range scan shows neither Ni nor Cr in the spectrum, due to the minute amount of MI. We therefore performed more careful scans in much narrower specific energy ranges: around 852 eV for Ni and 577 eV for Cr. The absence of any peaks in Fig. 14.3 (b) indicates no traceable amount of Ni atoms on the sample, while the spectrum in Fig. 14.3 (c), with a peak at ~577 eV, clearly demonstrates the presence of Cr in its oxidized form. Similar XPS measurements performed on multiple samples yielded consistent results, showing only Cr and no detectable Ni on the Pb films.

Figure 14.4 shows the $R_{\square}(T)$ data for a Pb film with an initial critical temperature (T_{c0}) of 1.22 K ($R_N = 3.1$ kΩ) at different normalized MI density. Since the MI

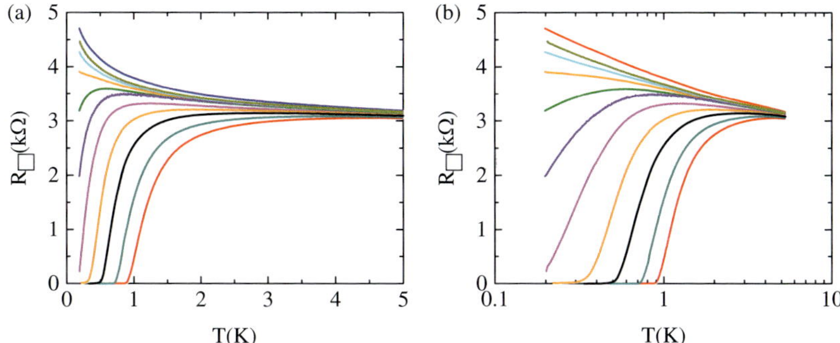

Fig. 14.4 Magnetic impurity-tuned SIT. (a) $R_\square(T)$ at increasing Cr impurity density (from bottom to top) for a homogeneous Pb film with initial T_c of 1.22 K and normal state sheet resistance of 3.1 kΩ. (b) The same set of data plotted on logarithmic temperature scale.

deposition rate was much below the resolution of the quartz crystal monitor, the absolute density of the MI is unknown. However, the normalized density, $\frac{n}{n_c}$ (n_c is the critical MI density at which T_c is suppressed to zero), is known precisely for all samples from the cumulative deposition time. As n increases, T_c is progressively suppressed while the resistive transitions remain well-defined. There is a distinct boundary between the superconducting phase and the weakly insulating normal phase. At n_c, superconductivity is completely destroyed and the film reaches a weakly insulating state, signified by an approximately log-T dependence of R in the entire temperature range (Fig. 14.4 (b)).

14.3.2 The model systems

14.3.2.1 d-tuned superconductor–insulator transition in granular films

Here we summarize a few distinct features of d-tuned SITs in granular films from published work. The SITs in granular films of a variety of materials have been studied in depth and in great detail. The most systematic data on d-tuned SITs have been obtained in quench-condensed films (Barber *et al.*, 1994; Jaeger *et al.*, 1989). It is well known that direct quench-condensation of a metal onto a cold substrate (without a Ge or Sb wetting layer) produces a granular film, and an insulator-to-superconductor transition results from an increase in the average thickness (decrease of R_N) of the film. Unlike the case of a homogeneous film, in which measurable conductance occurs at monolayer thickness and the film conductance increases linearly with its thickness, a granular film does not show any measurable conductance until a large critical thickness (e.g. $\sim$ 9 nm for Pb (Barber *et al.*, 1994)) is reached, beyond which the film conductance increases exponentially with the increase in the average thickness, often by orders of magnitude within a few Angstroms (Dynes *et al.*, 1978). These characteristics strongly suggest that the electron transport in the granular films is dominated by charge-tunneling between the grains.

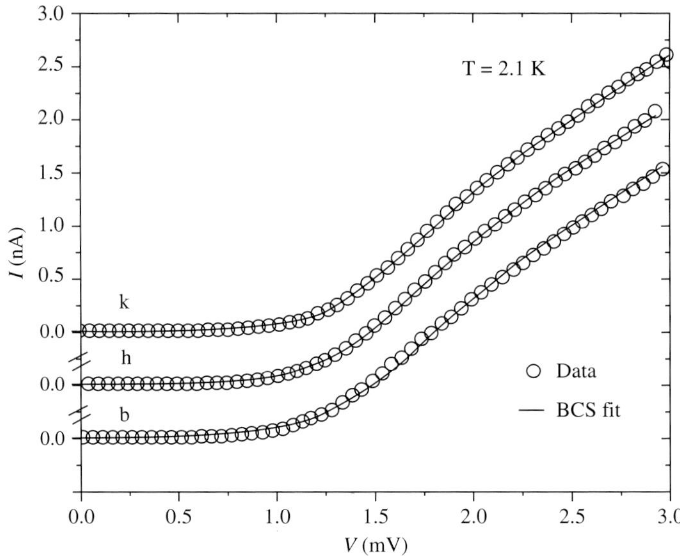

Fig. 14.5 The I–V characteristics ($T = 2.1$ K) for an Al-insulator–granular Pb tunnel junction at different Pb thicknesses. The spectra are almost identical, although in terms of transport h and k are superconducting and b is an insulator. Adapted from Barber *et al.* (1994).

A necessary signature of a bosonic SIT is the presence of robust superconducting pairing in the insulating state. This was revealed directly in the d-tuned SIT in granular films through electron tunneling (Barber *et al.*, 1994; Dynes *et al.*, 1978). Figure 14.5 shows the tunneling I–V curves for the same Al/AlOx/granular-Pb junction at three different $R_{\rm N}$ for the Pb film. These curves span the SIT. It is readily apparent that the three I–V curves are virtually identical; specifically, three important features in the data are worth noting (Barber *et al.*, 1994; Dynes *et al.*, 1978):

1. The superconducting energy gap is the same in the insulating state as in the superconducting state, and consistent with the bulk value, proving that the SOP amplitude is fully developed on individual grains even on the insulating side.
2. The high-bias slope of the I–V curves is the same, indicating that the normal-state electronic density of states at the Fermi level ($E_{\rm F}$), $D(E_{\rm F})$, does not change with $R_{\rm N}$, in contrast to homogeneous films, in which $D(E_{\rm F})$ is found to be suppressed concurrently with the film $T_{\rm c}$ with increasing $R_{\rm N}$ (Valles, Jr. *et al.*, 1989).
3. There is little inelastic broadening in the tunneling spectra, even deep in the insulating state. This, along with the robust energy gap, implies that there is negligible pair-breaking by the pronounced phase fluctuations, which are sufficient to induce the SIT.

Besides the direct signature for the presence of superconducting pairing in the insulating state as revealed by electron tunneling, there are several distinct characteristics in the transport and magnetotransport of the granular films that can be viewed as indirect signatures of a super-insulating state and a bosonic SIT (Barber *et al.*, 2006; Orr *et al.*, 1985):

- *Activated transport and large negative MR in the insulating state.* In the insulating state of the granular films, the film resistance rises rapidly with decreasing temperature. More interestingly, in a relatively broad range of normal-state resistances, the activated transport appears to commence at a temperature, which coincides with the onset of the resistive transitions in the superconducting state. These observations are consistent with the opening of the superconducting energy gap below (the same) T_c in the grains, which freezes out quasiparticle excitations and creates an energy barrier for quasiparticle tunneling between the grains. In fact, for both granular Sn (Orr *et al.*, 1985) and Pb (Barber *et al.*, 2006) films, $R_\square(T)$ in the insulating state can be well modeled with the temperature dependence of the quasiparticle tunneling resistance of a simple superconductor–insulator–superconductor tunnel junction. The application of a magnetic field suppresses the energy gap, thus reducing the energy barrier and dramatically decreasing the film resistance (as large as orders of magnitude at low temperatures), hence the large negative MR.
- *Reentrance and double reentrance near the SIT.* At the boundary of the d-tuned SIT in granular films, the film resistance exhibits pronounced non-monotonic dependence on temperature. Such non-monotonic behavior has been observed in a wide variety of granular systems (Barber *et al.*, 2006; Jaeger *et al.*, 1989; Orr *et al.*, 1985), but the complete picture of the characteristics and evolution of the reentrance behavior with R_N were revealed in great detail only recently, with a measurement of the d-tuned SIT in granular Bi (Parendo *et al.*, 2007). As shown in Fig. 14.6, at sufficiently low R_N the film resistance decreases with temperature below a certain characteristic temperature, reaches a minimum and then turns around and appears to approach infinite resistance at $T = 0$. At even lower R_N, however, there is a second turn-around for the resistance at a lower temperature, and the film appears to be headed for a zero-resistance state at $T = 0$. The complex temperature dependence in the transition region can be explained quite straightforwardly based on the physical picture of a film consisting of superconducting grains with varying degrees of electrical coupling. At high R_N the electrical transport is dominated by quasiparticle tunneling, which becomes thermally activated with the opening of the superconducting gap below T_c. At low R_N the Josephson coupling energy becomes sufficiently large to overcome thermal fluctuations; with decreasing temperature (thermal energy) the film resistance decreases continuously, leading eventually to a superconducting state. At intermediate R_N near the SIT, the competition between quasiparticle and pair tunneling results in the non-monotonic behavior: the initial decrease in the resistance corresponds to the grains turning superconducting and the subsequent upturn originates from the opening of the superconducting gap, which suppresses quasiparticle tunneling exponentially. Whether the film eventually approaches an insulating or superconducting state at $T = 0$ depends on whether the Josephson coupling is strong enough (as measured by R_N) to overcome the thermal

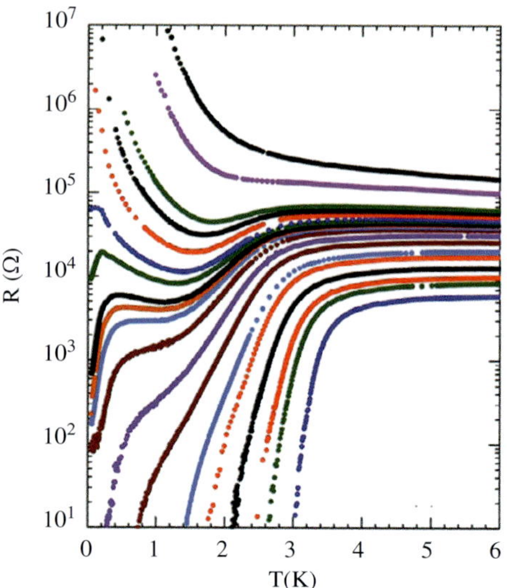

Fig. 14.6 Sheet resistance versus temperature for a granular Bi film at different thicknesses. The non-monotonic change in resistance and double reentrant behavior are typical characteristics of granular film. Adapted from Parendo *et al.* (2007).

fluctuations at a finite temperature. Such an evolution as a function of R_N is evident in Fig. 14.6; at sufficiently low R_N the pair tunneling becomes prevalent at low temperature, leading to the double reentrance.

Finally, we note that both of these indirect signatures are observed in the phase-fluctuation-driven SITs in Josephson junction arrays (van der Zant *et al.*, 1996). Therefore, these features in the readily accessible transport and magnetotransport measurements are strong indicators of a bosonic SIT with an electronically inhomogeneous insulating state having superconducting pairing.

14.3.2.2 MI-*tuned SIT in homogeneous Pb films*

On the central issue in the controversy surrounding the SITs in homogeneous 2D systems, i.e. the phase uniformity in the critical region and whether superconductivity persists into the insulating state, the MI-tuned transition provides a useful reference at the other end of the spectrum. The destruction of superconductivity by MI in general is well understood in the framework of Abrikosov and Gorkov (1960): increasing magnetic pair-breaking reduces T_c continuously, and beyond an n_{cr}, superconducting order is completely destroyed and the superconductor reverts to a normal electronic system. Figure 14.7 shows the $R_\Box(T)$ data for several films of different T_{c0} (T_{c0} being the T_c of the film before the deposition of any MI). In all cases, the film T_c is progressively suppressed by increasing MI concentration, while the resistive transitions remain sharp. Figure 14.8 shows the normalized T_c, T_c/T_{c0}, as a function of $\frac{n}{n_{cr}}$, for these films of different T_{c0}. All of the data collapse onto a single curve which is in good

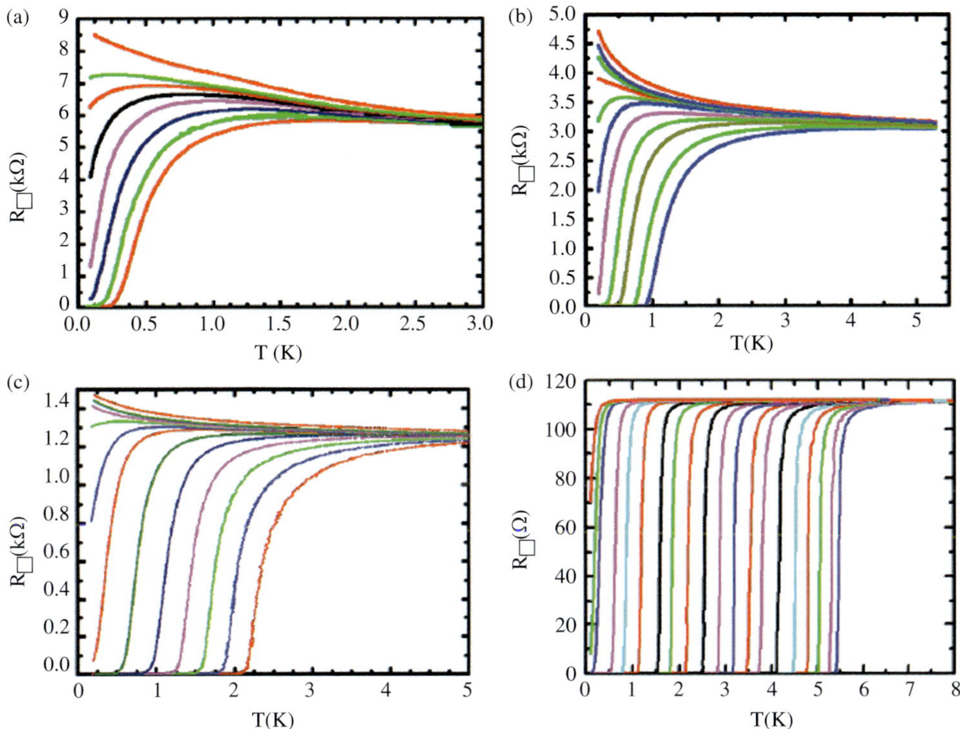

Fig. 14.7 Magnetic-impurity-tuned transitions for four different Pb films. (a)–(d) sheet resistance versus temperature at increasing Cr impurity densities for four films of starting T_c of 0.47 K, 1.22 K, 2.37 K and 5.44 K, respectively.

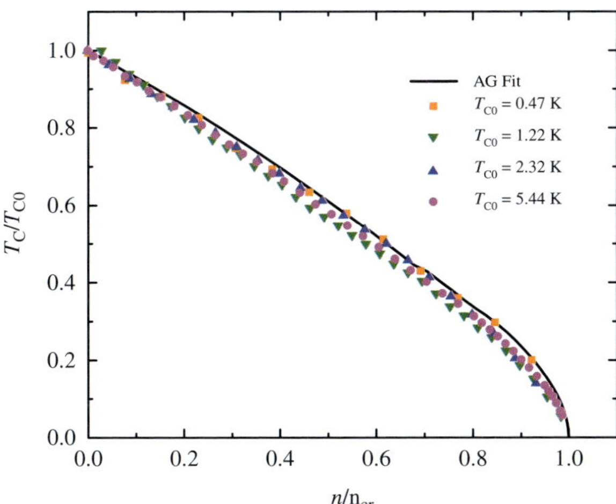

Fig. 14.8 Normalized critical temperature, T_c/T_{c0}, as a function of normalized magnetic impurity density, $\frac{n}{n_{cr}}$, for the four Pb films shown in Fig. 14.7.

agreement with theory (Abrikosov and Gorkov, 1960) as represented by the solid line. The microscopic nature of the MI pair-breaking makes it highly unlikely that there are any Cooper pairs beyond n_{cr} in the normal state. These results are consistent with the picture of a continuous 2D quantum phase transition (Ramazashvili and Coleman, 1997) between a BCS superconductor and a disordered fermionic conductor driven by MI pair-breaking. In our opinion, the MI-tuned transition establishes the behavior of a conventional fermionic SIT in three important aspects: the way superconductivity is suppressed in the superconducting phase (through continuous suppression of T_c by pair-breaking without affecting the long range phase coherence), the behavior in the vicinity of the transition (with a well-defined phase boundary), and the nature of the insulating state (a uniform disordered electronic system without any Cooper pairs).

The specific characteristics of the direct and indirect signatures discussed in the previous section for this model system are in stark contrast to those for the previous one (d-tuned SIT in granular films). Prior tunneling measurements (Reif and Woolf, 1964) in three-dimensional superconductors with paramagnetic impurities revealed progressive suppression of the energy gap and filling of the subgap states with increasing impurity concentration. In a very narrow range of impurity density before n_{cr}, the gap is reduced to zero although superconductivity persists (gapless superconductivity) (Reif and Woolf, 1964). Beyond n_{cr} the quasiparticle tunneling spectra show no sign of superconducting pairing.

Several key features in the transport data here are also in contrast to those in the d-tuned SIT in granular films:

1. There is a distinct boundary between the superconducting phase and the weakly insulating normal phase. No reentrance or double reentrance is present in the transition region.
2. In the insulating state, R diverges logarithmically, instead of exponentially, with decreasing temperature. This is seen more clearly in Fig. 14.4 (b) in which the data of Fig. 14.4 (a) are replotted in a semi-logarithmic fashion. We believe that the log-T dependence is an indicator of a disordered fermionic system with quantum conductance corrections.
3. There is no significant MR in the insulating state. Unlike the large (orders of magnitude) negative MR in the insulating state of the granular films, the MR in uniform films is very small (a few percent), and in the case of Pb films the MR is actually positive.

In a disordered conductor, certain quantum effects, including coherent backscattering (Bergmann, 1983) and electron–electron interaction (Altshuler *et al.*, 1980), lead to reductions in the electrical conductance. Both types of conductance corrections exhibit logarithmic divergence with decreasing temperature, resulting in an insulator (zero conductance/infinite resistance at 0 K). However, the temperature dependence of the resistance of such conductors is much weaker than those in which the electrons are intrinsically localized (e.g. hopping conductors). In many cases, at experimentally accessible temperatures the quantum conductance corrections are small and to the first order the sample resistance increases logarithmically with decreasing temperature. Such logarithmic temperature dependence has been observed in conventional disordered electronic materials (Dolan and Osheroff, 1979) as well as cuprate superconductors (Oh *et al.*, 2006; Ono *et al.*, 2000). In the cuprates, the temperature range and the magnitude of the effect make it unlikely that it originates from the quantum interference or interaction effect; more exotic mechanisms such as charge confinement onto stripes (Ono *et al.*, 2000) have been postulated. On the other hand, in conventional materials the most likely origins of the log-T resistance are weak localization effects. The conclusion is corroborated by the small positive MR in the normal state of the films, which is consistent with the dephasing of the anti-localization effect by the magnetic field in a disordered fermionic conductor with strong spin–orbit coupling. Therefore, we conclude that the MI-tuned transition is a continuous fermionic SIT, although it is debatable whether it is more appropriate to term the normal state here a canonical insulator or a metal with quantum conductance corrections.

In summary, in many aspects the MI-tuned SIT can serve as a model system in which the SIT is fermionic in nature, as manifested by a direct transition from a BCS superconductor to an electronically homogeneous fermionic insulator. Below we discuss the nature of the d-tuned and B-tuned SITs in homogeneous Pb films through direct comparisons with the two model systems, focusing on the key direct (tunneling) and indirect (transport) signatures outlined above.

14.3.3 *d*-tuned superconductor–insulator transition in homogeneous films

Experimentally, electron tunneling measurements on homogeneous Pb films have consistently pointed to a BCS superconducting gap that vanishes right at the SIT. Early experiments (Dynes *et al.*, 1986) deep into the superconducting state demonstrated that the BCS parameter $\frac{2\Delta}{k_B T_c}$ remains constant at a strong coupling value over a broad range of R_N (T_c). The extrapolation of this result implies that Δ should vanish with T_c at the SIT. Later experiments (Valles, Jr. *et al.*, 1992) in the close vicinity of the SIT boundary indeed revealed a vanishingly small energy gap on the superconducting side and no sign of superconducting pairing on the insulating side. Just as importantly, the tunneling measurements also revealed a progressive suppression of the normal-state quasiparticle density of states at E_F, $D(E_F)$, with increasing R_N, concomitant with the reduction of T_c (Valles, Jr. *et al.*, 1989). The close correlation between $D(E_F)$ suppression and T_c reduction strongly suggest a fermionic origin for the destruction of superconductivity in the homogeneous films. These results are in contrast to the constant Δ and $D(E_F)$ across the SIT in granular films (Fig. 14.5)(Barber *et al.*, 1994).

The overall transport characteristics of the d-tuned SIT (Fig. 14.1) were described in Section 14.3.1 The indirect signatures for a bosonic transition inferred from the d-tuned SIT in granular films are clearly absent here: there is no reentrance behavior near the SIT, and immediately on the insulating side, $R_\square(T)$ shows weakly localized behavior rather than activated transport. It is interesting to note that the SIT definitely occurs inside the weakly localized region of the Pb films. The crossover from weak localization to strong localization does not happen until a much higher R_N (~ 100 kΩ), as determined by a distinct switch in the sign of the MR (Hsu and Valles, Jr., 1995). In

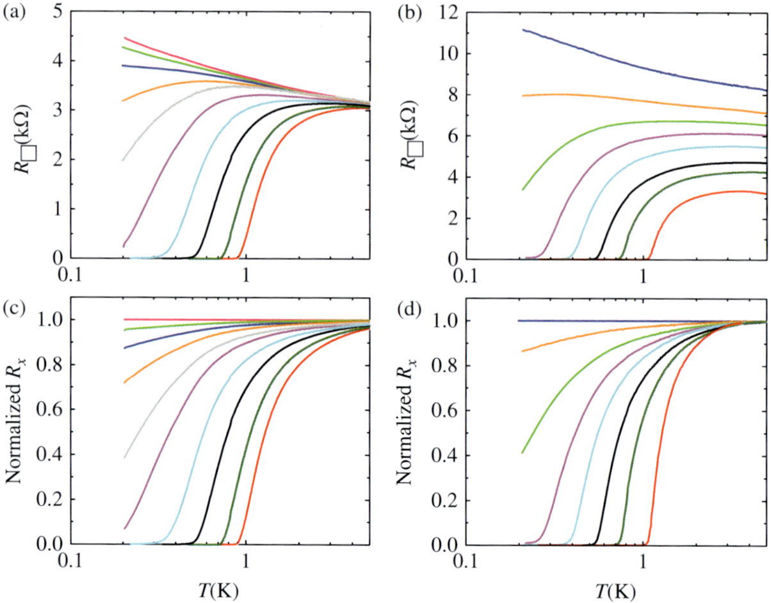

Fig. 14.9 (a) Semi-logarithmic plots of sheet resistance versus temperature for an MI-tuned SIT ($\frac{n}{n_{\rm cr}}$ = 0, 0.18, 0.39, 0.50, 0.64, 0.73, 0.82, 0.91, 1.00, and 1.09, from bottom to top). (b) Similar plots for a disorder-tuned SIT. (c) and (d) are the same data after the normalization of the logarithmic conductance corrections, for (a) and (b) respectively.

fact, in all aspects of the transport data, the d-tuned SIT bears a close resemblance to the other model system, the MI-tuned SIT. In Fig. 14.9 we directly compare the behavior of the MI-tuned and d-tuned transitions, as shown in Figs 14.9 (a) and (b), respectively. In both cases, there is a clear boundary separating the superconducting and normal phases. The qualitative resemblance of the d- and MI-tuned transitions becomes even more striking with the removal of the logarithmic contributions to the normal-state conductance using a procedure described in (Chervenak and Valles, Jr., 1999). The results are shown in Fig. 14.9 (c) and (d).

14.3.4 *B*-tuned superconductor–insulator transition in homogeneous films

In contrast to the well-defined MI- and d-tuned transitions, the B-tuned SIT exhibits much more complex behavior in the transport and in many aspects resembles the behavior of a d-tuned SIT in granular films (Parendo *et al.*, 2007). Figure 14.10 (a)–(e) shows the B-tuned SITs for a film at five different thicknesses. Besides the broad resistive transitions, a more notable feature is the absence of a distinct phase boundary separating the superconducting and insulating states. In fact, there exists a pronounced reentrant behavior in the critical region, which evolves systematically with the normal-state resistance, $R_{\rm N}$, of the film, as shown more clearly in the insets.

For high R_N, the resistance first decreases with decreasing T, reaches a minimum, and then rises rapidly and appear to diverge (Fig. 14.10 (a)–(c)). For lower R_N, at the lowest temperatures $R_\square(T)$ reaches a maximum and begins to decrease again, indicating a prevailing superconducting state (Fig. 14.10 (d) and 10(e)). Such non-monotonic $R_\square(T)$ is a distinguishing signature of the d-tuned SIT in granular films, as discussed earlier. Furthermore, immediately on the insulating side of the SIT, $R_\square(T)$ shows a rapid increase with decreasing T at low T. Although the temperature range is too narrow for us to precisely determine the T-dependence, it is close to activated transport and clearly much stronger than log T, as shown in Fig. 14.2 (b), in which the data in Fig. 14.2 (a) are replotted on logarithmic temperature scale. More interestingly, upon further increase in B, the T-dependence actually becomes weaker and eventually approaches a log-T dependence at 8 T. In some samples this results in crossing of $R_\square(T)$ at 8 T with those at lower fields and a peak in $R_\square(B)$ at low temperatures (Fig. 14.2 (c)).

Similar reentrant and double-reentrant behavior was reported in B-tuned SITs in ultra-thin TiN films (Hadacek *et al.*, 2004) and oxygen-doping controlled SITs in a high-T_c cuprate (Oh *et al.*, 2006). The latter results were attributed to collective electronic phase separation, which results in hole-rich (superconducting) regions coupled through a hole-poor (insulating) background. All of our observations can be understood with a similar picture of B-induced mesoscale phase inhomogeneities across the SIT. Near B_c, the film breaks up into localized superconducting regions with well-defined SOP in a weakly insulating background. The transport behavior in the critical region is determined by the coupling of the superconducting islands through the background: for high R_N the Josephson coupling between the superconducting islands is weak and the transport is dominated by quasiparticles. Hence the opening of the superconducting gap in the islands leads to activated transport at low temperatures across the film. Increases in the conductivity of the background (lower R_N) lead to stronger Josephson coupling between the islands, which prevails at lowest temperatures and results in a double reentrance into a superconducting state near B_c. In all cases, further increase in B eventually leads to the complete destruction of superconductivity in the film and a pure fermionic system with weakly insulating behavior in the entire temperature range. The weakening of the temperature dependence in $R_\square(T)$ leads to a negative MR at high field, hence a MR peak (Fig. 14.2 (b) and (c)). Much more pronounced MR peaks have been observed in thin films of compound superconductors of TiN (Baturina *et al.*, 2007) and InO_x (Gantmakher *et al.*, 1998; Paalanen *et al.*, 1992; Sambandamurthy, 2004; Steiner and Kapitulnik, 2005), probably reflecting an even stronger degree of phase inhomogeneity in the insulating state in those materials. Clearly, in this picture superconducting pairing persists well into the insulating state of the B-tuned SIT, and phase fluctuations dominate the critical region, meeting a necessary condition for the dirty-boson model (Jaeger *et al.*, 1989).

14.3.4.1 Theoretical implications

The origin of the B-induced phase separation is an intriguing open question. It is unlikely that it is caused by physical inhomogeneity. Our samples are made of

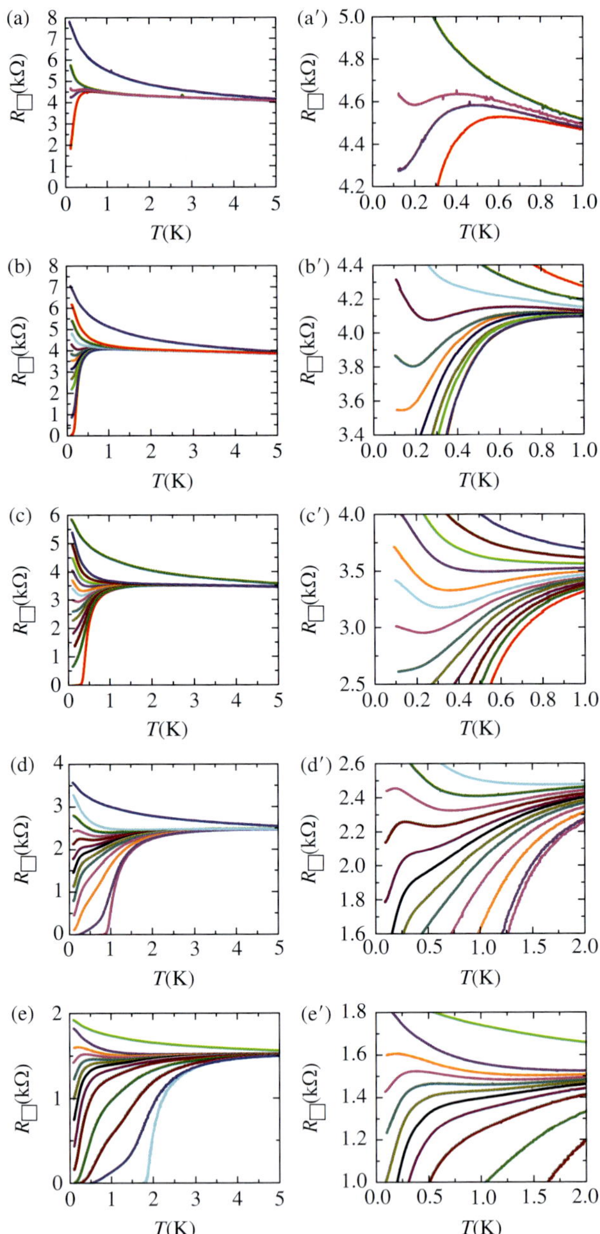

Fig. 14.10 Evolution of the reentrance and double-reentrance behavior in the magnetic field-tuned SITs in the same homogeneous Pb film at increasing film thicknesses. From (a) to (e) is the sheet resistance versus temperature at applied fields of (a) 0.00, 0.10, 0.12, 0.2, and 8.0 T; (b) 0.00, 0.02, 0.075, 0.10, 0.125, 0.15, 0.17, 0.20, 0.25, 0.34, 0.50, and 8.00 T; (c) 0.00, 0.10, 0.16, 0.21, 0.25, 0.28, 0.32, 0.39, 0.40, 0.44, 0.50, 0.575, 0.70, and 8.00 T; (d) 0.00, 0.13, 0.47, 0.67, 0.80, 0.88, 0.95, 1.02, 1.10, 1.25, 1.35, 1.45, and 8.00 T; (e) 0.00, 0.40, 1.00, 1.50, 1.90, 2.10, 2.22, 2.32, 2.41, 2.50, 2.60, 2.80, and 8.00 T, all from bottom to top. The corresponding close-ups of the critical region are presented (a to e) to clearly illustrate the reentrance and double-reentrance behavior.

elemental Pb and the growth method is known to produce electrically homogeneous films (Strongin, 1970). Moreover, the MI-tuned transition on the same film and the d-tuned SIT in even thinner films do not show any of the signatures indicative of phase inhomogeneity. The complex transport behavior is also inconsistent with vortex physics. The quasi-reentrance shown in Fig. 14.10 (a)–(c) and the negative MR at high field may be qualitatively described by theories based on fluctuation effects (Galitski and Larkin, 2001; Sambandamurthy, 2004). It was shown that in 2D dirty superconductors near 0 K and B_c, the three contributions to the fluctuation conductivity are of the same order but can have different signs, and the competition between them leads to non-monotonic $R_\square(T)$ and $R_\square(B)$. In these theories, however, the total conductivity correction at 0 K is always negative, and therefore cannot account for the double reentrance as shown in Fig. 14.10 (d) and (e).

A scenario that may apply to our case is based on disorder-induced enhancement of the upper critical field, H_{c2}, in dirty superconductors (Lopatin *et al.*, 2005; Spivak and Zhou, 1995). It was shown that near the transition point, mesoscopic effects could lead to local H_{c2} far exceeding the system-wide average, and these regions form superconducting islands even at fields above the macroscopic H_{c2}. Several numerical simulations (Dubi *et al.*, 2007; Ghosal *et al.*, 1998) uncovered the possibility that disorder or disorder/magnetic field could induce the formation of superconducting islands in an insulating background. Using the Bogoliubov–de Gennes approach, Ghosal *et al.* (1998) found that disorder (in zero field) induces strong fluctuations of the SOP amplitude in space, even in atomically homogeneous 2D systems. That is, with increasing disorder the system self-organizes into a nanoscale granular structure in terms of the local pairing amplitude and therefore superconductivity is destroyed in similar fashion to that of granular films. Following a similar approach, Dubi *et al.* (2007) demonstrated the formation of superconducting islands in homogeneously disordered films and their evolution with applied magnetic field, providing an explanation for the observed giant MR peak (Gantmakher *et al.*, 1998; Paalanen *et al.*, 1992; Sambandamurthy, 2004; Steiner and Kapitulnik, 2005). Our results indicate that disorder alone does not cause such phase separation, and a combination of high disorder and a magnetic field is required.

14.3.5 Other observations

14.3.5.1 B-tuned SITs in homogeneous Pb films with magnetic impurity

Figure 14.11 (a) shows the B-tuned SIT for a pure Pb film with $T_c = 550$ mK, which exhibits all the expected features (e.g. reentrance) described above. More Pb was deposited so that the film T_c was increased to 1.30 K; and the B-tuned SIT was measured again (Fig. 14.11 (b)). Finally, Cr impurities were deposited onto the film to reduce its T_c to close to the original value of the pure Pb film, i.e. 560 mK; and the B-tuned SIT measurements were performed on the film with magnetic impurities (Fig. 14.11 (c)). The experiment enabled us to directly compare the behavior of the B-tuned SITs (especially the transport signatures for a bosonic SIT) for samples with similar T_c and different R_N (Figs 14.11 (a) and(c)) as well as samples with the same R_N and different T_c (Figs 14.11 (b) and (c)). The evolution of the reentrance

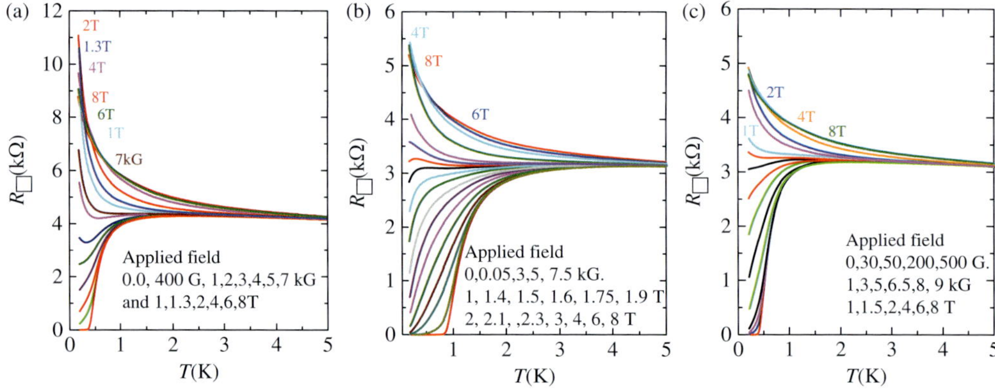

Fig. 14.11 Perpendicular magnetic-field-tuned SITs in Pb films without and with Cr impurities. (a) A pure Pb film with initial T_c of 0.55 K and normal-state sheet resistance of 4.1 kΩ. (b) The same sample with zero-field T_c of 1.22 K (increased by evaporating more Pb) and normal-state sheet resistance of 3.1 kΩ. (c) The same film with zero-field T_c of 0.56 K (reduced by adding Cr impurities) and normal-state sheet resistance 3.1 kΩ.

behavior in the two pure Pb films (from Fig. 14.11 (a) to Fig. 14.11 (b)) is consistent with results from the previous series (Fig. 14.10). The addition of MI diminishes the reentrance effect as can be inferred from a comparison of Figs 14.11 (c) and (b). The suppression is much stronger when compared to the pure Pb film with a similar T_c but higher R_N (Fig. 14.11 (a)), which attests to the important role the background conductivity plays in the reentrance behavior. Clearly, the presence of MI effectively curtails the ability of the magnetic field to produce an inhomogeneous insulating state with localized superconducting pairing; this can be readily attributed to the pair-breaking by the MI, which suppresses the superconductivity in the superconducting islands and produces a more homogeneous electronic state.

14.3.5.2 Parallel magnetic field

With a rotatable sample stage we were able to change the magnetic-field orientation *in situ*, and compare the effects of perpendicular and parallel magnetic fields. Figure 14.12 (a) and (b) shows the $R_\square(T)$ data on the same film in perpendicular and parallel magnetic fields up to 8 T respectively. The film resistance in the resistive transition region is extremely sensitive to any misalignment from the parallel orientation: a misalignment of much less than 0.2° leads to significant broadening of the transition. However, when near-perfect alignment is achieved through in- situ adjustment, a parallel magnetic field as large as 8 T results in no noticeable suppression of superconductivity. In fact, in a large range of field strengths (as high as 8 T for some samples) a parallel magnetic field actually enhances superconductivity (increases T_c) (Gardner *et al.*, 2011). It can be inferred from the data that the parallel critical field for the film is far greater than 8 T, while the perpendicular critical field is only 3.0 T.

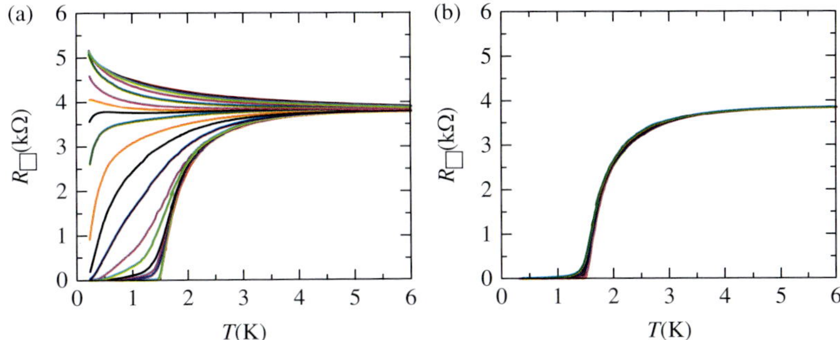

Fig. 14.12 Sheet resistance versus temperature for a homogeneous Pb film with a zero-field T_c of 1.8 K in: (a) perpendicular magnetic fields of 0.0, 0.005, 0.025, 0.050, 0.10, 0.20, 0.25, 0.50, 1.0, 2.0, 2.50, 2.80, 3.0, 3.30, 4.0, 5.0, 6.0, 7.0, 8.0 T, and (b) parallel magnetic fields from 0 to 8 T.

14.4 Summary

We have carried out a detailed systematic comparative study of 2D SITs tuned by disorder, paramagnetic impurities, and perpendicular magnetic fields in ultra-thin amorphous Pb films. We believe that the d-tuned SIT in granular films and the MI-tuned SIT in homogeneous films establish the behavior of two distinct classes of 2D SIT: the former is bosonic in nature with robust superconducting pairing in the insulating state and phase fluctuations driving the SIT, while the latter is a fermionic SIT where increasing magnetic pair-breaking leads to a continuous destruction of superconductivity and a transition to a purely fermionic weakly insulating normal state. From these two model systems we extract a set of direct (tunneling) and indirect (transport) signatures for bosonic and fermionic SITs, which we use to elucidate the origin of the d-tuned and B-tuned SITs in homogeneous films. In all aspects, the d-tuned and MI-tuned transitions in uniform Pb films are qualitatively similar, suggesting that the two quantum phase transitions have similar underlying mechanisms. The d-tuned SIT is therefore also fermionic in nature. In comparison, the B-tuned SIT appears to be in a qualitatively different class, and in many aspects resembles the d-tuned SIT in granular films. The disordered bosonic model may therefore be applicable. The complex transport behavior in the critical region of the B-tuned SIT suggests that the magnetic field induces mesoscale phase separation near the SIT and Cooper pairing persists well into the insulating state. Our picture of the d-tuned and B-tuned SITs in homogeneous films also imply that disorder alone does not induce such superconducting puddles in the insulating state. Instead, a combination of disorder and magnetic field is required.

Acknowledgements

We are indebted to Jeff Parker and Dan Read for their early contributions to the work. We thank Vlad Dobrosavljević, Lev Gorkov, Pedro Schlottmann, Oskar Vafek, and Jim Valles, Jr. for helpful discussions.

References

Abrikosov, A.A and Gorkov, L.P. (1960). *JETP*, **39**, 1781.

Altshuler, B.L., Aronov, A.G., and Lee, P.A. (1980). *Phys. Rev. Lett.*, **44**, 1288–1291.

Barber, R.P., Hsu, S.Y., Valles, Jr., J.M., Dynes, R.C., and Glover, R.E. (2006). *Phys. Rev. B*, **73**, 134516.

Barber, R.P., Merchant, L.M., Laporta, A., and Dynes, R.C. (1994). *Phys. Rev. B*, **49**, 3409–3412.

Baturina, T.I, Mironov, A.Y., Vinokur, V.M., Baklanov, M.R., and Strunk, C. (2007). *Phys. Rev. Lett.*, **99**, 257003.

Bergmann, G. (1983). *Phys. Rev. B*, **28**, 2914–2920.

Chervenak, J.A. and Valles, Jr., J.M. (1999). *Phys. Rev. B*, **59**, 11209–11212.

Dolan, G.J. and Osheroff, D.D. (1979). *Phys. Rev. Lett.*, **43**, 721–724.

Dubi, Y., Meir, Y., and Avishai, Y. (2007). *Nature*, **449**, 876–880.

Dynes, R., Garno, J., and Rowell, J. (1978). *Phys. Rev. Lett.*, **40**, 479–482.

Dynes, R.C., White, A.E., Graybeal, J.M., and Garno, J.P. (1986). *Phys. Rev. Lett.*, **57**, 2195–2198.

Finkelstein, A. (1994). *Physica B*, **197**, 636–648.

Fisher, M.P.A. (1990). *Phys. Rev. Lett.*, **65**, 923–926.

Fisher, M.P.A., Grinstein, G., and Girvin, S.M. (1990). *Phys. Rev. Lett.*, **64**, 587–590.

Fukuyama, H. (1984, Jan). *Physica B & C*, **126**, 306–313.

Galitski, V.M. and Larkin, A.I. (2001). *Phys. Rev. Lett.*, **87**, 087001.

Gantmakher, V.F., Golubkov, M.V., Dolgopolov, V.T., Tsydynzhapov, G.E., and Shashkin, A.A. (1998). *JETP Lett.*, **68**, 363–369.

Gardner, H.J., Kumar, A., Yu, L., Xiong, P., Warusawithana, M.P., Wang, L., Vafek, O., and Schlom, D.G. (2011). *Nature Phys.*. DOI: 10.1038/NPHY52075.

Ghosal, A., Randeria, M., and Trivedi, N. (1998). *Phys. Rev. Lett.*, **81**, 3940–3943.

Goldman, A. and Markovic, N. (1998). *Phys Today*, **51**, 39–44.

Hadacek, N., Sanquer, M., and Villegier, J.C. (2004). *Phys. Rev. B*, **69**, 024505.

Haviland, D.B., Liu, Y., and Goldman, A.M. (1989, Jan). *Phys. Rev. Lett.*, **62**, 2180–2183.

Hsu, S.Y. and Valles, Jr., J.M. (1995). *Phys. Rev. Lett.*, **74**, 2331–2334.

Jaeger, H., Haviland, D., Orr, B., and Goldman, A. (1989). *Phys. Rev. B*, **40**, 182–196.

Lopatin, A.V., Shah, N., and Vinokur, V.M. (2005). *Phys. Rev. Lett.*, **94**, 037003.

Merchant, L., Ostrick, J., Barber, R., and Dynes, R. (2001). *Phys. Rev. B*, **6313**, 134508.

Oh, S., Crane, T.A., Harlingen, D.J. Van, and Eckstein, J.N. (2006). *Phys. Rev. Lett.*, **96**, 107003.

Ono, S., Ando, Y., Murayama, T., Balakirev, F.F., Betts, J.B., and Boebinger, G.S. (2000). *Phys. Rev. Lett.*, **85**, 638–641.

Orr, B.G., Jaeger, H.M., and Goldman, A.M. (1985). *Phys. Rev. B*, **32**, 7586–7589.

Paalanen, M., Hebard, .A, and Ruel, R. (1992). *Phys. Rev. Lett.*, **69**, 1604–1607.

Parendo, K.A., Sarwa, K.H., Tan, B., and Goldman, A.M (2007). *Phys. Rev. B*, **76**, 100508.

Parker, J., Read, D., Kumar, A., and Xiong, P. (2006). *Europhys. Lett.*, **75**, 950–956.

Ramazashvili, R. and Coleman, P. (1997). *Phys. Rev. Lett.*, **79**, 3752–3754.
Reif, F. and Woolf, M.A. (1964). *Rev. Mod. Phys.*, **36**, 238.
Sachdev, S. (1999). *Quantum phase transitions.* Cambridge University Press, Cambridge.
Sambandamurthy, G. (2004). *Phys. Rev. Lett.*, **92**, 107005.
Sondhi, S. (1997). *Rev. Mod. Phys*, **69**, 315–333.
Spivak, B. and Zhou, F. (1995). *Phys. Rev. Lett.*, **74**, 2800–2803.
Steiner, M. and Kapitulnik, A. (2005). *Physica C*, **422**(1-2), 16–26.
Strongin, M. (1970). *Phys. Rev. B.*, **1**, 1078.
Valles, Jr., J.M., Dynes, R.C., and Garno, J.P. (1989). *Phys. Rev. B*, **40**, 6680–6683.
Valles, Jr., J.M., Dynes, R.C., and Garno, J.P. (1992). *Phys. Rev. Lett.*, **69**, 3567–3570.
van der Zant, H.S.J., Elion, W.J., Geerligs, L.J., and Mooij, J.E. (1996). *Phys. Rev. B*, **54**, 10081–10093.

15

Spin Effects Near the Superconductor–Insulator Transition

P.W. Adams

Department of Physics and Astronomy
Louisiana State University
Baton Rouge, LA 70803

15.1 Introduction

Ultra-thin metal films have proven to be extraordinarily fertile systems for studying a variety of quantum-scattering and interaction processes that ultimately serve to destroy the metallic state of their bulk counterparts (Bergmann, 1983; Dynes *et al.*, 1978). By the early 1980s it was recognized that coherent backscattering in moderately disordered films produces logarithmically insulating behavior at low temperature (Lee and Ramakrishnan, 1985). In addition, disorder tends to enhance the impact of electron–electron interactions, which manifest themselves as a logarithmic suppression of the density of states near the Fermi energy (Altshuler *et al.*, 1987). The theoretical description of weakly disordered two-dimensional systems has, in fact, been a great success, having given us a quantitative description of a wide spectrum of transport and tunneling density-of-states experiments (Abrahams *et al.*, 1979; Lee and Ramakrishnan, 1985). In contrast, the magnetotransport properties of highly disordered films, with sheet resistance R of the order of the quantum resistance $R_Q = h/e^2$, remain poorly understood (Butko and Adams, 2001). To date there is no clear consensus as to what roles film morphology (Epstein *et al.*, 1983), phase-coherent hopping (Entin-Wohlman *et al.*, 1989; Medina *et al.*, 1996; Nguyen *et al.*, 1985), Zeeman splitting (Butko *et al.*, 2000*a*; Hernandez *et al.*, 2003; Matveev *et al.*, 1995), and/or spin–orbit scattering (S-OS) (Hernandez and Sanquer, 1992; Pichard *et al.*, 1990; Shapir and Ovadyahu, 1989) play in producing the correlated insulator phase of ultra-thin metal films. Recently, however, investigators have recognized that new insights into the processes that contribute to the formation of the correlated insulator phase can be obtained through the study of metal films that undergo a superconductor-to-insulator (SIT) (Goldman and Markovic, 1998). The reason for this is obvious. On the one hand, superconductors are characterized by a macroscopic quantum state that exhibits long-range phase coherence and non-dissipative current flow. Insulators, on the other hand, have no long-range coherence of any sort and exhibit dissipative, glassy dynamics. The fact that this striking juxtaposition of electronic properties can be controlled via an external tuning parameter, such as film resistance, allows one to explore the emergence of the insulating phase from the perspective of the superconducting phase and its attendant fluctuations.

In practice, a superconducting film can be driven into the insulating phase by increasing its disorder beyond a specific threshold. Typically this is done by making the film thinner; once the normal-state sheet resistance is of the order of R_Q, the superconducting phase gives way to a highly insulating phase (Haviland *et al.*, 1989). Alternatively, if the system is close to the insulating threshold, a magnetic field can be used to tune the system through the SIT (Hebard and Paalanen, 1990). The SIT has been the subject of intense investigation for more than two decades now. The primary interest has been in those systems that are homogeneously disordered and, in particular, non-granular. It is generally believed that the disorder-driven SIT in these systems is mediated by electron–electron interaction effects (Valles, Jr. *et al.*, 1992). With increasing disorder, an otherwise perturbative depletion of quasiparticle states at the Fermi energy grows into a full-blown correlation gap when $R \gg R_Q$ (Butko and Adams, 2001; Butko *et al.*, 2000*b*). This has the effect of undermining the superconducting order-parameter amplitude, thereby suppressing the transition temperature T_c (Valles, Jr. *et al.*, 1992). However, the exact nature of the insulating state and its relation to the superconducting phase remain unclear. For instance, anomalously large, multi-fold, negative magnetoresistances have been reported in ultra-thin TiN films (Baturina *et al.*, 2007), in InO_x films (Gantmakher *et al.*, 2000, 1998; Steiner *et al.*, 2005) and in insulating Be films (Butko *et al.*, 2000*b*). The magnetoresistances of these films saturates at a weakly temperature-dependent resistance that is always near R_Q, i.e. the "quantum metal" phase (Baturina *et al.*, 2007; Butko and Adams, 2001). This observation suggests that the zero-field insulating ground state is distinctly different from the high-field ground state, which has led to speculation that the zero-field ground state has an incoherent superconducting component (Sambandamurthy *et al.*, 2004; Stewart *et al.*, 2007).

Although the SIT has been studied extensively over the last 30 years, there are, in fact, very few experiments that have directly probed the spin degrees of freedom of the system in the region near the zero-field transition. Several conditions must be met for one to probe spin effects in thin disordered films. First, the film must have a relatively low intrinsic S-OS rate. Otherwise, the spin–orbit interaction will mix the spin-up and spin-down eigenstates, and spin will no longer be a good quantum number. Second, one must find a way to couple to the spin moments without coupling to the orbital degrees of freedom. In practice, this is done through the Zeeman splitting from an applied magnetic field. If a film is very thin, i.e. thinner than the diffusion length and/or the superconducting coherence length, then the orbital response can be effectively suppressed by aligning the plane of the film parallel to the applied field. Third, one needs a microscopic probe that is sensitive to spin, such as electron tunneling.

Numerous spin studies of ultra-thin Be and Al films have shown that these light elements have a very low intrinsic S-OS rate (Adams, 2004; Adams *et al.*, 1998; Tedrow and Meservey, 1979) and are true spin-singlet superconductors. Here we will review the spin properties of these systems on either side of the zero-field SIT. We will begin with an overview of high-field superconductivity in thin, moderately disordered Al and Be films with an emphasis on their spin degrees of freedom. Next we will consider the magnetotransport properties of homogeneously disordered Be films that are on the insulating side of the SIT. Of course, our ultimate goal is to use spin to probe the quantum characteristics of the zero-field SIT.

15.2 The parallel field superconductor–insulator transition

15.2.1 The spin-paramagnetic transition

Magnetic fields generally have a detrimental effect on superconductors via two independent channels. The first is an orbital effect associated with the fact that cyclotron motion is incompatible with the formation of Cooper pairs and hence superconductivity; for the vast majority of superconducting systems the critical field transition is completely dominated by the orbital response of the conduction electrons. The second channel is the Zeeman coupling to the electron spin. This Zeeman splitting can be made the dominant pair-breaking mechanism by inhibiting the orbital response.

This is typically done by applying a magnetic field in the plane of a superconducting film whose thickness t is much less than the superconducting coherence length ξ and whose electron diffusivity is low (Fulde, 1973; Meservey *et al.*, 1970; Wu and Adams, 1994). Under these conditions the phase transition to the normal state is mediated by the spin polarization of the electrons, and near the critical field the electron Zeeman splitting is of the order of the superconducting gap energy. A first-order transition from the superconducting state to the paramagnetic normal state occurs at the Clogston–Chadrasekhar critical field (Chandrasekhar, 1962; Clogston, 1962)

$$H_{\mathrm{c}}^{\mathrm{CC}} = \frac{\Delta_0\sqrt{1+G^0}}{\sqrt{2}\mu_{\mathrm{B}}} \tag{15.1}$$

where Δ_0 is the zero-field, zero-temperature gap energy; μ_{B} is the Bohr magneton; and G^0 is the antisymmetric Fermi-liquid parameter. Note that, if Δ_0 is known, then by eqn (15.1), the parallel critical field of a thin film gives a direct measure of G^0, assuming spin–orbit effects are negligible. G^0 affects the spin response of the system and is related to the ratio of the spin-susceptibility density of states $N(\chi)$ to the heat capacity density of states $N(\gamma)$ by $G^0 = N(\gamma)/N(\chi) - 1$ (Baym and Pethick, 1991).

Though G^0 is a fundamentally important parameter in the many-body description of metals, there have been very few measurements of it reported in the literature. This is due, in part, to the fact that it is exceedingly difficult to measure it directly in bulk systems. To date, measurements of G^0 have been extracted from high-field tunneling density of states (DOS) and critical-field studies of low-atomic-mass superconducting films. Estimates of the normal-state value of G^0 have been reported via tunneling studies of the Zeeman splitting of the BCS density of states (DOS) in superconducting Al films, with values of $G^0 \sim 0.3$–0.4, (Alexander *et al.*, 1985; Tedrow *et al.*, 1984) in the intermediate-temperature regime where the superconducting order parameter is partly suppressed by thermal fluctuations. Low-temperature DOS measurements in amorphous Ga films, which is a strong-coupling superconductor, give a somewhat larger value of $G^0 \sim 0.81$ (Gibson *et al.*, 1989). Alternatively, G^0 can be extracted from parallel critical-field measurements. This method gives $G^0 \sim 0.23$ in TiN films (Suzuki *et al.*, 2000). The only direct measurements of G^0 have been obtained in the normal state via the field dependence of the pairing resonance (Aleiner and Altshuler, 1997). In Al films $G^0 \sim 0.17$ (Butko *et al.*, 1999) and in Be films $G^0 \sim 0.21$ (Adams and Butko, 2000). In all these systems the S-OS is quite low, thus spin remains a good quantum number.

15.2.2 Parallel-field studies of ultra-thin Al and Be films

Thin-film synthesis and characterization

In thin metal films the S-OS rate, $1/\tau_{\mathrm{so}}$, increases with increasing atomic mass Z as $\tau_{\mathrm{s}}/\tau_{\mathrm{so}} \sim Z^4$, where τ_{s} is the surface scattering time (Meservey and Tedrow, 1976), suggesting that light elemental films are the best candidates for purely spin-singlet superconductivity. Indeed, extensive studies of the parallel critical-field transition, commonly known as the spin–paramagnetic (S–P) transition, in Al and Be films have shown that the characteristic S-OS rate in these elements is very low. Aluminum films tend to form a granular microstructure, but very thin Be films can be made with a smooth, dense, homogeneously disordered microstructure. Al (Be) films used in S–P studies are typically made by e-beam deposition of 3–5 nm of 99.999% Al (99.5% Be) onto fire-polished glass microscope slides cooled to 84 K. Usual deposition rates are ~ 0.1 nm/s in a vacuum of ~ 0.5 μTorr. The Al (Be) films have a transition temperature $T_c \sim 2.7$ K ($T_c \sim 0.5$ K) and a parallel critical field $H_{c||} \approx 6.0$ T ($H_{c||} \approx 1.0$ T). Tunnel junctions are formed by exposing the films to atmosphere for 0.2–1 h in order to form a native oxide. Then a strip of metal is deposited on the upper surface of the film at 84 K to form a tunneling counter-electrode. The counter-electrode can either be a non-superconducting material such as Ag or a relatively thick ($t > 7$ nm) strip of Al, which will have a low parallel critical field. At low temperatures the tunneling conductance is proportional to the product $N_{ce}N_{film}$, where N_{film} is the quasiparticle DOS. Since the counter-electrode DOS is independent of energy, the tunneling conductance gives a direct measure of the superconducting film DOS. Alignment with the external field is crucial in these experiments, so alignment must be made with an *in situ* mechanical rotator.

S–P phase diagram

Low-temperature measurements of the tunneling conductance as a function of parallel magnetic field reveal that the tunneling spectrum of thin Al and Be films changes abruptly at the critical field and displays a surprisingly large hysteresis. The hysteresis is a consequence of the first-order nature of the critical field transition. The most complete studies of the hysteretic S–P transition have been made on Al films, which will be the primary focus of this section. In Fig. 15.1 the zero-bias tunnel junction conductance G(0) is plotted as a function of parallel field at the critical field transition of a 1 kΩ/sq Al film. The precipitous attenuation in G(0) as the field is lowered through the transition is due to the sudden opening of the superconducting gap in the single-particle DOS. As a consequence there is an exponential suppression of the zero-bias tunnelling conductance (Tinkham, 1996).

The hysteresis in the DOS indicates that the non-equilibrium aspects of the transition are intrinsic to the superconducting condensate itself. In Fig. 15.2 a phase diagram is produced by plotting the up-sweep and downsweep critical fields as a function of temperature, where the critical field is defined by the onset of a gap in zero bias tunneling conductance, G(0). Note that the hysteresis opens up below the tricritical point temperature $T_{Tri} \sim 0.6$ K. A similar phase diagram has been reported for Be films.

The dashed lines in Fig. 15.2 are provided to help the reader visualize the field and temperature cycles used to explore the phase diagram. Indeed, Fig. 15.2 represents the classic S–P phase diagram in which a high-temperature line of second-order phase transitions terminates into a line of first-order transitions at the tricritical point $T_{\rm tri} \sim 0.3T_c$. The low-temperature superconducting S and normal N phases are separated by a robust coexistence region in which the state of the system is solely determined by the system state prior to entering the region, i.e. the state memory SM region. If one were to start from the high-field, low-temperature point a in the normal state

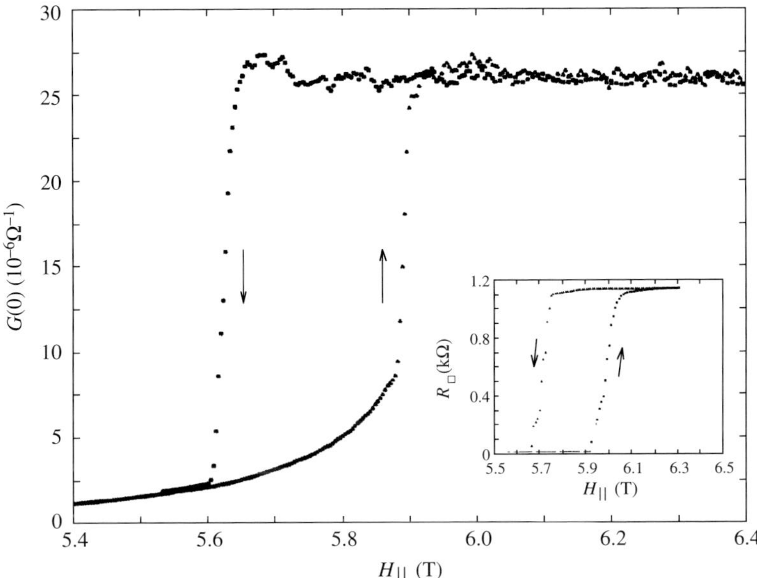

Fig. 15.1 Hysteresis in zero-bias tunneling conductance as a function of parallel field for a 3-nm-thick Al film at 30 mK. Arrows depict field sweep direction. Inset: corresponding hysteresis in film sheet resistance.

and then lower the field to point *c*, just above the lower critical field branch, then the system would be in metastable normal state. This "glassy" normal state can be melted by simply cutting across the diagram c $\rightarrow$ e, thereby inducing superconductivity via warming. Similarly, a metastable superconducting state can be formed by moving from d $\rightarrow$ b. The precise role disorder and microscopic morphology play in determining the details of the phase diagram are unknown. However, it is evident that the first-order S–P transition is intrinsically hysteretic, and films that are near the SM region exhibit glassy, non-equilibrium dynamics (Wu and Adams, 1995).

15.2.3 High-field spin-resolved tunneling

Superconducting phase

One of the most striking consequences of spin-rotation symmetry is the field-induced modification of the BCS DOS spectrum due to the Zeeman interaction. Under the same conditions that give rise to the first order S–P transition, one finds that, in a subcritical parallel field, the BCS quasiparticle coherence peaks are split into spin-up and spin-down subbands, as shown in Fig. 15.3. This effect was first reported by Tedrow and Meservey, who would later go on to use it as a spin-resolved tunneling probe of thin magnetic films. The BCS peaks are separated by the Zeeman energy. Spectra such as these provide irrevocable evidence that spin is a good quantum number in Al and Be films. Note that, with increasing field, the innermost peaks move toward

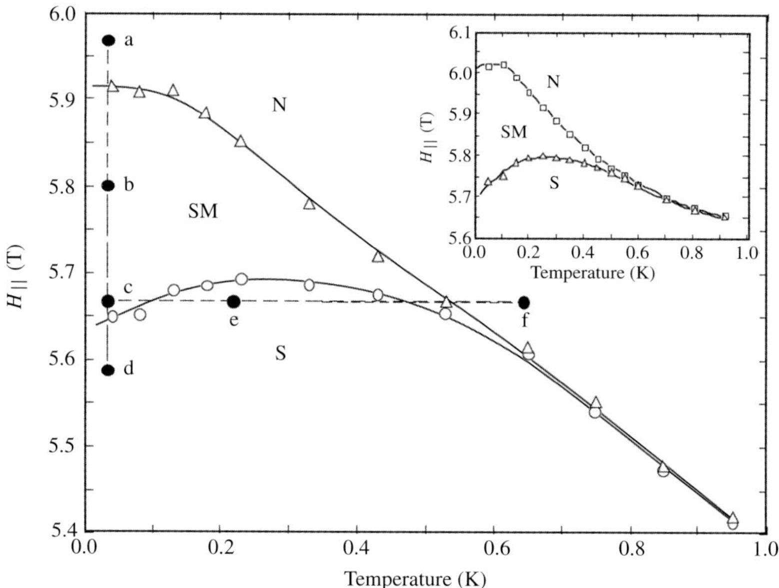

Fig. 15.2 Parallel critical fields as a function of temperature as measured by the zero-bias tunnel junction conductance. Triangles refer to up-sweep transitions and circles to down-sweep transitions. The letters and dotted lines are provided as a guide to the field and temperature cycles discussed in the text. S, superconducting phase; N, normal phase; SM, state memory region. Inset: Parallel critical fields as measured by resisitive transitions.

the Fermi energy (i.e. $V = 0$). However, by eqn (15.1) the critical field will be crossed before the spin-up and spin-down subbands overlap.

Pauli-limited normal state

The spin symmetry of ultra-light superconductors is not only manifested in the S–P transition itself but also in the Pauli-limited paramagnetic normal state, as well (i.e. $H > H_{c\|}$). In particular, an unexpected pairing resonance, clearly associated with virtual Cooper-pair formation, was discovered in the paramagnetic DOS spectrum of Al and Be films (Adams and Butko, 2000; Wu *et al.*, 1995) in the mid 1990s. Figure 15.4 shows typical 60-mK tunneling spectra of a 2.6-nm-thick Al film in several parallel fields above $H_{c\|}$. The spectra display a logarithmic depletion of states near the Fermi energy, commonly known as the zero-bias anomaly (ZBA). It is now well established that the logarithmic ZBA, which is independent of field, is associated with electron–electron interactions in two-dimensional (2D) disordered systems (Altshuler *et al.*, 1987). Indeed, the magnitude of the ZBA grows rapidly with increasing film resistance, and in films with $R \sim R_Q$ the depletion can be sufficiently extensive so as to undermine the formation of superconducting condensate.

The satellite features in Fig. 15.4 are due to the pairing resonance and represent a window into the processes that ultimately lead to the formation of stable Cooper pairs

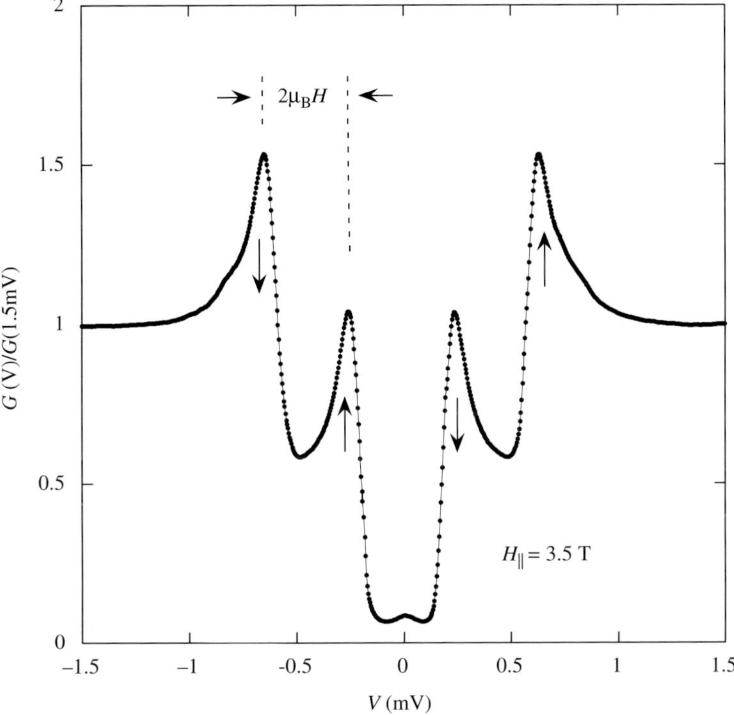

Fig. 15.3 Tunneling conductance of 26 Å superconducting Al in a parallel field of 3.5 T. Note the Zeeman splitting of the BCS DOS. The arrows denote the spin assignments of the peaks.

at lower fields. In the paramagnetic normal state there is a net polarization of spins near the Fermi energy. If a counter-electrode electron tunnels into this state with its spin aligned along the polarization direction, nothing of interest happens. However, if an electron with spin anti-parallel to the polarization direction tunnels into the film, it can form an evanescent Cooper pair with a local up-spin electron. This evanescent Cooper pair is energetically unbound, but it nevertheless can produce a resonance feature near the Zeeman energy. Indeed, the satellite dips in Fig. 15.4 result from the consumption of local single-particle states as the resonance attempts to open a gap near the Zeeman energy. The resonance can be observed in other low spin-orbit materials as well, Fig. 15.5 shows the pairing resonance in a 2.3-nm-thick Be film with $T_c \sim 0.6$ K. The fact that the resonance can be readily observed in high-field tunneling is due to the fact that it is a spin-singlet mode riding on top of a paramagnetic background. The spin structure of the resonance is depicted in Fig. 15.6, which shows an idealized tunnel junction between a non-superconducting paramagnetic counter-electrode and the Pauli-limited normal state of a superconducting film in a supercritical parallel magnetic field. The discrete momentum states near the Fermi energy are singly occupied, $E_Z = g_L \mu_B H$ is the Zeeman energy, and $g_L = 2/(1 + G^0)$ is the renormalized Landé g-factor.

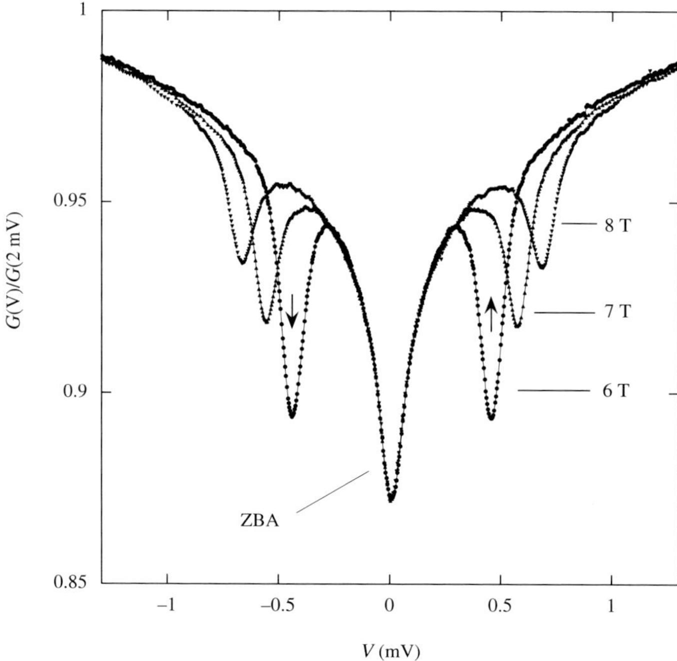

Fig. 15.4 The T = 70 mK normal-state tunneling conductance at several supercritical parallel magnetic fields in a 2.6-nm-thick Al film with sheet resistance $R \sim 1$ kΩ. The central feature is the zero-bias anomaly (ZBA), which is an electron–electron interaction effect. The satellite features represent the pairing resonance. The arrows denote the spin assignments of the occupied and unoccupied resonances.

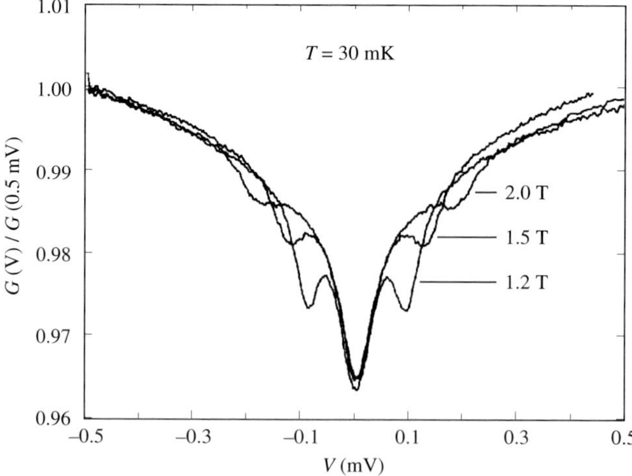

Fig. 15.5 The pairing resonance in a 2.3-nm-thick Be film with $T_c = 0.55$ K and $R = 260$ Ω.

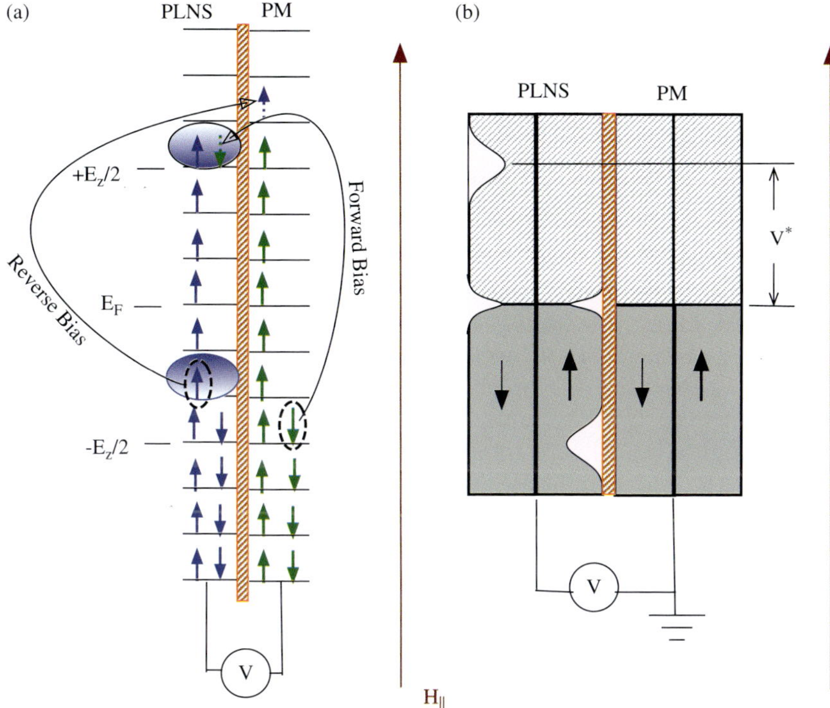

Fig. 15.6 (a) Schematic of a tunnel junction between the Pauli-limited normal state (PLNS) of a superconducting film in a supercritical parallel magnetic field and a paramagnetic (PM) non-superconducting metal. Note that the Zeeman splitting results in singly occupied orbitals near the Fermi energy. By applying a voltage, the states on either side of the junction can be shifted relative to each other. The diagram shows the tunneling channel that produces the relevant singlet-electron state in forward bias and the corresponding singlet-hole state in reverse bias. These singlet states resonantly mix with nearby empty states thereby forming evanescent Cooper pairs. (b) DOS profiles of a tunnel junction comprised of an Al film in the PLNS on one side and a paramagnetic film on the other. Note the depletion of states in the PLNS due to the pairing resonance and the ZBA.

The pairing resonance is associated with evanescent Cooper pairs formed via the two tunneling channels indicated by the forward and reverse bias arrows. In particular, at the appropriate bias a doubly occupied electron (hole) state is formed, which then mixes with the nearby unoccupied electron (hole) states lying above (below). This mixing results in a virtual Cooper pair, which in turn depletes single particle states in the vicinity, hence the dips in the spectra of Figs 15.4 and 15.5. Though the resonance lies near the Zeeman voltage $V_Z = E_Z/e$, its precise energy is determined, in part, by the potential energy of the Cooper pair . Using a non-perturbative approach, Aleiner and Altshuler (1997) showed that the resonance occurs at an energy that is universal for 0D (grain), 1D (wire), and 2D (film) systems,

$$V^* = \frac{1}{2}\left[V_Z + \sqrt{V_Z^2 - (\Delta_0/e)^2}\right], \tag{15.2}$$

where $V_Z = E_Z/e$. As can be seen in Fig. 15.7, eqn (15.2) is in reasonably good agreement with parallel-field measurements in Al and Be films. The solid lines are least-squares fits to eqn (15.2) in which only g_L was varied. Recently, it has been shown that by fitting the resonance profile to theory one can obtain quite accurate estimates of virtually all of the relevant superconducting microscopic parameters, Δ_0, the S-OS rate $1/\tau_{so}$, G^0, and the elastic scatter rate $1/\tau_0$.

Up to now the pairing resonance (PR) has been primarily studied in relatively low-resistance Al films, $R < R_Q/4$. Clearly, however, one would like to follow its evolution as the zero-field SIT is approached. It is now generally accepted that, for sufficiently strong disorder, the order parameter in a non-granular superconducting film will be destroyed by the repulsive electron–electron interactions that give rise to the ZBA. Insights into this process could, in principle, be obtained by comparing the superconducting-gap amplitude with the width of the pairing-resonance feature as one increases the resistance to values near R_Q. Since the resonance lies well above the Fermi energy, it may survive on the insulating side of the SIT by virtue of the fact that it is shifted away from the electron–electron correlation singularity at $V = 0$. Interestingly, in moderately disordered films the resonance deepens with increasing film resistance because resonant electron pairs with lower diffusivity spend more time interacting. Of course, the same is true in the repulsive interaction channel, which results in a deepening of the ZBA.

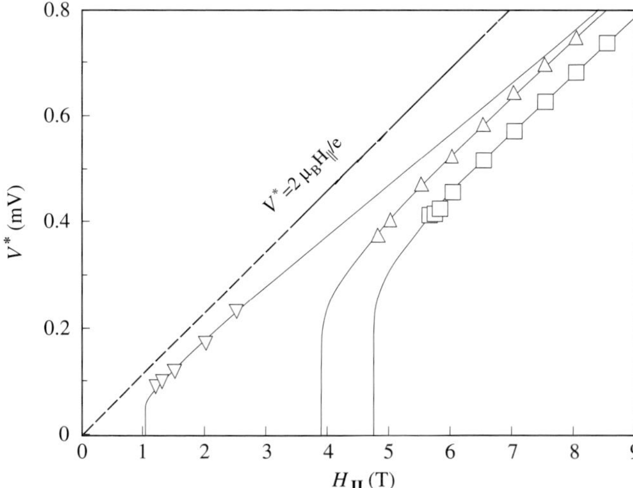

Fig. 15.7 Voltage position of the PR as a function of parallel magnetic field. Squares: $\Delta/e =$ 0.45-mV Al film; Up triangles: $\Delta/e = 0.38$-mV Al film; Down triangles: $\Delta/e = 0.1$-mV Be film. The dashed line is the Zeeman voltage with $g_L = 2$. The curved lines are least squares fits of eqn (15.2) to the data in which g_L was varied.

15.3 Spin effects in the insulating phase

15.3.1 Introduction and background

Here we will review the magneto-transport and tunneling DOS properties of ultra-thin, homogeneously disordered Be films that are on the insulating side of the zero-field SIT. Be films with $R \ll R_Q$ are typically superconducting with $T_c \sim 0.5 - 3$ K depending on the details of the deposition parameters. However, if one decreases the film thickness to the point where $R \sim R_Q$, then the superconducting phase gives way to a highly correlated, variable-range hopping phase. The relationship between the insulating ground state and the superconducting ground state on either side of the SIT remains the subject of intense experimental and theoretical work. A growing body of evidence suggests that localized Cooper pairs play an important role in the formation of the insulating ground state. For our purposes here it is best to begin with an overview of the theory of interaction effects in disordered electronic systems that neglects the possibility of localized Cooper pairs. From this vantage point we can consider recent experimental results within the context of the standard localization formalism in order to assess the need for more exotic models. Localization theory has for the most part been developed in two extreme limits. In the weak disorder/interaction 2D limit, it is well established that the primary effect of electron–electron interactions is to produce a logarithmic suppression of the DOS at the Fermi energy (Altshuler *et al.*, 1987), $\delta N \sim -\ln(V)$. This is manifest as the ZBA and has been well established in a number of different systems via tunneling measurements of the DOS (Imry and Ovadyahu, 1982; White *et al.*, 1985). The depletion of the DOS is perturbative and results in a weakly metallic ln-T transport conductivity. In the opposite limit, i.e. the strongly insulating regime, Efros and Shklovskii (1975; 1984) have shown that the Coulombic interactions can produce a non-perturbative gap in the DOS, which is commonly known as the Coulomb gap (Massey and Lee, 1995). Interestingly, the 2D Coulomb gap is expected to be linear in energy (Efros and Shklovskii, 1984),

$$N(eV) = \frac{\alpha(4\pi\epsilon_o\kappa)^2|eV|}{e^4} \tag{15.3}$$

where κ is the relative dielectric constant, ϵ_o is the permittivity of free space, and α is a constant of order unity. The Coulomb gap is usually associated with a modified variable-range hopping law of the form in the transport characteristics

$$R(T) = R_o \exp(T_o/T)^\nu \tag{15.4}$$

where R is the film sheet resistance and R_o is a constant. In the case of a flat DOS near the Fermi energy, the transport simply obeys Mott's variable-range hopping law with $\nu = \frac{1}{3}$ (Mott and Davis, 1979). If there is a simple gap in the DOS then T_o is the gap energy for fixed-range hopping and $\nu = 1$. Finally, if the DOS spectrum is given by eqn (15.3), then one expects $\nu = \frac{1}{2}$, R_o to be of the order of the quantum resistance R_Q (Efros and Shklovskii, 1984), and

$$T_o = \frac{2.8e^2}{4\pi k_B \epsilon_o \kappa \xi} \tag{15.5}$$

where k_B is the Boltzmann constant and ξ is the localization length. Thus, when the $\nu = \frac{1}{2}$ hopping form is observed, the DOS spectrum is simply assumed to be that of eqn (15.3). Below we examine this assumption through a systematic electron-tunneling DOS study of uniformly disordered Be films, whose transport properties range from that of weakly metallic to strongly insulating.

15.3.2 Insulating Be films

Beryllium forms smooth, dense, non-granular films when thermally evaporated onto glass. In fact, scanning force micrographs of the films' exposed oxide surfaces do not reveal any salient morphological features down to a resolution of 0.5 nm. This nongranular morphology is crucial in that it assures one that the measured resistance is representative of electron–electron correlation processes and not those associated with grain charging. In extremely high-resistance films any significant granularity will result in field-independent grain-charging (Coulomb-blockade) effects which preempt the many-body effects of interest. The Be films used in these studies ranged in thickness from 1.5–2.0 nm, with corresponding sheet resistances $R \sim 500\ \Omega$–$3\ \mathrm{M}\Omega$ at $T = 50$ mK. They were deposited by thermally evaporating 99.5% pure beryllium powder onto fire-polished glass substrates held at 84 K. The evaporations were made in a 0.4 μTorr vacuum at a rate of 0.30 nm/s. The film area was 1.5×4.5 mm. Transmission electron microstructural analysis of 15-nm-thick Be films deposited on cleaved NaCl crystals at 84 K revealed that the films were composed of an ultrafine base structure that was interspersed with 5–15 nm Be nanocrystallites. Electron diffraction measurements showed no diffraction from the metallic base structure suggesting that it was amorphous. Similarly the oxide (BeO) produced a broad, continuous diffraction ring indicating its grain size was less than 1 nm (Adams *et al.*, 1998).

15.3.3 Variable-range hopping and the Coulomb gap

Shown in the inset of Fig. 15.8 is the normal-state conductance of the 2600-Ω film as a function of $\ln T$. This sample had a superconducting transition temperature $T_c = 0.33$ K which was suppressed by the application of a magnetic field. The film exhibited the expected ln-T weakly insulating behavior, with some rounding below 100 mK. In stark contrast, the main body of Fig. 15.8 shows the activated-like behavior of a 2.6-MΩ film. The solid line is a least-squares fit to the data. The linearity of the data in Fig. 15.8 shows unequivocally that the hopping exponent is $\nu = \frac{1}{2}$. Furthermore the slope and intercept of the fit determine the parameters $T_o = 1.6$ K and $R_o \sim R_Q/2$ in eqn (15.4).

In principle we can use eqn (15.5) along with the measured value of T_o to calculate a localization length ξ. However, the relative dielectric constant κ in eqn (15.5) is unknown for a highly disordered metal film. Alternatively, if one instead assumes that $\xi \sim 1$ nm and takes $T_o = 1$ K, then eqn (15.5) gives $\kappa \sim 10^4$. This value seems reasonable in that it lies between the metallic and insulating limits of $\kappa \sim \infty$ and

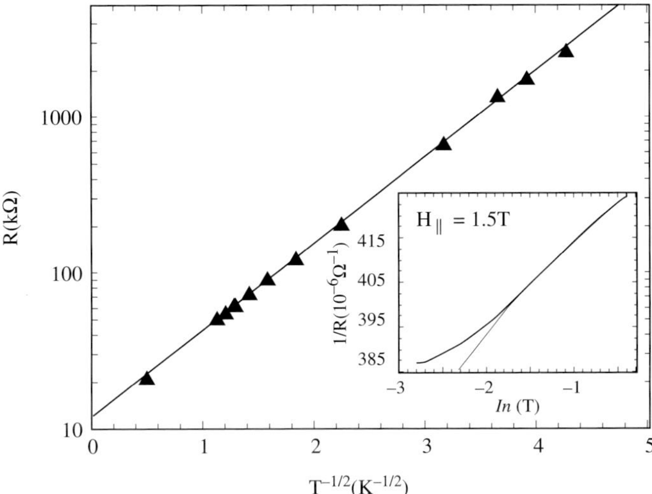

Fig. 15.8 Semi-log plot of the resistance of a 2.6-MΩ Be film as a function of $T^{-1/2}$. The solid line is a linear fit to the data for which we get $T_o = 1.6$ K and $R_o = h/2e^2$. Inset: conductance of a 2600-Ω film as a function of $\ln T$. The solid line is a guide to the eye.

$\kappa \sim 10$, respectively. This is not an issue in 3D semiconducting systems where the dielectric constant is well defined in the insulating phase. In any case, it is evident that Fig. 15.8 is consistent with the existence of the Coulomb gap described by eqn (15.4).

In Fig. 15.9 we show the evolution of the tunneling DOS of thin Be films as the sheet resistance is increased from a few hundred ohms to several megaohms. We note that the DOS spectrum of the lowest-resistance film in this plot displays a logarithmic ZBA, as expected. However, the magnitude of the ZBA feature grows rapidly with increasing disorder. Indeed, in films with $R > 10^4$ Ω the spectrum is no longer perturbative, and no longer logarithmic in form. Deep in the insulating phase, the energy dependence of the DOS spectrum becomes linear as can be seen in the 2.6-MΩ film. This linear dependence is predicted by eqn (15.3) and represents the 2D Coulomb gap.

15.3.4 Multi-fold magnetoconductance and quantum metallicity

Although the data in Figs. 15.8 and 15.9 are consistent with standard descriptions of modified variable-range hopping, the application of magnetic field shows that the underlying quantum state of high-resistance films is, in fact, anomalous. The magnetoconductance (MC) of a 3-MΩ Be film at 50 mK, with field applied perpendicular and parallel to the film surface, is plotted in Fig. 15.10. Below ~ 1 T the MC of the film is negative, with the conductance falling by about a factor of two. Above 1 T the MC is positive and linear for both field orientations. The magnitude of the MC is a function of the film disorder. In the inset of Fig. 15.10 we show a similar, but much smaller, MC in a 16-kΩ film.

The overall magnitude of the MC data in Fig. 15.10 is extremely large and cannot easily be explained within the framework of variable-range hopping theory. Assuming

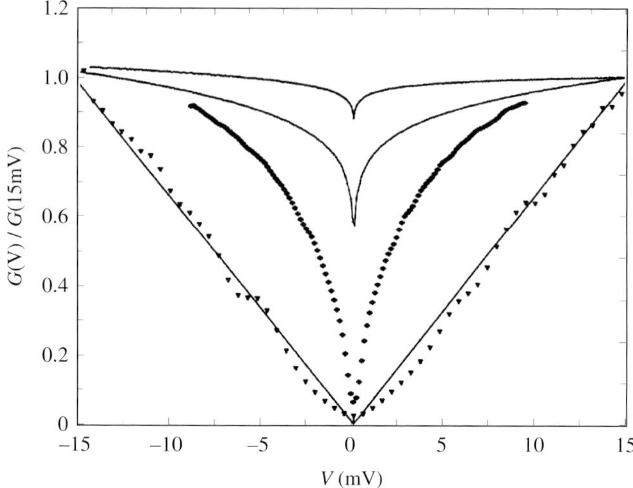

Fig. 15.9 Tunneling conductances normalized to $G = 15$ mV for Be films with $T = 50$ mK resistances of $R = 530\ \Omega$, 2.6 kΩ, 16 kΩ, and 2.6 MΩ (top to bottom). The solid lines are a best fit to the form $G(V) = b|V|$, where b is an adjustable parameter. The 2.6-MΩ data was taken at 700 mK; the other curves were measured at 50 mK.

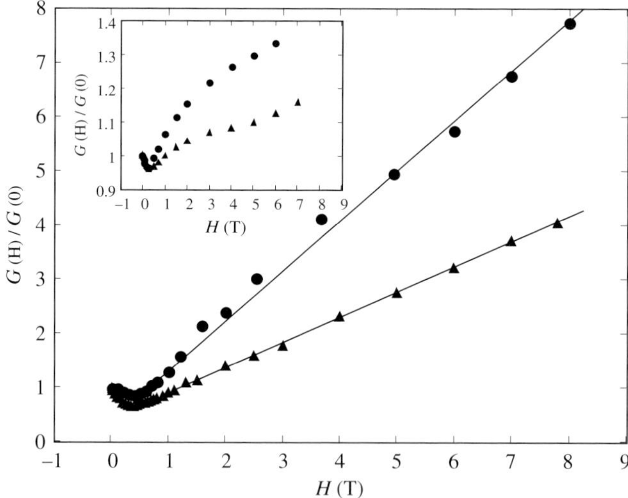

Fig. 15.10 Relative magnetoconductance of a 3-MΩ Be film at 50 mK. Circles, field perpendicular to film surface; triangles, field parallel to film surface. The solid lines are linear fits to the data above 1 T with slopes of 1/(1.1T) and 1/(2.2T) for the perpendicular and parallel data respectively. Inset: relative magnetoconductance of a 16-kΩ Be film.

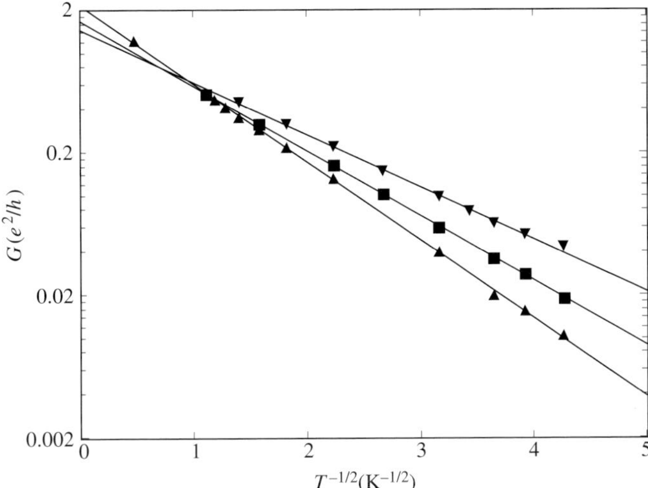

Fig. 15.11 Semi-log plot of the film conductance as function of $T^{-1/2}$ at three different parallel magnetic fields. Up triangles, $H_{||} = 0$; squares, $H_{||} = 3.0T$; down triangles, $H_{||} = 7.0T$. The solid lines are linear fits to the data from which $T_o(H_{||})$ were obtained via eqn (15.4).

that the parallel-field MC is entirely mediated by the Zeeman interaction, it is clear from the data that the orbital and spin contributions to the MC are comparable. In Fig. 15.11 we show that the $T^{-1/2}$ variable-range hopping (VRH) behavior is preserved in parallel field, and that the primary effect of the field is to diminish the correlation energy T_o. Again, this is not easily explained in the context of standard strong-localization theory.

The positive MC in the main panel of Fig. 15.10 shows no sign of saturation, up to the highest fields studied. Naturally, one would expect the MC to saturate on field scales in which the correlation energy T_o is driven to zero. In order to address this issue we present magnetoresistance (MR) data on Be films with substantially lower disorder than the films in Fig. 15.10. Figure 15.12 shows the temperature dependence of the film resistance normalized by R_{Q} for two critically disordered samples in zero field and at 8.4 T respectively. The zero-field data are strongly insulating, but the high-field data appear to be more metallic in character. Obviously the magnetic field is producing a significant suppression of the insulating phase. This is also evident in Fig. 15.13, where we show the low-temperature MR of the films. Note that the low-field MR is positive, but above 1 T the MR becomes strongly negative. The overall structure of the data in Fig. 15.13 is consistent with the MR behavior of highly insulating films of Fig. 15.10, with the notable exception of the provocative saturation of the MR to the quantum resistance.

MR peaks and anomalously large, multi-fold, negative magnetoresistances have also been reported in ultra-thin TiN films (Baturina *et al.*, 2007) and InO_x films (Gantmakher *et al.*, 2000, 1998; Steiner *et al.*, 2005). In a fashion similar to the data in Fig. 15.13 the MR of these films saturates at a weakly temperature-dependent

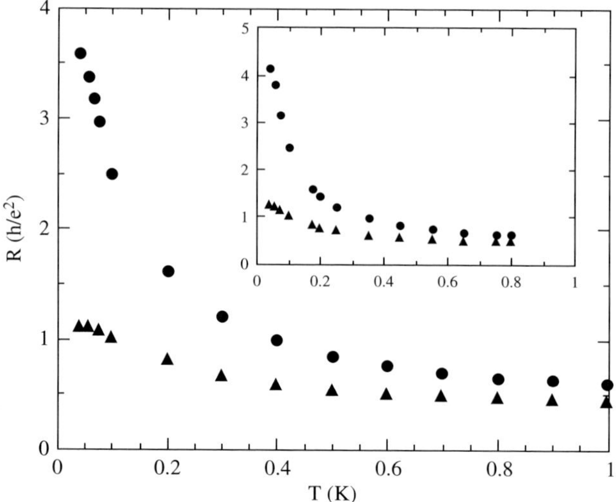

Fig. 15.12 Temperature dependence of the resistance of moderately insulating Be film in units of h/e^2 in zero field (circles) and in a perpendicular field of H = 8.4 T (triangles). Inset: temperature dependence of the resistance of a similar Be film, with axes the same as the main figure.

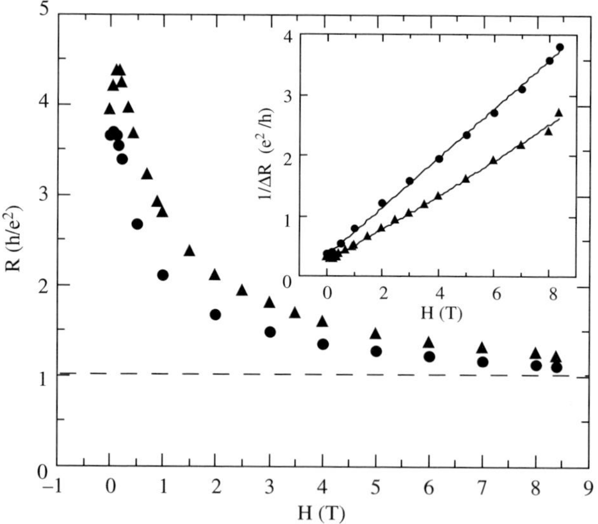

Fig. 15.13 Low-temperature magnetoresistance of the films in Fig. 15.12 in units of h/e^2 at T = 50 mK (circles on main panel data, triangles on inset data). Note the saturation at $R \sim R_Q$. Inset: linear behavior after subtracting a saturation resistance of $0.85R_Q$, $\Delta R = R - 0.85$. The solid lines are provided as a guide to the eye.

resistance that is always near R_Q, suggesting the existence of a high-field "quantum metal" phase (Baturina *et al.*, 2007; Butko and Adams, 2001). These experiments provide strong evidence that the zero-field insulating ground state is distinctly different from the high-field ground state. This has led to speculation that the zero-field ground state has an incoherent superconducting component (Sambandamurthy *et al.*, 2004; Stewart *et al.*, 2007) that mediates the insulating behavior.

15.3.5 Quantum metallicity and spin–orbit scattering

Interestingly, both Be and TiN films form dense homogeneously disordered, non-granular films with an intrinsic, clean limit, $T_c \sim 1$ K. As we discuss below, the fact that both of these systems have a well documented, low S-OS rate is crucial to the observation of the quantum metal phase. If the insulating side of the SIT in Be, TiN, and InO_x films is mediated by localized Cooper pairs, i.e. a Bose insulator, then one should be able to probe the local condensate with magnetic field. In particular, assuming that parallel magnetic field acts only on the electron spins, then the MR of the Be films can be attributed to the Zeeman splitting of the localized Cooper pairs. This is analogous to the S–P mechanism in the superconducting phase. Since the Zeeman-mediated critical field of low S-OS superconductors is simply proportional to the superconducting gap (Wu *et al.*, 2006), it is natural to assume that the local pair-breaking field in the insulating phase will also be proportional to the local energy gap. To account for the field range over which MR is significant, one must assume that there is a rather broad distribution of local pair-binding energies. In the presence of S-OS the Zeeman critical field of a superconductor can be much larger than that of the zero-S-OS case. Beryllium and Al films have a very low intrinsic S-OS rate (Adams, 2004; Adams *et al.*, 1998; Tedrow and Meservey, 1979), but a controllable amount of S-OS can be induced in these films by coating them with heavy noble metals. In particular, Tedrow and Meservey (1982) reported that, for each monolayer of Pt deposited on a 40-Å-thick Al film, the S-OS rate h/τ_{so} increased by 3.2 meV. Similarly, large S-OS rates can be induced in Be films by coating them with Au (Wu *et al.*, 2006). The primary effect of S-OS is to disrupt the spin-rotation symmetry of the system, so that spin is no longer a good quantum number. However, conventional BCS superconductivity does not require spin-rotation symmetry, therefore S-OS has little effect on the zero-field properties of the condensate. Nevertheless, the spin response of a superconductor, as probed by a parallel magnetic field, is much different in the presence of S-OS. The most apparent ramification of S-OS is a dramatic increase in the parallel critical field (Wu *et al.*, 2006). Thin superconducting Be films coated with 5 Å of Au can have a parallel critical field that is an order of magnitude larger that the Clogston–Chandresekhar limit. Naively, one would expect a similar S-OS effect on the parallel pair-breaking field of localized Cooper pairs.

In Fig. 15.14 we compare the perpendicular and parallel-field MR of the Be/Au bilayers. The bilayers were formed by depositing varying Au thicknesses onto Be films. All of the depositions were made on fire-polished glass substrates held at 84 K. First a Be film with thickness ~ 18 Å was deposited at a rate of 1.4 Å/s, then an Au overlayer was deposited at 0.1 Å/s without breaking the vacuum. The data, which

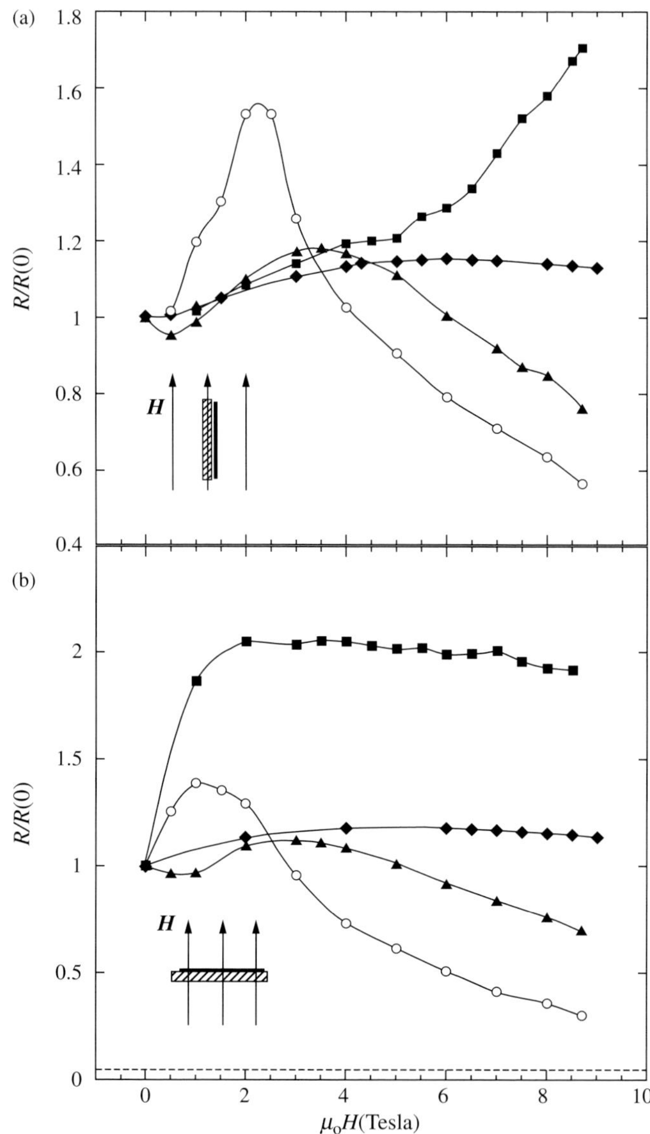

Fig. 15.14 The normalized resistance of Be/Au films with varying Au thickness as a function of parallel (a) and perpendicular magnetic (b) field. The Be thickness of each film is 18 Å. Circles, $d_{\mathrm{Au}} = 0.0$ Å; triangles, $d_{\mathrm{Au}} = 0.3$ Å; diamonds, $d_{\mathrm{Au}} = 0.6$ Å; squares, $d_{\mathrm{Au}} = 1.0$ Å. The $d_{\mathrm{Au}} = 0.0$ curves were taken at 500 mK. The other curves were taken at 400 mK.

have been normalized by the zero-field resistance, were taken at 400–500 mK in order to circumvent the long, non-exponential relaxations that hinder measurements below 100 mK. The open circle symbols correspond to the uncoated 18-Å Be film which displays the previously reported low-field positive MR followed by a multi-fold negative MR (Butko *et al.*, 2000*a*,*b*). The dashed line near the x-axis in Fig. 15.14 (b) corresponds to R_{Q} for the Be film. The MR appears to be asymptotic to R_{Q}, in accordance with the high-field quantum metal phase (Butko and Adams, 2001). In contrast, the overall scale of the MR for the $d_{\mathrm{Au}} = 0.3$ Å bilayer is somewhat diminished and almost completely quenched in the $d_{\mathrm{Au}} = 0.6$ Å bilayer. Both of these samples have low-temperature sheet resistances $R \gg R_{\mathrm{Q}}$ and correlation energies $T_o \gg T$, as is the case for the uncoated film. Because of this we conclude that the MR is being modified by the S-OS and not the lowering of T_o, for instance. Note that the MR peaks move to substantially higher fields with increasing Au coverage in the $d_{\mathrm{Au}} = 0.0$-, 0.3-, and 0.6-nm curves. The $d_{\mathrm{Au}} = 1.0$ curves display the largest MR anisotropy, but this may be a consequence of the fact that $T_o \sim T$ for this sample. Nevertheless, the high-field perpendicular MR is only weakly negative for the highest Au coverage bilayer, while the parallel MR maximum, if it exists, lies beyond 9 T.

15.4 Summary and zero-temperature phase diagram

The ground-state properties of the Be films are most effectively summarized using the schematic phase diagram in Fig. 15.15. The x-axis measures the amount of disorder, which we parameterize as $1/t$, where t is the film thickness. The y-axis is the applied magnetic field. In the limit of $1/t \to 0$ the films are 3D. As the thickness is lowered, however, the 3D-to-2D threshold is crossed, and the films become weakly metallic with a perturbative logarithmic increase in resistance with decreasing temperature and a perturbative logarithmic suppression of the DOS near the Fermi energy. Interestingly,

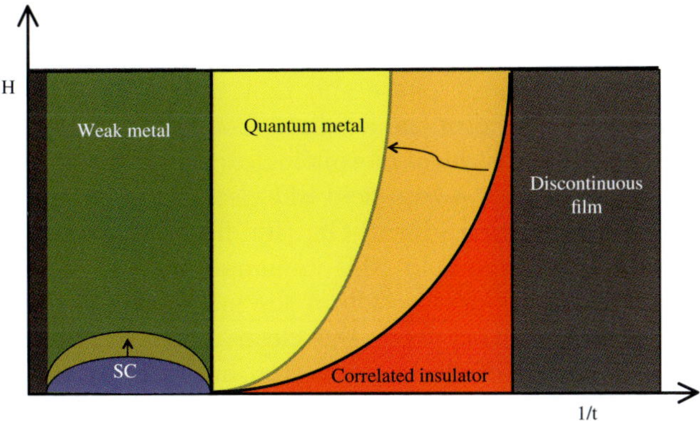

Fig. 15.15 Zero temperature phase diagram of a disordered thin film system with low S-OS. The arrows in the diagram show how the phase boundaries are shifted in the presence of the S-OS. SC, superconductor.

Be also has a locally enhanced superconducting phase in the weakly metallic regime, shown in blue. The S–P transition occurs at the boundary between the rightmost portion of the superconducting dome and the weak metal phase. In the vernacular of the S–P formalism, the weak metal phase above the dome is often referred to as the Pauli-limited normal state. The small arrow at the top of the dome shows how the phase boundary is modified in the presence of S-OS.

As the film thickness is decreased further the disorder increases and electron correlation effects begin to dominate the normal-state transport properties. In this regime $R \rightarrow R_{\mathrm{Q}}$ and states near the Fermi energy are rapidly depleted, signaling the emergence of the Coulomb gap and destruction of long-range superconducting phase coherence. This zero-field SIT occurs between the blue and red portions of the phase diagram. There is now good evidence that localized Cooper pairs participate in this transition. Furthermore, anomalous ultra-large MR peaks are also seen in this region. The latter observation suggests that the phase diagram in Fig. 15.15 may be oversimplified near the zero-field S–I boundary. Further decreases in thickness at zero field result in films with $R \gg R_{\mathrm{Q}}$. These films are correlated insulators, shown in red, and are characterized by a well-defined Coulomb gap and Efros–Shklovskii hopping transport. The deeper one moves to the right in the correlated insulator region of the diagram the greater the correlation energy T_o becomes, until one eventually reaches the percolation threshold of the film, at which point it is no longer electrically continuous. This occurs at $t \sim 1$ nm in Be films. We have shown that the application of a magnetic field lifts the Coulomb gap, thereby producing extremely large decreases in resistance at low temperatures. As the diagram in Fig. 15.15 indicates, we believe that this process eventually leads to a saturation in the resistance near R_{Q}, even for films deep in the correlated insulator regime. This weakly temperature-dependent, high-field state is depicted in yellow and essentially behaves as a weak metal. Thus the yellow region in Fig. 15.15 describes films of varying disorder but with resistance near R_{Q}. The boundary between the quantum metal phase and the correlated insulator phase is sensitive to S-OS in manner similar to that of the superconductor phase. The curved arrow depicts the effect of S-OS on the correlated insulator–quantum metal boundary. This sensitivity to S-OS suggests that localized Cooper pairs persist deep into the correlated insulator phase and that the primary effect of S-OS is to increase the Zeeman critical field of the localized Cooper pairs beyond the Clogston–Chandrashekar limit. The almost ideal morphological characteristics of Be films, along with the fact that Be undergoes a zero-field SIT, lead us to believe that the general features of Fig. 15.15 are generic to low S-OS, thin-film S–I systems.

Until very recently, the zero-field SIT in homogeneously disordered films was thought to be mediated by the destruction of the magnitude of the order parameter as one approached the transition from the clean limit. A general consensus had emerged that the convolution of electron–electron interactions and disorder reduced the available normal-state DOS to the point that a superconducting condensate was no longer sustainable, even at zero temperature, in critically disordered systems. However, as discussed in this and other chapters there is compelling new evidence that although the order parameter indeed goes to zero at the SIT, local Cooper pairing occurs well into the insulating phase. This somewhat counter-intuitive scenario is made even more plausible by the fact that incoherent Cooper pairing is clearly evident in the Pauli-limited normal state of moderately disordered superconducting films. Hopefully future experiments will be able elucidate the relationship between incoherent pairing on the superconducting side of the SIT with localized Cooper pairing on the insulating side. Quantitative progress on this issue is a crucial ingredient to our ultimate understanding of underlying quantum nature of the SIT.

References

Abrahams, E., Anderson, P.W., Licciardello, D.C, and Ramakrishnan, T.V. (1979). *Phys. Rev. Lett.*, **42**, 673–676.

Adams, P.W. (2004). *Phys. Rev. Lett.*, **92**, 067003.

Adams, P.W. and Butko, V.Yu. (2000). *Physica B*, **284**, 673–674.

Adams, P.W., Herron, P., and Meletis, E.I. (1998). *Phys. Rev. B*, **58**, R2952–R2955.

Aleiner, I.L. and Altshuler, B.L. (1997). *Phys. Rev. Lett.*, **79**, 4242–4245.

Alexander, J.A.X., Orlando, T.P., Rainer, D., and Tedrow, P.M. (1985). *Phys. Rev. B*, **31**, 5811–5825.

Altshuler, B.L., Aronov, A.G., Gershenson, M.E., and Sharvin, Yu.V. (1987). *Sov. Sci. Rev. A*, **9**, 223.

Baturina, T.I., Strunk, C., Baklanov, M.R., and Satta, A. (2007). *Phys. Rev. Lett.*, **98**, 127003.

Baym, G. and Pethick, C. (1991). *Landau Fermi-Liquid Theory: Concepts and Applications*. John Wiley & Sons, New York.

Bergmann, G. (1983). *Phys. Rev. B*, **28**, 2914.

Butko, V.Yu. and Adams, P.W. (2001). *Nature*, **409**, 161.

Butko, V.Y., Adams, P.W., and Aleiner, I.L. (1999). *Phys. Rev. Lett.*, **82**, 4284–4287.

Butko, V.Yu., DiTusa, J.F., and Adams, P.W. (2000*a*). *Phys. Rev. Lett.*, **85**, 162.

Butko, V.Yu., DiTusa, J.F., and Adams, P.W. (2000*b*). *Phys. Rev. Lett.*, **84**, 1543.

Chandrasekhar, B.S. (1962). *Appl. Phys. Lett.*, **1**, 7–8.

Clogston, A.M. (1962). *Phys. Rev. Lett.*, **9**, 266–267.

Dynes, R.C., Garno, J.P., and Rowell, J.M. (1978). *Phys. Rev. Lett.*, **40**, 479.

Efros, A.L. and Shklovskii, B.I. (1975). *J. Phys. C*, **8**, L49.

Efros, A.L. and Shklovskii, B.I. (1984). *Electronic Properties of Doped Semiconductors*. Springer, New York.

Entin-Wohlman, O., Imry, Y., and Sivan, U. (1989). *Phys. Rev. B*, **40**, 8342.

Epstein, K., Goldman, A.M., and Kadin, A.M. (1983). *Phys. Rev. B*, **27**, 6685.

Fulde, P. (1973). *Adv. Phys.*, **22**, 667.

Gantmakher, V.F., Golubkov, M.V., Dolgopolov, V.T., Shashkin, A.A., and Tsydynzhapov, G.E. (2000). *Pis'ma Zh. Eksp. Teor. Fiz.*, **71**, 693. (*JETP Lett.* **71**, 473 (2000)).

Gantmakher, V.F., Golubkov, M.V., Dolgopolov, V.T., Tsydynzhapov, G.E., and Shashkin, A.A. (1998). *Pis'ma Zh. Eksp. Teor. Fiz.*, **68**, 337. (*JETP Lett.* **68**, 363 (1998)).

Gibson, G.A., Tedrow, P.M., and Meservey, R. (1989). *Phys. Rev. B*, **40**, 137.

Goldman, A.M. and Markovic, N. (1998). *Phys. Today*, **51**(11), 39.

Haviland, D.B., Liu, Y., and Goldman, A.M. (1989). *Phys. Rev. Lett.*, **62**, 2180.

Hebard, A.F. and Paalanen, M.A. (1990). *Phys. Rev. Lett.*, **65**, 927.

Hernandez, L.M., Bhattacharya, A., Parendo, K.A., and Goldman, A.M. (2003). *Phys. Rev. Lett.*, **91**, 126801.

Hernandez, P. and Sanquer, M. (1992). *Phys. Rev. Lett.*, **68**, 1402.

Imry, Y. and Ovadyahu, Z. (1982). *Phys. Rev. Lett.*, **49**, 841.

Lee, P.A. and Ramakrishnan, T.V (1985). *Rev. Mod. Phys.*, **57**, 287–337.

Massey, J.G. and Lee, M. (1995). *Phys. Rev. Lett.*, **75**, 4266.

Matveev, K.A., Glazman, L.I., Clarke, P., Ephron, D., and Beasley, M.R. (1995). *Phys. Rev. B*, **52**, 5289.

Medina, E., Kardar, M., and Rangel, R. (1996). *Phys. Rev. B*, **53**, 7663.

Meservey, R. and Tedrow, P.M. (1976). *Phys. Lett. A*, **58**, 131–132.

Meservey, R., Tedrow, P.M., and Fulde, P. (1970). *Phys. Rev. Lett.*, **25**, 1270–&.

Mott, N.F. and Davis, G.A. (1979). *Electronic Processes in Non-Crystalline Materials.* Springer-Verlag, New York.

Nguyen, V.I., Spivak, B.Z., and Shklovskii, B.I. (1985). *Pis'ma Zh. Eksp. Teor. Fiz.*, **41**, 35. (*JETP Lett.* **41**, 42 (1985)).

Pichard, J.L., Sanquer, M., Slevin, K., and Debray, P. (1990). *Phys. Rev. Lett.*, **65**, 1990.

Sambandamurthy, G., Engel, L.W., Johansson, A., and Shahar, D. (2004). *Phys. Rev. Lett.*, **92**, 107005.

Shapir, Y. and Ovadyahu, Z. (1989). *Phys. Rev. B*, **40**, 12441.

Steiner, M.A., Boebinger, G., and Kapitulnik, A. (2005). *Phys. Rev. Lett.*, **94**, 107008.

Stewart, M.D., Yin, A., Xu, J.M., and Valles, Jr., J.M. (2007). *Science*, **318**, 1273.

Suzuki, T., Seguchi, Y., and Tsuboi, T. (2000). *J. Phys. Soc. Jpn.*, **69**, 1462–1471.

Tedrow, P.M., Kucera, J.T., Rainer, D., and Orlando, T.P. (1984). *Phys. Rev. Letts*, **52**, 1637.

Tedrow, P.M. and Meservey, R. (1979). *Phys. Rev. Lett.*, **43**, 384–387.

Tedrow, P.M. and Meservey, R. (1982). *Phys. Rev. B*, **25**, 171.

Tinkham, M. (1996). *Introduction to Superconductivity.* McGraw-Hill, New York.

Valles, Jr., J.M., Dynes, R.C., and Garno, J.P. (1992). *Phys. Rev. Lett.*, **69**, 3567.

White, A.E., Dynes, R.C., and Garno, J.P. (1985). *Phys. Rev. B*, **31**, 1174.

Wu, Wenhao and Adams, P.W. (1994). *Phys. Rev. Lett.*, **73**, 1412–1415.

Wu, Wenhao and Adams, P.W. (1995). *Phys. Rev. Lett.*, **74**, 610–613.

Wu, Wenhao, Goodrich, R.G., and Adams, P.W. (1995). *Phys. Rev. B*, **51**, 1378–1380.

Wu, X.S., Adams, P.W., and Catelani, G. (2006). *Phys. Rev. B*, **74**, 144519.

16

Spectroscopic Imaging STM Studies of Electronic Structure in Both the Superconducting and Pseudogap Phases of Underdoped Cuprates

K. Fujita[1,2,3], A.R. Schmidt[1,2,4], E.-A. Kim[1], M.J. Lawler[1,5], H. Eisaki[6], S. Uchida[3], D.-H. Lee[4] and J.C. Davis[1,2,7]

[1] LASSP, Department of Physics, Cornell University, Ithaca, NY 14853, USA
[2] CMPMS Department, Brookhaven National Laboratory, Upton, NY 11973, USA
[3] Department of Physics, University of Tokyo, Bunkyo-ku, Tokyo 113-0033, Japan
[4] Department of Physics, University of California, Berkeley, California 94720, USA.
[5] Department of Physics and Astronomy, Binghamton University, Binghamton, NY 13902-6000, USA
[6] Institute of Advanced Industrial Science and Technology, Tsukuba, Ibaraki 305-8568, Japan.
[7] School of Physics and Astronomy, University of St. Andrews, St. Andrews, Fife KY16 9SS, Scotland

16.1 Introduction to electronic structure of hole-doped cuprates

16.1.1 Electronic structure of the superconducting and pseudogap phases

The CuO_2 plane electronic structure is dominated by Cu $3d$ and O $2p$ orbitals (Zaanen *et al.*, 1985). Each Cu d_{x2-y2} orbital is split into upper and lower states by on-site Coulomb interactions, with the O p-states intervening energetically. The resulting "charge-transfer" (Zaanen *et al.*, 1985) Mott insulator is strongly antiferromagnetic (Anderson, 1950, 1987). Removing electrons from the O atoms (Chen *et al.*, 1991) (so-called hole-doping) results in the highest-temperature superconductivity known. The electronic phase diagram (Orenstein and Millis, 2000) is shown schematically in Fig. 1 (a) with p being the number of holes per CuO_2 plaquette. The antiferromagnetism persists for $p<$ 2–5%, the d-wave superconductivity (dSC) appears in the range 5–10% $<p<$ 25%, and a Fermi-liquid state is found at $p >$ 25–30%. The highest superconducting critical temperature T_c always occurs near $p\sim$16% which is referred to as "optimal" doping and the superconducting state always exhibits d-wave symmetry. As p is reduced, an excitation spectrum with energy scale $E = \Delta_1$ and which is anisotropic in **k**-space appears far above the superconducting T_c (Campuzano *et al.*, 2004; Damascelli *et al.*, 2003; Hufner *et al.*, 2008; Norman and Pepin, 2003; Orenstein and Millis, 2000; Timusk and Statt, 1999) This phase is referred to as the "pseudogap" (PG) phase because of the existence of the excitation energy-scale Δ_1 which could be the energy gap of another ordered phase (Campuzano *et al.*, 2004; Damascelli *et al.*, 2003; Hufner *et al.*, 2008; Timusk and Statt, 1999). Possible explanations for the PG phase include, for example, effects of hole-doping an antiferromagnetic Mott insulator (Anderson, 1987; Anderson *et al.*, 2004; K-Y. *et al.*, 2006; Kotliar, 1988; Paramekanti *et al.*, 2001; Zhang *et al.*, 1988) or that it is a d-wave superconductor without phase coherence (Berg and Altman, 2007; Carlson, Kivelson, Emery and Manousakis, Carlson *et al.*; Emery and Kivelson, 1995; Franz and Millis, 1998; Kwon *et al.*, 2001; Randeria *et al.*, 1992) or that it is an electronic ordered phase (Chakravarty *et al.*, 2001; Emery *et al.*, 1999; Fradkin *et al.*, 2010; Honerkamp *et al.*, 2007; Kim *et al.*, 2008; Kivelson *et al.*, 2003, 1998; Lee *et al.*, 2006*b*; Newns and Tsuei, 2007; Sachdev, 2003; Varma, 2006; Vojta, 2009; Vojta and Sachdev, 1999; White and Scalapino, 1998; Zaanen and Gunnarsson, 1989) due to the breaking of one or more crystal or electronic symmetries unrelated to superconductivity. Some combination of these three types of explanation is also an obvious possibility. A central challenge in cuprate high-temperature superconductivity studies is therefore to achieve a clear and widely accepted understanding of the electronic structure of the PG phase, and to determine its relationship to the superconductivity.

16.1.2 Two characteristic types of electronic excited states as p→0

The two energy scales Δ_1 and Δ_0 associated with two distinct types of electronic excited states are detected by a wide variety of different spectroscopies in underdoped cuprates (Damascelli *et al.*, 2003; Deutscher, 1999; Gedik *et al.*, 2003; Hufner *et al.*, 2008; Khasanov *et al.*, 2010; Le Tacon *et al.*, 2006; Orenstein and Millis, 2000; Timusk and Statt, 1999). These two excitation energies diverge with diminishing p as shown in Fig. 16.1 (b) (reproduced from Hufner *et al.* (2008)). Angle-resolved photoemission (ARPES) first revealed that in the PG phase, excitations at an energy $E{\sim}\Delta_1$ and largely independent of temperature occur in the regions of momentum space near $\mathbf{k} \cong (\pi/a_0,0)$, $(0,\pi/a_0)$. Moreover, $\Delta_1(p)$ increases rapidly as $p \to 0$ approaching the Mott insulator (Campuzano *et al.*, 2004; Damascelli *et al.*, 2003; Hufner *et al.*, 2008; Timusk and Statt, 1999). By contrast, the "nodal" region of **k**-space supports an ungapped "Fermi arc" (Norman *et al.*, 1998) of excitations in the PG phase, upon which a momentum- and temperature-dependent energy gap opens only in the dSC phase (H.-B. Yang and Kidd; Kanigel *et al.*, 2007, 2006; Kondo *et al.*, 2009; Norman *et al.*, 1998; Shen *et al.*, 2005; Tanaka *et al.*, 2006). Results from a wide variety of other spectroscopic techniques appear to be in agreement with this picture. For example, optical transient grating spectroscopy finds that the higher-energy excitations near Δ_1 are spatially exclusive of the superfluid and propagate exceedingly slowly without recombination to form Cooper pairs, whereas lower-energy excitations surrounding the $E=0$ nodes propagate and reform delocalized Cooper pairs as expected (Gedik *et al.*, 2003). Andreev tunneling exhibits two diverging excitation energy scales as $p \to 0$, the first being the PG energy Δ_1 and the second, lower-scale Δ_0, viewed as the maximum energy of delocalized Cooper-pair binding (Deutscher, 1999). Similarly Raman spectroscopy reveals that scattering near the node, but not the anti-node, is consistent with delocalized Cooper pairing (Le Tacon *et al.*, 2006). Finally, muon spin rotation studies show that that the superfluid density evolution is inconsistent with the carriers over the whole Fermi surface being available for condensation as if anti-nodal regions cannot develop delocalized Cooper pairs (Khasanov *et al.*, 2010).

Non-imaging tunneling spectroscopy has established an energetically particle–hole-symmetric excitation energy $E = \pm\Delta_1$, which is indistinguishable in magnitude in the PG and dSC phases (Fischer *et al.*, 2007; Renner *et al.*, 1998). The evolution of spatially-averaged differential tunnelling conductance (Alldredge *et al.*, 2008; McElroy *et al.*, 2005*a*,*b*) $g(E)$ for $Bi_2Sr_2CaCu_2O_{8+\delta}$gn Fig. 16.2 shows the p-dependence of this PG energy $E = \pm\Delta_1$ as indicated with blue dashed curve (see Sections 16.3, 16.5 and 16.7). The approximate p-dependence of Δ_0 as determined from several different tunnelling techniques (Section 16.3) is shown by red dashed curves.

Spectroscopic imaging scanning tunneling microscopy (SI-STM) studies of underdoped cuprates have revealed the rich spatial complexity of these two classes of electronic state. For excitation energies below the slowly doping-dependent (Alldredge *et al.*, 2008; Kohsaka *et al.*, 2008) and lower-scale $E{\sim}\Delta_0$, dispersive Bogoliubov quasiparticles of a spatially homogeneous superconductor (Fig. 16.2 (a)) are observed (Hanaguri *et al.*, 2007; Hoffman *et al.*, 2002; Kohsaka *et al.*, 2007, 2008; Lee *et al.*, 2009; McElroy *et al.*, 2003; Wise *et al.*, 2008). In contrast, the states near $E{\sim}\Delta_1$ are spatially disordered on the 3-nm scale (Alldredge *et al.*, 2008; Boyer *et al.*, 2007; Cren *et al.*, 2001; Gomes *et al.*, 2007; Howald *et al.*, 2001; Lang *et al.*, 2002; Matsuda *et al.*, 2003; McElroy *et al.*, 2005*a*,*b*; Pan *et al.*, 2001; Pushp *et al.*, 2009). More importantly, when the spatial structure of these non-dispersive states surrounding $E{\sim}\Delta_1$ is imaged

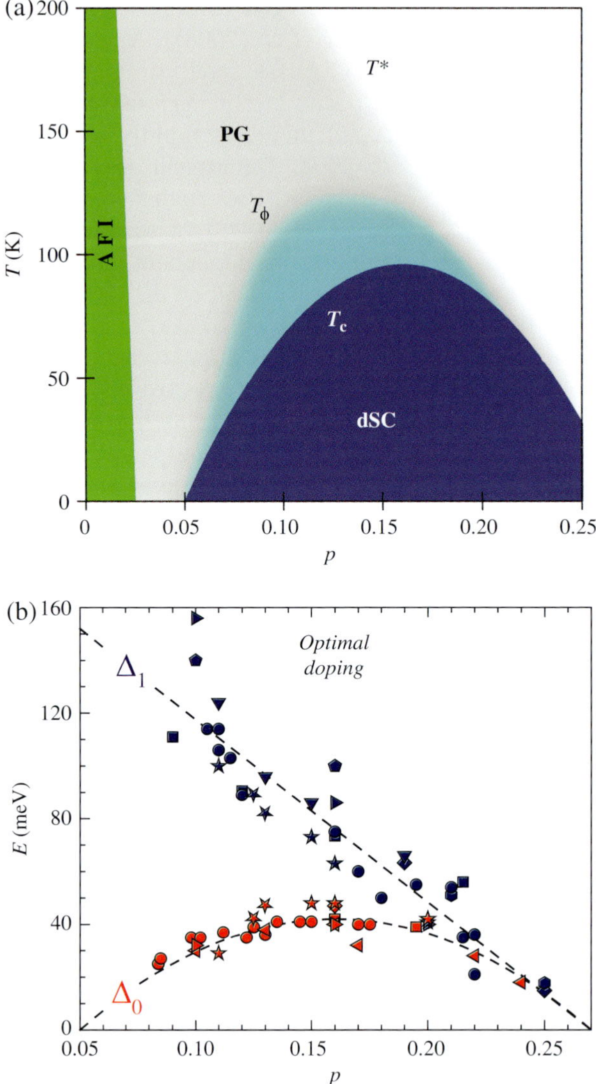

Fig. 16.1 (a) Schematic copper-oxide phase diagram. Here T_c is the critical temperature circumscribing a "dome" of superconductivity, T_ϕ is the maximum temperature at which phase fluctuations are detectable within the pseudogap phase, and T^* is the approximate temperature at which the pseudogap phenomenology first appears. (b) Two classes of electronic excitation in copper oxides as $p \to 0$. The separation between the energy scales associated with excitations of the superconducting state (dSC) and those of the pseudogap state (PG) increases as p decreases (reproduced from Hufner *et al.* (2008)). The different symbols correspond to the use of different experimental techniques.

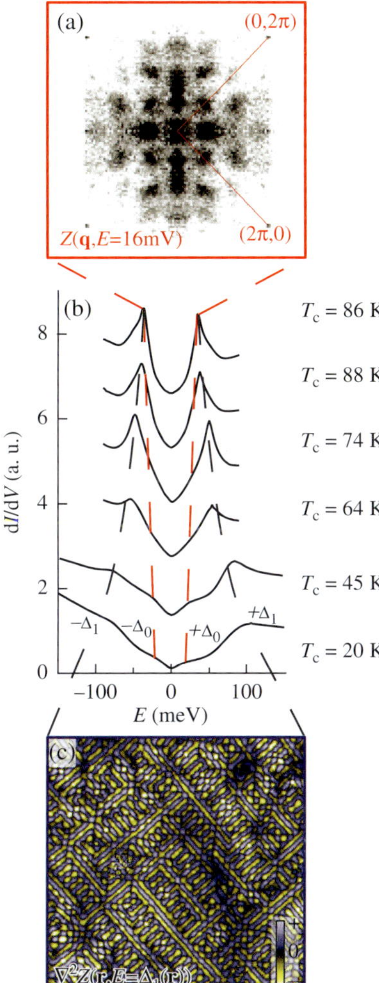

Fig. 16.2 (a) Fourier transform of the conductance ratio map $Z(\mathbf{r}, E)$ at a representative energy below Δ_0 for $T_c = 45$ K $Bi_2Sr_2Dy_{0.2}Ca_{0.8}Cu_2O_{8+\delta}$, which only exhibits the patterns characteristic of homogenous d-wave superconducting quasiparticle interference. (b) Evolution of the spatially averaged tunneling spectra of $Bi_2Sr_2CaCu_2O_{8+\delta}$ with diminishing p, here characterized by $T_c(p)$. The energies $\Delta_1(p)$ (blue dashed line) are easily detected as the pseudogap edge while the energies $\Delta_0(p)$ (red dashed line) are more subtle but can be identified by the correspondence of the "kink" energy with the extinction energy of Bogoliubov quasiparticles, following the procedures in Alldredge *et al.* (2008) and Kohsaka *et al.* (2008). (c) Laplacian of the conductance ratio map $Z(\mathbf{r})$ at the pseudogap energy $E = \Delta_1$, emphasizing the local symmetry-breaking of these electronic states for strongly underdoped $Ca_{1.88}Na_{0.12}CuO_2Cl_2$.

with sub-angstrom precision, several distinct broken symmetries are revealed (Kohsaka *et al.*, 2007, 2008; Lawler, 2010; Lee *et al.*, 2009; McElroy *et al.*, 2005*a*) (Fig. 16.2 (c)). These two classes of excitation exhibit increasing energy segregation in SI-STM data as $p \to 0$.

16.2 Spectroscopic imaging scanning tunneling microscopy

16.2.1 SI-STM techniques

Spectroscopic imaging scanning tunneling microscopy (SI-STM) techniques allow imaging of the differential tunneling conductance $dI/dV\,(\mathbf{r}, E = eV) \equiv g\,(\mathbf{r}, E = eV)$ with atomic resolution as a function of both location $\mathbf{r}$ and electron energy E, at cuprate crystal surfaces. This technique is distinguished from other electron spectroscopies in that it can access simultaneously the real space ($\mathbf{r}$-space) and momentum space ($\mathbf{k}$-space) electronic structure for both filled and empty states. But great care must be taken to avoid the serious systematic errors that are endemic to it.

The first of these occurs because the STM tip-sample tunneling current is given by

$$I(\mathbf{r}, z, V) = f(\mathbf{r}, z) \int_0^{eV} N(\mathbf{r}, E)\, dE, \tag{16.1}$$

where z is the tip-surface distance, V the tip-sample bias voltage, $N\,(\mathbf{r}, E)$ the sample's local density of electronic states, while $f(\mathbf{r}, z)$ contains both effects of tip elevation and of spatially-dependent tunneling matrix elements. The $g\,(\mathbf{r}, E)$ data are then related to $N\,(\mathbf{r}, E)$ by (Hanaguri *et al.*, 2007; Kohsaka *et al.*, 2007, 2008; Lee *et al.*, 2009; Pan *et al.*, 2001) as,

$$g(\mathbf{r}, E = eV) = \frac{eI_S}{\int_0^{eV_S} N(\mathbf{r}, E')\, dE'} N(\mathbf{r}, E), \tag{16.2}$$

where V_S and I_S are the junction stabilization bias voltage and current respectively. Thus when $\int_0^{eV_S} N(\mathbf{r}, E')\, dE'$ is strongly heterogeneous at the atomic scale (as it typically is in underdoped cuprates; (Alldredge *et al.*, 2008; Boyer *et al.*, 2007; Cren *et al.*, 2001; Gomes *et al.*, 2007; Howald *et al.*, 2001; Kohsaka *et al.*, 2007, 2008; Lang *et al.*, 2002; Lawler, 2010; Lee *et al.*, 2009; Matsuda *et al.*, 2003; McElroy *et al.*, 2005*a,b*; Pan *et al.*, 2001; Pushp *et al.*, 2009; Wise *et al.*, 2008)) $g\,(\mathbf{r}, E = eV)$ cannot be used to measure $N\,(\mathbf{r}, E)$. However, as eqn. 2 shows, these potentially severe systematic errors can be cancelled (Hanaguri *et al.*, 2007; Kohsaka *et al.*, 2007, 2008; Lee *et al.*, 2009) by using the observable (Hanaguri *et al.*, 2007)

$$Z(\mathbf{r}, E) \equiv \frac{g(\mathbf{r}, E = +eV)}{g(\mathbf{r}, E = -eV)} = \frac{N(\mathbf{r}, +E)}{N(\mathbf{r}, -E)}. \tag{16.3}$$

$Z(\mathbf{r}, E)$ measures the ratio of the probability of electron injection to the probability of extraction at a given $\mathbf{r}$ and E. A related observable that also avoids these systematic errors (but lacks energy resolution) is (Kohsaka *et al.*, 2007)

$$R(\mathbf{r}) \equiv \frac{I(\mathbf{r}, E = +eV)}{I(\mathbf{r}, E = -eV)}. \tag{16.4}$$

Another challenge is the random nanoscale variation (Kohsaka *et al.*, 2007, 2008) of $\Delta_1(\mathbf{r})$, which causes the $E{\sim}\Delta_1$ PG states to be detected at different locations for different bias voltages (Fig. 16.3 (a)). This effect can be mitigated (Kohsaka *et al.*, 2008; Lawler, 2010) by scaling the tunnel-bias energy scale $E= eV$ at each $\mathbf{r}$ by the PG magnitude $\Delta_1(\mathbf{r})$ at the same location, thus defining a reduced energy scale $e = E/\Delta_1(\mathbf{r})$ such that

$$Z(\mathbf{r}, e) \equiv Z(\mathbf{r}, E/\Delta_1(\mathbf{r})) \tag{16.5}$$

In such $Z(\mathbf{r}, e)$ images, the $\mathrm{E}{\sim}\Delta_1$ PG states all occur together at $E{=}1$ (Kohsaka *et al.*, 2008).

A final systematic error concerns $g(\mathbf{q}, E)$ and $Z(\mathbf{q}, E)$, the Fourier transforms of $g(\mathbf{r}, E)$ and $Z(\mathbf{r}, E)$ respectively. These are used to distinguish the dispersive wavevectors $\boldsymbol{q}_i(E)$ due to quantum interference patterns of delocalized states from any non-dispersive wavevectors $\boldsymbol{Q}^*$ indicative of an electronic ordered phase. To achieve adequate precision in $|\boldsymbol{q}_i(E)|$ for such discrimination requires

i. that $g(\mathbf{r}, E)$ or $Z(\mathbf{r}, E)$ be measured in large fields-of-view (FOV) while maintaining atomic resolution and register
ii. that the energy resolution be at or below ~2 meV

Using any smaller FOV or poorer energy resolution when measuring $g(\mathbf{r}, E)$ creates unavoidably the erroneous impression of non-dispersive modulations. For the case of $Bi_2Sr_2CaCu_2O_{8+\delta}$, we have demonstrated that no deductions distinguishing between dispersive and non-dispersive excitations can be made using Fourier-transformed $g(\mathbf{r},E)$ data from a FOV smaller than ~45 nm-square (Lee *et al.*, 2009; McElroy *et al.*, 2003)

16.2.2 Materials and methods

We have applied these techniques during the sequence of studies summarized herein by measuring $g(\mathbf{r},V)$ in ~ 45±5 nm square fields of view in $Bi_2Sr_2CaCu_2O_{8+\delta}$ samples with p (holes/Cu)≅ 0.19, 0.17, 0.14, 0.12, 0.10, 0.08, and 0.06 or with T_c(K) = 86, 88, 74, 64, 45, 37, and 20 respectively. Several of these samples were studied in both the dSC and PG phases. Each sample is inserted into the cryogenic ultra-high vacuum of the SI-STM system, cleaved to reveal an atomically clean BiO surface, and all $g(\mathbf{r},V)$ measurements were made between 1.9 K and 65 K. Three cryogenic SI-STMs (optimized for different purposes) are used throughout these studies. The resulting data set, acquired over approximately a decade, consists of $>10^8$ atomically resolved and registered tunneling spectra.

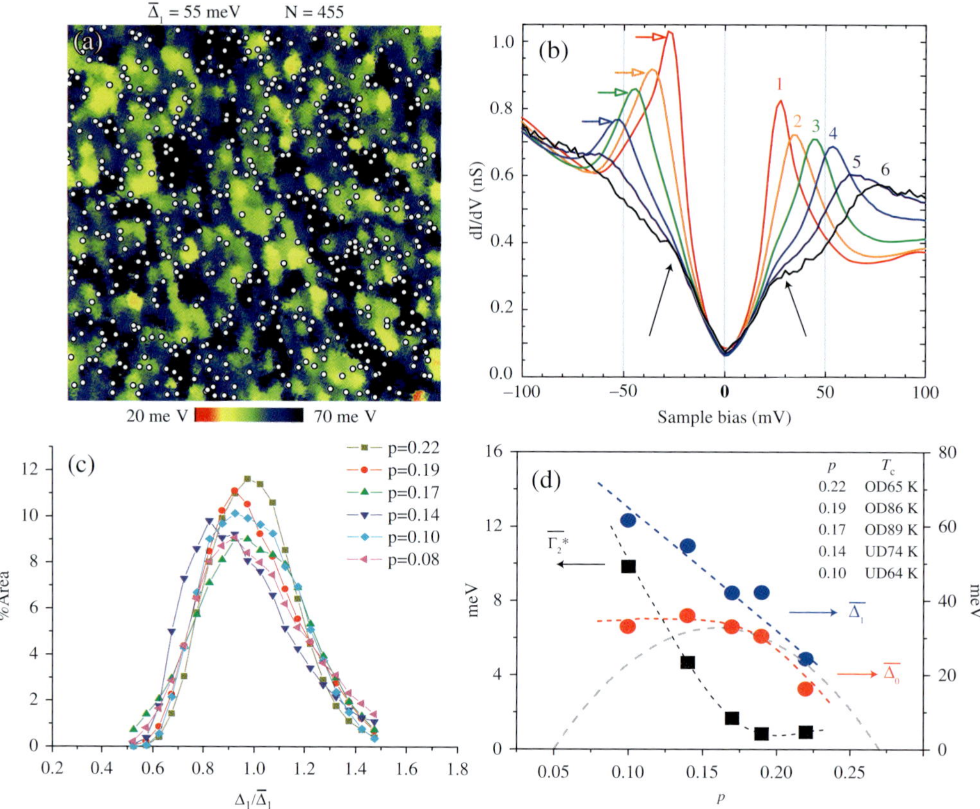

Fig. 16.3 (a) Map of the local energy scale $\Delta_1(\mathbf{r})$ from a 49-nm field of view (corresponding to $\sim$16,000 CuO_2 plaquettes) measured on a sample with $T_c = 74$ K. Average gap magnitude Δ_1 is at the top, together with the values of N, the total number of dopant impurity states (shown as white circles) detected in the local spectra. (b) The average tunneling spectrum, $g(E)$, associated with each gap value in the field of view in (a). The arrows locate the "kinks" whose energy is Δ_0. (c) Histograms of equivalent $\Delta_1(\mathbf{r})$ maps from samples with $p = 0.08$, 0.10, 0.14, 0.17, 0.19, and 0.22 normalized to the average Δ_1 in each map. Obviously, these distributions are statistically highly similar. (d) The doping dependence of the average Δ_1 (blue circles), average Δ_0 (red circles), and average anti-nodal scattering rate $\Gamma_2{}^*$ (black squares), each set interconnected by dashed guides to the eye. The higher-scale Δ_1 evolves along the PG line whereas the lower-scale Δ_0 represents segregation in energy between homogeneous and heterogeneous electronic structure.

16.3 Doping dependence of nanoscale electronic disorder in $Bi_2Sr_2CaCu_2O_{8+\delta}$

16.3.1 Nanoscale electronic disorder of the $E \sim \Delta_1$ pseudogap states

Nanoscale electronic disorder is seen universally (Alldredge *et al.*, 2008; Boyer *et al.*, 2007; Cren *et al.*, 2001; Gomes *et al.*, 2007; Howald *et al.*, 2001; Kohsaka *et al.*, 2007, 2008; Lang *et al.*, 2002; Lawler, 2010; Lee *et al.*, 2009; Matsuda *et al.*, 2003; McElroy *et al.*, 2005*a*,*b*, 2003; Pan *et al.*, 2001; Pushp *et al.*, 2009) in images of $\Delta_1(\mathbf{r})$ in $Bi_2Sr_2CaCu_2O_{8+\delta}$, with $|\Delta_1|$ ranging from above 130 meV to below 10 meV as the hole density p ranges from 0.06 to 0.22. Virtually identical nanoscale electronic disorder is seen in $Bi_2Sr_2Cu_1O_{6+\delta}$ (Boyer *et al.*, 2007; Wise *et al.*, 2008) and in $Bi_2Sr_2Ca_2Cu_3O_{10+\delta}$ (Jenkins *et al.*, 2009). Figure 16.3 (a) shows a typical $Bi_2Sr_2CaCu_2O_{8+\delta}$ $\Delta_1(\mathbf{r})$ image, with the sites of the non-stoichiometric oxygen dopant ions overlaid (McElroy *et al.*, 2005*b*). Figure 16.3 (b) shows the typical $g(E)$ spectrum associated with each value of $\pm\Delta_1$ in such an image (McElroy *et al.*, 2005*a*); it also shows how the electronic structure becomes homogeneous (Alldredge *et al.*, 2008; Hanaguri *et al.*, 2007; Kohsaka *et al.*, 2008; Lee *et al.*, 2009; McElroy *et al.*, 2005*a*,*b*; Wise *et al.*, 2008) below a lower energy scale $E = \pm\Delta_0$ as indicated by the arrows. Similar effects are seen in $Bi_2Sr_2CuO_{6+\delta}$ and $Bi_2Sr_2Ca_2Cu_3O_{10+\delta}$ samples (Boyer *et al.*, 2007; Jenkins *et al.*, 2009; Wise *et al.*, 2008). Figure 16.3 (c) shows the distributions of $|\Delta_1|$ measured in units of the spatially-averaged $\bar{\Delta}_1$ from six samples with varying hole densities. These normalized distributions are virtually independent of p, indicating that the microscopic trigger for the random Δ_1 disorder is universal. Imaging in the PG phase reveals highly similar (Boyer *et al.*, 2007; Gomes *et al.*, 2007; Lee *et al.*, 2009; Matsuda *et al.*, 2003) nanoscale electronic disorder in $\Delta_1(\mathbf{r})$. Explaining these Δ_1 disorder phenomena has been an interesting challenge for cuprate studies.

16.3.2 Effect of interstitial oxygen dopant atoms

A key component of the explanation lies in the fact that electron-acceptor atoms must be introduced (Maekawa, 2004) in order to generate hole-doped superconductivity from the Mott-insulating "parent" phase. This process almost always creates random distributions of differently charged dopant atoms near the CuO_2 planes (Eisaki *et al.*, 2004). In $Bi_2Sr_2CaCu_2O_{8+d}$ the dopant atoms are -2e charged interstitial oxygen ions, which could conceivably cause a variety of different local effects. For example, electrostatic screening of each ion could accumulate holes at those locations thereby reducing the energy-gap values nearby (Martin and Balatsky, 2001; Wang *et al.*, 2002). Another possibility is that the dopant ions could generate local crystalline stress/strain at the nanoscale (He *et al.*, 2006; Kaneshita *et al.*, 2004; Mori *et al.*, 2008; Nunner *et al.*, 2005; Zhu *et al.*, 2003) thus disordering various hopping matrix elements and electron–electron interactions within the unit cell. The locations of interstitial O^{2-} dopant ions can be identified in $Bi_2Sr_2CaCu_2O_8$ because an atomic-scale impurity state occurs at $E=$ -0.96 V whose density scales with their density (McElroy *et al.*, 2005*b*) (Fig. 16.3 (a)). Strong spatial correlations are found between the distribution of these impurity states and $\Delta_1(\mathbf{r})$ maps, implying that dopant disorder is responsible for much of the $\Delta_1(\mathbf{r})$ electronic disorder. The primary effect near each dopant ion is a shift of spectral weight from low to high energy so that the Δ_1g excitation energy is increased strongly. Simultaneous imaging of the dopant ion locations and $g(\mathbf{r},E<\Delta_0)$ reveals that the dispersive $g(\mathbf{r},E)$ modulations are well correlated with dopant ion locations for $E<\Delta_0$. Therefore the dopant ions are a key source of the scattering generating the observed Bogoliubov quasiparticle interference effects (Hanaguri *et al.*, 2007; Hoffman *et al.*, 2002; Kohsaka *et al.*, 2008; Lee *et al.*, 2009; McElroy *et al.*, 2005*a*,*b*, 2003; Wise *et al.*, 2008) (Sections 16.6 and 16.7). The direct demonstration that it is the chemical doping process itself that both disorders Δ_1 and causes strong quasiparticle scattering is of great practical significance because similar nanoscale electronic disorder phenomena are then likely to be common (although with different intensities) in all non-stoichiometric high-T_c cuprates.

16.3.3 Microscopic mechanism of nanoscale disorder in $E{\sim}\Delta_1$ pseudogap states

The microscopic mechanism of the Δ_1 disorder is, however, not understood. It does not appear to be due to hole accumulation surrounding O^{2-} dopant ions because (i) the modulations in integrated density of filled states are observed to be weak (McElroy *et al.*, 2005*b*) and (ii) Δ_1 is increased nearby the dopant ions (McElroy *et al.*, 2005*b*) a situation diametrically opposite to the expected effect from hole-accumulation there. Random isovalent atomic substitution at the Sr site is known to suppress superconductivity strongly (Eisaki *et al.*, 2004), possibly by distortions of the unit cell and associated changes in the hopping matrix elements. It has therefore been proposed that the interstitial dopant ions might act similarly, perhaps by displacing the Sr or apical oxygen atoms (Eisaki *et al.*, 2004; Kaneshita *et al.*, 2004; Zhu *et al.*, 2003) and thereby distorting the unit-cell geometry. Direct support for this point of view comes from the observation that quasi-periodic geometrical distortions of the crystal unit-cell geometry yield virtually identical perturbations in $g(E)$ and $\Delta_1(\mathbf{r})$ but now unrelated to the dopant ions (Slezak *et al.*, 2008). Thus it seems that the Δ_1 disorder is not caused primarily by carrier-density modulations but by geometrical distortions to the unit-cell dimensions with resulting strong but local changes in its high-energy electronic structure.

16.3.4 "Kinks" in N(E) separating homogeneous and heterogeneous states

Finally, we discuss the "kinks" that have been reported ubiquitously (Alldredge *et al.*, 2008; Boyer *et al.*, 2007; Cren *et al.*, 2001; Gomes *et al.*, 2007; Hanaguri *et al.*, 2007; Howald *et al.*, 2001; Kohsaka *et al.*, 2008; Lang *et al.*, 2002; McElroy *et al.*, 2005*a*,*b*; Pan *et al.*, 2001; Pushp *et al.*, 2009; Wise *et al.*, 2008) in cuprate $g(E)$ spectra. In general, these kinks are weak perturbations to $N(E)$ near optimal doping, becoming more clear and increasing in number as p is strongly diminished (Alldredge *et al.*, 2008; McElroy *et al.*, 2005*a*). Representative Δ_1-sorted spectra in Fig. 16.3 (b) show that for $\Delta_1>50$ meV the kinks become very obvious (Alldredge *et al.*, 2008; McElroy *et al.*, 2005*a*) Each kink can be identified and its energy is labelled $\Delta(\mathbf{r})$. By determining

$\bar{\Delta}_0$ (the spatial average of $\Delta_0(\mathbf{r})$) as a function of p, we find that this energy $\bar{\Delta}_0$ always divides the electronic structure into two categories (Alldredge *et al.*, 2008). For E $<\bar{\Delta}_0$ the excitations are homogenous in $\boldsymbol{r}$-space and are welldefined Bogoliubov quasiparticle eigenstates in $\boldsymbol{k}$-space (Section 16.6). By contrast the PG excitations with $E{\sim}\Delta_1$ are heterogeneous in $\boldsymbol{r}$-space and ill-defined in $\boldsymbol{k}$-space (Section 16.7). Figure 16.3 (c) provides a summary of the evolution of $\bar{\Delta}_0$ and $\bar{\Delta}_1$ with p.

16.3.5 Summary

The nanoscale Δ_1 disorder in $Bi_2Sr_2CaCu_2O_{8+d}$ is strongly influenced by the random distribution of dopant ions (McElroy *et al.*, 2005*b*) through an electronic process probably involving geometrical distortions of the crystal unit cell (He *et al.*, 2006; Mori *et al.*, 2008; Nunner *et al.*, 2005; Slezak *et al.*, 2008) and which appears local at the 3-nm scale. This disorder only exists for energies near the PG energy $E{\sim}\Delta_1$ while the electronic excitations with E$<\Delta_0$ are only influenced by the dopant ions via scattering and are otherwise relatively homogeneous when studied using Bogoliubov quasiparticle interference (QPI) or by direct imaging (Alldredge *et al.*, 2008; Lang *et al.*, 2002; McElroy *et al.*, 2005*a*,*b*) As the equivalent $\Delta_1(\mathbf{r})$ disorder exists in the PG phase (Boyer *et al.*, 2007; Gomes *et al.*, 2007; Lawler, 2010; Lee *et al.*, 2009; Matsuda *et al.*, 2003), an intuitively appealing idea has been that these $\Delta_1(\mathbf{r})$ arrangements (Fig. 16.3 (a)) could represent images of nanoscale superconducting “grains”. However, when the superconducting energy gap $\Delta(\mathbf{k})$ is imaged using Bogoliubov QPI (Hanaguri *et al.*, 2007; Hoffman *et al.*, 2002; Kohsaka *et al.*, 2008; Lee *et al.*, 2009; McElroy *et al.*, 2005*a*, 2003; Wise *et al.*, 2008) it is found to be spatially homogeneous (Section 16.6). Moreover, the $E{\sim}\Delta_1$ PG states are also involved in a classic oxygen isotope effect, which indicates a strong localized electron–lattice interaction (Lee *et al.*, 2006*a*). Finally, atomic resolution imaging of the $E{\sim}\Delta_1$ states shows them to be energetically non-dispersive and to break several spatial symmetries locally (Kohsaka *et al.*, 2007, 2008; Lawler, 2010; Lee *et al.*, 2009) (Section 16.7). Since none of these latter phenomena are expected characteristics of d-wave Bogoliubov quasiparticles within a superconducting grain it appears implausible at present that $\Delta_1(\mathbf{r})$ represents merely the image of a d-wave granular superconductor.

16.4 Bogoliubov quasiparticle interference imaging

16.4.1 d-wave Bogoliubov quasiparticle interference

Bogoliubov quasiparticles are the excitations generated by breaking Cooper pairs. Bogoliubov QPI occurs when scattering centers break translational invariance causing quasiparticles to undergo quantum interference. In a d-wave cuprate superconductor with a single hole-like band within the CuO_2 Brillouin zone, the Bogoliubov quasiparticle dispersion $E(\mathbf{k})$ has “banana-shaped” constant energy contours. The d-symmetry superconducting energy gap causes strong maxima to appear in the joint-density-of-states at the eight tips $\mathbf{k}_j(E)$; j=1, 2,..., 8 of these “bananas”. Elastic scattering between the $\mathbf{k}_j(E)$ then produces $\mathbf{r}$-space interference patterns in the local density of states $N(\mathbf{r}, E)$. The resulting energy-dispersive $g(\mathbf{r}, \mathbf{E})$ modulations detectable by

SI-STM should exhibit $16\pm\mathbf{q}$ pairs of dispersive wavevectors in $g(\mathbf{q}, E)$. The minimal set of these wavevectors that is specifically characteristic of *d*-wave superconductivity consists of seven: $\mathbf{q}_i(E)i$=1,...,7 with $\mathbf{q}_i(+E) = \mathbf{q}_i(+E)$. In this so-called "octet model" (Capriotti *et al.*, 2003; Nunner *et al.*, 2006; Wang and Lee, 2003) the relationship between the measured $\mathbf{q}_i(E)$ and the locus of the Bogoliubov band-minima along the trajectory of the normal-state Fermi surface $\mathbf{k}_B(E) = (k_x(E),k_y(E))$, is given by:

$$\begin{array}{lll} \mathbf{q}_1 = (2k_x, 0) & \mathbf{q}_4 = (2k_x, 2k_y) & \mathbf{q}_7 = (k_x - k_y,\, k_y - k_x) \\ \mathbf{q}_2 = (k_x + k_y, k_y - k_x) & \mathbf{q}_5 = (0, 2k_y) & \\ \mathbf{q}_3 = (k_x + k_y, k_y + k_x) & \mathbf{q}_6 = (k_x - k_y, k_y + k_x) & \end{array} \tag{16.6}$$

When this set of $\mathbf{q}_i(E)$ are measured from $Z(\mathbf{q}, E)$, the Fourier transform of spatial modulations seen in $g(\mathbf{r}, \mathbf{E})$ (see Fig. 16.2 (a) for example), the $\mathbf{k}_B(E)$ can be determined by using eqn. 6 with the requirement that all its independent solutions be consistent at all energies. The superconductor's Cooper-pairing energy-gap structure $\Delta(\mathbf{k})$ is then determined directly by inverting $\mathbf{k}_B(E)$. In $Bi_2Sr_2CaCu_2O_{8+\delta}$ near optimal doping, measurements from QPI of $\mathbf{k}_B(E)$ and the superconducting $\Delta(\mathbf{k})$ (Fig. 16.4 (a)) are consistent with ARPES (McElroy *et al.*, 2003; Mesot *et al.*, 1999). In both $Ca_{2-x}Na_xCuO_2Cl_2$ and $Bi_2Sr_2CaCu_2O_{8+\delta}$ the octet model and QPI yields $\mathbf{k}_B(E)$ and $\Delta(\mathbf{k})$ equally well (Hanaguri *et al.*, 2007; Wise *et al.*, 2008). Moreover, the correctness of the fundamental **k**-space phenomenology behind this *d*-wave QPI "octet" model has been demonstrated directly by ARPES studies (Chatterjee *et al.*, 2006, 2007; McElroy *et al.*, 2006).

16.4.2 Summary

Fourier transformation of $Z(\mathbf{r}, E)$ plus the octet model of *d*-wave Bogoliubov QPI, yields the two branches of the Bogoliubov excitation spectrum $\mathbf{k}_B(\pm E)$ and the superconducting energy gap magnitude $\pm\Delta(\mathbf{k})$ for both filled and empty states in a single experiment. As only the Bogoliubov states of a *d*-wave superconductor could exhibit this set of 16 pairs of interference wavevectors with $\mathbf{q}_i$(-E)=$\mathbf{q}_i$(+E) and all dispersions internally consistent within the octet model, the energy gap $\pm\Delta(\mathbf{k})$ determined by these procedures is definitely that of the delocalized Cooper-pairs. Moreover, because $\Delta(\mathbf{k})$ must be translationally invariant for Bogoliubov QPI to generate such long-range interference patterns, cuprate superconductivity is always spatially homogeneous when they are detected (Hanaguri *et al.*, 2007; Hoffman *et al.*, 2002; Kohsaka *et al.*, 2008; Lawler, 2010; Lee *et al.*, 2009; McElroy *et al.*, 2003; Wise *et al.*, 2008).

16.5 Low energy excitations in the superconducting phase

16.5.1 Bogoliubov quasiparticle interference in the dSC phase

Bogoliubov QPI imaging techniques have been used to study the evolution of **k**-space electronic structure with falling p in $Bi_2Sr_2CaCu_2O_{8+\delta}$. In the superconducting phase, the expected 16 pairs of **q**-vectors are always observed in $Z(\mathbf{q}, E)$ and are found consistent with each other within the octet model (Fig. 16.2 (a)). However, the dispersion of octet model **q**-vectors always stops at the same weakly doping-dependent (Kohsaka *et al.*, 2008; McElroy *et al.*, 2005*a*; Wise *et al.*, 2008) excitation energy Δ_0. For E> Δ_0 we find three non-dispersive **q**-vectors: the reciprocal lattice vector **Q** along with $\mathbf{q}_1^*$ and $\mathbf{q}_5^*$ (see Fig. 16.4 (b)). The equivalent pair of non-dispersive wavevectors to $\mathbf{q}_1^*$ and $\mathbf{q}_5^*$ has also been detected by ARPES in optimally doped $Bi_2Sr_2CaCu_2O_{8+\delta}$ (Chatterjee *et al.*, 2007) and in underdoped $Ca_{2-x}Na_xCuO_2Cl_2$ (Shen *et al.*, 2005).

By using the QPI imaging techniques described in Section 16.4, we show in Fig. 16.4 (c) the locus of Bogoliubov quasiparticle states $\mathbf{k}_B(E)$ determined as a function of p. Here we see that when the Bogoliubov QPI patterns disappear at Δ_0, the **k**-states are near the diagonal lines between **k**=(0,π/a$_0$) and **k**=(π/a$_0$,0) within the CuO_2 Brillouin zone. These **k**-space Bogoliubov arc tips are defined by both the change from dispersive to non-dispersive characteristics and by the disappearance of the $\mathbf{q}_2$, $\mathbf{q}_3$, $\mathbf{q}_6$, and $\mathbf{q}_7$ modulations (see Fig. 16.4 (b)). Thus, the complete QPI signature of delocalized Cooper pairs is confined to an arc (fine solid lines in Fig. 16.4 (c)) and this arc shrinks with falling p (Kohsaka *et al.*, 2008). This observation is supported by angle resolved photoemission studies (Chatterjee *et al.*, 2006, 2007), by QPI studies of $Bi_2Sr_2CuO_{6+\delta}$ (Wise *et al.*, 2008) and by analyses of g(E) when fitted to a multi-parameter model for **k**-space structure plus a dSC energy gap (Pushp *et al.*, 2009).

The minima (maxima) of the Bogoliubov bands $\mathbf{k}_B(\pm E)$ should occur at the **k**-space location of the Fermi surface of the non-superconducting state. One can then ask if the carrier-density count satisfies Luttinger's theorem, which states that twice the **k**-space area enclosed by the Fermi surface, measured in units of the area of the first Brillouin zone, equals the number of electrons per unit cell, n. In Fig. 16.4 (c) we show as fine solid lines hole-like Fermi surfaces fitted to our measured $\mathbf{k}_B(E)$. Using Luttinger's theorem with these **k**-space contours extended to the zone face would result in a calculated hole-density p for comparison with the estimated hole density in the samples. These data are shown by filled symbols in the inset to Fig. 16.4 (c). We see that the Luttinger theorem is strongly violated at all doping below $p{\sim}10\%$. However, the Luttinger theorem can be amended in a doped Mott insulator (Kohsaka *et al.*, 2008) so that the zero-energy contours bounding the region representing carriers are defined, not only by poles in the Green's functions, but also by their zeros. The locus of zeros of these Green's functions may be expected to occur at the lines joining **k**=(0,π/a$_0$) to **k**=(π/a$_0$,0). In that situation, the hole density is related quantitatively to the area between the **k**=(0,π/a$_0$) - **k**=(π/a$_0$,0) lines and the arcs. The carrier densities calculated in this fashion are shown by open symbols in the inset to Fig. 16.4 (c) and are obviously in much better agreement with the chemical hole density.

Figure 16.5 provides a doping-dependence analysis of the locations of the ends of the arc-tips at which the Bogoliubov QPI signature disappears. Figure 16.5 (a) shows a typical $Z(\mathbf{q},E)$ for which Δ_0 <E<Δ_1. The vectors $\mathbf{q}_1{}^*$ and $\mathbf{q}_5{}^*$ (see Fig. 16.4 (b)) are labeled along with the Bragg vectors $\mathbf{Q}_x$ and $\mathbf{Q}_y$. Figure 16.5 (b) shows a schematic representation of the arc of the **k**-space supporting Bogoliubov QPI in blue. Its termination points on the lines linking **k** =±(0,π_1 a$_0$) and **k** =±(π_1a$_0$,0) directly link the $\mathbf{q}_1{}^*$ and $\mathbf{q}_5{}^*$ wavevectors to the Brillouin zone size (as shown in red). Equivalent phenomena have also been reported for $Bi_2Sr_2CuO_{6+\delta}$ (Wise *et al.*, 2008).

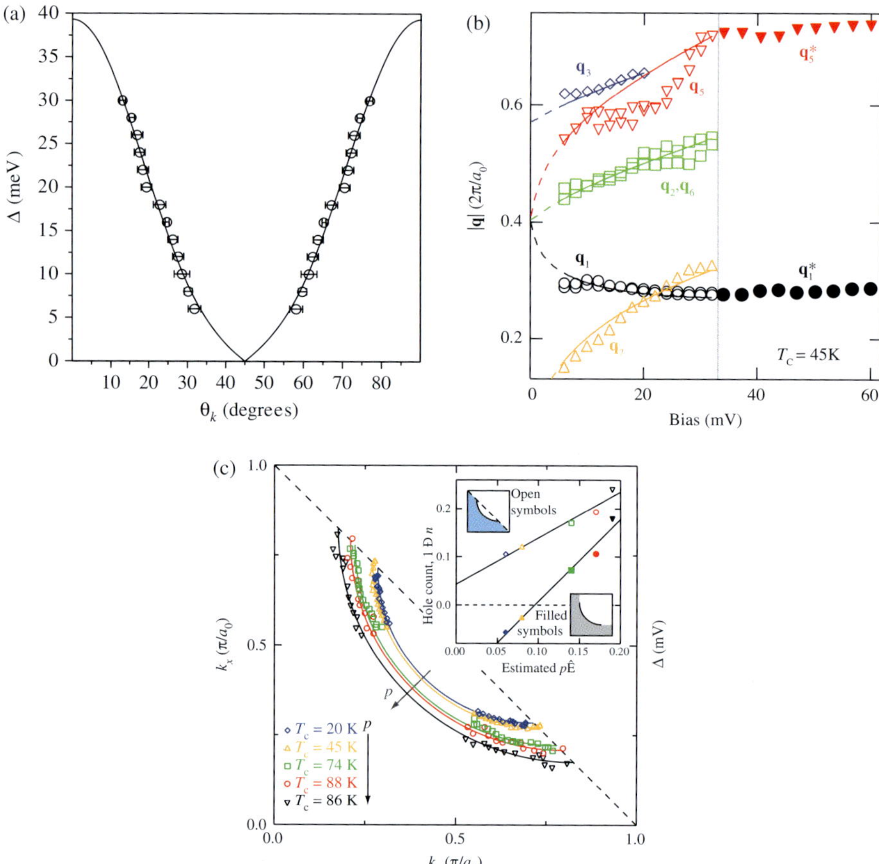

Fig. 16.4 (a) A plot of the superconducting energy gap $\Delta(\theta_k)$ determined from octet model inversion of quasiparticle interference measurements, shown as open circles (McElroy *et al.*, 2003). These were extracted using the measured position of scattering vectors $\mathbf{q}_1$ through $\mathbf{q}_7$ (excluding $\mathbf{q}_4$) in (a). The solid line is a fit to the data. The mean value of Δ_1 for this overdoped $T_c = 86$ K sample was 39 meV. (b) The magnitude of various extracted QPI vectors, plotted as a function of energy. Whereas the expected energy dispersion of the octet vectors $\mathbf{q}_i(E)$ is apparent for $|E| < 32$ mV, the peaks that avoid extinction ($\mathbf{q}_1$* and $\mathbf{q}_5$*) always become non-dispersive above Δ_0 (vertical grey line). (c) Locus of the Bogoliubov band minimum $\mathbf{k}_B(E)$ found from extracted QPI peak locations $\mathbf{q}_i(E)$, in five independent $Bi_2Sr_2CaCu_2O_{8+\delta}$ samples with decreasing hole density. Fits to quarter-circles are shown and, as p decreases, these curves enclose a progressively smaller area. The BQP interference patterns disappear near the perimeter of a $\mathbf{k}$-space region bounded by the lines joining $\mathbf{k} = (0, \pm\pi/a_0)$ and $\mathbf{k} = (\pm\pi/a_0, 0)$. The spectral weights of $\mathbf{q}_2$, $\mathbf{q}_3$, $\mathbf{q}_6$ and $\mathbf{q}_7$ vanish at the same place (dashed line; see also Kohsaka *et al.* (2008). Filled symbols in the inset represent the hole count p = 1 - n derived using the simple Luttinger theorem, with the fits to a large, hole-like Fermi surface indicated schematically here in grey. Open symbols in the inset are the hole counts calculated using the area enclosed by the Bogoliubov arc and the lines joining $\mathbf{k} = (0, \pm\pi/a_0)$ and $\mathbf{k} = (\pm\pi/a_0, 0)$, and are indicated schematically in blue.

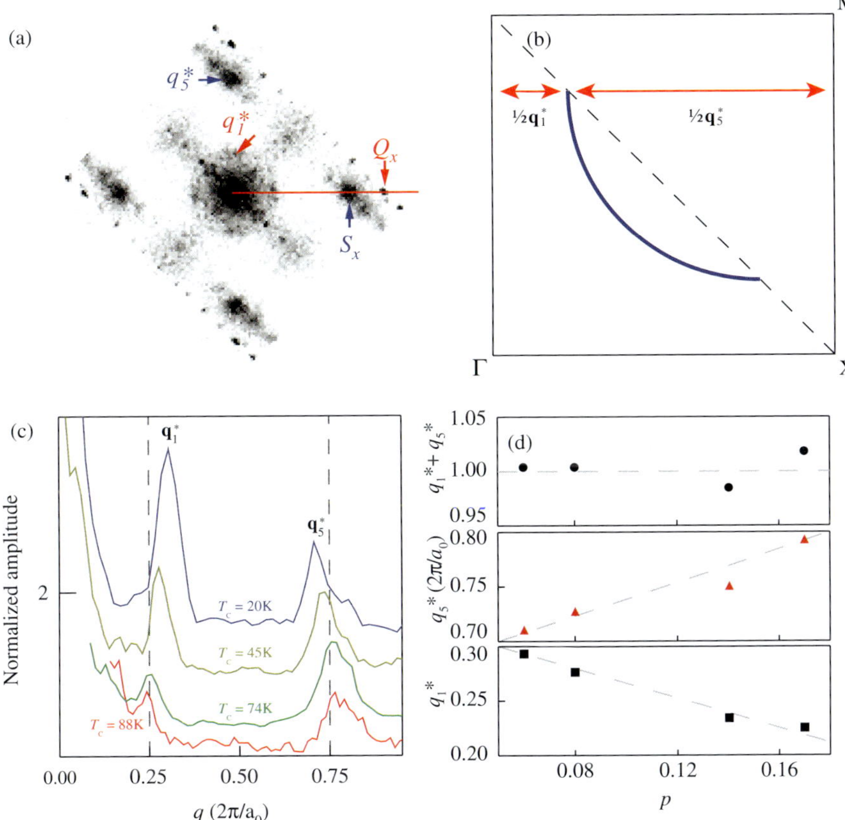

Fig. 16.5 (a) Fourier transform of the conductance ratio $Z(\mathbf{q}, E= 48$ meV) at a representative energy between Δ_0 and Δ_1 for underdoped T_C=74 K $Bi_2Sr_2CaCu_2O_{8+\delta}$. The red line schematically indicates the source of the data in **c**. The arrows label the location of the wavevectors $\mathbf{q}_1$*, $\mathbf{q}_5$*, $\boldsymbol{S}_x$, and $\boldsymbol{Q}_x$ described in the text. (b) Schematic diagram of the Brillouin zone illustrating the relationship of non-dispersive $\mathbf{q}_1$* and $\mathbf{q}_5$* to the ends of the Bogoliubov arc. (c) Doping dependence of line-cuts of $Z(\mathbf{q}, E= 48$ meV) extracted along the Cu–O bond direction $\boldsymbol{Q}_x$. The vertical dashed lines demonstrate that the non-dispersive **q**-vectors at energies between Δ_0 and Δ_1 are not commensurate harmonics of a $4a_0$ periodic modulation, but instead evolve in a fashion directly related to the extinction point of the Fermi arc. The data in **c**. have been normalized to the peak amplitude of $\mathbf{q}_5$* and offset vertically for clarity. (d) $\mathbf{q}_1$*, $\mathbf{q}_5$*, and their sum $\mathbf{q}_1$* + $\mathbf{q}_5$* as a function of p demonstrating that individually these modulations evolve with doping while their sum does not change and is equal to the reciprocal lattice vector defining the first Brillouin zone.

Figure 16.5 (c) shows the doping dependence for $Bi_2Sr_2CaCu_2O_{8+\delta}$ of the location of both $\mathbf{q}_1{}^*$ and $\mathbf{q}_5{}^*$ from $Z(\mathbf{q}, E)$ (Kohsaka *et al.*, 2008). The measured magnitude of $\mathbf{q}_1{}^*$ and $\mathbf{q}_5{}^*$ versus p are then shown in Fig. 16.5 (d) along with the sum $\mathbf{q}_1{}^*+\mathbf{q}_5{}^*$, which is always equal to 2π. This demonstrates that, as the Bogoliubov QPI extinction point travels along the line from $\mathbf{k}=(0,\pi_1 a_0)$ and $\mathbf{k}=(\pi_1 a_0,0)$ (Kohsaka *et al.* (2008) and Figs 16.4 (c) and 16.5 (b), the wavelengths of incommensurate modulations $\mathbf{q}_1{}^*$ and $\mathbf{q}_5{}^*$ are controlled by its **k**-space location (Kohsaka *et al.*, 2008).

16.5.2 Summary

Because the superconducting $\Delta(\mathbf{k})$ must be translationally invariant for Bogoliubov QPI to exhibit the observed long-range interference patterns, cuprate superconductivity is found to be rather spatially homogeneous (as implied also by direct $g(E)$ spectra studies). When p is reduced, the Bogoliubov QPI signature of which **k**-space states contribute to Cooper pairing is confined to a shrinking arc (Kohsaka *et al.*, 2008; McElroy *et al.*, 2005*a*, 2003; Wise *et al.*, 2008) in **k**-space. The arc tips lie near the diagonal lines connecting $\mathbf{k}=(0,\pm\pi/a_0)$ and $\mathbf{k}=(\pm\pi/a_0,0)$ and occur at a weakly doping-dependent (Kohsaka *et al.*, 2008; McElroy *et al.*, 2005*a*) energy $E = \Delta_0$ that is indistinguishable from (i) where the $g(E)$ kinks occur (McElroy *et al.*, 2005*a*) and (ii) where electronic homogeneity is lost (Kohsaka *et al.*, 2008; Lang *et al.*, 2002; Lee *et al.*, 2009; McElroy *et al.*, 2005*a*; Pan *et al.*, 2001; Wise *et al.*, 2008). The shrinking of this arc with decreasing hole-density could satisfy Luttinger's theorem if it is actually the front side of a hole-pocket bounded behind by the $\mathbf{k}=(0,\pm\pi/a_0) - \mathbf{k}=(\pm\pi/a_0,0)$ lines. Finally the gap energy at the arc tip Δ_0 is associated with the disappearance of the QPI signature of delocalized Cooper pairs for $E\geq\Delta_0$, while the upper energy Δ_1 is associated with a quite distinct **r**-space electronic structure of the $E{\sim}\Delta_1$ PG excitations (Section 16.7).

16.6 Low-energy excitations in the pseudogap phase

16.6.1 QPI in a phase-fluctuating d-wave superconductor

Because cuprate superconductivity is quasi-two-dimensional, the superfluid density increases from zero approximately linearly with p, and the superconducting energy gap $\Delta(\mathbf{k})$ exhibits four **k**-space nodes, fluctuations of the quantum phase $\phi(\mathbf{r}, t)$ of the superconducting order parameter $\Psi = \Delta(\mathbf{k})e^{i\phi(\mathbf{r},t)}$ could have strong effects on the superconductivity at low hole density (Berg and Altman, 2007; Carlson, Kivelson, Emery and Manousakis, Carlson *et al.*; Emery and Kivelson, 1995; Franz and Millis, 1998; Kwon *et al.*, 2001; Randeria *et al.*, 1992). Phenomena indicative of phase-fluctuating superconductivity are detectable for cuprates in particular regions of the phase diagram (Bergeal *et al.*, 2008; Corson *et al.*, 1999; Li *et al.*, 2005; Wang *et al.*, 2005, 2006; Xu *et al.*, 2000) as indicated by the region $T_c < T < T_\phi$ (Fig. 16.1 (a)). The techniques involved include terahertz transport studies (Corson *et al.*, 1999), the Nernst effect (Wang *et al.*, 2006; Xu *et al.*, 2000), torque-magnetometry measurements (Wang *et al.*, 2005), field dependence of the diamagnetism (Li *et al.*, 2005), and zero-bias conductance enhancement (Bergeal *et al.*, 2008).

A spectroscopic signature of phase-incoherent d-wave superconductivity in the PG phase could be the continued existence of the Bogoliubov QPI octet described in the previous two sections. This is because, if the quantum phase $\phi(\mathbf{r}, t)$ is fluctuating while the energy-gap magnitude $\Delta(\mathbf{k})$ remains largely unchanged, the particle-hole symmetric octet of high joint-density-of-states regions generating the QPI should continue to exist (Misra *et al.*, 2004; Pereg-Barnea and Franz, 2003; Wulin *et al.*, 2009). However, any gapped **k**-space regions supporting Bogoliubov QPI in the PG phase must then occur beyond the tips of the ungapped Fermi arc.

16.6.2 Bogoliubov quasiparticle interference in the PG phase

The temperature evolution of the Bogoliubov octet in $Z(\mathbf{q}, E)$ was studied as a function of increasing temperature from the dSC phase into the PG phase using a 48 nm square FOV and with subatomic resolution. Representative $Z(\mathbf{q}, E)$ for six temperatures are shown in Fig. 16.6; the $\mathbf{q}_i(E)$ (i=1,2…7) characteristic of the superconducting octet model are observed to remain unchanged upon passing above T_c to at least T$\sim$1.5T_c. This demonstrates that the Bogoliubov QPI octet phenomenology exists in the cuprate PG phase (although it is generated by different regions of **k**-space, and thus different $\Delta(\mathbf{k})$, than in the same sample in the superconducting phase). Thus for the low-energy (E<35 mV) excitations in the underdoped PG phase, the $\boldsymbol{q}_i(E)$ (i=1,2…7) characteristic of the octet model are preserved unchanged upon passing above T_c. Importantly, all seven $\boldsymbol{q}_i(E)$ (i=1,2…7) modulation wavevectors that are dispersive in the dSC phase remain dispersive into the PG phase, which is still consistent with the octet model (Lee *et al.*, 2009). The octet wavevectors also retain their particle-hole symmetry $\boldsymbol{q}_i$(+E) $=\boldsymbol{q}_i$(-E) in the PG phase and the $g(\mathbf{r}, E)$ modulations occur in the same energy range and emanate from the same contour in **k**-space as those observed at lowest temperatures (Lee *et al.*, 2009). However the particle—hole-symmetric energy gap $|\Delta(\mathbf{k})|$ moves away from the nodes with increasing T leaving behind a growing Fermi arc of gapless excitations.

16.6.3 Summary

All the Bogoliubov QPI signatures detectable in the dSC phase survive virtually unchanged into the underdoped PG phase—up to at least T$\sim$1.5T$_c$ for strongly underdoped $Bi_2Sr_2CaCu_2O_{8+\delta}$ samples. Moreover, for E<Δ_0 all seven dispersive $\boldsymbol{q}_i(E)$ modulations characteristic of the octet model in the dSC phase remain dispersive in the PG phase. These observations rule out the existence for all E $\leq\Delta_0$ of non-dispersive $g(E)$ modulations at finite ordering wavevector $\mathbf{Q}^*$ which would be indicative of a static electronic order. This conclusion is in agreement with the results of ARPES studies (Chatterjee *et al.*, 2006, 2007). Instead, the observed excitations are indistinguishable from the dispersive **k**-space eigenstates of a phase-incoherent d-wave superconductor (Lee *et al.*, 2009). Thus the SI-STM picture of electronic structure in the strongly underdoped PG phase actually contains three elements:

1. the ungapped Fermi arc (Norman *et al.*, 1998)
2. the particle—hole-symmetric gap $\Delta(\mathbf{k})$ of a phase-disordered superconductor (Lee *et al.*, 2009)

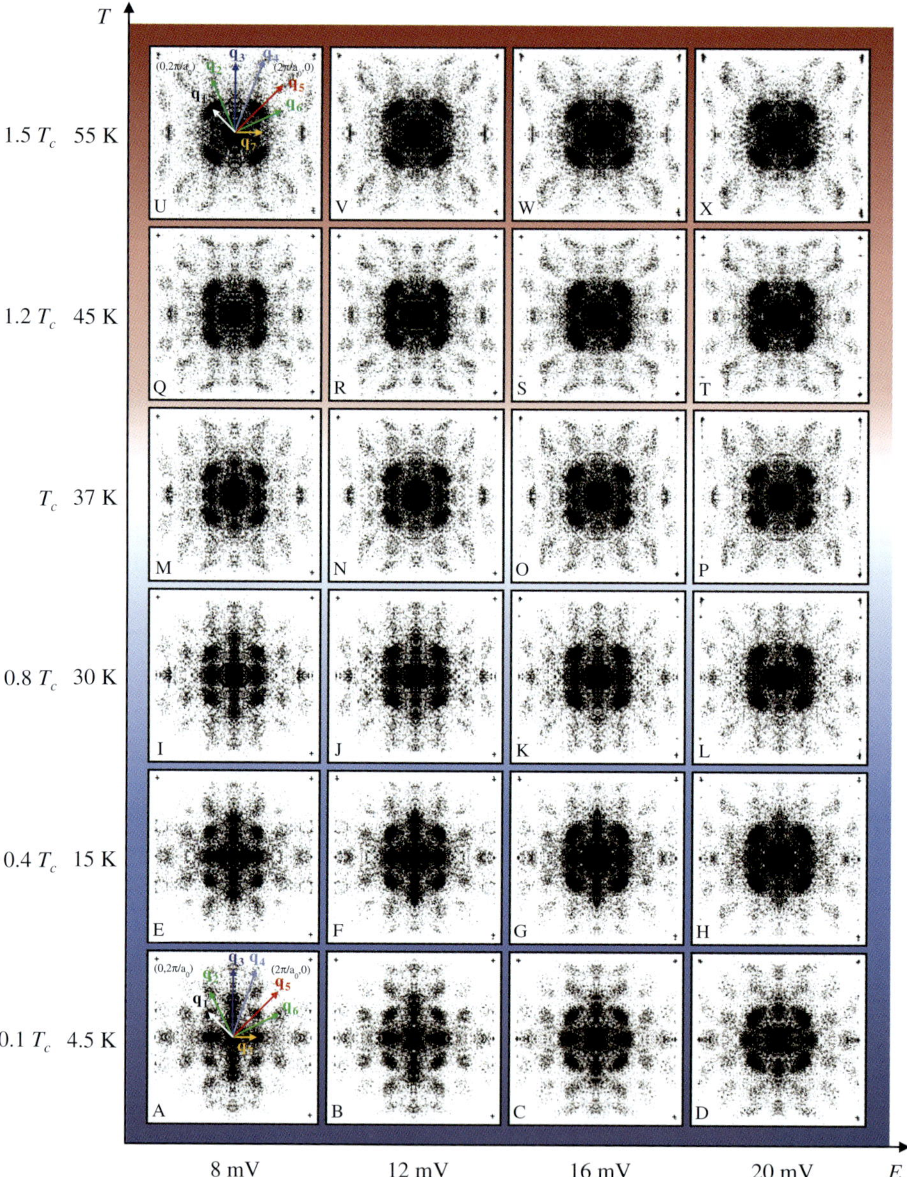

Fig. 16.6 Differential conductance maps $g(\mathbf{r},E)$ were obtained on the same sample in an atomically resolved and registered FOV $> 45 \times 45$ nm^2 at six temperatures. Each panel shown is the Fourier transform $Z(\mathbf{q}, E)$ of $Z(\mathbf{q}, E) \equiv \mathrm{g}(\mathbf{r},+E)/\mathrm{g}(\mathbf{r},-E)$ for a given energy and temperature. The QPI signals evolve dispersively with energy along the horizontal energy axis. The temperature dependence of QPI for a given energy evolves along the vertical axis. The octet-model set of QPI wave vectors is observed for every E and T, as seen, for example, by comparing (a) and (u), each of which has the labeled octet vectors. Within the basic octet QPI phenomenology, there is no particular indication in these data of where the superconducting transport T_c occurs.

3. the non-dispersive and locally symmetry breaking excitations at the $E{\sim}\Delta_1$ energy scale (Kohsaka *et al.*, 2007, 2008; Lawler, 2010; Lee *et al.*, 2009; McElroy *et al.*, 2005*a*) (which remain completely unaltered upon the transition between the dSC and the PG phases (Lawler, 2010; Lee *et al.*, 2009)).

16.7 Broken spatial symmetries of $E \sim \Delta_1$ states in both the dSC and PG phases

16.7.1 Atomic-scale imaging of the $E{\sim}\Delta_1$ pseudogap states

As described in Sections 5 and 6, underdoped cuprates exhibit an octet of dispersive Bogoliubov QPI wavevectors $\mathbf{q}_i(E)$ which always disappear above $E \cong \Delta_0$ to be replaced by a spectrum of non-dispersive states (Kohsaka *et al.*, 2007, 2008; Lawler, 2010; Lee *et al.*, 2009; McElroy *et al.*, 2005*a*) surrounding $E{\sim}\Delta_1$ (Fig. 16.4 (b)). The $Z(\mathbf{q}$,E$> \Delta_0)$ modulations exhibit the two non-dispersive $\mathbf{q}$-vectors, $\mathbf{q}_1^*$ and $\mathbf{q}_5^*$, which evolve with p as shown in Fig. 16.5. The $\mathbf{q}_1^*$ modulations appear as the energy transitions from below to above Δ_0 but then they disappear quickly, leaving only two primary electronic structure elements of the PG-energy electronic structure in $Z(\mathbf{r}, E \cong \Delta_1)$. These occur at $\mathbf{Q}_x$=(1,0)$2\pi/\mathrm{a}_0$ and $\mathbf{Q}_y$=(0,1)$2\pi/\mathrm{a}_0$, which are the Bragg peaks representing the periodicity of the unit cell, and $\mathbf{S}_x \equiv$(~3/4,0)$2\pi/\mathrm{a}_0$, $\mathbf{S}_y \equiv$(0,~3/4)$2\pi/\mathrm{a}_0$, which are due to the local breaking of lattice translation symmetry at the nanoscale. The doping evolution of $|\mathbf{S}_x| = |\mathbf{S}_y|$ (which is by definition that of $\mathbf{q}_5$*—see Fig. 16.5), as shown in Fig. 16.5 (d), indicates that these incommensurate modulations are linked to the doping-dependence of the extinction point of the arc of Bogoliubov QPI.

Atomically resolved $\mathbf{r}$-space images of the static phenomena in $Z(\mathbf{r}, E)$ show highly similar spatial patterns at all energies near Δ_1 but with variations of intensity due to the Δ_1 disorder (Fig. 16.3 (a)). By changing to reduced energy variables $e(\mathbf{r}) = E/\Delta_1(\mathbf{r})$ and imaging $Z(\mathbf{r},e)$ it becomes clear that these modulations exhibit a strong maximum in intensity at $e=1$. This is demonstrated directly in Fig. 16.7 where the relative intensity of the modulations (all in same units and contrast scales) exhibits a strong maximum at e=1 (Kohsaka *et al.*, 2008). Thus the PG states of underdoped cuprates locally break translational symmetry, and reduce the 90°-rotational (C_4) symmetry of states within the unit cell to a 180°-rotational symmetry (C_2) in a fashion which is centered on the planar oxygen sites (Kohsaka *et al.*, 2007, 2008; Lee *et al.*, 2009).

16.7.2 Universality of the broken symmetries of the $E{\sim}\Delta_1$ states

Theoretical concerns have been advanced about such spatial imaging of the cuprate PG states including the possibility of spurious rotation-symmetry-breaking due to the dopant atoms (Y. *et al.*, 2006). To address such issues, we carried out a sequence of identical experiments on two radically different cuprates at the same p: strongly underdoped $Ca_{1.88}Na_{0.12}CuO_2Cl_2$ (Na-CCOC; $T_c\sim$ 21 K) and $Bi_2Sr_2Dy_{0.2}Ca_{0.8}Cu_2O_{8+\delta}$ (Dy-Bi2212; $T_c{\sim}$45 K). These materials have completely different crystallographic

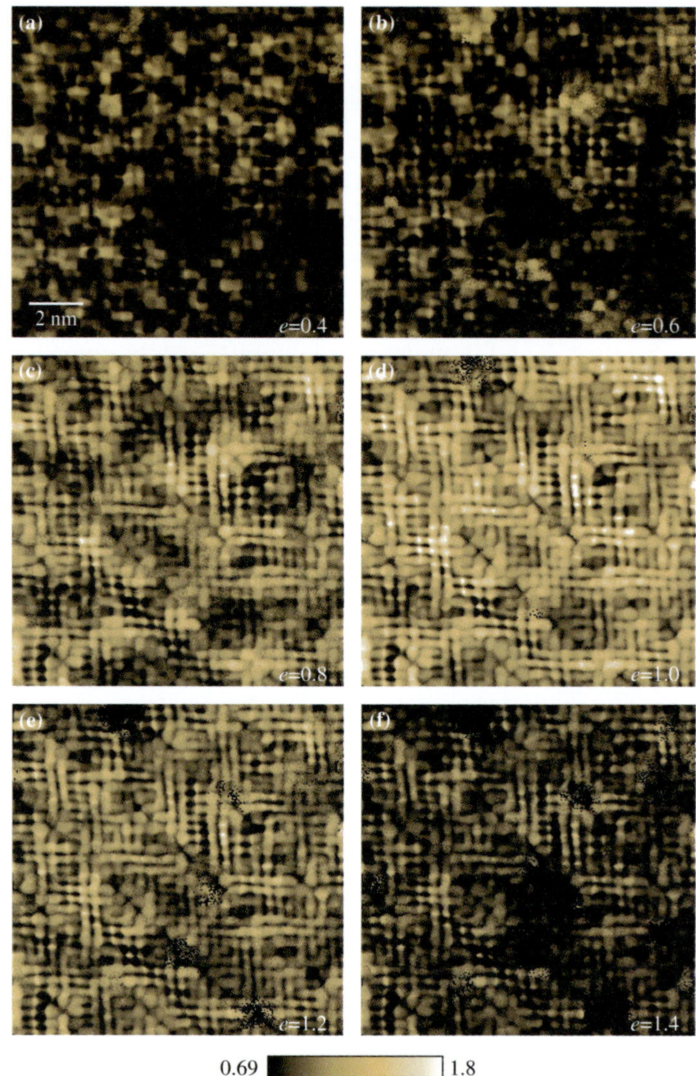

Fig. 16.7 A series of images displaying the real-space conductance ratio $Z(\mathbf{r}, E)$ as a function of energy, rescaled to the local pseudogap value, $e = E/\Delta_1(\mathbf{r})$. Each pixel location was rescaled independently of the others. The common color scale illustrates that the broken electronic symmetry patterns appear strongest in Z exactly at $E = \Delta_1(\mathbf{r})$, the local pseudogap energy.

structures, chemical constituents, dopant-ion species, and inequivalent dopant-ion sites within the crystal-termination layers lying between the CuO_2 plane and the STM tip (Kohsaka *et al.*, 2007). However, images of the $E\sim\Delta_1$ PG states for these two systems demonstrate virtually indistinguishable electronic structure arrangements (Kohsaka *et al.*, 2007). Obviously these symmetry-breaking effects within every CuO_2 unit cell cannot be governed by individual dopant ions because there is only a single such ion for every ~20 planar oxygen atoms in Dy-Bi2212. Moreover, the dopant ions occur at quite different locations (substitutional or interstitial respectively) in the unit cells of Na-CCOC and Dy-Bi2212. Thus the virtually identical phenomena in images of the atomic-scale broken symmetries $E\sim\Delta_1$ PG states in Na-CCOC and Dy-Bi2212 must occur because of the only common characteristic of these two radically different materials. Therefore $Z(\mathbf{r},e=1)$ images of the spatial structure of the cuprate PG states (Kohsaka *et al.*, 2007, 2008; Lawler, 2010; Lee *et al.*, 2009) should be ascribed to the intrinsic electronic structure of the CuO_2 plane.

16.7.3 Imaging the distinct broken spatial symmetries of the pseudogap $E\sim\Delta_1$ states

To explore which spatial symmetries are actually broken by the cuprate PG states, we use sub-unit-cell resolution $Z(\mathbf{r},e)$ imaging performed on multiple different under-doped $Bi_2Sr_2CaCu_2O_{8+\delta}$ samples with T_cs between 20 K and 55 K. The necessary registry of the Cu sites in each $Z(\mathbf{r},e)$ is achieved by a picometer-scale transformation that renders the topographic image $T(\mathbf{r})$ perfectly a_0-periodic. The same transformation is then applied to the simultaneously acquired $Z(\mathbf{r},e)$ to register all the electronic structure data to this ideal lattice. The topograph $T(\mathbf{r})$ is shown in Fig. 16.8 (a); the inset compares the Bragg peaks of its real (in-phase) Fourier components $\mathrm{Re}\,T(\mathrm{Q}_x)$, $\mathrm{Re}\,T(\mathrm{Q}_y)$ and demonstrates that $\mathrm{Re}\,T(\mathrm{Q}_x)/\,\mathrm{Re}\,T(\mathrm{Q}_y)=1$. Therefore $T(\mathbf{r})$ preserves the C_4 symmetry of the crystal lattice. In contrast, Fig. 16.8 (b) shows that the $Z(\mathbf{r},e=1)$ determined simultaneously with Fig. 16.8 (a) breaks various crystal symmetries (Kohsaka *et al.*, 2007, 2008; Lee *et al.*, 2009). The inset shows that since $\mathrm{Re}\,\tilde{Z}(\mathrm{Q}_x,e=1)\big/\mathrm{Re}\,\tilde{Z}(\mathrm{Q}_y,e=1)\neq 1$, the PG states break C_4 symmetry on the average throughout Fig. 16.8 (b). We defined a normalized order parameter over the entire FOV for intra-unit-cell electronic nematicity as a function of e:

$$O_N^Q(e) \equiv \frac{\mathrm{Re}\,\tilde{Z}(\mathbf{Q}_y,e)-\mathrm{Re}\,\tilde{Z}(\mathbf{Q}_x,e)}{\bar{Z}(e)} \tag{16.7}$$

where the spatial average of $Z(\mathbf{r},e)$, $\bar{Z}(e)$ ensures that any nematicity is measured relative to overall intensity. The plot of $O_N^Q(e)$ in Fig. 16.8 (c) shows that the magnitude of $O_N^Q(e)$ is low for $e << \omega/\Delta_0$, begins to grow near $e\sim\omega/\Delta_0$, and becomes well defined as $e\sim 1$ or $E\sim\Delta_1$. Thus the observed electronic nematicity is specific to the PG states.

To determine the source of these effects at atomic scale within the CuO_2 unit cell, we study $Z(\mathbf{r},e)$ with sub-unit-cell resolution. Figure 16.8 (d) shows the topographic image of a representative region from Fig. 16.8 (a); the locations of each Cu site $\mathbf{R}$ and of the two O atoms within its unit cell are indicated. Figure 16.8 (e) shows $Z(\mathbf{r},e)$

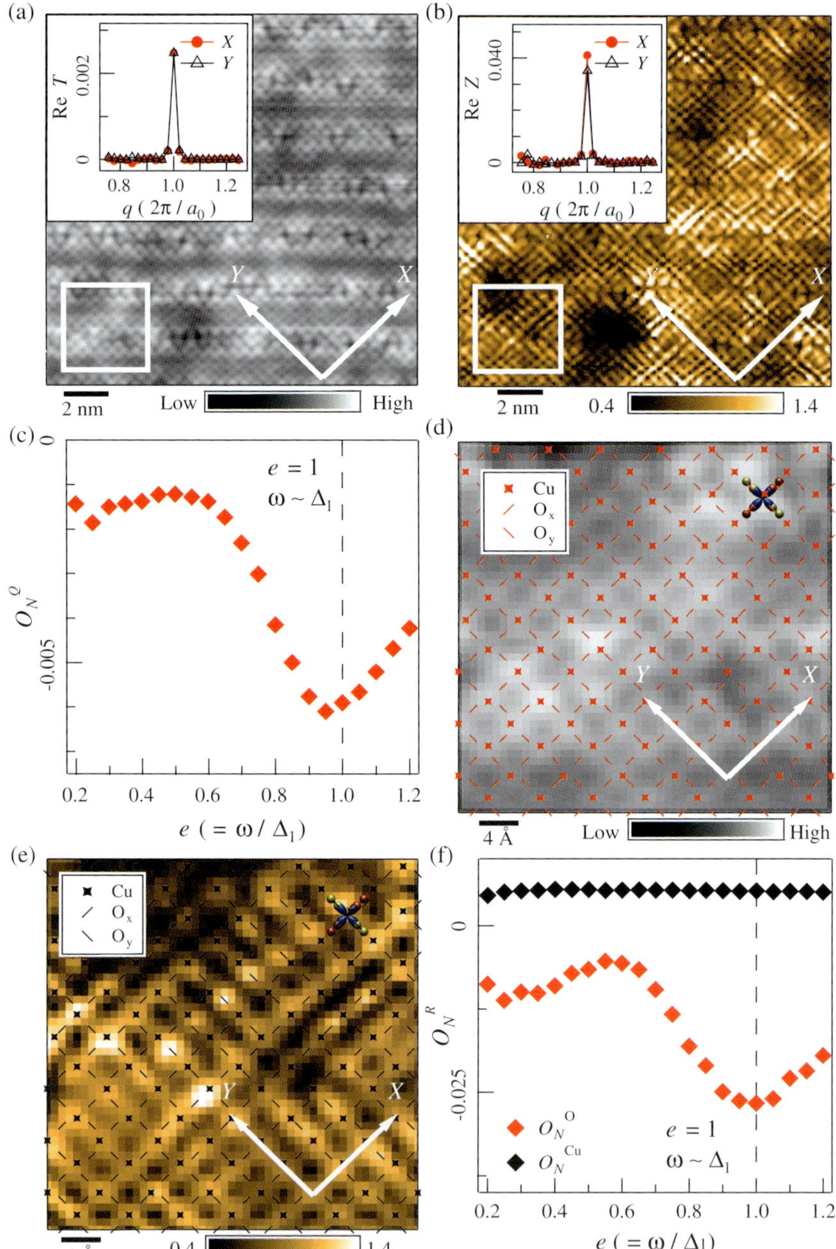

Fig. 16.8 (a) Topographic image $T(\mathbf{r})$ of the $Bi_2Sr_2CaCu_2O_{8+\delta}$ surface. The inset shows that the real part of its Fourier transform $\mathrm{Re}T(\mathbf{q})$ does not break C_4 symmetry at its Bragg points because plots of $T(\mathbf{q})$ show its values to be indistinguishable at $\boldsymbol{Q}_x = (1, 0)2\pi/a_0$ and $\boldsymbol{Q}_y = (0, 1)2\pi/a_0$. Importantly, this means that neither the crystal nor the tip used to image it

measured simultaneously with Fig. 16.8 (d) with same Cu and O site labels. Next we define

$$O_{\mathrm{N}}^{\mathrm{R}}(e) = \sum_{\mathbf{R}} \frac{Z_x(\mathbf{R}, e) - Z_y(\mathbf{R}, e)}{\bar{Z}(e)N} \tag{16.8}$$

where $Z_x(\mathbf{R}, e)$ is the magnitude of $Z(\mathbf{r}, e)$ at the O site $a_0/2$ along the x-axis from $\mathbf{R}$ while $Z_y(\mathbf{R}, e)$ is the equivalent along the y-axis, and N is the number of unit cells. This is the $\boldsymbol{r}$-space equivalent of $O_N^Q(e)$ in eqn. 7 but counting only O-site contributions. Figure 16.8 (e) contains the calculated value of $O_N^R(e)$ from the same FOV as Fig. 16.8 (a) and (b), revealing the good agreement with$O_N^Q(e)$. Thus the electronic nematicity of the cuprate PG states derives primarily from the inequivalence of electronic structure at the two O sites within each unit cell.

16.7.4 Separating Q=0 nematic from Q=$\mathbf{S}_x$, $\mathbf{S}_y$ smectic contributions to the $\boldsymbol{E}\sim\boldsymbol{\Delta}_1$ electronic structure

The smectic contributions to the $E\sim\Delta_1$ electronic structure can be examined by defining a measure analogous to eqn 7 of C_4 symmetry-breaking, but now in the modulations with $\mathbf{S_x}$, $\mathbf{S_y}$:

(and its $Z(\mathbf{r}, E)$ simultaneously) exhibits C_2 symmetry. (b) The $Z(\mathbf{r}, e = 1)$ image measured simultaneously with $T(\mathbf{r})$ in (a). The inset shows that the Fourier transform $Z(\mathbf{q}, e = 1)$ does break C_4 symmetry at its Bragg points because $\mathrm{Re}Z(\boldsymbol{Q}_x, e{\sim}1) \neq \mathrm{Re}Z(\boldsymbol{Q}_y, e{\sim}1)$. This means that, on average throughout the FOV of (a) and (b), the modulations of $Z(\mathbf{r}, E< \Delta_1)$ that are periodic with the lattice have different intensities along the $x-$axis and along the $y-$axis. (c) The value of $O_N^Q(e)$ defined in eqn. 7 computed from $Z(\mathbf{r}, e)$ data measured in the same FOV as (a) and (b). Its magnitude is low for all $e<\Delta_0$ and then rises rapidly to become well established near $e< 1$ or $E<\Delta_1$. Thus the quantitative measure of intra-unit-cell electronic nematicity reveals that the PG states in this FOV of a strongly underdoped $Bi_2Sr_2CaCu_2O_{8+\delta}$ sample are nematic. (d) Topographic image $T(\mathbf{r})$ from the region identified by a small white box in (a). It is labeled with the locations of the Cu atom plus both the O atoms within each CuO_2 unit cell (labels shown in the inset). Overlaid is the location and orientation of a Cu and four surrounding O atoms. The simultaneous $Z(\mathbf{r}, e = 1)$ image in the same FOV as (d) (the region identified by small white box in (b)) showing the same Cu- and O- site labels within each unit cell (see inset). Thus the physical locations at which the nematic measure $O_N^R(e)$ of eqn 8 is evaluated are labeled by the dashes. Overlaid is the location and orientation of a Cu atom and four surrounding O atoms. The value of O_N^R (e) is computed from $Z(\mathbf{r}, e)$ data measured in the same FOV as (a) and (b). As in (c), its magnitude is low for all $E<\Delta_0$ and then rises rapidly to become well established at $e< 1$ or $E<\Delta_1$.

$$O_S^Q(e) = \frac{\mathrm{Re}\,\tilde{Z}(\mathrm{S}_y, e) - \mathrm{Re}\,\tilde{Z}(\mathrm{S}_x, e)}{\bar{Z}(e)} \tag{16.9}$$

For all samples studied, the low values found for $\left|O_S^Q(e)\right|$ at low e occur because these states are dispersive Bogoliubov quasiparticles (Hanaguri *et al.*, 2007; Hoffman *et al.*, 2002; Lee *et al.*, 2009; McElroy *et al.*, 2003; Wise *et al.*, 2008), and cannot be analyzed in terms of any static electronic structure, smectic or otherwise. More importantly $\left|O_S^Q(e)\right|$ shows no tendency to become well established at the PG or any other energy (Lawler, 2010).

To visualize the separate broken symmetries in the $E{\sim}\Delta_1$ electronic structure, we consider Z($\mathbf{q}$, e=1) in Fig. 16.9 (a); this is the Fourier-space representation of electronic structure of the $E{\sim}\Delta_1$ states. Taking into account only the Bragg peaks at $\mathbf{Q}_x$, $\mathbf{Q}_y$ (red circles in Fig. 16.9 (a)) the C_2 symmetry of $\mathbf{Q}$=0 intra-unit-cell electronic structure is revealed as shown, for example, by Fig. 16.9 (b). In contrast, if one focuses upon the incommensurate modulations $\mathbf{S}_x$, $\mathbf{S}_y$ (blue circles in Fig. 16.9 (a)), we find a disordered electronic structure with incommensurate modulations, which break both C_2 and translational symmetry locally as shown, for example, in Fig. 16.9 (c). Although these two types of electronic phenomena represent completely different broken symmetries, they coexist in the $E{\sim}\Delta_1$ PG electronic structure of underdoped cuprates.

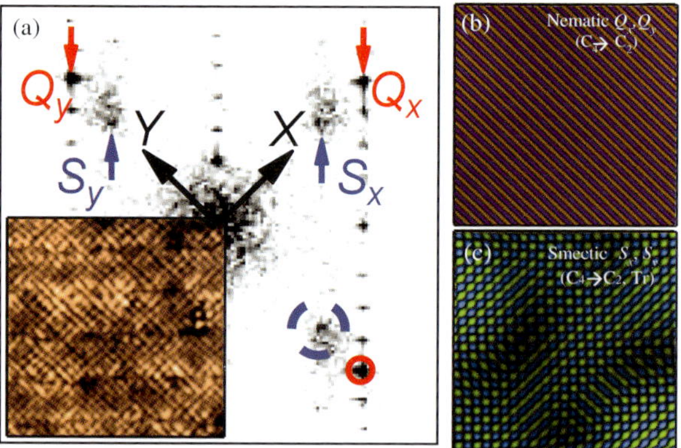

Fig. 16.9 (a) The Fourier transform Z($\mathbf{q}$,e= 1) of a typical image Z($\mathbf{r}$,e= 1) of the spatial structure of the pseudogap states in underdoped $Bi_2Sr_2CaCu_2O_{8+\delta}$. The Bragg peaks are identified by red circles and $\mathbf{Q}_x$, $\mathbf{Q}_y$ labels. The wavevectors of the smectic modulations in electron structure are identified by blue circles and $\mathbf{S}_x$, $\mathbf{S}_y$ labels. (b) The spatial information in the Bragg peaks $\mathbf{Q}_x$, $\mathbf{Q}_y$ alone reveals the intra unit-cell C_2-symmetric electronic structure. (c) The spatial information in the $\mathbf{S}_x$, $\mathbf{S}_y$ wavevectors alone reveals the disordered breaking of both rotational and translational symmetry.

16.7.5 Summary

When SI-STM images in $Z(\mathbf{r}, E)$ of the intra-unit-cell states in underdoped $Bi_2Sr_2CaCu_2O_{8+\delta}$ are analyzed using two independent techniques, compelling evidence for intra-unit-cell (or $\mathbf{Q}$=0) electronic nematicity specifically of the states at the $E{\sim}\Delta_1$ PG energy is detected. This broken electronic symmetry occurs due to electronic inequivalence at the two oxygen sites within each CuO_2 unit cell. Moreover, this $\mathbf{Q}$=0 electronic nematicity coexists with finite $\mathbf{Q}$=$\mathbf{S}$ smectic electronic modulations, but they can be analyzed separately by using Fourier filtration techniques. The wavevector of smectic electronic modulations is controlled by the point in $\mathbf{k}$-space where the Bogoliubov interference signature disappears when the arc supporting delocalized Cooper pairing approaches the lines between $\mathbf{k}$=$\pm(0,\pi_0\, a_1)$ and $\mathbf{k}$ =$\pm(\pi_0 a_1,0)$ (see Fig. 16.5 (b) and (d). This appears to indicate that the $\mathbf{Q}$=$\mathbf{S}$ smectic effects are dominated by the same $\mathbf{k}$-space phenomena that restrict the regions of Cooper pairing (Kohsaka *et al.*, 2008).

16.8 Overview, conclusions, and future

16.8.1 Bipartite electronic structure of underdoped cuprates derived from SI-STM

A clearer picture of the fundamentally bipartite electronic structure of strongly underdoped cuprates approaching the Mott insulator emerges from these SI-STM studies. This is summarized in Fig. 16.10. In the dSC phase (Fig. 16.10 (a)–(c)) the Bogoliubov QPI signature of delocalized Cooper pairs (Section 16.5 and Fig. 16.10 (c)) exists upon the arc in $\mathbf{k}$-space labeled by region II in Fig. 16.10 (a). The Bogoliubov QPI disappears near the lines connecting $\mathbf{k}$=$(0,\pm\pi/a_0)$ to $\mathbf{k}$=$(\pm\pi/a_0,0)$, thus defining a $\mathbf{k}$-space arc which supports the delocalized Cooper pairing. This arc shrinks rapidly towards the $\mathbf{k}$=$(\pm\pi/2a_0,\pm\pi/2a_0)$ points with falling hole-density, in a fashion that would satisfy Luttinger's theorem if it is actually a hole-pocket bounded behind by the $\mathbf{k}$=$\pm(\pi_1 a_0,0)$ - $\mathbf{k}$=$\pm$ $(0,\pi_1 a_0)$ lines. The $E{\sim}\Delta_1$ PG excitations (Section 16.7) are labeled by region I in Fig. 16.10 (a) and exhibit a radically different $\mathbf{r}$-space phenomenology locally breaking the C_4 symmetry down to C_2 within each CuO_2 unit cell (Fig. 16.10 (b)). In the PG phase (Fig. 16.10 (d)–(f)), the Bogoliubov QPI signature (Section 16.6 and Fig. 16.10 (f)) exists upon part of the same arc in $\mathbf{k}$-space as it did in the dSC phase. This is labeled as region II in Fig. 16.10 (d). Here, however, since the ungapped Fermi arc (region III) predominates, the gapped region supporting *d*-wave QPI has shrink into a narrow sliver near a line connecting $\mathbf{k}$=$(\pi_1 a_0,0)$ and $\mathbf{k}$=$(0,\pi_1/a_0)$ (Fig. 16.10 (d)). The $E{\sim}\Delta_1$ excitations in the PG phase (Section 16.7) are again labeled by region I in Fig. 16.10 (d) and exhibit locally broken symmetries indistinguishable from those in the dSC phase (Fig. 16.10 (e)).

16.8.2 Microscopic mechanism of Q=0 intra-unit-cell electronic nematicity

The microscopic source of the $\mathbf{Q}$=0 intra-unit-cell electronic nematicity in $E{\sim}\Delta_1$ states (Fig. 16.8) is unknown at present. One important point to consider is the

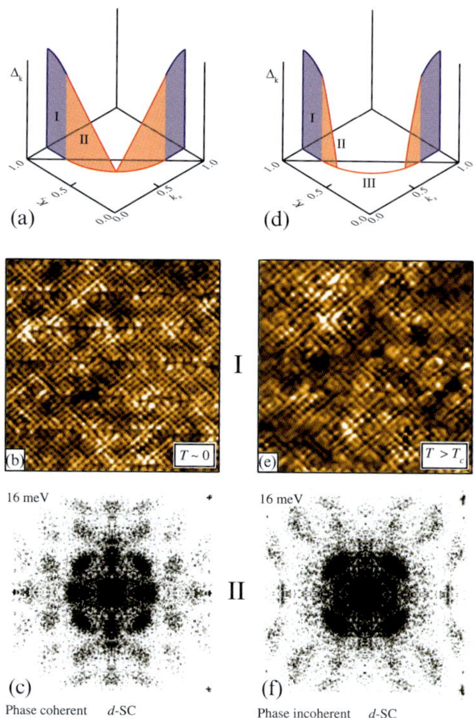

Fig. 16.10 (a) A schematic representation of the electronic structure in one quarter of the Brillouin zone at lowest temperatures in the dSC phase. The region marked II in front of the line joining $\mathbf{k}=(\pi/a_0,0)$ and $\mathbf{k}=(0,\pi/a_0)$ is the locus of the Bogoliubov QPI signature of delocalized Cooper pairs. (b) An example of the broken spatial symmetries that are concentrated upon pseudogap energy $E\sim\Delta_1$, as measured at lowest temperatures. (c) An example of the characteristic Bogoliubov QPI signature of 16 pairs of interference wavevectors, all dispersive and internally consistent with the octet model as well as particle—hole-symmetric $\mathbf{q}_i(+\mathrm{E})=\mathbf{q}_i(-\mathrm{E})$, here measured at lowest temperatures. (d) A schematic representation of the electronic structure in one quarter of the Brillouin zone at $T\sim1.5T_c$ in the PG phase. The region marked III is the Fermi arc, which is seen in QPI studies as a set of interference wavevectors $\mathbf{q}_i(E=0)$, which indicate that there is no gap-node at $E=0$. Region II in front of the line joining $\mathbf{k}=(\pi/a_0,0)$ and $\mathbf{k}=(0,\pi/a_0)$ is the locus of the Bogoliubov QPI signature. Here all wavevectors of the octet model are detected and found to be dispersive. Thus although the sample is not a long-range phase-coherent superconductor, it does give strong clear QPI signatures of *d*-wave pairing. (e) An example of the broken spatial symmetries that are concentrated upon pseudogap energy $E\sim\Delta_1$ as measured in the PG phase; they are indistinguishable from measurements at $T\sim0$. (f) An example of the characteristic Bogoliubov QPI signature of sixteen pairs of interference wavevectors, all dispersive and internally consistent with the octet model as well as particle—hole-symmetric $\mathbf{q}_i(+\mathrm{E})=\mathbf{q}_i(-\mathrm{E})$, here measured at $T\sim1.5T_c$.

relationship between ARPES, elastic neutron scattering (NS) and SI-STM studies of broken electronic symmetries of the PG phase. ARPES reveals spontaneous dichroism of antinodal states (Kaminski *et al.*, 2002) which exhibit C_2 symmetry because the opposite sign of the effect occurs at $\mathbf{k}=(\pi/a_0,0)$ and $\mathbf{k}=(0,\pi/a_0)$ (Kaminski *et al.*, 2002). The $\mathbf{Q}=0$ magnetic order detected by NS at the Bragg peak (Fauqué *et al.*, 2006; Li *et al.*, 2008) consists of intra-unitcell, apparently antiferromagnetic, C_2 symmetric states in both $YBa_2Cu_3O_{6+x}$ and $HgBa_2CuO_{4+\delta}$. The SI-STM studies described in this review also reveal intra-unit-cell, C_2 symmetric states at the PG energy and show that these effects are associated primarily with electronic inequivalence at the two O sites within the CuO_2 unit cell (Section 16.7). With such commonality between the results from such disparate techniques it is not implausible they are detecting different characteristics of the same broken symmetry states. In that case, one immediate implication is that a three-band model using both the $d_{x^2-y^2}$ states at Cu and the $2p_\sigma$ states of O may be necessary to accurately describe the electronic structure of underdoped cuprates and to explain the relationship (if it exists) between the ARPES, NS and SI-STM observations of $\mathbf{Q}=0$ C_2 symmetry electronic structure.

16.8.3 Relationship between the nematic and smectic electronic structures and to superconductivity

Both nematic and smectic broken symmetries have been reported in the electronic structure of different cuprate compounds (Gu *et al.*, 2008; J.M. *et al.*, 1995; Tranquada *et al.*, 2004) A spin/charge smectic broken-symmetry phase (stripes) exists in $La_{2-x}Sr_xCuO_4$ and $La_{2-x}Ba_xCuO_4$ when $p{\sim}0.125$. Nematic broken symmetry has been reported in underdoped $YBa_2Cu_3O_{6+\delta}$ (Fauqué *et al.*, 2006) underdoped $Bi_2Sr_2CaCu_2O_{8+\delta}$ and underdoped $HgBa_2CuO_{4+x}$ (Li *et al.*, 2008). To understand how both these broken symmetry states can coexist, and the form of their interactions, will be important in unraveling the mystery of the cuprate phase diagram. That both types of broken symmetry coexist at the nanoscale in $Bi_2Sr_2CaCu_2O_{8+\delta}$ (Lawler (2010) and Fig. 16.9) represents an important new opportunity to address their interactions. Should that be possible, the next challenge for SI-STM would then be to demonstrate directly the relationship between the superconductivity and the two broken symmetries of the $E{\sim}\Delta_1$ PG states with the (ambitious) view towards a complete Ginzburg–Landau understanding of the cuprate phase diagram.

16.8.4 Electronic structure of the cuprate pseudogap phase

Among the explanations for the PG phase is that it is a spin liquid created by hole-doping an antiferromagnetic Mott insulator, or that it is a *d*-wave superconductor without phase coherence, or that it is an electronic ordered phase with additional broken symmetries. SI-STM reveals that the basic particle–hole-symmetric, dispersive, octet phenomenology is consistent with theoretical predictions for the QPI characteristics of a phase-incoherent *d*-wave superconductor (Fig. 16.10 (f)). Furthermore, since all the $\mathbf{q}_i(E)$ i=1,...7 disperse internally consistently with the octet-model, they cannot represent the signature of any static ordered state of fixed wavevector $\mathbf{Q}^*$. Thus the low-energy $E<\Delta_0$ electronic structure of the PG phase (which is what is probed by transport and thermodynamics) is indeed that of a phase-incoherent d-wave superconductor. Nevertheless, the high-energy electronic states at the PG energy scale $E{\sim}\Delta_1$ exhibit strongly broken symmetries, including $\mathbf{Q}$=0 nematicity and finite-$\mathbf{Q}$ smectic modulations (Fig. 16.9). Finally, the truncated arc of Bogoliubov QPI seen below T_c, which remains unchanged in the PG phase except for the appearance of an ungapped portion, can be indicative of spin-liquid phenomenology (see below). Thus the characteristics of the PG phase from SI-STM contain some elements of all three theoretical approaches to the electronic structure of hole-doped CuO_2 approaching the Mott insulator.

16.8.5 Fundamental electronic structure of the hole-doped CuO_2 Mott insulator from SISTM

The overall cuprate electronic structure derived from SI-STM studies (Fig. 16.10) raises a number of questions. Why, with increasing energy, does the Bogoliubov QPI signature of delocalized Cooper pairs disappear (Kohsaka *et al.*, 2008) near the $\mathbf{k}=(0,\pm\pi/a_0)$ – $\mathbf{k}=(\pm\pi/a_0,0)$ connecting lines? And why do the PG states $E{\sim}\Delta_1$ exhibit such dramatically different symmetries (Kohsaka *et al.*, 2007; Lawler, 2010) to the Bogoliubov quasiparticles at $E<\Delta_0$? One reason could be that the $\mathbf{r}$-space electronic structure has undergone a $\sqrt{2}\times\sqrt{2}$ reconstruction due to the appearance of a coexisting long-range ordered state. The arcs supporting Cooper pairing would then represent one side of a hole-pocket within a reduced Brillouin zone. But neither antiferromagnetism nor other long-range ordered electronic phases (Chakravarty *et al.*, 2001; Lee *et al.*, 2006*b*) necessary for such a reconstruction are detected in $Bi_2Sr_2CaCu_2O_{8+\delta}$. A related explanation could be inelastic scattering of the quasiparticles by spin fluctuations (Graser *et al.*, 2008; Lee *et al.*, 2008) at $\mathbf{Q}=(\pi/a_0,\pi/a_0)$ or by fluctuations of other ordered states that would exhibit a $\sqrt{2}\times\sqrt{2}$ reconstruction if stabilized. Neither of these approaches explains the broken spatial symmetries of the $E{\sim}\Delta_1$ PG states, however. Another type of explanation could be a spin-charge stripe glass (Kohsaka *et al.*, 2007) coexisting with superconductivity (Balents *et al.*, 2005; Granath, 2008; M. and O., 2008; Tešanović, 2008). This could explain the loss of translational symmetry and reduction of C_4 to C_2 symmetry within the $E{\sim}\Delta_1$ PG states, but it does not (yet) explain the $\mathbf{Q}$=0 electronic nematicity nor the disappearance of quasiparticle interference along the $\mathbf{k}=(0,\pm\pi/a_0)$ – $\mathbf{k}=(\pm\pi/a_0,0)$ lines, although smectic fluctuations may provide an explanation for the latter (Xu *et al.*, 2000). Yet another proposal, that orbital charge currents exist within each CuO_2 unit-cell (Varma, 2006), receives support from NS experiments (Fauqué *et al.*, 2006; Li *et al.*, 2008) and may provide an explanation for the $\mathbf{Q}$=0 nematicity discussed here. But it does not (yet) explain the finite-$\mathbf{Q}$ smectic modulations or the disappearance of Bogoliubov QPI near $\mathbf{k}=(0,\pm\pi/a_0)$ – $\mathbf{k}=(\pm\pi/a_0,0)$ lines. Yet another possibility, which is revealed by the fact that the Luttinger theorem can be satisfied by using the region bounded the Bogoliubov QPI arcs and the $\mathbf{k}=(0,\pm\pi/a_0)$ – $\mathbf{k}=(\pm\pi/a_0,0)$ lines (Kohsaka *et al.*, 2008), is that many of the effects summarized in Fig. 16.10 are to be expected as the basic properties of a hole-doped spin liquid (K-Y. *et al.*, 2006). This approach can explain (at least phenomenologically) the Bogoliubov arc termination as due to zeros of Green's-function poles along the $\mathbf{k}=(0,\pm\pi/a_0) - \mathbf{k}=(\pm\pi/a_0,0)$ lines (K-Y. *et al.*, 2006; Kohsaka *et al.*, 2008), how the Luttinger theorem can be satisfied given the exotic **k**-space structure observed (K-Y. *et al.*, 2006; Kohsaka *et al.*, 2008), and possibly the cause of smectic finite-**Q** non-dispersive modulations (E. and B., 2008). However it does not (yet) appear to explain $\mathbf{Q}=0$ electronic nematicity.

Under these circumstances, a key recognition necessary for progress may be that there are several distinct electronic phenomena, generated by different microscopic influences, which are coexisting with each other at low hole-density in the CuO2 phase diagram. The key challenge then becomes the development of a quantitative understanding of these coexisting and competing electronic phenomena in underdoped cuprates. For example, by discovering 2π topological defects within the phase-fluctuating smectic states and that they are associated with the spatial fluctuations of the robust intra-unit-cell nematicity, we have demonstrated empirically a coupling between these two locally broken electronic symmetries of the cuprate pseudogap states (Mesaros *et al.*, 2011). This allowed identification of a Ginzburg–Landau functional that explains how these phenomena coexist and predicts their interplay at the atomic scale. If such a tendency for an intra-unit-cell nematicity to coexist with a disordered electronic smectic is ubiquitous to underdoped cuprates, which broken symmetry manifests at the macroscopic scale then depends on the coefficients in the relevant GL functional and on other material specific aspects such as crystal symmetry.

Acknowledgements

We acknowledge and thank our collaborators J.W. Alldredge, T. Hanaguri, P.J. Hirschfeld, J.E. Hoffmann, E.W. Hudson, Chung-Koo Kim, Y. Kohsaka, K.M. Lang, Jhinhwan Lee, Jinho Lee, V. Madhavan, K. McElroy, S.H. Pan, J. Slezak, J. Sethna, H. Takagi, C. Taylor, P. Wahl, and M. Wang. This project was supported by the Center for Emergent Superconductivity, an Energy Frontier Research Center funded by the U.S. Department of Energy, Office of Basic Energy Sciences under Award Number DE-2009-BNL-PM015.

References

Alldredge, J.W., Lee, J., McElroy, K., Wang, M., Fujita, K., Kohsaka, Y., Taylor, C., Eisaki, H., Uchida, S., and Hirschfeld, P.J. (2008). *Nat. Phys.*, **4**, 319–326.

Anderson, P.W. (1950). *Phys. Rev.*, **79**, 350.

Anderson, P.W. (1987). *Science*, **6**, 1196.

Anderson, P.W., Lee, P.A., Randeria, M., Rice, T.M., Trivedi, N., and Zhang, F.C. (2004). *J. Phys. Cond. Matt.*, **16**, R755–R769.

Balents, L., Bartosch, L., Burkov, A., Sachdev, S., and Sengupta, K. (2005). *Phys. Rev. B*, **71**, 144509.

Berg, E. and Altman, E. (2007). *Phys. Rev. Lett.*, **99**, 247001.

Bergeal, N., Lesueur, J., Aprili, M., Faini, G., Contour, J.P., and Leridon, B. (2008). *Nature Phys.*, **4**, 608–611.

Boyer, M.C., Wise, W.D., Chatterjee, K., Yi, M., Kondo, T., Takeuchi, T., Ikuta, H., and Hudson, E.W. (2007). *Nat. Phys.*, **3**, 802–806.

Campuzano, J.C., Norman, M.R., and Randeria, M. (2004). *The Physics of Superconductors II.* Springer, New York.

Capriotti, L., Scalapino, D.J., and Sedgewick, R.D. (2003). *Phys. Rev. B*, **68**, 014508.

Carlson, E.W., Kivelson, S.A., Emery, V.J., and Manousakis, E. *Phys. Rev. Lett.*, **83**, 612–615.

Chakravarty, S., Laughlin, R.B., Morr, D.K., and Nayak, C. (2001). *Phys. Rev. B*, **63**, 094503.

Chatterjee, U., Shi, M., Kaminski, A., Kanigel, A., Fretwell, H.M., Terashima, K., Takahashi, T., Rosenkranz, S., Li, Z.Z., and Raffy, H. (2006). *Phys. Rev. Lett.*, **96**, 107006.

Chatterjee, U., Shi, M., Kaminski, A., Kanigel, A., Fretwell, H.M., Terashima, K., Takahashi, T., Rosenkranz, S., Li, Z.Z., and Raffy, H. (2007). *Phys. Rev. B*, **76**, 012504.

Chen, C.T., Sette, F., Ma, Y., Hybertsen, M.S., Stechel, E.B., Foulkes, W.M.C., Schulter, M., Cheong, S.-W., Cooper, A.S., Rupp, L.W.Jr, Batlogg, B., Soo, Y.L., Ming, Z.H., Krol, A., and Kao, Y.H. (1991). *Phys. Rev. Lett.*, **66**, 104.

Corson, J., Mallozzi, R., Orenstein, J., Eckstein, J.N., and Bozovic, I. (1999). *Nature*, **398**, 221–223.

Cren, T., Rditchev, D., Sacks, W., and Lkein, J. (2001). *Euro. Phys. Lett.*, **54**, 84–90.

Damascelli, A., Hussain, Z., and Shen, Z.-X. (2003). *Rev. Mod. Phys.*, **75**, 473.

Deutscher, G. (1999). *Nature*, **397**, 410–412.

E., Bascones and B., Valenzuela (2008). *Phys. Rev. B*, **77**, 024527.

Eisaki, H., Kaneko, N., Feng, D.L., Damascelli, A., Mang, P.K., Shen, K.M., Shen, Z.X., and Greven, M. (2004). *Phys. Rev. B*, **69**, 064512.

Emery, V.J. and Kivelson, S.A. (1995). *Nature*, **374**, 434–437.

Emery, V.J., Kivelson, S.A., and Tranquada, J.M. (1999). *Proc. Natl. Acad. Sci. USA*, **96**, 8814–8817.

Fauqué, B., Sidis, Y., Hinkov, V., Pailhes, S., Lin, C.T., Chaud, X., and Bourges, P. (2006). *Phys. Rev. Lett.*, **96**, 197001.

Fischer, O., Kugler, M., Maggio-Aprile, I., Berthod, C., and Renner, Ch. (2007). *Rev. Mod. Phys.*, **79**, 353–419.

Fradkin, E., Kivelson, S.A., Lawler, M.J., Eisenstein, J.P., and Mackenzie, A.P. (2010). *Annu. Rev. Condens. Matter Phys.*, **1**, 153–178.

Franz, M. and Millis, A.J. (1998). *Phys. Rev. B*, **58**, 14572–14580.

Gedik, N., Orenstein, J., Liang, R., Bonn, D.A., and Hardy, W.N. (2003). *Science*, **300**, 1410–1412.

Gomes, K.K., Pasupathy, A.N., Pushp, A., Ono, S., Ando, Y., and Yazdani, A. (2007). *Nature*, **447**, 569–572.

Granath, M. (2008). *Phys. Rev. B*, **77**, 165128.

Graser, S., Hirschfeld, P.J., and Scalapino, D.J. (2008). *Phys. Rev. B*, **77**, 184504.

Gu, Kim Y.-J., G.D., Gog, T., and Casa, D. (2008). *Phys. Rev. B*, **77**, 064520.

H.-B. Yang, J.D. Ramaeu, Z.-H. Pan G.D. Gu P.D. Johnson R.H. Claus D.G. Hinks and Kidd, T.E. arXiv 1008.3121.

Hanaguri, T., Kohsaka, Y., Davis, J.C., Lupien, C., Yamada, I., Azuma, M., Takano, M., Ohishi, K., Ono, M., and Takagi, H. (2007). *Nat. Phys.*, **3**, 865–871.

He, Y., Nunner, T.S., Hirschfeld, P.J., and Cheng, H.-P. (2006). *Phys. Rev. Lett.*, **96**, 197002.

Hoffman, J.E., McElroy, K., Lee, D.H., Lang, K.M., Eisaki, H., Uchida, S., and Davis, J.C. (2002). *Science*, **297**, 1148–1151.

Honerkamp, C., Fu, H.C., and Lee, D.-H. (2007). *Phys. Rev. B*, **75**, 014503.

Howald, C., Fournier, P., and Kapitulnik, A. (2001). *Phys. Rev. B*, **64**, 100504.

Hufner, S., Hossain, M.A., Damascelli, A., and Sawatzky, G.A. (2008). *Prog. Phys.*, **71**, 062501.

Jenkins, N., Fasano, Y., Berthod, C., Maggio-Aprile, I., Piriou, A., Giannini, E., Hoogenboom, B.W., Hess, C., Cren, T., and Fischer, Ø. (2009). *Phys. Rev. Lett.*, **103**, 227001.

J.M., Tranquada, Sternlieb, B.J., Axe, J.D., Nakamura, Y., and Uchida, S. (1995). *Nature*, **375**, 561–563.

K-Y., Yang, T.M., Rice, and F.-C., Zhang (2006). *Phys. Rev. B*, **73**, 174501.

Kaminski, A., Rosenkranz, S., Fretwell, H.M., Campuzano, J.C., Li, Z., Raffy, H., Cullen, W.G., You, H., Olson, C.G., and Varma, C.M. (2002). *Nature*, **416**, 610–613.

Kaneshita, E., Martin, I., and Bishop, A.R. (2004). *Phys. Soc. Japan Lett.*, **73**, 3223–3226.

Kanigel, A., Chatterjee, U., Randeria, M., Norman, M.R., Souma, S., Shi, M., Li, Z.Z., Raffy, H., and Campuzano, J.C. (2007). *Phys. Rev. Lett.*, **99**, 157001.

Kanigel, A., Norman, M.R., Randeria, M., Chatterjee, U., Souma, S., Kaminski, A., Fretwell, H.M., Rosenkranz, S., Shi, M., and Sato, T. (2006). *Nat. Phys.*, **2**, 447–451.

Khasanov, R., Kondo, T., Bendele, M., Hamaya, Y., Kaminski, A., Lee, S.L., Ray, S.J., and Takeuchi, T. (2010). *Phys. Rev. B*, **82**, 020511(R).

Kim, E.-A., Lawler, M.J., Oreto, P., Sachdev, S., Fradkin, E., and Kivelson, S.A. (2008). *Phys. Rev. B*, **77**, 184514.

Kivelson, S.A., Bindloss, I.P., Fradkin, E., Oganesyan, V., Tranquada, J.M., Kapitulnik, A., and Howald, C. (2003). *Rev. Mod. Phys.*, **75**, 1201–1241.

Kivelson, S.A., Fradkin, E., and Emery, V.J. (1998). *Nature*, **393**, 550–553.

Kohsaka, Y., Taylor, C., Fujita, K., Schmidt, A., Lupien, C., Hanaguri, T., Azuma, M., Takano, M., Eisaki, H., and Takagi, H. (2007). *Science*, **315**, 1380–1385.

Kohsaka, Y., Taylor, C., Wahl, P., Schmidt, A., Lee, J., Fujita, K., Alldredge, J.W., McElroy, K., Lee, J., and Eisaki, H. (2008). *Nature*, **454**, 1072–1078.

Kondo, T., Khasanov, R., Takeuchi, T., Schmalian, J., and Kaminski, A. (2009). *Nature*, **457**, 296–300.

Kotliar, G. (1988). *Phys. Rev. B*, **37**, 3664–3666.

Kwon, H.-J., Dorsey, A.T., and Hirschfeld, P.J. (2001). *Phys. Rev. Lett.*, **86**, 3875–3878.

Lang, K.M., Madhavan, V., Hoffman, J.E., Hudson, E.W., Eisaki, H., Uchida, S., and Davis, J.C. (2002). *Nature*, **415**, 412–416.

Lawler, M.J. et al. (2010). *Nature*, **466**, 374–351.

Le Tacon, M., Sacuto, A., Georges, A., Kotliar, G., Gallais, Y., Colson, D., and Forget, A. (2006). *Nature Phys.*, **2**, 537–543.

Lee, J., Fujita, K., McElroy, K., Slezak, J.A., Wang, M., Aiura, Y., Bando, H., Ishikado, M., Masui, T., and Zhu, J.X. (2006*a*). *Nature*, **442**, 546–550.

Lee, J., Fujita, K., Schmidt, A.R., Kim, C.K., Eisaki, H., Uchida, S., and Davis, J.C. (2009). *Science*, **325**, 1099–1103.

Lee, P.A., Nagaosa, N., and Wen, X.-G. (2006*b*). *Rev. Mod. Phys.*, **78**, 17–85.

Lee, W. C., Sinova, J., Burkov, A.A, Joglekar, Y., and MacDonald, A.H. (2008). *Phys. Rev. B*, **77**, 214518.

Li, L., Wang, Y., Naughton, M.J., Ono, S., Ando, Y., and Ong, N.P. (2005). *Europhys. Lett.*, **72**, 451–457.

Li, Y., V Balédent, N.B., Cho, Y., Fauqué, B., Sidis, Y., Yu, G., Zhao, X., and P Bourges, M.G. (2008). *Nature*, **455**, 372–375.

M., Vojta and O., Rsch (2008). *Phys. Rev. B*, **77**, 094504.

Maekawa, S. (2004). *Physics of Transition Metal Oxides.* Springer-Verlag, Berlin.

Martin, I. and Balatsky, A.V. (2001). *Physica C*, **357**, 46–48.

Matsuda, A., Fujii, T., and Watanabe, T. (2003). *Physica C*, **388**, 207–208.

McElroy, K., Gweon, G.H., Zhou, S.Y., Graf, J., Uchida, S., Eisaki, H., Takagi, H., Sasagawa, T., Lee, D.H., and Lanzara, A. (2006). *Phys. Rev. Lett.*, **96**, 067005.

McElroy, K., Lee, D.H., Hoffman, J.E., Lang, K.M., Lee, J., Hudson, E.W., Eisaki, H., Uchida, S., and Davis, J.C. (2005*a*). *Phys. Rev. Lett.*, **94**, 197005.

McElroy, K., Lee, J., Slezak, J.A., Lee, D.H., Eisaki, H., Uchida, S., and Davis, J.C. (2005*b*). *Science*, **309**, 1048–1052.

McElroy, K., Simmonds, R.W., Hoffman, J.E., Lee, D.H., Orenstein, J., Eisaki, H., Uchida, S., and Davis, J.C. (2003). *Nature*, **422**, 592–596.

Mesaros, A., Fujita, K., Eisaki, H., Uchida, S., Davis, J.C., Sachdev, S., Zaanen, J., Lawler, M.J., and Kim, E.A. (2011). *Science*, **333**, 426.

Mesot, J., Norman, M.R., Ding, H., Randeria, M., Campuzano, J.C., Paramekanti, A., Fretwell, H.M., Kaminski, A., Takeuchi, T., and Yokoya, T. (1999). *Phys. Rev. Lett.*, **83**, 840–843.

Misra, S., Vershinin, M., Phillips, P., and Yazdani, A. (2004). *Phys. Rev. B*, **70**, 220503.

Mori, M., Khaliullin, G., Tohyama, T., and Maekawa, S. (2008). *Phys. Rev. Lett.*, **101**, 247003.

Newns, D.M. and Tsuei, C.C. (2007). *Nature Phys.*, **3**, 184–191.

Norman, M.R., Ding, H., Randeria, M., Campuzano, J.C., Yokoya, T., Takeuchi, T., Takahashi, T., Mochiku, T., Kadowaki, K., and Guptasarma, P. (1998). *Nature*, **392**, 157–160.

Norman, M.R. and Pepin, C. (2003). *Rep. Prog. Phys.*, **66**, 1547.

Nunner, T.S., Andersen, B.M., Melikyan, A., and Hirschfeld, P.J. (2005). *Phys. Rev. Lett.*, **95**, 177003.

Nunner, T.S., Chen, W., Andersen, B.M., Melikyan, A., and Hirschfeld, P.J. (2006). *Phys. Rev. B*, **73**, 104511.

Orenstein, J. and Millis, A.J. (2000). *Science*, **288**, 468.

Pan, S.H., O'Neal, J.P., Badzey, R.L., Chamon, C., Ding, H., Engelbrecht, J.R., Wang, Z., Elsaki, H., Uchida, S., and Gupta, A.K. (2001). *Nature*, **413**, 282–285.

Paramekanti, A., Randeria, M., and Trivedi, N. (2001). *Phys. Rev. Lett.*, **87**, 217002.

Pereg-Barnea, T. and Franz, M. (2003). *Phys. Rev. B*, **68**, 180506.

Pushp, A., Parker, C.V., Pasupathy, A.N., Gomes, K.K., Ono, S., Wen, J., Xu, Z., Gu, G., and Yazdani, A. (2009). *Science*, **324**, 1689–1693.

Randeria, M., Trivedi, N., Moreo, A., and Scalettar, R.T. (1992). *Phys. Rev. Lett.*, **69**, 2001–2004.

Renner, Ch., Revaz, B., Genoud, J.-Y., Kadowaki, K., and Fischer, . (1998). *Phys. Rev. Lett.*, **80**, 149–152.

Sachdev, S. (2003). *Rev. Mod. Phys.*, **75**, 913–932.

Shen, K.M., Ronning, F., Lu, D.H., Baumberger, F., Ingle, N.J.C., Lee, W.S., Meevasana, W., Kohsaka, Y., Azuma, M., and Takano, M. (2005). *Science*, **307**, 901–904.

Slezak, J.A., Lee, J., Wang, M., McElroy, K., Fujita, K., Andersen, B.M., Hirschfeld, P.J., Eisaki, H., Uchida, S., and Davis, J.C. (2008). *Proc. Nat'l Acad. Sci.*, **105**, 3203.

Tanaka, K., Lee, W.S., Lu, D.H., Fujimori, A., Fujii, T., Terasaki, I., Scalapino, D.J., Devereaux, T.P., Hussain, Z., and Shen, Z.X. (2006). *Science*, **314**, 1910–1913.

Tešanović, Z. (2008). *Nat. Phys.*, **4**, 408–414.

Timusk, T. and Statt, B. (1999). *Rep. Prog. Phys.*, **62**, 61.

Tranquada, J.M., Woo, H., Perring, T.G., Goka, H., Gu, G.D., Xu, G., Fujita, M., and Yamada, K. (2004). *Nature*, **429**, 534–538.

Varma, C.M. (2006). *Phys. Rev. B*, **73**, 155113.

Vojta, M. (2009). *Adv. Phys.*, **58**, 699–820.

Vojta, M. and Sachdev, S. (1999). *Phys. Rev. Lett.*, **83**, 3916–3919.

Wang, Q.-H., Han, J.H., and Lee, D.-H. (2002). *Phys. Rev. B*, **65**, 054501.

Wang, Q.-H. and Lee, D.-H. (2003). *Phys. Rev. B*, **67**, 020511.

Wang, Y., Li, L., Naughton, M.J., Gu, G.D., Uchida, S., and Ong, N.P. (2005). *Phys. Rev. Lett.*, **95**, 247002.

Wang, Y., Li, L., and Ong, N.P. (2006). *Phys. Rev. B*, **73**, 024510.

White, S.R. and Scalapino, D.J. (1998). *Phys. Rev. Lett.*, **80**, 1272–1275.

Wise, W.D., Boyer, M.C., Chatterjee, K., Kondo, T., Takeuchi, T., Ikuta, H., Wang, Y., and Hudson, E.W. (2008). *Nat. Phys.*, **4**, 696–699.

Wulin, D., He, Y., Chien, C.-C., Morr, D.K., and Levin, K. (2009). *Phys. Rev. B*, **80**, 134504.

Xu, Z.Z., Ong, N. P, Wang Y., Kakeshita, T., and Uchida, S. (2000). *Nature*, **406**, 486–488.

Y., Chen, Rice, T.M., and F.C., Zhang (2006). *Rev. Lett.*, **97**, 237004.

Zaanen, J. and Gunnarsson, O. (1989). *Phys. Rev. B*, **40**, 7391–7394.

Zaanen, J., Sawatzky, G.A., and Allen, J.W. (1985). *Phys. Rev. Lett.*, **55**, 418.

Zhang, F.C., Gros, C., Rice, T.M., and Shiba, H. (1988). *Super. Sci. & Tech.*, **1**, 36–46.

Zhu, J.-X., Ahn, K.H., Nussinov, Z., Lookman, T., Balatsky, A.V., and Bishop, A.R. (2003). *Phys. Rev. Lett.*, **91**, 057004.

17
Theoretical Studies of Superconductor–Insulator Transitions

Yen LEE LOH[1] and Nandini TRIVEDI[2]

[1]Department of Physics and Astrophysics, University of North Dakota, Grand Forks, ND 58202

[2]Department of Physics, The Ohio State University, 191 W Woodruff Avenue, Columbus, OH 43210

17.1 Introduction

The superconductor–insulator transition (SIT) is an intriguing example of a quantum phase transition in a fermionic system. It is a theme that runs across boundaries between disciplines—strongly correlated electrons, helium, cold atoms, and nuclear physics—and is highly relevant to some of the biggest unsolved problems in superconductivity. For a broad perspective, see Chapter 10 and the review by Gantmakher and Dolgopolov (2010).

In this article we study SITs within the general framework of an attractive Hubbard model. This is a well-defined model of s-wave superconductivity, which permits use of different tuning parameters (disorder and field). Furthermore, it allows a comparison of various analytical and computational approaches in order to gain a complete understanding of the various effects of amplitude and phase fluctuations. We present a systematic pedagogical approach, aiming to equip the "lay" reader with enough apparatus to be able to understand the numerical calculations, reproduce some of the simpler results, and be able to tackle future problems related to inhomogeneous phases. We go into considerable detail on mean-field theory (MFT) and the Bogoliubov–de Gennes (BdG) approach, as these are a first line of attack, which can capture much of the physics. However, we also outline cases where this fails to capture phase fluctuations and more sophisticated quantum Monte Carlo (QMC) calculations are necessary. We discuss the behavior of many observables, including densities of states, superfluid stiffness, and dynamical conductivity, for the disorder-tuned SIT.

The general Hamiltonian is

$$H = -\sum_{ij\sigma} t_{ij} c^{\dagger}_{i\sigma} c_{j\sigma} - \sum_{i} U n_{i\uparrow} n_{i\downarrow} - \sum_{i\sigma} \mu_{i\sigma} n_{i\sigma}, \tag{17.1}$$

where i and j are site indices, $\sigma = \pm 1$ are fermion spin indices, t_{ij} is the hopping from site i to site j, $c^{\dagger}_{i\sigma}$ and $c_{i\sigma}$ are fermion creation and annihilation operators, U is the on-site attractive interaction, $\mu_{i\sigma} = \mu - v_i - h\sigma$ is the site-dependent chemical potential, μ is the average chemical potential, v_i are disorder potentials on each site drawn independently from a uniform distribution $[-V, +V]$, h is a uniform Zeeman field, and $n_{i\sigma} = c^{\dagger}_{i\sigma} c_{i\sigma}$ is the local density for spin σ. In particular, we will consider various combinations of the parameters t, U, μ, h, disorder strength V, and temperature T. If necessary, the Hubbard model parameters t, V, U can be connected to physical observables such as Δ_0, k_{F}, E_{F}, the mean free path l, and the disorder energy scale $\hbar/\tau$. We shall focus on transport and spectra, which are the properties most directly affected by the SIT (rather than thermodynamic properties).

This model covers many situations of interest. We begin with the tight-binding model (with hopping t) and introduce the other ingredients one by one (disorder V, attraction U, and Zeeman field h). Adding V alone leads to the Anderson model of localization and the metal–insulator transition in a non-interacting system (Section 17.2). This simple model provides an opportunity for us to introduce various basic concepts. Conversely, with U alone, the model describes a superconductor in which the size of the pairs can be controlled by the strength of the attraction and allows one to access the beautiful physics of the crossover from the Bardeen–Cooper–Schrieffer (BCS) regime of large overlapping Cooper pairs to the Bose–Einstein condensation (BEC) regime of tightly bound bosons (Section 17.3.1).

The interplay between U and V produces a disorder-tuned SIT. With analytical and numerical treatments of increasing sophistication, culminating in QMC simulations, we elucidate the bosonic nature of this transition and its consequences (Section 17.3).

Magnetic-field-tuned SITs are quite different due to the pairbreaking nature of magnetism. The competition between U and h leads to a parallel-field-tuned SIT from a fully paired superconductor to a fully polarized Fermi liquid via a Fulde–Ferrell–Larkin–Ovchinnikov (FFLO) state, in which the pairing and magnetism form a self-organized layered microstructure—an example of microscale phase separation (Section 17.4.2). Finally, when U, h, and V are all present, a disordered version of the FFLO state ("dLO") appears to survive at weak disorder (Section 17.4.3). In dLO phases, Andreev bound states in domain walls contribute mid-gap weight in the density of states (DOS), which may be an explanation for the anomalous zero-bias tunneling conductance observed in purely parallel-field-tuned SITs.

17.2 Free fermions in a random potential

In this chapter we introduce the tight-binding model and obtain the dispersion relation and DOS. We then discuss the properties of the Anderson Hamiltonian for on-site random potentials. We calculate single-particle properties (DOS, localization length, and participation ratio) and two-particle properties (conductivity), which serve as a basis for comparing with later results on disordered superconductors.

17.2.1 Tight-binding model on square lattice

In this article we will focus on two-dimensional (2D) systems. The starting point for our study is the tight-binding model Hamiltonian for spinless fermions on a square lattice:

$$H = -\sum_{ij} t_{ij} c_i^\dagger c_j - \mu \sum_i n_i, \qquad (17.2)$$

where i and j are site indices, t_{ij} are hopping amplitudes (such that $t_{ij} = t$ for nearest neighbors and 0 otherwise), $c_i^\dagger$ and c_i are fermion creation and annihilation operators, μ is the chemical potential, and $n_i = c_i^\dagger c_i$ are number operators. Since H is translationally invariant, it can be diagonalized by Fourier transforming into momentum space:

$$H = \sum_{\mathbf{k}} (\varepsilon_{\mathbf{k}} - \mu) c_{\mathbf{k}}^\dagger c_{\mathbf{k}} \qquad (17.3)$$

where the dispersion relation is

$$\varepsilon_{\mathbf{k}} = -2t(\cos k_x + \cos k_y). \qquad (17.4)$$

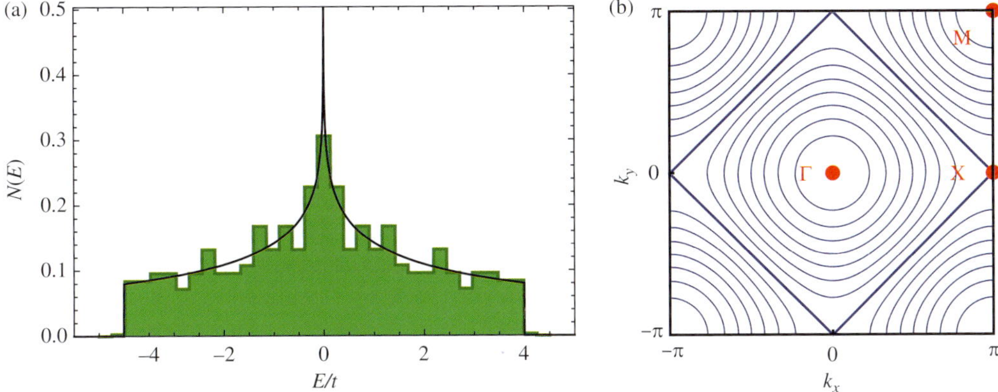

Fig. 17.1 Properties of the square lattice tight-binding model. (a) density of states. (b) Constant-energy contours of dispersion relation from eqn (17.4). The thick diamond indicates the contour at zero energy, which would be the Fermi surface for a half-filled system. The solid line is the analytic formula of eqn (17.6). The histogram is for a 36×36 square lattice with periodic boundary conditions.

This dispersion relation has a bandwidth of $8t$, and it has stationary points at the Γ, X, and M symmetry points of the Brillouin zone. The DOS (per site) is

$$N_{\mathrm{TB}}(E) = \frac{1}{N}\sum_{\mathbf{k}} \delta(E - \varepsilon_{\mathbf{k}}) = \int_{BZ} \frac{d^2k}{(2\pi)^2}\delta(E - \varepsilon_{\mathbf{k}}), \tag{17.5}$$

where N is the number of sites in the system. This can be written in terms of the complete elliptic integral K:[1]

$$N_{\mathrm{TB}}(E) = \left| \frac{2}{\pi^2 E} \operatorname{Im} K\left(\frac{16}{E^2}\right) \right|. \tag{17.6}$$

The DOS has step singularities at $E = -4t$ and $E = 4t$ and a logarithmic singularity at $E = 0$, as illustrated in Fig. 17.1 (a). These van Hove singularities correspond to the Γ, M, and X symmetry points in Fig. 17.1 (b). For the finite systems treated in this article, the van Hove singularities are quite severely smeared out.

When the chemical potential lies within the band, this model is a Fermi liquid. Its thermodynamic properties correspond to those of a Fermi liquid. For example, the specific heat is linear in temperature. However, the transport properties of this model are pathological: it has zero superfluid stiffness but infinite DC conductivity (with finite charge stiffness). These concepts will be discussed in Section 17.3. In order to model real metals, one has to introduce some amount of disorder, as we do in the next section.

[1]There are several definitions for elliptic integrals; ours is the `EllipticK` function in Mathematica.

17.2.2 Anderson Hamiltonian

Consider a tight-binding model with a disorder potential v_i at each site, picked independently from a uniform distribution on $[-V, +V]$, where V is the disorder strength.[2] This is known as the "Anderson model" of localization:

$$H = -\sum_{ij} t_{ij} c_i^\dagger c_j + \sum_i (v_i - \mu) n_i. \tag{17.7}$$

The hopping alone would produce plane-wave eigenstates with a bandwidth of $8t$, whereas the disorder potential alone would produce site-localized eigenstates with a bandwidth of $2V$. The competition between hopping and disorder makes this a non-trivial problem, and in some situations it can give rise to a metal–insulator transition.

We wish to calculate thermodynamic and transport properties averaged over disorder configurations. Many methods have been developed for studying such problems, each with their own advantages: exact diagonalization, Lanczos methods, Chebyshev methods, transfer matrices, mapping to non-linear sigma models, and so on. In this article we will focus on exact diagonalization,[3] to set the stage for the formalism of the disordered superconductor in Section 17.3.

Although it is convenient to write the Hamiltonian in operator form, for computational purposes it is necessary to construct the Hamiltonian matrix elements explicitly, such that $\hat{H} = \sum_{ij} H_{ij} c_i^\dagger c_j$:

$$H_{ij} = -t_{ij} + (v_i - \mu)\delta_{ij}. \tag{17.8}$$

For example, for a 3×3 square lattice with open boundaries, the Hamiltonian matrix is a sparse matrix with only five non-zero diagonals:

$$\mathbf{H} = \begin{pmatrix} v_{11} & -t & & -t & & & & & \\ -t & v_{12} & -t & & -t & & & & \\ & -t & v_{13} & & & -t & & & \\ -t & & & v_{21} & -t & & -t & & \\ & -t & & -t & v_{22} & -t & & -t & \\ & & -t & & -t & v_{23} & & & -t \\ & & & -t & & & v_{31} & -t & \\ & & & & -t & & -t & v_{32} & -t \\ & & & & & -t & & -t & v_{33} \end{pmatrix}. \tag{17.9}$$

The next step is to diagonalize the Hamiltonian matrix to find the eigenvalues E_α and unitary eigenvector matrix $\phi_{i\alpha}$ such that $\sum_j H_{ij}\phi_{j\alpha} = E_\alpha \phi_{i\alpha}$, where α labels the eigenmodes. Then, $\gamma^\dagger$, and γ are operators that create and annihilate fermions in eigenmodes, where $c_i = \sum_\alpha \phi_{i\alpha}\gamma_\alpha$ and $\gamma_\alpha = \sum_i \phi_{i\alpha}^* c_i$. Figure 17.2 shows localized and "extended" eigenstates for a single disorder realization.

[2] In the literature, the disorder bandwidth $W = 2V$ is often used as the disorder-strength parameter.

[3] The word "exact" is misleading, because the diagonalization is usually performed numerically; for example, LAPACK 2003 uses Householder tridiagonalization and the "relatively robust representation" algorithm. Some authors prefer to write "direct diagonalization," although this is also a misnomer, as most diagonalization methods involve an "indirect" stage (iteration to convergence).

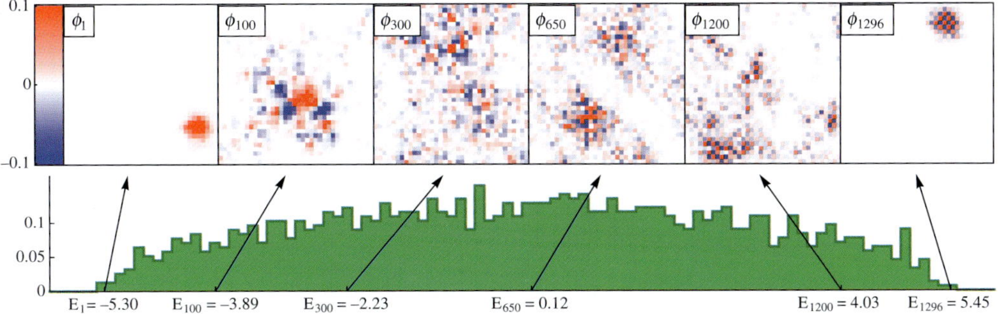

Fig. 17.2 Six eigenstates of the Anderson model on a 36×36 square lattice for a single disorder realization ($\mu = 0, V = 3t$). Red and blue colors indicate signs of eigenfunctions $\psi_{i\alpha}$. The states in the band tail are localized, whereas the states in the band center are quasi-extended over the size of the system.

17.2.3 Participation ratio and localization length

The Anderson model of localization has been extensively studied in the context of metal–insulator transitions. The nature of the eigenstates is now well understood in terms of the scaling theory of localization (Abrahams *et al.*, 1979) and the theory of weak localization (Altshuler *et al.*, 1980). See Chapter 3 for more on the topic of Anderson localization.

For three-dimensional (3D) systems, eigenstates can be extended or localized. Extended states occupy the center of each band, whereas localized states exist near the edges of the band (the tails in the DOS). The boundaries between regions of extended and localized states are called mobility edges. The loci of the mobility edges form a curve in the (E, V) plane (Bułka *et al.*, 1985), as illustrated in Fig. 17.3. Sweeping the chemical potential across a mobility edge produces a metal–insulator transition, which is an example of a quantum phase transition. Figure 17.3 also serves as a phase diagram in the (μ, V) plane.

In 1D and 2D systems, an infinitesimal amount of disorder is sufficient to make all states localized. However, localized states appear to be extended if their localization length is greater than the box size, as is evident in Fig. 17.2.

One way to quantify the spatial extent of an eigenstate $\phi_{i\alpha}$ is via its participation number

$$p_\alpha = 1 \Big/ \sum_i |\phi_{i\alpha}|^4 , \tag{17.10}$$

which describes the number of sites over which the wavefunction has an appreciable magnitude.[4] Sometimes $R_\alpha = (p_\alpha)^{1/d}$ is identified as an "eigenmode radius," an indicator of the spatial extent of eigenmode α.

[4]In the literature one often encounters the participation ratio p_α/N, which is the fraction of sites over which the wavefunction has an appreciable magnitude.

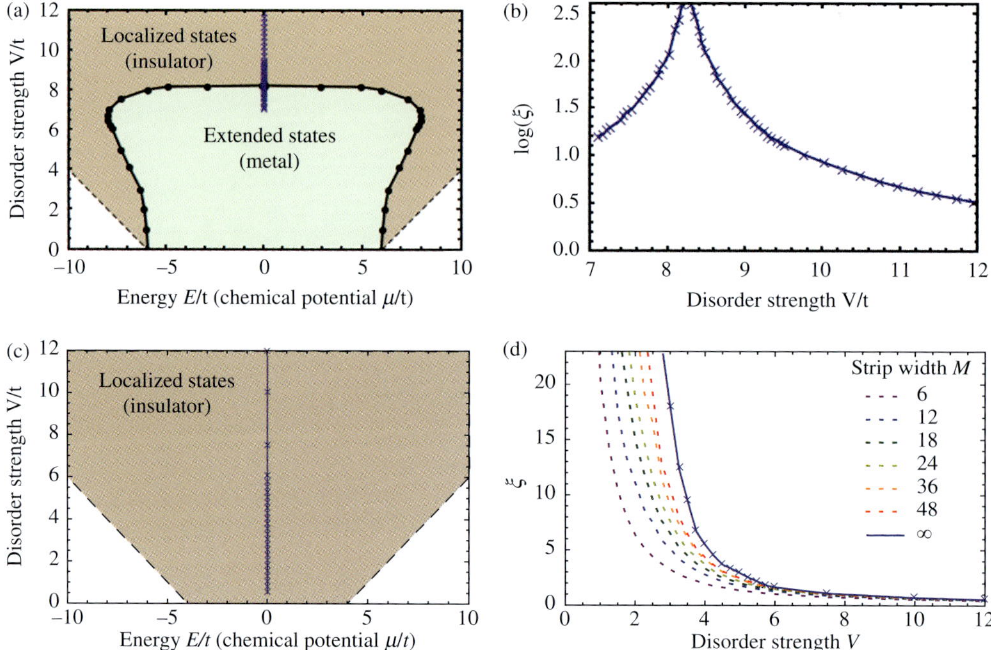

Fig. 17.3 Properties of the Anderson model. 3D figures are adapted from the review by Kramer and MacKinnon. (a) Mobility edge trajectories (solid), for the cubic lattice (3D) Anderson model with box disorder, $v_i \in [-V, +V]$, as well as band edge trajectories (dashed) for comparison. The blue crosses correspond to data points in (b). (b) Divergence of the localization length ξ at the critical disorder strength V_c. For $V < V_c$, ξ is to be interpreted as a scaling length instead. (c) Phase diagram of the square lattice (2D) Anderson model with box disorder. (d) Localization length ξ_{loc} on strips of width M with cylindrical boundary conditions, calculated using the Green's function method, eqn (17.11). The thick blue curve is the infinite-size limit, extracted using a rudimentary form of finite-size scaling. ξ_{loc} is finite for all $V > 0$, but it diverges strongly as $V \to 0$.

Another important concept is the localization length ξ_{loc}. Qualitatively, this is the length scale for the exponential decay of an eigenfunction far from its center of mass, but for practical calculations, ξ_{loc} is usually defined as the decay length of the transmission coefficient along a long strip, and can be calculated using transfer matrix methods or Green's function methods (MacKinnon and Kramer, 1981). The Green's function $\mathbf{G}_l$ is the result of evolving the Schrödinger equation l layers away from a source at the zeroth layer (see Fig. 17.4 for an illustration). It can be calculated using the recursion relation given by

$$\begin{aligned}
\mathbf{A}_{-1} &= \mathbf{0}, \\
\mathbf{A}_0 &= \mathbf{1}, \\
\mathbf{A}_l &= (E\mathbf{1} - \mathbf{H}_l)\mathbf{A}_{l-1} - \mathbf{A}_{l-2}, \qquad l = 1, \ldots, L,
\end{aligned}$$

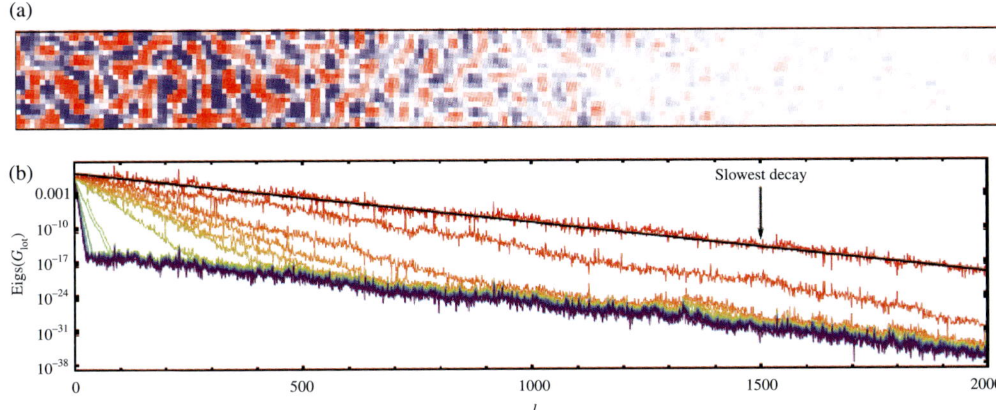

Fig. 17.4 Illustration of the Green's function and localization length on a strip. (a) Response of a 200×20 strip of the Anderson model to a source at one edge, calculated by a brute-force linear system solution. Periodic boundary conditions were used in the short direction. Parameters were $E = -3$, $V = 1.5$, $t = 1$. (b) Eigenvalues of $\mathbf{G}_l$ as a function of strip length l. The largest eigenvalue of $\mathbf{G}_l$ decays exponentially with l. The length scale for this decay gives the localization length.

$$\mathbf{G}_l = (\mathbf{A}_l)^{-1},$$

$$\xi_{\rm loc}{}^{-1} = \lim_{l\to\infty} \frac{1}{2(l-1)} \ln \operatorname{tr}(\mathbf{G}_l)^2, \tag{17.11}$$

where $\mathbf{H}_L$ is the $M \times M$ Hamiltonian for the Lth slice of the strip. One must take special precautions against numerical instabilities, for example, by periodically restarting the recursion. Typically, $\xi_{\rm loc}$ is calculated as a function of strip width M and then finite-size scaling is used to extrapolate to $M \to \infty$. See MacKinnon and Kramer (1981), Bułka *et al.* (1985), and Kramer and MacKinnon (1993) for details.

Figure 17.3 (b), adapted from Kramer and MacKinnon (1993), shows $\xi_{\rm loc}$ as a function of box disorder strength for the cubic-lattice Anderson model, clearly showing a metal–insulator transition. Similar results have been obtained for other disorder distributions, indicating universal behavior for the critical exponents at the metal–insulator transition.

Figure 17.3 (d) shows $\xi_{\rm loc}(V)$ for the square lattice. Because of weak localization in two dimensions, $\xi_{\rm loc}$ is finite for all $V > 0$. Nevertheless, $\xi_{\rm loc}$ diverges strongly as $V \to 0$, and for $V \lesssim 2t$, it greatly exceeds the dimensions of the systems we will be simulating.

In principle, the average "eigenmode radius" at a given energy $R(E)$, and the localization length, $\xi_{\rm loc}(E)$, are distinct quantities, as the former refers to the bulk of the eigenmodes whereas the latter refers to the tails. Nevertheless, both quantities diverge at the mobility edge, and for the purposes of this article we will loosely refer to both as $\xi_{\rm loc}$.

17.2.4 Single-particle properties

Many quantities can be computed by using the eigenmode representation. The (retarded) single-particle Green's function is

$$G_{ij}(E) = \sum_\alpha \phi_{i\alpha}\phi^*_{j\alpha} \frac{1}{E - E_\alpha + i0^+} \tag{17.12}$$

where the infinitesimal shift in the denominator imposes causality. From the Green's function one can obtain single-particle spectral properties, such as the local density of states (LDOS) $N_i(E)$, spectral function $A(\mathbf{k}, E)$, and total DOS $N(E)$:[5]

$$N_i(E) = -\tfrac{1}{\pi}\,\mathrm{Im}\,G_{ii}(E) = \sum_\alpha \delta(E - E_\alpha)\left|\phi_{i\alpha}\right|^2, \tag{17.13}$$

$$A(\mathbf{k}, E) = -\tfrac{1}{\pi}\,\mathrm{Im}\,G_{\mathbf{kk}}(E) = \sum_\alpha \delta(E - E_\alpha)\left|\phi_{\mathbf{k}\alpha}\right|^2, \tag{17.14}$$

$$N(E) = \sum_i N_i(E) = \sum_{\mathbf{k}} A(\mathbf{k}, E) = \sum_\alpha \delta(E - E_\alpha), \tag{17.15}$$

where $\phi_{\mathbf{k}\alpha} = \frac{1}{N}\sum_i e^{-i\mathbf{k}\cdot\mathbf{r}_i}\phi_{i\alpha}$ are the eigenfunctions in the momentum representation. Note that sums of Dirac delta functions are most efficiently calculated by accumulating weights in bins. Furthermore, one can obtain static properties such as the local number density $\langle n_i \rangle$ and momentum distribution $\langle n(\mathbf{k}) \rangle$,

$$\langle n_i \rangle = \left\langle c_i^\dagger c_i \right\rangle = \int dE\ f(E) N_i(E) = \sum_\alpha f_\alpha \left|\phi_{i\alpha}\right|^2, \tag{17.16}$$

$$\langle n(\mathbf{k}) \rangle = \left\langle c_{\mathbf{k}}^\dagger c_{\mathbf{k}} \right\rangle = \int dE\ f(E) A(\mathbf{k}, E) = \sum_\alpha f_\alpha \left|\phi_{\mathbf{k}\alpha}\right|^2. \tag{17.17}$$

where f_α are the Fermi occupation factors of the eigenmodes,

$$f_\alpha = f(E_\alpha) = \frac{1}{e^{\beta(E_\alpha - \mu)} + 1} = \tfrac{1}{2} - \tfrac{1}{2}\tanh\tfrac{\beta(E_\alpha - \mu)}{2}. \tag{17.18}$$

Figure 17.5 (a) shows the DOS for different disorder strengths. The single-particle DOS is gapless. For weak disorder ($V = 1t$) it resembles the tight-binding DOS, whereas for strong disorder ($V = 12t$) it approaches the uniform distribution of the disorder potential.

[5]The random potential breaks translational invariance, so that momentum is not a good quantum number; particles can scatter from $\mathbf{k}$ to $\mathbf{k} + \mathbf{q}$, and hence the spectral function and momentum distribution are also functions of the total momentum $\mathbf{q}$. Averaging over many disorder realizations recovers translational invariance, such that only the $\mathbf{q} = \mathbf{0}$ term remains.

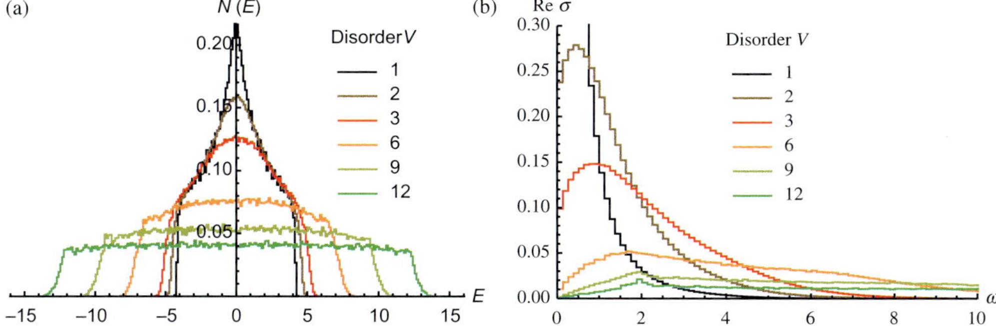

Fig. 17.5 Properties of the Anderson model with box disorder on a 36×36 square lattice. (a) The DOS $N(E)$. As disorder increases, the DOS gets flatter and more box-like, but no gaps appear. (b) The real part of the conductivity, $\mathrm{Re}\,\sigma(\omega)$, for chemical potential $\mu = 0$. At weak disorder, for the finite systems studied here, the dynamical conductivity has a large Drude-like peak. At strong disorder, the conductivity is suppressed, and it develops a "soft gap" (presumably due to Mott variable-range hopping), even though the DOS remains gapless. Note the suppression of $\sigma(\omega)$ at low frequency ω. For an infinite 2D system, the DC conductivity would be suppressed all the way to zero.

17.2.5 Electromagnetic response (DC and AC conductivity)

For the purposes of this article the most important two-particle property is the electromagnetic response tensor, $\Upsilon_{\mu\nu}(\mathbf{q},\omega) = -\frac{dj_\nu(\mathbf{q},\omega)}{dA_\mu(\mathbf{q},\omega)}$, which is the current response in direction ν to an applied vector potential with wavevector $\mathbf{q}$, frequency ω, and polarization μ. This can be calculated in the eigenbasis using the Kubo formula (see Section 17.8 for derivation),

$$\frac{\mathrm{Im}\,\Upsilon_{\mu\nu\mathbf{q}\omega}}{\pi} = \sum_{\alpha\beta} \Gamma_{\alpha\beta\mu\mathbf{q}}\Gamma_{\beta\alpha\nu\bar{\mathbf{q}}}(f_\beta - f_\alpha)\,\delta(E_\alpha - E_\beta - \omega), \tag{17.19}$$

$$\mathrm{Re}\,\Upsilon_{\mu\nu\mathbf{q}\omega} = \langle -k_{\mu\nu}\rangle + \mathcal{P}\int_{-\infty}^{\infty}\frac{d\omega'}{\pi}\,\frac{\mathrm{Im}\,\Upsilon_{\mu\nu\mathbf{q}\omega'}}{\omega-\omega'}, \tag{17.20}$$

where

$$\langle -k_{\mu\nu}\rangle = \sum_{\alpha ij} r_{ij\mu} r_{ij\nu} t_{ij} \phi^*_{i\alpha}\phi_{j\alpha} f_\alpha,$$

$$\Gamma_{\alpha\beta\mu\mathbf{q}} = \sum_{ij} e^{-i\mathbf{q}\cdot\mathbf{r}_i} r_{ij\mu} i t_{ij} \phi^*_{i\alpha}\phi_{j\beta}. \tag{17.21}$$

Here, $\langle -k_{\mu\nu}\rangle$ is the diamagnetic response, which is related to the mean kinetic energy, and $r_{ij\mu}$ is the μth Cartesian component of the displacement vector from site i to site j.

The dynamical conductivity (typically the optical or microwave conductivity) is $\sigma(\omega) = \frac{\Upsilon(\omega,q=0)}{i\omega}$. Figure 17.5 (b) shows the behavior of $\text{Re}\,\sigma(\omega)$ for the Anderson model at various disorder strengths. In an infinite 2D system, the DC conductivity $\sigma(\omega = 0)$ is zero for all $V > 0$. However, because the calculations were done on finite 2D systems, there is a small finite DC conductivity. Nevertheless, it can be seen that the AC conductivity is strongly suppressed at low frequencies. At weak disorder, this can be interpreted in terms of weak localization. At strong disorder, this can be understood according to the theory of Mott variable-range hopping (VRH), which predicts the following temperature and frequency dependence of the conductivity:

$$\sigma(\omega = 0, T) \propto \exp\left[-(T_0/T)^{1/(d+1)}\right], \tag{17.22}$$

$$\text{Re}\,\sigma(\omega, T = 0) \propto \omega^2 \left[\ln(\text{const}/\omega)\right]^4, \tag{17.23}$$

(where $d = 2$ is the dimensionality).

17.2.6 Summary

In this section we have seen that the 2D Anderson model is generically an insulator. For Anderson insulators, the single-particle spectrum is gapless, but the conductivity has a “soft gap” due to weak localization or variable-range hopping. Due to the presence of low-lying fermionic excitations, an Anderson insulator can be classified as a Fermi insulator.

In the next section we will see that including attraction leads to a different type of insulator—a Bose insulator—in which the single-particle spectrum has a hard gap.

17.3 Disorder-tuned superconductor–insulator transition

17.3.1 Clean superconductor

We now proceed to the case of a clean superconductor modeled by the attractive Hubbard model on a square lattice. The Hamiltonian is

$$H = -\sum_{ij\sigma} t_{ij} c^\dagger_{i\sigma} c_{j\sigma} - \mu \sum_{i\sigma} n_{i\sigma} - \sum_i U c^\dagger_{i\uparrow} c^\dagger_{i\downarrow} c_{i\downarrow} c_{i\uparrow}. \tag{17.24}$$

17.3.1.1 BCS mean-field theory

Most of the properties of this model can be understood within BCS mean-field theory (Bardeen *et al.*, 1957; de Gennes, 1966; Tinkham, 1996). Here, in order to obtain quantitative results and to set the stage for the treatment of disorder, we explicitly take into account the lattice densities of states and Hartree corrections.

The quartic interaction is decoupled in terms of a uniform pairing potential $\Delta = U\left\langle c_{i\downarrow}c_{i\uparrow}\right\rangle$ and a Hartree chemical potential $\mu^{\mathrm{H}} = U\left\langle n\right\rangle$. Up to a constant,[6]

$$H_{\mathrm{MF}} = -\sum_{ij\sigma} t_{ij}c^{\dagger}_{i\sigma}c_{j\sigma} - (\mu+\mu^{\mathrm{H}})\sum_{i\sigma} n_{i\sigma} - \sum_{i}(\Delta^* c_{i\downarrow}c_{i\uparrow} + \Delta c^{\dagger}_{i\uparrow}c^{\dagger}_{i\downarrow}). \qquad (17.25)$$

Since the Hamiltonian has translational symmetry, it can be diagonalized by Fourier transforming to momentum space:

$$H_{\mathrm{MF}} = \sum_{\mathbf{k}}\left(\sum_{\sigma}\xi_{\mathbf{k}}c^{\dagger}_{\mathbf{k}\sigma}c_{\mathbf{k}\sigma} - \Delta^* c_{-\mathbf{k}\downarrow}c_{\mathbf{k}\uparrow} - \Delta c^{\dagger}_{\mathbf{k}\uparrow}c^{\dagger}_{-\mathbf{k}\downarrow}\right), \qquad (17.26)$$

where $\xi_{\mathbf{k}} = \varepsilon_{\mathbf{k}} - \mu - \mu^{\mathrm{H}}$ and $\varepsilon_{\mathbf{k}} = -2t(\cos k_x + \cos k_y)$ as defined earlier. This can be written in a 2×2 matrix form. Up to a constant,

$$H_{\mathrm{MF}} = \sum_{\mathbf{k}}\begin{pmatrix} c^{\dagger}_{\mathbf{k}\uparrow} & c_{-\mathbf{k}\downarrow}\end{pmatrix}\begin{pmatrix}\xi_{\mathbf{k}} & -\Delta \\ -\Delta & -\xi_{\mathbf{k}}\end{pmatrix}\begin{pmatrix} c_{\mathbf{k}\uparrow} \\ c^{\dagger}_{-\mathbf{k}\downarrow}\end{pmatrix}. \qquad (17.27)$$

The matrix can be further diagonalized by a Bogoliubov transformation of the fermion operators to bogolon creation and annihilation operators γ,

$$\begin{pmatrix} c_{\mathbf{k}\uparrow} \\ c^{\dagger}_{-\mathbf{k}\downarrow}\end{pmatrix} = \begin{pmatrix} u_{\mathbf{k}} & v_{\mathbf{k}} \\ -v_{\mathbf{k}} & u_{\mathbf{k}}\end{pmatrix}\begin{pmatrix} \gamma_{\mathbf{k}\uparrow} \\ \gamma^{\dagger}_{-\mathbf{k}\downarrow}\end{pmatrix}, \qquad (17.28)$$

where $u_{\mathbf{k}} = \cos\theta_{\mathbf{k}}$, $v_{\mathbf{k}} = \sin\theta_{\mathbf{k}}$, $E_{\mathbf{k}} = \sqrt{{\xi_{\mathbf{k}}}^2 + \Delta^2}$, and $\tan 2\theta_{\mathbf{k}} = \frac{\Delta}{E_{\mathbf{k}}}$. The Hamiltonian is bilinear in the bogolon operators:

$$H_{\mathrm{MF}} = \sum_{\mathbf{k}}\begin{pmatrix} \gamma^{\dagger}_{\mathbf{k}\uparrow} & \gamma_{-\mathbf{k}\downarrow}\end{pmatrix}\begin{pmatrix} E_{\mathbf{k}} & 0 \\ 0 & -E_{\mathbf{k}}\end{pmatrix}\begin{pmatrix} \gamma_{\mathbf{k}\uparrow} \\ \gamma^{\dagger}_{-\mathbf{k}\downarrow}\end{pmatrix} = \sum_{\mathbf{k}\sigma} E_{\mathbf{k}}\gamma^{\dagger}_{\mathbf{k}\sigma}\gamma_{\mathbf{k}\sigma} \qquad (17.29)$$

(up to a constant). Therefore, expectations of bogolon operators are simply Fermi occupation factors: $\left\langle \gamma^{\dagger}_{\mathbf{k}\sigma}\gamma_{\mathbf{k}\sigma}\right\rangle = f_{\mathbf{k}} = f(E_{\mathbf{k}}) = \frac{1}{2} - \frac{1}{2}\tanh\frac{\beta}{2}E_{\mathbf{k}}$.

[6] In textbooks, BCS mean-field decoupling is often performed by writing $c_{i\downarrow}c_{i\uparrow} = \frac{\Delta}{|U|} + (c_{i\downarrow}c_{i\uparrow} - \frac{\Delta}{|U|})$, expanding $c^{\dagger}_{i\uparrow}c^{\dagger}_{i\downarrow}c_{i\downarrow}c_{i\uparrow}$ using the binomial theorem, and expanding up to first order in the term in parentheses. However, for decoupling in two or more channels, this method over-counts the Hubbard interaction, and it is not clear what the constant term in the Hamiltonian should be. The rigorous variational formalism in Section (17.9) resolves these problems.

The "pair density" (or anomalous Green's function) on each site is

$$\begin{aligned}
F &= \left\langle c_{i\downarrow} c_{i\uparrow} \right\rangle = \frac{1}{N} \sum_{\mathbf{k}} \left\langle c_{-\mathbf{k}\downarrow} c_{\mathbf{k}\uparrow} \right\rangle \\
&= \int_{\mathbf{k}} \left\langle (-v_{\mathbf{k}} \gamma^{\dagger}_{\mathbf{k}\uparrow} + u_{\mathbf{k}} \gamma_{-\mathbf{k}\downarrow})(u_{\mathbf{k}} \gamma_{\mathbf{k}\uparrow} + v_{\mathbf{k}} \gamma^{\dagger}_{-\mathbf{k}\downarrow}) \right\rangle \\
&= \int_{\mathbf{k}} u_{\mathbf{k}} v_{\mathbf{k}} \left\langle -\gamma^{\dagger}_{\mathbf{k}\uparrow} \gamma_{\mathbf{k}\uparrow} + \gamma_{\mathbf{k}\downarrow} \gamma^{\dagger}_{\mathbf{k}\downarrow} \right\rangle = \int_{\mathbf{k}} u_{\mathbf{k}} v_{\mathbf{k}} (1 - 2 f_{\mathbf{k}}) \\
&= \int_{\mathbf{k}} \frac{\Delta}{2E_{\mathbf{k}}} \tanh \frac{E_{\mathbf{k}}}{2T}
\end{aligned} \tag{17.30}$$

where $\int_{\mathbf{k}} \equiv \int_{\mathrm{BZ}} \frac{d^2 k}{(2\pi)^2}$. The pairing amplitude, or order parameter, is self-consistently determined by $\Delta = |U|\, F$, leading to the order-parameter equation,[7]

$$\frac{1}{|U|} = \int_{\mathbf{k}} \frac{1}{2E_{\mathbf{k}}} \tanh \frac{E_{\mathbf{k}}}{2T}. \tag{17.31}$$

The zero-temperature order parameter Δ_0 and the mean-field critical temperature $T_{\mathrm{c}}^{\mathrm{MF}}$ are given by setting $T = 0$ and $\Delta = 0$ respectively in the order-parameter equation:

$$\frac{1}{|U|} = \int_{\mathbf{k}} \frac{1}{2\sqrt{\xi_{\mathbf{k}}{}^2 + \Delta_0{}^2}} = \int_{\mathbf{k}} \frac{1}{2\xi_{\mathbf{k}}} \tanh \frac{\xi_{\mathbf{k}}}{2T_{\mathrm{c}}^{\mathrm{MF}}}. \tag{17.32}$$

The number density on each site is

$$\begin{aligned}
n &= \left\langle c^{\dagger}_{i\uparrow} c_{i\uparrow} + c^{\dagger}_{i\downarrow} c_{i\downarrow} \right\rangle = \frac{1}{N} \sum_{\mathbf{k}} \left\langle c^{\dagger}_{\mathbf{k}\uparrow} c_{\mathbf{k}\uparrow} + c^{\dagger}_{\mathbf{k}\downarrow} c_{\mathbf{k}\downarrow} \right\rangle \\
&= \int_{\mathbf{k}} \sum_{\sigma} \left(u_{\mathbf{k}}{}^2 \left\langle \gamma^{\dagger}_{\mathbf{k}\sigma} \gamma_{\mathbf{k}\sigma} \right\rangle + v_{\mathbf{k}}{}^2 \left\langle \gamma_{\mathbf{k}\sigma} \gamma^{\dagger}_{\mathbf{k}\sigma} \right\rangle \right) \\
&= 1 + \int_{\mathbf{k}} \frac{\xi_{\mathbf{k}}}{E_{\mathbf{k}}} \tanh \frac{E_{\mathbf{k}}}{2T}
\end{aligned} \tag{17.33}$$

(after some algebra). This depends (through ξ) on the chemical potential μ and Hartree potential μ^{H}. The order-parameter equation and number equation can be iterated to self-consistency.

Figure 17.6 shows quantities as functions of attraction, $|U|/t$, at a general filling $n = 0.875$. This illustrates the important dichotomy between amplitude physics (pairing) and phase physics (coherence). Superconductivity requires both pair formation, which occurs below the mean-field critical temperature $T_{\mathrm{c}}^{\mathrm{MF}}$, and phase coherence,

[7] In the literature, Δ is often called the "gap", and the self-consistent equation for Δ is called the "gap equation." We will avoid this terminology, because, as we shall see, in dirty superconductors near the SIT, the gap E_{g} and the order parameter Δ are completely distinct quantities.

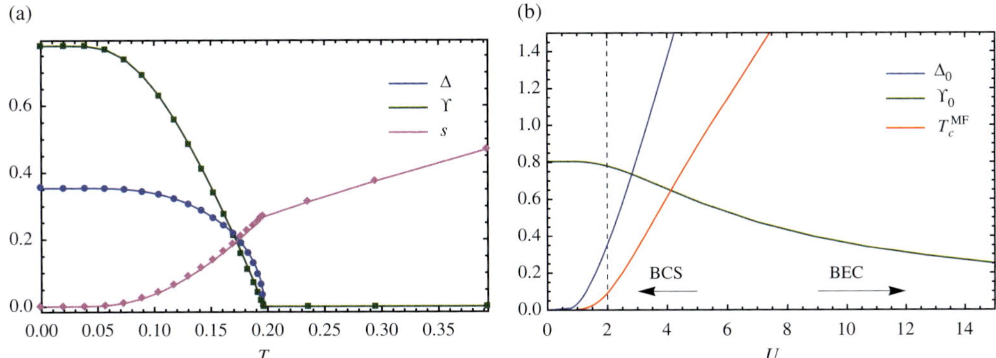

Fig. 17.6 Properties of the attractive Hubbard model on a square lattice within mean-field theory. All energies are in units of the hopping amplitude t. (a) Properties as a function of temperature for $n = 0.875$ and $|U| = 2t$, corresponding to the dashed line in Fig. 17.6 (b). All quantities have an Arrhenius behavior ($e^{-\Delta_0/T}$) at low temperature. At T_c, Δ has a square-root singularity, whereas Υ goes linearly to zero. The entropy per site s has a kink, corresponding to a jump in the specific heat c. (b) Properties as a function of on-site attraction U for $n = 0.875$, calculated within mean-field theory. The blue curve is the zero-temperature order parameter Δ_0. The red curve is the mean-field critical temperature T_c^{MF}. The green curve is the zero-temperature superfluid stiffness $\Upsilon_0 \sim \rho_s$, which is proportional to the phase fluctuation temperature T_θ. The true T_c (not shown) is bounded above by T_c^{MF} and T_θ.

which is governed by the phase stiffness Υ (or ρ_s). The superconducting critical temperature T_c is determined by the lower of these two energy scales. In the weak-coupling limit, $T_c^{\mathrm{MF}} \ll \Upsilon$, and the transition is well described by BCS mean-field theory. In the strong-coupling limit, $T_c^{\mathrm{MF}} \gg \Upsilon$, so the transition is dominated by phase fluctuations. In that case, T_c^{MF} is a pseudogap temperature corresponding to pair formation. This is the BEC limit, in which composite bosons become superfluid at low temperatures. For the purposes of this chapter we shall work in the BCS limit, which is relevant to the materials used in SIT experiments. However, we will see that phase fluctuations nevertheless become important near the SIT.

Figure 17.6 (a) shows the temperature dependence of the order parameter at $|U| = 2t$ (corresponding to the dashed line in Fig. 17.6 (b)). The "strong-coupling ratio" $2\Delta/T_c$ is about 4. For comparison, BCS theory in the continuum gives the universal number $2\Delta/T_c = \frac{2\pi}{e^\gamma} \approx 3.53$, where γ is the Euler–Mascheroni constant.

17.3.1.2 Single-particle spectrum

We have already derived Δ and n above. Now we consider other single-particle properties. The single-particle Green's function is

$$G_{\mathbf{k}}(\tau) = \langle c_{\alpha\tau} c_\alpha^\dagger \rangle = u_{\mathbf{k}}{}^2 \left\langle \gamma_{\mathbf{k}\tau}^\dagger \gamma_{\mathbf{k}} \right\rangle + v_{\mathbf{k}}{}^2 \left\langle \gamma_{\mathbf{k}\tau} \gamma_{\mathbf{k}}^\dagger \right\rangle = u_{\mathbf{k}}{}^2 G_{\alpha\tau} - v_{\mathbf{k}}{}^2 G_{\alpha,-\tau}, \qquad (17.34)$$

where $G_{\alpha\tau}$ is the standard Green's function for a fermionic eigenmode as described in Section 17.6 and $0 < \tau < \beta$. Fourier transforming, analytically continuing to real frequencies, and taking the imaginary part gives the spectral function,

$$A_{\mathbf{k}}(E) = u_{\mathbf{k}}{}^2\delta(E - E_\alpha) + v_{\mathbf{k}}{}^2\delta(E + E_\alpha). \tag{17.35}$$

The spectral function has a pole of strength $u_{\mathbf{k}}{}^2$ at positive frequencies, along the gapped bogolon dispersion relation, and a pole of strength $v_{\mathbf{k}}{}^2$ at negative frequencies. This structure is illustrated in Fig. 17.7 (b).

The DOS can be obtained by integrating $A_{\mathbf{k}}(E)$ over wavevectors $\mathbf{k}$. Reducing the 2D integral to a 1D integral over the tight-binding DOS and using identities for Dirac delta functions leads to

$$N(E) = \frac{\operatorname{sgn} E}{2}\left[\left(\frac{E}{\sqrt{E^2 - \Delta^2}} + 1\right) N_{\mathrm{TB}}\left(\mu + \sqrt{E^2 - \Delta^2}\right) + \left(\frac{E}{\sqrt{E^2 - \Delta^2}} - 1\right) N_{\mathrm{TB}}\left(\mu - \sqrt{E^2 - \Delta^2}\right)\right] \tag{17.36}$$

where $N_{\mathrm{TB}}(\varepsilon)$ is the tight-binding DOS given in eqn (17.6). The resulting DOS has BCS coherence peaks (inverse square-root divergences) at $E = \pm\Delta$ as well as van Hove singularities at $E = \sqrt{\mu^2 + \Delta^2}$ and $E = \sqrt{(\mu \pm 4t)^2 + \Delta^2}$, as illustrated in Fig. 17.7 (a). There is a clearly-defined gap in the spectrum E_{g}, equal to Δ. Hence, it is

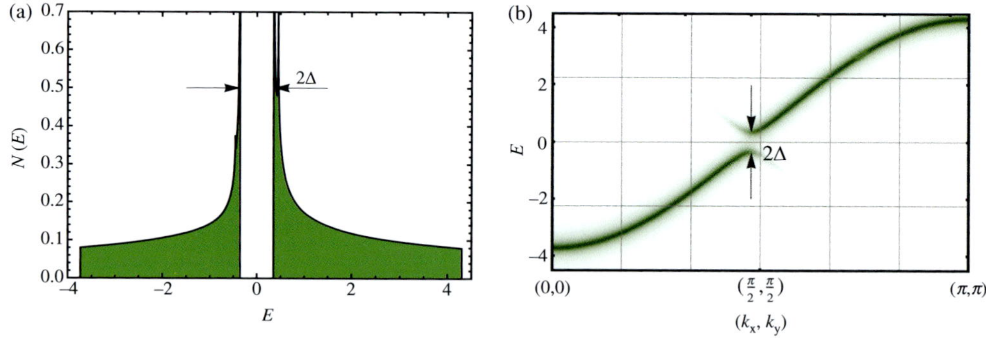

Fig. 17.7 Spectral properties of a clean superconductor within mean-field theory for $U = -2$, $n = 0.875$, and $T = 0$, for which $\Delta = 0.354$. All energies are in units of the hopping amplitude t. (a) Density of states, showing coherence peaks at $E = \pm\Delta$. In this case, the inverse-square root singularities are further enhanced by the nearby logarithmic van Hove singularity. (b) Spectral function $A(\mathbf{k}, \omega)$, showing back-bending due to particle-hole mixing.

common practice to use the terms "gap" E_g and "order parameter Δ" interchangeably. However, as we will see, this is very misleading when discussing the SIT.[8]

17.3.1.3 *Electromagnetic response*

One of the most important properties of a superconductor is its superfluid stiffness or phase stiffness Υ. For a clean SC on a square lattice, the Kubo formula (see Section 17.8.1) gives

$$\Upsilon = \Upsilon_{xx} = \langle -k_{xx} \rangle - \Lambda_{xx}, \tag{17.37}$$

$$\langle k_{\mu\nu} \rangle = \sum_{\mathbf{k}} \frac{\partial^2 \varepsilon_{\mathbf{k}}}{\partial k_\mu \partial k_\nu} \frac{\xi_{\mathbf{k}}}{E_{\mathbf{k}}} (2 f_{\mathbf{k}} - 1), \tag{17.38}$$

$$\Lambda_{xx} = -\frac{8}{N} \sum_{\mathbf{p}} \sin^2 p_x \frac{\partial f(E_{\mathbf{p}})}{\partial E_{\mathbf{p}}}, \tag{17.39}$$

where $\frac{\partial f}{\partial E_{\mathrm{p}}} = -\frac{\beta}{4} \operatorname{sech}^2 \frac{\beta}{2} E_{\mathbf{p}}$. The terms $\langle -k_{xx} \rangle$ and Λ_{xx} are often referred to as "diamagnetic" and "paramagnetic" contributions respectively. At $T = 0$, $\Lambda_{xx} = 0$ vanishes, whereas for $T > T_c$, $\langle -k_{xx} \rangle = \Lambda_{xx}$ and $\Upsilon = 0$ as required. The behavior of Υ is illustrated in Fig. 17.6 (a) and (b).

We refer the reader to Scalapino *et al.* (1993) for a discussion of the finite-frequency conductivity, pausing only to mention that a perfectly clean Hubbard-model superconductor has pathological properties.

With the exception of Λ_{xx}, all of the 2D integrals appearing in this section can be reduced to 1D integrals by writing $\int_{\mathrm{BZ}} \frac{d^2 k}{(2\pi)^2} f(\xi_{\mathbf{k}}) \equiv \int_{-\infty}^{\infty} d\xi \; N_{\mathrm{TB}}(\xi + \mu) f(\xi)$, where $N_{\mathrm{TB}}(E)$ is the square-lattice tight-binding DOS defined in eqn (17.6).

Rather than dwelling on more sophisticated treatments, we will now go on to the problem of a disordered superconductor.

17.3.2 Dirty superconductor

We represent a dirty superconductor by an attractive Hubbard model with a disorder potential,

$$H = -\sum_{ij\sigma} t_{ij} c_{i\sigma}^\dagger c_{j\sigma} - \sum_i \mu_i n_{i\sigma} - U \sum_i c_{i\uparrow}^\dagger c_{i\downarrow}^\dagger c_{i\downarrow} c_{i\uparrow}, \tag{17.40}$$

where $\mu_i = v_i - \mu$ and where the disorder potential at each site v_i is picked independently from a uniform distribution on $[-V, +V]$, as before.

[8]Another place where it is misleading is in Abrikosov–Gor'kov theory, where a large concentration of magnetic impurities can produce gapless superconductivity with a finite order parameter but zero spectral gap.

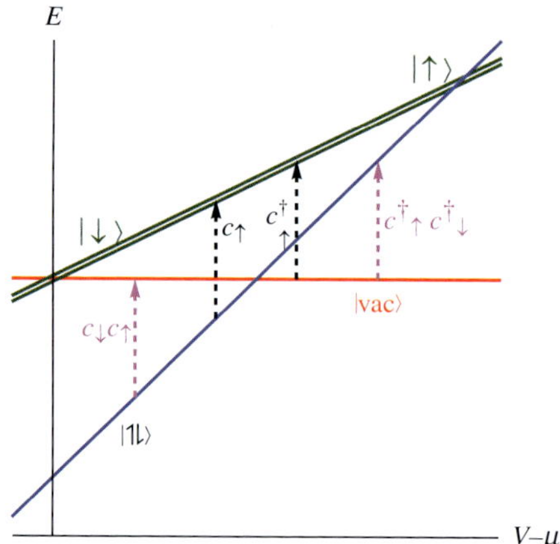

Fig. 17.8 Energy levels E and single-particle and two-particle transition energies for a single site with potential V, chemical potential μ, and attraction U (the atomic limit of the Hubbard model).

17.3.3 Atomic limit

In the limit of extreme disorder, the hopping can be neglected, and the system then reduces to an ensemble of single-site Hubbard models, each with the Hamiltonian

$$H = Un_\uparrow n_\downarrow + (V - \mu)(n_\uparrow + n_\downarrow). \tag{17.41}$$

This system has just four Fock states. The energies of these states are $E_0 = 0$, $E_\uparrow = E_\downarrow = V - \mu$, and $E_\updownarrow = U + 2(V - \mu)$. The four states occur with relative Boltzmann weights $\exp(-\beta E_n)$. The spectral function (the DOS for single-particle excitations) can be obtained by considering transitions between these four Fock states (amplitudes and energies). This is illustrated in Fig. 17.8. Regardless of the on-site potential V, single-particle transitions (black arrows) always cost at least $|U|/2$, and therefore the spectrum is always gapped.[9] Pair excitations (purple arrows), however, may cost zero energy if V is just right. This is understood in the literature in terms of the "parity gap."

We have generalized the above calculation to exact diagonalization of the many-body Hubbard Hamiltonian on small clusters of a few sites, which leads to the same conclusions: single-particle excitations are gapped whereas two-particle excitations can be gapless.

We know that a clean s-wave superconductor ($V = 0$) has a gap $E_\text{g} = \Delta$ given by the BCS gap equation. We have just found that in the limit of extreme disorder, $V \gg (U, t)$, the gap is finite and large: $E_\text{g} = U/2$. We will see that the gap remains finite between these two extremes.

[9]This is in contrast to the repulsive Hubbard model in the atomic limit, for which the spectrum is only gapped if $U > 2V$.

17.3.4 Pairing of exact eigenstates (PoEE)

The above Hamiltonian contains three terms: hopping, disorder, and attraction. In typical s-wave superconductors, the first two terms have the largest energy scales. Thus, it makes sense to solve the non-interacting problem first by direct diagonalization, to find the disorder eigenvalues and eigenstates ξ_α and $\phi_{i\alpha}$ (as in Section 17.2.2), and then examine the effect of U. This method, known as the "pairing of exact eigenstates" (PoEE) approximation, is very much in the spirit of Anderson's original derivation of his theorem.

In the basis of exact eigenstates, the Hamiltonian is

$$H = \sum_{\alpha}^{\sigma} \xi_\alpha \gamma^\dagger_{\alpha\sigma}\gamma_{\alpha\sigma} - U \sum_{\alpha\beta\gamma\delta i} \phi_{i\alpha}\phi_{i\beta}\phi^*_{i\gamma}\phi^*_{i\delta} c^\dagger_{\alpha\uparrow} c^\dagger_{\beta\downarrow} c_{\gamma\downarrow} c_{\delta\uparrow}. \tag{17.42}$$

Following Anderson's suggestion, let us assume that instead of pairing between $\mathbf{k}$ and $-\mathbf{k}$, we have pairing between time-reversed eigenstates α and $\bar{\alpha}$ (i.e. complex conjugate eigenfunctions). Retaining only those terms in the Hamiltonian that connect such eigenstates,

$$\begin{aligned} H_{\text{PoEE}} &= \sum_{\alpha}^{\sigma} \xi_\alpha \beta^\dagger_{\alpha\sigma}\beta_{\alpha\sigma} - U \sum_{\alpha\beta i} \phi_{i\alpha}\phi_{i\bar{\alpha}}\phi^*_{i\bar{\beta}}\phi^*_{i\beta} c^\dagger_{\alpha\uparrow} c^\dagger_{\bar{\alpha}\downarrow} c_{\bar{\beta}\downarrow} c_{\beta\uparrow} \\ &= \sum_{\alpha}^{\sigma} \xi_\alpha \beta^\dagger_{\alpha\sigma}\beta_{\alpha\sigma} - \sum_{\alpha\beta} M_{\alpha\beta} c^\dagger_{\alpha\uparrow} c^\dagger_{\bar{\alpha}\downarrow} c_{\bar{\beta}\downarrow} c_{\beta\uparrow}, \end{aligned} \tag{17.43}$$

where $M_{\alpha\beta} = U \sum_i |\phi_{i\alpha}|^2 |\phi_{i\beta}|^2$. Approximating this by a mean-field Hamiltonian,

$$H_{\text{MF}} = \sum_{\alpha}^{\sigma} \xi_\alpha \beta^\dagger_{\alpha\sigma}\beta_{\alpha\sigma} - \sum_{\beta} (\Delta^*_\beta c_{\bar{\beta}\downarrow} c_{\beta\uparrow} + h.c.) \tag{17.44}$$

(up to a constant), where the order parameter is

$$\Delta^*_\beta = U \sum_\alpha M_{\alpha\beta} \left\langle c^\dagger_{\alpha\uparrow} c^\dagger_{\bar{\alpha}\downarrow} \right\rangle, \tag{17.45}$$

(assuming that ξ_α have been redefined in this step to include Hartree shifts).

The gap equation then works out to be

$$\Delta_\alpha = U \sum_\beta M_{\alpha\beta} \frac{\Delta_\beta}{2E_\beta} \tanh \frac{E_\beta}{2T}, \tag{17.46}$$

where $E_\beta = \sqrt{\xi_\beta{}^2 + \Delta_\beta{}^2}$, and the chemical potential is determined by the number equation

$$\langle n \rangle = \frac{1}{N} \sum_\alpha \left(1 - \frac{\xi_\alpha}{E_\alpha} \right). \tag{17.47}$$

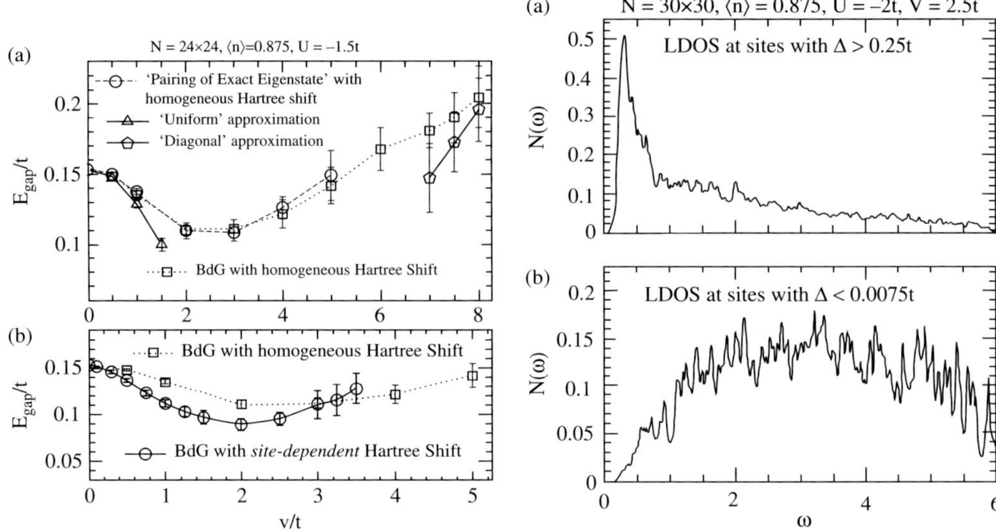

Fig. 17.9 Left panels, top (a) and bottom (b) showing single-particle gap from pairing of exact eigenstates compared with BdG. Right panels, top (a) and bottom (b) showing local density of statesobtained from BdG.

The PoEE theory can be used in the above form, or one can perform further approximations as follows. In the low-disorder regime, the disorder eigenstates $\phi_{i\alpha}$ are extended on the scale of the system, so that $M_{\alpha\beta} \approx 1/N$ independent of α and β. In this limit Anderson's theorem applies—the gap equation takes the simple BCS form and Δ is spatially uniform. In the high-disorder regime, on the other hand, the disorder eigenstates are strongly localized with localization lengths $\xi_\alpha^{\rm loc}$, and the M matrix is approximately diagonal, $M_{\alpha\beta} \approx \delta_{\alpha\beta} \sum_i |\phi_{i\alpha}|^4 \approx \delta_{\alpha\beta}/(\xi_\alpha^{\rm loc})^2$.

The results of the PoEE theory implemented numerically, and in the low-disorder and high-disorder approximations, are compared in Fig. 17.9 (a). Surprisingly, *the gap is finite for all values of disorder* $0 < V < \infty$. At large disorder,

$$E_{\rm g} = \frac{U}{2{\xi_{\rm loc}}^2}, \tag{17.48}$$

where $\xi_{\rm loc}$ is the localization length at the chemical potential. At extremely high disorder one recovers the atomic limit described in Section 17.3.3.

The PoEE approach is useful for understanding the robustness of the gap, but it does not give the full story. It predicts the the BCS coherence peaks in the DOS survive up to infinite disorder, whereas more accurate calculations show that they do not. Furthermore, being a mean-field theory, PoEE fails to capture the destruction of phase coherence at the SIT due to quantum phase fluctuations. We now proceed to more sophisticated treatments.

17.3.5 Bogoliubov-de Gennes

In the Bogoliubov–de Gennes (BdG) formalism the Hubbard interaction is decoupled in terms of mean fields that are allowed to take arbitrary values at each site. The most general decoupling involves six fields at every site (see Section 17.9):

$$\begin{aligned} -U c_{i\uparrow}^\dagger c_{i\downarrow}^\dagger c_{i\downarrow} c_{i\uparrow} \longrightarrow \;& \Delta_i^* c_{i\downarrow} c_{i\uparrow} + \Delta_i c_{i\uparrow}^\dagger c_{i\downarrow}^\dagger \quad \text{(Bogoliubov)} \\ & + \mu_{i\uparrow}^{\mathrm{H}} c_{i\uparrow}^\dagger c_{i\uparrow} + \mu_{i\downarrow}^{\mathrm{H}} c_{i\downarrow}^\dagger c_{i\downarrow} \quad \text{(Hartree)} \\ & + h_i^+ c_{i\downarrow}^\dagger c_{i\uparrow} + h_i^- c_{i\uparrow}^\dagger c_{i\downarrow} \quad \text{(Fock)}. \end{aligned} \tag{17.49}$$

Although in this section only pairing and density channels are required, we nevertheless present a decoupling in three channels (pairing, density, and z-magnetization) which we use later in Section 17.4.3. In this formalism the mean-field Hamiltonian is

$$H_{\mathrm{BdG}} = -\sum_{ij\sigma} t_{ij} c_{i\sigma}^\dagger c_{j\sigma} - \sum_{i\sigma} (\mu + \mu_{i\sigma}^{\mathrm{H}}) n_{i\sigma} - \sum_i (\Delta_i^* c_{i\downarrow} c_{i\uparrow} + \Delta_i c_{i\uparrow}^\dagger c_{i\downarrow}^\dagger), \tag{17.50}$$

where $\mu_{i\uparrow}^{\mathrm{H}}$, $\mu_{i\downarrow}^{\mathrm{H}}$, and Δ_i are $3N$ parameters to be determined self-consistently. This can be written in a $2N \times 2N$ matrix form:

$$H_{\mathrm{BdG}} = \sum_{ij} \begin{pmatrix} c_{i\uparrow}^\dagger & c_{i\downarrow} \end{pmatrix} \underbrace{\begin{pmatrix} -t_{ij} - \tilde{\mu}_{i\uparrow}\delta_{ij} & -\Delta_i \delta_{ij} \\ -\Delta_i \delta_{ij} & t_{ij} + \tilde{\mu}_{i\downarrow}\delta_{ij} \end{pmatrix}}_{\mathbf{H}} \begin{pmatrix} c_{j\uparrow} \\ c_{j\downarrow}^\dagger \end{pmatrix} \tag{17.51}$$

(up to a constant). This is a real symmetric matrix, so its eigenvalues E_α and eigenvectors ϕ_α are real. It will be convenient to split the eigenvectors into particle parts $u_{i\alpha} = \phi_{i1;\alpha}$ and hole parts $v_{i\alpha} = \phi_{i2;\alpha}$ (note that the u and v in our three-channel formalism are different from the u and v variables in traditional BdG). Then, the fermion operators can be expressed in terms of bogolon operators γ_α as

$$c_{i\uparrow} = \sum_\alpha u_{i\alpha} \gamma_\alpha, \qquad c_{i\downarrow} = \sum_\alpha v_{i\alpha} \gamma_\alpha^\dagger. \tag{17.52}$$

The Hamiltonian is bilinear in the bogolon operators:

$$H_{\mathrm{BdG}} = \sum_\alpha E_\alpha \gamma_\alpha^\dagger \gamma_\alpha \tag{17.53}$$

(up to a constant), so expectations of bogolon operators are easy to calculate, e.g. $\langle \gamma_\alpha^\dagger \gamma_\alpha \rangle = f_\alpha = 1/(\exp \beta E_\alpha + 1)$. Expectations of fermion operators may be calculated by transforming to the bogolon basis; the derivations are quite similar to those in Section 17.2.4. The basic recipe for a BdG calculation is as follows:

1. Make initial guesses for the internal fields (the Hartree potentials $\mu_{i\sigma}^{\mathrm{H}}$ and the self-consistent pairing field Δ_i).
2. Find the total (effective) fields at site i by combining the external (applied) fields with the internal fields: $\tilde{\mu}_i = \mu_i + \mu_i^{\mathrm{H}}$ and $\tilde{h}_i = h_i + h_i^{\mathrm{H}}$.

3. Construct the $2N \times 2N$ Hamiltonian matrix $\mathbf{H}$.
4. Find the eigenvalues E_α, eigenvectors $(u_{i\alpha}, v_{i\alpha})$, and occupation numbers f_α.
5. Compute the number densities $n_{i\sigma}$ and the pairing density F_i at every site i:

$$n_{i\uparrow} = \sum_\alpha f_\alpha {u_{i\alpha}}^2, \tag{17.54}$$

$$n_{i\downarrow} = \sum_\alpha (1 - f_\alpha) {v_{i\alpha}}^2, \tag{17.55}$$

$$F_i = \sum_\alpha (f_\alpha - 1/2) u_{i\alpha} v_{i\alpha}. \tag{17.56}$$

6. Recompute the internal fields $\Delta_i := UF_i$, $\mu^{\mathrm{H}}_{i\uparrow} := Un_{i\downarrow}$, and $\mu^{\mathrm{H}}_{i\downarrow} := Un_{i\uparrow}$.
7. Go back to step 2. Repeat till convergence.

There are several possible improvements to the above scheme. The scheme, as described, uses fixed-point iteration to approach self-consistency. This may converge slowly or it may become unstable; convergence can be improved by introducing an empirically determined linear-mixing factor γ:

$$\Delta_i := \Delta_i + \gamma(UF_i - \Delta^i_i), \tag{17.57}$$

$$\mu^{\mathrm{H}}_{i\uparrow} := \mu^{\mathrm{H}}_{i\uparrow} + \gamma(Un_{i\downarrow} - \mu^{\mathrm{H}}_{i\uparrow}), \tag{17.58}$$

$$\mu^{\mathrm{H}}_{i\downarrow} := \mu^{\mathrm{H}}_{i\downarrow} + \gamma(Un_{i\uparrow} - \mu^{\mathrm{H}}_{i\downarrow}). \tag{17.59}$$

One can go even further and use the Broyden method for multi-dimensional root-finding, which converges superlinearly sufficiently close to the solution.

For a given set of internal fields, one can calculate the variational free energy Ω (see Section 17.9 for a derivation),

$$\Omega = -T\sum_\alpha \ln(2\cosh\tfrac{1}{2}\beta E_\alpha) + \sum_i U(F_i^2 + x_i^2 - m_i^2) + \sum_i 2(\Delta_i F_i + \mu^{\mathrm{H}}_i x_i + h^H_i m_i), \tag{17.60}$$

where $x_i = \frac{1}{2}n_i - \frac{1}{2}$ and $m_i = \frac{1}{2}(n_{i\uparrow} - n_{i\downarrow})$. It is good practice to track the value of Ω, to verify that the iteration is converging to a minimum and not to a saddle-point or maximum.

To study systems with fixed average densities $n^{\mathrm{target}}_\sigma$, one can include Lagrange multipliers $\mu^{\mathrm{Lag}}_\sigma$ as two additional variables in the self-consistency iteration. However, in that case, the root-finding problem can no longer be rephrased as the problem of minimizing Ω.

17.3.5.1 Eigenstates

Figure 17.10 shows some of the BdG eigenstates (bogolon modes). The lowest-energy modes are concentrated in the same locations as the superconducting puddles. In contrast, the higher-energy excitations correspond to breaking of localized pairs.

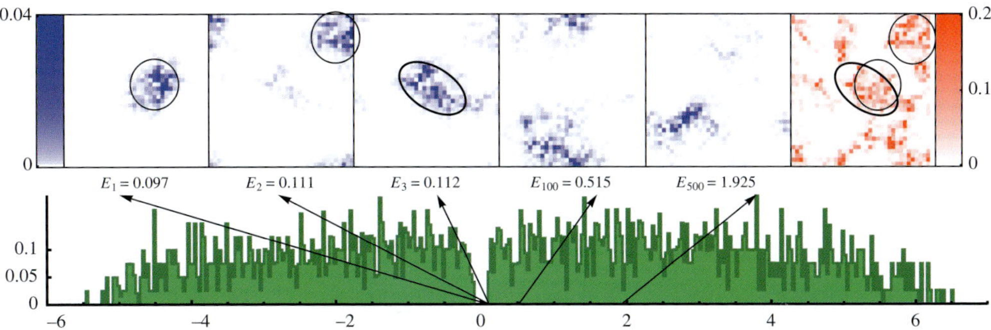

Fig. 17.10 The first five panels show the magnitude of five BdG eigenstates (bogolon wavefunctions), $|u_i|^2 + |v_i|^2$. The last panel (red) is a map of the local pairing amplitude Δ_i. The low- and high-energy eigenstates are localized, whereas the intermediate-energy eigenstates are quasi-extended. In particular, the lowest eigenstates correspond to the locations of the superconducting puddles (where Δ_i is large). The parameters were $U = -1.5t$, $n = 0.875$, $N = 36 \times 36$, $V = 3t$.

17.3.5.2 Single-particle spectrum

After the iteration has converged one may calculate further quantities of interest. In the same manner as in Section 17.2.4, the single-particle Green's functions in the Matsubara frequency domain are

$$G_{ij\uparrow}(i\varepsilon_n) = \langle \psi_{i\uparrow}(i\varepsilon_n)\bar{\psi}_{j\uparrow}(i\varepsilon_n)\rangle = \sum_{\alpha\beta} u_{i\alpha}u_{j\beta}\,\langle \gamma_\alpha(i\varepsilon_n)\bar{\gamma}_\beta(i\varepsilon_n)\rangle$$

$$= \sum_\alpha u_{i\alpha}u_{j\alpha}\frac{1}{i\varepsilon_n - E_\alpha}, \tag{17.61}$$

$$G_{ij\downarrow}(i\varepsilon_n) = \sum_\alpha v_{i\alpha}v_{j\alpha}\frac{1}{i\varepsilon_n + E_\alpha}, \tag{17.62}$$

so the densities of states for up and down electrons are

$$N_{i\uparrow}(E) = \sum_\alpha \delta(E - E_\alpha){u_{i\alpha}}^2, \tag{17.63}$$

$$N_{i\downarrow}(E) = \sum_\alpha \delta(E + E_\alpha){v_{i\alpha}}^2. \tag{17.64}$$

Figure 17.11 shows the number density, pairing amplitude, and DOS from BdG calculations on 36×36 lattices at various disorder strengths. At weak disorder, the number density and pairing amplitude are uniform, and the DOS has coherence peaks. At strong disorder, most sites are empty ($n_i = 0$) or doubly occupied ($n_i = 2$) due to the on-site attraction. However, there are still locally superconducting puddles with intermediate occupation numbers n_i and finite pairing amplitudes Δ_i. In fact, the sites

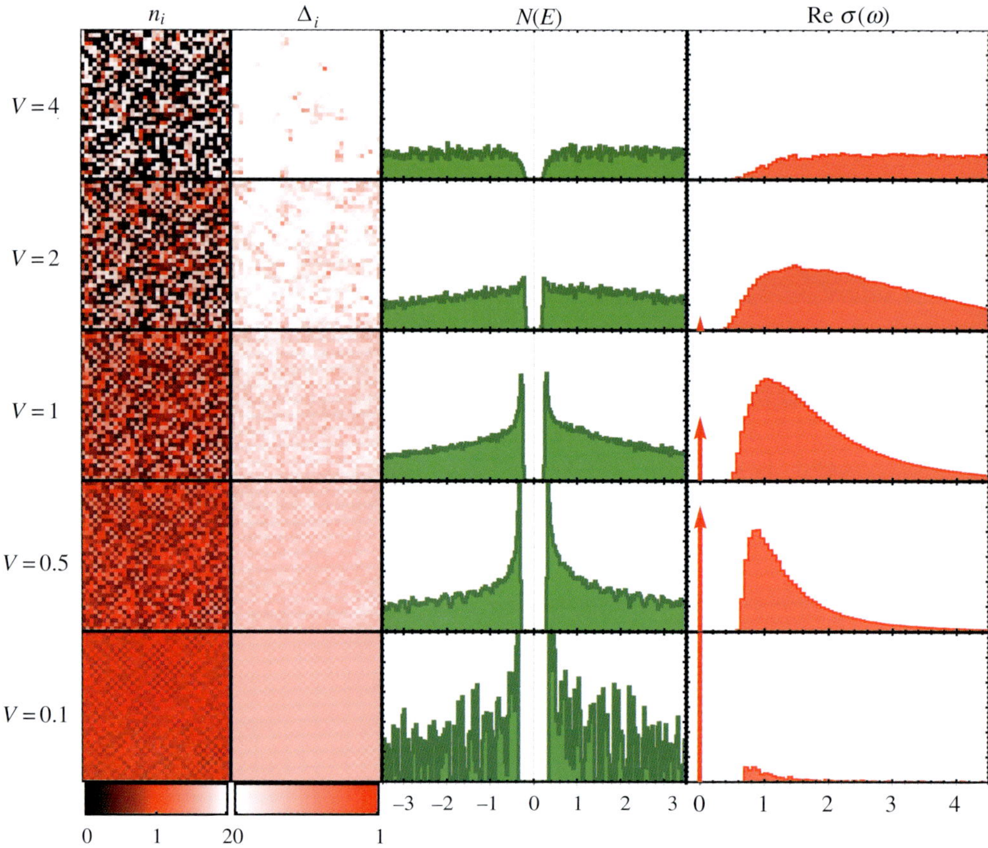

Fig. 17.11 The four columns show the number density, pairing density, single-particle spectrum, and optical conductivity of the attractive Hubbard model with a disorder potential, on a 36×36 square lattice with $U = -2t$, $\langle n \rangle = 0.875$, and $T = 0$, within the BdG approximation. $N(E)$ and σ are averaged over 10 disorder configurations.

with the smallest values of Δ_i tend to be the ones with the largest values of the local spectral gap E_g.

The BdG calculation shows that the spectral gap persists at all disorder strengths, and the size of the gap is approximately in agreement with the prediction from PoEE (see Fig. 17.9). However, unlike in PoEE, the coherence peaks are suppressed by disorder and vanish for $V \geq 2t$ (see Fig. 17.11).

17.3.5.3 Electromagnetic response

The electromagnetic response function can be calculated using the Kubo formula (see Section 17.8.2 for derivation):

$$\frac{\mathrm{Im}\,\Upsilon_{\mu\nu\mathbf{q}\omega}}{\pi} = \sum_{\alpha\beta} \Gamma_{\alpha\beta\mu\mathbf{q}}\Gamma_{\beta\alpha\nu\bar{\mathbf{q}}}(f_\beta - f_\alpha)\;\delta(E_\alpha - E_\beta - \omega), \tag{17.65}$$

$$\mathrm{Re}\,\Upsilon_{\mu\nu\mathbf{q}\omega} = \langle -k_{\mu\nu}\rangle + \mathcal{P}\int_{-\infty}^{\infty}\frac{d\omega'}{\pi}\;\frac{\mathrm{Im}\,\Upsilon_{\mu\nu\mathbf{q}\omega'}}{\omega-\omega'}, \tag{17.66}$$

where $\langle -k\rangle$ and Γ now involve both "spin components" of the eigenvectors (eqn (17.67)) and the eigenmode indices α and β now run from 1 to $2N$:

$$\langle k_{\mu\nu}\rangle = -\sum_{\alpha ij} r_{ij\mu}r_{ij\nu}t_{ij}\,(u_{i\alpha}u_{j\alpha} - v_{j\alpha}v_{i\alpha})\,f_\alpha,$$

$$\Gamma_{\alpha\beta\mu\mathbf{q}} = \sum_{ij} e^{-i\mathbf{q}\cdot\mathbf{r}_i} r_{ij\mu} i t_{ij}\,(u_{i\alpha}u_{j\beta} - v_{j\alpha}v_{i\beta})\,. \tag{17.67}$$

First consider the superfluid stiffness, $\Upsilon \equiv \Upsilon(\omega = 0)$, which is plotted in Fig. 17.12. As disorder increases, Υ decreases, but it never falls to zero. Even at large disorder, there are rare regions that are relatively disorder-free and have a large order parameter. These contribute to a finite stiffness within BdG theory. This means that *BdG, by itself, does not capture the SIT.*

Now consider the finite-frequency response. The real part of the dynamical conductivity from BdG calculations is of the form

$$\mathrm{Re}\,\sigma(\omega) = \Upsilon_{xx}\pi\delta(\omega) - \frac{1}{\omega}\,\mathrm{Im}\,\Lambda_{xx}(\omega). \tag{17.68}$$

This quantity is shown in the rightmost column of Fig 17.11. The heights of the red arrows indicate the relative weights of the delta functions. The weight of the delta

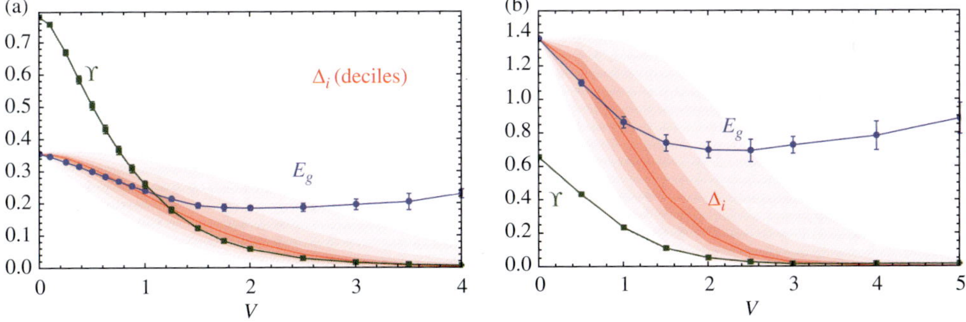

Fig. 17.12 BdG results at $T = 0$ on a 36×36 lattice (a) U = 2t and (b) U = 4t. The single-particle gap E_g remains finite, whereas the stiffness Υ falls to very small values for $V > V_\mathrm{c}$. The pairing amplitude Δ_i varies from site to site. The distribution, $P(\Delta_i)$, is visualized in terms of deciles; that is, successive red-colored bands represent the lowest 10% of Δ values, the next 10%, and so on up to the highest value of Δ. The central red line is the median value of Δ. All energy scales are in units of the hopping amplitude t.

function is the charge stiffness, which is equal to the superfluid stiffness Υ plotted in Fig. 17.12. [10] At weak coupling, the behavior is roughly in accord with Mattis-Bardeen theory. Within the BdG approximation, $\sigma(\omega)$ has a hard gap that is twice the single-particle gap: $\omega_g = 2E_{\rm g}$. The weight above the gap grows as the disorder strength increases. However, this is not the whole story. Physically, if the SIT is a continuous phase transition due to long-range phase coherence between superconducting puddles, one would expect that near the transition there should be low-frequency weight due to charge sloshing around between distant puddles. Indeed, near a quantum critical point, one might expect $\sigma(\omega, T)$ to follow a universal scaling form possibly with low-frequency weight. Thus the BdG results are questionable. The true behavior of the conductivity is a topic of further research.

The BdG method gives a lot of insight into the nature of the SIT and of the "Cooper pair insulator" at large V, but it only models the amplitude of the order parameter. The fact that the pairing amplitude becomes highly inhomogeneous gives a picture of superconducting islands weakly coupled by Josephson tunneling, and points to the importance of phase fluctuations in driving the SIT, as illustrated in Fig. 17.13. Generally, it appears that phase fluctuations are less important for single-particle properties, but more important for two-particle properties. To gain a full understanding of the latter, it is necessary to include quantum phase fluctuations.

17.3.6 Self-consistent harmonic approximation

There are several ways to treat phase fluctuations beyond BdG. In Ghosal *et al.* (2001), a 2D quantum XY action in imaginary time was used to describe the dynamics of the phase variables $\theta(\mathbf{r}, \tau)$ defined on a coarse-grained square lattice of spacing ξ:

$$S_\theta = \frac{\kappa\xi^2}{8}\int_0^\beta d\tau \sum_{\mathbf{r}} \left(\frac{\partial\theta(\mathbf{r},\tau)}{\partial\tau}\right)^2 + \frac{D_{\rm s}^0}{4}\int_0^\beta d\tau \sum_{\mathbf{r}\boldsymbol{\delta}} \{1 - \cos[\theta(\mathbf{r},\tau) - \theta(\mathbf{r}+\boldsymbol{\delta},\tau)]\}, \tag{17.69}$$

where κ was the mean-field static uniform compressibility and $D_{\rm s}^0$ was the mean-field phase stiffness (referred to in this article as Υ, up to a constant factor). The authors made the approximation of ignoring the spatial variations of κ and $D_{\rm s}^0$. Then, they performed a self-consistent harmonic approximation (SCHA) by choosing the optimal Gaussian action to minimize the free energy,

$$S_\theta = \frac{\kappa\xi^2}{8}\int_0^\beta d\tau \sum_{\mathbf{r}} \left(\frac{\partial\theta(\mathbf{r},\tau)}{\partial\tau}\right)^2 + \frac{D_{\rm s}}{8}\int_0^\beta d\tau \sum_{\mathbf{r}\boldsymbol{\delta}} \Big[\theta(\mathbf{r},\tau) - \theta(\mathbf{r}+\boldsymbol{\delta},\tau)\Big]^2, \tag{17.70}$$

[10] In general, one has to be very careful when taking limits of the electromagnetic response function $\Upsilon_{xx}(\omega, q_x, q_y)$. Different limits give three quantities—charge stiffness D, superfluid stiffness $D_{\rm s}$ ($\equiv \Upsilon$), and longitudinal response $\frac{ne^2}{m}$—which are, in general, different. In this case, because the single-particle spectrum is gapped, it has been proven that charge stiffness and superfluid stiffness are equal (Scalapino *et al.*, 1993).

where the renormalized stiffness was

$$D_s = D_s^0 \exp(-\langle \theta_{ij}^2 \rangle_0 /2). \tag{17.71}$$

Here $\langle \theta_{ij}^2 \rangle_0$ is the mean square fluctuation of the nearest-neighbor phase difference

$$\langle \theta_{ij}^2 \rangle_0 = \frac{2}{\xi} \int_{\mathbf{Q}} \left(\frac{\varepsilon_{\mathbf{Q}}}{D_s \kappa} \right)^{1/2} \tag{17.72}$$

where $\varepsilon_{\mathbf{Q}} = 2(2 - \cos Q_x - \cos Q_y)$ and $\int_{\mathbf{Q}}$ is the average over the Brillouin zone. Defining the renormalization factor $X = D_s/D_s^0$ and

$$\sqrt{\alpha} = \frac{1}{\xi \sqrt{D_s^0 \kappa}} \int_{\mathbf{Q}} \varepsilon_{\mathbf{Q}}{}^{1/2}, \tag{17.73}$$

the SCHA equation becomes

$$X = \exp(-\sqrt{\alpha/X}). \tag{17.74}$$

Solving eqn (17.74) gives the renormalized stiffness D_s as shown in Fig. 17.13. We see that the BdG+SCHA approach *successfully predicts a SIT*—the renormalized

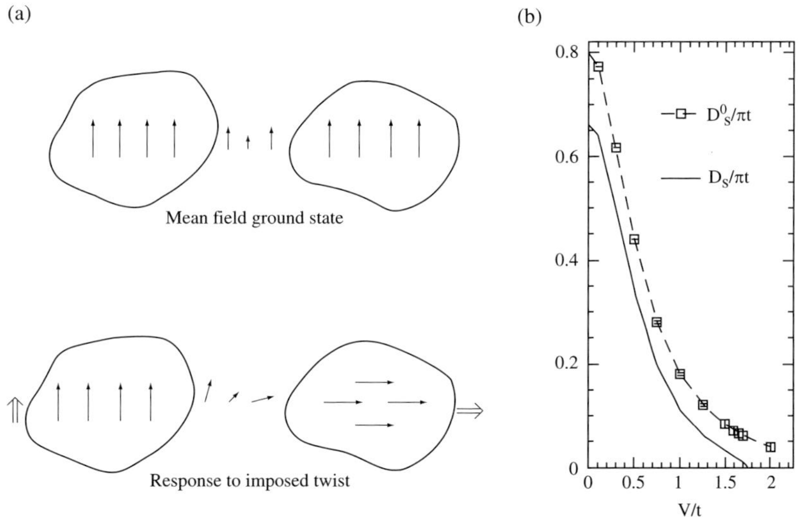

Fig. 17.13 (a) Schematic of a disordered superconductor in which the non-uniform amplitude results in the formation of superconducting islands. The length and direction of each arrow represent the amplitude and phase of $\Delta(\mathbf{r})$. The phases on different islands are weakly coupled, so that the global phase stiffness D_s^0 is greatly reduced from that in a clean superconductor. (b) Renormalization of the BdG phase stiffness D_s^0 to the SCHA value D_s, which vanishes for $V > V_c \approx 1.75t$. Figures are from Ghosal et al. (2001) and parameters were $|U| = 1.5t$ and $\langle n \rangle = 0.875$. The BdG was performed on a 24×24 lattice.

stiffness at $V_c = 1.75t$ for the parameters being considered. The critical disorder obtained from such a calculation is in reasonable agreement with quantum Monte Carlo results for parameter values ($|U| = 4t$) for which such a comparison can be made.

Despite the above success, the SCHA is ultimately a theory of Gaussian fluctuations. It predicts a transition at $\alpha_{\text{crit}} = 4e^{-2}$, with a jump discontinuity of e^{-2} in the value of X, which is probably an artifact. The true transition is expected to be a continuous quantum phase transition in the universality class of the disordered Bose–Hubbard model.[11]

17.3.7 Quantum Monte Carlo

Compared to the PoEE and BdG approaches, determinant quantum Monte Carlo (DQMC) is a computationally intensive but even more powerful technique (Hirsch and Fye, 1986). It includes thermal and quantum fluctuations of the amplitude and phase, and for the model being considered it is free of the sign problem. For a description of the DQMC algorithm as applied to repulsive Hubbard models, please refer to Chapter 5.

- *Superfluid stiffness.* Using DQMC, it was found that the superfluid stiffness D_s does indeed fall to zero at a critical disorder $V = V_c$, as shown in Fig. 17.14 (a). Furthermore, the DC conductivity shows a crossing point between superconducting and insulating behavior, confirming that the model does indeed contain a SIT. See Trivedi *et al.* (1996) and Scalettar *et al.* (1999) for details.

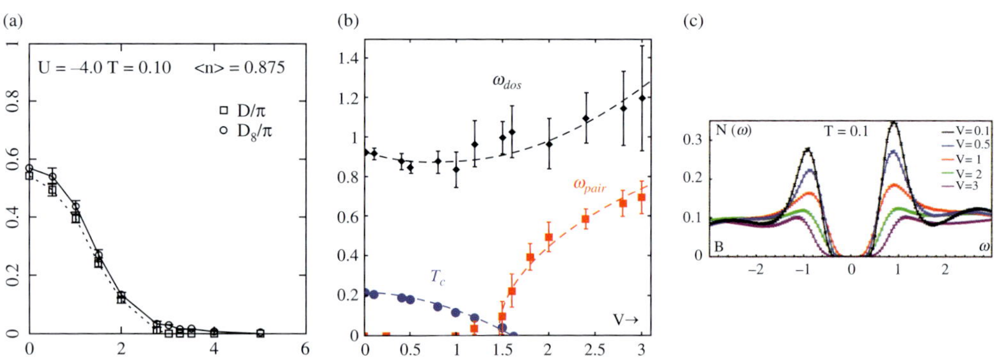

Fig. 17.14 (a) DQMC results for the low-temperature superfluid stiffness D_s as a function of disorder $2V$ (Trivedi et al. (1996)). D_s falls to zero at $V_c \approx 1.6$. (b) DQMC+MEM results for the single-particle gap E_g ($\equiv \omega_{\text{DOS}}$), which remains finite for all V (Bouadim et al. (2011)). (c) DQMC+MEM results for the DOS $N(\omega)$. With increasing V the gap remains robust, but the coherence peaks disappear beyond $V_c \approx 1.6$. All energies are in units of hopping t.

[11]The actual universality class and critical exponents depend on whether particle-hole symmetry is satisfied exactly, on average, or not at all. See Chapter 11 for a discussion of experimentally observed scaling.

- *Single-particle spectrum.* In a more recent DQMC study (Bouadim *et al.*, 2011), dynamical quantities were obtained by analytic continuation of imaginary-time correlation functions using the maximum entropy method (MEM). It was found that the single-particle gap is robust against disorder, and the coherence peaks are more or less correlated with the existence of long-range order (see Fig. 17.14 (b) and 17.14 (c)). This is in agreement with the results of BdG+SCHA.
  The implication is that the SIT in this model occurs via a *bosonic* mechanism, involving the delocalization/localization of *bound* pairs, rather than via a fermionic mechanism—the state at large V is a "Cooper pair insulator." Indeed, recent experiments (Sacépé *et al.*, 2008) provide evidence for such a bosonic SIT in InO_x films.
- *Two-particle spectrum.* A quantum phase transition is typically characterized by energy and temperature scales that vanish as the transition is approached from either side. On the superconducting side, we have seen that D_s and T_c go to zero (Figs. 17.14 (a) and (b)). What are the relevant energy scales on the insulating side? QMC results suggest that there is a peak in the spectrum of *two-particle excitations* in the insulator, and the location of this peak moves to zero energy as the SIT is approached (Bouadim *et al.*, 2011) because pairs can be inserted into a superconductor at no cost. However, this two-particle peak is not fully understood and is a topic of further research.
- *Dynamical conductivity.* Whereas BdG predicts that the conductivity $\mathrm{Re}\,\sigma(\omega)$ has a hard gap $\omega_\sigma^{\mathrm{BdG}} = 2E_\mathrm{g}^{\mathrm{BdG}}$, preliminary results from DQMC+MEM suggest that there is weight within this gap arising from phase fluctuations, i.e. vertex corrections beyond BdG. In particular, the integrated low-energy weight $I = \int_0^{2E_\mathrm{g}^{\mathrm{QMC}}} d\omega\ \mathrm{Re}\,\sigma(\omega)$ was found to be finite, and it had a peak at $V \sim V_\mathrm{c}$, suggesting quantum critical fluctuations near the SIT. However, there are complications from finite-temperature effects and from the difficulty of analytic continuation, and the study of $\sigma(\omega)$ is still a work in progress.

17.3.8 Finite temperature

The SIT is a quantum phase transition that occurs at zero temperature, so we have concentrated our effort on the zero-temperature behavior of the model. The zero-temperature phase diagram nevertheless has important ramifications for properties at finite temperature. For example, we have found that the spectral gap E_g at $T = 0$ persists even on the insulating side of the SIT. What is the effect of temperature? The only sensible scenario is that at finite temperature $0 < T \lesssim E_\mathrm{g}$, the gap must fill up gradually to form a pseudogap. This expectation is borne out by actual BdG calculations at finite temperature, as illustrated in Fig. 17.15.

By examining the behavior of the DOS as a function of disorder V and temperature T, as in Fig. 17.15, one can identify crossover temperature scales T_cp and T_pg for features in the DOS (coherence peaks and pseudogaps), as shown in Figs. 17.16 (a) and (b). Although the critical temperature T_c for spontaneous order within BdG theory is largely unaffected by disorder (because it is controlled by the existence of

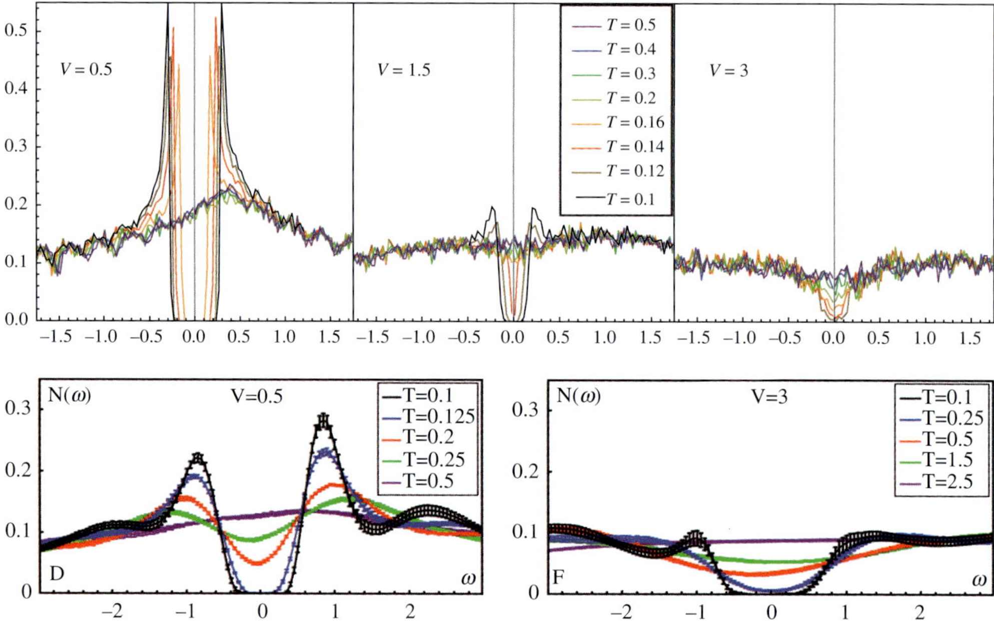

Fig. 17.15 BdG results for (a) U = 2t and (b) U = 4t showing the the evolution of the DOS $N(\omega)$ with disorder V and temperature T. For weak disorder ($V = 0.5$), the coherence peaks remain while the superconducting gap closes. For strong disorder ($V = 3$), the insulating gap fills up slowly, forming a pseudogap (a suppression of the DOS near the Fermi level). Bottom: DQMC results for $|U| = 4$. There is an interaction-induced pseudogap at weak disorder, but otherwise the physics is essentially the same. All energies are in units of the hopping amplitude t.

rare clean regions), the coherence peaks seem to be an indication of whether the system ultimately has phase coherence or not—even though BdG neglects phase fluctuations.

Figure 17.15 shows DQMC+MEM results for the DOS at finite temperature (Bouadim *et al.*, 2011). Due to the relatively large coupling ($|U| = 4$) there is already an interaction-induced pseudogap at weak disorder. Nevertheless, it was found that the size of the pseudogap temperature range increased with disorder, in agreement with BdG. Recent experiments on TiN and NbN films do indeed see a pseudogap up to many times T_c (Mondal *et al.*, 2011; Sacépé *et al.*, 2010).

It is common practice to characterize superconductors by the so-called strong-coupling ratio $2E_g/T_c$, which is the ratio of two experimentally measurable quantities, the zero-temperature gap E_{g0} (from tunneling) and the critical temperature T_c (from transport). BCS MFT predicts that $2E_g/T_c = 2\pi/e^\gamma \approx 3.52775$. In the present situation, DQMC+MEM predicts that E_g is robust against disorder whereas T_c is suppressed to zero, so that the ratio $2E_g/T_c$ tends to infinity (see Fig. 17.14). This is a strong deviation from the BCS result. Experiments on InO_x do indeed see a divergence of this ratio as the SIT is approached (Sacépé *et al.*, 2008).

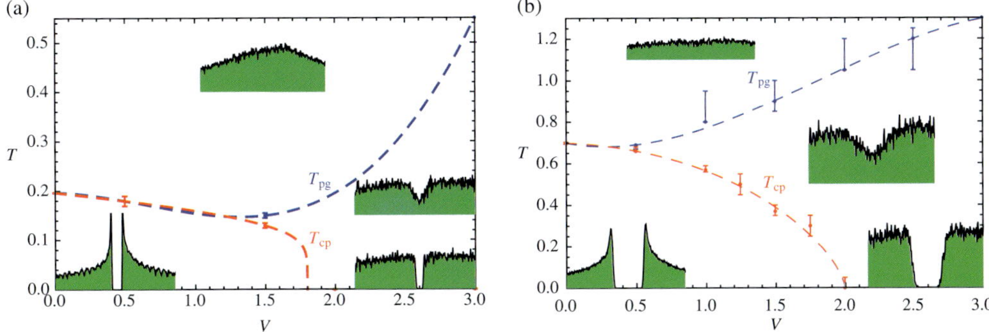

Fig. 17.16 BdG results showing the temperature T_{cp} below which the DOS has coherence peaks, and the temperature T_{pg} above which the pseudogap disappears. These are crossover temperatures, but nevertheless, a clear qualitative trend is visible. All energy scales are in units of the hopping amplitude t.

17.3.9 Emerging picture of the superconductor–insulator transition

A summary of the SIT, within the model of this section, is presented in Fig. 17.17. For a weak-coupling superconductor like Al it is known that the superfluid density ρ_s and energy gap E_g decrease with increasing temperature T and vanish at T_c as seen in Fig. 17.17 (a). The scale for ρ_s is set by the Fermi energy, whereas the scale for E_g is exponentially suppressed from the Fermi energy in weak coupling. The behavior of these two quantities as a function of disorder at $T = 0$ is markedly different. While ρ_s decreases as expected with increasing disorder and vanishes at a critical disorder strength V_c, the gap in the spectrum remains a hard gap for all values of the disorder (see Fig. 17.17 (b)). The behavior at finite disorder as a function of T is quite distinct from the behavior at zero disorder. As seen in Fig. 17.17 (c), ρ_s vanishes at $T_c(V)$ where $T_c(V) < T_c(0)$. However, the energy gap, which started as a hard gap at $T = 0$, starts filling up and finally approaches the normal-state value at a temperature $T^*(V) > T_c(V)$. Thus, for $T_c < T < T^*$, there is a pseudogap: the DOS at the Fermi level $N(0)$ is suppressed relative to its normal-state value. This is a separation between the temperatures for pairing and long-range phase coherence occurring even in a weak-coupling superconductor, this time produced by the combined effects of interaction and disorder. Figs. 17.17 (d) and 17.17 (f) illustrate this picture. Although the simulations in this section were performed at finite interaction U, it can be argued (Ghosal *et al.*, 2001) that the conclusions remain valid in the limit of infinitesimal interaction (Fig. 17.17 (e)).

17.3.10 Summary

The results of this section are summarized below:

- The pairing-of-exact-eigenstates (PoEE) approximation finds that the single-particle gap remains finite for all values of disorder, but fails to describe the vanishing of the coherence peaks.

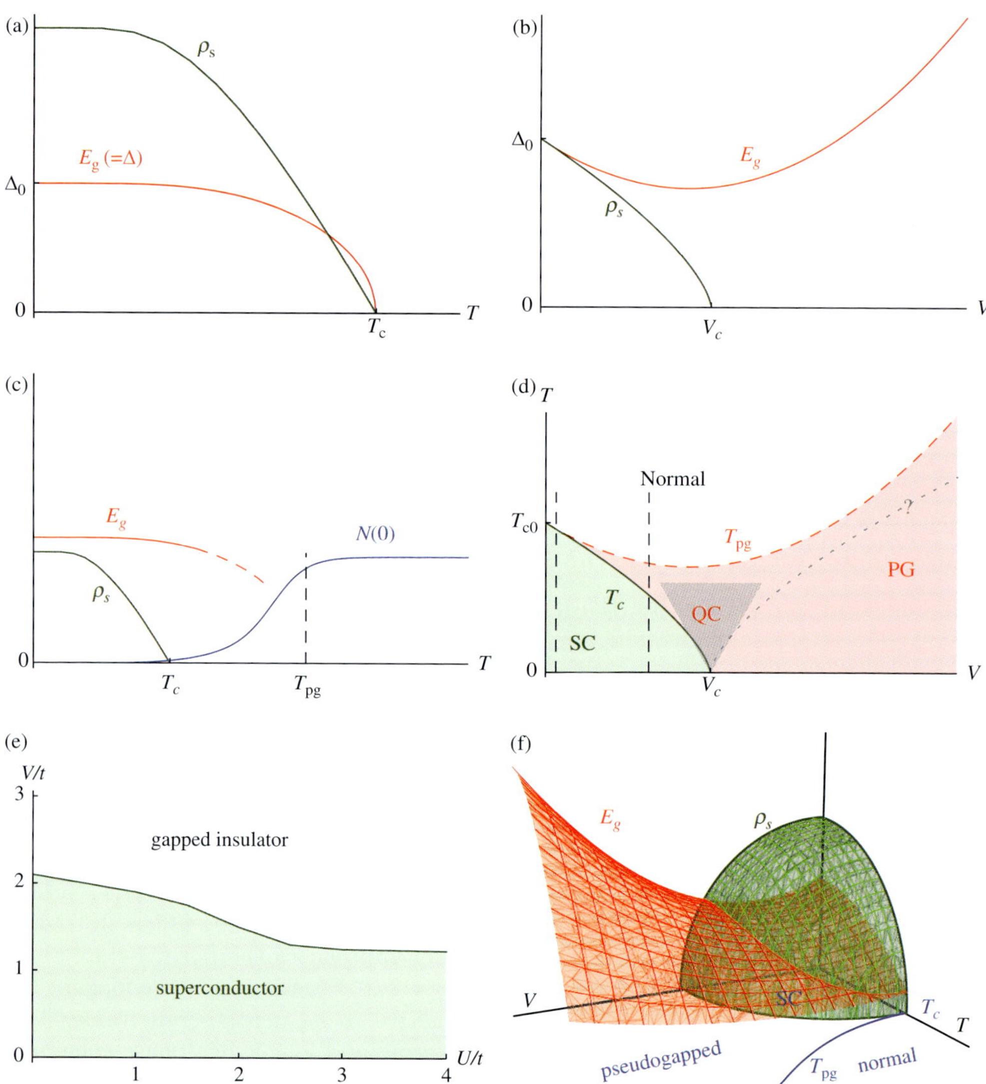

Fig. 17.17 Schematic diagrams illustrating the physics of the disorder-tuned SIT in two dimensions. (a) Temperature dependence in a clean weak-coupling superconductor. (b) Superfluid stiffness ρ_s and single-particle gap E_g as a function of disorder strength V. (c) Temperature dependence in a disordered superconductor near the SIT. (d) Phase diagram, showing normal state, superconducting state (SC), pseudogap state (PG), and quantum critical fan (QC). Dashed vertical lines correspond to Figs. 17.17 (a) and 17.17 (c). (e) Estimated phase diagram of the attractive Hubbard model at filling $n = 0.875$ as a function of attraction U and disorder strength V, adapted from Ghosal *et al.* (2001). (f) Three-dimensional visualization of various quantities as functions of V and T.

- The Bogoliubov–de Gennes (BdG) approach finds that the pairing amplitude becomes extremely inhomogeneous with increasing disorder. It predicts that coherence peaks disappear and that the phase stiffness drops precipitously, but it does not by itself explain the SIT. At finite temperature, BdG predicts the existence of a disorder-induced pseudogap.
- The self-consistent harmonic approximation (SCHA) predicts that quantum phase fluctuations suppress the phase stiffness (and hence the critical temperature) to zero beyond a certain critical disorder, thus capturing the SIT.
- Determinant quantum Monte Carlo (DQMC), which includes all amplitude and phase fluctuations, confirms the above predictions for the gap, coherence peaks, pseudogap, and stiffness.
- The attractive Hubbard model undergoes a bosonic SIT as a function of disorder strength, which is due to localization of Cooper pairs by disorder. This scenario is in agreement with experiments on InOx, TiN, and NbN.

The present model includes attraction and disorder, but ignores the Coulomb repulsion between electrons. Is it possible to include the Finkel'stein mechanism (suppression of uniform pairing amplitude by Coulomb and disorder) together with the physics of amplitude inhomogeneity and phase fluctuations, and thereby obtain a quantitative explanation of experiments across all materials and parameter ranges? This and many other questions (such as the behavior of two-particle spectra and dynamical conductivity) are still unanswered and hopefully will be addressed in future research.

17.4 Parallel field-tuned superconductor–insulator transition

17.4.1 Introduction

The superconducting state is characterized by a large diamagnetic susceptibility due to the Meissner effect and a vanishing paramagnetic susceptibility due to the binding of spins into Cooper pairs. Conversely, applying a magnetic field to a superconductor raises its free energy by inducing diamagnetic currents and by tending to align the electron spins. When an external magnetic field is applied to a superconductor, it suppresses superconductivity via both the orbital effect and the Zeeman effect. We shall consider only parallel fields on thin films, so that the Zeeman effect dominates.[12]

17.4.2 Clean superconductor

The problem of superconductivity in a Zeeman field has been studied since the 1960s. The simplest theories assume a uniform order parameter. With this restriction, there is a first-order transition from a superconductor to a high-field normal metal at the Chandrasekhar–Clogston critical field $h_{\text{CC}} = \Delta_0/\sqrt{2} \approx 0.71\Delta_0$, where Δ_0 is the

[12] In real materials there are complications arising from perpendicular field components, spin-orbit interactions, magnetic impurities, and disorder; there has, however, been some progress toward realizations of the pure Zeeman physics.

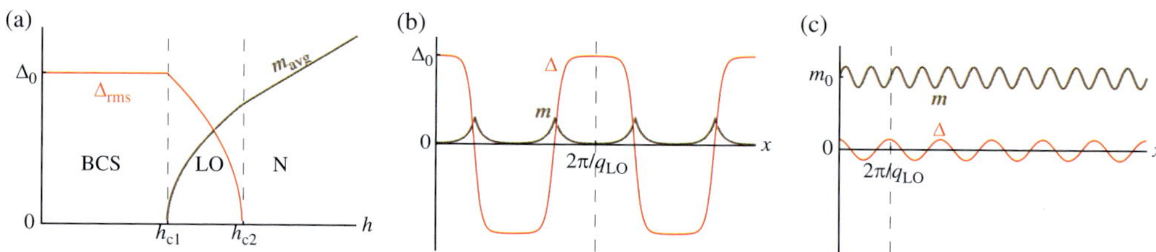

Fig. 17.18 Schematic illustration of Larkin–Ovchinnikov physics. (a) shows the root-mean-square pairing amplitude $\Delta_{\rm rms}$ and average magnetization $m_{\rm avg}$ in the ground state of a superconductor as a function of Zeeman field h, within a mean-field picture. At low field the ground state is a uniform BCS superconductor. As the field is increased beyond a lower critical field h_{c1}, it eventually becomes energetically favorable for magnetization to penetrate the system in the form of domain walls at which the order parameter changes sign, as illustrated in (b). The magnetization (i.e. the excess majority spins) occupies Andreev bound states whose wavefunctions are localized in the domain walls. The spatial periodicity $\lambda_{\rm LO} = \frac{2\pi}{q_{\rm LO}}$ is related to the LO wavevector, $\mathbf{q}_{\rm LO}$, which is the best nesting vector for the Fermi surfaces of the given numbers of up and down spins *in the absence of pairing*. As the field increases further, m increases and $\lambda_{\rm LO}$ decreases, so that the domain walls begin to overlap and the modulation becomes small and sinusoidal. Finally, beyond an upper critical field h_{c2}, pairing is completely destroyed and a uniform magnetization prevails everywhere.

zero-temperature zero-field gap (Chandrasekhar, 1962; Clogston, 1962; Gor'kov and Rusinov, 1964; Sarma, 1963). However, allowing the mean-field order parameter to vary in space reveals that in an intermediate range of fields the system can lower its free energy by forming periodic patterns known as Fulde–Ferrell–Larkin–Ovchinnikov (FFLO) states. Fulde and Ferrell (1964) used the ansatz $\Delta(\mathbf{r}) \propto e^{i\mathbf{Q}\cdot\mathbf{r}}$, which breaks time-reversal symmetry, whereas Larkin and Ovchinnikov (1964) studied additional modulation patterns of the form $\Delta(\mathbf{r}) \propto \sum_{\mathbf{Q}} \cos \mathbf{Q}\cdot\mathbf{r}$, which break translational symmetry. It is generally found that LO states are favored over FF states, so we will henceforth simply refer to LO states. A general LO state consists of regions of positive and negative pairing amplitude $\Delta(\mathbf{r})$ separated by a regular array of domain walls where the majority fermions are concentrated; it exemplifies the emergent phenomenon of "microscale phase separation" (Radzihovsky and Vishwanath, 2009). A proper treatment of the LO state requires some amount of numerical work (Burkhardt and Rainer, 1994; Koponen *et al.*, 2007; Loh and Trivedi, 2010; Machida and Nakanishi, 1984; Yoshida and Yip, 2007); see Casalbuoni and Nardulli (2004) for a review. Figure 17.18 illustrates one version of the story.

In the 3D continuum, FFLO only occupies a tiny sliver of the mean-field phase diagram, between $h_{c1} = 0.665$ and $h_{c2} = 1$(Sheehy and Radzihovsky, 2006). Furthermore, quantum and thermal fluctuations destroy even this sliver (Radzihovsky and Vishwanath, 2009). In 2D, the mean-field FFLO region is larger ($h_{c2} = 1$), as shown in Fig. 17.19, but fluctuations are even more severe. Thus, it is not surprising that FFLO order has not been observed except in some reports on layered organic and heavy-fermion superconductors (Radovan *et al.*, 2003). We have attempted to summarize the

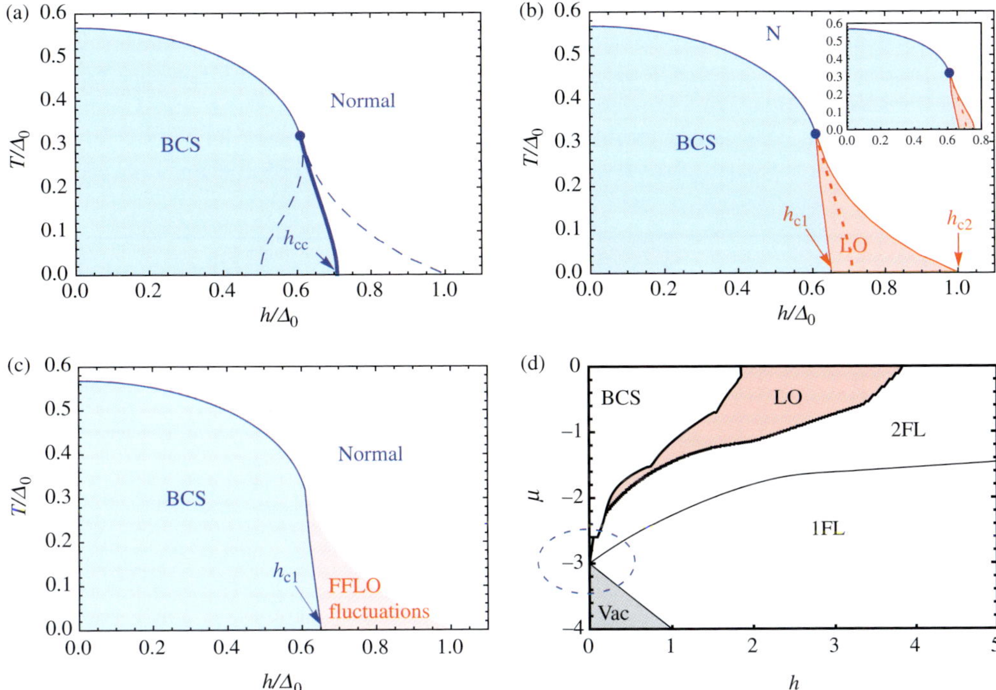

Fig. 17.19 Approximate phase diagrams of a clean s-wave superconductor, at weak coupling and in the continuum, as a function of Zeeman field h and temperature T in units of the zero-temperature zero-field pairing amplitude Δ. (a) Phase diagram assuming a uniform mean-field pairing amplitude Δ. At low temperatures T and fields h the superconductor is stable. At high T or strong h the system becomes a polarized normal Fermi liquid. The thin solid curve is a second-order boundary. The thick solid curve is the first-order boundary at which $\Omega_s = \Omega_n$. The dashed curves are the limits of metastability of the normal state in the superconductor and vice versa. (b) Phase diagram allowing inhomogeneous pairing $\Delta(\mathbf{r})$. The dotted curve is the hypothetical first-order BCS-normal boundary, corresponding to the solid curve in Fig. 17.19. At the lower critical field h_{c1} (shown schematically), magnetization begins to penetrate the superconductor in domain walls at which the pairing amplitude changes sign. At the upper critical field $h_{c2} = \Delta_0$ (for the case of a 2D continuum), the normal polarized Fermi liquid becomes unstable to FFLO pairing. The inset shows the analogous phase diagram in three dimensions, for which $h_{c2} \approx 0.755\Delta_0$. (c) Fluctuations destroy the continuum LO state in two and three dimensions, as illustrated here schematically. (d) BdG phase diagram of the cubic lattice (3D) Hubbard model as a function of field h and chemical potential μ, in units of the hopping amplitude t. The LO region is considerably larger than for the 3D continuum. Furthermore, the lattice suppresses translation and rotation of the LO pattern, so quantum fluctuations should be less severe.

above discussion in Fig. 17.19 and 17.20, although it should be acknowledged that the phase diagram is exquisitely sensitive to strong-coupling corrections, material properties, and calculational methods, and that many aspects of FFLO physics continue to be debated (Bulgac and Forbes, 2008; Matsuo *et al.*, 1998; Mora and Combescot, 2005; Shimahara, 1998). For example, some authors find first-order transitions to a crystalline LO state with minority spins localized in a superconducting background.

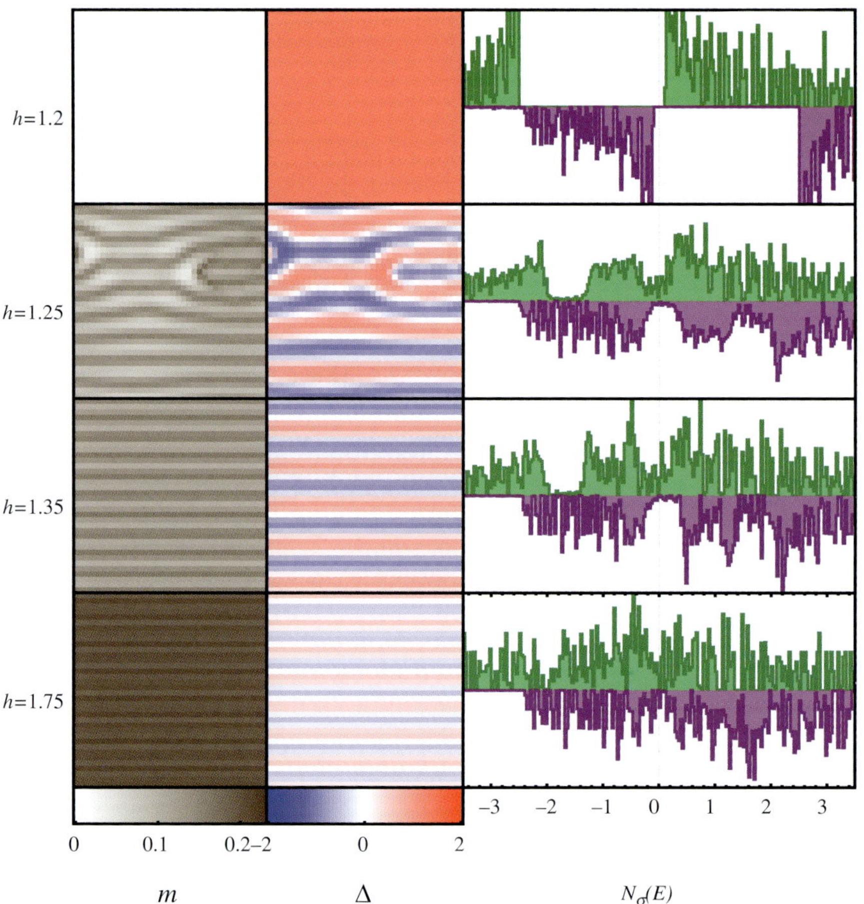

Fig. 17.20 Evolution of a clean superconductor with increasing Zeeman field within BdG (the disorder strength was set to a negligible value, $V = 0.02$). The panels show the results of independent BdG calculations starting from random configurations. Because of symmetry, some runs converged to vertical patterns instead of horizontal patterns; these are not shown. For $h = 1.25$, the smectic LO pattern is disrupted by a pair of dislocations. In this case, the BdG procedure had become trapped in a local minimum of the free energy, and was unable to find the global minimum. This does, however, reflect actual physics that may happen in experiments on condensed matter or cold atoms. All energy scales are in units of the hopping amplitude t.

In contrast, cold Fermi gases in optical lattices are a promising arena in which to search for FFLO physics (Koponen *et al.*, 2007; Loh and Trivedi, 2010; Parish *et al.*, 2007; Zhao and Liu, 2008). For example, the cubic lattice phase diagram Fig. 17.19 (d)) has a much larger LO region than the 3D continuum phase diagram (inset of Fig. 17.19 (b)). The most favorable systems appear to be coupled tubes or anisotropic lattices at weak-to-intermediate coupling. Since this chapter is about the field-tuned SIT, we shall not dwell on this topic; we will just present some BdG pictures in order to make a connection with the next section, which includes disorder.

The full BdG calculation follows the formalism described in the previous chapter.[13] BdG calculations on lattices have a strong tendency to give LO states in suitable parameter regimes: if the system is initialized in a BCS-like state, it may go through many iterations, exploring the free-energy landscape, before settling into a reasonable LO state.

17.4.3 Dirty superconductor

In the presence of disorder potential as well as a parallel field, diagrammatic calculations by Zhou and Spivak (1998) suggest that the superconductor becomes like an "XY glass", with random positive and negative Josephson couplings, so that there are many metastable states, in which the superconducting order parameter is positive and negative in different places.

Cui and Yang (2008) did BdG calculations of the kind described in this article. They found that an LO-like ground state still survives at weak disorder (see Fig. 17.21). Although the orientation of the LO stripes is disrupted, there is still some signs of periodicity.

Dubi, Meir, and Avishai (2007; 2008) did BdG calculations, as well as Monte Carlo calculations in which the BdG pairing amplitude was treated as a fluctuating auxiliary field. They found that the superconducting islands are destroyed with the application

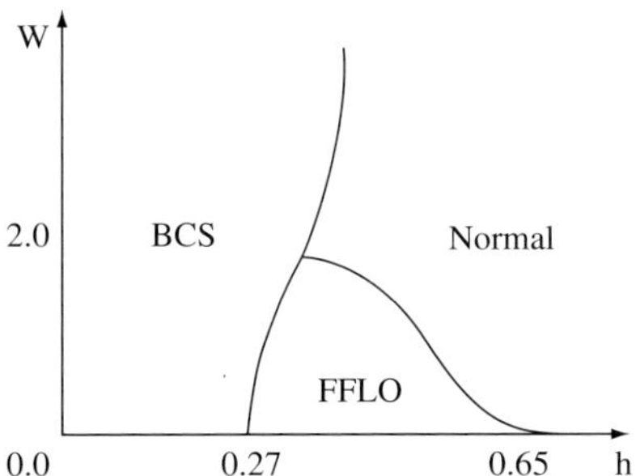

Fig. 17.21 Approximate phase diagram of a superconductor in a Zeeman field h and a random potential of strength W (Cui et al. (2008)). A disordered (FF)LO ground state exists over a significant field range $h_{c1} < h < h_{c2}$, provided that disorder is not too large.

[13]Due to the breaking of spin symmetry, all $2N$ eigenvalues and eigenvectors are now independent, and certain optimizations are no longer possible.

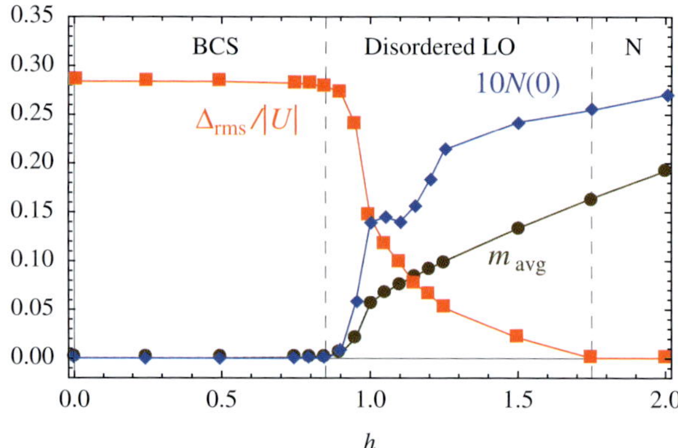

Fig. 17.22 Disordered superconductor in Zeeman field. Root-mean-square pairing $\Delta_{\rm rms}$ (red squares), average magnetization $m_{\rm avg}$ (brown circles), and Fermi-level DOS $N(0)$ (blue diamonds), as functions of Zeeman field h. Between the two dashed lines there is a disordered LO state with coexistent pairing and magnetization, in which the gap is partially filled in. These are results of BdG simulations on a 36×36 square-lattice Hubbard model with attraction $|U|/t = 4$, disorder strength $W/t = 1$, and temperature $T/t = 0.1$, for different values of the Zeeman field h.

of a field. However, they did not report sign changes of the order parameter and the connection to FFLO physics.

Figure 17.22 shows the coexistence of pairing and magnetization as a function of Zeeman field h. As the Zeeman field is increased beyond a critical field, a finite DOS develops within the gap, due to the formation of a disordered Larkin–Ovchinnikov (dLO) state with bound states in domain walls (Loh *et al.*, 2011). This is illustrated in Fig. 17.22.

Figure 17.23 illustrates the order parameter $\Delta(\mathbf{r})$, the magnetization density $m(\mathbf{r}) = \frac{1}{2}\left[n_\uparrow(\mathbf{r}) - n_\downarrow(\mathbf{r})\right]$, and the spatially averaged densities of states of up and down spins $A_\sigma(E)$, for various values of field. The disorder strength is $W = 1$, which corresponds to a normal-state sheet resistance $R_\square$ of the order of $0.3R_{\rm Q}$ (the sheet resistance in zero field at temperatures somewhat above $T_{\rm c}$), where $R_{\rm Q} = \frac{h}{4e^2} \approx 6.4\ {\rm k}\Omega$ is the quantum resistance appropriate to Cooper-paired systems.

At low fields the system is a BCS superconductor with a nearly uniform order parameter $\Delta(\mathbf{r}) \approx \Delta_0$, whose DOS contains coherence peaks at $\pm\Delta \pm h$ that are slightly broadened by inhomogeneous Hartree shifts (Ghosal *et al.*, 1998, 2001). At high fields the system is a normal metal with nearly uniform magnetization.[14] However, at intermediate fields, the simulations find inhomogeneous states where the order parameter $\Delta(\mathbf{r})$ has sign changes and there are patches of finite magnetization.

[14]In principle, weak localization corrections in two dimensions would cause the system to be an insulator; such effects are not visible on the scale of the simulated systems. Also, in real films, the interplay of disorder and Coulomb repulsion (which is absent from our model) produces Altshuler–Aronov corrections, which suppress the Fermi-level DOS.

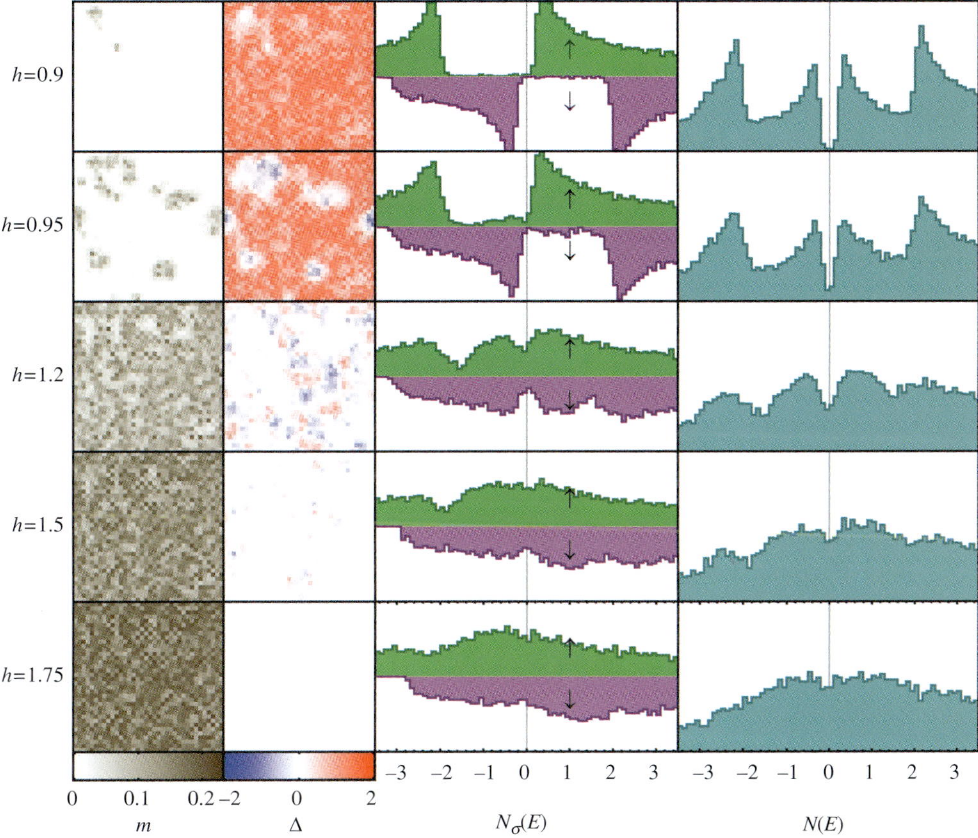

Fig. 17.23 The first two columns show spatial maps of the order parameter Δ and the magnetization m. The last two columns show the DOSs of up and down electrons $N_\sigma(E)$ as well as the total DOS $N(E)$. For intermediate fields the system exhibits disordered LO states with domain walls at which m is finite and the sign of $\Delta(\mathbf{r})$ varies in space (as indicated by the red and blue patches). The appearance of such states is associated with the filling-in of the gap in the DOS. Parameters are as in Fig. 17.22.

These can be viewed as disordered FFLO states (Cui and Yang, 2008), in which the order-parameter oscillations at wavevector $q_{\mathrm{FFLO}} \approx 2k_{\mathrm{F}}$ are partially disrupted by the disorder potential. Since the lowest-energy solutions always have a real order parameter $\Delta(\mathbf{r})$, we will refer to these as disordered Larkin–Ovchinnikov (LO) states, to emphasize the contrast with higher-energy Fulde–Ferrell (FF) states with a more restrictive complex-valued ansatz for $\Delta(\mathbf{r})$. This is in contrast with the traditional view of a first-order transition directly from a BCS state with a hard gap to a gapless Fermi liquid.

To gain some insight into the sign changes in the order parameter, it is interesting to consider a different approach studied by Zhou and Spivak (1998): the effective functional describing the order parameter $\Delta(\mathbf{r})$ takes the form of an XY spin-glass Hamiltonian. The ground state of such a spin glass has real $\Delta(\mathbf{r})$, with positive and negative signs. Zhou and Spivak did not, however, make the connection between XY spin glasses and LO physics.

It should be noted that the DOS in the high-field state exhibits a very interesting many-body effect called the pairing resonance, which was predicted by Aleiner and Altshuler (1997) and Kee *et al.* (1998), and discovered by Butko *et al.* (1999). Unfortunately, a full discussion of this phenomenon is beyond the scope of this chapter.

Even if an ordered FFLO state is prevented by quantum fluctuations, fluctuating FFLO physics may nevertheless be important near the BCS-normal transition. In particular, the DOS is a probe of local physics at all energies, not of global order, so it is likely that some basic features of the LO state (such as the "soft gap") may remain even when the system does not possess long-range order.

17.4.4 Summary

At intermediate fields, the pairing and magnetization coexist in a disordered pattern with some remnant periodicity (as found by Cui and Yang), with spontaneously occurring "π-junctions" where the magnetization is concentrated. Furthermore, such dirty LO states are associated with peculiar "soft gaps" at the Fermi energy in the minority spin DOS (Loh *et al.*, 2011).

17.5 Conclusions

In this article we have examined two types of SITs using various analytical and numerical methods (PoEE, BdG, SCHA, and DQMC). In the disorder-tuned SIT, the Cooper pairs become localized by the disorder potential, leading to an insulating "Cooper pair glass" (expected to be adiabatically connected to the Bose glass). The single-particle gap *remains finite* across the SIT. The theory predicts a pseudogap at finite temperature, which is indeed observed in experiments. Many other predictions remain to be tested.

The parallel-field-tuned SIT, from the viewpoint of BdG simulations and with some experimental evidence, proceeds via a disordered LO state. The gap gradually *fills up* due to Andreev bound states. The pair-breaking effect of the field ultimately leads to a fermionic Anderson insulator.

Clearly, the "I" in "SIT" can stand for many different types of insulator. Our studies lead to a classification of electronic ground states based on transport as well as spectral properties, as shown in Table 17.1.

In this article we have ignored the effects of Coulomb repulsion. Sufficiently near the SIT, it is likely that the Coulomb effects renormalize away and the phonon-mediated attraction is the dominant effect (as in the theory of ordinary superconductors). However, the Coulomb amplitude suppression mechanism may play an important role in determining the overall shape of the phase diagram (see Chapter 10).

Table 17.1 Comparison of various fermionic ground states, characterized by transport (metallic/superconducting/insulating), single-particle DOS $N(\omega)$, and dynamical conductivity $\sigma(\omega)$.

State	Description	$N(\omega)$	$\sigma(\omega)$
Fermi liquid	Fermi sea filled up to k_{F}	Gapless	Gapless
Band insulator	Fermions localized by Pauli exc.	Hard gap	Hard gap
Anderson insulator	Fermions localized by disorder	Gapless	Soft gap (WL/MVRH)
Fermi MI	Fermions localized by Hubbard repulsion	Soft gap	Soft gap
Electron glass	Fermions localized by disorder and Coulomb repulsion	Soft gap (ES)	Soft gap (ES VRH)
Superconductor	Mobile Cooper pairs	Hard gap + coh peaks	$\delta(\omega)$ + hard gap
Cooper pair MI	CPs localized by Coulomb blockade	Hard gap	Hard gap
Cooper pair glass	CPs localized by disorder	Hard gap	Unknown

Abbreviations: MI, Mott insulator; CP, Cooper pair; WL, weak localization; (M)VRH, (Mott) variable-range hopping; ES, Efros-Sh'klovskii; coh, coherence.

17.6 Conventions and standard formulas for imaginary-time Green's functions

Consider the Hamiltonian for a single eigenmode, $H = E_\alpha c_\alpha^\dagger c_\alpha$. We will define the fermionic imaginary-time Green's function as the amplitude for inserting a fermion at zero time and removing it at time τ, $G_{\alpha\tau} = \left\langle \mathcal{T} c_{\alpha\tau} c_{\alpha 0}^\dagger \right\rangle = \langle \psi_{\alpha\tau} \bar{\psi}_{\alpha 0} \rangle$, where ψ and $\bar{\psi}$ are the Grassmann fields in the coherent-state path-integral formalism. This Green's function has an exponential behavior and is antiperiodic with period β such that

$$G_{\alpha\tau} = \begin{cases} -\beta < \tau < 0 & f_\alpha e^{-\tau E_\alpha} \\ 0 < \tau < \beta & (f_\alpha - 1) e^{-\tau E_\alpha} \end{cases} \tag{17.75}$$

where $f_\alpha = f(\beta E_\alpha) = \frac{1}{e^{\beta E_\alpha} - 1} = \frac{1}{2} - \frac{1}{2} \tanh \frac{\beta}{2} E_\alpha$, as illustrated in Fig. 17.24. In the Matsubara frequency domain this Green's function becomes

$$G_\alpha(i\varepsilon_n) = \frac{1}{i\varepsilon_n - E_\alpha}. \tag{17.76}$$

Analytically continuing to real frequencies and taking the imaginary part gives $-\frac{1}{\pi} \operatorname{Im} G(E + i0^+) = \delta_{E - E_\alpha}$.

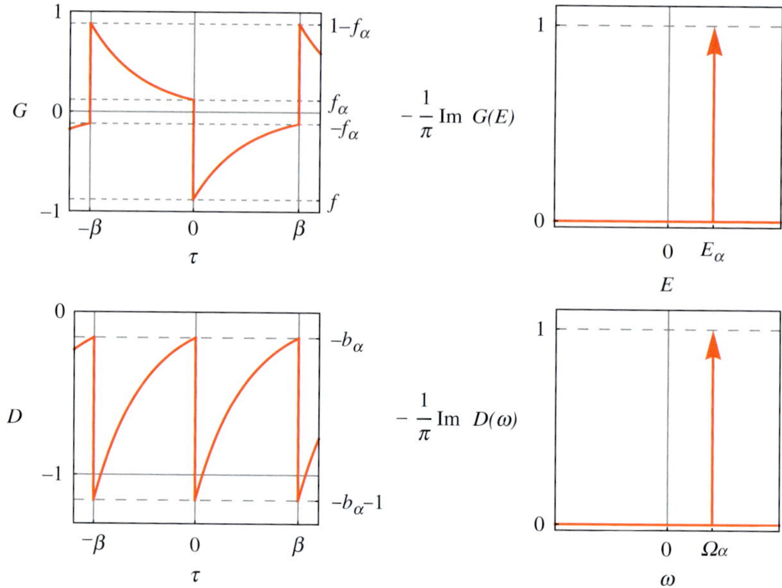

Fig. 17.24 Imaginary-time fermion Green's function $G_{\alpha\tau}$ and boson Green's function $D_{\alpha\tau}$ corresponding to delta-function spectra (single eigenmodes).

Similarly, the standard bosonic Green's function for a simple harmonic oscillator, $H = \Omega a^\dagger a$, is

$$D_{\alpha\tau} = \begin{cases} -\beta < \tau < 0 & -b_\alpha e^{-\tau\Omega_\alpha} \\ 0 < \tau < \beta & -(b_\alpha + 1)e^{-\tau\Omega_\alpha} \end{cases} \tag{17.77}$$

where $b_\alpha = b(\beta\Omega_\alpha) = \frac{1}{2}\coth\frac{\beta}{2}\Omega_\alpha - \frac{1}{2} = \frac{1}{e^{\beta\Omega_\alpha}-1}$. This corresponds to

$$D_\alpha(i\omega_l) = \frac{1}{i\omega_l - \Omega_\alpha}. \tag{17.78}$$

Analytically continuing to real frequencies and taking the imaginary part gives $-\frac{1}{\pi}\operatorname{Im} D(E + i0^+) = \delta_{\omega-\Omega_\alpha}$.

Now consider the pairing bubble diagram corresponding to the creation of two fermions in eigenmodes α and β, as shown in Fig. 17.25:

$$\begin{aligned} P_\tau &= G_{\alpha\tau}G_{\beta\tau} \\ &= (1 - f_\alpha)(1 - f_\beta)e^{-\tau(E_\alpha + E_\beta)} \quad (\text{for } 0 < \tau < \beta). \end{aligned}$$

Fourier transforming to Matsubara frequencies gives

$$P(i\omega_l) = \frac{f_\alpha + f_\beta - 1}{i\omega_l - E_\alpha - E_\beta} \tag{17.79}$$

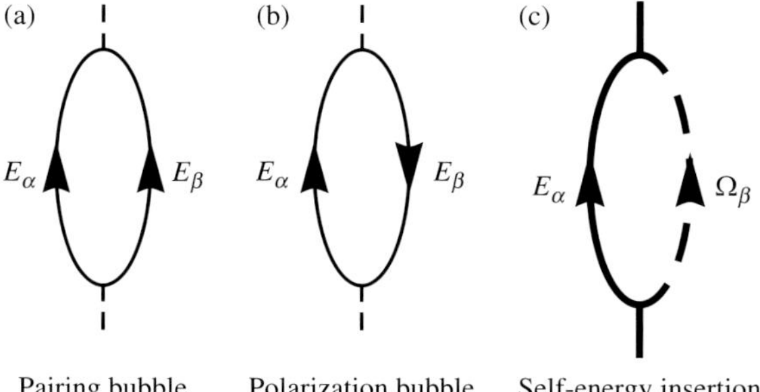

Fig. 17.25 Two-particle diagrams involving fermion lines (solid) and boson lines (dashed).

after some algebra.

This is a very useful formula, as it tells us the spectrum of a bosonic response in terms of the spectra (eigenenergies) of the fermionic states. The physical meaning is that the amplitude for creating a fermion pair in modes α and β depends on the occupation of those two modes. The formula automatically satisfies the requirement that $-\frac{1}{\pi}\operatorname{Im} P(\omega)$ has the same sign as ω.

Note that $P(\omega = 0)$, like $D(\omega = 0)$, is negative with these conventions. Response functions like compressibility are traditionally defined to be positive at zero frequency; therefore, they correspond to the *negatives* of bosonic propagators.

The formula for a polarization (particle-hole) bubble, Fig. 17.25, can be derived simply by changing the signs of E_α and E_β:

$$\kappa_\tau = G_{\alpha\tau} G_{\beta,-\tau}, \tag{17.80}$$

$$\kappa(i\omega_l) = \frac{f_\alpha - f_\beta}{i\omega_l - E_\alpha + E_\beta}. \tag{17.81}$$

Analogous formulas can be derived for the self-energy of a fermion due to emission of a boson (Fig. 17.25):

$$\Sigma_\tau = G_{\alpha\tau} D_{\beta\tau}, \tag{17.82}$$

$$\Sigma(i\varepsilon_n) = \frac{-(1 - f_\alpha + b_\beta)}{i\varepsilon_n - E_\alpha - \Omega_\beta}. \tag{17.83}$$

17.7 Derivation of Kubo formulas for electromagnetic response

17.7.1 General derivation

Here we derive the Kubo formula for the electromagnetic response of the Anderson model, a BCS superconductor, and a dirty superconductor.

The response to an electromagnetic field (superfluid response, conductivity, and dielectric polarizability) is an important quantity characterizing the SIT. We now derive the Kubo formula for the electromagnetic response tensor (Scalapino *et al.*, 1993). For simplicity we will omit spin indices. We will work in units where $e = \hbar = 1$. In this system, the unit of magnetic flux is $\hbar/e$, and the unit of conductivity is $e^2/\hbar$. Thus (for a Cooper-paired system) the flux quantum is $\Phi_Q = h/e = 2\pi$ and the conductance quantum is $G_Q = e^2/4h = 2/\pi$.

For non-relativistic charged particles in the continuum, the electromagnetic vector potential $\mathbf{A}(\mathbf{r})$ couples to the matter fields in a gauge-invariant way in the kinetic energy term of the Hamiltonian

$$H_{\text{kin}} = \int d^d r \, \tfrac{1}{2m} c^\dagger(\mathbf{r})(-i\nabla - \mathbf{A}(\mathbf{r})) \cdot (i\nabla - \mathbf{A}(\mathbf{r}))\mathbf{c}(\mathbf{r}).$$

For charged particles on a lattice the analogous term is

$$H_{\text{kin}} = -\sum_{ij} t_{ij} e^{iA_{ij}} c_i^\dagger c_j. \tag{17.84}$$

where the lattice vector potential is the line integral of the continuum vector potential, $A_{ij} = \int_{\mathbf{r}_i}^{\mathbf{r}_j} \mathbf{dr} \cdot \mathbf{A}(\mathbf{r})$. The units of A are V s m^{-1}, whereas the units of A_{ij} are V s, so $\frac{e}{\hbar} A_{ij}$ is dimensionless as required by eqn (17.84).

Following the usual Kubo procedure, we compute the response to a general vector potential that depends on imaginary time and space, $A_{\tau b}$. It will be convenient to perform this derivation in the coherent-state path-integral formalism, where fermion operators c are replaced by Grassmann fields ψ. The kinetic energy term in the action is

$$S_{\text{kin}}[\psi, \bar{\psi}, A] = \int_\tau \sum_{ij} t_{ij} e^{iA_{ij\tau}} \bar{\psi}_{i\tau} \psi_{j\tau}. \tag{17.85}$$

The full action S may contain other terms that do not depend on A.

The current field (corresponding to the current operator $\hat{j}$ in the operator formalism) can be identified as the field that couples linearly to A:

$$j_{ij\tau}[\psi, \bar{\psi}, A] = \frac{\partial S}{\partial A_{ij}} = i t_{ij} e^{iA_{ij\tau}} \bar{\psi}_{i\tau} \psi_{j\tau}, \tag{17.86}$$

where the subscripts i are site indices whereas those in the prefactors refer to $\sqrt{-1}$. (For simplicity in later derivations, we are treating A_{ij} and A_{ji} as independent fields, so in the above equation, j_{ij} only includes the forward current from site i to site j.)

For later convenience, let k be the second derivative of the action,

$$k_{ij\tau}[\psi, \bar{\psi}, A] = \frac{\partial^2 S}{\partial A_{ij}^2} = -t_{ij} e^{iA_{ij\tau}} \bar{\psi}_{i\tau} \psi_{j\tau}. \tag{17.87}$$

The partition function, or generating functional, is

$$Z[A] = \int [d\psi \; d\bar{\psi}] \exp S[\psi, \bar{\psi}, A]. \tag{17.88}$$

The electromagnetic response function is the second derivative of the free energy:

$$\begin{aligned} \Upsilon_{ijkl\tau\tau'} &= -\frac{\partial \langle j_{ij\tau} \rangle}{\partial A_{kl\tau'}} = -T \frac{\partial^2 \ln Z}{\partial A_{ij\tau} \partial A_{kl\tau'}} \\ &= -T \frac{\partial^2}{\partial A_{ij\tau} \partial A_{kl\tau'}} \ln \int [d\psi \; d\bar{\psi}] \exp S[\psi, \bar{\psi}, A]. \end{aligned} \tag{17.89}$$

Differentiating the ln and exp functions produces three terms:

$$\begin{aligned} \beta \Upsilon_{ijkl\tau\tau'} = &- \left\langle \frac{\partial^2 S}{\partial A_{ij\tau} \; \partial A_{kl\tau'}} \right\rangle - \left\langle \frac{\partial S}{\partial A_{ij\tau}} \; \frac{\partial S}{\partial A_{kl\tau'}} \right\rangle \\ &+ \left\langle \frac{\partial S}{\partial A_{ij\tau}} \right\rangle \left\langle \frac{\partial S}{\partial A_{kl\tau'}} \right\rangle . \end{aligned} \tag{17.90}$$

We shall be studying the linear response to an infinitesimal perturbation, so we set $A_{ij} = 0$. Then, the last term in eqn (17.90) vanishes. Also, the response function depends only upon $\tau - \tau'$, so we can set $\tau' = 0$ without loss of generality. In terms of k and j,

$$\beta \Upsilon_{ijkl\tau} = \delta_{ik} \delta_{jl} \delta(\tau) \langle -k_{ij} \rangle - \langle j_{ij\tau} j_{kl} \rangle , \tag{17.91}$$

where $k_{ij} = -t_{ij} \bar{\psi}_i \psi_j$ is the bond kinetic energy and $j_{ij\tau} = i t_{ij} \bar{\psi}_{i\tau} \psi_{j\tau}$ is the current field. (We are using the convention that omitted τ indices mean $\tau = 0$, in other words $\psi_i \equiv \psi_i(\tau = 0)$, just as in the operator formalism, where $c_i \equiv c_i(\tau = 0)$.)

Ultimately, we wish to find the electromagnetic (EM) response tensor $\Upsilon_{\mu\nu\mathbf{q}\omega} = -\partial j_{\nu\mathbf{q}\omega} / \partial A_{\mu\mathbf{q}\omega}$, which is the current response to an applied vector potential with wavevector $\mathbf{q}$, frequency ω, and polarization μ. Here we consider the transformations from the lattice basis ij to the polarization-wavevector basis $\mu\mathbf{q}$ (omitting τ for the moment):

$$A_{ij} = \sum_{\mu\mathbf{q}} e^{i\mathbf{q}\cdot\mathbf{r}_i} r_{ij\mu} A_{\mu\mathbf{q}}, \tag{17.92}$$

$$j_{\mu\mathbf{q}} = \sum_{ij} e^{-i\mathbf{q}\cdot\mathbf{r}_i} r_{ij\mu} j_{ij} = \sum_{ij} e^{-i\mathbf{q}\cdot\mathbf{r}_i} r_{ij\mu} i t_{ij} \bar{\psi}_i \psi_j, \tag{17.93}$$

$$-k_{\mu\nu} = -\sum_{ij} r_{ij\mu} r_{ij\nu} k_{ij} = \sum_{ij} r_{ij\mu} r_{ij\nu} t_{ij} \bar{\psi}_i \psi_j, \tag{17.94}$$

where $\mathbf{r}_i$ is the position vector of site i and $\mathbf{r}_{ij} = \mathbf{r}_j - \mathbf{r}_i$ is the displacement vector from site i to site j. In terms of these quantities,

$$\Upsilon_{\mu\nu\mathbf{q}\tau} = \delta(\tau) \langle -k_{\mu\nu} \rangle - \langle j_{\mu\mathbf{q}\tau} j_{\nu\bar{\mathbf{q}}} \rangle . \tag{17.95}$$

Equation (17.95) is a Kubo formula for the electromagnetic linear response function in terms of the current–current correlation. This is an example of the fluctuation-dissipation theorem. The first term is the kinetic energy for each bond weighted by a geometrical factor depending on the hopping distance, and it gives a diamagnetic response (the induced $\mathbf{j}$ is opposite to the applied $\mathbf{A}$). The second term is the current–current correlation function, which gives a paramagnetic contribution.

The derivation of eqn (17.95) is very general. It can easily be adapted for XY models instead of fermions, for example, by considering an XY model action coupled to an A field, $S[\theta, A] = \sum_{ij} \beta J_{ij} \cos(\theta_i - \theta_j + A_{ij})$, and carrying out the differentiations to obtain j, k, and Υ.

17.8 Electromagnetic response of Anderson model

Now let us derive the Kubo formula explicitly for the Anderson model.

Return to the operator formalism (replace $\bar{\psi}_i \psi_j$ by $c_i^\dagger c_j$), and transform the fermion operators c_i from the site basis to the eigenmode basis, γ_α:

$$
\begin{aligned}
c_i &= \sum_\alpha \phi_{i\alpha} \gamma_\alpha, \\
k_{\mu\nu} &= -\sum_{\alpha\beta ij} r_{ij\mu} r_{ij\nu} t_{ij} \phi_{i\alpha}^* \phi_{j\beta} \gamma_\alpha^\dagger \gamma_\beta, \\
j_{\mu\mathbf{q}\tau} &= \sum_{\alpha\beta} \sum_{ij} e^{-i\mathbf{q}\cdot\mathbf{r}_i} r_{ij\mu} i t_{ij} \phi_{i\alpha}^* \phi_{j\beta} \gamma_{\alpha\tau}^\dagger \gamma_{\beta\tau} = \sum_{\alpha\beta} \Gamma_{\alpha\beta\mu\mathbf{q}} \gamma_{\alpha\tau}^\dagger \gamma_{\beta\tau},
\end{aligned} \tag{17.96}
$$

where

$$
\Gamma_{\alpha\beta\mu\mathbf{q}} = \sum_{ij} e^{-i\mathbf{q}\cdot\mathbf{r}_i} r_{ij\mu} i t_{ij} \phi_{i\alpha}^* \phi_{j\beta} \tag{17.97}
$$

are matrix elements for the coupling between electromagnetic plane waves and particle-hole excitations in disorder eigenstates. Since the Hamiltonian is bilinear and diagonal in the eigenbasis, expectations of products of γs can be conveniently reduced to products of Green functions using Wick's theorem. Using $\langle \gamma_\alpha^\dagger \gamma_\beta \rangle = f_\alpha \delta_{\alpha\beta}$, $\left\langle \gamma_{\alpha\tau} \gamma_\beta^\dagger \right\rangle = G_{\alpha\tau} \delta_{\alpha\beta}$, etc., we obtain

$$
\begin{aligned}
\langle -k_{\mu\nu} \rangle &= \sum_{\alpha ij} r_{ij\mu} r_{ij\nu} t_{ij} \phi_{i\alpha}^* \phi_{j\alpha} f_\alpha, \\
\langle j_{\mu\mathbf{q}\tau} j_{\nu\bar{\mathbf{q}}} \rangle &= -\sum_{\alpha\beta} \Gamma_{\alpha\beta\mu\mathbf{q}} \Gamma_{\beta\alpha\nu\bar{\mathbf{q}}} G_{\alpha,-\tau} G_{\beta\tau},
\end{aligned} \tag{17.98}
$$

where $G_{\alpha\tau}$ is the Green function for inserting a fermion into an eigenmode α with energy E_α (see Section 17.6).[15] Transforming to the Matsubara frequency domain and analytically continuing to real frequencies gives

$$\Upsilon_{\mu\nu\mathbf{q}\omega} = \langle -k_{\mu\nu} \rangle + \sum_{\alpha\beta} \Gamma_{\alpha\beta\mu\mathbf{q}} \Gamma_{\beta\alpha\nu\bar{\mathbf{q}}} \frac{f_\beta - f_\alpha}{E_\alpha - E_\beta - \omega}. \tag{17.99}$$

Taking the imaginary part leads to an expression for the spectral (dissipative) part of the electromagnetic response, which can be computed efficiently by accumulating delta function weights in bins:

$$\frac{\operatorname{Im} \Upsilon_{\mu\nu\mathbf{q}\omega}}{\pi} = \sum_{\alpha\beta} \Gamma_{\alpha\beta\mu\mathbf{q}} \Gamma_{\beta\alpha\nu\bar{\mathbf{q}}} (f_\beta - f_\alpha)\, \delta(E_\alpha - E_\beta - \omega). \tag{17.100}$$

This equation is visualized diagrammatically in Fig. 17.26. The vertex factors $\Gamma_{\alpha\beta\mu\mathbf{q}}$ are matrix elements connecting plane electromagnetic waves with disorder eigenstates, the delta function imposes energy conservation, and the Fermi occupation factors affect the amplitude of creating a particle and hole in disorder eigenstates.

To obtain the reactive response as a function of frequency, it is most efficient to infer it using the Kramers–Kronig relation:

$$\operatorname{Re} \Upsilon_{\mu\nu\mathbf{q}\omega} = \langle -k_{\mu\nu} \rangle + \mathcal{P} \int_{-\infty}^{\infty} \frac{d\omega'}{\pi} \frac{\operatorname{Im} \Upsilon_{\mu\nu\mathbf{q}\omega'}}{\omega - \omega'}. \tag{17.101}$$

Equations 17.100 and 17.101 provide an efficient way to calculate the electromagnetic response.

To make contact with the notation in the literature (Scalapino *et al.*, 1993), observe that for a square lattice with lattice spacing a and nearest-neighbor hopping t, the

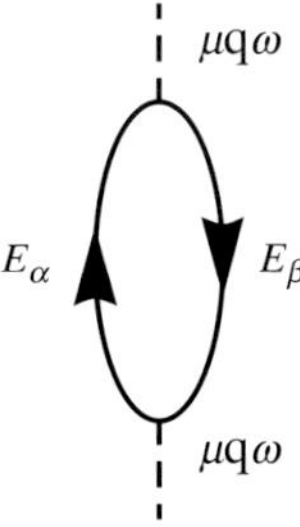

Fig. 17.26 Diagrammatic visualization of the Kubo formula for the electromagnetic response, eqn (17.100).

[15]In general, one has to be careful with infinitesimal time shifts in the path-integral formalism that arise from anticommutation relations in the operator formalism. In this particular working, however, the 1s arising from commutation relations drop out.

equations for k and Γ can be written

$$\langle k_x \rangle = ta^2 \sum_{i\alpha} f_\alpha (\phi^*_{i\alpha} \phi_{i+\hat{x},\alpha} + h.c.), \tag{17.102}$$

$$\Gamma_{\alpha\beta x\mathbf{q}} = ita \sum_{i} e^{i\mathbf{q}\cdot\mathbf{r}_i} (\phi^*_{i\alpha} \phi_{i+\hat{x},\beta} - h.c.), \tag{17.103}$$

where $i + \hat{x}$ refers to the neighbor of site i in the x direction.

17.8.1 Electromagnetic response of clean superconductor

The electromagnetic response is rather complicated to derive, so we have made an entire section for it. The diamagnetic term k (which involves bond kinetic energies) and bond current j now include a sum over both spin species,

$$k_{ij} = -\sum_{\sigma} t_{ij} c^\dagger_{i\sigma} c_{j\sigma}, \tag{17.104}$$

$$j_{ij} = \sum_{\sigma} it_{ij} c^\dagger_{i\sigma} c_{j\sigma}. \tag{17.105}$$

Transform this into the wavevector-polarization basis:

$$k_{\mu\nu} = -\sum_{ij\sigma} r_{ij\mu} r_{ij\nu} t_{ij} c^\dagger_{i\sigma} c_{j\sigma}, \tag{17.106}$$

$$j_{\mu\mathbf{q}} = \sum_{ij\sigma} e^{-i\mathbf{q}\cdot\mathbf{r}_i} r_{ij\mu} it_{ij} c^\dagger_{i\sigma} c_{j\sigma}. \tag{17.107}$$

Using the Fourier relations,

$$c_{j\sigma} = \sum_{\mathbf{k}} e^{-i\mathbf{k}\cdot\mathbf{r}_j} c_{\mathbf{k}\sigma}, \tag{17.108}$$

$$t_{ij} = \sum_{\mathbf{k}} e^{i\mathbf{k}\cdot(\mathbf{r}_j - \mathbf{r}_i)} \varepsilon_{\mathbf{k}}, \tag{17.109}$$

$$r_{ij\mu} it_{ij} = \sum_{\mathbf{k}} e^{i\mathbf{k}\cdot(\mathbf{r}_j - \mathbf{r}_i)} \frac{\partial \varepsilon_{\mathbf{k}}}{\partial k_\mu}, \tag{17.110}$$

and one obtains

$$k_{\mu\nu} = \sum_{\mathbf{k}} \frac{\partial^2 \varepsilon_{\mathbf{k}}}{\partial k_\mu \partial k_\nu} \left(c^\dagger_{\mathbf{k}\uparrow} c_{\mathbf{k}\uparrow} + c^\dagger_{\mathbf{k}\downarrow} c_{\mathbf{k}\downarrow} \right), \tag{17.111}$$

$$j_{\mu\mathbf{q}} = \sum_{\mathbf{k}} \frac{\partial \varepsilon_{\mathbf{k}}}{\partial k_\mu} \left(c^\dagger_{\mathbf{k}\uparrow} c_{\mathbf{p}\uparrow} + c^\dagger_{\mathbf{k}\downarrow} c_{\mathbf{p}\downarrow} \right) \tag{17.112}$$

where $\mathbf{p}$ is shorthand for $\mathbf{p} = \mathbf{k} + \mathbf{q}$. Now, transform the fermion fields into the bogolon basis using

$$c_{\mathbf{k}\uparrow} = u_{\mathbf{k}}\gamma_{\mathbf{k}\uparrow} + v_{\mathbf{k}}\gamma^{\dagger}_{-\mathbf{k}\downarrow} \tag{17.113}$$

$$c_{\mathbf{k}\downarrow} = u_{\mathbf{k}}\gamma^{\dagger}_{\mathbf{k}\downarrow} - v_{\mathbf{k}}\gamma_{-\mathbf{k}\uparrow} \tag{17.114}$$

(assuming $u_{\mathbf{k}} = u_{-\mathbf{k}}$). This gives

$$k_{\mu\nu} = \sum_{\mathbf{k}} \frac{\partial^2 \varepsilon_{\mathbf{k}}}{\partial k_\mu \partial k_\nu} \Big[(u_{\mathbf{k}}\gamma^{\dagger}_{\mathbf{k}\uparrow} + v_{\mathbf{k}}\gamma_{-\mathbf{k}\downarrow})(u_{\mathbf{k}}\gamma_{\mathbf{k}\uparrow} + v_{\mathbf{k}}\gamma^{\dagger}_{-\mathbf{k}\downarrow}) +$$
$$(u_{\mathbf{k}}\gamma_{\mathbf{k}\downarrow} - v_{\mathbf{k}}\gamma^{\dagger}_{-\mathbf{k}\uparrow})(u_{\mathbf{k}}\gamma^{\dagger}_{\mathbf{k}\downarrow} - v_{\mathbf{k}}\gamma_{-\mathbf{k}\uparrow}) \Big],$$

$$j_{\mu\mathbf{q}} = \sum_{\mathbf{k}} \frac{\partial \varepsilon_{\mathbf{k}}}{\partial k_\mu} \Big[(u_{\mathbf{k}}\gamma^{\dagger}_{\mathbf{k}\uparrow} + v_{\mathbf{k}}\gamma_{-\mathbf{k}\downarrow})(u_{\mathbf{p}}\gamma_{\mathbf{p}\uparrow} + v_{\mathbf{p}}\gamma^{\dagger}_{-\mathbf{p}\downarrow}) +$$
$$(u_{\mathbf{k}}\gamma_{\mathbf{k}\downarrow} - v_{\mathbf{k}}\gamma^{\dagger}_{-\mathbf{k}\uparrow})(u_{\mathbf{p}}\gamma^{\dagger}_{\mathbf{p}\downarrow} - v_{\mathbf{p}}\gamma_{-\mathbf{p}\uparrow}) \Big].$$

Now, the prefactor $\frac{\partial \varepsilon_{\mathbf{k}}}{\partial k_\mu k_\nu}$ is even in $\mathbf{k}$ (unchanged under the transformation $\mathbf{k} \to -\mathbf{k}$), whereas $\frac{\partial \varepsilon_{\mathbf{k}}}{\partial k_\mu}$ is odd in $\mathbf{k}$. Carefully collecting terms and simplifying by substituting dummy indices, being mindful of the above-mentioned symmetries and the anticommutation relations, leads to

$$k_{\mu\nu} = \sum_{\mathbf{k}} \frac{\partial^2 \varepsilon_{\mathbf{k}}}{\partial k_\mu \partial k_\nu} \Big[({u_{\mathbf{k}}}^2 - {v_{\mathbf{k}}}^2)(\gamma^{\dagger}_{\mathbf{k}\uparrow}\gamma_{\mathbf{k}\uparrow} + \gamma_{\mathbf{k}\downarrow}\gamma^{\dagger}_{\mathbf{k}\downarrow}) + (\text{terms involving } \gamma\gamma \text{ and } \gamma^\dagger\gamma^\dagger) \Big],$$

$$j_{\mu\mathbf{q}} = \sum_{\mathbf{k}} \frac{\partial \varepsilon_{\mathbf{k}}}{\partial k_\mu} \Big[(u_{\mathbf{k}}u_{\mathbf{p}} + v_{\mathbf{k}}v_{\mathbf{p}})(\gamma^{\dagger}_{\mathbf{k}\uparrow}\gamma_{\mathbf{p}\uparrow} + \gamma^{\dagger}_{\mathbf{k}\downarrow}\gamma_{\mathbf{p}\downarrow}) + (u_{\mathbf{k}}v_{\mathbf{p}} - v_{\mathbf{k}}u_{\mathbf{p}})(\gamma^{\dagger}_{\mathbf{k}\uparrow}\gamma^{\dagger}_{\mathbf{p}\downarrow} + \gamma_{\mathbf{k}\uparrow}\gamma_{\mathbf{p}\downarrow}) \Big].$$

The expression for j contains combinations of u and v known as "Case II coherence factors" (Tinkham, 1996). We can now take the expectation to obtain the diamagnetic term in the Kubo formula,

$$\langle k_{\mu\nu} \rangle = \sum_{\mathbf{k}} \frac{\partial^2 \varepsilon_{\mathbf{k}}}{\partial k_\mu \partial k_\nu} \frac{\xi_{\mathbf{k}}}{E_{\mathbf{k}}} (2 f_{\mathbf{k}} - 1). \tag{17.115}$$

The factor $\frac{\xi_{\mathbf{k}}}{E_{\mathbf{k}}}$ is due to the binding of electrons into Cooper pairs, which lowers their potential energy at the expense of a gain in kinetic energy. (For continuum electrodynamics, the diamagnetic term is simply proportional to the electron density, and this factor is absent.) For the square lattice, putting in the explicit forms of the dispersion relation gives

$$\langle k_{xx} \rangle = \int_{\mathbf{k}} (2 \cos k_x) \frac{\xi_{\mathbf{k}}}{E_{\mathbf{k}}} (2 f_{\mathbf{k}} - 1). \tag{17.116}$$

For the paramagnetic term, Wick contraction of γ and $\gamma^\dagger$ operators leads to

$$\langle j_{\mu\mathbf{q}\tau} j_{\nu\bar{\mathbf{q}}} \rangle = \sum_{\mathbf{k}} \frac{\partial \varepsilon_{\mathbf{k}}}{\partial k_\mu} \frac{\partial \varepsilon_{\mathbf{k}}}{\partial k_\nu} \Big[(u_{\mathbf{k}} u_{\mathbf{p}} + v_{\mathbf{k}} v_{\mathbf{p}})^2 (G_{\mathbf{k}\tau} G_{\mathbf{p}\bar{\tau}} + G_{\mathbf{k}\tau} G_{\mathbf{p}\bar{\tau}}) + (u_{\mathbf{k}} v_{\mathbf{p}} - v_{\mathbf{k}} u_{\mathbf{p}})^2 (G_{\mathbf{k}\tau} G_{\mathbf{p}\bar{\tau}} + G_{\mathbf{k}\tau} G_{\mathbf{p}\bar{\tau}}) \Big]. \tag{17.117}$$

Fourier-transforming to Matsubara frequencies and using the explicit forms of $\varepsilon_{\mathbf{k}}$, $u_{\mathbf{k}}$, and $v_{\mathbf{k}}$ gives the result from Scalapino *et al.* (1993):

$$\Lambda_{xx}(q_y, \omega_m) = \tfrac{4}{N} \sum_{\mathbf{p}} \sin^2 p_x \Bigg\{ \left[\tfrac{1}{2} \left(1 - \tfrac{\xi_{\mathbf{p}} \xi_{\mathbf{p}+\mathbf{q}} + \Delta^2}{E_{\mathbf{p}} E_{\mathbf{p}+\mathbf{q}}} \right) \right] \left[\tfrac{1 - f_{\mathbf{p}} - f_{\mathbf{p}+\mathbf{q}}}{E_{\mathbf{p}} + E_{\mathbf{p}+\mathbf{q}} + i\omega_m} + \tfrac{1 - f_{\mathbf{p}} - f_{\mathbf{p}+\mathbf{q}}}{E_{\mathbf{p}} + E_{\mathbf{p}+\mathbf{q}} - i\omega_m} \right] + \left[\tfrac{1}{2} \left(1 + \tfrac{\xi_{\mathbf{p}} \xi_{\mathbf{p}+\mathbf{q}} + \Delta^2}{E_{\mathbf{p}} E_{\mathbf{p}+\mathbf{q}}} \right) \right] \left[\tfrac{f_{\mathbf{p}+\mathbf{q}} - f_{\mathbf{p}}}{E_{\mathbf{p}} - E_{\mathbf{p}+\mathbf{q}} + i\omega_m} + \tfrac{f_{\mathbf{p}+\mathbf{q}} - f_{\mathbf{p}}}{E_{\mathbf{p}} - E_{\mathbf{p}+\mathbf{q}} - i\omega_m} \right] \Bigg\}.$$

Finally, combining diamagnetic and paramagnetic contributions gives

$$\Upsilon_{xx\mathbf{q}\omega} = \langle -k_{xx} \rangle - \Lambda_{xx\mathbf{q}\omega}. \tag{17.118}$$

The above Kubo formula can be visualized roughly in terms of a two-fluid picture, where the superfluid exhibits a diamagnetic response, whereas the normal fluid gives a paramagnetic contribution arising from quasiparticle excitations. The static uniform superfluid stiffness is obtained in the limit $\omega_m = 0, \mathbf{q} \to 0$:

$$\Upsilon_{xx} = \langle -k_{xx} \rangle - \Lambda_{xx}, \quad \text{where} \quad \Lambda_{xx} = -\frac{8}{N} \sum_{\mathbf{p}} \sin^2 p_x \frac{\partial f(E_{\mathbf{p}})}{\partial E_{\mathbf{p}}}, \tag{17.119}$$

where $\frac{\partial f}{\partial E_{\mathbf{p}}} = -\frac{\beta}{4} \operatorname{sech}^2 \frac{\beta}{2} E_{\mathbf{p}}$.

17.8.2 Electromagnetic response of a dirty superconductor

Let us follow the derivation of eqn (17.95), but including spin indices. For later convenience, begin by anticommuting the down-spin fermion operators:

$$j_{ij} = it_{ij} \left(c_{i\uparrow}^\dagger c_{j\uparrow} - c_{j\downarrow} c_{i\downarrow}^\dagger \right), \tag{17.120}$$

$$k_{ij} = -t_{ij} \left(c_{i\uparrow}^\dagger c_{j\uparrow} - c_{j\downarrow} c_{i\downarrow}^\dagger \right). \tag{17.121}$$

Transforming the fermion fields from the site basis, $c_{i\tau}$, to the eigenmode basis, $\gamma_{\alpha\tau}$ using

$$c_{i\uparrow} = \sum_\alpha u_{i\alpha} \gamma_\alpha, \quad c_{i\downarrow} = \sum_\alpha v_{i\alpha} \gamma_\alpha^\dagger, \tag{17.122}$$

and interchanging some dummy indices, gives

$$j_{ij} = it_{ij} \sum_{\alpha\beta} (u_{i\alpha} u_{j\beta} - v_{j\alpha} v_{i\beta}) \gamma^\dagger_\alpha \gamma_\beta, \tag{17.123}$$

$$k_{ij} = -t_{ij} \sum_{\alpha\beta} (u_{i\alpha} u_{j\beta} - v_{j\alpha} v_{i\beta}) \gamma^\dagger_\alpha \gamma_\beta. \tag{17.124}$$

Carrying out the rest of the derivation, in analogy with the Anderson case, leads to the same form for the Kubo formula,

$$\frac{\operatorname{Im} \Upsilon_{\mu\nu\mathbf{q}\omega}}{\pi} = \sum_{\alpha\beta} \Gamma_{\alpha\beta\mu\mathbf{q}} \Gamma_{\beta\alpha\nu\bar{\mathbf{q}}} (f_\beta - f_\alpha)\, \delta(E_\alpha - E_\beta - \omega), \tag{17.125}$$

$$\operatorname{Re} \Upsilon_{\mu\nu\mathbf{q}\omega} = \langle -k_{\mu\nu} \rangle + \mathcal{P} \int_{-\infty}^{\infty} \frac{d\omega'}{\pi}\, \frac{\operatorname{Im} \Upsilon_{\mu\nu\mathbf{q}\omega'}}{\omega - \omega'}, \tag{17.126}$$

except that $\langle -k \rangle$ and Γ now involve both "spin components" of the eigenvectors and the eigenmode indices α and β now run from 1 to $2N$:

$$\langle k_{\mu\nu} \rangle = -\sum_{\alpha ij} r_{ij\mu} r_{ij\nu} t_{ij} \left(u_{i\alpha} u_{j\alpha} - v_{j\alpha} v_{i\alpha} \right) f_\alpha,$$

$$\Gamma_{\alpha\beta\mu\mathbf{q}} = \sum_{ij} e^{-i\mathbf{q}\cdot\mathbf{r}_i} r_{ij\mu} i t_{ij} \left(u_{i\alpha} u_{j\beta} - v_{j\alpha} v_{i\beta} \right). \tag{17.127}$$

17.9 Variational Bogoliubov–de Gennes formalism

In this appendix we describe a variational mean-field treatment of the Hubbard model. This is the only way to decouple the Hubbard interaction in multiple channels without overcounting it.

17.9.1 Variational method in statistical mechanics

For a system described by Hamiltonian $\hat{H}$ at temperature $T = 1/\beta$, the partition function is

$$Z = \operatorname{Tr} e^{-\beta \hat{H}}. \tag{17.128}$$

Let $\hat{\rho}$ be an arbitrary density matrix. Then, formally, we may write

$$Z = \operatorname{Tr} \hat{\rho} e^{-\beta \hat{H} - \ln \hat{\rho}} = \left\langle e^{-\beta \hat{H} - \ln \hat{\rho}} \right\rangle_{\hat{\rho}}. \tag{17.129}$$

For a classical random variable X and any convex function f, Jensen's inequality states that $\langle f(X) \rangle \geq f(\langle X \rangle)$. This theorem is easily extended to expectations of a Hermitian operator with respect to a density matrix, i.e. $\left\langle f(\hat{X})\right\rangle_{\hat{\rho}} \geq f(\left\langle \hat{X}\right\rangle_{\hat{\rho}})$. Thus,

$$Z \geq \exp\left\langle -\beta\hat{H} - \ln\hat{\rho}\right\rangle_{\hat{\rho}}. \tag{17.130}$$

Taking logs of both sides gives

$$\Omega = -T\ln Z \leq \left\langle \hat{H} + T\ln\hat{\rho}\right\rangle_{\hat{\rho}}. \tag{17.131}$$

The variational free energy $\Omega_{\text{var}} = \left\langle \hat{H} + T\ln\hat{\rho}\right\rangle_{\hat{\rho}}$ with respect to *any* density matrix is an upper bound on the true free energy Ω. This is the variational principle of quantum statistical mechanics. It is exploited in the variational method, in which Ω_{var} is optimized with respect to the trial density matrix ρ to obtain a *least* upper bound to Ω.[16]

17.9.2 Variational mean-field theory with a trial Hamiltonian

Chaikin and Lubensky (2000) describe a general formalism in which the trial density matrix $\hat{\rho}$ is written as the product of arbitrary matrices at each site. However, in the context of this article, it is most convenient to use the exact density matrix of a solved trial Hamiltonian $\hat{H}_t$,

$$\hat{\rho}_t = \frac{1}{Z_t}e^{-\beta\hat{H}_t} = e^{\beta\Omega_t - \beta\hat{H}_t}, \tag{17.132}$$

where $\Omega_t = -T\ln\operatorname{Tr}e^{-\beta\hat{H}_t}$ is the trial free energy. The task is then to optimize the variational free energy

$$\Omega_{\text{var}} = \left\langle \hat{H} + T(\beta\Omega_t - \beta\hat{H}_t)\right\rangle_{\hat{\rho}_t} = \left\langle \hat{H} - \hat{H}_t\right\rangle_{\hat{\rho}_t} + \Omega_t \tag{17.133}$$

with respect to the trial Hamiltonian $\hat{H}_t$. To explore the consequences of this decision we must explicitly introduce a model for $\hat{H}_t$.

17.9.3 Hubbard model: two-channel decoupling

Let us illustrate the procedure with a clean Hubbard model with an applied chemical potential μ^a,

$$\hat{H} = \underbrace{-\sum_{ij\sigma} t_{ij}c^\dagger_{i\sigma}c_{j\sigma}}_{\hat{t}} \underbrace{- g\sum_i c^\dagger_{i\uparrow}c^\dagger_{i\downarrow}c_{i\downarrow}c_{i\uparrow}}_{\hat{g}} \underbrace{- \mu^a\sum_{i\sigma} c^\dagger_{i\sigma}c_{i\sigma}}_{\hat{\mu}^a}. \tag{17.134}$$

[16] One also hopes that the optimal trial density matrix is *close* to the true density matrix, but this is not guaranteed. All variational calculations suffer from the inevitable bias involved in choosing a reasonably simple form for the trial density matrix (or wavefunction).

We can construct a suitable trial Hamiltonian by decoupling the interaction in terms of a pairing amplitude Δ_i and a Hartree potential μ_i^{H} at every site, which serve as variational parameters:

$$\hat{H}_t(\{\Delta_i, \mu_i^{\mathrm{H}}\}) = -\sum_{ij\sigma} t_{ij} c_{i\sigma}^\dagger c_{j\sigma} - \underbrace{\sum_i \Delta_i (c_{i\uparrow}^\dagger c_{i\downarrow}^\dagger + c_{i\downarrow} c_{i\uparrow})}_{\hat{\Delta}} - \underbrace{\sum_i \mu_i (c_{i\uparrow}^\dagger c_{i\uparrow} + c_{i\downarrow}^\dagger c_{i\downarrow})}_{\hat{\mu}},$$

where $\mu_i = \mu^a + \mu_i^{\mathrm{H}}$. This bilinear trial Hamiltonian can be solved by diagonalizing it, that is, by calculating the eigenvalues E_α, eigenvectors u_α and v_α, and occupation numbers f_α (see Section 17.3.5). In the BdG basis we simply have $\hat{H}_t = \sum_\alpha E_\alpha (\gamma_\alpha^\dagger \gamma_\alpha - \gamma_\alpha \gamma_\alpha^\dagger)$. The resulting trial density matrix,

$$\hat{\rho}_t = \frac{1}{Z_t} \exp\left[-\beta \sum_\alpha E_\alpha (\gamma_\alpha^\dagger \gamma_\alpha - \gamma_\alpha \gamma_\alpha^\dagger) \right], \tag{17.135}$$

is Gaussian, and so the trial free energy can be evaluated straightforwardly,

$$\Omega_t = -T \ln \operatorname{Tr} e^{-\beta \hat{H}_t} = -T \sum_\alpha \ln \left(2 \cosh \frac{\beta E_\alpha}{2} \right). \tag{17.136}$$

Now substituting eqns (17.134) and (17.136) into eqn (17.133), the first term gives

$$\begin{aligned}
\left\langle \hat{H} - \hat{H}_t \right\rangle_{\hat{\rho}_t} &= \left\langle -\hat{t} - \hat{g} - \hat{\mu}^a - (-\hat{t} - \hat{\Delta} - \hat{\mu}) \right\rangle_{\hat{\rho}_t} \\
&= \left\langle -\hat{g} + \hat{\Delta} + \hat{\mu}^a - \hat{\mu} \right\rangle \\
&= -g \sum_i \left\langle c_{i\uparrow}^\dagger c_{i\downarrow}^\dagger c_{i\downarrow} c_{i\uparrow} \right\rangle + \sum_i \Delta_i \left\langle c_{i\uparrow}^\dagger c_{i\downarrow}^\dagger + c_{i\downarrow} c_{i\uparrow} \right\rangle + \sum_i (\mu_i - \mu^a) \langle n_i \rangle ,
\end{aligned}$$

where all the expectations are taken with respect to the trial density matrix $\hat{\rho}_t$. Since $\hat{\rho}_t$ is Gaussian, the quartic term reduces to a sum of Wick contractions. The final expression for the variational free energy is

$$\Omega_{\mathrm{var}}(\{\Delta_i, \mu_i^{\mathrm{H}}\}) = \Omega_t + \sum_i \Big[-g({F_i}^2 + {n_i}^2) + 2\Delta_i F_i + 2(\mu_i - \mu^a) n_i \Big], \tag{17.137}$$

where Ω_t is given by eqn (17.136) and F_i and n_i are

$$\begin{aligned}
F_i &= \frac{1}{2} \left\langle c_{i\uparrow}^\dagger c_{i\downarrow}^\dagger + c_{i\downarrow} c_{i\uparrow} \right\rangle = \sum_\alpha (f_\alpha - 1/2) u_{i\alpha} v_{i\alpha}, \\
n_i &= \frac{1}{2} \left\langle c_{i\uparrow}^\dagger c_{i\uparrow} + c_{i\downarrow}^\dagger c_{i\downarrow} \right\rangle = \sum_\alpha f_\alpha {u_{i\alpha}}^2 + \sum_\alpha (1 - f_\alpha) {v_{i\alpha}}^2.
\end{aligned} \tag{17.138}$$

Here, $n_i \in [0, 1]$ is the average density per spin species.

It can be verified that, upon minimizing eqn (17.137) with respect to the variational parameters, one recovers the gap and number equations, $\Delta_i = gF_i$ and $\mu_i^{\mathrm{H}} = gn_i$, which are familiar from the traditional BdG formalism.

17.9.4 Six-channel decoupling

For completeness we now present the most general mean-field theory for the Hubbard model, in which the Hubbard interaction is decoupled in all six channels. It is convenient to adopt a $4N \times 4N$ version of the Nambu–Gor'kov matrix formalism and write the Hubbard Hamiltonian as

$$\hat{H}_{\text{true}} = \underbrace{-\sum_{ijs} \tfrac{1}{2} t_{ij} \eta_{6ss'} c_{is}^\dagger c_{js}}_{\hat{t}} \underbrace{- \sum_{ilss'} \tfrac{1}{2} \Sigma_{il}^a \eta_{lss'} c_{is}^\dagger c_{is'}}_{\hat{\Sigma}^a} + \underbrace{\sum_i U_i x_{i\uparrow} x_{i\downarrow}}_{\hat{U}} . \tag{17.139}$$

The notation is as follows: t_{ij} and U_i are hopping and on-site repulsion and $s, s' = 1, 2, 3, 4$ are superspin indices that distinguish between the four fermionic degrees of freedom at each site (up-spin and down-spin particles and holes),

$$c_{is} = \left(c_{i\uparrow} \; c_{i\downarrow} \; c_{i\uparrow}^\dagger \; c_{i\downarrow}^\dagger \right)_s . \tag{17.140}$$

The factors of $\frac{1}{2}$ in the Hamiltonian compensate for particle–hole doubling. The operator $x_{i\sigma} = \frac{1}{2}\left(\bar{\psi}_{i\sigma}\psi_{i\sigma} - \psi_{i\sigma}\bar{\psi}_{i\sigma}\right)$ is the local density measured with respect to half-filling. This definition has the advantage of being particle–hole symmetric. The applied potentials (such as a disorder potential or a Zeeman field) are represented by the self-energy fields Σ_{il}^a on sites $i = 1, \ldots, N$ in channels $l = 1, 2, 3, 4, 5, 6$. The six channels correspond to the three components of the Zeeman field $\mathbf{h}$, the real and imaginary parts of the pairing potential Δ, and the chemical potential μ:

$$\begin{aligned} &\Sigma_1 \equiv h_X \qquad \Sigma_2 \equiv h_Y \qquad \Sigma_3 \equiv h_Z, \\ &\Sigma_4 \equiv \Delta_{\mathrm{R}} \qquad \Sigma_5 \equiv \Delta_{\mathrm{I}} \qquad \Sigma_6 \equiv \mu. \end{aligned} \tag{17.141}$$

The basis matrices $\boldsymbol{\eta}_l$ are

$$\boldsymbol{\eta}_1 = \begin{pmatrix} 0 & 1 & 0 & 0 \\ 1 & 0 & 0 & 0 \\ 0 & 0 & 0 & -1 \\ 0 & 0 & -1 & 0 \end{pmatrix} \qquad \boldsymbol{\eta}_2 = \begin{pmatrix} 0 & -i & 0 & 0 \\ i & 0 & 0 & 0 \\ 0 & 0 & 0 & i \\ 0 & 0 & -i & 0 \end{pmatrix} \qquad \boldsymbol{\eta}_3 = \begin{pmatrix} 1 & 0 & 0 & 0 \\ 0 & -1 & 0 & 0 \\ 0 & 0 & -1 & 0 \\ 0 & 0 & 0 & 1 \end{pmatrix},$$

$$\boldsymbol{\eta}_4 = \begin{pmatrix} 0 & 0 & 0 & 1 \\ 0 & 0 & -1 & 0 \\ 0 & -1 & 0 & 0 \\ 1 & 0 & 0 & 0 \end{pmatrix} \qquad \boldsymbol{\eta}_5 = \begin{pmatrix} 0 & 0 & 0 & i \\ 0 & 0 & -i & 0 \\ 0 & i & 0 & 0 \\ -i & 0 & 0 & 0 \end{pmatrix} \qquad \boldsymbol{\eta}_6 = \begin{pmatrix} 1 & 0 & 0 & 0 \\ 0 & 1 & 0 & 0 \\ 0 & 0 & -1 & 0 \\ 0 & 0 & 0 & -1 \end{pmatrix}. \tag{17.142}$$

The $\boldsymbol{\eta}_l$ are Hermitian, particle–hole symmetric, mutually orthogonal, and normalized such that $\operatorname{tr} \boldsymbol{\eta}_l^\dagger \boldsymbol{\eta}_l = 4$. The self-energy term can be written out explicitly as

$$\hat{\Sigma}^a = \begin{pmatrix} c_\uparrow \\ c_\downarrow \\ c_\uparrow^\dagger \\ c_\downarrow^\dagger \end{pmatrix}^\dagger \begin{pmatrix} \mu + h_Z & -h_X + ih_Y & 0 & \Delta_\mathrm{R} + i\Delta_\mathrm{I} \\ -h_X - ih_Y & \mu - h_Z & -\Delta_\mathrm{R} - i\Delta_\mathrm{I} & 0 \\ 0 & -\Delta_\mathrm{R} + i\Delta_\mathrm{I} & -\mu - h_Z & h_X - ih_Y \\ \Delta_\mathrm{R} - i\Delta_\mathrm{I} & 0 & h_X + ih_Y & -\mu + h_Z \end{pmatrix} \begin{pmatrix} c_\uparrow \\ c_\downarrow \\ c_\uparrow^\dagger \\ c_\downarrow^\dagger \end{pmatrix} \tag{17.143}$$

(where we have omitted site indices i and superscripts a for clarity). The self-energy matrix is a Hermitian matrix with particle–hole symmetry; these symmetries constrain the 16 complex matrix elements, so that six real numbers are sufficient to parametrize the self-energy. Six is the number of generators of the group $SU(2) \times SU(2)$; the six parameters transform into each other under suitable rotations in spin space or particle–hole space.

17.9.4.1 Trial Hamiltonian and trial density matrix

Following the procedure illustrated earlier, we then construct a trial Hamiltonian by decoupling $\hat{U}$ in six channels,

$$\hat{H}_t = -\sum_{ijs} \tfrac{1}{2} t_{ij} \eta_{6ss'} c_{is}^\dagger c_{is'} - \underbrace{\sum_{ilss'} \tfrac{1}{2} \Sigma_{il} \eta_{lss'} c_{is}^\dagger c_{is'}}_{\hat{\Sigma}}, \tag{17.144}$$

where the total (effective) self-energy $\Sigma_{il} = \Sigma_{il}^a + \Sigma_{il}^H$ is the applied (external) self-energy plus the internal (Hartree/Fock/Bogoliubov) self-energy arising from the decoupling of the interaction. This bilinear trial Hamiltonian can be constructed explicitly as a $4N \times 4N$ matrix and diagonalized to give eigenvalues $E_\alpha (\alpha = 1, \ldots, 4N)$ and eigenvectors Φ_{is}^α. The trial free energy is

$$\Omega_t = -T \ln \operatorname{Tr} e^{-\beta \hat{H}_\mathrm{trial}} = -\tfrac{T}{2} \sum_\alpha \ln \left(2 \cosh \tfrac{\beta}{2} E_\alpha \right). \tag{17.145}$$

The variational free energy works out to be

$$\begin{aligned} \Omega_\mathrm{var} = \Omega_t &+ \sum_i U_i \left(-m_Z^2 - m_X^2 - m_Y^2 + F_\mathrm{R}^2 + F_\mathrm{I}^2 + x^2 \right)_i \\ &+ 2 \sum_i \left(h_X^H m_X + h_Y^H m_Y + h_Z^H m_Z + \Delta_\mathrm{R}^H F_\mathrm{R} + \Delta_\mathrm{I}^H F_\mathrm{I} + \mu^\mathrm{H} x \right)_i. \end{aligned} \tag{17.146}$$

In eqn (17.146), there are six densities at every site: (m_X, m_Y, m_Z) are the three components of magnetization, $(F_\mathrm{R}, F_\mathrm{I})$ are the real and imaginary parts of the anomalous Green's function, and x is the average density per spin species measured with respect to half-filling. Each of these quantities lies in the interval $[-\frac{1}{2}, +\frac{1}{2}]$. Explicitly,

$$G_1 = m_X = \tfrac{1}{2}\left\langle \bar{\psi}_\uparrow \psi_\downarrow + \bar{\psi}_\downarrow \psi_\uparrow \right\rangle G_2 = m_Y = \tfrac{1}{2i}\left\langle \bar{\psi}_\uparrow \psi_\downarrow - \bar{\psi}_\downarrow \psi_\uparrow \right\rangle G_3 = m_Z = \tfrac{1}{2}(x_\uparrow - x_\downarrow),$$
$$G_4 = F_\mathrm{R} = \tfrac{1}{2}\left\langle \psi_\downarrow \psi_\uparrow + \bar{\psi}_\uparrow \bar{\psi}_\downarrow \right\rangle G_5 = F_\mathrm{I} = \tfrac{1}{2i}\left\langle \psi_\downarrow \psi_\uparrow - \bar{\psi}_\uparrow \bar{\psi}_\downarrow \right\rangle G_6 = x = \tfrac{1}{2}(x_\uparrow + x_\downarrow).$$

These quantities can be calculated from knowledge of the occupation numbers and eigenvectors:

$$G_{il} = \tfrac{1}{4}\sum_{ss'} \eta_{lss'} G_{iss'}$$
$$\text{where} \quad G_{iss'} = \sum_\alpha (f_\alpha - \tfrac{1}{2}) \Phi_{is}^{\alpha\,*} \Phi_{is'}^{\alpha}. \tag{17.147}$$

Minimizing eqn (17.146) with respect to the $6N$ variational parameters Σ_{il}^a gives the self-consistency conditions at each site,

$$\mathbf{h}_i^H = +U_i m_i, \qquad \Delta_i^H = -U_i F_i, \qquad \mu_i^\mathrm{H} = -U_i x_i. \tag{17.148}$$

This makes physical sense: a repulsive interaction $U > 0$ produces positive feedback in the spin channel, which tends to produce spontaneous magnetic order, whereas an attractive interaction $U < 0$ produces a tendency towards pairing, charge separation, and charge ordering.

As remarked in Section 17.3.5, the variational mean-field calculation can be performed using Broyden-type methods to solve the self-consistency equations (eqn (17.148)), while monitoring the variational free energy (eqn (17.146)) to ensure that the iteration is converging to a minimum of Ω_t and not a saddle-point or maximum.

References

Abrahams, E. Anderson, P.W., Licciardello, D.C., and Ramakrishnan, T.V. (1979). *Phys. Rev. Lett.*, **42**, 673–676.

Aleiner, I.L. and Altshuler, B.L. (1997). *Phys. Rev. Lett.*, **79**, 4242–4245.

Altshuler, B.L., Khmel'nitzkii, D., Larkin, A.I., and Lee, P.A. (1980). *Phys. Rev. B*, **22**, 5142–5153.

Bardeen, J., Cooper, L.N., and Schrieffer, J.R. (1957). *Phys. Rev.*, **108**, 1175–1204.

Bouadim, K., Loh, Y.L., Randeria, M., and Trivedi, N. (2011). arxiv:1011.3275; accepted in *Nature Physics* (2011).

Bulgac, A. and Forbes, M.M. (2008). *Phys. Rev. Lett.*, **101**, 215301.

Bułka, B.R., Kramer, B., and MacKinnon, A. (1985). *Z. Phys. B*, **60**, 13–17.

Burkhardt, H. and Rainer, D. (1994). *Ann. Physik*, **3**, 181.

Butko, V.Y., Adams, P.W., and Aleiner, I.L. (1999). *Phys. Rev. Lett.*, **82**, 4284–4287.

Casalbuoni, R. and Nardulli, G. (2004). *Rev. Mod. Phys.*, **76**, 263–320.

Chaikin, P.M. and Lubensky, T.C. (2000). *Principles of Condensed Matter Physics*. Cambridge University Press, Cambridge.

Chandrasekhar, B.S. (1962). *Appl. Phys. Lett.*, **1**, 7–8.

Clogston, A.M. (1962). *Phys. Rev. Lett.*, **9**, 266–267.
Cui, Q.H. and Yang, K. (2008). *Phys. Rev. B (Condensed Matter and Materials Physics)*, **78**, 054501.
de Gennes, P.G. (1966). *Superconductivity in Metals and Alloys.* Benjamin, New York.
Dubi, Y., Meir, Y., and Avishai, Y. (2007). *Nature*, **449**, 876–880.
Dubi, Y., Meir, Y., and Avishai, Y. (2008). *Phys. Rev. B*, **78**, 024502.
Fulde, P. and Ferrell, R.A. (1964). *Phys. Rev.*, **135**, A550.
Gantmakher, V.F. and Dolgopolov, V.T. (2010). *Physics-Uspekhi*, **53**, 3.
Ghosal, A., Randeria, M., and Trivedi, N. (1998). *Phys. Rev. Lett.*, **81**, 3940.
Ghosal, A., Randeria, M., and Trivedi, N. (2001). *Phys. Rev. B*, **65**, 014501.
Gor'kov, L.P. and Rusinov, A.I. (1964). *Sov. Phys. JETP*, **19**, 922.
Hirsch, J.E. and Fye, R.M. (1986). *Phys. Rev. Lett.*, **56**, 2521–2524.
Kee, H.Y., Aleiner, I.L., and Altshuler, B.L. (1998). *Phys. Rev. B*, **58**, 5757–5776.
Koponen, T.K., Paananen, T., Martikainen, J.-P., and Törmä, P. (2007). *Phys. Rev. Lett.*, **99**, 120403.
Kramer, B and MacKinnon, A (1993). *Rep. Prog. Phys*, **56**, 1469.
Larkin, A.I. and Ovchinnikov, Y.N. (1964). *Zh. Eksp. Teor. Fiz.*, **47**, 1136. [Sov. Phys. JETP 20, 762 (1965)].
Loh, Y.L. and Trivedi, N. (2010). *Phys. Rev. Lett.*, **104**, 165302.
Loh, Y.L., Trivedi, N., Xiong, Y.M., Adams, P.W., and Catelani, G. (2011). arxiv:1102.3889. accepted in *Phys. Rev. Lett.* (2011).
Machida, K. and Nakanishi, H. (1984). *Phys. Rev. B*, **30**, 122–133.
MacKinnon, A. and Kramer, B. (1981). *Phys. Rev. Lett.*, **47**, 1546.
Matsuo, S., Higashitani, S., Nagato, Y., and Nagai, K. (1998). *J. Phys. Soc. Jpn.*, **67**, 280–289.
Mondal, M., Kamlapure, A., Chand, M., Saraswat, G., Kumar, S., Jesudasan, J., Benfatto, L., Tripathi, V., and Raychaudhuri, P. (2011). *Phys. Rev. Lett.*, **106**, 047001.
Mora, C. and Combescot, R. (2005). *Phys. Rev. B*, **71**, 214504.
Parish, M.M., Baur, S.K., Mueller, E.J., and Huse, D.A. (2007). *Phys. Rev. Lett.*, **99**, 250403.
Radovan, H. A, Fortune, N. A, Murphy, T.P., Hannahs, S.T., Palm, E.C., Tozer, S.W., and Hall, D. (2003). *Nature*, **425**, 51–55.
Radzihovsky, L. and Vishwanath, A. (2009). *Phys. Rev. Lett.*, **103**, 010404.
Sacépé, B., Chapelier, C., Baturina, T.I., Vinokur, V.M., Baklanov, M.R., and Sanquer, M. (2008). *Phys. Rev. Lett.*, **101**, 157006.
Sacépé, B., Chapelier, C., Baturina, T.I., Vinokur, V.M., Baklanov, M.R., and Sanquer, Marc (2010). *Nat. Commun.*, **1**, 140.
Sarma, G. (1963). *J. Phys. Chem. Solids*, **24**, 1029–32.
Scalapino, D.J., White, S.R., and Zhang, S. (1993). *Phys. Rev. B*, **47**, 7995–8007.
Scalettar, R.T., Trivedi, N., and Huscroft, C. (1999, February). *Phys. Rev. B*, **59**, 4364.
Sheehy, D.E. and Radzihovsky, L. (2006). *Phys. Rev. Lett.*, **96**, 060401.
Shimahara, H. (1998). *J. Phys. Soc. Jpn.*, **67**, 736–739.

Tinkham, M. (1996). *Introduction to Superconductivity.* McGraw-Hill, NY.
Trivedi, N., Scalettar, R.T., and Randeria, M. (1996). *Phys. Rev. B*, **54**, R3756.
Yoshida, N. and Yip, S.-K. (2007). *Phys. Rev. A*, **75**, 063601.
Zhao, E. and Liu, W.V. (2008). *Phys. Rev. A*, **78**, 063605.
Zhou, F. and Spivak, B. (1998). *Phys. Rev. Lett.*, **80**, 5647–5650.

18

Suppression of Tunneling of Superconducting Vortices Caused by a Remote Gate: Anderson's Orthogonality Catastrophe and Localization

---

K. MICHAELI[1] and A.M. FINKEL'STEIN[2]

[1] Department of Condensed Matter Physics, The Weizmann Institute of Science, Rehovot 76100, Israel
[2] Department of Condensed Matter Physics, The Weizmann Institute of Science, Rehovot 76100, Israel, and
Department of Physics, Texas A&M University, College Station, TX 77843 – 4242, USA

We address the experimental observation of a strong decrease of the resistance of a superconducting film when a remote unbiased gate is placed above it. The resistance in the experiment discussed was measured at various magnetic fields, both with and without a gate. Here we explain the experimental finding as a suppression of the vortex tunneling by the normal electrons inside the gate. The gate and the film are coupled by the magnetic field of the vortices, which pierces the gate. We use two different approaches to describe the effect of the gate on the vortex tunneling rate: Anderson's orthogonality catastrophe and tunneling with dissipation. We interpret the change in the resistance of the film as a delocalization–localization transition in the system of vortices induced by the gate. We show that in this system Anderson's orthogonality catastrophe promotes Anderson's localization.

18.1 Introduction

The vortex motion in a superconductor is a source of energy losses (Anderson, 1966), destroying the perfect conductivity of the superconductor. The dissipation is caused by non-superconducting electrons located inside the vortex core (Bardeen and Stephen, 1965). The pinning potential created by impurities opposes the motion of vortices. The pinning potential has minima that are typically separated by a distance of the order of the coherence length ξ (Larkin, 1970; Larkin and Ovchinnikov, 1971). Vortices may change their positions either by thermal activation (Anderson and Kim, 1964) or by quantum tunneling between the potential minima at sufficiently low temperatures (Ephron *et al.*, 1996; Liu *et al.*, 1992; Tafuri *et al.*, 2006; van Oudenaarden *et al.*, 1996), see Fig. 18.1. In this context, the observation of a strong decrease of the resistance when

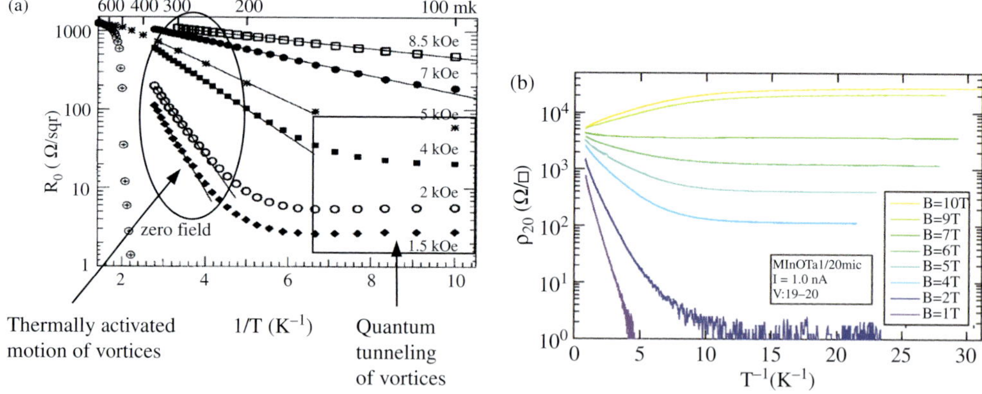

Fig. 18.1 Measurements of the resistance as a function of the inverse temperature of (a) $Mo_{43}Ge_{57}$ film, adapted from Ephron *et al.* (1996), (b) InO_x film (courtesy of Shahar and Sambandamurthy, pers. comm.). The resistance is measured at various values of magnetic fields. Close to the transition temperature, $T \lesssim T_c$, the resistance shows a thermal activation behavior, $R_\square \propto e^{-U/T}$. The saturation of the resistance at low temperatures indicates that vortices move by quantum tunneling.

an unbiased metallic gate is placed above an amorphous superconducting film (Mason and Kapitulnik, 2002) is of great interest. Here we show that the electrons inside the gate suppress the vortices' tunneling rate due to Anderson's orthogonality catastrophe (Anderson, 1967). In other words, the electrons oppose the tunneling and can even localize the vortices, restoring the perfect superconducting properties of the film at low temperatures. We interpret the change in the resistance of the superconducting film as a delocalization–localization transition of the vortices induced by normal electrons inside the gate.

The resistance in the experiment of Mason and Kapitulnik (2002) was measured at various magnetic fields, both with and without a gate (see Fig. 18.2). In the absence of the gate, in magnetic fields lower than the critical one, the resistance initially decreases with lowering of the temperature, but eventually saturates at a finite value. The saturation indicates the possibility of vortex tunneling. However, when an unbiased metallic gate is placed above the superconducting film the resistance reduces significantly and does not saturate at low T. Remarkably, the resistance drop induced by the gate begins at the temperatures where the resistance of the ungated film starts to saturate. In the experiment discussed the film thickness is $a \approx 30$ Å and the gate thickness is $d \approx 400$ Å. The gate is separated from the film by an oxide layer of 160 Å. Therefore, the film is thermally isolated from the gate,

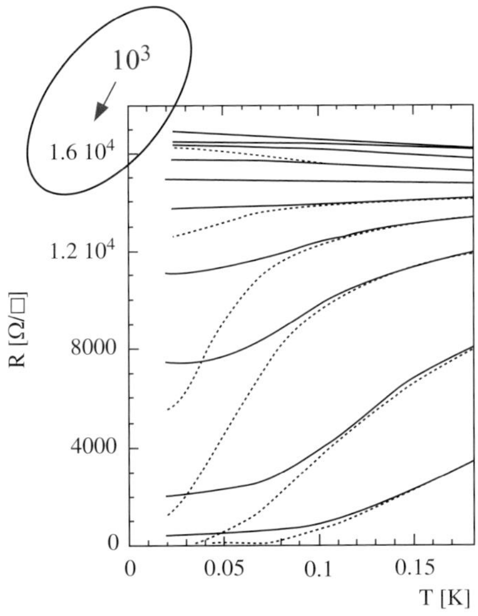

Fig. 18.2 Resistance versus temperature curves for the ungated (solid lines) and gated (dashed lines) films of $Mo_{43}Ge_{57}$; adapted from (Mason and Kapitulnik, 2002). The resistance is measured for magnetic field $H = 1.5$, 1.25, 1.18, 1.1, 1.0, 0.8, and 0.6 T from the top; $H_{C2} = 1.4$ T. The deviation between the curves of the gated and ungated films begins when the resistance of the ungated films starts to saturate. Note that the actual resistance scale is ten times smaller than the one shown in the plot.

ruling out the possibility that the saturation of the resistance in the ungated film can be attributed to heating. Another indication that the gate and the film are well separated is that the superconducting transition temperature, T_c, as well as the critical magnetic field, H_c, are practically unchanged by adding the gate. Since the gate does not affect the superconductivity at $T \approx T_c$ when it is the weakest, its influence on the superconducting properties, such as the energy gap, can be ignored at lower temperatures.

The use of double-layer system to study superconductors is well known (Giaever, 1964; Kruithof *et al.*, 1991; Rimberg *et al.*, 1997). One usually considers the capacitive coupling between the superconducting film and the normal gate. In the case of the Josephson junction arrays (or granular superconductors) the capacitive coupling reduces the fluctuations of the phase of the superconducting order parameter (Rimberg *et al.*, 1997; Wagenblast *et al.*, 1997). As a result, the system may undergo a transition from an insulating to a superconducting state. However, for the homogenous film with a relatively small resistance $\sim 1.5k\Omega/\Box$ used by Mason and Kapitulnik (2002)[1] the phase fluctuations at low temperature are not very effective (Ramakrishnan, 1989). This is confirmed by the observed insensitivity of the critical magnetic field H_c to the presence of the gate. In view of the above, we searched for an alternative explanation of the experiment. Instead of the capacitive coupling, we studied the magnetic coupling. The gate and the film are coupled by the magnetic field of the vortices, which pierces through the gate (see Fig. 18.3).

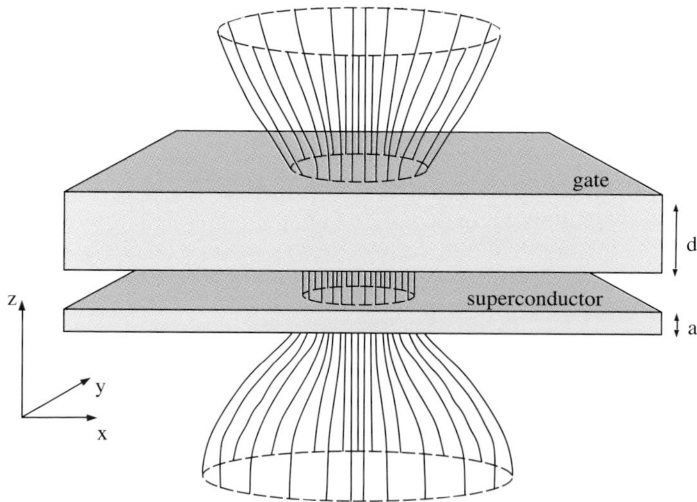

Fig. 18.3 Superconducting film magnetically coupled to a metallic gate. The magnetic field of the vortex penetrates the gate.

[1]In Figs. 3 and 4 of Mason and Kapitulnik (2002), the indicated resistance is one order of magnitude larger than the actual one.

Since little is known about the motion of the vortex density at low temperatures,[2] we employ the following strategy. We accept the tunneling of vortices at low temperatures as an established experimental fact. We do not try to calculate the tunneling rate of the vortices. Instead, we concentrate on how the response of the electrons inside the gate to a change in the vortex position suppresses the tunneling rate. With this question in mind, in the complex problem of vortex tunneling we wish to isolate the effect induced by the gate. For studying the role of the gate, it is sufficient to assume that the change in the vortex position is a rare tunneling event. This can be a tunneling of a single vortex, a bundle of vortices, or topological defects such as dislocation pairs in the case of a vortex lattice (or a glassy state). Phenomenologically, the change in the vortex position can be described by the hopping Hamiltonian

$$H = \sum_i \varepsilon_i a_i^+ a_i + \sum_{\langle i,j \rangle} \left(\Gamma_{ij} a_i^+ a_j + h.c. \right) . \tag{18.1}$$

According to the standard criterion of the metal–insulator transition (Anderson, 1958), the ratio of the variation of the potential minima $\varepsilon = \langle \varepsilon_i \rangle$ to the typical value of the tunneling rates $\Gamma = \langle \Gamma_{ij} \rangle$ specifies whether the vortices are itinerant or localized. The finite resistance at low temperature in the absence of the gate indicates that the vortices are mobile, i.e. the tunneling rate in the ungated film $\Gamma_{\text{unG}} > \varepsilon$. The dissipationless nature of the superconducting film is revived when the tunneling rate in the gated film is reduced to $\Gamma_{\text{G}} < \varepsilon$; see Fig. 18.4, which illustrates the two cases. Thus we interpret the experiment (Mason and Kapitulnik, 2002) as a transition from delocalized to localized states of the system of tunneling vortices induced by the gate. Let us emphasize that the interactions between the vortices are not ignored in eqn 18.1. They are incorporated into the parameters ε and Γ_{ij} in the phenomenological hopping Hamiltonian describing rare tunneling events.

The fact that the tunneling of the vortices can be blocked by placing a gate above the film indicates that the tunneling event gives rise to a dramatic response of the electrons inside the gate. This response can be analyzed in terms of low-energy

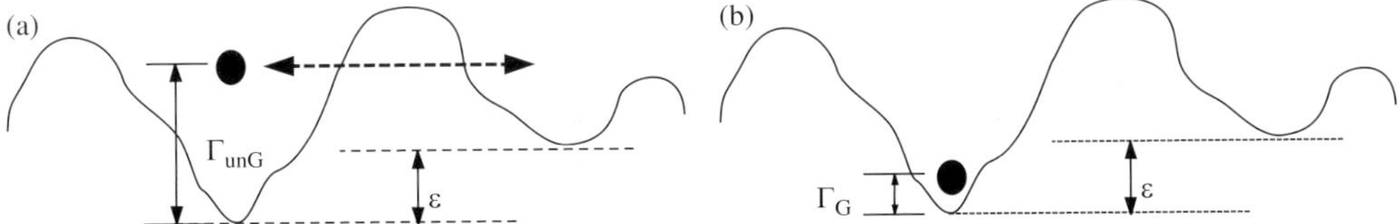

Fig. 18.4 The delocalization–localization transition. The vortex in the effective potential landscape is represented by a hopping "particle." (a) The tunneling rate in the ungated film Γ_{unG} exceeds the energy difference between the potential minima ε; the vortices are in a delocalized phase. (b) The tunneling rate in the gated film is reduced to $\Gamma_{\text{G}} < \varepsilon$, and the system of vortices becomes localized; the dissipationless property of the superconducting film is restored.

[2]The quantum tunneling of vortices leading to the variable-range hopping resistivity in superconducting films was discussed in (Fisher *et al.*, 1991).

electron-hole excitations "decorating" the tunneling vortex. One may consider the cloud of virtual excitations as a part of the tunneling process that lasts long after the change in the vortex positions occurs. One may better understand the specifics of tunneling with the participation of low-energy degrees of freedom, i.e. in the presence of a dissipative environment, using the picture given by Iordanskii and one of the authors (Iordanskii and Finkel'stein, 1973). This describes the quantum formation of a nucleation center in the decay of a metastable macroscopic state,[3] but it can also provide a general perspective. Quantum nucleation is an example of a tunneling process in which a large number of degrees of freedom participate. Alternatively, this kind of process can be treated as a tunneling of an artificial "particle" in a multi-dimensional space. When low-energy degrees of freedom are involved in the process of quantum nucleation, the nucleation develops in two stages (Iordanskii and Finkel'stein, 1973): the motion of the "particle" in this multi-dimensional space along the trajectory minimizing the imaginary time action consists of fast and slow stages. A cartoon illustrating the separation into the fast and slow stages is given in Fig. 18.5.

The slow stage appears because of the long time needed for the slow (low-energy) degrees of freedom to adjust themselves to the new state of the fast degrees of freedom.[4] This time is much longer than needed for the fast degrees of freedom to complete the tunneling. This is why the tunneling has to develop in two stages. It has been shown in Iordanskii and Finkel'stein (1973) that despite the fact that the low-energy degrees of freedom yield only a small contribution to the energy of the barrier, their participation in the tunneling process increases parametrically the overall tunneling time. This results in a large increase of the action and, correspondingly, in the strong suppression of the tunneling rate.

This picture of changing a quantum state in the presence of low-energy degrees of freedom is typical for condensed-matter systems. In the course of the fast stage, a quantum-mechanical object changes its state (position, spin projection, phase of

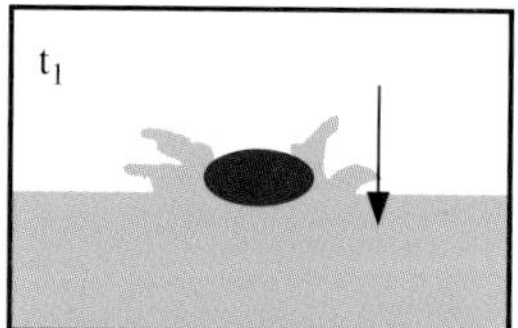

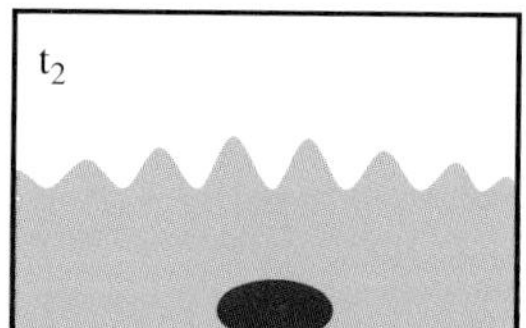

Fig. 18.5 A stone falling into water as an illustration of the fast and slow stages of tunneling in an environment. The stone represents a tunneling object and the water is the environment. The time interval $t_1 < t < t_2$ corresponds to the fast stage of the tunneling when the falling stone creates a splash. The ripples, which last long after the stone reaches its final position, correspond to the slow stage that follows the fast one at $t > t_2$.

[3]A similar phenomenon was later rediscovered in the context of high-energy physics as the "fate of the false vacuum".

[4]In the decay of a metastable phase considered by Iordanskii and Finkel'stein (1973) the slow stage precedes the fast one, thereby preparing the environment for subsequent motion of the fast degrees of freedom. In fact, the tunneling process can also develop in the reverse order.

the Josephson junction, etc). The accompanying slow degrees of freedom act as an environment for the fast degrees of freedom. In the problem of tunneling in the gated superconducting film, the tunneling of the vortex from one potential minimum to another corresponds to the fast stage. During the slow stage, the electrons inside the gate adjust their state to the new position of the vortex. The ensemble of electron–hole pairs in the gate represents the low-energy degrees of freedom of the environment. As we have already mentioned, we concentrate only on the slow stage, which develops when the tunneling of the vortex degrees of freedom is mostly accomplished, and we do not specify how the fast stage develops.

In the following sections we apply two different approaches to describe the response of the electrons inside the gate to the change of the vortex position. In Section 18.2 we describe the elastic scattering of the gate electrons on the vector potential of the magnetic field of the vortex. The zero overlap between the states of the electrons before and after the change of the scattering potential is known generally as Anderson's orthogonality catastrophe (OC) (Anderson, 1967). We show that the OC caused by the change in the vortex position effectively suppresses vortex tunneling. The novel element here (Michaeli and Finkel'stein, 2006) is that the OC is connected to the Aharonov–Bohm effect (Aharonov *et al.*, 1984; Aharonov and Bohm, 1959). In Section 18.3 we consider the dissipative eddy (Foucault) currents induced in the gate, which continue long after the vortex has changed its position (Michaeli and Finkel'stein, 2007). This allows us to examine the vortex tunneling in the gated superconducting film in the context of the well-known problem of tunneling in the presence of a dissipative environment (Caldeira and Leggett, 1981). We conclude with a discussion of the relationship between the two approaches and the peculiarities of the tunneling of an extended object, such as a vortex in a thin superconducting film.

18.2 Anderson's orthogonality catastrophe

Here we analyze the response of electrons inside the metallic gate to a sudden change in the vortex position in terms of Anderson's orthogonality catastrophe (OC) (Anderson, 1967). The magnetic flux of a vortex penetrating inside the gate scatters the electrons in a way similar to Aharonov–Bohm (A–B) scattering (Aharonov *et al.*, 1984; Aharonov and Bohm, 1959). The OC manifests itself in the vanishing overlap $\langle\Psi_f|\Psi_i\rangle$ of the two wave functions describing the macroscopic electron system before and after the change in the scattering potential. As has been already said, we assume that the gate does not change the superconducting properties (such as T_c), but rather affects only the motion of vortices by renormalizing the tunneling rate:

$$\Gamma_G = \Gamma_{unG}\langle\Psi_f|\Psi_i\rangle. \tag{18.2}$$

The overlap $\langle\Psi_f|\Psi_i\rangle$ accounts for the response of the electron gas. Formally, the overlap factor in eqn (18.2) can be presented in terms of the operators $\hat{S}_i$ and $\hat{S}_f$ describing the scattering of the electrons by the magnetic field of the tunneling vortices in their initial and final positions (Yamada and Yosida, 1982):

$$|\langle\Psi_f|\Psi_i\rangle| = N^{-K_{OC}};\ K_{OC} = -\frac{1}{8\pi^2}Tr\left\{\ln^2(\hat{S}_f\hat{S}_i^{-1})\right\}. \tag{18.3}$$

Here, N is the number of electrons and hence the overlap factor vanishes unless there is a mechanism that limits the effectiveness of the OC. At finite temperatures (Kagan and Prokof'ev, 1986; Yamada and Yosida, 1982), the large parameter N should be substituted by another large parameter $1/(\max\{T, \Gamma_{\rm G}\}\tau_{\rm tun})$. Here $\tau_{\rm tun}$ is the time of vortex tunneling in the absence of the gate; $\tau_{\rm tun}^{-1}$ acts as the high-energy cutoff because only slow excitations that cannot follow the tunneling particle adiabatically reduce the overlap factor. In contrast, the temperature enters as a low-energy cutoff because the excitations with energy smaller than T, being thermally activated, do not contribute to the orthogonality of the initial and final states (it is assumed that $\tau_{\rm tun}^{-1} \gg T$). In addition, electrons with energies smaller than $\Gamma_{\rm G}$ cannot react to the tunneling events, which are too frequent for them. This is why the tunneling rate determines its own renormalization in a self-consistent way (like in the anisotropic Kondo problem (Anderson *et al.*, 1970)):

$$\Gamma_{\rm G}(T) = \Gamma_{\rm unG}(\max\{T, \Gamma_{\rm G}\}\tau_{\rm tun})^{K_{\rm OC}}. \tag{18.4}$$

For $K_{\rm OC} < 1$, the tunneling rate remains finite at low temperatures $\Gamma_{\rm G}(T \to 0) = \Gamma_{\rm unG}(\Gamma_{\rm unG}\tau_{\rm tun})^{K_{\rm OC}/(1-K_{\rm OC})}$, while for $K_{\rm OC} > 1$ the tunneling rate $\Gamma_{\rm G}(T)$ goes to zero with the temperature as $T^{K_{\rm OC}}$.

The renormalization of $\Gamma_{\rm G}$ induced by the gate can explain the localization of vortices (i.e. a transition from $\Gamma_{\rm unG} > \varepsilon$ to $\Gamma_{\rm G} < \varepsilon$) if the exponent $K_{\rm OC}$ is comparable to or larger than 1. In the following part of this section we show that for a superconducting film magnetically coupled to a metallic gate the exponent $K_{\rm OC}$ is

$$K_{\rm OC} = \varsigma(\alpha\delta r)^2\frac{(k_{\rm F}^{\rm gate})^2 d}{64R}. \tag{18.5}$$

Here δr is the typical distance that vortices have to tunnel, which is approximately the coherence length $\delta r \sim \xi$ (Larkin and Ovchinnikov, 1971). The parameter α is the total flux that moves a distance δr as a result of a tunneling event, measured in units $\Phi_0 = 2\pi\hbar c/e$. For a single vortex, $\alpha = \frac{1}{2}$; the same holds for interstitials or vacancies. When a bundle (or a pair of dislocations) is tunneling, α should be multiplied by the number of vortices. The result is valid as long as $\delta r \ll R$, where R is the radius of the area inside the gate occupied by the magnetic field of a vortex, which in the case of Mason and Kapitulnik (2002) is about the superconducting penetration depth, $R \approx \lambda$. The expression for λ is determined by the vortex solution specific for thin-film superconductors, known as the Pearl vortex (Abrikosov, 1988; Pearl, 1964). Other factors determining $K_{\rm OC}$ are the gate thickness d and the Fermi momentum of the electrons in the gate $k_{\rm F}^{\rm gate}$. The prefactor ς is evaluated numerically as ≈ 0.4.

Using the expressions for λ and ξ in disordered thin films the exponent can be rewritten as ($\hbar = 1$):

$$K_{\rm OC} \sim \varsigma\frac{\alpha^2}{48\pi}\left(\frac{e^2}{c}\right)^2\frac{v_{\rm F}^{\rm sc}}{e^2}(k_{\rm F}^{\rm gate}d)(k_{\rm F}^{\rm gate}a)(k_{\rm F}^{\rm sc}l^{\rm sc})^2. \tag{18.6}$$

The index sc refers to the electrons in the superconducting film: $l^{\rm sc}$ is their mean-free path (in the normal state) and $v_{\rm F}^{\rm sc}$ is the Fermi velocity. Interestingly, $T_{\rm c}$ drops out from $K_{\rm OC}$ so that it depends only on the geometrical factors and the non-superconducting properties of the electrons. We see that the value of the exponent $K_{\rm OC}$ is determined by a small factor $\sim 10^{-7}$, which is opposed by a product of a few large factors. The condition for the vortex localization can be easily fulfilled for a not-too-thin gate and not-too-disordered superconducting film.

Next, we briefly sketch the steps in the derivation of eqn (18.5) starting from eqn (18.3). Let us consider the OC in response to a single vortex tunneling. The cylindrical symmetry of the vortex allows us to analyze the scattering of electrons using the basis of cylindrical waves, $|\ell, q, k_z\rangle$; here ℓ is the angular momentum along the z-axis, while q and k_z are the magnitudes of the in-plane and z components of the momentum. In this basis the scattering operator is diagonal and can be described in terms of the phase shifts

$$\langle \ell, q, k_z|S|\ell', q', k_z'\rangle = e^{2i\delta_\ell}\delta_{\ell,\ell'}\delta_{q,q'}\delta_{k_z,k_z'} . \tag{18.7}$$

Since the exponent in eqn (18.3) contains a product of two scattering operators corresponding to the different vortex positions shifted by δr, we have to use the transformation matrix between the two bases of cylindrical waves centered at these positions: ${}_f\langle \ell, q, k_z|\ell' q', k_z'\rangle_i = J_{(\ell-\ell')}(q\delta r)\delta_{q,q'}\delta_{k_z,k_z'}$, where $J_\nu(z)$ is the Bessel function. The elements of the matrix $S_f S_i^{-1}$ can be easily calculated as

$${}_f\langle \ell|S_f S_i^{-1}|\ell'\rangle_f = \sum_n e^{2i\delta_\ell - 2i\delta_{n+\ell}} J_n(q\delta r) J_{n-\ell'+\ell}(q\delta r) . \tag{18.8}$$

To proceed further, we need to find the phase shifts specific for the scattering by the vortex. An analogy to classical scattering, where the angular momentum is related to the impact parameter $b = |\ell|/q$, helps elucidate the behavior of the phase shift as a function of ℓ. For $b \gg R$, the scattering by the vortex is similar to the A–B scattering by a flux $\alpha\Phi_0$. In A–B scattering (Aharonov *et al.*, 1984; Aharonov and Bohm, 1959), electrons acquire the phase $\delta_\ell^{\rm A-B} = \frac{\pi}{2}(|\ell| - |\ell - \alpha|)$. The uniqueness of this scattering is in its infinite range: δ_ℓ does not vanish when $|\ell| \to \infty$. For scattering by the vortex, the jump in the A–B phase shifts is smeared out, but the infinite-range character of this scattering is preserved. Hence, δ_ℓ varies monotonically as a function of ℓ between the two limits:

$$\delta_\ell \xrightarrow[\ell \gg qR]{} \alpha\frac{\pi}{2} sgn\ \ell . \tag{18.9}$$

Naturally, for $qR \gg 1$ the phase shift depends on b and R only through the dimensionless combination $b/R = \ell/qR$ such that $\delta_\ell = \frac{\alpha\pi}{2} g\left(\ell/qR\right)$; see Fig. 18.6 for illustration.

We now notice that the sum determining the elements of ${}_f\langle \ell|S_f S_i^{-1}|\ell'\rangle_f$ is accumulated at $-q\delta r \lesssim n \lesssim q\delta r$. This is because the Bessel functions $J_\nu(z)$ decay exponentially with their order when $\nu > z$. Since for thin superconductors $\delta r/R \sim \xi/\lambda \ll 1$, the phase-shift difference in eqn (18.8) can be approximated as:

$$\delta_\ell - \delta_{n+\ell} \xrightarrow[n \ll qR]{} -n\delta'_\ell; \qquad \delta'_\ell \approx \frac{\alpha\pi}{2qR} g'\left(\frac{\ell}{qR}\right) \ll 1. \tag{18.10}$$

The final step of the calculation is to expand the logarithm in eqn (18.3) in $\delta r/R$, and take the trace over ℓ and the momentum on the Fermi surface. The outcome of the calculation is given in eqn (18.5). The gate thickness d appears here as a result of taking the trace. The specifics of the vortex solution enter only through $g(x)$, with the integral yielding $\varsigma = \int dx(dg/dx)^2 \approx 0.4$.

The expression in eqn (18.5) can be applied for any bundle with a total flux $\alpha\Phi_0$ that moves a distance δr as a result of a tunneling event. This is because the magnetic field of the tunneling vortices extends over a large distance, so that the exact configurations of vortices participating in a tunneling event is not important. The only relevant quantity is the product $\alpha\delta r$.

Finally, note that although we invoke the expansion in terms of $\delta r/R \ll 1$, we get $K_{\mathrm{OC}} \propto (\delta r)^2/R$. This is a typical feature of the OC in the case of an extended scattering potential (Matveev and Larkin, 1992) when a large number of harmonics is involved. This can be understood from the following arguments. It was shown that the OC in the discussed problem is determined by $\sum_\ell(\delta_\ell - \delta_{\ell+1})^2 \approx \sum_\ell(\delta'_\ell)^2$. Since the phase shifts approach the limit $\pm\alpha\frac{\pi}{2}$ asymptotically, then the sum

$$\sum_\ell(\delta_\ell - \delta_{\ell+1}) \approx \sum_\ell \delta'_\ell = \pi\alpha. \tag{18.11}$$

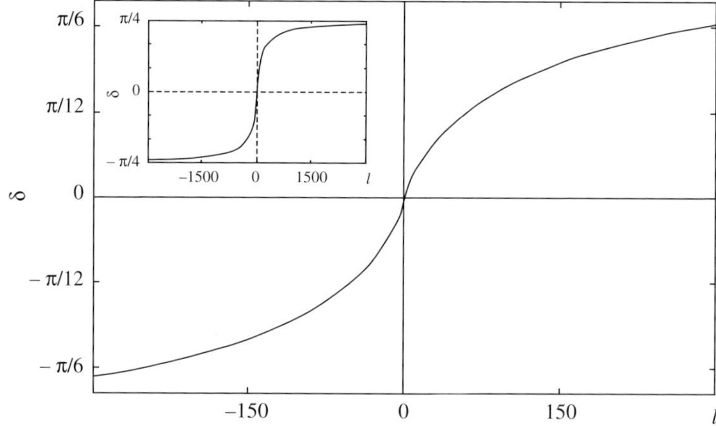

Fig. 18.6 The phase shift for an electron scattered by the magnetic field of a superconducting vortex as a function of the angular momentum. The phase shift is calculated for $qR = 50$, 100, and 150. The three curves are rescaled to $qR = 100$ to demonstrate the universality of the scattering. The inset shows the asymptotic behavior of the phase shift at large angular momenta, $\delta_\ell \to \pm\frac{\pi}{4}$.

Therefore, the result obtained for the exponent $K_{\rm OC}$ corresponds to the differences $(\delta_\ell - \delta_{\ell+1})$ that are distributed almost equally between $L \sim qR$ channels:

$$\sum_\ell (\delta_\ell - \delta_{\ell+1})^2 \sim L\left(\frac{\pi\alpha}{L}\right)^2 \sim \frac{(\pi\alpha)^2}{qR}. \tag{18.12}$$

The above arguments helps to explain the effect of disorder in the gate on the OC. One may expect that the randomization of the phase shifts due to the disorder can only increase the value of the exponent $K_{\rm OC}$. In the general case, ℓ should be substituted by an index i of the scattering channel (i.e. the index of the states diagonalizing the scattering matrix). The scattering by impurities leads to the randomization of the phase differences, while the asymptotic limits of the phase shifts remain the same, $\pm\alpha\frac{\pi}{2}$. Therefore, the value of the exponent $K_{\rm OC}$, which is determined by the squares of the phase differences, should increase in the presence of disorder. Under the condition of eqn (18.11), the obtained exponent $K_{\rm OC} \propto (\delta r)^2/R$ is close to the minimum possible value which is at equal phase differences. Our conclusion that disorder increases the effect of the gate in suppressing the tunneling rate $\Gamma_{\rm G}$ is in accordance with the existing theoretical results regarding the enhancement of the OC exponent by not-too-strong disorder (Chen and Kroha, 1992; Gefen *et al.*, 2002).

18.3 Tunneling with dissipative eddy currents

In this section we describe another approach for analyzing the tunneling of a vortex in the presence of large number of low-energy states, which are the electron–hole excitations located in the metallic gate. For this purpose, we write the action describing a vortex magnetically coupled to the gate. In general, when the action describing the tunneling process is formulated in terms of a single fast degree of freedom (i.e. after integrating out the environmental degrees of freedom), the long-time response of the slow degrees of freedom reveals itself as a term that is non-local in time (Leggett *et al.* 1987):

$$S_{\rm env} = \frac{1}{2}\int d\tau \int d\tau' M(\tau - \tau')\left[\mathbf{R}(\tau) - \mathbf{R}(\tau')\right]^2. \tag{18.13}$$

Here, $\mathbf{R}(\tau)$ is the coordinate of the fast degree of freedom at time τ; in our problem $\mathbf{R}(\tau)$ is the position of the vortex. In the case of an ohmic dissipative environment (Caldeira and Leggett, 1981), $M(\tau - \tau') \propto 1/(\tau - \tau')^2$. At small time differences one can substitute $[\mathbf{R}(\tau) - \mathbf{R}(\tau')]/(\tau - \tau')$ with the velocity $\dot{\mathbf{R}}$. Then, the integrand in eqn (18.13) is reminiscent of the Rayleigh dissipation function. The Rayleigh function is used in the Euler–Lagrange equation to include dissipation (Goldstein, 1980):

$$\frac{d}{dt}\left(\frac{\partial\mathcal{L}}{\partial\dot{R}}\right) - \frac{\partial\mathcal{L}}{\partial R} + \frac{\partial\mathcal{F}}{\partial\dot{R}} = 0, \tag{18.14}$$

where $\mathcal{F} = \eta\dot{R}^2/2$ is the Rayleigh function with a friction coefficient η. Since the Rayleigh function enters the equation of motion without a time derivative (unlike the

Lagrangian), the inclusion of dissipation into the equation requires an additional time integration. Therefore, the corresponding term must be non-local in time.

The environment produces the most significant effect on tunneling at large time differences, when the slow degrees of freedom have sufficient time to develop. Naturally, the effect is the strongest when the tunneling occurs between states that are almost degenerate. In the latter case the long-time response can fully develop and considerably reduce or even block the tunneling (this statement is often formulated in the context of the dissipative quantum phase transition (Herrero and Zaikin, 2002; Schmid, 1983)).

Next, we write the action describing the tunneling vortex and the eddy currents induced inside the gate as a response to the change in the vortex position. We integrate out the environmental degrees of freedom inside the gate, and obtain the dissipative term in the effective action of the tunneling vortex. We start with the imaginary time action:

$$S = S_{\rm sc} + S_{\rm gate} + S_{\rm int}. \tag{18.15}$$

Here $S_{\rm sc}$ is an action of the superconducting film in the absence of the gate. Since we are not trying to solve the problem of vortex tunneling in full scale, but are interested only in the effect of the gate on the tunneling rate, this part of the action is not specified.

In writing $S_{\rm gate}$, which describes the dynamics of the electrons in the gate, one should keep in mind the following argument. The charge and current densities that are relevant for the motion of the vortex in the film are characterized by large length and time scales. Therefore, their dynamics can be described macroscopically. Since the deviations of the charge and current densities from their equilibrium values are small, the action that describes the dynamics of their fluctuations should be consistent with the fluctuation-dissipation theorem. (Examples of such an approach can be found in the calculation of the dephasing time of the cooperons due to electromagnetic fluctuations (Altshuler *et al.*, 1982), and in a macroscopic calculation of the zero bias anomaly (Levitov and Shytov, 1997).)

The current in the gate has two contributions. One is the ohmic response to the electric field, while the other one is the diffusive current from the gradients of the density; $J(r,\tau) = J_{\rm ohmic}(r,\tau) - D\boldsymbol{\nabla}\rho(r,\tau)$. The fluctuations of the charge and current densities can be expressed through the correlation function of the ohmic part of the current, written in terms of the Matsubara frequency as follows:

$$\begin{aligned}\hat{K}^{-1}_{i,j}(\mathbf{k}, i\omega_n) &= \langle J^i_{\rm ohmic}(\mathbf{k}, i\omega_n) J^j_{\rm ohmic}(-\mathbf{k}, -i\omega_n)\rangle \\ &= \sigma_{i,j}(\mathbf{k},\omega_n)|\omega_n| + \sigma_{i,i'} D_{j,j'}(\mathbf{k},\omega_n) k_{i'} k_{j'}.\end{aligned} \tag{18.16}$$

Here the diffusion constants tensor $\hat{D}$ and the conductivity $\hat{\sigma}$ are connected through the Einstein relation $\hat{\sigma} = e^2(dn/d\mu)\hat{D}$. The gate, being a simple homogeneous metal, is adequately described by the Drude formula; we ignore the Hall conductivity in the gate. Since we are interested only in the low-frequency and long-wavelength behavior, we take the conductivity σ to be constant.

Following the above arguments, one can write the gate action S_{gate} for the charge and current densities inside the gate in the Matsubara time ($\beta = \frac{1}{T}$):

$$S_{\text{gate}} = \frac{1}{2}\int_0^\beta d\tau_1 d\tau_2 \int d\mathbf{r}_1 d\mathbf{r}_2 \mathbf{J}_{\text{ohmic}}(\mathbf{r}_1, \tau_1)\hat{K}(\mathbf{r}_2 - \mathbf{r}_1, \tau_2 - \tau_1)\mathbf{J}_{\text{ohmic}}(\mathbf{r}_2, \tau_2) + S_{\text{Maxwell}} \tag{18.17}$$

The first term in the above action reproduces the charge and current density fluctuations, in accordance with the fluctuation–dissipation theorem. The term S_{Maxwell} describes the interaction of the electromagnetic fields and the charge and current densities in the gate in a way that reproduces the Maxwell equations. (Since the fluctuations of the current and density are not independent, the charge continuity condition has to be imposed in eqn (18.17); for more details see (Michaeli and Finkel'stein, 2007).)

The last term in the action, S_{int}, describes the connection between the superconducting film and the gate. The current density in the gate interacts with the vector potential created by the superconducting film:

$$S_{\text{int}} = i\int_0^\beta d\tau \int d\mathbf{r}\frac{1}{c}\mathbf{J}(\mathbf{r}, \tau) \cdot \mathbf{A}_{sc}(\mathbf{r}, \tau). \tag{18.18}$$

The factor i appears because the action is written in imaginary time, as usual for tunneling problems. We are interested in the limited problem of the long-tail response of electrons inside the gate that develops when the tunneling of the vortex degrees of freedom is mostly accomplished. Therefore, we consider τ that is sufficiently large so that one can assume that the deformation of the field $\mathbf{A}_{sc}(r, \tau)$ appearing during the process of vortex tunneling is already relaxed—by deformation we mean the deviation of $\mathbf{A}_{sc}(r, \tau)$ from the field of a static vortex centered at $\mathbf{R}(\tau)$(Abrikosov, 1988). With this in mind, we put in eqn (18.18)

$$\mathbf{A}_{sc}(\mathbf{r}, \tau) = \mathbf{A}_{vor}(\mathbf{r}; \mathbf{R}(\tau)), \tag{18.19}$$

where $\mathbf{A}_{vor}(\mathbf{r}; \mathbf{R}(\tau))$ is the standard solution for the vector potential of a vortex in thin films (Abrikosov, 1988; Carneiro and Brandt, 2000; Pearl, 1964).

Since the action is quadratic in the charge and current densities, it is possible to integrate them out (Michaeli and Finkel'stein, 2007). As a result we obtain the term S_{env} describing the effect of the environment on the tunneling vortex:

$$S_{\text{env}} = -\frac{\pi\sigma}{2\beta^2 c^2}\int d\tau_1 d\tau_2 \int \frac{d^2q}{(2\pi)^2}dz \frac{e^{i\mathbf{q}(\mathbf{R}(\tau_1)-\mathbf{R}(\tau_2))}}{\sin^2\left(\frac{\pi}{\beta}(\tau_1 - \tau_2)\right)} \frac{B^z_{vor}(-\mathbf{q}, z)B^z_{vor}(\mathbf{q}, z)}{q^2}. \tag{18.20}$$

Here $\mathbf{q}$ is the in-plane momentum of the charge and current densities inside the gate. For the purpose of illustration, let us compare this action with the action of Caldeira and Leggett (1981) for a mechanical particle moving in a dissipative environment:

$$S_{\rm CL} = \frac{1}{4\pi}\int d\tau_1 d\tau_2 \eta \frac{(\mathbf{R}(\tau_1)-\mathbf{R}(\tau_2))^2}{(\tau_1-\tau_2)^2}. \tag{18.21}$$

If $R(\tau)$ describes a particle moving with a constant velocity, this action becomes $1/2\pi \int d\tau_1 d\tau_2 \eta \dot{\mathbf{R}}^2/2$. Under the same approximation of constant velocity, the action $S_{\rm env}$ can be written in the coordinate representation as:

$$S_{\rm env} = \frac{1}{4\pi}\int d^2r dz \int d\tau_1 d\tau_2 \frac{\sigma}{c^2}\left[\mathbf{v}_{\rm vor}\times\mathbf{B}_{\rm vor}(\mathbf{r},z)\right]^2. \tag{18.22}$$

The combination $1/c\,[\mathbf{v}_{\rm vor}\times\mathbf{B}_{\rm vor}(\mathbf{r})]$ is an electric field created by a moving vortex. The integrand in the action $S_{\rm env}$ is then merely the energy dissipation rate in the gate caused by the vortex motion, $S_{\rm env} = 1/2\pi\int d^2rdz\int d\tau_1 d\tau_2 \sigma E^2_{vor}(r,z)/2$. This expression is in full correspondence with the one obtained for a constant motion of a particle in the presence of friction η.

We return to eqn (18.20) and analyze it for the case of tunneling between two minima separated by a distance δr. After integration over the z coordinate and momentum $\mathbf{q}$ we obtain:

$$S_{\rm env} = \frac{\alpha^2\sigma d}{8e^2\lambda^2}\int d\tau_1 d\tau_2 \frac{-\mathbf{R}(\tau_1)\mathbf{R}(\tau_2)}{(\tau_1-\tau_2)^2}\ln\left(\frac{d}{\lambda}+\frac{8\pi^2\sigma d^2}{|\tau_1-\tau_2|c^2}\right)^{-1}. \tag{18.23}$$

The appearance of the log factor is very natural if one recalls that we integrate the square of the magnetic field, and it is well known that $B_{\rm vor}\sim 1/r$ at a distance r from the center of the vortex, when $r<\lambda$. The time dependence of the logarithm results from the fact that the integration over the momenta is limited to the ohmic regime, $|\omega_n| \ll c^2 q/(2\pi\sigma d)$. The time dependence of the logarithm is important because in thin superconducting films the ratio d/λ can be very small. At large time differences, which are essential for low temperatures, the logarithmic factor in the action becomes $\ln\lambda/d$.

Following the standard procedure (Leggett *et al.*, 1987) one gets the result that the modified tunneling rate $\Gamma_{\rm G}$ is:

$$\Gamma_{\rm G}(T) = \Gamma_{\rm unG}\left(max\{T,\Gamma_G\}\tau_{\rm tun}\right)^{K_{\rm E}}; \tag{18.24a}$$

$$K_{\rm E} = \sigma d\frac{(\alpha\delta r)^2}{8e^2\lambda^2}\ln\left(\frac{\lambda}{d}\right). \tag{18.24b}$$

As discussed in the previous section, low-frequency components of the eddy current with frequencies smaller than $\Gamma_{\rm G}$ cannot develop when the tunneling events are too frequent. This is why the tunneling rate determines its own renormalization in a self-consistent way. The full calculation of the renormalized tunneling rate includes a number of subtleties that have been described by Michaeli and Finkel'stein (2007).

Let us discuss the result obtained for the exponent $K_{\rm E}$. Naturally, the action $S_{\rm env}$ is proportional to the square of the magnetic field. After integration over the coordinates one obtains the factor $(\alpha/\lambda)^2$. From the general structure of eqn (18.13) it follows that $K_{\rm E}$ is proportional to the squared distance of the change of vortex position, which is typically of order ξ. Altogether, the exponent K_{E} acquires the factor $(\xi/\lambda)^2 \ll 1$. This smallness is opposed by the sheet conductance of the gate, $\sigma^{2D} = \sigma d$, which in the case of a thick well-conducting gate gives a large factor $h\sigma^{2D}/e^2$. As usual, albeit paradoxically, the effect of dissipation is stronger for cleaner systems.

18.4 Comparison between the orthogonality catastrophe and tunneling with dissipation

We considered two approaches to describe the response of the electrons inside the gate on the tunneling event:

(i) the OC caused by the A–B scattering of electrons inside the gate
(ii) the eddy current in the gate generated by the change in the magnetic flux.

The exponent determining the renormalized tunneling rate $\Gamma_{\mathrm{G}}(T) = \Gamma_{\mathrm{unG}}\left(T\tau_{\mathrm{tun}}\right)^{K}$ is given by eqn (18.5) for the OC and by eqn (18.24b) for the effect of the Eddy current. We find that for the experimental setup of Mason and Kapitulnik (2002) the effect of the OC provides an exponent sufficient for a substantial suppression of the tunneling rate of the vortices, $K_{\mathrm{OC}} \sim 1$. The two expressions for the exponent, K_{OC} and K_{E}, share the same dependence on $\alpha\delta r$ and d. Therefore, the ratio between them is "universal":

$$\frac{K_{\mathrm{E}}}{K_{\mathrm{OC}}} = \frac{\sigma \ln(\lambda/d)}{e^2(k_{\mathrm{F}}^{\mathrm{gate}})^2\lambda} \sim \frac{l_{\mathrm{gate}}}{\lambda}, \tag{18.25}$$

where l_{gate} is the mean free path of the electrons in the gate. Since in a thin superconducting film the penetration depth λ is very large, under the conditions of the experiment (Mason and Kapitulnik, 2002), the OC is dominant.

The expressions in eqns (18.24b) and (18.5) were obtained assuming that only one vortex participates in each tunneling event. As has been already mentioned, vortices can tunnel as a bundle or as topological defects. Still, the calculation remains valid as long as $\delta r \ll \lambda$. This is because the magnetic field of the tunneling vortices extends over a large distance, so that the exact configurations of vortices participating in a tunneling event is not important.

Although the two approaches to analyze tunneling in the presence of slow (low-energy) degrees of freedom (i.e. the scattering and functional-integral formalisms) give rather different results for K, we still believe that they both describe the same mechanism of suppression of the tunneling rate. Therefore, we have to explain the reasons why K_{OC} appears to be significantly larger than K_{E}. Basically, there are two reasons. First, the calculation scheme of the tunneling with dissipation for the eddy currents corresponds to the OC expression calculated in the perturbation theory in the second order with respect to the change of the vector potential. (It merely corresponds to finding the elements of the scattering matrix in the Born approximation.) However, the A–B effect is a non-perturbative phenomenon, and the asymptotic limits of the phase shifts were not captured in the tunneling-with-dissipation calculation. The other more general reason is related to the extended size of the tunneling vortex.

In the original formulation of tunneling with dissipation described by Caldeira and Leggett (1981), the environment was substituted by one-dimensional phonons in order to imitate the constant density of states of the electron–hole excitations. The disadvantage of this effective model is that it capture the scattering only in the s-wave channel. Since the tunneling of an extended object creates excitations in many channels of the environment, the usual formulation of tunneling with dissipation appears to be incomplete. Thus, while the sum over the channels (harmonics) enters in a natural way into the exponent describing the effect of the OC, such a summation is absent in the calculation of the tunneling with dissipation.

18.5 Summary

We address our analysis to an experiment (Mason and Kapitulnik, 2002) in which the resistance of a superconducting film was measured in a magnetic field both with and without a gate. We interpret the result of the experiment by assuming that the origin of the finite resistance at low temperatures is the tunneling of vortices. From our point of view, the difference in the resistance of the gated and ungated films indicates that the gate reduces the tunneling rate of the vortices making them localized. Indeed, we show here that adding a gate may effectively suppress the vortex tunneling. The gated system discussed here can be used as an effective experimental tool for investigating the vortex motion at low temperatures.[5] The gated system provides a unique opportunity to study the vortex tunneling in thin superconducting films by such simple means as varying the characteristics of the gate, in particular the gate thickness and/or the sheet conductance of the gate.

In this work we showed that Anderson's orthogonality catastrophe (Anderson, 1967) promotes Anderson localization (Anderson, 1958) of tunneling vortices. Besides these studies, we used Anderson's work on the flux flow (Anderson, 1966; Anderson and Kim, 1964) and the Kondo problem (Anderson *et al.*, 1970). However, these remarkable papers are only fraction of P. W. Anderson's seminal contributions to contemporary physics.

Acknowledgements

We thank A. Kapitulnik, T. M. Klapwijk, J. E. Mooij, and B. Spivak for useful discussions. This work is supported by the US–Israel BSF Foundation.

[5]For example, it was suggested that at low temperatures the heat released as a result of a single vortex motion initiates the motion of the surrounding vortices. In other words, the finite resistance is caused not by tunneling of vortices, but by a chemical reaction-like activation motion. A mechanism of this kind was observed in magnetic deflagration Suzuki *et al.* (2005). Since in the experimental setup described in Mason and Kapitulnik (2002) the remote gate is well isolated from the superconducting film, we believe that this experiment rules out such a scenario for the motion of vortices.

References

Abrikosov, A.A. (1988). *Fundamentals of the Theory of Metals*. Elsevier, Amsterdam.

Aharonov, Y., Au, C.K., Lerner, E.C., and Liang, J.Q. (1984). *Phys. Rev. D*, **29**, 2396–2398.

Aharonov, Y. and Bohm, D. (1959). *Phys. Rev*, **115**, 485–491.

Altshuler, B.L., Aronov, A.G., and Khmel'nitskii, D.E. (1982). *J. Phys. C*, **15**, 7367–7386.

Anderson, P.W. (1958). *Phys. Rev*, **109**, 1492–1505.

Anderson, P.W. (1966). *Rev. Mod. Phys.*, **38**, 298–310.

Anderson, P.W. (1967). *Phys. Rev. Lett*, **18**, 1049–1051.

Anderson, P.W. and Kim, Y.B. (1964). *Rev. Mod. Phys*, **36**, 39–43.

Anderson, P.W., Yuval, G., and Hamann, D.R. (1970). *Phys. Rev. B*, **1**, 4464–4473.

Bardeen, J. and Stephen, M.J. (1965). *Phys Rev*, **140**, A1197–A1207.

Caldeira, A.O. and Leggett, A.J. (1981). *Phys Rev. Lett*, **46**, 211–214.

Carneiro, G. and Brandt, E.H. (2000). *Phys. Rev. B*, **61**, 6370–6376.

Chen, Y. and Kroha, J. (1992). *Phys. Rev. B*, **46**, 1332–1337.

Ephron, D., Yazdani, A., Kapitulnik, A., and Beasley, M.R. (1996). *Phys. Rev. Lett*, **76**, 1529–1531.

Fisher, M.P.A., Tokuyasu, T.A., and Young, A.P. (1991). *Phys. Rev. Lett*, **66**, 2931–2934.

Gefen, Y., Berkovits, R., Lerner, I.V., and Altshuler, B.L. (2002). *Phys. Rev. B*, **65**, 081106.

Giaever, I. (1964). *Phys. Rev. Lett*, **15**, 825–827.

Goldstein, H. (1980). *Classical Mechanics*. Addison-Wesley, Massachusetts.

Herrero, C.P. and Zaikin, A.D. (2002). *Phys. Rev. B.*, **65**, 104516.

Iordanskii, S.V. and Finkel'stein, A.M. (1973). *J. Low. Temp. Phys*, **10**, 423–447.

Kagan, Y. and Prokof'ev, N.V. (1986). *Zh. Eksp. Teor. Fiz*, **90**, 2176–2195. (*Sov. Phys. JETP*, **63**, 1276–1286).

Kruithof, G.H., van Son, P.C., and Klapwijk, T.M. (1991). *Phys. Rev. Lett*, **67**, 2725–2728.

Larkin, A.I. (1970). *Zh. Eksp. Teor. Fiz*, **58**, 1466–1470. (*Sov. Phys. JETP.* (1970), **31**, 784–786).

Larkin, A.I. and Ovchinnikov, Yu.N. (1971). *Zh. Eksp. Teor. Fiz*, **61**, 1221–1230. (*Sov. Phys. JETP.*, **34**, 651–655).

Leggett, A.J., Chakravarty, S., Dorsey, A.T., Fisher, M.P.A, Garg, A., and Zwerger, W. (1987). *Rev. Mod. Phys.*, **59**, 1–85. See also references therein.

Levitov, L.S. and Shytov, A.V. (1997). *Pis'ma Zh. Eksp. Teor. Fiz*, **66**, 200–205.

Liu, Y., Haviland, D.B., Glazman, L.I., and Goldman, A.M. (1992). *Phys. Rev. Lett*, **68**, 2224–2227.

Mason, N. and Kapitulnik, A. (2002). *Phys. Rev. B.*, **65**, 220505(R).

Matveev, K.A. and Larkin, A.I. (1992). *Phys. Rev. B*, **46**, 15337–15347.

Michaeli, K. and Finkel'stein, A.M. (2006). *Phys. Rev. Lett.*, **97**, 117004.

Michaeli, K. and Finkel'stein, A.M. (2007). *Phys Rev. B*, **76**, 064506.

Pearl, J. (1964). *App Phys Lett*, **5**, 65–66.

Ramakrishnan, T.V. (1989). *Physica Scripta*, **T27**, 24–30.

Rimberg, A.J., Ho, T.R., Kurdak, C., Clarke, J., Campman, K.L., and Gossard, A.C. (1997). *Phys. Rev. Lett*, **78**, 2632–2635.

Schmid, A. (1983). *Phys. Rev. Lett*, **51**, 1506. The correct form of the phase diagram is given by Bulgadaev, S.A. (1984). Phase diagram of a dissipative quantum system. *Pis'ma Zh. Eksp. Teor. Fiz.*; 264–267 (*Sov. Phys. JETP Lett.*) **39**, 315–319).

Suzuki, Y., Sarachik, M.P., Chudnovsky, E.M., McHugh, S., Gonzalez-Rubio, R., Avraham, N., Myasoedov, Y., Zeldov, E., Shtrikman, H., Chakov, N.E., and Christou, G. (2005). *Phys. Rev. Lett.*, **95**, 147201.

Tafuri, F., Kirtley, J.R., Born, D., Stornaiuolo, D., Medaglia, P.G., Orgiani, P., Balestrino, G., and Kogan, V.G. (2006). *Europhys. Lett.*, **73**, 948–954.

van Oudenaarden, A., Vardy, S.J.K., and Mooij, J.E. (1996). *Phys. Rev. Lett*, **77**, 4257–4260.

Wagenblast, K.H., van Otterlo, A., Schon, G., and Zimanyi, G.T. (1997). *Phys. Rev. Lett.*, **79**, 2730–2733.

Yamada, K. and Yosida, K. (1982). *Prog Theor. Phys.*, **68**, 1504–1517.